Advances in Intelligent Systems and Computing

Volume 1076

The series “Advances in Intelligent Systems and Computing” contains publications on theory, applications, and design methods of Intelligent Systems and Intelligent Computing. Virtually all disciplines such as engineering, natural sciences, computer and information science, ICT, economics, business, e-commerce, environment, healthcare, life science are covered. The list of topics spans all the areas of modern intelligent systems and computing such as: computational intelligence, soft computing including neural networks, fuzzy systems, evolutionary computing and the fusion of these paradigms, social intelligence, ambient intelligence, computational neuroscience, artificial life, virtual worlds and society, cognitive science and systems, Perception and Vision, DNA and immune based systems, self-organizing and adaptive systems, e-Learning and teaching, human-centered and human-centric computing, recommender systems, intelligent control, robotics and mechatronics including human-machine teaming, knowledge-based paradigms, learning paradigms, machine ethics, intelligent data analysis, knowledge management, intelligent agents, intelligent decision making and support, intelligent network security, trust management, interactive entertainment, Web intelligence and multimedia.

The publications within “Advances in Intelligent Systems and Computing” are primarily proceedings of important conferences, symposia and congresses. They cover significant recent developments in the field, both of a foundational and applicable character. An important characteristic feature of the series is the short publication time and world-wide distribution. This permits a rapid and broad dissemination of research results.

**** Indexing: The books of this series are submitted to ISI Proceedings, EI-Compendex, DBLP, SCOPUS, Google Scholar and Springerlink ****

More information about this series at http://www.springer.com/series/11156

Vikrant Bhateja · Suresh Chandra Satapathy ·
Hassan Satori
Editors

Embedded Systems and Artificial Intelligence

Proceedings of ESAI 2019, Fez, Morocco

Editors
Vikrant Bhateja
Department of Electronics
and Communication Engineering
Shri Ramswaroop Memorial Group
of Professional Colleges (SRMGPC)
Lucknow, Uttar Pradesh, India

Dr. A.P.J. Abdul Kalam
Technical University
Lucknow, Uttar Pradesh, India

Hassan Satori
Department of Computer Sciences
Faculty of Sciences Dhar Mahraz
Sidi Mohammed Ben Abbdallah University
Fez, Morocco

Suresh Chandra Satapathy
School of Computer Engineering
Kalinga Institute of Industrial Technology
(KIIT)
Bhubaneswar, Odisha, India

ISSN 2194-5357 ISSN 2194-5365 (electronic)
Advances in Intelligent Systems and Computing
ISBN 978-981-15-0946-9 ISBN 978-981-15-0947-6 (eBook)
https://doi.org/10.1007/978-981-15-0947-6

This Springer imprint is published by the registered company Springer Nature Singapore Pte Ltd.
The registered company address is: 152 Beach Road, #21-01/04 Gateway East, Singapore 189721, Singapore

Conference Organization Committees

Honorary Committee

Pr. Radouane Mrabet, President of Sidi Mohamed Ben Abdellah University, Fez, Morocco
Pr. Khadija Essafi, Vice-President of Sidi Mohamed Ben Abdellah University, Fez, Morocco
Pr. Mohammed Benlemilh, Dean of Faculty of Sciences Dhar elmahraz, Fez, Morocco
Pr. Abderrahim Lahrach, Director of ENSA, Fez, Morocco
Pr. Driss Chennouni, Director of ENS, SMBA University, Fez, Morocco
Pr. Sidi Adil Ibrahimi, Dean of Faculty of Medicine and Pharmacy, Fez, Morocco

Conference Chair

Hassan Satori, FSDM Fez, Morocco

Publication Chair

Dr. Suresh Chandra Satapathy, KIIT, Bhubaneswar, Odisha, India

Technical Program Committee Chair

Dr. Vikrant Bhateja, SRMGPC, Lucknow, Uttar Pradesh, India

Advisory Commitee

Abdelaziz Bensrhair, National Institute of Applied Sciences (INSA) of Rouen, France
Abdelhakim Senhaji Hafid, University of Montreal, Canada
Bansidhar Majhi, IIITDM Kancheepuram, Tamil Nadu, India
Christophe Grova, Faculty of Engineering and Computer Science, Concordia University, Canada
Ganpati Panda, IIT Bhubaneswar, Odisha, India
Habib Benali, Faculty of Engineering and Computer Science, Concordia University, Canada
Hamid Mcheick, University of Quebec at Chicoutimi, Quebec, Canada
Jagdish Chand Bansal, South Asian University, New Delhi, India
João Manuel R. S. Tavares, Universidade do Porto (FEUP), Porto, Portugal
Le Hoang Son, Vietnam National University, Hanoi, Vietnam
Mohamed Masmoudi, National Engineering School of Sfax (ENIS), Tunisia
Mohammad Abushariah, King Abdullah II School of IT, Jordan
Mourad Fakhfakh, ENET, Sfax University, Tunisia
Naeem Hanoon, Multimedia University, Cyberjaya, Malaysia
Pietro Pala, Florence University, Italy
Roman Senkerik,Tomas Bata University in Zlin, Czech Republic
Siba K. Udgata, University of Hyderabad, Telangana, India
Swagatam Das, Indian Statistical Institute, Kolkatta, India
Thierry Bouwmans, University of La Rochelle, France
Tai Kang, Nanyang Technological University, Singapore
Valentina Balas, Aurel Vlaicu University of Arad, Romania
Yu-Dong Zhang, University of Leicester, UK

Local Organizing Committee

Abdellatif El Abderrahmani, FP Larache, Morocco
Ali Yahyaouy, FSDM Fez, Morocco
Bachir Benhala, FS Meknes, Morocco
Chakir Loqman, FSDM Fez, Morocco
Elhabib Nfaoui, FSDM Fez, Morocco
El mehdi Mellouli, ENSA Fez, Morocco
Hamid Tairi, FSDM Fez, Morocco
Hassane Moustabchir, ENSA Fez, Morocco
Hassan Satori, FSDM Fez, Morocco
Imane Satauri, ENSAM Casablanca, Morocco
Jamal Riffi, FSDM Fez, Morocco
Jaouad Boumhidi, FSDM Fez, Morocco

Karim El Moutaouakil, ENSA Al Hoceima, Morocco
Khalid Haddouch, ENSA Al Hoceima, Morocco
Khalid Satori, FSDM Fez, Morocco
Lahcen Oughdir, ENSA Fez, Morocco
Mohammed adnane MAHRAZ, FSDM Fez, Morocco
Mohammed Alfidi, ENSA Fez, Morocco
Mohammed Berrada, ENSA Fez, Morocco
Mohammed Chakib Sosse Alaoui CRMEF Fez, Morocco
Mohammed Zouiten, FP Taza, Morocco
Mostafa Merras, EST Meknes, Morocco
My Abdelouahed Sabri, FSDM Fez, Morocco
Nabil El Akkad, ENSA Fez, Morocco
Naouar Laaidi, FSDM Fez, Morocco
Noura Aherrahrou, FSDM Fez, Morocco
Noureddine El Makhfi, FST Al Hoceima, Morocco
Noureddine Ennahnahi, FSDM Fez, Morocco
Omar El Beqqali, FSDM Fez, Morocco
Zakaria Chalh, ENSA Fez, Morocco

Program Committee

Abdelali Boushaba, FST Fez, Morocco
Abdelaziz Bouras, Qatar University, Qatar
Abdelghni Lakehal, FP Larache, Morocco
Abdelhak Boulaalam, FP Taza, Morocco
Abdelhakim Senhaji Hafid, University of Montreal, Canada
Abdelhamid Zouhair, ENSA Al Hoceima, Morocco
Abdelkhalak Bahri, ENSA Al Hoceima, Morocco
Abdelkader Doudou, FP Nador, Morocco
Abdellatif El Abderrahmani, FP Larache, Morocco
Abdellatif El Afia, ENSIAS Rabat, Morocco
Abdellatif Ezzouhairi, ENSA Fez, Morocco
Abdelmajid El Bakkali, FST Errachidia, Morocco
Abdelmjid Saka, ENSA Fez, Morocco
Abdelouahed Sabri, FSDM Fez, Morocco
Abdeslam El Akkad, CRMEF Fez, Morocco
Abdessalam Ait Madi, ENSA Kenitra, Morocco
Aberrahim Saaidi, FP Taza, Morocco
Adil BEN Abbou, FST Fez, Morocco
Adil Benhdich, EST Meknes, Morocco
Adil Jeghal, FSDM Fez, Morocco
Adil Kenzi, ENSA Fez, Morocco
Ahmad Taher Azar, Benha University, Egypt

Ahmed Aberqi, ENSA Fez, Morocco
Ahmed Boutejdar, DFG, Braunschweig-Bonn, Germany
Ahmed Faize, FP Nador, Morocco
Ahmed Khoumsi, University Of Sherbrooke, Canada
Aissam Khaled, ENSA Al Hoceima, Morocco
Akram Halli, FSJES Meknes, Morocco
Ali Yahyaouy, FSDM Fez, Morocco
Amal Zouaq, University of Ottawa, Canada
Andrews Sobral, University of La Rochelle, France
Aris M. Ouksel, University of Illinois, Chicago, USA
Arsalane Zarghili, FST Fez, Morocco
Aziz Baataoui, FST Errachidia, Morocco
Aghoutane Badraddine, FST Errachidia, Morocco
Bachir Benhala, FS Meknes, Morocco
Belkacem Ould Bouamama, University of lille, France
Benaissa Bellach, ENSA Oujda, Morocco
Chaabane Lamiche, University of M'sila, Algeria
Chakir Loqman, FSDM Fez, Morocco
Christine Fernandez-Maloigne, Poitiers University, France
Anouar Darif, FS Beni Mellal, Morocco
Driss Achemlal, FP Taza, Morocco
Eduardo Sousa de Cursi, INSA Rouen, France
El mehdi Mellouli, ENSA Fez, Morocco
Elhabib Nfaoui, FSDM Fez, Morocco
Elmiloud Chaabelasri, FS Oujda, Morocco
El Mustapha Mouaddib, University of Picardie Jules Verne, France
Et-touhami Es-Sebbar, Paul Scherrer Institute, Switzerland
Fadoua Ataa Allah, IRCAM, Rabat Morocco
Fadoua Yakine, ENSA Fez, Morocco
Habib Benali, Concordia University, Canada
Hamid Gualous, University of Caen, France
Hamid Tairi, FSDM Fez, Morocco
Hanaa Hachimi, ENSA Kenitra, Morocco
Hassan Satori, FSDM Fez, Morocco
Hassane Moustabchir, ENSA Fez, Morocco
Hicham ELMOUBTAHIJ, EST Agadir, Morocco
Hicham Touil, FP Larache, Morocco
Houcine Chafouk, Normandy University of Rouen, France
Imane Satauri, ENSAM Casablanca, Morocco
Imad Badi, ENSA Al Hoceima, Morocco
Jamal Riffi, FSDM Fez, Morocco
Jamal Zbitou, FST Settat, Morocco
Jaouad Boumhidi, FSDM Fez, Morocco
Jean-yves Dieulot, University of Lille, France
Karima Aissaoui, ENSA Fez, Morocco

Karim El Moutaouakil, ENSA Al Hoceima, Morocco
Khalid El Afazazy, FPO Agadir, Morocco
Khalifa Mansouri, ENSET Mohammedia, Morocco
Khalid Haddouch, ENSA Al Hoceima, Morocco
Khalid Satori, FSDM Fez, Morocco
Lahcen Oughdir, ENSA Fez, Morocco
Mohamed Sayyouri, ENSA El Jadida, Morocco
Mohamed Abd El Azziz Khamis, E-JUST, Alexandria, Egypt
Mohammad Abushariah, King Abdullah II School of IT, Jordan
Mohamed Ben Halima, Gabes University, Tunisia
Mohamed Benslimane, EST Fez, Morocco
Mohamed Chrayah, ENSAT Tetouan, Morocco
Mohamed Lazaar, ENSIAS, Mohammed V University, Rabat, Morocco
Mohamed Naiel, Concordia University, Montreal, Canada
Mohamed Ouahi, ENSA Fez, Morocco
Mohamed Wassim JMAL, ISSAT, Gafsa, Tunisia
Mohammed adnane Mahraz, FSDM Fez, Morocco
Mohammed Ayoub Alaoui Mhamdi, Bishop's University, Canada
Mohammed Alfidi, ENSA Fez, Morocco
Mohammed Berrada, ENSA Fez, Morocco
Mohammed Chakib Sosse Alaoui Crmef Fez, Morocco
Mohammed Ouanan, FS Meknes, Morocco
Mohammed Oudghiri Bentaie, ENSA Fez, Morocco
Mohammed Zouiten, FP Taza, Morocco
Mostafa Merras, EST Meknes, Morocco
Mostafa El Mallahi, ENS Fez, Morocco
Mounir Kriraa, ENSA Safi, Morocco
Mourad Fakhfakh, ENET, Université de Sfax, Tunisia
My Abdelouahed Sabri, FSDM Fez, Morocco
Nabil El Akkad, ENSA Fez, Morocco
Naouar Laaidi, FSDM Fez, Morocco
Noreddine Chaibi, FSDM Fez, Morocco
Noura Aherrahrou, FSDM Fez, Morocco
Noureddine El Makhfi, FST Al Hoceima, Morocco
Noureddine ENNAHNAHI, FSDM Fez, Morocco
Nour El Houda Chaoui, ENSA Fez, Morocco
Omar El Beqqali, FSDM Fez, Morocco
Omar El harrouss, FSDM Fez, Morocco
Pereira Pedro, University NOVA of Lisbon, Portugal
Pradorn Sureephong, Chiang Mai Univesity, Thailand
Rachid Ben Abbou, FST Fez, Morocco
Rachid Saadane, EHTP Casablanca, Morocco
Rafik Lasri, FP Larache, Morocco
Richard Chbeir, University of Pay and Adour Countries, France
Said El Kafhali, FST Settat, Morocco

Said Idrissi, FP Safi, Morocco
Said Mazer, ENSA Fez, Morocco
Salim Hariri, University of Arizona, USA
Sebastian Kurtek, The Ohio State University, USA
Soulaiman El Hazzat, FSDM Fez, Morocco
Souinida Laaidi, FST ERRACHIDIA, Morocco
Sudipta Das Nityanandapur, Chandipur, India
Thierry Bouwmans, University of La Rochelle, France
Vikrant Bhateja, SRMGPC, Lucknow (Uttar Pradesh), India
Wolfgang Borutzky, Bonn Rhein-Sieg University, Germany
Yasser El Madani El Alami, ENSIAS Rabat, Morocco
Younes Bennani, Paris 13 University, France
Younes Balboul, ENSA Fez, Morocco
Younes Dhassi, FSDM Fez, Morocco
Youssef Ghanou, EST Meknes, Morocco
Youssouf El Allioui, FP Khouribga, Morocco
Zakaria Chalh, ENSA Fez, Morocco

PHD Student Committee

Mohammed Es-Sabry, FSDM Fez, Morocco
Ouissam Zealouk, FSDM Fez, Morocco
Mohamed Hamidi, FSDM Fez, Morocco
Ilham Addarrazi, FSDM Fez, Morocco

Preface

This book is a collection of high-quality peer-reviewed research papers presented at the First International Conference on Embedded Systems and Artificial Intelligence (ESAI'19), May 02–03, 2019, held at Faculty of Sciences, Dhar El Mahraz, Sidi Mohammed Ben Abdellah University, Fez, Morocco.

Artificial intelligence (AI) has driven a revolution, not only in the domain of computer science, communications and information technology but also in diverse engineering applications. With the growing power of AI—new computing platforms–embedded systems are necessary to support the emerging AI algorithms and applications from system level to circuit level. ESAI-2019 conference promoted the transfer of recent AI research into practical applications and implementations, i.e., developing intelligent ICT systems and pursuing artificial intelligence applications for contents, platforms, networks and devices. The goal of this conference was to bring together researchers and practitioners from academia and industry to focus on understanding recent research in the said domain and establishing new collaborations in these areas.

ESAI-2019 had received a number of submissions from the field of embedded systems, artificial intelligence, intelligent computing and its prospective applications in different spheres of engineering. The papers received have undergone a rigorous peer-review process with the help of the technical program committee members of the conference from various parts of country as well as abroad. The review process has been crucial with minimum two–three reviews each along with due checks on similarity and content overlap.

The book consists of 80+ quality papers segregated into two major modules of embedded system and artificial intelligence. There were seven parallel tracks (of oral presentation sessions chaired by esteemed professors from premier institutes of the country), and they are compiled in this volume for publication. The conference featured many distinguished keynote addresses by eminent speakers from various regions across the globe like: Prof. Abdelhakim Senhaji Hafid, University of Montreal, Canada; Dr. Vikrant Bhateja, Shri Ramswaroop Memorial Group of Professional Colleges (SRMGPC), Lucknow, Uttar Pradesh, India; Prof. Jamal Zbitou, University of Hassan 1st, Morocco; Prof. Mohammad A. M. Abushariah,

King Abdullah II School of Information Technology, The University of Jordan, Jordan; Prof. Suresh Chandra Satapathy, KIIT (Deemed to be University), Bhubaneshwar, Odisha, India; and Prof. Assal A. M. Alqudah, Nuance Communication Services Ireland Ltd., Dubai Branch, UAE. These keynote lectures/talks embraced a huge toll of audience including students, faculties, budding researchers as well as delegates.

The editors are thankful to Faculty of Sciences, Dhar El Mahraz, Sidi Mohammed Ben Abdellah University, Fez, Morocco, for their whole-hearted support in organizing the ESAI conference. The editorial board takes this opportunity to thank authors of all the submitted papers for their hard work, adherence to the deadlines and patience during the review process.

Lucknow, India Dr. Vikrant Bhateja
Bhubaneswar, India Dr. Suresh Chandra Satapathy
Fez, Morocco Dr. Hassan Satori

Contents

About the Editors

Vikrant Bhateja is Associate Professor, Department of Electronics & Communication Engineering, Shri Ramswaroop Memorial Group of Professional Colleges (SRMGPC), Lucknow, and also the Head (Academics & Quality Control) in the same college. He is doctorate in Biomedical Imaging & Signal Processing and has a total academic teaching experience of 16 years with around 125 publications in reputed international conferences, journals, and online book chapter contributions. His areas of research include digital image and video processing, computer vision, medical imaging, machine learning, pattern analysis and recognition. Dr. Vikrant has edited 15 proceeding books/editorial volumes with Springer Nature. He is Editor-in-Chief of IGI Global—International Journal of Natural Computing and Research (IJNCR). He is also Associate Editor in International Journal of Synthetic Emotions (IJSE) and International Journal of Ambient Computing and Intelligence (IJACI) under IGI Global Press. He is Guest Editor in Special Issues in reputed Scopus/SCIE indexed journals under Springer Nature: "Evolutionary Intelligence" and "Arabian Journal of Science and Engineering".

Suresh Chandra Satapathy holds a Ph.D. in Computer Science, currently working as Professor at KIIT (Deemed to be University), Bhubaneshwar, Odisha, India. He held the position of the National Chairman Div-V (Educational and Research) of Computer Society of India and is also a senior member of IEEE. He has been instrumental in organizing more than 20 international conferences in India as Organizing Chair and edited more than 30 book volumes from Springer LNCS, AISC, LNEE, and SIST Series as Corresponding Editor. He is quite active in research in the areas of Swarm Intelligence, Machine Learning, and Data Mining. He has developed a new optimization algorithm known as Social Group Optimization (SGO) published in Springer Journal. He has delivered a number of keynote address and tutorials in his areas of expertise in various events in India. He has more than 100 publications in reputed journals and conference proceedings. Dr. Suresh is in editorial board of IGI Global, Inderscience, Growing Science journals and also Guest Editor for Arabian Journal of Science and Engineering published by Springer.

Dr. Hassan Satori is Associate Professor, Department of Computer Science, Faculty of Sciences, Dhar El Mahraz, Sidi Mohamed Ben Abdellah University, Fez, Morocco. He holds Ph.D. in Speech Recognition and Signal Processing in 2009. He received M.Sc. in Physics from the Mohamed Premier University and Ph.D. in nanotechnology and computer simulation, also from the same university, in 2001. From 2001 to 2002, he was a postdoctoral fellow in physics, mathematical modeling, and computer simulation at the Chemnitz University of Technology, Germany, and was then a postdoctoral fellow in computer simulation in the Department of Interface Chemistry and Surface Engineering, Max-Planck-Institut, Düsseldorf, Germany, (2003–2005). Dr. Hassan has large academic teaching and research experience in the fields of the mathematical modeling, speech and signal processing machine learning. His research interests include areas like: nanotechnology, computer simulation, speech, signal processing, speech analysis, speech recognition, natural language processing, and artificial intelligence.

Embedded Computing and Applications

Real-Time Implementation of Artificial Neural Network in FPGA Platform

Mohamed Atibi, Mohamed Boussaa, Abdellatif Bennis and Issam Atouf

Abstract In this paper, we present the implementation of artificial neural networks in the FPGA embedded platform. The implementation is done by two different methods: a hardware implementation and a softcore implementation, in order to compare their performances and to choose the one that best approaches real-time systems and processes. For this, we have exploited the tools of this platform such as blocks Megafunctions and softcore NIOS II processor. The results obtained in terms of execution time have shown that the hardware implementation is much more efficient than that based on the NIOS II softcore.

Keywords Artificial neural networks · FPGA · Hardware implementation · Softcore implementation · Execution time

1 Introduction

Artificial intelligence is an area in which machines try to interpret the content of the information (image or signal) close to human capabilities [1, 2]. It is a very broad field of research, and its applications are very diverse and varied [3].

This is a field that aims to design intelligent systems that can evolve over time by interacting with their environments by using several disciplines such as computer science, mathematics, neuroscience, embedded systems, etc.

One of the widely used algorithms in this field is artificial neural networks, which offer satisfactory solutions for a wide variety of real-world applications, in various fields such as industrial engineering, robotics and automotive.

M. Atibi (✉) · M. Boussaa · A. Bennis · I. Atouf
LTI Lab. Faculty of Science Ben Msik, Hassan II University of Casablanca,
El Harti B.P 7955, Sidi Othmane, Casablanca, Morocco
e-mail: atibi.simo@gmail.com

M. Atibi
Pluridisciplinary Laboratory of Research and Innovation (LPRI), EMSI Casablanca,
Casablanca, Morocco

V. Bhateja et al. (eds.), *Embedded Systems and Artificial Intelligence*,
Advances in Intelligent Systems and Computing 1076,
https://doi.org/10.1007/978-981-15-0947-6_1

An artificial neural network is a network whose neurons are interconnected according to an architecture, and the links are configured through a learning process for a specific application, as pattern recognition or classification. However, as in all biological nervous systems, this learning process requires the adjustment of the connections (synaptic weights) between the neurons [4]. The importance of these decision systems is increasing because data are available in several fields of application [5]. Artificial neural networks can be used in various applications such as smart cars [6, 7], medicine [8–10] and robotic vision.

However, the properties of the artificial neural network require a powerful hardware implementation circuit, hence the use of a platform capable of implementing this ANN and using parallelism in processing and reconfiguration. Thus, several researchers have tried to implement this network in embedded platforms, such as graphics processors [11] and application-specific integrated circuits ASIC [12]. But, these systems have major drawbacks related to application areas, such as the speed of calculation which is a paramount parameter in the field of road safety.

To overcome these difficulties, researchers have tried to implement artificial neural networks in the programmable gate arrays FPGA. Field-programmable gate array FPGA or programmable logic networks are fully reconfigurable components, which allow them to be reprogrammed at will to accelerate particular phases of calculations [13]. The advantage of using this kind of circuit is its great flexibility that allows reuse in different algorithms in a very short time. These benefits are suitable for calculating ANN.

Several implementations have been proposed in the literature, but it has always been found that there is a compromise to be made between the hardware implementation of the neural network and their computational efficiency (implementation of the activation function without the use of approximations) [14].

For this, the document will be dedicated to the implementation of artificial neural networks in the FPGA platform of Cyclone II Altera, according to two types of implementations. The first is a hardware implementation, and the second is a softcore implementation based on a NIOS II integrated processor. The document will present the comparative results of these two implementations, in order to show the advantages of each implementation in front of the other, in terms of speed and in terms of material resources consumed.

2 ANN et FPGA

2.1 ANN

Artificial neural networks are networks that model the architecture of natural neural networks. They consist of two essential parts, the neuron model (base unit) and the connection model between the neurons [15, 16].

There are several neuron models, the most used in the literature are based on the model developed by McCulloch and Pitts. This model receives two vectors at its input; a parameter vector dedicated to the application made and the other connection weight vector called synaptic weight, so that it performs a weighted summation between the two vectors (1). The result of this operation is transferred to the output through a transfer function (activation function)[8]. The values of these synaptic weights are already set during the learning phase.

$$V = \sum_{i=0}^{n} W_i * X_i \quad (1)$$

With :
$X_1 X_n$: the object vector of the neuron.
$W_1 W_n$: the synaptic weights contained in the neuron.
$\sum$: a function that calculates the sum of the multiplications between the object vector and the synaptic weights according to equation (1).
The formal neuron model uses the sigmoid function as an activation function, which has several advantages over other functions such as threshold function and Gaussian. Among its advantages are the differentiability and continuity of the function.
The topology of an artificial neural network is defined by the choice of architecture and the choice of synaptic weight adjustment algorithm.
Most artificial neural networks use the multilayer perceptron architecture, and this architecture uses in addition to the input layer and the output layer an intermediate layer called hidden layer (one or more). The architecture of the network is described by the number of layers and the number of neurons in each layer.
Most multilayer perceptron networks use an appropriate algorithm for adjusting synaptic weights. Among these algorithms, the most used is the error propagation algorithm (Back pro) that simulates the phenomenon of learning by error. The goal of this algorithm is to minimize the overall error by using the gradient descent method.

2.2 FPGA

The implementation of artificial neural networks represents a task that requires powerful and efficient embedded platforms in terms of computing power, speed and abundance of hardware resources.
In the literature, several integrated circuits can give solutions to these difficulties, such as ASIC, ISP and DSP circuits. But, the FPGA platform remains the most suitable platform for this type of system. This is due to the advantages offered by this circuit, such as programming flexibility and prototyping of complex functions.
FPGAs are integrated circuits that offer inexpensive, easy and flexible solutions [17]. The FPGA is used for the implementation and design of complex circuits and complete digital systems on a single chip [18].

The FPGA offers several advantages of parallel processing [19], speed and flexibility. This will be useful in signal and image processing applications, as in the case of driver assistance systems that require speed and parallelism.
The FPGA consists of three basic blocks that are configurable logic blocks, input–output blocks and interconnects.
The implementation of an algorithm in the FPGA platform is done in two methods:

– Hardware Implementation.

The hardware implementation is based on a hardware description language VHDL Very High Speed Hardware Integrated Circuit Description language or Verilog; this language consists of programming the application physically. These languages are based on the description and programming of the electronic behavior of a given system. This type of design requires a compiler and a simulator, for Cyclone II, Quartus II software acts as a compiler and simulator.

– Softcore Implementation.

The current development, complexity and reliability of FPGA chips have given birth to discipline very close to system on chip (SOC) called system on programmable chip (SOPC). The SOPC consists in designing a system on chip either from existing intellectual property (IP) or from a description made from A to Z in HDL language generally with the VHDL or Verilog standards and to implement them on FPGA.
The basis of any modern digital system is the use of one or more microprocessors. In the field of SOPCs, these microprocessors are descriptions in HDL language (VHDL or Verilog) or compilations of these descriptions called “Netlists” hence the name “SOFTCORE” [20].
These “softcore” microprocessors have the advantage of being more flexible compared to “hardcore” microprocessors that are in the form of integrated circuits.
NIOS II is a 32-bit softcore microprocessor developed by Altera. NIOS was introduced in 2001 to be the first processors marketed and used in the industry. Its creation was specifically for the design of embedded systems in FPGAs. Its compact and advantageous architecture makes it possible to consume fewer FPGA hardware resources comparable to 8-bit microprocessors. It is specially designed for Spartan-3, Virtex and cyclone [21, 22].

3 Implementation of Artificial Neural Networks

This section describes in detail the different modules needed for the hardware and softcore implementation of the ANN.

3.1 Hardware Implementation

Among the problems of implementing the ANN in an FPGA platform with the VHDL language is the non-synthesis of real numbers. In order to work around this problem, we had to design an architecture that manipulates this data into floating-point representations. This has allowed great efficiency in terms of calculation accuracy.

For this, we used blocks named Megafunctions which are blocks offered by the manufacturers of the FPGA, these blocks are written in VHDL in order to implement complex arithmetic operations with a floating-point representation (32 or 64 bits), and these blocks are useful to guarantee the precision of calculation of the ANN.

The implementation of the ANN-type MLP requires a set of formal neurons interconnected with each other according to a given architecture. This number of neurons increases progressively with the complexity of the task to be performed. This has an impact on the material resources needed.

To remedy this difficulty, our implementation is based on the exploitation of only one formal neuron, which will play all the neurons of the architecture.

– Implementation of the formal neuron.

The formal neuron is constituted of the Megafunctions of multiplication, addition, exponential and division. The following diagram (Fig. 1) illustrates the different blocks of this formal neuron:

The implementation is divided into two parts:

– The first part concerns the implementation of internal activation.
– The second part concerns the implementation of the sigmoid activation function.

a. Implmentation of internal activation: The formal neuron receives as input an object vector $X = (X_1, X_2, \ldots, X_n)$ and a synaptic weight vector $W = (W_1, W_2, \ldots, W_n)$ which represent the connection between the neuron and its ith input .

The work of the neuron consists in first calculating the weighted sum of its inputs. The output of the sum is called the internal activation of the neuron (1).

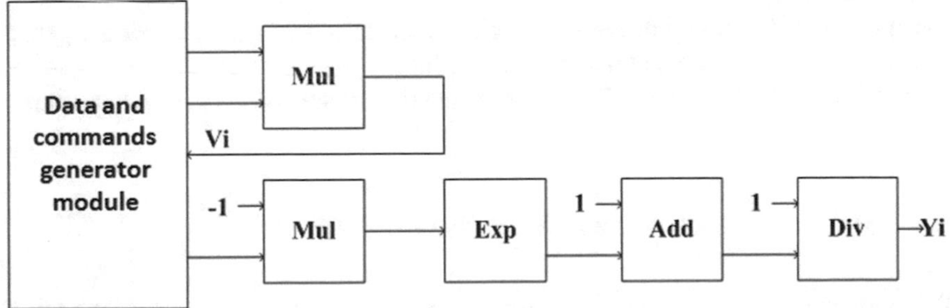

Fig. 1 Synoptic diagram of the computational chain of the formal neuron

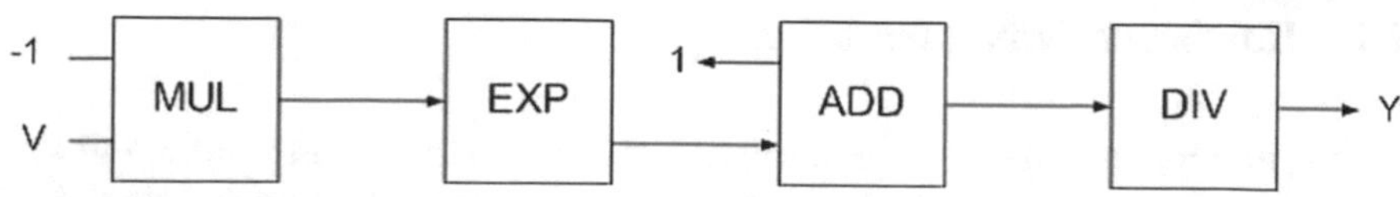

Fig. 2 Hardware implementation of the sigmoid function

A module for generating data and commands sends the Xi and Wi data to the multiplication block Megafunction, and the calculated product will be cumulated and stored in the internal memory of this module.
multiplication Megafunction block:
The multiplication block used is a Megafunction block that implements the functions of the multipliers [15]. It follows the IEEE-754 standard for representations of single-precision floating-point numbers, double precision and extended precision. In addition, it allows the representation of special values like zero and infinity.

b. The sigmoid function implementation: The second block is a transfer function called activation function. It makes it possible to limit the exit of the neuron in an interval [0,1]. The most used function is sigmoid.
The implementation of this function (Fig. 2) requires a number of complex operations such as division and exponential. It requires the use of the Megafunctions of exponential and division, and these blocks Megafunctions have the same capacities of calculation of the multiplication.

– ANN implementation.

The implementation of the ANN consists of implementing a single formal neuron controlled by the data and command generator module, in order to perform the function of the ANN.
This neuron calculates the output Y11 of the first neuron of the first layer and stores it in the internal memory of the data and command generator module. Then, the same neuron calculates the Y12 output of the second neuron of the first layer and stores it in the same memory. This process is repeated until the last neuron of the output layer is calculated.
During this calculation process, the synchronization and data commands are generated by the data and command generator module.
The following RTL scheme (Fig. 3) illustrates the module of the ANN algorithm:

3.2 *Softcore Implementation*

The implementation of the MLP network in the FPGA platform based on the NIOS II can improve efficiency and reduce design complexity on the FPGA platform; this implementation requires two essential steps:

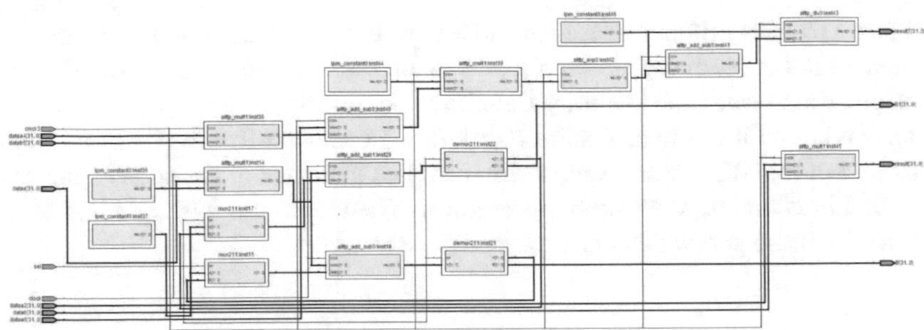

Fig. 3 RTL diagram of the artificial neural network

Material part This part is devoted to the implementation of the on chip system based on the NIOS II microprocessor; the hardware implementation of NIOS II requires two development tools QUARTUS and SOPC Builder.

Altera's Quartus II design software provides a comprehensive, multiplatform design environment that easily adapts to specific design needs. It is a complete environment for the design of SOPCs. Quartus II software includes solutions for all design phases in FPGA and CPLD.

SOPC is a new concept and approach proposed by Altera Corporation. It has features like EDA, SOC, DSP and IP combination programmable logic elements that help with flexible design, ability to adapt the application, extension, scalability and software and hardware programming for system development. The evolution of SOPC design technology is the product of several factors: modern computer-aided design technology, EDA technology and the huge development of large-scale integrated circuit technology.

As shown in the figure, the SOPC Builder simulates and then generates all the necessary hardware for a given application such as a NIOS II microprocessor, memories (ROM, RAM, SDRAM, DRAM and FLASH), LED PIO, LCD and an Avalon bus (Fig. 4). After complete compilation and generation, the SOPC Builder loads the result as a single module (Fig. 5) that adds to the FPGA component library. This module (NIOS II and its peripherals) is the equivalent of a simple microcontroller that has its own set of instructions and can be programmed with advanced languages such as C language.

Software part The module's programming tool is the NIOS IDE, which is based on Eclipse, to which Altera added plug-ins and obviously a compiler and assembler for the NIOS II processor.

The NIOS IDE allows programming with three languages that are C, C++ and assembler.

The NIOS II EDS provides two separate development streams and includes many proprietary and open-source tools for creating NIOS II programs. This platform is suitable for the implementation of entire project software, since all developments, debugging and state diagrams can be performed in a single window [22]. Eclipse

IDE also provides different software models or error messages and functions as a terminal when executing code on a NIOS II processor. The terminal displays any message printed using the C library functions like printf ().

After having built the final system (hardware part and software part), the implementation of the MLP or any application requires the entry of the code of the MLP described in C language in the Eclipse editor. Then, compile this code and load it into the hardware part which takes care of the execution.

3.3 Result and Discussion

In this section, the two hardware and softcore implementations of artificial neural networks will be simulated and tested in the same DE2_70 FPGA card. This section will be devoted to taking the execution time and hardware resources used by the two implementations in order to compare their performance.

Regarding hardware resources, Table 1 shows that the two implementations have undeniable advantages in terms of hardware resources, and these two architectures remain optimized architectures that do not consume a lot of hardware resources.

With regard to execution time, which is a key parameter in road safety and robotics applications.

The hardware implementation has undeniable advantages in terms of speed of calculation, which reaches 0.0000001 s, and this is due to the use of Megafunctions blocks.

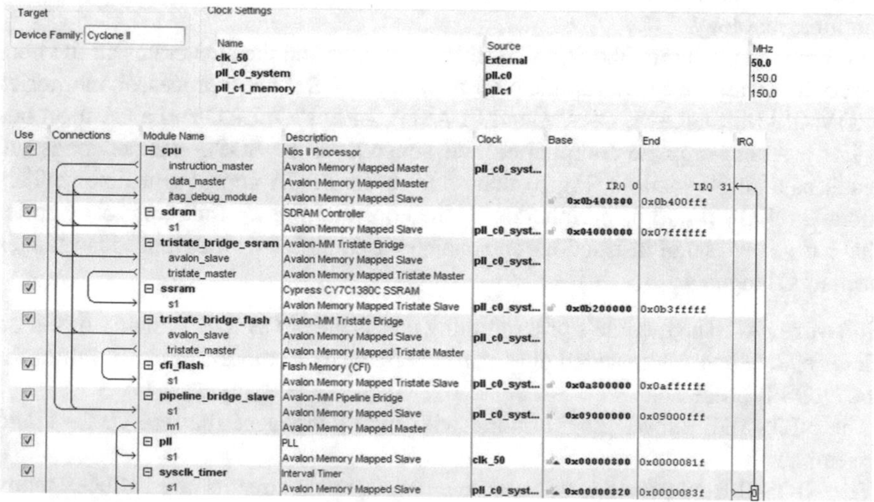

Fig. 4 Global system generated by the SOPC Builder

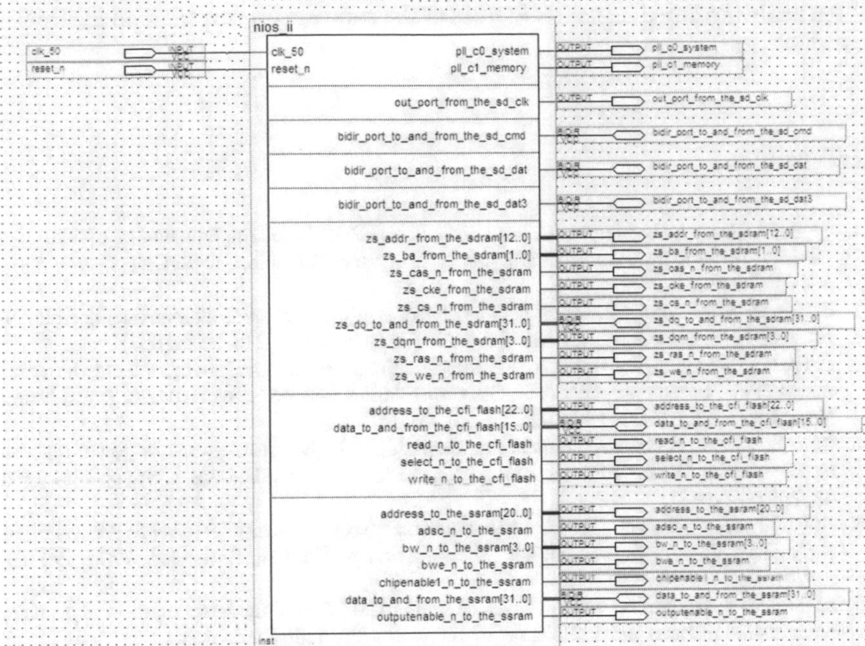

Fig. 5 Microcontroller generated

Table 1 Comparison between the hardware resources used by the two implementations

	Hardware implementation	NIOS II implementation
Number of element	5321 (8%)	5214 (8%)
Quantity of bit memory	69,568 (6%)	69,568 (6%)

On the other hand, the NIOS II-based implementation has consumed up to 0.003 s, due to the implementation of the peripherals necessary for the NIOS II to work.

In summary, the hardware architecture has remarkable advantages in terms of speed of computation compared to the software architecture (NIOS II) which requires a time 10^4 times greater.

4 Conclusion

The aim of this document was first to implement artificial neural networks in the FPGA DE2_70 platform by two different methods (hardware and software) and secondly to compare the performances of these two implementations in terms of execution time and material resources used. This comparison highlighted the interest of the hardware implementation in front of the software implementation in terms of execution time.

References

1. Mellit, A., Mekki, H., Messai, A., Kalogirou, S.A.: FPGA-based implementation of intelligent predictor for global solar irradiation, part I: theory and simulation. Expert. Syst Appl. **38**(3), 2668–2685 (2011)
2. Savich, A.W., Moussa, M., Areibi, S.: The impact of arithmetic representation on implementing MLP-BP on FPGAs: a study. IEEE Trans. Neural Netw. **18**(1), 240–252 (2007)
3. Shah, N., Chaudhari, P., Varghese, K.: Runtime programmable and memory bandwidth optimized FPGA-based coprocessor for deep convolutional neural network. IEEE Trans. Neural Netw. Learn. Syst. **99**, 1–13 (2018)
4. Sun, Y., Cheng, A.C.: Machine learning on-a-chip: a high-performance low-power reusable neuron architecture for artificial neural networks in ECG classifications. Comput. Biol. Med. **42**, 751757 (2012)
5. Roy Chowdhury, S., Saha, H.: Development of a FPGA based fuzzy neural network system for early diagnosis of critical health condition of a patient. Comput. Biol. Med. **40**(2), 190–200 (2010)
6. Mohamed, A., Issam, A., Mohamed, B., Abdellatif, B.: Real-time detection of vehicles using the haar-like features and artificial neuron networks. Procedia Comput. Sci. **73**, 24–31 (2015)
7. Atibi, M., Atouf, I., Boussaa, M., Bennis, A.: Comparison between the MFCC and DWT applied to the roadway classification. In: 2016 7th International Conference on Computer Science and Information Technology (CSIT), pp. 1–6. IEEE (2016)
8. Mohamed, B., Issam, A., Mohamed, A., Abdellatif, B.: ECG image classification in real time based on the haar-like features and artificial neural networks. Procedia Comput. Sci. **73**, 32–39 (2015)
9. Boussaa, M., Atouf, I., Atibi, M., Bennis, A.: Comparison of MFCC and DWT features extractors applied to PCG classification. In: 2016 11th International Conference on Intelligent Systems: Theories and Applications (SITA), pp. 1–5. IEEE (2016)
10. Boussaa, M., Atouf, I., Atibi, M., Bennis, A.: ECG signals classification using MFCC coefficients and ANN classifier. In: 2016 International Conference on Electrical and Information Technologies (ICEIT), pp. 480–484. IEEE (2016)
11. Yadan, O., Adams, K., Taigman, Y., Ranzato, M.A.: Multi-gpu training of convnets. arXiv preprint arXiv:1312.5853 (2013)
12. Chen, Y.H., Krishna, T., Emer, J.S., Sze, V.: Eyeriss: an energy-efficient reconfigurable accelerator for deep convolutional neural networks. IEEE J. Solid-State Circuits **52**(1), 127–138 (2017)
13. Zhang, D., Wang, Q., Xu, C.: Data acquisition system of a self-balancing robot based on FPGA. J. Theor. Appl. Inf. Technol. **47**(2) (2013)
14. Lin, C.J., Tsai, H.M.: FPGA implementation of a wavelet neural network with particle swarm optimization learning. Math. Comput. Model. **47**(9), 982–996 (2008)
15. Atibi, M., Bennis, A., Boussaa, M.: Precise calculation unit based on a hardware implementation of a formal neuron in a FPGA platform. Int. J. Adv. Eng. Technol. **7**(3) (2014)
16. Boussaa, M., Bennis, A., Atibi M.: Comparison between two hardware implementations of a formal neuron on FPGA platform. Int. J. Innov. Technol. Explor. Eng. **4**(1) (2014)
17. Alin, M., Pehlivan, I., Koyuncu, I.: Hardware design and implementation of a novel ANN-based chaotic generator in FPGA. Optik (Stuttg) **127**(13), 5500–5505 (2016)
18. Mohammed, H.K., Ali, E.Z.: Hardware implementation of artificial neural network using field programmable gate array. Int. J. Comput. Theory Eng. **5**(5), 780 (2013)

19. Atibi, M., Atouf, I., Boussaa, M., Bennis, A.: Parallel and mixed hardware implementation of artificial neuron network on the FPGA platform. Int. J. Eng. Technol. **6**(5), 0975–4024 (2014)
20. http://www.altera.com
21. Pokale, M.S.M., Kulkarni, M.K., Rode, S.V.: NIOS II processor implementation in FPGA: an application of data logging system. Int. J. Sci. Technol. Res. **1**(11) (2012)
22. Mukadam, M.S.B., Titarmare, A.S.: Design of power optimization using C2H hardware accelerator and NIOS II processor (2014)

Design of a 0.18 μm CMOS Inductor-Based Bandpass Microwave Active Filter

Mariem Jarjar and Nabih Pr. El Ouazzani

Abstract This work proposes the use of an active inductor to construct an RF bandpass filter. The main characteristic of the circuit is the incorporation of active inductors that rely on the gyrator-C topology according to the 0.18 μm CMOS TSMC technology. Such a new configuration enables to build an OTA circuit using simple current mirrors. On the basis of the frequency transformation method, a C-coupled bandpass filter is designed. Simulation results, obtained by means of the PSPICE software, show good performances in terms of the scattering parameters and the noise figure demonstrating the effectiveness of the new topology.

Keywords Bandpass filter · Active inductor · OTA · Current mirror · 0.18 μm CMOS · J-inverter · Chebyshev response

1 Introduction

Active microwave bandpass filters are widely used in wireless communication systems covering several aspects that deal with data transmission and signal processing [1].

As well known, important properties are offered by active filters such as small occupied area, low insertion losses and sharp rejection [2]. Several configurations have been proposed and also much work has been done in order to bring more improvements related to the key parameters [3, 2].

The major task consists of overcoming the issue of losses and the complexity of designing inductances of low values as well. As an alternative solution, a large number of active inductors' topologies have been developed and have become increasingly suitable for many microwave applications leading to high quality performances [4]. Much effort has been made to optimize size and power consumption along with the enhancement of frequency responses.

M. Jarjar (✉) · N. Pr. El Ouazzani
Faculty of Sciences and Technology (FST), Signals, Systems and Components Laboratory (LSSC), Fez, Morocco
e-mail: jarjarmariem@gmail.com

V. Bhateja et al. (eds.), *Embedded Systems and Artificial Intelligence*, Advances in Intelligent Systems and Computing 1076,
https://doi.org/10.1007/978-981-15-0947-6_2

Hence, this paper focuses on designing a third-order microwave Chebyshev bandpass filter using active inductors. The synthesis of the proposed active inductors is based on the gyrator-C topology involving two operational transconductance amplifiers (OTA). The novelty comes from the fact that those circuits are made from CMOS simple current mirrors under the 0.18 μm technology.

Thanks to the frequency transformation, the proposed filter's elements are derived taking into account the insertion of J-inverters between different parallel LC resonators [5]. The whole circuit presents a C-coupled structure that helps reduce the number of inductors.

The remainder of this paper is organized as follows. The synthesis of a grounded CMOS active inductor is presented in Sect. 2. Section 3 is dedicated to the proposed process of designing the active bandpass filter with Chebyshev response. Simulation results, through the PSPICE software, are given and analyzed in Sect. 4. Section 5 concludes the work.

2 Synthesis of a Grounded CMOS Active Inductor

There are several configurations of the OTA circuits depending on many parameters and constraints. The proposed OTA is based on the use of a simple CMOS current mirror (M1, M2) connected to a differential pair (M3, M4) [6]. Another current mirror (M5, M6) stands for the current source and allows the biasing of the whole circuit. We also insert a symmetrical power supply to reduce the initial voltage of 5 V to the minimum value of 1.8 V according to the 0.18 μm technology.

The following figure shows the OTA circuit (Fig. 1):

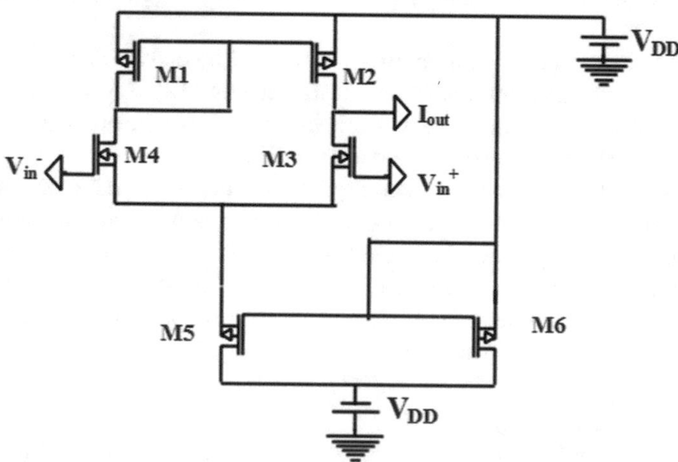

Fig. 1 Proposed current mirror-based OTA circuit

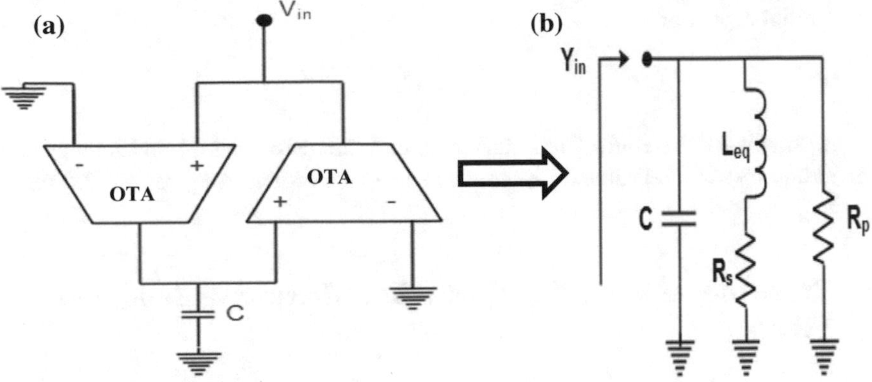

Fig. 2 **a** Active inductor structure, **b** Equivalent resonant circuit

Figure 2 presents the proposed active inductor structure depending on the gyrator-C topology. It consists of two OTA circuits that are connected back to back in a negative feedback with a grounded capacitor which is represented by the grid–source capacitance C_{GS} in the real structure.

Considering the small signal assumption, the input admittance Y_{in} of the active inductor is given by (1); it obviously demonstrates that the inductor is equivalent to an RLC resonator

$$Y_{\text{in}} = \frac{I_{\text{in}}}{V_{\text{in}}} = G_m + \frac{G_m^2}{j\omega C_{\text{gs}} + G_{\text{ds}}} + j\omega C_{\text{gs}} \tag{1}$$

where C_{gs} and C_{ds} are the grid–source and the drain–source capacitances, respectively. G_m is the transconductance of the OTA.

From Eq. (1), we can identify different elements of the equivalent resonant circuit, shown in Fig. 2b [6], that are:

Inductance

$$L_{\text{eq}} = \frac{C_{\text{gs}}}{G_m^2} \tag{2}$$

Series resistance

$$R_s = \frac{G_{\text{ds}}}{G_m^2} \tag{3}$$

Parallel resistance

$$R_p = \frac{1}{G_m} \tag{4}$$

Parallel capacitor

$$C_p = C_{gs} \tag{5}$$

L_{eq} stands for the equivalent inductance, and the resistances R_s and R_p represent the circuit losses. The value of the active inductance proposed is around 47.8 pF.

3 Procedure of Designing Chebyshev Microwave Bandpass Filters

The developed bandpass filter consists mainly in connecting parallel and series LC resonators according to the frequency transformation technique. In order to have lower insertion losses, we must reduce the inductors' number involved in the circuit [7]. As an alternative, admittance J-inverters converts load admittance to its inverse according to Eq. (4). Consequently, they are inserted between shunt parallel LC resonators as shown in Fig. 3:

$$Y' = \frac{J^2}{Y_L} \tag{7}$$

Amongst several available J-inverters, we have chosen those based on capacitor pi-networks and thus, the proposed bandpass filter presents only shunt LC resonators separated by capacitors. As a result, we obtain the simplified topology shown in Fig. 4.

The design procedure is carried out according to the following steps. First, the center frequency f_0 and the bandwidth BW of the filter are defined. Second, the normalized parameters $g_{(i)}$ are obtained from the Chebyshev prototype depending on the chosen order n. Finally, the algorithm made of (8), (9), (10), (11), (12), (13) and (14), is applied for a predefined value of L. Thus, different capacitors of the C-coupled structure are known [7].

$$C_{(n,n+1)} = \frac{J(n)}{2\pi f_0 \sqrt{(1 - (J(n) \times Z_0)^2}} \tag{8}$$

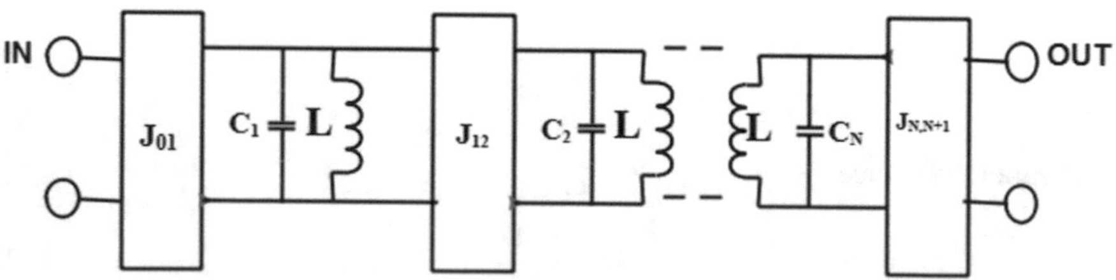

Fig. 3 Bandpass filter structure

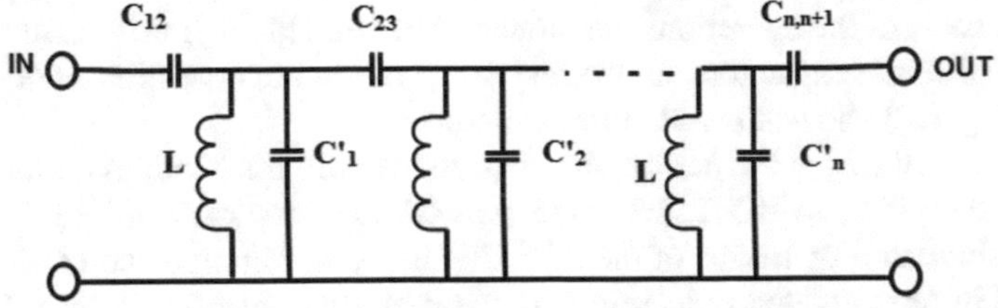

Fig. 4 Final structure

$$C'_{(n)} = C_0 - C_{(n,n+1)} - C_{(n+1,n+2)} \tag{9}$$

where

$$J(1) = \sqrt{\frac{2\pi f_0 \times C_0 \times \Delta}{Z_0 \times 1 \times g_{(1)}}} \tag{10}$$

$$J(n) = \sqrt{\frac{2\pi f_0 \times C_0 \times \Delta}{Z_0 \times g_{(n-1)} \times g_{(n)}}} \tag{11}$$

For $1 < i < n$

$$J(i) = \frac{2\pi f_0 \times C_0 \times \Delta}{\sqrt{g_{(i-1)} \times g_{(i)}}} \tag{12}$$

where

$$\Delta = \frac{BW}{f_0} \tag{13}$$

And

$$C_0 = \frac{1}{(2\pi f_0)^2 L} \tag{14}$$

where Z_0 is the characteristic impedance which is equal to 50 Ω.

4 Example of a Third-Order Chebyshev Bandpass Filter

In this section, we present a third-order 1-dB ripple Chebyshev bandpass filter in which active inductors are incorporated to stand for the chosen value of $L_{eq} = L = 47.8$ pH. The targeted response has the following parameters:

BW = 7 MHz; $f_0 = 1.13$ GHz.

On the basis of the operations mentioned above and taking into account the association of different capacitors, all the necessary values are determined and shown in Table 1. Figure 5 shows the final filter structure.

In order to illustrate the performances of the circuit, the bandpass filter has been designed according to the TSMC 0.18 μm CMOS process with ±1.8 V supply and then simulated by means of the PSPICE software. Figure 6 shows high quality responses in terms of the reflection parameter and the insertion losses. Indeed, at the center frequency $f_0 = 1.13$ GHz the reflection coefficient S_{11} presents a value of −43 dB and S_{21} equals 0 dB.

Moreover, in the context of frequency performances, we propose a noise figure (NF) analysis as shown in Fig. 7 from which we obviously notice that the NF is less than 5 dB in the bandwidth.

The following table summarizes some major parameters which illustrate the good quality of the proposed filter (Table 2).

Table 1 Capacitors' values

C12 = C45 (pF)	C23 (pF)	C34 (pF)	C′1 = C′3 (pF)	C′2 (pF)
11.3	1.91	1.91	425	432.72

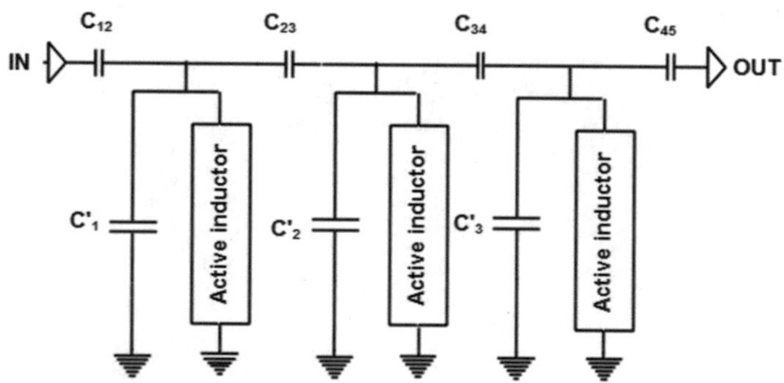

Fig. 5 Proposed bandpass filter

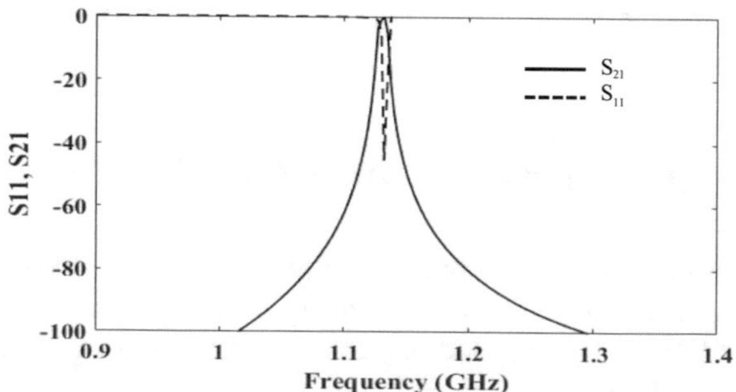

Fig. 6 S_{11} and S_{21} responses

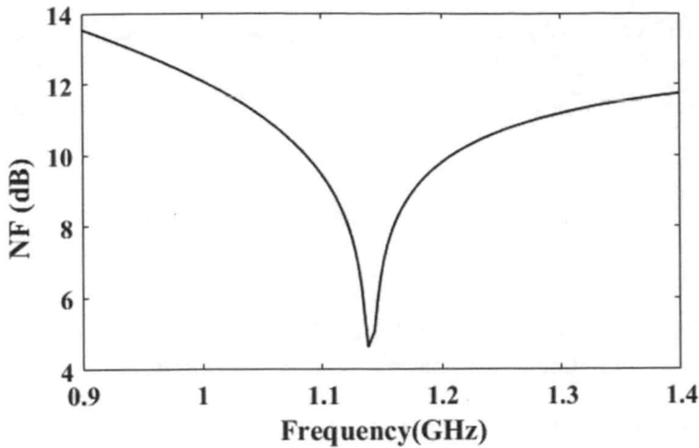

Fig. 7 Noise figure response

Table 2 Filter parameters

BW at −3 dB (MHZ)	BW at −15 dB (MHZ)	NF at f_0 (dB)	NF at f_{C1} (dB)	NF at f_{C2}(dB)
7.3	11.3	4.3	4.8	5.1

Table 3 Comparison of the present results with the previous works

Reference	NF (dB)	Central frequency (GHz)
[8]	15	3.6
[9]	18	1.92
[10]	4.8	2
This work	4.3	1.13

With regard to the most important performances, Table 3 shows a comparison between several results including the proposed ones according to the same technology (CMOS 0.18 μm). It is obvious that our contribution produces better values in terms of noise figure.

5 Conclusion

An OTA based on a simple current mirror and made under the CMOS 0.18 μm TSMC technology has been used as the main block of an active inductor with respect to the gyrator-C topology. The novelty of this work comes from the use of 0.18 μm CMOS components as fundamental elements in the design.

As an application we have been interested in the synthesis of an active bandpass microwave filter. Indeed, the proposed Chebyshev third-order bandpass filter is designed using those active inductors according to the C-coupled structure. The simulation process of the whole circuit, through the PSPICE software, provides excellent responses of the active filter in complete agreement with the desired performances.

References

1. Sassi, Z., Darfeuille, S., Barelaud, B., Billonnet, L., Jarry, B., Marie, H., Le, N.T.L., Gamand, P.: 2 GHz Tuneable Integrated Differential Active Bandpass Filter on Silicon, pp. 90–93. GeMiC (2005)
2. Sabaghi, M., Rahnama, M., Lahiji, M.N., Miri, M.S., Rezakhani, Sh.: Design and simulation of differential active inductor with 0.18 μm CMOS technology. Can. J. Electr. Electron. Eng. **2**(9) (2011)
3. Córdova, D., Cruz, J.D.L., Silva, C.: A 2.3-GHz CMOS high-Q bandpass filter design using an active inductor. In: XV Workshop Iberchip, Buenos Aires, pp. 25–27 Argentina, de Marzo de (2009)
4. Manjula, J., Malarvizhi, S.: Performance analysis of active inductor based tunable band pass filter for multiband RF front end. Int. J. Eng. Technol. (IJET) **5**(3) (2013, June–July)
5. Filter design by the insertion loss method 389
6. Jarjar, M., El Ouazzani, N.: Synthesis of a microwave CMOS active inductor using simple current mirror-based operational Transconductance Amplifier (OTA). Int. J. Sci. Eng. Res. **7**(8) (2016, August). ISSN 2229–5518
7. David M.P.: Microwave Engineering, 4th ed. TK7876.P69 (2011)
8. Dinh, A., Ge, J.: A Q-enhanced 3.6 GHz, tunable, sixth-order bandpass filter using 0.18 μm CMOS. Research Article ID 84650 (2007)
9. Gao, Z., Ma, J., Yu, M., Ye, Y.: Fully integrated CMOS active bandpass filter for multiband rf front-ends. IEEE Trans. Circuits Syst. II Express Briefs **55**(8), 718–722 (2008)
10. Weng, R.M., Kuo, R.C.: An ω0-Q tunable CMOS active inductor for RF bandpass filters. Int. Symp. Signals Syst. Electron. 571–574 (2007)

Improvement of the Authentication on In-Vehicle Controller Area Networks

Asmae Zniti and Nabih El Ouazzani

Abstract This paper deals with the issue of the Controller Area Network (CAN) bus security in terms of node authentication and preventing attacks. Several techniques have been developed to overcome this problem, but unfortunately, there are still some threats to be addressed. The proposed concept is based upon the security centralized authentication system involving a new approach related to the process of generation and receiving data frames and digests along with falsified message monitoring in real time.

Keywords Security · Automobile · Attack · Authentication · Controller Area Network

1 Introduction

Nowadays, CAN bus security is a major problem in automotive manufacturing industries. Vehicles are increasingly vulnerable and subject to a wide range of attacks locally as well as remotely [1, 2]. The current technological tendency towards cars, overwhelmingly connected, makes it even worse. The main issue lies on the fact that various messages can be sent over the bus network by either a malicious node or compromised calculators, and subsequently are regarded as legitimate by the other nodes. Much work has been achieved aiming to improve CAN security by means of encryption and message authentication in a context where the risk of increasing system response time is present due to the constraints and limitations of computing power and storage capacity of the electronic control units (ECUs).

A. Zniti (✉) · N. El Ouazzani
Faculty of Sciences and Technology, Signals, Systems and Components Laboratory (LSSC), Fez, Morocco
e-mail: znitiasmae@gmail.com

N. El Ouazzani
e-mail: elouazzani.nabih@gmail.com

V. Bhateja et al. (eds.), *Embedded Systems and Artificial Intelligence*, Advances in Intelligent Systems and Computing 1076,
https://doi.org/10.1007/978-981-15-0947-6_3

Over the past years, several techniques that have been developed in order to make the CAN bus capable of identifying messages from a trusted node and those from a malicious one. Amongst these techniques we have:

- CANAuth [3]: based on the idea of sending authentication data via an out-of-band channel using the CAN+ protocol, allowing additional bits to be inserted between the sampling points of a CAN bus interface without disturbing the smooth operation of the CAN bus controllers.
- LIBRACAN [4]: an authentication mechanism, in which only a few bits of the message authentication code (MAC) are added in each packet to each verifier, and each part of the MAC, can be verified by more than one receiver.
- Centralized Authentication System [5]: includes the following two phases: Node authentication and key delivery phase, falsified message monitoring phase and overwriting the message with a real-time error frame.
- Lightweight Authentication Framework [6]: enables a secure and efficient distribution of cryptographic keys between ECUs without pre-shared secrets. This system combines symmetric and asymmetric cryptographic methods to implement two-phase authentication of ECUs and message flows.
- LEIA [7]: is designed to run under the stringent time and bandwidth constraints of automotive applications and is compatible with the existing vehicle infrastructure. The protocol is suitable for implementation using lightweight cryptographic primitives yet providing appropriate security levels by limiting the usage of every key in the system.
- Vatican [8]: allows recipients of a message to verify its authenticity via HMACs (keyed hash message authentication code), while not changing CAN messages for legacy, non-critical components.
- Vulcan [9]: involves the integrity and authenticity of critical messages circulating on the unreliable vehicular network, as well as during processing by isolated software components on participating calculators.

The comparison between the techniques that have been developed in order to secure the CAN bus shows that some methods are of more interest than others, in terms of security and acceptable cost. The analysis of the protection tools used by each method against several attack scenarios suggests that the centralized authentication system [5] is the least expensive technique able to authenticate message streams taking into account different threats with minimal overheads. In addition, it does not require a significant change in hardware; the only additional component is the monitoring node.

This paper proposes a relevant approach which considers different attack scenarios and makes it simple, convenient, efficient and lightweight, so that its implementation within CAN automotive networks will be as easy as effective. In line with the overall objective of the above techniques and considering various attack scenarios, we are strongly interested in the centralized authentication system. Weaknesses will be highlighted and improvements will be included accordingly.

The remainder of this paper is organized as follows. Section 2 is dedicated to the proposed authentication method with the emphasis on the proprieties of the security monitoring system and the improvements we have brought in. The hash algorithm used to calculate digests is described in Sect. 3. Finally, the work is concluded in Sect. 4.

2 Development of an Authentication Method

2.1 Principles

The proposed security system involves a monitoring node including a particular CAN controller in charge of authenticating each calculator and verifying the MAC assigned to the messages.

The security monitoring system is implemented according to the following two phases [5]:

- Phase 1: (1) Node authentication and key delivery
  We use a program code as pre-shared information.

$$\text{AUTHKEY } I = \text{hash function(MSG} \mid\mid \text{NONCE)} \tag{1}$$

where:

AUTHKEY I the authentication code of the transmission node I.
MSG the program code of the transmission node I.
NONCE a random seed

- Phase 2: (2) Falsified message monitoring
  In the security monitoring system, thc remaining part of AUTHKEY described above is used as a MAC generation/verification key. As a result, a 128-bit MAC generation key has been mounted.

$$\text{MAC}_i = \text{hash function } (\text{ID}_i,\ \text{D}_i,\ \text{FC}_i,\ \text{AUTHKEY } I) \tag{2}$$

where:

Id_i CAN-ID.
D_i message i data.
FC_i Complete monotones counter for the message i.
AUTHKEY I the encryption key for the transmission node I.

2.2 Current Technique

According to the centralized authentication system [5], the length of the MAC is set to 1 byte, that is the attacker is required to do 2^8 attempts to carry out an attack. So, the probability of successful spoofing by random attack is $1/2^8$, which means that 2^8 messages must be transmitted until the attack succeeds. On the other hand, it acts on the length of the frame payload and thereafter the data field becomes limited to 7 bytes as indicated in Fig. 1.

The approach that we have chosen uses an error frame to overwrite spoofed messages in real time. As soon as the monitoring node receives a message, it calculates and checks the MAC immediately, while nodes still have not transmitted an ACK signal to the CAN bus, Fig. 2.

However, if the monitoring node detects a MAC error, it overwrites the falsified message using the error frame as shown in Fig. 3.

Start of Frame	…	Data Field		…	End of Frame
1	…	0-64 bits		…	7
		0-7 bytes : Payload	1 byte : MAC		

Fig. 1 CAN frame of security monitoring system

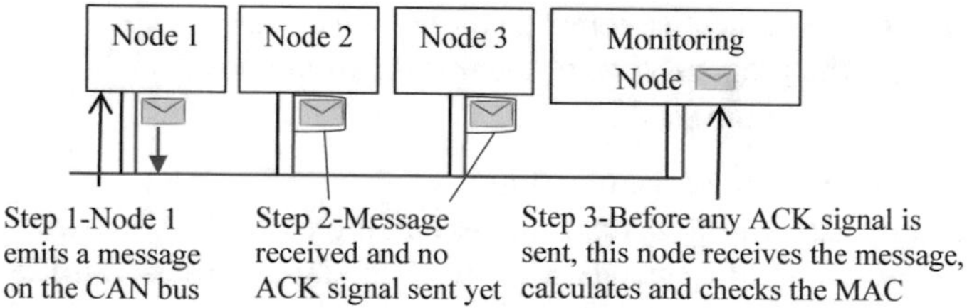

Fig. 2 Centralized authentication system (Transmission protocol)

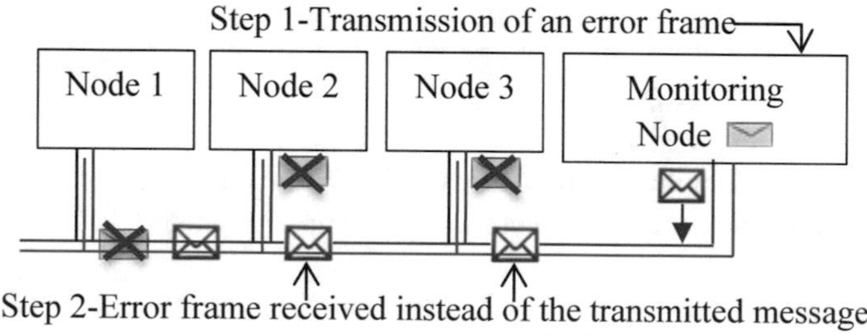

Fig. 3 Centralized authentication system (Error authentication)

2.3 Proposed Improvements

We should point out that the previous approach implies that the payload must be reduced. This constraint turns out detrimental to the transmission quality in the current context.

As an alternative and for better functioning, the proposed technique allows the enabled node to generate two frames, one after another. First, the data frame is sent and then the authentication data. Figure 4 illustrates the procedure.

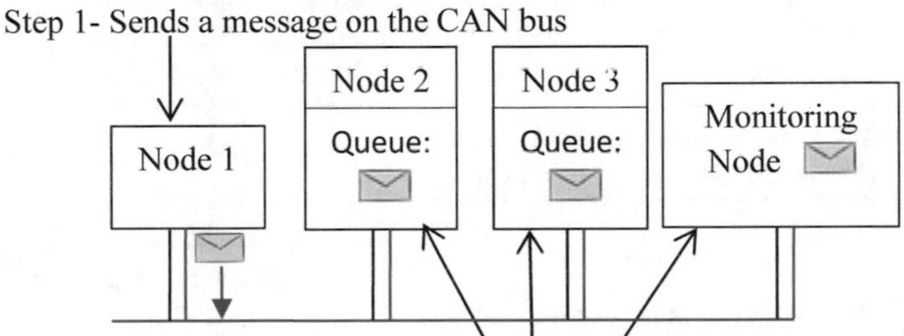
Start of Frame
…
Data Field 0-8 byte : the payload
…
End of Frame
Start of Frame
…
Data Field 0-8 byte : the MAC
…
End of Frame

Fig. 4 CAN frames of our proposed security system

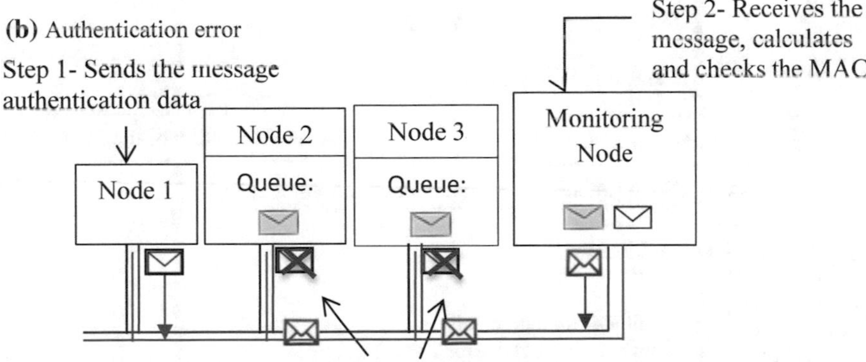
(a) Message transmission
Step 1- Sends a message on the CAN bus
Node 1
Node 2
Queue:
Node 3
Queue:
Monitoring Node
Step 2- The message is received by different nodes
(b) Authentication error
Step 1- Sends the message authentication data
Step 2- Receives the message, calculates and checks the MAC
Node 1
Node 2
Queue:
Node 3
Queue:
Monitoring Node
Step 3- The participating nodes wait for the arrival of either the authentication frame to ensure the legitimacy of the message in the queue or an error frame

Fig. 5 Proposed security system

According to our new technique all the nodes are programmed not to start the data processing and wait for the confirmation of the message legitimacy issued by the monitoring node. Using the ATMega328p, the total time between sending a message and receiving the authentication data is less than 4.5 ms which is an acceptable range. In the case of an attack, the monitoring node destroys the message by means of an error frame as can be seen in Fig. 5.

Key Delivery Cycle This phase allows not only to guarantee the sharing of keys between the monitor node and the other participating nodes but also to ensure the software integrity.

Since an attacker can act on the ECU firmware taking advantage of the numerous existing weaknesses, this phase should then be performed in a periodic fashion. The cyclic functioning takes into account the following cases:

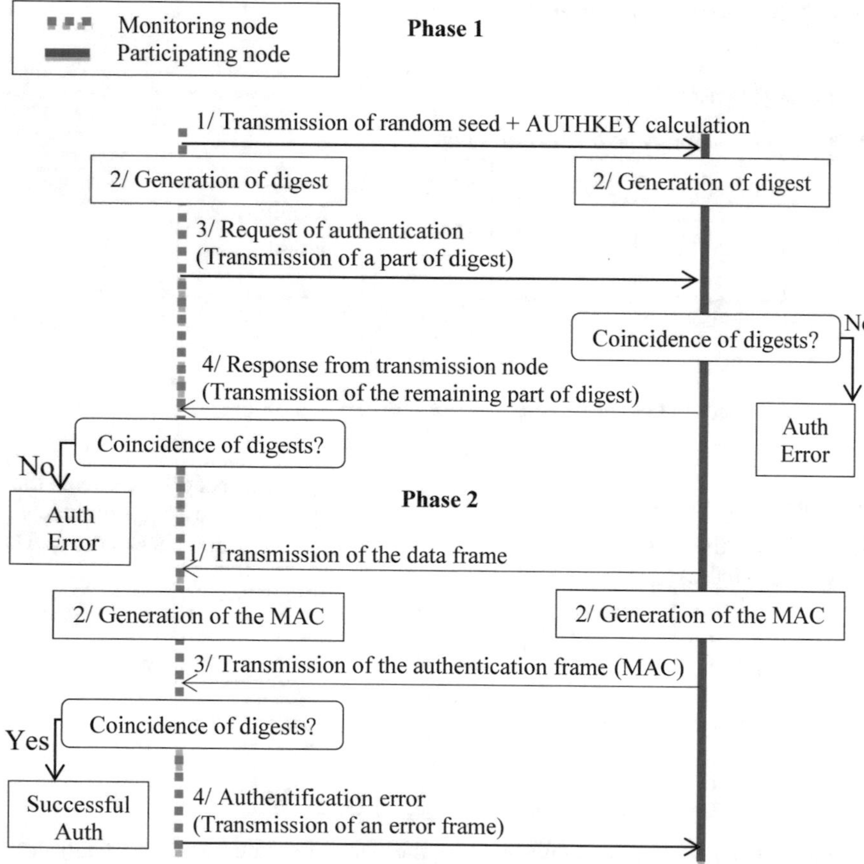

Fig. 6 Proposed security system protocol

- Switching on the engine.
- Receiving an error frame.

The overall procedure is explained through the flow graph of Fig. 6.

3 Hash Algorithm

A hash algorithm will be used to calculate two types of digest considering two different entries:

- The code implemented on the CPUs participating in the CAN communication.
- The message sent.

To make the best choice, the findings indicated by the report XBX Benchmarking Results January 2012 [10] are taken into consideration, whereas the comparison of several algorithms run on different microcontrollers is made and the following conclusion reached (Table 1):

BLAKE is the main candidate which produces the best performances among all the SHA-3 cryptographic hash algorithms.

The proposed protocol can be implemented on a field programmable gate array (FPGA) board so that the hash algorithms confrontation based on FPGAs [11, 12, 13] is available.

Table 2 gives a comparison related to the throughput over area ratio of three different implementations of the SHA-3 finalists algorithm "Grøstl, Keccak, JH, Skein and BLAKE" for long messages on FPGAs, i.e. the messages that have a length higher than one message block. The throughput area measure, which is derived from the area and the throughput all together, yields the necessary assessment:

- The area of FPGA implementations is usually measured in the total number of slices required for an implementation.
- The throughput is measured in Mbit/s. For hash functions, the throughput changes with respect to the message length.

We obviously notice that the most suitable algorithms, with the best throughput over area on Xilinx Virtex-5 device, are BLAKE and Grøstl. The following ones are Keccak and JH. Finally, Skein trails far behind.

Table 1 Overall SHA-3 candidate ranking

Candidate	Ranked 1st	Ranked 2nd	Ranked 3rd	Score
BLAKE	7	2	0	25
Keccak	1	2	4	11
Grostl	1	1	3	8
JH	0	0	1	1
Skein	2	4	1	15

Table 2 Results for the SHA-3 finalists (Virtex-5)

Algorithm	BLAKE			Grøstl			JH			Keccak			Skein		
FPGA	Virtex-5														
References	[11]	[12]	[13]	[11]	[12]	[13]	[11]	[12]	[13]	[11]	[12]	[13]	[11]	[12]	[13]
TP-area [(MBits/s)/Slice]	1.76	1.65	1.86	2.71	1.33	1.50	2.12	0.88	0.99	0.99	0.43	0.48	0.49	0.15	0.17

As a result, we propose to test BLAKE algorithms in our system and examine its performances in terms of speed, storage and security as well as the feasibility and validity.

4 Conclusion

We have proposed an enhanced authentication method for automotive embedded networks based on the centralized authentication technique. The emphasis has been on the monitoring system function that involves mainly the generation of two frames.

The first handles the transmission of the application data, whereas the second caries out the authentication and potentially the elimination of unauthorized frames. The whole operation is based on the BLAKE algorithm thanks to its performances in terms of storage memory and calculation speed.

The advantage of this improved technique comes from the fact that it can easily be implemented with no need for modifying the built-in hardware nor for complicated procedures. Consequently, the next step is the implementation of the enhanced platform and the confirmation of the expected results.

References

1. Checkoway, S., McCoy, D., Kantor, B., Anderson, D., Shacham, H., Savage, S.,Koscher, K., Czeskis, A., Roesner, F., Kohno, T., et al.: Comprehensive experimental analyses of automotive attack surfaces. In: 20th USENIX Security Symposium. San Francisco (2011)
2. Miller, C., Valasek, C.: Remote Exploitation of an Unaltered Passenger Vehicle (2015)
3. Van Herrewege, A., Singelee, D., Verbauwhede, I.: CANAuth—a simple, backward compatible broadcast authentication protocol for CAN bus. In: ECRYPT Workshop on Lightweight Cryptography 2011 (2011)
4. Groza, B., Murvay, S., Van Herrewege, A., Verbauwhede, I.: LiBrA-CAN: a lightweight broadcast authentication protocol for controller area networks. In: Cryptology and Network Security. pp. 185–200. Springer (2012)
5. Ueda, H., Kurachi, R., Takada, H., Mizutani, T., Inoue, M., Horihata, S.: Security authentication system for in-vehicle network. SEI, Oaks, PA, USA, Tech (2015)
6. Mundhenk, S., Steinhorst, M., Lukasiewycz, S., Fahmy, A., Chakraborty, S.: Lightweight authentication for secure automotive networks. In: Proceedings of the Conference on Design, Automation and Test in Europe (DATE 2015)
7. Radu, A.-I., Garcia, F.D., Leia: a lightweight authentication protocol for can. In: 21st European Symposium on Research in Computer Security (ESORICS 2016) (2016)
8. Nurnberger, S., Rossow, C.: Vatican–vetted, authenticated can bus. In: International Conference on Cryptographic Hardware and Embedded Systems. Springer (2016)
9. Van Bulck, J., Mühlberg, J.T., Piessens, F.: VulCAN: efficient component authentication and software isolation for automotive control networks. In: Proceedings of the 33th Annual Computer Security Applications Conference (ACSAC'17), ACM (2017)
10. Wenzel-Benner, C., Gräf, J., Pham, J., Kaps, J.P.: XBX Benchmarking Results January 2012. In: Third SHA-3 Candidate Conference (2012)
11. Jungk, B.: FPGA-based Evaluation of Cryptographic Algorithms-Ph.D. Dissertation (2016)

12. Kaps, J.-P., Yalla, P., Surapathi, K.K., Habib, B., Vadlamudi, S., Gurung, S., Pham, J.: Lightweight implementations of SHA3 finalists on FPGAs. Submission to NIST (Round 3) (2011). http://csrc.nist.gov/groups/ST/hash/sha-3/Round3/March2012/documents/papers/KAPS_paper.pdf
13. Kaps, J.-P., Yalla, P., Surapathi, K.K., Habib, B., Vad-lamudi, S., Gurung, S.: Lightweight implementations of SHA-3 finalists on FPGAs. In: The Third SHA-3 Candidate Conference. Cite-seer (2012)

Comparative Evaluation of Speech Recognition Systems Based on Different Toolkits

Fatima Barkani, Hassan Satori, Mohamed Hamidi, Ouissam Zealouk and Naouar Laaidi

Abstract Speech recognition is a method that allows machines to convert the incoming speech signals into text commands. This paper presents a brief survey on automatic speech recognition systems based on HTK, Julius, MATLAB, Sphinx and Kaldi. A description of the mentioned speech recognition systems is discussed, and the structure and performance of these different systems are presented.

Keywords Speech recognition · HMMs · CMU Sphinx · HTK · Julius · Kaldi

1 Introduction

Automatic speech recognition (ASR) or computer speech recognition is a computer technology that allows to transcribe an oral message and extracts linguistic information from an audio signal through an algorithm which is implemented as a computer program. Also, it permits software to interpret a natural human language. The automatic speech recognition world has become very flexible and practical thanks to the progress of the toolboxes and their availabilities which allowed the researchers to advance very quickly these last years. The speech recognition evolution passed through five generations that are being formulated based on the research themes [1].

Generation 1: Use of ad hoc methods to recognize sounds or small vocabularies of isolated words.
Generation 2: Use of acoustic–phonetic approaches to recognize phonemes, phones or digit vocabularies.
Generation 3: Use of pattern recognition approaches to speech recognition of small or medium-sized vocabularies of isolated and connected word sequences. Also, it includes linear predictive coding, dynamic programming methods and vector quantization.

F. Barkani (✉) · H. Satori · M. Hamidi · O. Zealouk · N. Laaidi
LIIAN Laboratory, Faculty of Sciences Dhar Mahraz, Sidi Mohammed Ben Abbdallah University, Fez, Morocco

V. Bhateja et al. (eds.), *Embedded Systems and Artificial Intelligence*,
Advances in Intelligent Systems and Computing 1076,
https://doi.org/10.1007/978-981-15-0947-6_4

Generation4: Use of hidden Markov model statistical methods for modeling speech dynamics and statistics in a continuous speech recognition system, use of forward–backward, Viterbi alignment methods, maximum likelihood and various other performance criteria, Introduction of neural network methods, use of adaptation methods that modify the parameters associated with either the speech signal or the statistical model.

Generation 5: Use of parallel processing methods to increase recognition decision reliability, combinations of hidden Markov models (HMMs) and acoustic–phonetic approaches, increased robustness for speech recognition in noise, machine learning of optimal combination of models, ASR applications such as dictation, command recognition, interactive and voice response. Also, it may be particularly pleasant to handicapped people when the speech methods are not supported.

In this work, we are interested in the study of the different construction tools of ASR systems based on HMMs to see their ability to manage large vocabulary systems.

This paper is organized as follows. Some of the related works are presented in Sect. 2. Section 3 presents the architecture and functioning of the automatic speech recognition system. Section 4 represents the hidden Markov models. Speech recognition libraries are given in Sect. 5. Section 6 compares the ASR systems. Finally, the conclusion is in Sect. 7.

2 Related Works

This section presents some of the reported works proving ASR system for other languages.

Satori and El Haoussi [2] have described the development of a speaker-independent continuous automatic amazing speech recognition system-based HMMs. In their implementation, the hidden Markov models have been used to develop the system, and Mel frequency cepstral coefficients (MFCCs) have been used to extract the feature. In [3], an HMM automatic speech recognition system was designed to detect smoker speaker. In their project, the amazing language is used to compare the voice of normal persons to smokers one. Their obtained results present that the created system makes the diagnostic between smokers and non-smokers.

In [4, 5], Hamidi et al. provide an insight into the theoretical and implementation details of an amazing speech recognition system over interactive voice response server (IVR). Their works describe how an amazing speech recognition system can be displayed over the Asterisk server. Zealouk et al. [6] present a technical overview of the speech recognition systems and their recent improvement concerning feature extraction techniques, speech classifiers and performance assessment. The authors in [7] talk about the implementation of an isolated word automatic speech recognition system for a Punjabi language using the HTK toolkit. Their system is trained for 115 distinct Punjabi words by collecting data from eight speakers and then is tested by using samples from six speakers in real-time environments. The experiential results show that the overall system performance is 95.63 and 94.08%. In [8], the authors have constructed the system for amazing recognition system based on the hidden Markov model and HTK toolkit. They used the HMM to model the phonetic units corresponding to words taken from the training base. The results obtained are very encouraging given the size of the training set and the number of people taken to the registration. In [11], Addarrazi et al. describe the recording and evaluation of an audiovisual database which is as far as they know the first audiovisual corpus uses amazing language. This database is called AmDigit_AVSR (Amazigh Digit _ Audiovisual speech recognition system) which contains 4000 video and audio files of 40 speakers utter the 10 first Amazigh digits.

Medennikov and Prudniko [10] used the Kaldi speech recognition toolkit for experiments-based Russian speech recognition. The resulting system achieves WER of 16.4%, with an absolute reduction of 8.7% and relative reduction of 34.7% compared to our previous system result on this test set. In [9], authors use Kaldi for speech recognition. They proposed a two-pronged strategy to reduce the performance gap in far-field ASR systems, when using alignments from close-talk microphone (IHM) and distant microphone (SDM/MDM) audio using a lattice-free MMI objective function which is tolerant to minor mis-alignment errors and a data filtering technique based on lattice oracle WER. They reduced the relative change in WER, on using IHM alignments, from 7.2 to 1.3%, on average. Mittal and Kaur [12] have implemented the system for Punjabi language recognition system using the HTK toolkit along with Julius toolkit. The experiential results show that the accuracy of the system comes out to be 57.54%. In [13], the authors presented frequency analysis of Urdu numbers. The data was acquired in moderate noisy environment by word utterances of 15 different speakers. Fast Fourier transform algorithm was used in MATLAB to analyze the data. As expected, they found high correlation among frequency contents of the same word, when spoken by many different speakers. In [14], authors examined the Swahili language structure and sound synthesis processes to building a Swahili speech recognizer using Sphinx4. They built 40 words Swahili acoustic model based on the observed language and sound structures using CMU SphinxTrain and associated tools. Their recognition rates were lower compared to other researches. In [15], the authors implemented an Arabic automatic speech recognition system using the HTK. The experimental results showed that the overall system performance was 90.62%, 98.01% and 97.99% for sentence correction, word correction and word accuracy, respectively. The researchers in [16] aim to construct a connected words speech recognition system for Hindi language using HTK. The training data has been collected from 12 speakers including both males and females. The test data used for evaluating the system performance has been collected from the five speakers. The experimental results show that the presented system provides the overall word accuracy of 87.01%, word error rate of 12.99% and word correction rate of 90.93%, respectively. The authors in [17] describe their experience to design a secure telephony spoken system over the VoIP network by combining both HMM automatic speech recognition and IVR technology based on the open-source platform. The created system permits the user to control the distance of the security

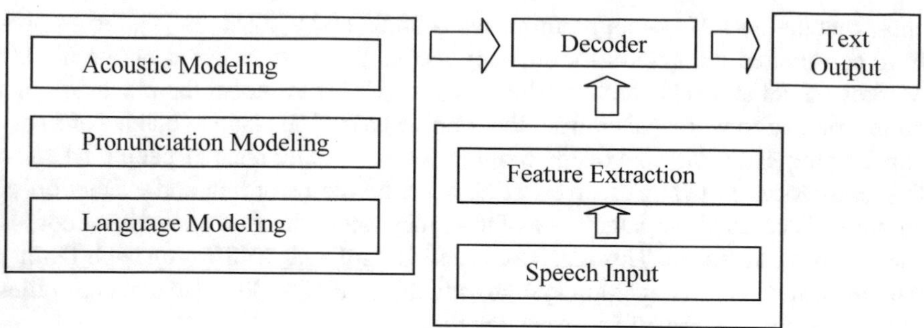

Fig. 1 Setups involved in ASR system

network and data backup by using his voice for network tasks administration managing and his biological voiceprint for security identification. Their obtained results show that the recognition rates were all above 80% for admin voice and less than 6% for non-admin voice.

3 ASR System Architecture

ASR system is shown in Fig. 1. The steps of speech recognition can be stated as follows: A microphone perceives a waveform input that is converted into a set of observations, through a process called "feature extraction" by which the decoder seeks for a sequence of words [18].

4 Hidden Markov Models (HMMs)

In artificial intelligence, the hidden Markov models represent a statistical tool used to model stochastic phenomena. HMMs were first described in a series of statistical papers by Baum and other authors in the 1960s [19]. Jelinek and colleagues at IBM [20] are generally known as the first researchers to apply hidden Markov models to speech recognition. The HMMs have become the most successful technique in speech recognition after the pioneering work in the 1970s and 1980s. Also, HMMs began to be applied to the analysis of biological sequences such as DNA in the second half of the 1980s.

5 Speech Recognition Libraries

There are several libraries for the field of ASR; among these libraries, we find HTK, Sphinx, Kaldi, MATLAB ASR, Java Speech and ISIP. This section provides an overview of the structure of different types of speech recognition systems: CMU Sphinx, HTK, JULIUS, Kaldi and MATLAB.

5.1 CMU Sphinx

Sphinx is a free downloadable speech recognition library, with the ability to modify the source code, and it can implement systems with a wide vocabulary, independent of the speaker. The first version of Sphinx (1, 2 and 3) is written in C language, but the recent version of Sphinx4 is written in Java. Sphinx 1, 2, 3 and 4 are decoders, and the learning tool is called SphinxTrain. The tool for creating SphinxTrain acoustic models is written in C language, and the Sphinx4 recognition tool is written in Java [21].

5.2 HTK

Hidden Markov model toolbox (HTK) is implemented in C language, and it runs on the command line; it is capable of implementing considerable vocabulary, independently of the speaker and is applicable on any languages. Such toolbox is dedicated to hidden Markov models and mainly used for the recognition of speech [22].

5.3 Julius

It is of high performance, based on the technique "n-grams" and dependent on the HMM context. Main research techniques are fully integrated, such as lexicon, the manipulation of context-dependent crosswords and Gaussian selection [23]. It is also carefully modulated to be independent of model structures and the different types of HMM that are supported.

5.4 Kaldi

Kaldi is a speech recognition toolkit, written in C++ and licensed under the Apache License v2.0. The aim of Kaldi is to have a modern and a flexible code that is easy to understand, modify and extend. Kaldi depends on two external libraries that are also freely available: One is OpenFst for the finite state framework, and the other is numerical algebra libraries [24].

5.5 MATLAB

MATLAB includes a toolbox including artificial learning algorithms based on hidden Markov models and detection algorithms [25].

6 ASR Systems Comparison

A comparative study between different speech recognition softwares-based features is shown in Table 1.

After a detailed study of speech recognition tools, we choose the three most used softwares which are CMU Sphinx, Julius and HTK. In [26], the authors have referenced their evaluation in their research results that we have illustrated in Table 2. Their comparison between the mentioned softwares illustrates that Julius has good recognition rates and works very fast where a higher score means better performance. Their proposed solution is generating the acoustic model by using HTK toolkit in a

Table 1 Comparative table between HTK, Sphinx and Julius

Toolkit	Sphinx	HTK	Julius
Programming languages	Java, C, Python, others	C, Python	C, Python
Open-source software	Yes	Yes	Yes
Input formats	WAV MP3	WAV, TIMIT, NIST, OIG, AIFI	WAV
Gross platform	Yes	Yes	Yes
Large vocabulary	Yes	Yes	Yes

Table 2 Comparison between state-of-the-art speech recognition software

Criteria	SPHINX4	JULIUS	HTK
Recognition rate (60% of the total score)	6.5	8	6
Platform independence (15%)	8	9	8
Cost (5%)	10	10	7
Modularity (15%)	10	10	10
Actuality (5%)	7	7	6
Total score	7.45	8.5	6.35

Table 3 Word error rates comparison

Recognizer	Verbmobil 1	Wall street journal 1
Julius	27.2	23.1
PocketSphinx	23.9	21.4
Sphinx4	26.9	22.7
Kaldi	12.7	6.5

way that it suits Julius. Therefore, they chose Sphinx for their recognition task and started to create a speech corpus by using their lecture videos in order to increase the recognition rate.

In addition, an investigation of a large-scale evaluation of open-source speech recognition toolkits (Julius, Sphinx4, PocketSphinx and Kaldi) is down in [27]. The authors have generated the language and acoustic models for Kaldi, PocketSphinx, Julius, Sphinx4 and HDecode. They have tuned the recognition tests on the German Verbmobil 1 and the English Wall Street Journal 1 corpus. Their essential contributions are the Kaldi inclusion to a global evaluation of open-source ASR systems. The tests' results are shown in Table 3. In their experiments, the toolkits evaluated order is presented as follows based on the obtained performance:

- Kaldi surpasses all the other recognition toolkits, in the training and decoding cases including the most advanced techniques. This gives the best results in a short time but has the highest computational cost.
- The Sphinx is able to generate good results, but not advanced as Kaldi toolkit.
- HTK is the most difficult toolkit. Setting up the system intended the training pipeline development which was time-consuming and error-prone. The obtained results are like Sphinx, but the effort to get these is quite larger.

Next to a comparative study between HTK and Sphinx according to [28, 29] (on Chinese Mandarin, English and Hindish), we have mentioned the following:

- The acoustic models which are trained by Sphinx were better than the ones for HTK [28].
- Sphinx3 performs better than HTK for continuous speech recognition, but both are comparable in the speech recognition of isolated words [29].
- Sphinx3 and HDecode afford similar levels of WER and xRT (the real-time factor) [30].

7 Conclusion

Research in the speech recognition field has become less difficult thanks to the availability of free and open-source libraries. In this paper, we have seen the differences between the most used libraries in the ASR. We have studied the state-of-the-art speech recognition software, the differences between the most used ASR libraries and the proposed solution for a speech recognition system to perform the best results.

In our next work, we will do a comparative study of Amazigh speech recognition using the three systems: JULIUS, CMU Sphinx, HTK, Kaldi and MATLAB.

References

1. Karpagavalli, S., Deepika, R., Kokila, P., Usha Rani, K., Chandra, E.: Automatic speech recognition: architecture, methodologies and challenges-a review. Int. J. Adv. Res. Comput. Sci. **2**(6) (2011)
2. Satori, H., ElHaoussi, F.: Investigation Amazing speech recognition using CMU tools. Int. J. Speech Technol. **17**(3), 235–243 (2014)
3. Satori, H., Zealouk, O., Satori, K., ElHaoussi, F.: Voice comparison between smokers and non-smokers using HMM speech recognition system. Int. J. Speech Technol. **20**(4), 771–777 (2017)
4. Hamidi, M., Satori, H., Satori, K.: Implementing a voice interface in VOIP network with IVR server using Amazing digits. Int. J. Multi. Sci. **2**, 38–43 (2016)
5. Hamidi, M., Satori, H., Zealouk, O., Satori, K.: Speech coding effect on amazing alphabet speech recognition performance. J. Adv. Res. Dyn. Control Syst. **11**(2), 1392–1400 (2019)
6. Zealouk, O., Satori, H., Hamidi, M., Satori, K.: Speech recognition for Moroccan dialects: feature extraction and classification methods. J. Adv. Res. Dyn. Control Syst. **11**(2), 1401–1408 (2019)
7. Dua, M., Aggarwal, R.K., Kadyan, V., Dua, S.: Punjabi automatic speech recognition using HTK. Int. J. Comput. Sci. Issues (IJCSI) **9**(4), 359 (2012)
8. El Ghazi, A., Daoui, C., Idrissi, N., Fakir, M., Bouikhalene, B.: Speech recognition system based on Hidden Markov Model concerning the Moroccan dialect DARIJA. Global J. Comput. Sci. Technol (2011)
9. Ilham, A., Hassan, S., Khalid, S.: Building a first amazing database for automatic audiovisual speech recognition system. In: Proceedings of the 2nd International Conference on Smart Digital Environment, pp. 94–99. ACM (2018, October)
10. Medennikov, I., Prudnikov, A.: Advances in STC Russian spontaneous speech recognition system. In: International Conference on Speech and Computer, pp. 116–123. Springer, Cham (2016, August)
11. Peddinti, V., Manohar, V., Wang, Y., Povey, D., Khudanpur, S.: Far-field ASR without parallel data. In: Interspeech, pp. 1996–2000 (2016, September)
12. Mittal, S., Kaur, R.: Implementation of word level speech recognition system for Punjabi language. Int. J. Comput. Appl. **146**(3) (2016)
13. Husnain, S.K., Beg, A., Awan, M.S.: Frequency analysis of spoken Urdu numbers using MATLAB and Simulink. PAF KIET J. Eng. Sci. **1**, 5 (2007)
14. Kimutai, S.K., Milgo, E., Gichoya, D.: Isolated Swahili words recognition using Sphinx4. Int. J. Emerg. Sci. Eng. **2**(2), 2319–6378 (2013)
15. Al-Qatab, B.A., Ainon, R.N.: Arabic speech recognition using hidden Markov model toolkit (HTK). In: 2010 International Symposium in Information Technology (ITSim), Vol. 2, pp. 557–562. IEEE (2010, June)
16. Kumar, K., Aggarwal, R.K., Jain, A.: A Hindi speech recognition system for connected words using HTK. Int. J. Comput. Syst. Eng. **1**(1), 25–32 (2012)
17. Mohamed, H., Hassan, S., Ouissam, Z., Khalid, S., Naouar, L.: Interactive voice response server voice network administration using hidden Markov model speech recognition system. In: 2018 Second World Conference on Smart Trends in Systems, Security and Sustainability (WorldS4), pp. 16–21. IEEE (2018, October)

18. Kraleva, R., Kralev, V.: On model architecture for a children's speech recognition interactive dialog system. (2016). arXiv preprint arXiv:1605.07733
19. Hayes, B.: First links in the Markov chain. Am. Sci. **101**(2), 252 (2013)
20. Rabiner, L.R., Juang, B.H.: An introduction to hidden Markov models. IEEE ASSP Mag. **3**(1), 4–16 (1986)
21. CMUSphinx, Open Source Toolkit For Speech Recognition, Project By CMU, "Sphinx-4 Application Programmer's Guide"
22. Young, S., Evermann, G., Gales, M., Hain, T., Kershaw, D., Liu, X., Valtchev, V.: The HTK Book, 3rd edn, p. 175. Cambridge University Engineering Department, Cambridge (2002)
23. Lee, A., Kawahara, T.: Recent development of open-source speech recognition engine julius. In: Proceedings: APSIPA ASC 2009: Asia-Pacific signal and information processing association, 2009 annual summit and conference. Asia-Pacific Signal and Information Processing Association, 2009 Annual Summit and Conference, International Organizing Committee, pp. 131–137 (2009)
24. Povey, D., Ghoshal, A., Boulianne, G., Burget, L., Glembek, O., Goel, N., Hannemann, M., Motlicek, P., Qian, Y., Schwarz, P. and Silovsky, J.: The Kaldi speech recognition toolkit. In: IEEE 2011 Workshop on Automatic Speech Recognition and Understanding (No. EPFL-CONF-192584). IEEE Signal Processing Society (2011)
25. Campbell, D., Palomaki, K., Brown, G.: A MATLAB simulation of "shoebox" room acoustics for use in research and teaching. Comput. Inf. Syst. **9**(3), 48 (2005)
26. Yang, H., Oehlke, C., Meinel, C.: German speech recognition: a solution for the analysis and processing of lecture recordings. In: IEEE/ACIS 10th International Conference on Computer and Information Science (ICIS), 2011, pp. 201–206. IEEE (2011, May)
27. Gaida, C., Lange, P., Petrick, R., Proba, P., Malatawy, A., Suendermann-Oeft, D.: Comparing Open-source Speech Recognition Toolkits. Technical Report. DHBW Stuttgart, Stuttgart (2014)
28. Samudravijaya, K., Barot, M.: A comparison of public domain software tools for speech recognition. In: WSLP2003, pp. 125–131 (2003)
29. Ma, G., Zhou, W., Zheng, J., You, X. Ye, W. A Comparison between HTK and SPHINX on Chinese Mandarin. In: 2009 International Joint Conference on Artificial Intelligence, pp. 394–397. IEEE (2009)
30. Vertanen, K.: Baseline WSJ acoustic models for HTK and Sphinx: Training recipes and recognition experiments. Technical report). Cavendish Laboratory, Cambridge, United Kingdom (2006)

Stability Analysis and H_∞ Performance for Nonlinear Fuzzy Networked Control Systems with Time-Varying Delay

Ahmed Ech-charqy, Said Idrissi, Mohamed Ouahi, El Houssaine Tissir and Ghali Naami

Abstract This paper studies the problem of delay-dependent stability and H_∞ performance for nonlinear fuzzy networked control systems with time-varying delay. Based on Lyapunov–Krasovskii theorem, a less conservative sufficient condition is obtained in terms of linear matrix inequalities (LMIs) in such a way that it can be easily solved by using standard numerical package. The main novelty of the proposed method is to use the information about the time delays in sensor and actuator. Finally, numerical examples are given to show the advantage and usefulness of this work.

Keywords Networked control systems · Fuzzy systems · Time-varying delay · Linear matrix inequalities (lmis)

1 Introduction

The fuzzy model approach [1] is a very powerful and effective tool to control nonlinear systems due to the fact that it can describe complex nonlinear systems by using fuzzy sets and the membership functions. Therefore, great attention has been paid to stability analysis and controller synthesis of T-S fuzzy systems [2–5].

Time delays are common phenomena in many practical, industrial and engineering systems. They are regarded as an important sources of performance degradation or dynamic system instability. During the past three decades, the stability analysis

A. Ech-charqy (✉) · E. H. Tissir
LESSI, Department of Physics Faculty of Sciences Dhar El Mehraz,
B.P. 1796 Fes-Atlas, Morocco
e-mail: ahmed.echcharqy1@usmba.ac.ma

S. Idrissi
Department of Physics, Faculty of Polydisciplinary Safi University Cadi Ayyad,
Marakech, Morocco

M. Ouahi · G. Naami
National school of Applied Sciences, LISA,
Avenue My Abdallah km 5 Route d'Imouzzer, FP 72 Fes, Morocco

V. Bhateja et al. (eds.), *Embedded Systems and Artificial Intelligence*,
Advances in Intelligent Systems and Computing 1076,
https://doi.org/10.1007/978-981-15-0947-6_5

problem of time-delay systems has been received a great important interest [6–8] and the references therein. Generally speaking, the stability conditions for time-delay systems can be classified into two categories: One called delay-dependent stability criteria and the other one is delay-independent stability criteria. It is well known that delay-dependent stability criteria are usually less conservatives than delay-independent stability criteria especially when the time delay is small.

On the other hand, the network control systems (NCSs) play an increasingly significant role in the infrastructure of society, such as artificial intelligence systems, chemical systems and public transportation systems [9–11]. In recent years, many researchers pay their attention to NCSs and get a lot of important results. For example, the author in [12] studied the problem of robust H_∞ design for networked control systems with uncertainties. In [13], the authors study the problem of output feedback control for network control systems. In [14], the authors have investigated the event-triggered controller design for network control systems. However, to the best of our knowledge, no result has been reported on the problem of delay-dependent control for uncertain fuzzy network control systems with nonlinear parameters. This motivates the study in this work.

Then, we focus in this paper on the problem of analysis for nonlinear networked control systems(NNCSs) which can be represented by a T-S fuzzy model. First, a delay-dependent stability criterion is established for the considered systems without disturbance in terms of LMIs. Next, we analysis the H_∞ performance under disturbance. Moreover, a numerical examples are given to express the potential of the proposed theories.

2 Problem Formulation

Consider a fuzzy network control system with time-varying delay. The rule of the model has the following form:
Rule i: IF α_1 is F_1^i and α_2 is F_2^i ... and α_p is F_p^i THEN.

$$\begin{cases} \dot{x}(t) = A_i x(t) + B_i u(t) + B_{fi} f(x(t)) + B_{wi} w(t) \\ z(t) = C_i x(t) + D_i u(t) \\ x(t) = \phi(t), \quad t \in [-d, 0] \end{cases} \tag{1}$$

where $A_i, B_i, B_{fi}, B_{wi}, C_i$ and D_i are constant matrices with appropriate dimensions. $F_j^i (j = 1, \ldots, p)$ are fuzzy sets, $\alpha = [\alpha_1, \ldots, \alpha_p]$ is the premise variable vector. It is assumed that the premise variables do not depend on the input variables $u(t)$, which is needed to avoid a complicated defuzzification process of fuzzy controllers. $x(t) \in R^n$ is the state vector, $u(t) \in R^m$ is the control input, $w(t)$ is the disturbance signal, $z(t) \in R^s$ is the controlled output, and $f(t)$ is a nonlinear function vector; we define $f(x(t)) = \left[f_1(x_1(t)) \; f_1(x_1(t)) \; \ldots \; f_m(x_m(t))\right]$, and $f_j(\bullet)$ satisfy $f_j(\bullet) \in H_j[0, h_j] = \{f_j(x_j) | f_j(0) = 0, \; 0 \leq x_j f_j(x_j) \leq h_j x_j^2\}$ for $0 < h_j < +\infty$.

By fuzzy blending, the overall fuzzy model is inferred as follows

$$\begin{cases} \dot{x}(t) = Ax(t) + Bu(t) + B_f f(x(t)) + Bw(t) \\ z(t) = Cx(t) + Du(t) \\ x(t) = \phi(t), \end{cases} \tag{2}$$

where

$$A = \sum_{i=1}^{r} \lambda_i(\alpha) A_i,\ B = \sum_{i=1}^{r} \lambda_i(\alpha) B_i,\ B_f = \sum_{i=1}^{r} \lambda_i(\alpha) B_{fi},\ B_w = \sum_{i=1}^{r} \lambda_i(\alpha) B_{wi}$$

$$C = \sum_{i=1}^{r} \lambda_i(\alpha) C_i,\ D = \sum_{i=1}^{r} \lambda_i(\alpha) D_i.$$

and

$$\lambda_i(\alpha) = \frac{\prod_{j=1}^{p} F_j^i(\alpha_j)}{\sum_{i=1}^{r} \prod_{j-1}^{p} F_j^i(\alpha_j)}$$

with $F_j^i(\alpha_j)$ represents the membership degrees of α_j in the fuzzy set F_j^i. Note that the normalized weights $\lambda_i(\alpha)$ satisfy

$$\lambda_i(\alpha) \geq 0,\ i = 1, 2, \ldots, r\ \sum_{i=1}^{r} \lambda_i(\alpha) = 1$$

we assume that the state variable is fully observable and the controller time delay is neglected, then we have

$$u_c(t) = Kx(t - \tau_{sc}) \tag{3}$$

with $K = \sum_{j=1}^{r} \lambda_i(\alpha) K_j$.
For the controller of the system could be expressed as

$$\begin{aligned} u(t) = u_c(t) &= Kx(t - \tau_{sc}) \\ &= Kx(t - (\tau_{sc}(t) + \tau_{ca}(t))) \\ &= Kx(t - \tau(t)) \end{aligned} \tag{4}$$

where $u_c(t)$ is the output of sensor, $u(t)$ is the output of actuator. $\tau_{ca}(t)$ is the transmission time from the controller to the executor, and $\tau_{sc}(t)$ is the transmission time from the sensor to controller. These time-varying delays are satisfying:

$$\begin{aligned} 0 \leq \tau_{ca}(t) \leq \bar{\tau}_{ca},\quad \dot{\tau}_{ca}(t) \leq d_1 \\ 0 \leq \tau_{sc}(t) \leq \bar{\tau}_{sc},\quad \dot{\tau}_{sc}(t) \leq d_2 \end{aligned} \tag{5}$$

and

$$\begin{aligned} \tau(t) &= \tau_{ca}(t) + \tau_{sc}(t) \\ \bar{\tau} &= \bar{\tau}_{ca} + \bar{\tau}_{sc} \\ d &= d_1 + d_2 \end{aligned} \tag{6}$$

Now we can rewrite the fuzzy network control system as

$$\begin{cases} \dot{x}(t) = Ax(t) + BKx(t - \tau_{ca}(t) - \tau_{sc}(t)) + B_f f(x(t)) + Bw(t) \\ z(t) = Cx(t) + DKx(t - \tau_{ca}(t) - \tau_{sc}(t)) \\ x(t) = \phi(t), \end{cases} \tag{7}$$

In the sequel, the following lemma is needed.

Lemma 1 *[15]. For a positive definite matrix $Q > 0$, and an integrable function $z(\lambda)|\ \lambda \in [r_2, r_1]$, the following inequalities hold:*

$$\int_{r_1}^{r_2} z(\lambda)^T Q z(\lambda) d\lambda \geq \frac{1}{r_2 - r_1} \left(\int_{r_1}^{r_2} z(\lambda) d\lambda \right)^T Q \left(\int_{r_1}^{r_2} z(\lambda) d\lambda \right)$$

3 Main Result

3.1 Delay-Dependent Stability

In this section, a delay-dependent condition for the system (7) without disturbance to be stable will be proposed.

Theorem 1 *For given scalars $\bar{\tau}_{ca}$, $\bar{\tau}_{sc}$ d_1 and d_2, the system (7) with $w(t) = 0$ is asymptotically stable if there exist matrices $P > 0$, $Q_1 > 0$, $Q_2 > 0$, $R_1 > 0$, $R_2 > 0$, $T = diag(t_1, t_2, \ldots, t_r) > 0$, $\lambda = diag(\lambda_1, \lambda_2, \ldots, \lambda_r) > 0$, $H = diag(h_1 t_1, h_2 t_2, \ldots, h_r t_r) > 0$ and $F_l, l = 1, 2, 3$ such that the following set of LMIs holds:*

$$\Psi^{ij} = \begin{bmatrix} \Psi_{11} & \Psi_{12} & \Psi_{13} & \Psi_{14} & \Psi_{15} \\ * & \Psi_{22} & \Psi_{23} & 0 & 0 \\ * & * & \Psi_{33} & \Psi_{34} & \Psi_{35} \\ * & * & * & \Psi_{44} & \Psi_{45} \\ * & * & * & * & \Psi_{55} \end{bmatrix} < 0, \quad for\ \ 1 \leq i \leq j \leq r. \tag{8}$$

where

$$\Psi_{11} = -\frac{1}{\bar{\tau}_{sc}} Q_1 + R_1 + F_1 A_i + A_i^T F_1^T, \quad \Psi_{12} = \frac{1}{\bar{\tau}_{sc}} Q_1, \quad \Psi_{13} = F_1 B_i K_j + A_i^T F_2^T$$
$$\Psi_{14} = P - F_1 + A_i^T F_3^T, \quad \Psi_{15} = H + F_1 B_f$$

$$\Psi_{22} = -\frac{1}{\bar{\tau}_{sc}} Q_1 - \frac{1}{\bar{\tau}_{ca}} Q_2 - (1 - d_1)(R_1 - R_2), \quad \Psi_{23} = \frac{1}{\bar{\tau}_{ca}} Q_2$$

$$\Psi_{33} = -\frac{1}{\bar{\tau}_{ca}} Q_2 - (1 - d_1 - d_2) R_2 + F_2 B_i K_j + K_j^T B_i^T F_2^T$$

$$\Psi_{34} = -F_2 + K_j^T B_i^T F_3^T, \quad \Psi_{35} = F_2 B_f, \quad \Psi_{44} = \bar{\tau}_{sc} Q_1 + \bar{\tau}_{ca} Q_2 - F_3 - F_3^T$$

$$\Psi_{45} = \lambda + F_3 B_w, \quad \Psi_{55} = -T - T^T$$

Proof. Consider the Lyapunov-Krasovskii functional candidate:

$$V(x_t) = \sum_{i=1}^{4} V_i(x_t) \tag{9}$$

where

$$V_1(x_t) = x^T(t) P x(t) \tag{10}$$

$$V_2(x_t) = 2 \sum_{j=1}^{m} \lambda_j \int_0^{x_j} f_j(s) \mathrm{d}s \tag{11}$$

$$V_3(x_t) = \int_{t-\tau_{sc}(t)}^{t} x^T(s) R_1 x(s) \mathrm{d}s + \int_{t-\tau_{sc}(t)-\tau_{ca}(t)}^{t-\tau_{sc}(t)} x^T(s) R_2 x(s) \mathrm{d}s \tag{12}$$

$$V_4(x_t) = \int_{\bar{\tau}_{sc}}^{0} \int_{t+\theta}^{t} \dot{x}^T(s) Q_1 \dot{x}(s) \mathrm{d}s + \int_{-\bar{\tau}_{sc}-\bar{\tau}_{ca}}^{-\bar{\tau}_{sc}} \int_{t+\theta}^{t} \dot{x}^T(s) Q_2 \dot{x}(s) \mathrm{d}s \tag{13}$$

From (7), for any appropriately dimensional matrices $F_i; i = 1, 2, 3$, we may construct the following null equation:

$$\begin{aligned} 2\left[x^T(t) F_1 + x^T(t - \tau(t)) F_2 + \dot{x}^T(t) F_3\right]&[-\dot{x}(t) + Ax(t) + BKx(t - \tau(t)) \\ &+ B_f f(t)] = 0 \end{aligned} \tag{14}$$

Taking the time derivative of (9) and using lemma 1 and considering (14) yield

$$\dot{V}(x(t)) \le \xi^T \bar{\Psi}^{ij} \xi \tag{15}$$

where

$$\xi(t) = \left[x^T(t)\ x^T(t-\tau_{sc}(t))\ x^T(t-\tau(t))\ \dot{x}^T(t)\ f^T(t)\right]^T$$

$$\bar{\Psi}^{ij} = \begin{bmatrix} \Psi_{11} & \Psi_{12} & \Psi_{13} & \Psi_{14} & F_1 B_f \\ * & \Psi_{22} & \Psi_{23} & 0 & 0 \\ * & * & \Psi_{33} & \Psi_{34} & \Psi_{35} \\ * & * & * & \Psi_{44} & \Psi_{45} \\ * & * & * & * & 0 \end{bmatrix}$$

Then, we consider the nonlinear term, if there exists $T = diag(t_1, t_2, \ldots, t_m) > 0$, if

$$\dot{V}(x(t)) - 2\sum_{j=1}^{m} t_j f_j(x_j(t))\left[f_j(x_j(t)) - h_j x(t)\right] \leq 0 \tag{16}$$

holds, then $\dot{V}(x(t)) < 0$, which ensure that the system (7) is asymptotically stable. From the Shur complement, (16) holds if $\Psi^{ij} < 0$, this completes the proof.

3.2 H_∞ Performance Analysis

In this subsection, we analysis the H_∞ performance under disturbance.

Theorem 2 *For given scalars $\bar{\tau}_{ca}$, $\bar{\tau}_{sc}$ d_1 and d_2, the system* (7) *is asymptotically stable with H_∞ norm bound γ if there exist matrices $P > 0$, $Q_1 > 0$, $Q_2 > 0$, $R_1 > 0$, $R_2 > 0$, $T = diag(t_1, t_2, \ldots, t_r) > 0$, $\lambda = diag(\lambda_1, \lambda_2, \ldots, \lambda_r) > 0$, $H = diag(h_1 t_1, h_2 t_2, \ldots, h_r t_r) > 0$ and $F_l, l = 1, 2, 3$ such that the following set of LMIs holds:*

$$\Phi^{ij} = \begin{bmatrix} \Phi_{11} & \Phi_{12} & \Phi_{13} & \Phi_{14} & \Phi_{15} & \Phi_{16} \\ * & \Phi_{22} & \Phi_{23} & 0 & 0 & 0 \\ * & * & \Phi_{33} & \Phi_{34} & \Phi_{35} & \Phi_{36} \\ * & * & * & \Phi_{44} & \Phi_{45} & \Phi_{46} \\ * & * & * & * & \Phi_{55} & \Phi_{56} \\ * & * & * & * & * & \Phi_{56} \end{bmatrix} < 0, \quad for\ \ 1 \leq i \leq j \leq r. \tag{17}$$

where

$$\Phi_{11} = -\frac{1}{\bar{\tau}_{sc}} Q_1 + R_1 + C_i^T C_i + F_1 A_i + A_i^T F_1^T, \quad \Phi_{12} = \frac{1}{\bar{\tau}_{sc}} Q_1$$
$$\Phi_{13} = C_i^T D_i K_j + F_1 B_i K_j + A_i^T F_2^T, \quad \Phi_{14} = P - F_1 + A_i^T F_3^T, \quad \Phi_{15} = H + F_1 B_f$$
$$\Phi_{16} = F_1 B_w, \quad \Phi_{22} = -\frac{1}{\bar{\tau}_{sc}} Q_1 - \frac{1}{\bar{\tau}_{ca}} Q_2 - (1 - d_1)(R_1 - R_2), \quad \Phi_{23} = \frac{1}{\bar{\tau}_{ca}} Q_2$$

$$\Phi_{33} = -\frac{1}{\bar{\tau}_{ca}} Q_2 - (1 - d_1 - d_2)R_2 + K_j^T D_i^T D_i K_j + F_2 B_i K_j + K_j^T B_i^T F_2^T$$
$$\Phi_{34} = -F_2 + K_j^T B_i^T F_3^T, \quad \Phi_{35} = F_2 B_f, \quad \Phi_{36} = F_2 B_w$$
$$\Phi_{44} = \bar{\tau}_{sc} Q_1 + \bar{\tau}_{ca} Q_2 - F_3 - F_3^T, \quad \Phi_{45} = \lambda + F_3 B_w$$
$$\Phi_{46} = F_3 B_w, \quad \Phi_{55} = -T - T^T, \quad \Phi_{66} = -\gamma^2 I$$

Proof. For a given scalar $\gamma > 0$, $w(t) \in L[0, +\infty)$ under the condition $x(t) = 0,\ \exists t \in [-h, 0]$, the H_∞ performance index is defined to be

$$J_{zw} = \int_0^\infty \left(z^T(t)z(t) - \gamma^2 w^T(t)w(t)\right) \mathrm{d}t \tag{18}$$

Note that

$$J_{zw} = \int_0^\infty \left(z^T(t)z(t) - \gamma^2 w^T(t)w(t) + \dot{V}(x(t))\right) \mathrm{d}t - \lim_{t\to\infty} V(x(t)) + V(x(0)) \tag{19}$$

Under zero initial condition, we have

$$J_{zw} \le \int_0^\infty \left(z^T(t)z(t) - \gamma^2 w^T(t)w(t) + \dot{V}(x(t))\right) \mathrm{d}t \tag{20}$$

From (7), the following equation holds:

$$\begin{aligned} 2\left[x^T(t)F_1 + x^T(t-\tau(t))F_2 + \dot{x}^T(t)F_3\right]&[-\dot{x}(t) + Ax(t) + BKx(t-\tau(t)) \\ &+B_f f(t) + B_w w(t)\big] = 0 \end{aligned} \tag{21}$$

From (15) and (21), it can be established that

$$J_{zw} \le \zeta^T(t)\bar{\Phi}^{ij}\zeta(t) \tag{22}$$

where $\zeta(t) = \left[\xi^T(t)\ w^T(t)\right]^T$ and

$$\bar{\Phi}^{ij} = \begin{bmatrix} \Phi_{11} & \Phi_{12} & \Phi_{13} & \Phi_{14} & F_1 B_f & \Phi_{16} \\ * & \Phi_{22} & \Phi_{23} & 0 & 0 & 0 \\ * & * & \Phi_{33} & \Phi_{34} & \Phi_{35} & \Phi_{36} \\ * & * & * & \Phi_{44} & \Phi_{45} & \Phi_{46} \\ * & * & * & * & 0 & \Phi_{56} \\ * & * & * & * & * & \Phi_{66} \end{bmatrix}, \quad \text{for } 1 \le i \le j \le r. \tag{23}$$

Then, we consider the nonlinear term, if $\Phi^{ij} < 0$, then

$$J_{zw} - 2\sum_{j=1}^{m} t_j f_j(x_j(t)) \left[f_j(x_j(t)) - h_j x(t)\right] < 0 \tag{24}$$

And from (24), we can conclude that $J_{zw} < 0$, then the system is asymptotically stable, this complete the proof .

4 Numerical Examples

Example 1 Consider the following system:

$$\begin{aligned} Rule1 &: If \alpha_1 is \pm \tfrac{\pi}{2}, then \dot{x}(t) = A_1 x(t) + B_1 x(t-\tau(t)) \\ Rule2 &: If \alpha_2 is \pm 0, then \dot{x}(t) = A_2 x(t) + B_2 x(t-\tau(t)) \end{aligned} \tag{25}$$

where

$$A_1 = \begin{bmatrix} -1 & 0 \\ 0 & -0.9 \end{bmatrix}, \; B_1 = \begin{bmatrix} -1 & 0 \\ -1 & -1 \end{bmatrix}, \; A_2 = \begin{bmatrix} -1 & 0.5 \\ 0 & -1 \end{bmatrix}, \; B_2 = \begin{bmatrix} -1 & 0 \\ 0.1 & -1 \end{bmatrix}.$$

By applying Theorem 1 to the above system, the maximum value of $\bar{\tau}$ ensuring the asymptotic stability is computed as 1.8970. For comparison, Table 1 lists the upper bounds obtained from the criteria in [4, 5]. It is seen that our method in Theorem 1 leads to better results.

Example 2 Consider system (7) with the following parameter:

$$\begin{aligned} &A = \begin{bmatrix} -2.7 & 0 \\ 1.6 & -2.4 \end{bmatrix}, \; B = \begin{bmatrix} -2 & 1 \\ 0.1 & 0.7 \end{bmatrix}, \; B_f = \begin{bmatrix} 1 & 2 \\ 2 & -2 \end{bmatrix}, \; B_w = \begin{bmatrix} -2 & 1 \\ 1.2 & 1.55 \end{bmatrix}, \\ &C = \begin{bmatrix} 0.1 & 0 \\ 0.2 & 0.1 \end{bmatrix}, \; D = \begin{bmatrix} -0.2 & -0.1 \\ 0.1 & -0.3 \end{bmatrix} \text{ and } K = \begin{bmatrix} 1 & 0 \\ 0 & 1 \end{bmatrix}. \end{aligned} \tag{26}$$

Let $\gamma = 0.9$. Then from Theorem 2, we can obtain the following Table 2.

Table 1 Comparisons of maximum allowed delay $\bar{\tau}$ For $d = 0$

Methods	[4]	[5]	Theorem 1
$\bar{\tau}$	1.5974	1.6609	1.8970

Table 2 Allowable upper bound of $\bar{\tau} = \bar{\tau}_{sc} + \bar{\tau}_{ca}$ with $d = 0.1$

Upper bound $\bar{\tau}_{ca}$ for given $\bar{\tau}_{sc}$			Upper bound $\bar{\tau}_{sc}$ for given $\bar{\tau}_{ca}$		
$\bar{\tau}_{sc} = 1$	$\bar{\tau}_{sc} = 1.2$	$\bar{\tau}_{sc} = 1.5$	$\bar{\tau}_{ca} = 0.2$	$\bar{\tau}_{ca} = 0.3$	$\bar{\tau}_{ca} = 0.5$
2.259	2.139	1.912	2.483	2.463	2.413

5 Conclusion

In this paper, the stability analysis of nonlinear networked control systems described by Takagi–Sugeno fuzzy systems has been investigated. We also considered the H_∞ stability problem. By employing Lyapunov approach and LMI technique, sufficient conditions for stability and H_∞ performance of the concerned systems are presented in terms of LMIs. Finally, numerical examples are provided to show the effectiveness of our proposed conditions.

References

1. Takagi, T., Sugeno, M.: Fuzzy identification of systems and its application to modeling and control. IEEE Trans. Syst. Man Cybernet **15**, 116–132 (1985)
2. Lamrabet, O., Ech-charqy, A., Tissir, E.H., El Haoussi, F.: Sampled data control for Takagi-Sugeno fuzzy systems with actuator saturation. Second. Int. Conf. Intell. Comput. Data Sci. **148**, 448–454 (2019)
3. Idrissi, S., Tissir, E.H., Boumhidi, I., Chaibi, N.: New delay dependent robust stability criteria for TS fuzzy systems with constant delay. Int. J. Control. Autom. Syst. **11**, 885–892 (2013)
4. Peng, C., Yue, D., Tian, Y.-C.: New approach on robust delay-dependent H_∞ control for uncertain T-S fuzzy systems with interval time-varying delay. IEEE Trans. Fuzzy Syst. **17**, 890–900 (2009)
5. Kwon, O.M., Park, M.J., Lee, S.M., Park, J.H.: Augmented Lyapunov-Krasovskii functional approaches to robust stability criteria for uncertain Takagi-Sugeno fuzzy systems with time-varying delays. Fuzzy Set Syst. **201**, 1–19 (2012)
6. Tissir, E.H.: Delay dependent robust stabilization for systems with uncertain time varying delays. J. Dyn. Syst. Meas. Control. **132**(5), 054504 (2010)
7. Ech-charqy, A., Ouahi, M., Tissir, E.H.: Robust observer-based control design for singular systems with delays in states. Int. J. Autom. Smart Technol. **8**(3), 127–137 (2018)
8. Ech-charqy, A., Ouahi, M., Tissir, E.H.: Delay dependent robust stability criteria for singular time-delay systems by delay partitioning approach. Int. J. Syst. Sci. **49**(14), 2957–2967 (2018)
9. Liu, S., Liu, P.X., Saddik, A.E.: A stochastic security game for Kalman filtering in networked control systems (NCSs) under denial of service (DoS) attacks. Ifac Proc. Vol. **46**, 106–111 (2013)
10. El Fezazi, N., El Haoussi, F., Tissir, E.H., Alvarez, T.: Design of robust H_∞ controller for congestion control in data networks. J. Frankl. Inst. **357**, 7828–7845 (2017)
11. Wu, Z.G., Xu, Y., Pan, Y.J., Su, H., Tang, Y.: Event-triggered control for consensus problem in multi-agent systems With quantized relative state measurements and external disturbance. IEEE Trans. Circuits Syst. I Regul. Pap. **1–11** (2018)
12. Zhang, H., Yang, J.: T-S fuzzy-model-based robust H_∞ design for networked control systems with uncertainties. IEEE Trans Ind. Inform. **3**, 289–301 (2007)
13. Niu, Y., Jia, T., Wang, X., Yang, F.: Output-feedback control design for NCSs subject to quantization and dropout. Inf. Sci. **179**, 3804–3813 (2009)
14. Liu, J., Yue, D.: Event-triggering in networked systems with probabilistic sensor and actuator faults. Inf. Sci. **240**, 145–160 (2013)
15. Gu, K., Kharitonov, V.L., Chen, J.: Stability of Time-Delay Systems. Birkhauser, Basel (2003)

Driver Behavior Assistance in Road Intersections

Safaa Dafrallah, Aouatif Amine, Stéphane Mousset and Abdelaziz Bensrhair

Abstract One of the main reasons of intersection accidents is related to drivers' behavior. Differences on age, gender or even personality affected drivers speed and level of aggressiveness which may cause huge problems and fatal crashes in intersections. Those differences assembled with others, such us vehicles type and the approach speed, may lead to different reactions while facing the yellow signal, which cause dilemma zone in signalized intersections. In this paper, we will report different authors views concerning drivers' behavior in both signalized and unsignalized intersections with a focus on the major predictors of aggressiveness.

Keywords Drivers' behavior · Intersections · Dilemma zone · Yellow light signal

1 Introduction

Thousands of people are dying every year due to road accidents, it becomes the first cause of death for those who are under 30. The main reasons for that accidents are human and drivers behavior. A wrong behavior in a wrong time and a wrong place can make a fatal accident. According to the Department of Transportation (DoT) in United States of America (USA), around 45% of injury crashes and 22% of roadway fatalities in the USA are intersection related. In a signalized intersection, each driver reacts differently at the onset of the yellow signal, the majority of them get confused about the way they should respond to that signal hesitated between decelerating and stopping or accelerating and running for crossing the intersection before the red phase signal. A wrong decision in such a time can have serious consequences, a decision of stopping can involve a rear end crashing and a decision of passing may cause a right angle accident, this zone of confusion is called the dilemma zone (DZ). The

S. Dafrallah · A. Amine
National School of Applied Sciences, Kenitra, Morocco

S. Dafrallah (✉) · S. Mousset · A. Bensrhair
National Institute of Applied Sciences, Rouen, France
e-mail: dafrallah.safaa@uit.ac.ma

V. Bhateja et al. (eds.), *Embedded Systems and Artificial Intelligence*,
Advances in Intelligent Systems and Computing 1076,
https://doi.org/10.1007/978-981-15-0947-6_6

remainder of this paper is organized as follows: Related works concerning dilemma zone are exposed in Sect. 2, in Sect. 3 we will analyze the different drivers' behaviors according to several authors, then we cited some of the most relevant predictors of aggressiveness indicated by different researchers.

This work is supported by the Moroccan Ministry of Environment, Transport, Logistics and Water (METLE), in collaboration with the National Center for the Scientific and Technical Research (CNRST).

2 Related Works

Most of the accidents in signalized intersections are related to the wrong management of the drivers for the relation between space and time while facing the yellow signal which lead to a zone of confusion where the driver ignored the well action to do, this zone is named the dilemma zone. Two types of dilemma zone exist: The first is caused by a short duration of the yellow light, while the second is due to drivers' behavior. The dilemma zone (DZ) was first introduced by Gazis et.al [1] as an area where the driver could neither stop safely nor cross the intersection before the end of the yellow signal [2]. The author defined the first type of the dilemma zone as a problem that is generally caused by the timing of the yellow signal that should be long enough to allow the driver either to stop or to pass the intersection safely [3]. So an incorrect choice in the length of the yellow signal leads to a dilemma zone. If a driver chooses to stop in an intersection, his vehicle must be at the onset of the yellow signal in a distance which is superior from the minimum stopping distance named X_c; however, in the case of crossing the vehicle must be in a distance which is inferior from the maximum yellow passing distance named X_0. A vehicle in a distance from the intersection which is smaller than X_c cannot stop safely, while a vehicle in a distance greater than X_0 cannot cross the intersection before the red light without accelerating. Then if the driver finds himself in a region between X_c and X_0 where $X_c > X_0$ so his in a region called the dilemma zone type 1 as shown in Fig. 1.

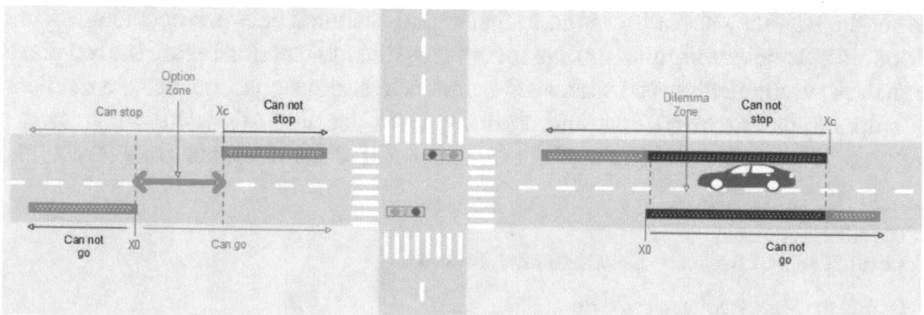

Fig. 1 Dilemma and option zones

2.1 Dilemma Zone Type I

According to Gazis [1], a driver can avoid being in a dilemma zone if his approach speed is less than the speed limit, because the critical distance X_c decreases rapidly as the approach speed decreases. So in this case, if the $X_c < X_0$ the driver would be in a zone which is called the option zone where he can both stop or cross the intersection safely. The stopping distance is the distance that the vehicle travels before it comes to stop, the minimum safe stopping distance depends on many factors such as the approach speed of the vehicle (V_0), the deceleration rate (a_{stop}), as well as the drivers ability to react at stopping which is known by the drivers perception reaction time to stop (δ_{stop})
The formula is:

$$X_c = V_0.\delta_{\text{stop}} + \frac{V_0^2}{2.a_{\text{stop}}} \tag{1}$$

The maximum yellow passing distance is the required distance for a vehicle to cross the intersection before the onset of red light with an acceptable acceleration. This maximum distance depends also on different factors such as the approach speed (V_0), the length of the yellow signal (τ), the acceleration rate (a_{run}), the perception reaction time of the driver to pass (δ_{run}), the width of the intersection (W) and the length of the vehicle (L). The formula is:

$$X_0 = V_0.\tau + \frac{1}{2}a_2(\tau - \delta_2)^2 - (W + L) \tag{2}$$

So the dilemma zone type I is (1)–(2):

$$X_c - X_0 = V_0.\delta_{\text{stop}} + \frac{V_0^2}{2.a_{\text{stop}}} - [V_0.\tau + \frac{1}{2}a_{\text{run}}(\tau - \delta_{\text{run}})^2 - (W + L)] \tag{3}$$

According to the Gazis, Herman and Maradudin (GHM) model the parameters of DZ type I have constant values where the approach speed (V_0) always equal the speed limit, as well as the perception reaction time for passing always equal the one for stopping ($\delta_{\text{stop}} = \delta_{\text{run}}$). It is also assumed that the DZ type I can be eliminated if the yellow signal duration complies with the one recommended by the Institute of Transportation Engineers (ITE) handbook. In addition to that affecting constant values to the parameter of the dilemma zone can cause inexact calculation of it, because those values could be different with the difference of the driver behaviors, which leads to the second type of the dilemma zone.

2.2 Dilemma Zone Type II

The second type of the dilemma zone was defined by Zegeer and Deen [4], by defining its boundaries:

- In terms of distance from the area, where 90% of drivers would stop at the onset of the yellow signal to the area where only 10% of drivers would stop.
- In terms of time, its boundaries are between 5.5 and 2.5 s from the stop bar, this time is equal to the decision of the 90th and 10th percentile of drivers, respectively.

3 Analysis of Drivers' Behavior

Some researchers have classified the drivers' behavior in two categories:

- Compliant
- Aggressive.

While others have grouped it into three categories:

- Conservative
- Normal
- Aggressive.

Aoude et al. [5] defined the aggressive driver as the one who does not stop before the stop bar or at most 3 m after the stop bar, while those who do stop are compliant.

Liu et al. [6] classified the drivers' behavior in three categories:

- The conservative driver: the one who stops even if the distance to the stop bar (ds) is smaller than the critical distance of stopping (X_c), which can cause a rear-end collision.
- The normal driver: the one who stops when the distance to the stop bar is greater than the critical distance of stopping and cross the intersection when it is smaller.
- The aggressive driver: the one who crosses the intersection regardless of the differences between the distance to the stop bar and the critical distance of stopping.

Papaioannou et al. [7] classified also the driver behavior in three categories but this time by assembling two criteria:

- The approach speed: drivers who exceed the limit speed by a certain percentage are considered as aggressive.
- Drivers reaction at the onset of the yellow signal (stopping or crossing the intersection).

For the case of an unsignalized intersection, Kaysi et al. [8] defined the aggressiveness of a vehicle is modeling by the gap acceptance between two vehicles. An aggressive driver is the one who accepts a gap which is smaller than the critical gap.

4 Predictors of Aggressiveness

The most common predictors used in different studies to classify the drivers' behavior are:

- Approach speed
- Driver characteristics (gender, age, personality)
- Vehicles type
- Road surface
- Duration of the yellow light
- Distance to the intersection
- Time to the intersection.

In his research, Papaioannou [7] found that female drivers are less aggressive than male ones, according to him, 90% of women drivers stop after the onset of the yellow signal, while only 50% do stop from males. This difference in behavior is usually caused by the difference in speed. As he also found that the majority of drivers who stopped, faced the yellow signal at a distance of 50 m, while those who choose to cross the intersection was at 40 m from it at the onset of the yellow light, and at 49 m about 50% of drivers choose to stop and 50% choose to cross it.

In [9], Koll et al. studied the behavior of drivers with and without the flashing green, concluding that the flashing green produces a large option zone, which gives the opportunity for the driver to choose either to stop or to cross the intersection which consequently reduces the dilemma zone.

Liu [6] concluded that the approach speed of the aggressive driver is about 10–20% higher than the mean speed of the traffic, while the approach speed of the conservative one is lower for about 10–15% from the mean speed.

In his paper [10], Elhenawy (2015) introduced a new predictor to detect the driver aggressiveness, which is based on the computation of the number of accelerations that the driver does when the Time To Intersection (TTI) is smaller than the duration of the yellow signal as well as if its speed is greater or equal to the limit speed. When the value of this new predictor is near to 1, it means that the driver stops rarely so his an aggressive one. Rhodes et al. in [11] studied the effect of the differences between ages and gender in risky driving, by using a phone survey for teens between 16 and 20 years and adults from 25 to 45 years for both genders. While comparing the values of the risk perception and the positive effects for both genders and ages, they have concluded that the risk perception is a higher predictor for the risky driving for females while the positive effect is for males. Concerning ages, they have found that the risk perception is a higher predictor for adults while the positive effects are for teenagers.

Poó et al. [12] compared between personality and driving styles using the Multidimensional Driving Style Inventory (MDSI) proposed by Taubman-Ben-Ari et al [13]. This MDSI evaluates five driving styles:

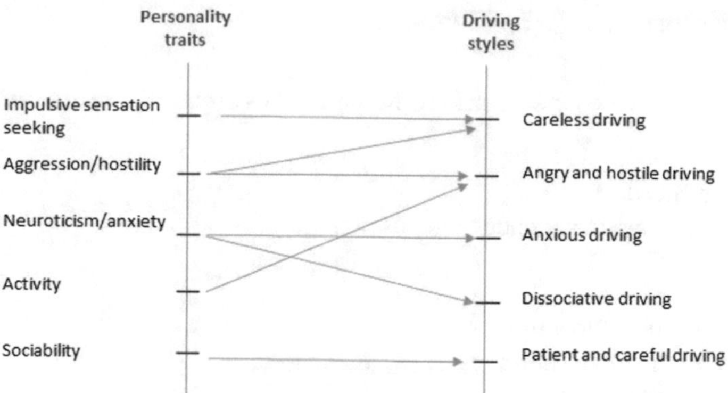

Fig. 2 Relationship between personality traits and driving styles

- Careless driving
- Anxious driving
- Angry and hostile driving
- Patient and careful driving
- Dissociative driving.

Each driving style was affected to a personality trait following the Zuckerman five-factor model [14] which are:

- Impulsive sensation seeking
- Aggression-hostility
- Neuroticism-anxiety
- Activity
- Sociability.

Concluding that each driving style is related to a personality trait as shown in Fig. 2.

What concern the unsignalized intersections, the research done by Kaysi et al. [8], shows that the aggressiveness increases if the driver is young (under 26 years) and driving a sport car, moreover driving a sport car increase the probability of doing an aggressive act two times more than a young age do.

5 Used Methods

Most of the studies have used the binary regression logit model to explain the driver choice in a signalized intersection since the driver has just two choices to stop or to run [7, 8, 10, 15, 16]. As well as to estimate the critical distance of the intersection [6]. To calculate the brake response time and the deceleration rate, Gates et al. [15] have used the multiple regression model, this method was also used by [10–12] to predict,

respectively, the driver decision, evaluate the effect of the personality trait in each driving style and to evaluate the effect of different predictors on drivers depending on their ages and gender.
The Support Vector Machine (SVM) model has been used for the prediction of the driver behavior and to classify it as a compliant or a violent one [5, 10], while the Hidden Markov Model (HMM) was used for the prediction of the driver behavior.

The log-likelihood was used to determine the importance of each parameter on the influence of the driver behavior [8, 16]. Also the t-test was used to explain the aggressive behavior in terms of predictors to find the most significant predictor and also to compare the evolution of the classification and the false positive rate before and after the new predictor as used in [8]. In [6], Liu et al. have also used the t-test to examine the difference of speed between drivers' behavior characteristics.

6 Conclusion

As any innovation, the car industrialization has its benefits but contains also perilous drawbacks that increases mortality in the worldwide, which leads authorities and researchers to make road safety one of their priorities. Many procedures have been applied to reduce road accidents, however the drivers' behavior remains without significant contributions to make it more compliant, thus our future works would be in this sense.

Acknowledgements This research work is supported by the "SAFEROAD Meta-plateforme pour la Sécurité Routière (MSR)" project under contract No:24/2017.

References

1. Gazis, D. Herman, R., Maradudin, A.: The problem of the amber signal light in traffic flow. Oper. Res. Inc. no. August 2015 (1960)
2. Zhang, Y., Fu, C.: Yellow light dilemma zone researches: a review. Traffic Transp. Eng. (2014)
3. Urbanik, T.: The dilemma with dilemma zones. In: Proceedings of Institute of Transportation Engineering (ITE) District (2007)
4. Zegeer, C.V., Deen, R.C.: Green-extension systems at high-speed intersections. ITE J. **48**(11), 19–24 (1978)
5. Aoude, G.S., Desaraju, V.R., Stephens, L.H., How, J.P.: Behavior classification algorithms at intersections and validation using naturalistic data. IEEE Intell. Veh. Symp. (IV) (Iv), 601–606 (2011)
6. Liu, Y., Chang, G.-L., Tao, R., Hicks, T., Tabacek, E.: Empirical observations of dynamic dilemma zones at signalized intersections. J. Transp. Res. Board **2035**(2035), 122–133 (2007)
7. Papaioannou, P.: Driver behaviour, dilemma zone and safety effects at urban signalised intersections in Greece. Accid. Anal. Prev. **39**, 147–158 (2007)
8. Kaysi, I.A., Abbany, A.S.: Modeling aggressive driver behavior at unsignalized intersections. Accid. Anal. Prev. **39**, 671–678 (2007)

9. Köll, H., Bader, M., Axhausen, K.W.: Driver behaviour during flashing green before amber : a comparative study. Accid. Anal. Prev. **36**, 273–280 (2004)
10. Elhenawy, M., Jahangiri, A., Rakha, H.A., El-shawarby, I.: Modeling driver stop / run behavior at the onset of a yellow indication considering driver run tendency and roadway surface conditions. Accid. Anal. Prev. **83**, 90–100 (2015)
11. Rhodes, N.: Age and gender differences in risky driving : The roles of positive affect and risk perception. Accid. Anal. Prev. (May 2011) (2017)
12. Poó, F., Nacional, U., Mar, D.: A study on the relationship between personality and driving styles. Traffic Inj. Prev. (2013)
13. Taubman-Ben-Ari, O., Mikulincer, M., Gillath, O.: The multidimensional driving style inventory scale construct and validation. Accid. Anal. Prev. **36**(3), 323–332 (2004)
14. Zuckerman, M.: Zuckerman-kuhlman personality questionnaire (zkpq): an alternative five-factorial model. Big Five Assess. 377–396 (2002)
15. Gates, T.J., Noyce, D.A., Laracuente, L., Nordheim, E.V.: Analysis of driver behavior in dilemma zones at signalized intersections. J. Transp. Res. Board (2030), 29–39
16. Savolainen, P.T.: Class logit models. Accid. Anal. Prev. 1–8 (2016)

BER Performance of CE-OFDM System: Over AWGN Channel and Frequency-Selective Channel Using MMSE Equalization

J. Mestoui, M. El Ghzaoui and K. El Yassini

Abstract The fifth generation (5G) wireless network technology is to be standardized by 2020, where main goals are to improve capacity, reliability, and energy efficiency. Constant envelope orthogonal frequency division multiplexing (CE-OFDM) is a technique to modify the OFDM signal with high peak-to-average power ratio (PAPR) to a constant envelope zero decibel PAPR waveform, in order to minimize energy consumption in future wireless communication systems (5G). The conventional CE-OFDM uses inverse discrete Fourier transform (IDFT) to calculate the signal time samples before feeding them to the phase modulator. In this paper, the performances of the CE-OFDM are studied in terms of the BER over AWGN channel and frequency-selective fading channel.

Keywords OFDM · CE-OFDM · PAPR · Phase modulator · BER · Frequency-selective fading channel

1 Introduction

The goal of 5G is to evolve 4G networks to meet different complementary needs, increase bit rates for mobile services, support low-power and long-range communications for connected objects, and finally, real-time communications for critical applications. All while targeting specific uses presented in Fig. 1, such as connected and autonomous vehicles, health and connected objects. 5G technology is designed

J. Mestoui (✉) · M. El Ghzaoui · K. El Yassini
Moulay Ismail University, BP 11201 Zitoune, Meknès, Morocco
e-mail: mestoui@hotmail.com

M. El Ghzaoui
e-mail: m.elghzaoui@fpe.umi.ac.ma

K. El Yassini
e-mail: khalid.elyassini@gmail.com

M. El Ghzaoui
Moulay Ismail University, BP 512 Boutalamine, Meknès, Morocco

V. Bhateja et al. (eds.), *Embedded Systems and Artificial Intelligence*,
Advances in Intelligent Systems and Computing 1076,
https://doi.org/10.1007/978-981-15-0947-6_7

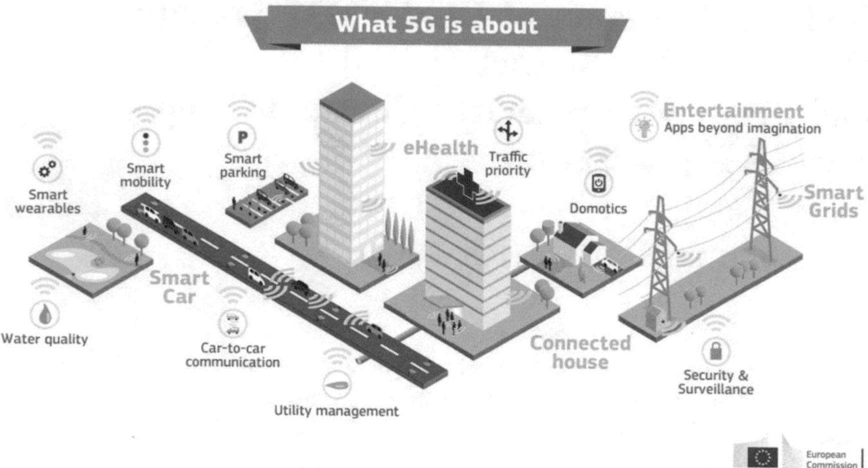

Fig. 1 5G technology applications

to support data rates of several Gbps [1, 2]. To achieve this goal, one of the difficult tasks of wireless transceivers is to generate a millimeter wave (mmW) band above 24 GHz [3, 4] with a reduced PAPR waveform.

Unfortunately, one of the limitations with the mm-Wave wireless communications is the increased link loss due to the reduced wavelength [5]. In order to compensate relatively large propagation attenuation at mm-Wave frequencies, the antenna array with the ability to generate directional beams would be essential in 5G mm-Wave applications [6–8]. The switch to 5G would be a significant departure from 4G systems, which are based on OFDM, exploiting its flexible resource allocation and simplified equalization. An important approach that eliminates PAPR and still allows one to maintain most of the advantages and functional blocks of OFDM is the constant envelope OFDM approach (CE-OFDM).

Orthogonal frequency division multiplexing (OFDM) uses its attractive properties such as (i) good spectral efficiency, (ii) capacity to minimize inter-symbol interference effects (ISI), and (iii) multi-path distortion resistance to support high-speed data links over wireless channels characterized by severe multi-path fading [9–14]. OFDM has been chosen for that matter in several telecommunication standards (ADSL, Wi-Max, IEEE 802.11a/g/n, LTE, DVB,…).

However, the main disadvantage of OFDM is that the modulated waveform has high amplitude fluctuations that produce large peak-to-average power ratios (PAPR) [15–17]. The high PAPR increases the OFDM's sensitivity to nonlinear distortion caused by the transmitter's power amplifier (PA) [18]. In the absence of sufficient power reduction, the system suffers from spectral broadening, intermodulation distortion, and, therefore, performance degradation [19–22]. These problems can be reduced by increasing the back off, but these result in reduced PA efficiency [23]. For battery-powered mobile devices, this is particularly problematic due to limited energy resources [24–26]. For future systems such as 5G, efficient amplification is therefore essential [27].

To reduce problems related to nonlinear amplification, the joint approaches to PAPR reduction and linearization of PA have been developed [28]. These techniques may include increased complexity, reduced spectral efficiency, and performance degradation. As a result, OFDM is considered inefficient in terms of power, which is undesirable especially for battery-powered wireless systems. An alternative approach to mitigate the PAPR problem is based on signal transformation. This technique involves a transformation of the signal before the amplification, then a reverse transformation at the receiver before the demodulation [29]. The advantage of the transform of the phase modulator is that the transformed signal has the lowest possible PAPR, 0 dB.

The constant envelope OFDM approach studied in [29] proposes to use the OFDM waveform to phase modulate the carrier. Such a signal can be amplified with a minimum power reduction, which (1) maximizes the efficiency of the power amplifier and (2) potentially increases the range of the system because more power of the signal is radiated into the channel. Therefore, these constant envelope signals can be amplified by highly nonlinear, inexpensive, and highly efficient PAs such as those of class D, E, and F [30].

The rest of this article is organized as follows: The model of the CE-OFDM system is described and characterized in Sect. 2. Section 3 presents a study of frequency-selective channels. Section 4 presents the results of the simulation. Section 5 concludes the work and indicates future work.

2 System Model

In CE-OFDM system, the OFDM signal is used to phase modulate the carrier. This is in contrast to conventional OFDM in which amplitude modulates the carrier. CE-OFDM can also be thought of as a transformation technique, as shown in Fig. 2. At the transmitter, the high PAPR OFDM signal is transformed into a low PAPR signal prior to the power amplifier. At the receiver, the inverse transformation is performed prior to demodulation.

The baseband OFDM waveform

$$m(t) = \sum_{i} \sum_{k=1}^{N} I_{i,k} q_k(t - iT_B) \tag{1}$$

where $\{I_{i,k}\}$ are the data symbols, and $\{q_k(t)\}$ are the orthogonal subcarriers. For CE-OFDM, before amplification $m(t)$ is passed through a phase modulator.

The baseband CE-OFDM signal is

$$s(t) = Ae^{j\phi(t)} \tag{2}$$

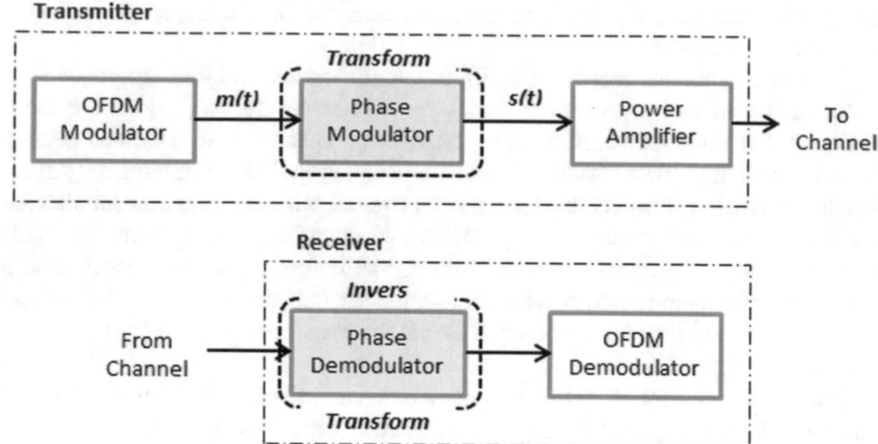

Fig. 2 Basic concept of CE-OFDM

where A is the signal amplitude, and $\phi(t)$ is the information bearing phase signal. The advantage of the CE waveform is that the instantaneous power is constant $|s(t)|^2 = A^2$. Consequently, the PAPR is 0 dB, and the required back off is 0 dB. The PA can therefore operate at the optimum (saturation) point, maximizing average transmit power (good for range) and maximizing PA efficiency (good for battery life).

The phase signal during ith block is written as

$$\phi(t) = \theta_i + 2\pi h C_N \sum_{k=1}^{N} I_{i,k} q_k(t - iT_B), \quad iT_B \le t < (i+1)T_B \tag{3}$$

where h is referred as the modulation index, and θ_i is a memory term. The normalizing constant, C_N, is set to

$$C_N = \sqrt{\frac{2}{N\sigma_I^2}} \tag{4}$$

where σ_I^2 is the data symbol variance:

$$\sigma_I^2 = E\left\{\left|I_{i,k}\right|^2\right\} = \frac{M^2 - 1}{3} \tag{5}$$

which is only a function of the modulation index. The signal energy is

$$\varepsilon_s = \int_{iT_B}^{(i+1)T_B} |s(t)|^2 \mathrm{d}t = A^2 T_B \tag{6}$$

and the bit energy is

$$\varepsilon_b = \frac{\varepsilon_s}{N\log_2 M} = \frac{A^2 T_B}{N\log_2 M} \tag{7}$$

CE-OFDM requires a real-valued OFDM message signal. Therefore, the data symbols in (1) are real-valued as follows:

$$I_{i,k}\in\{\pm 1, \pm 3, \ldots, \pm(M-1)\} \tag{8}$$

This one-dimensional constellation is known as pulse-amplitude modulation (PAM). The subcarriers $\{q_k(t)\}$ must also be real-valued. Three possibilities are considered [29]: half-wave cosines, half-wave sines, and full-wave cosines and sines. For $k = 1, 2, \ldots, N$, full-wave cosines and sines are as follows:

$$q_k(t) = \begin{cases} \cos 2\pi kt/T_B & 0 \le t < T_B;\ k \le N/2 \\ \sin 2\pi(k - N/2)/T_B & 0 \le t < T_B;\ k > N/2 \\ 0 & \text{othewise} \end{cases} \tag{9}$$

In terms of implementation, it can be computed with a discrete cosine transform (DCT), with a discrete sine transform (DST), and by taking the real part of a discrete Fourier transform (DFT), or equivalently by taking a 2N-point DFT of a conjugate symmetric data vector as processed in [31].

The smoothest phase results from continuous phase-DCT (CP-DCT) which, unlike DST and CP-DFT, has a first derivative equal to zero at the boundary times $t = iT_B$. Consequently, the CP-DCT is the most spectrally contained [31]. In this paper, the discrete cosines and sines transform is used for the simulation system.

In Fig. 3, the CE-OFDM system, described by Steve Thompson in [31], shares many of the same functional blocks as conventional OFDM. Therefore, an existing OFDM system can provide an additional CE-OFDM mode with relative ease, particularly for the case of software defined platforms.

In the block diagram of an uncoded CE-OFDM system, the source bits generate an information sequence, $X[k]$, which is passed through an inverse discrete Fourier transform (IDFT) block and then phase modulated to generate the sequence $s[n]$. A cyclic prefix (CP) sequence is added as in standard OFDM (the high PAPR OFDM

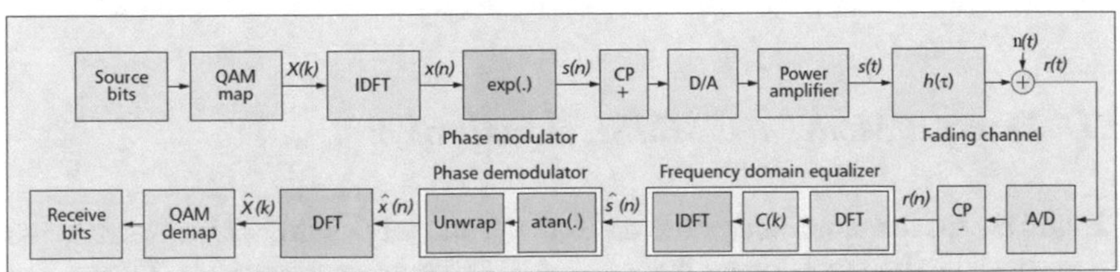

Fig. 3 Baseband block diagram of a CE-OFDM system [31]

signal is transformed into a low PAPR signal), and the signal is passed through the power amplifier before being transmitted. At the receiver, the CP is first removed, the DFT of the signal is generated, and equalization in the frequency domain is carried out. In order to undo the effect of the phase modulator, an oversampled IDFT, an arctan(.) operation, phase unwrapping, and a DFT are performed. In essence, the OFDM waveform X[k] is used to phase modulate the carrier.

In the discrete-time phase demodulator, a finite impulse response (FIR) filter is optional, but has been found effective at improving performance; arctan(·) simply calculates the arctangent of its argument, and the phase unwrapper is used to minimize the effect of phase ambiguities. As will be shown, the unwrapper makes the receiver insensitive to phase offsets caused by the channel and/or by the memory terms. Notice that a well-known characteristic of phase demodulator receivers is at low carrier-to-noise ratio (CNR), below a threshold value, the system performance degrades drastically. A commonly accepted threshold CNR for analog FM systems is 10 dB [29].

3 Equalization and Channel Models

In this section, the performance of CE-OFDM in frequency-selective channels is studied. The channel is time dispersive having an impulse response $h(\tau)$ that can be nonzero over $0 \leq \tau < \tau_{\max}$, where $\tau_{\max}$ is the channel's maximum propagation delay. The received signal is

$$\begin{aligned} r(t) &= \int_{-\infty}^{+\infty} h(\tau)s(t-\tau)\mathrm{d}\tau + n(t) \\ &= \int_{0}^{\tau_{\max}} h(\tau)s(t-\tau)\mathrm{d}\tau + n(t) \end{aligned} \tag{10}$$

where $s(t)$ is the CE-OFDM signal, and $n(t)$ is the complex Gaussian noise term. The lower bound of integration in (10) is due to the law of causality: $h(\tau) = 0$ for $\tau < 0$. The upperbound is $\tau_{\max}$ since, by definition of the maximum propagation delay, $h(\tau) = 0$ for $\tau > \tau_{\max}$.

3.1 Discrete Model and MMSE Equalization

CE-OFDM has the same block structure as conventional OFDM, with a block period, T_B, designed to be much longer than $\tau_{\max}$. A guard interval of duration $T_g \geq \tau_{\max}$ is inserted between successive CE-OFDM blocks to avoid interblock interference. At the receiver, $r(t)$ is sampled at the rate Fsa = 1/Tsa samp/s, the guard time samples are discarded, and the block time samples are processed. Using the discrete-time model outlined in [29], the processed samples are

$$r[i] = \sum_{m=0}^{N_C-1} h[m]s[i-m] + n[i], \quad i = 0, \ldots, N_B - 1 \tag{11}$$

Note that the eliminated samples are $\{r[i]\}_{i=-N_g}^{-1}$. Transmitting a cyclic prefix during the guard interval makes the linear convolution with the channel equivalent to circular convolution. Thus,

$$r[i] = \frac{1}{N_{\mathrm{DFT}}} \sum_{k=0}^{N_{\mathrm{DFT}}-1} H[k]S[k]\mathrm{e}^{j2\pi ik/N_{\mathrm{DFT}}}, \quad i = 0, \ldots, N_B - 1 \tag{12}$$

where $\{H[k]\}$ is the DFT of $\{h[i]\}$, and $\{S[k]\}$ is the DFT of $\{s[i]\}$.

$$\hat{s}[i] = \frac{1}{N_{\mathrm{DFT}}} \sum_{k=0}^{N_{\mathrm{DFT}}-1} R[k]C[k]\mathrm{e}^{j2\pi ik/N_{\mathrm{DFT}}}, \quad i = 0, \ldots, N_B - 1 \tag{13}$$

where $\{R[k]\}$ is the DFT of the processed samples, and $\{C[k]\}$ is the equalizer correction terms, which are computed in [32].

The MMSE criterion takes into account the signal-to-noise ratio, making an optimum trade between channel inversion and noise enhancement. Notice that the MMSE and ZF are equivalent at high SNR as follows:

$$\begin{aligned} \lim_{\varepsilon_b/N_0 \to \infty} C[k]|_{\mathrm{MMSE}} &= \lim_{\varepsilon_b/N_0 \to \infty} \frac{H^*[k]}{|H[k]|^2 + (\varepsilon_b/N_0)^{-1}} \\ &= \frac{H[k]}{|H[k]|^2} = \frac{1}{H[k]} = C[k]|_{ZF} \end{aligned} \tag{14}$$

3.2 Channel Models

Mathematically, linear channels with stochastic temporal variations are described by Bello [33]. The widely used assumption of WSSUS (broadly stationary uncorrelated scattering) is applied. In addition, it is assumed that the channel is composed of discrete paths, each having an associated gain and a discrete delay. The impulse response of the channel is

$$h(\tau) = \sum_{l=0}^{L-1} a_l \delta(\tau - \tau_l) \tag{15}$$

where a_l is the complex channel gain, and τ_l is the discrete propagation delay of the lth path; the total number of paths is represented by L. The delay of the 0th path is defined as $\tau_0 \equiv 0$, thus

$$\tau_l = lT_{sa}, \quad l = 0, 1, \ldots, L-1 \tag{16}$$

For each simulation trial, the set of path gains $\{a_l\}_{l=0}^{L-1}$ are generated randomly. Each gain is complex valued and has a zero mean and a variance

$$\sigma_{a_l}^2 = E\left\{|a_l|^2\right\}, \quad l = 0, 1, \ldots, L-1 \tag{17}$$

Both the real and imaginary parts of the path gains are Gaussian distributed; thus, the envelope $|a_l|^2$ is Rayleigh distributed. Also, the channels are normalized such that

$$\sum_{l=0}^{L-1} \sigma_{a_l}^2 = 1 \tag{18}$$

CE-OFDM is simulated over two channel models. The first channel is AWGN, but the second is a frequency-selective channel.

In the case of an AWGN channel, only the Gaussian additive noise $n(t)$ is taken into account.

Frequency-selective fading channel has an exponential delay power spectral density as follows:

$$\sigma_{a_l,C}^2 = \begin{cases} C_{Ch}\mathrm{e}^{-\tau_l/2\mu s}, & 0 \le \tau_l \le 125\,\mathrm{ns} \\ 0, & \text{otherwise} \end{cases} \tag{19}$$

where

$$C_{Ch} = 1 \Big/ \sum_{l=0}^{255} \exp(-\tau_l/2e-6)$$

is the normalizing constant used to guarantee (18). Note that the maximum propagation delay is 125 ns.

4 Simulation Results

The simulation results of this study are presented over three figures: Fig. 4 compares the performance of a CE-OFDM system, with fixed oversampling factor J and modulation index h, but varying M. Figure 5 compares the performance of a CE-OFDM system with fixed M and h, but varying J, and Fig. 6 compares the performance of constant envelope OFDM system with fixed M and J, but varying h over two channel models. The simulation results over the multi-path channel model are labeled with circles. For each case, the number of subcarriers is $N = 128$. The sampling factor J and the modulation index h are defined in [29]. The parameters of the CE-OFDM system considered are presented in Table 1. The samples $\{h[i]\}$, $\{s[i]\}$, and $\{n[i]\}$ are generated then used to calculate the received samples (11) which are then processed by the FDE and the demodulator.

Figure 4 compares the performance of an EC-OFDM system of $N = 128$ subcarrier on an AWGN channel and on a frequency-selective channel. The system is computer simulated with a sampling rate fsa = JN/TB, where $J = 8$ is the oversampling factor. For $M = 4$, $E_b/N_0 = 10$ dB, and $2\pi h = 1.0$, the system has a better performance of BER $= 10^{-3}$ over AWGN channel. For a frequency-selective channel, we need a $E_b/N_0 = 24$ dB to have the same BER value. However, in seeking to increase the flow rate, it is found that the BER is increased.

Figure 5 compares the performance of $N = 128$ subcarrier CE-OFDM system on an AWGN channel and a frequency-selective channel. The system is computer

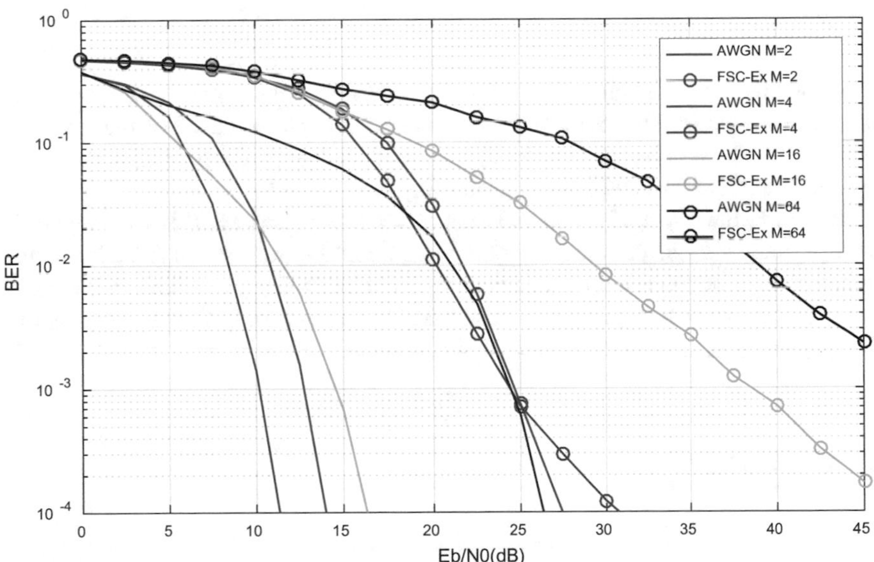

Fig. 4 BER performance of CE-OFDM system with varying values of M ($J = 8$, $2\pi h = 1.0$ and $N = 128$)

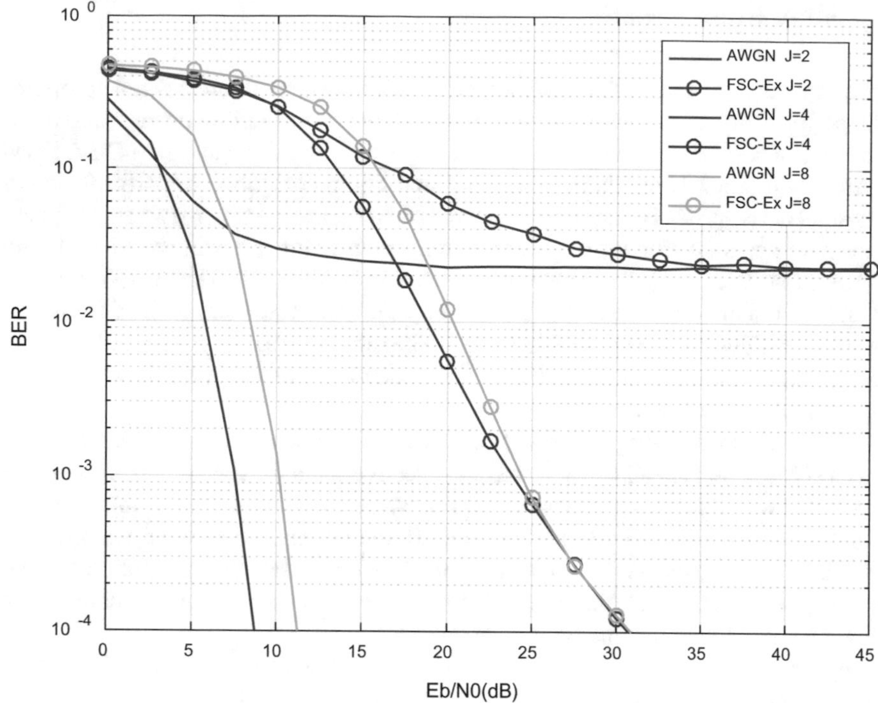

Fig. 5 BER performance of CE-OFDM system with varying values of J ($M = 4$, $2\pi h = 1.0$ and $N = 128$)

simulated with a modulation order $M = 4$ and $2\pi h = 1.0$ where h is the modulation index. From the results obtained, it is found that the performance of the system is better when the number of samples is increased; however, the performances of the system take their limits with an oversampling factor $J = 8$.

In Fig. 6, the effect of the modulation index h is presented. It is shown that for small values of h, the performance of the CE-OFDM system on an AWGN channel is better than that on a frequency-selective channel. In the case where h increases, the performance of the system in terms of BER increases for both types of channels. The system is at its limits for values of $2\pi h \geq 1.2$.

5 Conclusion

In this paper, a transformation technique that eliminates the PAPR problem associated with OFDM is presented. The phase modulation transformation produces 0 dB steady-state PAPR signals, which are well suited for efficient nonlinear amplification. For frequency-selective attenuation channels, a frequency domain equalizer is used.

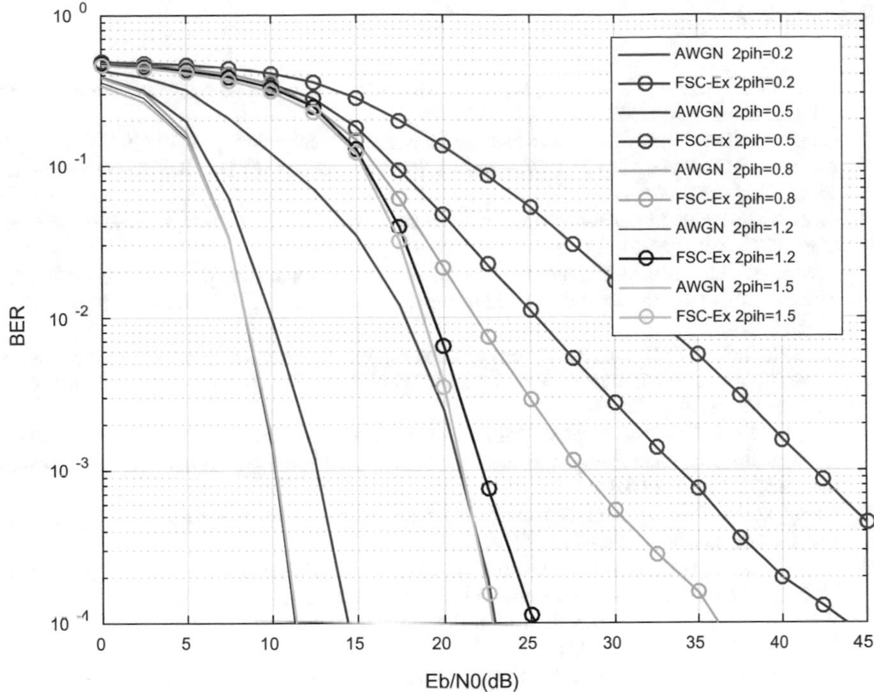

Fig. 6 BER performance of CE-OFDM system with varying values of **h** ($M = 4$, $J = 8$ and $N = 128$)

Table 1 System parameters

Parameter	Value
Block period (T_B)	500.10^{-9} s
Number of subcarrier (N)	128
Guard period (T_g)	125.10^{-9} s
Blocks/channel realization (victories)	MMSE
Equalizer correction terms	10
FIR filter length	11

As with traditional OFDM, a cyclic prefix is used to simplify the equalization. The results of the simulation demonstrate a limitation of the phase demodulator receiver for a high modulation index and a low signal-to-noise ratio. In future work, a study of the system will be made for millimeter waves to evaluate the performances of the CE-OFDM waveform for 5G technologies, and improvement of the system's capacity must be studied by combining spatial diversity from the massive MIMO technique to the CE-OFDM technique.

References

1. Rappaport, T.S., Sun, S., Mayzus, R., et al.: Millimeter wave mobile communications for 5G cellular: it will work. IEEE Access **1**, 335–349 (2013)
2. Ghosh, A.: The 5G mmWave radio revolution. Microw. J. **59**(9 Part I), 3–10 (2016)
3. Andrews, J.G., Buzzi, S., Choi, W., et al.: What will 5G be? IEEE J. Sel. Areas Commun. **32**(6), 1065–1082 (2014)
4. Pi, Z., Khan, F.: An introduction to millimeter-wave mobile broadband systems. IEEE Commun. Mag. **49**(6), 101–107 (2011)
5. Sulyman, A.I., Nassar, A.T., Samimi, M.K., et al.: Radio propagation path loss models for 5G cellular networks in the 28 and 38 GHz millimeter-wave bands. IEEE Commun. Mag. **52**(9), 78–86 (2014)
6. Kim, T., Park, J., Seol, J.Y., et al.: Tens of gbps support with mmWave beamforming systems for next generation communications. In: IEEE GlobeCom 13, pp. 3790–3795. Atlanta, GA, USA, 9–13 Dec 2013 (2013)
7. Roh, W., Seol, J.Y., Park, J., et al.: Millimeter-wave beamforming as an enabling technology for 5G cellular communications: theoretical feasibility and prototype results. IEEE Commun. Mag. **52**(2), 106–113 (2014)
8. Zhang, J., Ge, X., Li, Q., et al.: 5G millimeter-wave antenna array: design and challenges. IEEE Wirel. Commun. **24**(2), 106–112 (2017)
9. Tan, J., Stuber, G.L.: Constant envelope multi-carrier modulation. In: Proceedings of IEEE Military Communications Conference, vol. 1, pp. 607–611, Anaheim (2002)
10. Yesodha, P.: VLSI implementation of fast fourier transform used in OFDM. Elixir Int. J. 1246012462, 7 Jan 2013 (2013)
11. van Nee, R., Awater, G., Morikura, M., Takanashi, H., Webster, M., Halford, K.W.: New high-rate wireless LAN standards. IEEE Commun. Mag. 82–88 (1999)
12. Moose, P.H., Roderick, D., North, R., Geile, M.: A COFDM-based radio for HDR LOS networked communications. In: Proceedings IEEE ICC, vol. 1, pp. 187–192 June (1999)
13. Koffman, I., Roman, V.: Broadband wireless access solutions based on OFDM access in IEEE 802.16. IEEE Commun. Mag. 96–103, Apr (2002)
14. Yang, H.: A road to future broadband wireless access: MIMO-OFDM based air interface. IEEE Commun. Mag. 53–60, Jan (2005)
15. Shepherd, S., Orriss, J., Barton, S.: Asymptotic limits in peak envelope power reduction by redundant coding in orthogonal frequency division multiplex modulation. IEEE Trans. Commun. **46**(1), 5–10 (1998)
16. Ochiai, H., Imai, H.: On the distribution of the peak-to-average power ratio in OFDM signals. IEEE Trans. Commun. **49**(2), 282–289 (2001)
17. Thompson, S.C.: Constant envelope OFDM phase modulation. Ph.D. dissertation, University of California, San Diego (2005). [Online]. Available: http://elsteve.com/thesis/
18. Banelli, P.: Theoretical analysis and performance of OFDM signals in nonlinear fading channels. IEEE Trans. Wireless Commun. **2**(2), 284–293 (2003)
19. Cariolaro, G., Michieletto, G., Stivanello, G., Vangelista, L.: Spectral analysis at the output of a TWT driven by an OFDM signal. In: Proceedings ICCS, vol. 2, pp. 653–657, Singapore, Nov (1994)
20. Costa, E., Midrio, M., Pupolin, S.: Impact of amplifier nonlinearities on OFDM transmission system performance. IEEE Commun. Lett. **3**(2), 37–39 (1999)
21. Dardari, D., Tralli, V., Vaccari, A.: A theoretical characterization of nonlinear distortion effects in OFDM systems. IEEE Trans. Commun. **48**(10), 1755–1764 (2000)
22. Banelli, P., Baruffa, G., Cacopardi, S.: Effects of HPA non linearity on frequency multiplexed OFDM signals. IEEE Trans. Broadcast. **47**(2), 123–136 (2001)
23. Raab, F.H., Asbeck, P., Cripps, S., Kenington, P.B., Popovic, Z.B., Pothecary, N., Sevic, J.F., Sokal, N.O.: Power amplifiers and transmitters for RF and microwave. IEEE Trans. Microwave Theory Tech. **50**(3), 814–826 (2002)

24. Miller, S.L., O'Dea, R.J.: Peak power and bandwidth efficient linear modulation. IEEE Trans. Commun. **46**(12), 1639–1648 (1998)
25. Ochiai, H.: Power efficiency comparison of OFDM and single-carrier signals. In: Proceedings IEEE VTC, vol. 2. pp. 899–903, Sept (2002)
26. Wulich, D.: Definition of efficient PAPR in OFDM. IEEE Commun. Lett. **9**(9), 832–834 (2005)
27. Kiviranta, M., Mammela, A., Cabric, D., Sobel, D.A., Brodersen, R.W.: Constant envelope multicarrier modulation: performance evaluation in AWGN and fading channels. In: Proceedings IEEE Milcom, vol. 2, pp. 807–813 Oct (2005)
28. Abel Gouba, O.: Approche conjointe de la réduction du facteur de crête et de la linéarisation dans le contexte OFDM. May (2014)
29. Thompson, S.C.: Constant OFDM envelope phase modulation. Ph.D. dissertation, University of California, San Diego (2005)
30. Raab, F.H., Asbeck, P., Cripps, S., Kenington, P.B., Popovic, Z.B., Pothecary, N., Sevic, J.F., Sokal, N.O.: Power amplifiers and transmitters for RF and microwave. IEEE Trans. Microw. Theory Techn. **50**(3), 814–826 (2005)
31. Thompson, S.C., et al.: Constant envelope OFDM. IEEE Trans. Commun. **56**(8), 1300–1312 (2008)
32. Sari, H., Karam, G., Jeanclaude, I.: Transmission techniques for digital terrestrial TV broadcasting. IEEE Commun. Mag. 100–109 (1995)
33. Bello, P.A.: Characterization of randomly time-variant linear channels. IEEE Trans. Commun. 360–393 (1963)

Towards a Dynamical Adaptive Core for Urban Flows Simulations

Hind Talbi, El Miloud Chaabelasri, Najim Salhi and Imad Elmahi

Abstract In river hydraulic, the classical shallow water equations are governed by the Saint-Venant system, are widely used to model water flow in rivers, lakes, reservoirs, coastal areas. As their solutions are typically non-smooth and even discontinuous, among all techniques numerical, the finite volume schemes are well suited to problems conservative, hyperbolic and nonlinear such as the model of Saint-Venant and also among the numerical schemes are among the most popular tools. In this study, we promote a finite volume method, on an unstructured grid, with shock-capturing capabilities by an automatic algorithm to adapt grids near higher gradient. The performance of the numerical method has been investigated by applying it on a problem of urban flow including an oblique hydraulic jump.

Keywords Shallow water equations · Finite volume method · Shock-capturing method · Grid adaptation

1 Introduction

The free surface flows are described by Saint-Venant equations obtained by Adhémar Barre Saint-Venant in 1871 and precise set 1888. The shallow water equations are frequently used to describe flow in natural (river) and artificial (irrigation, sanitation) channels and for the simulation of many phenomena such as the maritime domain, meteorology, climatology, sedimentology, studies of flood. The physical model governed by shallow water equation is derived from depth-integrating the Navier–Stokes equations, in the case where the horizontal length scale is much greater than the vertical length scale. It is represented by a set of hyperbolic partial differential describing the mass and momentum conservation. At present, there are several approaches to the

H. Talbi (✉) · E. M. Chaabelasri · N. Salhi
LME, Faculté Des Sciences, 717, 60000 Oujda, B.P., Maroc
e-mail: h.talbi@ump.ac.ma

I. Elmahi
ENSAO Complexe Universitaire, 669, 60000 Oujda, B.P., Maroc

V. Bhateja et al. (eds.), *Embedded Systems and Artificial Intelligence*, Advances in Intelligent Systems and Computing 1076,
https://doi.org/10.1007/978-981-15-0947-6_8

numerical resolution of hyperbolic fluid conservation law, systems of fluid mechanics especially the system of shallow water. Schematically, the numerical simulations associated with the equations of shallow water coming can be obtained by the following three methods: Finite difference methods (M.D.F) (Fennema et Chaudry (1990)), Finite element methods (M.E.F) (Akambi et Katopodes (1988)) and Finite volume methods (M.V.F) (Zhao et Al (1994, 1996)).

In recent years, the finite volume method was among the first to reach an advanced stage of development in the context of urban hydraulics and for stationary and non-stationary flow calculations. This method has attracted wide attention from the community of the numericians and has achieved a series of successes. Among the scheme of the finite volumes frequently encountered in the literature, mention may be made of the upwind scheme based on approximate Riemann solvers [1].

In the literature, many authors have worked on development of robust and efficient schemes for conservation laws. Among others, Van Leer [2], Harten [3], Osher and Solomon [4], and later Glaister [5], the first ones, for the resolution of the one-dimensional shallow water equations applied to dam-break problems, and the other, for a two-dimensional discretization of the shallow water equations. A two-dimensional upwind scheme for the convective part of the shallow water equations has been studied and used to simulate a hydraulic jump by Paillere et al. [6]. However, the source terms taken into account in the system of the shallow water equations, that describe the friction and the bed slope, require a special treatment and must be studied with caution. Furthermore, the major difficulty for numerical simulation of water flow is to build a so called balanced scheme [16] which allow preserving the balance between convective flow and that the source term.

The numerical scheme constituted by Van leer and Roe which generates natural upwind discretization of the source terms has been improved by Bermudez and Vazquez [7] to approach the source terms of the inhomogeneous part of the hyperbolic system of Saint-Venant, with the concept of balancing the gradient flow with source terms. The resulting scheme, that preserving the exact conservation property, has been checking under several practical shallow water flow problems [8] for several types of practical shallow water flow problems has applied. Leveque [9] presented a Riemann solver, which balanced the source terms and the flux gradient, in each cell of a regular computing mesh by a different approach to adopt in the wave propagation method, quasi-stationary. Hubbard and Garcia-Navarro [10] illustrated the source-term processing studied by Vazquez to a second-order decreasing total rate (TDV) scheme. Alcrudo and Garcia-Navarro [11] proposed a Godunov type for numerical solution of shallow water equation. Kurgnov and Levy [12] have developed a technique based on central-upwind scheme exploits the surface elevation instead of the water depth. Alcrudo and Benkhaldoun [13] have improved the exact solutions for the Riemann problem at the interface with a variation in the topography. In the work of Zhou et al. [14], developed numerical methods based on surface gradient techniques were applied to equations in shallow water. Benkhaldoun et al. [15] proposed a new finite volume method for flux-gradient and source-term balancing of shallow water equation on non-flat topography. Furthermore, Greenberg and Leroux [16] have considered a numerical scheme, which retains a balance between the source terms and the internal forces caused by the presence of shock wave.

In this work, we describe a new finite volume scheme non-homogeneous Riemann solver (SRNH) for numerical solution of shallow water equation. Benkhaldoun [17] is the first to propose the scheme (SRNH) for free surface. Sahmim and Benkhaldoun [18] examined this scheme numerically for free surface flow in one-dimensional and two-dimensional cases on structured meshes. Benkhaldoun et al. [19] extend the SNRH scheme for the simulation of the hydrodynamic and pollutant transport in in the Strait of Gibralta on an adaptive mesh. The efficient results presented in this paper prove the high shock capturing capabilities of the proposed scheme.

This paper is organized as follows. The mathematical model with the basic equations is briefly presented in Sect. 2. The finite volume method is formulated in Sect. 3. This section includes a description of the dynamic mesh adaptation process. Numerical simulations results are illustrated and presented in Sect. 4.

2 Shallow Water Equations

The general conservative formulation for a system in 2D is

$$\frac{\partial U}{\partial t} + \frac{\partial F(U)}{\partial x} + \frac{\partial G(U)}{\partial y} = S. \tag{1}$$

Or

$$\frac{\partial U}{\partial t} + \nabla \cdot E = S. \tag{2}$$

Being $E = (F, G)$

The existence of a Jacobian matrix of the system is the basis of the upwind numerical discretization, and it is defined by

$$J_n = \frac{\partial E \cdot n}{\partial U}. \tag{3}$$

the normal flux to a direction is given by the projection on the normal unit vector of a cell $n = (n_x, n_y)$.

The numerical model is developed under the hypothesis that the problem is dominated by advection and is strongly determined by the source terms.

The 2D shallow water equation conforms a hyperbolic nonlinear system that can be written as in (1) where:

$$U = (h, q_x, q_y)^T, q_x = hu, q_y = hv \tag{4}$$

$$F = \left(q_x, \frac{q_x^2}{h} + \frac{1}{2}gh^2, \frac{q_x q_y}{h}\right)^T, G = \left(q_y, \frac{q_x q_y}{h}, \frac{q_y^2}{h} + \frac{1}{2}gh^2\right)^T. \quad (5)$$

$$S = \left(0, gh(S_{0x} - S_{fx}), gh(S_{0y} - S_{fx})\right)^T. \quad (6)$$

With h representing the water depth and (u, v), the depth-averaged components of the velocity vector along the (x, y) coordinates.
The slop terms are

$$S_{Ox} = -\frac{\partial z}{\partial x}, S_{0y} = -\frac{\partial z}{\partial y}. \quad (7)$$

And the friction-associated loses in terms of the manning roughness number n are

$$S_{fx} = \frac{n^2 u\sqrt{u^2 + v^2}}{h^{4/3}}, S_{fy} = \frac{n^2 v\sqrt{u^2 + v^2}}{h^{4/3}}. \quad (8)$$

The Jacobian matrix of the flux in the normal-pointing direction is

$$J_n = \frac{\partial E \cdot n}{\partial U} \quad (9)$$

where

$$J_n = \begin{pmatrix} 0 & n_x & n_y \\ -u(u \cdot n) + c^2 n_x & u \cdot n + u n_x & u n_y \\ -v(u \cdot n) + c^2 n_y & v n_x & u \cdot n + v n_y \end{pmatrix}. \quad (10)$$

The eigenvalues and eigenvectors are given by

$$\lambda^1 = u \cdot n - c, \lambda^2 = u \cdot n, \lambda^3 = u \cdot n + c \quad (11)$$

$$e^1 = \begin{pmatrix} 1 \\ u - c n_x \\ v - c n_y \end{pmatrix}, e^2 = \begin{pmatrix} 0 \\ -c n_y \\ c n_x \end{pmatrix}, e^3 = \begin{pmatrix} 1 \\ u + c n_x \\ v + c n_y \end{pmatrix}. \quad (12)$$

The matrices that diagonalize the Jacobian are

$$P = (e^1, e^2, e^3) = \begin{pmatrix} 1 & 0 & 1 \\ u - c n_x & -c n_y & u + c n_x \\ v - c n_y & c n_x & v + c n_y \end{pmatrix}. \quad (13)$$

$$P^{-1} = -\frac{1}{2c}\begin{pmatrix} -u.n - c & n_x & n_y \\ 2(v n_x - u n_y) & 2n_y & -2n_x \\ u.n - c & -n_x & -n_y \end{pmatrix}. \quad (14)$$

$$J_n = P\Lambda P^{-1}, P^{-1}J_nP = \Lambda, \Lambda = \begin{pmatrix} \lambda^1 & 0 & 0 \\ 0 & \lambda^2 & 0 \\ 0 & 0 & \lambda^3 \end{pmatrix}. \tag{15}$$

The approximate Jacobian $J_{n,w}$ of the system together with the source and wave strengths is constructed in terms of the averaged variables, corresponding to Roe's approximate Riemann solver:

$$\widetilde{u}_w = \frac{u_i\sqrt{h_i} + u_j\sqrt{h_j}}{\sqrt{h_i} + \sqrt{h_j}}, \widetilde{v}_w = \frac{v_i\sqrt{h_i} + v_j\sqrt{h_j}}{\sqrt{h_i} + \sqrt{h_j}}, \widetilde{c}_w = \sqrt{g\frac{h_i + h_j}{2}} \tag{16}$$

This results in the next eigenvalues and eigenvectors:

$$\widetilde{\lambda}^1_w = (\widetilde{u} \cdot n + \widetilde{c})_w, \widetilde{\lambda}^2_w = (\widetilde{u} \cdot n)_w, \widetilde{\lambda}^3_w = (\widetilde{u} \cdot n - \widetilde{c})_w \tag{17}$$

$$\widetilde{e}^1_w = \begin{pmatrix} 1 \\ \widetilde{u} + \widetilde{c}n_x \\ \widetilde{v} + \widetilde{c}n_y \end{pmatrix}_w, \widetilde{e}^2_w = \begin{pmatrix} 0 \\ -\widetilde{c}n_y \\ \widetilde{c}n_x \end{pmatrix}_w, \widetilde{e}^3_w = \begin{pmatrix} 1 \\ \widetilde{u} - \widetilde{c}n_x \\ \widetilde{v} - \widetilde{c}n_y \end{pmatrix}_w \tag{18}$$

3 Numerical Method Formulation

3.1 Finite Volumes Formalism

The finite volume method is based on two main steps: the discretization of the studied domain into a finite number of control volumes and the integration of the system of equations considered on each control volume.

We start by integrating the equation [1] over a control volume Ω written as:

$$\int_\Omega \frac{\partial U}{\partial t} d\Omega + \int_\Omega \frac{\partial}{\partial x} F(U) d\Omega + \int_\Omega \frac{\partial}{\partial y} G(U) d\Omega = 0. \tag{19}$$

Invoking Gauss theorem on each control volume T_i area A_i, the variational formulation is written:

$$A_i \frac{\partial U_i}{\partial t} + \sum_{j \in K_i} \int_{\Gamma_{ij}} F(U, \vec{n}_{ij}) d\Gamma = 0. \tag{20}$$

With:

U_i is the average value of the solution on the cell T_i, k_i is the set of indices of triangles having a common edge with the triangles T_i, Γ_{ij} is the interface separating the triangle T_i ti from the triangle T_j, $\vec{n}_{ij} = (n_x, n_y)$ unitary outside the cell T_i and directed to T_j or the flow function defined by

$$F(U, n) = F(U)n_x + G(U)n_y \tag{21}$$

3.2 Discretization of Gradient Flows: The SRNH Scheme (Non-homogeneous Riemann Solver)

In this section, we formulate the SRNH method used to solve the equation [19].

The schema (SRNH) consists of two steps, the first, of which called predictor step, consists in writing at each interface between two cells, an off-axis approximation of the approximate solution involving a scalar control parameter, and the second, said corrective step, consists in updating the states at time $n + 1$ to form the predicted states.

The domain Ω is approached by a domain Ω_h consisting of a finite number of triangular volumes T_i such that:

$$\Omega_h = \bigcup_{i=1}^{N_T} T_i \tag{22}$$

where N_e is the total number of triangles. Each triangle represents a control volume, and the physical variables are calculated at the centre of gravity of each cell. The cell-centred finite volume formulation (which calculates the gradient on each node of each mesh using the main unknown at the centre of the meshes and auxiliary unknowns defined on the edges of the mesh) is adopted in this work. All variables are represented by a constant state of the following form:

$$U_i = \frac{1}{|T_i|} \int\limits_{T_i} U dV \tag{23}$$

where $|T_i|$ is the surface of the triangles T_i. Let us divide the time interval $[0, t]$ in sub-intervals $\left[t^n, t^{n+1}\right]$ with time step size Δt.

The discrete finite volumes form of the equations is obtained by applying the Gauss theorem on the shallow water system [20] on an unstructured mesh.

The final equations are written in the following form:

$$U_i^{n+1} = U_i^n - \frac{\Delta t}{|T_i|} \sum_{j \in N(i)} \int\limits_{\Gamma_{ij}} F(U^n, n) d\Gamma + \frac{\Delta t}{|T_i|} \int\limits_{T_i} S(U^n) dV. \tag{24}$$

where $N(i)$ is the set of neighbouring triangle of the cell T_i and U_i^n is the average value of U in the cell T_i at time t_n. Note that, the spatial discretization is complete when numerical reconstruction of flux in are selected.

The convective part is approached to a numerical convective flow as follow:

$$\int_{\Gamma_{ij}} F(U, n)d\Gamma = \Phi(U_i, U_j, n_{ij})\left|\Gamma_{ij}\right|. \tag{25}$$

where Φ is the numerical flow, U_i and U_j are, respectively, the values of U on the cells T$_i$ and T$_j$, $\left|\Gamma_{ij}\right|$ the length of the edge Γ_{ij}. By construction of the SRNH schemas, the numerical flow is defined by

$$\Phi(U_i, U_j, n_{ij}) = F(U_{ij}^n, n_{ij}). \tag{26}$$

Or U_{ij}^n is determined in a first step called prediction step:

$$U_{ij}^n = \frac{1}{2}(U_i^n + U_j^n) - \frac{1}{2}\text{sgn}\left[\nabla F(\bar{U}_{ij}^n, \vec{n}_{ij})\right](U_j^n - U_i^n). \tag{27}$$

U_{ij} is the average of roe between states U_i^n and U_j^n, $\bar{U}_{ij}^n$ at the moment t^n, $\text{sgn}\left[\nabla F(\bar{U}_{ij}^n, \vec{n}_{ij})\right]$ represents the sign matrix of $\nabla F(\bar{U}_{ij}^n, \vec{n}_{ij})$. We refer the reader to the Ref. [20], which gives an overview on the sign matrix.

Finally, the solution at time t^{n+1} t is determined at the correction step by

$$U_i^{n+1} = U_i^n - \frac{\Delta t}{|T_i|}\sum_{j\in N(i)} \Phi(U_{ij}^n, n_{ij})\left|\Gamma_{ij}\right| + \Delta t S_i^n. \tag{28}$$

3.3 Procedure for Mesh Adaptation

In the context, improve the efficiency and accuracy of our scheme in finite volume. We realized a mesh adaptation to develop an almost optimal mesh able to capture the discontinuities and to increase the precision of the numerical solution. A refinement and unrefinement process is introduced which dynamically follows the solution of the physical problem. The idea is to start with a coarse mesh that covers the field of computation. We establish a criterion function that will identify regions where refinement is required. A list S of elements to be refined, their degree of refinement, and those to be unrefined is then established. This is accomplished by filling, and an integer array denoted for example by *IADIV* for all triangles of the coarse mesh. At time $t = t^n$ and for a macro-element T_i, we set $IADIV(T_i) = m$ which means that the element T_i has to be divided into 4^m triangles. Thus, starting from a mesh level

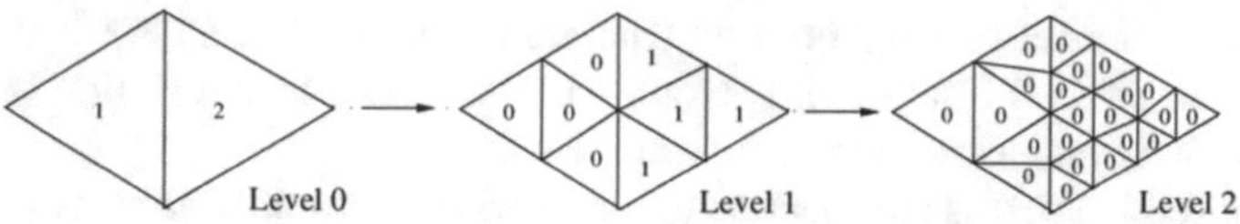

Fig. 1 Example of a two-level refining for triangular elements

l, made of $N^{(l)}$ cells, the next mesh level contains $N^{(l+1)} = 4 \times N^{(l)}$ cells. Clearly, this process can be repeated as long as $l \langle m_{\max}$ with $m_{\max}$ being the number of refinement levels. In order to obtain a mesh, which is not too distorted, the algorithm decides to divide into two equal parts some additional edges. An illustration of the procedure is shown in Fig. 1.

4 Numerical Results

4.1 *Oblique Hydraulic Jump*

The main purpose of this test is to first check the validity of the code used. This code is based on a finite volume method on an unstructured grid, with shock-capturing capabilities by an automatic algorithm to adapt grids near higher gradient. This test will allow us to compare the results for the three meshes, coarse mesh, refined mesh and adaptive mesh. We consider an oblique hydraulic jump. It is an interaction between a supercritical flow and a converging wall inclined through the angle $\theta = 8.95°$. The water flows in a channel of width $x = 30$ m. The initial and upstream conditions are $h = 1$ m, $u = 8.75\,m/s$ and $v = 0$ m/s and a number of Froude $Fr = 2.74$. The shock wave is formed with an angle $\beta = 30°$. As boundary conditions, we impose inflow fixed conditions on the left-hand side, on the right side, we have used free conditions, and on the upper and lower sides, slip boundary conditions are used. The hydraulic jump is achieved with a water height of 1.5m, velocity of 7.955m/s and a Froud number equal to 2.0055 (see Fig. 3). The reader can refer to the work of Alcrudo and Garcia-Navarro [11] for details of the exact solution. $h = 1.5$ m, $u = 7.955$ m/s and $Fr = 2.0055$.

In order to show the performance of the mesh adaptation procedure in the SRNH scheme, we present in Fig. 2 the using coarse refined and adaptive mesh, and in Figs. 3 and 4 the obtained results. We remark that the computed results on the adaptive mesh compare qualitatively well to those obtained on the refined mesh with a considerable differences on the CPU times [coarse (CPU Time = 1077, 7797), refined (CPU Time = 8692, 1410), adaptive (CPU Time = 1210, 0217)]. Note that because of the grid adaptation, the final mesh at the steady state consists of 14,315 elements only compared to other mesh (coarse, refined). This results in a significant reduction of computational costs. It is clear that the numerical diffusion is more pronounced in the results on the coarse mesh than the other results. Moreover, we remark the

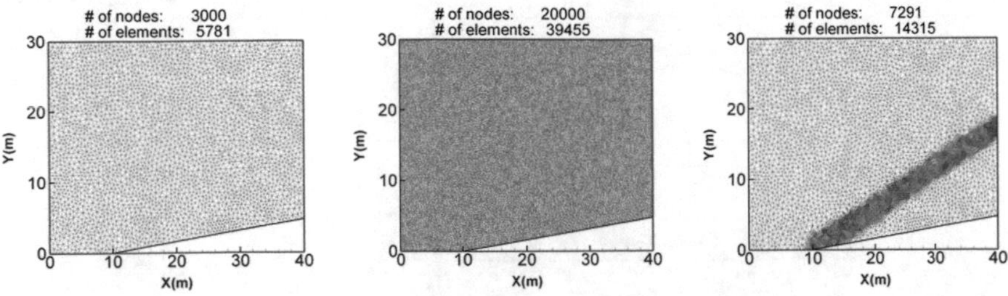

Fig. 2 Geometry and calculation mesh for the oblique hydraulic jump test for an unstructured triangulated mesh with three cases from left to right: coarse, refined and adaptive

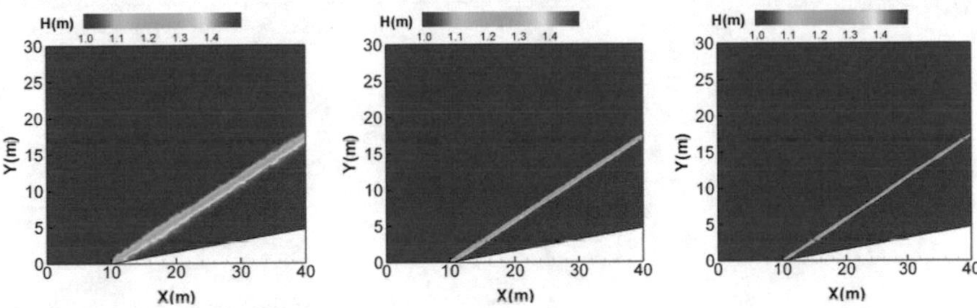

Fig. 3 Oblique hydraulic jump: water depth contours (2D) for the three cases from left to right (coarse, refined and adaptive mesh)

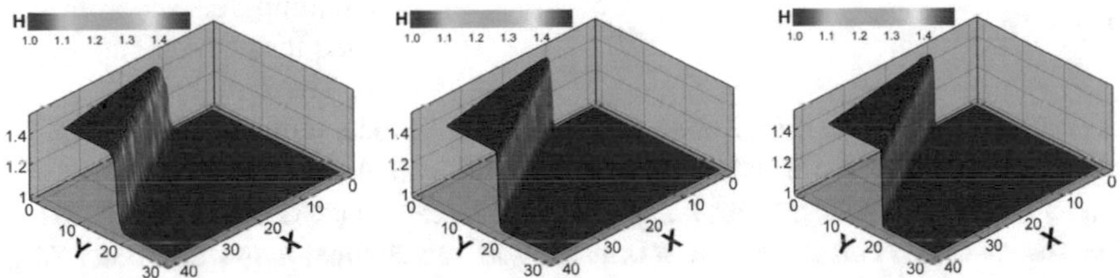

Fig. 4 Oblique hydraulic jump: water depth contours (3D) for the three cases from left to right (coarse, refined and adaptive mesh)

results obtained by the mesh adaptive at the levels of shocks and discontinuities are slightly higher than those obtained by the refined and coarse mesh. The propagation of the shock wave is clearly observed. We display in Fig. 5 the water profile cross section at the channel downstream $x = 40$ m. The resultants obtained for the SRNH using coarse mesh and fine mesh and adaptive mesh. It is obvious that the adaptive SRNH scheme produces the best results relative to coarse mesh and refined are more dissipative. Finally, Fig. 6 presents the history of convergence. The three cases converge at 1.388E-16, 2.585E-8 and 1.8512E-8 in relative error.

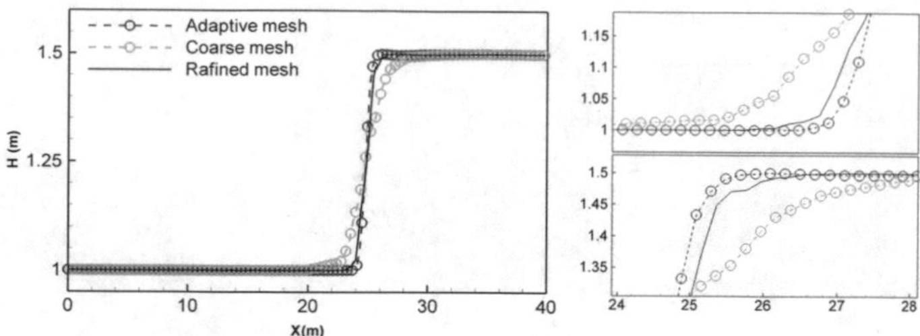

Fig. 5 Water depth profiles at x = 40 m for the three cases

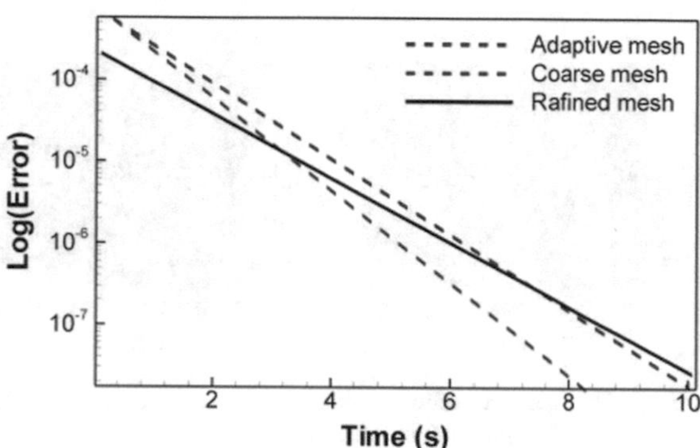

Fig. 6 Error for the three cases

4.2 *Hypothetical Urban Flood*

In this test case, we consider the urban flood propagation problem. The domain is a 17.7 m long and 3.6 m wide channel with horizontal bed. The city consists of 3 × 3 buildings, each of them being a square block with a side length of 0.50 m. The distance between the blocks is 0.50 m, with the channel axis was fixed 3.75 m downstream from the gate. The dam break is constructed by opening a 1 m gate (Their configurations are sketched in Fig. 7).with a courant number of 0.25. The numerical execution is done with a courant number of 0.25 and a Manning coefficient of 0.01 s/m$^{\frac{1}{3}}$. In this work, we studied the effects of nine buildings on dam-break wave propagation. Initially, the water was set at rest in the 0.4 m-deep reservoir behind

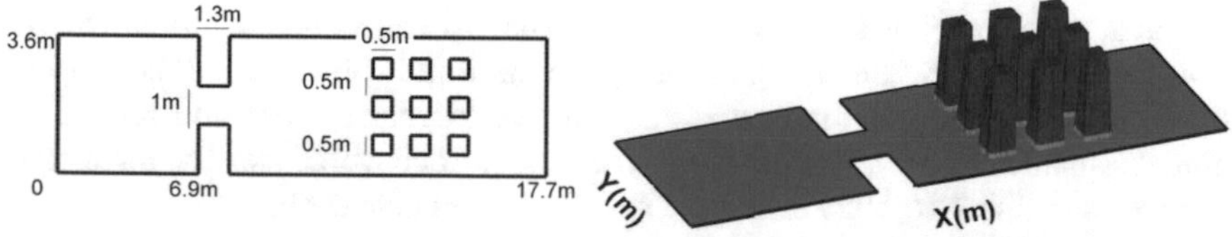

Fig. 7 Configuration of the urban flood propagation problem, dimension and 3D view

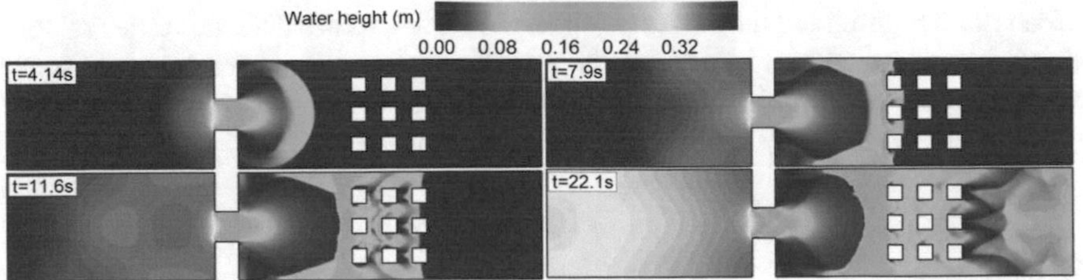

Fig. 8 Water height distribution inside urban space

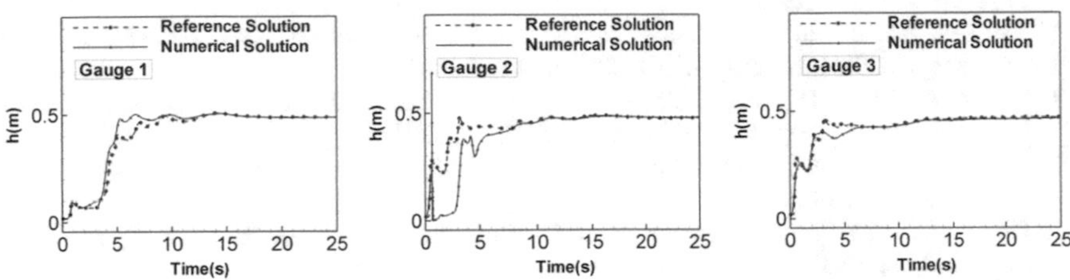

Fig. 9 Time evolution of water height at three gauges

the gate, whilst a 0.01 m-thin water layer was set in the downstream reach. The gate was pulled up in approximately 0.25 s. The duration of the experiment was 30 s. A complete description of the laboratory experiment and available measurements can be found in Soares-Frazao and Zech [21].

Figure 8 shows snapshots of the simulated water height inside urban space. After the gate opening, the flooding wave freely spreads over the initially dry downstream area (Fig. 8, $t = 4.14$ s). Whatever the configuration of the city, the urban district induced a reduction of the available flow section, which resulted in the formation of strong hydraulic jumps upstream of the buildings and in complex flow features (e.g. wake zones behind the buildings) inside the urban district (Fig. 8, $t = 7.9$ s and 11.6 s). As water progressed further downstream into the city, the flow became much more uniform, and the hydraulic jumps propagated in the upstream direction until their intensity decreased with time according to the inflow discharge (Fig. 8, $t = 22.1$ s). As shown in Fig. 9, the time histories of water level predicted by the model, at three gauges, match well with those obtained on the reference solution.

5 Conclusion

In this work, we have presented a finite volume method for flux gradient of shallow water. A finite volume scheme has been coupled with a dynamic mesh de-refinement refinement process. The simulated hydrodynamic problem, the hydraulic jump and urban flood highlighted the capacity of the model to capture discontinuities of solution without oscillation. Moreover, the use of the process mesh adaptation to the simulation of these problems has significantly increased the accuracy and precision of the results with a very significant simulation time saving.

References

1. Roe, P.L.: Approximate Riemann solvers, parameter vectors and difference schemes. J. Comput. Phys. **43**, 357–372 (1981)
2. Van Leer B.: Towards the ultimate conservation difference scheme a second order sequel to Godunov's method. J. Comput. Phy. **23**, 101–136(1979)
3. Harten, A., Lax, P.D., Leer, B.V.: On upstream differencing and Godunov type schemes for hyperbolic conservation laws. SIAM Rev. **25**, 35–61 (2015)
4. Ocher, S., Solomon, F.: Upwind difference schemes for hyperbolic conservation laws. Math. Comp. **38,** 339–374 (1982)
5. Glaister, P.: Approximate Riemann solutions of the shallow-water equations. J. Hydraul. Res. **26**, 239–306 (1988)
6. Paillere, H., Degrez, G., Deconick, H.: Multidimensional upwind schemes for then shallow water equations. Int. J. Numer. Meth. Fluids **26**, 987–1000 (1998)
7. Bermúdez, A., Vázquez, M.E.: Upwind methods for hyperbolic conservation laws with source terms. Comput. Fluids **23**, 1049–1071 (1994)
8. Vázquez-Cendón, M.E.: Improved treatment of source terms in upwind schemes for the shallow water equations in channels with irregular geometry. J. Comput. Phys. **148**, 497–526 (1999)
9. Leveque, R.J.: Balancing source terms and flux gradients in high resolution Godunov methods: the quasi-steady wave-propagation algorithm. J. Comput. Phys. **146**, 346–365 (1998)
10. Hubbard, M.E., Garcia-Navarro, P.: Flux Difference splitting and the balancing of source terms and flux gradients. J. Comput. Phys. **165**, 89–125 (2000)
11. Alcrudo, F., Garcia-Navarro, P.: A high resolution Godunov-type scheme in finite volumes for the 2D shallow water equation. Int. J. Numer. Methods Fluids. **16**, 489–505 (1993)
12. Kurganov, A., Levy, D.: Central-upwind schemes for the Saint-Venant system. Math. Model. Numer. Anal. **36**, 397–425 (2002)
13. Alcrudo, F., Benkhaldoun, F.: Exact solutions to the Riemann problem of the shallow water equations with a bottom step. Comput. Fluids **30**, 643–671 (2001)
14. Zhou, J.G., Causon, D.M., Mingham, C.G., Ingram, D.M.: The surface gradient method for the treatment of source terms in the shallow water equations. J. Comp. Phys. **20**, 1–25 (2001)
15. Benkhaldoun, F., Elmahi, I., Seaïd, M.: A new finite volume method for flux gradient and source-term balancing in shallow water equations. Comput. Method Appl. Mech. Eng **199**, 3324–335 (2010)
16. Greenberg, J.M., LeRoux, A.Y.: Well-balanced scheme for the numerical processing of source terms in hyperbolic. SINUM **33**, 1–16 (1996)
17. Benkhaldoun, F., Quivy, L.: A non homogeneous Riemann solver for shallow water. Flow Turbulence Combust **76**, 391–402 (2006)
18. Sahmim, S., Benkhaldoun, F.: Schéma SRNHS: Analyse et appmication d'un schema aux volumes finis dedié aux systems non homogènes. ARIMA **5**, 303–316 (2006)
19. Benkhaldoun, F., Elmahi, I., Seaid, M.: Well-balanced finite volume schemes for pollutant transport by shallow water equations on unstructured meshes. J. Comput. Phys. **226**(1), 180–203 (2007)

20. Benkhaldoun, F., Elmahi, I., Seaïd, M.A.: new finite volume method for flux-gradient and source-term balancing in shallow water equations. Comput. Methods Appl. Mech. Eng. **199**, 3324–3335 (2010)
21. Soares-Frazao, S., Zech, Y.: Experimental study of dam-break flow against an isolated obstacle. J. Hydraul. Res. **45**, 27–36 (2007)

An Efficiency Study of Adaptive Median Filtering for Image Denoising, Based on a Hardware Implementation

Hasnae El Khoukhi, Faiz Mohammed Idriss, Ali Yahyaouy and My Abdelouahed Sabri

Abstract Generally, in digital image processing applications and real-time signal, some unwanted signals capture the image which is termed as noise. It is desirable to be able to perform some image filtering techniques frequently used to reduce or eliminate noise on an image or signal. The two-dimensional spatial median filter is the most commonly used filter for image denoising. The median of a given sequence can be determined by sorting all values in the sequence and by choosing the middle value in the sorted sequence. Commonly, image filters are following up software approach in systems. But hardware implementation prefers in comparison with software implementation for better processing speed. In this paper, we propose an Adaptive Median Filter hardware (AMFh) implementation. Indeed, the $3 \times 3, 5 \times 5$ and 7×7 windows techniques are implemented for removal of Salt-Pepper and Impulse Noises from images and simulated using ModelSim (the Verilog language was utilized) and Matlab softwares. We are conducting a study on the effectiveness of the implemented filter, based on two main parameters, the peak signal-to-noise ratio and windows size. PSNR values for a set of benchmark images are calculated to quantify the impact of the proposed AMFh algorithm on the PSNR performance for treated windows size.

Keywords Digital image processing · Image denoising · Median filter · Verilog · Impulse noise (Salt & pepper noise)

1 Introduction

The digital image processing (DIP) has become a focus of widespread interest of several researchers in many areas ([1–4]), where this field helps to analyse, infer and make decisions. Generally, in DIP applications and real-time signal, some unwanted signals capture the image which is termed as noise. Indeed, digital images that play an

H. El Khoukhi (✉) · F. M. Idriss · A. Yahyaouy · M. A. Sabri
LIIAN, Department of Computer Science, Faculty of Sciences Dhar-Mahraz, Sidi Mohamed Ben Abdellah University Fez, Fez, Morocco
e-mail: hasnae.elkhoukhi@usmba.ac.ma

V. Bhateja et al. (eds.), *Embedded Systems and Artificial Intelligence*, Advances in Intelligent Systems and Computing 1076,
https://doi.org/10.1007/978-981-15-0947-6_9

important role both in daily life applications and in areas of research and technology are often corrupted by impulse noise due to errors generated in noisy sensor or transmission of images. It is related to the types of noise introduced to the image. Some examples of noise are: Gaussian or White, Rayleigh, Shot or Impulse, periodic, sinusoidal or coherent, uncorrelated and granular. Thus, Images Denoising (ID) is often a necessary and the first step to be taken before the images data is analysed. It is indispensable to apply an efficient denoising technique to compensate for such data corruption. Several linear and non-linear filters are proposed in the literature for ID. Although non-linear filters are more complex than linear filters, they are more commonly used for ID because they reduce the smoothing and preserve the image edges. The two-dimensional spatial median filter is the most commonly used non-linear filter for ID ([5–7]). It is a non-linear sorting-based filter. It sorts the pixels in the given window, determines the median value, and replaces the pixel in the middle of the given window with this median value. Therefore, it has high computational complexity.

Commonly, image filters are following up software approach in systems. But hardware implementation ([8–11]) prefers in comparison with software implementation for better processing speed. In this paper, we aim to study the efficiency of a proposed Adaptive Median Filtering hardware (AMFh) for image denoising, based on a hardware implementation. Indeed, the 3×3, 5×5 and 7×7 windows techniques are implemented for removal of Salt-Pepper and Impulse Noises from images, and simulated using ModelSim (the Verilog language was utilized) and Matlab softwares. We conducted an experimental study based on two main parameters, the peak signal-to-noise ratio and windows size.

The application of median filter has been investigated. As an advanced method compared with standard median filtering, the AMFh performs spatial processing to preserve detail and smooth non-impulsive noise. A prime benefit to this adaptive approach to median filtering is that repeated applications of this AMFh do not erode away edges or other small structure in the image.

The rest of the paper is organized as follows. In Sect. 2, the proposed Adaptive Median Filter hardware is explained and some implementation details are specified. The implementation results are also given. Section 3 presents the conclusion and some future directions.

2 Proposed Adaptive Median Filter Hardware and Implementation Details

Generally, the median of a given sequence can be determined by sorting all values in the sequence and by choosing the middle value in the sorted sequence. The median is just the middle value of all the values of the pixels in the neighbourhood. This is not the same as the average (or mean); instead, the median has half the values in the neighbourhood larger and half smaller. It presents a stronger "central indicator"

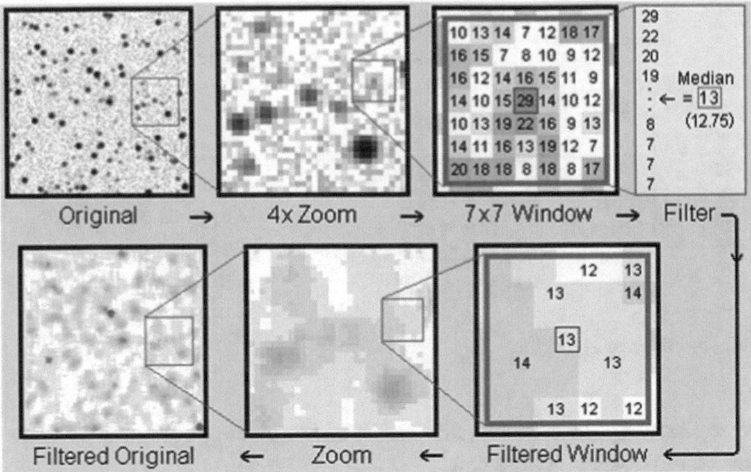

Fig. 1 7 × 7 window median filtering example

than the average. In particular, it is hardly affected by a small number of discrepant values among the pixels in the neighbourhood. Consequently, median filtering is very effective at removing various kinds of noise. The standard two-dimensional spatial median filter algorithm consists of sorting the pixels in the given window (several size of windows can be studied), determines the median value, and replaces the pixel in the middle of the given window with this median value (see Fig. 1).

2.1 *Proposed Methodology*

The proposed AMFh algorithm is an adaptive version of the standard one based on a hardware implementation. The performed algorithm is designed with the Verilog language and simulated using the Matlab and ModelSim tools environments. It can be adapted for several windows sizes. Indeed, in our study, the 3 × 3, 5 × 5 and 7 × 7 windows techniques are implemented for removal of Salt-Pepper and Impulse Noises from images.

The median is calculated by first sorting all the pixel values from the surrounding neighbourhood (it depends on the size of the window) into numerical order as a first step; and then replacing the pixel being considered with the middle pixel value as a second step (see Fig. 2). If the neighbourhood under consideration contains an even number of pixels, the average of the two middle pixel values is used. The size of the neighbourhood is adjustable, as well as the threshold for the comparison. A pixel that is different from a majority of its neighbours, as well as being not structurally aligned with those pixels to which it is similar, is labelled as impulse noise. These noise pixels

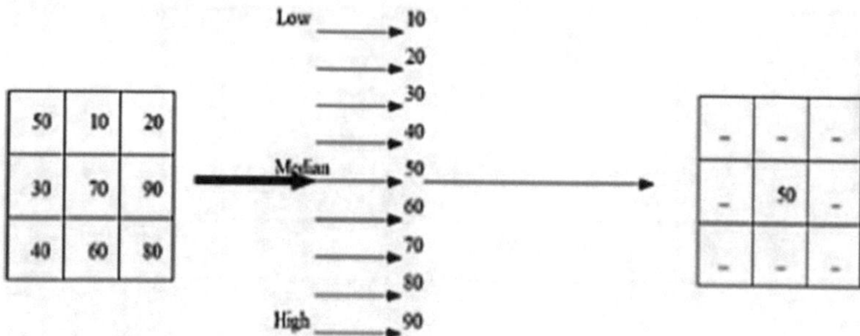

Fig. 2 Median filtering algorithm

are then replaced by the median pixel value of the pixels in the neighbourhood that have passed the noise labelling test.

We are conducting an experimental study on the effectiveness of the implemented filter, based on two main parameters, the peak signal-to-noise ratio and windows size.

The peak signal-to-noise ratio (PSNR) is one of the best-known techniques for assessing the amount of noise that an image is polluted with and, for that matter, the amount of noise left in a filtered image. The peak signal-to-noise criterion is adopted to measure the performance ([12–14]) of various digital filtering techniques quantitatively. It is the measure of the peak error.

PSNR values for a set of benchmark images are calculated to quantify the impact of the proposed AMFh algorithm on the PSNR performance for treated windows size.

2.2 *Experimental Results*

Aiming to evaluate the performance of the proposed design, computational experimentation has been carried out on a PC Intel(R) Core(TM) i5-7200U CPU @ 2.50 GHz 2.70 GHz, 4Go RAM, under Windows 10 Professional.

The proposed AMFh algorithm is implemented using Verilog HDL on ModelSim-INTEL FPGA STARTER EDITION 10.5b, and tested using two types of input images (Input Image 1 and Input Image 2) (see Fig. 3), in order to show its effectiveness. During the execution process, a logical link is established between ModelSim software and Matlab R2016a (64-bit).

By varying the peak signal-to-noise ratio from 2%, 10%, 20%, 30% and 40%, the simulation results associated to 3×3, 5×5 and 7×7 window sizes for Input Image 1 and Input Image 2 are presented in the following sections.

The proposed AMFh implemented using Verilog HDL on ModelSim software (AMFh_ModelSim) and the standard median filter (SMF) implemented using Matlab software (SMF_Matlab) are compared, based on two main parameters, the peak

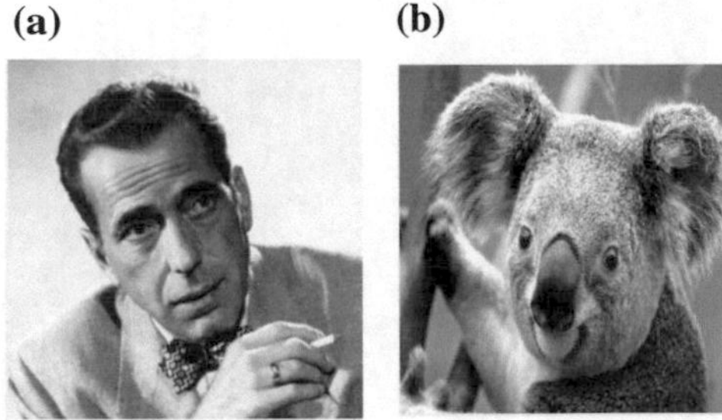

Fig. 3 Original greyscale Image 1 (**a**) and Image 2 (**b**)

signal-to-noise ratio (in our case, we use 2%, 10%, 20%, 30% and 40%) and windows size (in our case, we use 3×3, 5×5 and 7×7 window sizes). PSNR values for the set of benchmark images (Image 1 and Image 2) are calculated to quantify the impact of the proposed AMFh algorithm on the PSNR performance for treated windows size.

2.2.1 Obtained Results for Image 1 and Image 2

This section illustrates the results we obtained during the process of our algorithm applied to the first example of image (Image 1) and the second example of image (Image 2). Experimental results for the AMFh_ModelSim and the SMF_Matlab algorithms are presented explicitly below (see Tables 1 and 2).

Salt and pepper noise is added to original images. Then, we filtered the images with Matlab standard median filter function (medfilt2), and with the proposed AMFh algorithm. Varying the peak signal-to-noise ratio from 2%, 10%, 20%, 30% and 40%, we quantified the PSNR performance for all 3×3, 5×5 and 7×7 window sizes, which values are resumed in Tables 1 and 2. These results show well that the proposed AMFh_ModelSim algorithm achieves better results and produces higher quality filtered images than SMF_Matlab in all studied configurations (see Table 3).

2.2.2 Analysis of Obtained Results

In order to test the performance rate of our proposed algorithm compared to the standard one, we have performed experiments at a set of noise levels ranging from 10 to 40%, including the case of 2% on two input images. Each image is a two-dimensional 8-bit greyscale one. Impulse noise of different percentages is added. The outcome of the performances of the filtering operations is demonstrated by the output of the images. We study also the influence of windows size by treating tree cases the 3×3, 5×5 and 7×7 window sizes.

For both examples of images, we can remark, from the performed two tests, that when the impulse noise in the image is increased the AMFh_ModelSim algorithm achieves more enhanced images than the SMF_Matlab. Like the other median filters, the performance of the AMFh filter deteriorates as the noise factor increases.

Table 1 Peak signal-to-noise ratio for image 1, according to window size

Filter methods	Window size	2% implulse noise	10% implulse noise		20% implulse noise		30% implulse noise		40% implulse noise	
SMF_Matlab	3 × 3	**31.3738**[a]	3 × 3	**28.5958**	3 × 3	**24.6549**	3 × 3	**21.1206**	3 × 3	**17.1317**
	5 × 5	**25.2114**	5 × 5	**24.3471**	5 × 5	**22.3621**	5 × 5	**20.6603**	5 × 5	**19.4557**
	7 × 7	**23.6951**	7 × 7	**22.9577**	7 × 7	**20.9499**	7 × 7	**19.3997**	7 × 7	**18.3228**
AMFh_ModelSim	3 × 3	**33.2371**	3 × 3	**31.5042**	3 × 3	**27.1617**	3 × 3	**22.1996**	3 × 3	**17.7592**
	5 × 5	**28.1138**	5 × 5	**27.4944**	5 × 5	**26.8326**	5 × 5	**26.2211**	5 × 5	**24.6705**
	7 × 7	**25.7850**	7 × 7	**25.5614**	7 × 7	**25.1642**	7 × 7	**24.7412**	7 × 7	**23.7759**

Table 2 Peak signal-to-noise ratio for image 2, according to window size

Filter methods	Window size	2% implulse noise	10% implulse noise		20% implulse noise		30% implulse noise		40% implulse noise	
SMF_Matlab	3 × 3	**24.7205**	3 × 3	**24.4659**	3 × 3	**23.6402**	3 × 3	**22.5650**	3 × 3	**17.5521**
	5 × 5	**24.7018**	5 × 5	**24.4031**	5 × 5	**23.5298**	5 × 5	**22.6069**	5 × 5	**21.5491**
	7 × 7	**23.5076**	7 × 7	**23.3117**	7 × 7	**22.4533**	7 × 7	**21.5991**	7 × 7	**20.7449**
AMFh_ModelSim	3 × 3	**27.4219**	3 × 3	**26.8296**	3 × 3	**26.8182**	3 × 3	**26.1767**	3 × 3	**18.1865**
	5 × 5	**24.9243**	5 × 5	**24.7447**	5 × 5	**24.4705**	5 × 5	**24.0692**	5 × 5	**23.1936**
	7 × 7	**23.7883**	7 × 7	**23.6939**	7 × 7	**23.5268**	7 × 7	**23.3312**	7 × 7	**22.8998**

[a]The values in bold correspond to the PSNR values for the two benchmark images

Table 3 Visual results 3 × 3 window size from Image 1

Noise density (%)	Noisy image	Filtered image (SMF_Matlab)	Filtered image (AMFh_ModelSim)
2			
10			
20			
30			
40			

Logically, a higher value of PSNR is good because it means that the ratio of signal to noise is higher. And this is independently of the size of the window. But, we can see clearly that in 3 × 3 window size, we obtain the best values of PSNR compared to the 5 × 5 and 7 × 7 ones.

The results obtained from the peak signal-to-noise ratio computations show that at low to moderate noise levels, the performance of the AMFh stands out.

3 Conclusions and Future Directions

In this paper, we have proposed an AMFh algorithm. An experimental study on the effectiveness of the implemented filters AMFh on ModelSim software and SMF on Matlab software, based on two main parameters, the peak signal-to-noise ratio and windows size, is conducted. PSNR values for a set of two benchmark images are calculated to quantify the impact of the proposed AMFh algorithm on the PSNR performance for treated windows size. The results of the AMFh algorithm are better as compared to the results of the SMF one tested on two images with different percentages of impulse noise introduced in these images ranging from 2% and 10%

to 40%. And this is for all considered widows sizes 3 × 3, 5 × 5 and 7 × 7. More precisely, the 3 × 3 case gives the best values for PSNR compared to 5 × 5 and 7 × 7 windows size.

As future directions of this work, a large part of improvement is possible in the algorithm himself, we aim to propose an improved AMFh algorithm. In addition, we target developing a FPGA hardware implementation platform of the proposed AMFh and improved AMFh algorithms.

References

1. El Khoukhi, H., Sabri, M.A.: Comparative study between HDLs simulation and Matlab for image processing. In: IEEE 2018 International Conference on Intelligent System and computer Vision (ISCV) (2018)
2. Gonzalez, R.C., Woods, R.E.: Digital Image Processing. Prentice Hall, pp. 1–142. (2002). ISBN 0-13-094659
3. Maini, R., Aggarwal, H.: A comprehensive review of image enhancement techniques. J. Comput. **2**(3), 269–300 (2010)
4. Vanaparthy, P., Sahitya, G., Sree, K., Naidu, C.D.: FPGA implementation of image enhancement algorithms for biomedical image processing. Int. J. Adv. Res. Electri. Electron. Instrumentation Eng. **2**(11), 5747–5753 (2013)
5. Eric, A.: FPGA implementation of median filter using an improved algorithm for image processing. Int. J. Innovative Res. Sci. Technol. **1**(12), 25–30 (2015)
6. Kalali, E., Hamzaoglu, I.: A low energy 2D adaptive median filter hardware. In: Design, Automation & Test in Europe Conference & Exhibition (DATE) (2015)
7. Wei, W., Bing, Y.: The design and implementation of fast median filtering algorithm based on FPGA. Electr. Compon. Appl. **10**(1), 1–57 (2008)
8. Bisht, R., Vijay, R.: Hardware implementation of real time window based switching median filter. Int. J. Eng. Sci. Res. Technol. **6**(7), 732–740 (2017)
9. Chiuchisan, I., Cerlinca, M., Potorac, A.-D., Graur, A.: Image enhancement methods approach using Verilog hardware description language. In: International Conference On Development And Application Systems (2012)
10. Nausheen, N., Seal, A., Khanna, P., Halder, S.: A FPGA based implementation of Sobel edge detection. Microprocess. Microsyst. **56**, 84–91 (2018)
11. Neeraja, G., Deepika, S.: FPGA based area efficient median filtering for removal of salt-pepper and impulse noises. Int. J. Sci. Eng. Technol. Res. **2**(16), 1795–1804 (2013)
12. Sankur, B., Sayood, K., Avcıbas, I.: Statistical evaluation of Image quality measure. J. Electron. Imaging **11**(2), 206–223 (2002)
13. Filali, Y., Ennouni, A., Sabri, M.A., Aarab, A.: A study of lesion skin segmentation, features selection and classification approaches. In: IEEE 2018 International Conference on Intelligent System and computer Vision (ISCV). Fez, Morocco (2018)
14. Sabri, M.A., Ennouni, A., Aarab, A.: Automatic estimation of clusters number for K-means. 4th IEEE Int. Colloquium Inf. Sci. Technol. (CiSt), 450–454 (2016). https://doi.org/10.1109/cist.2016.7805089. Electronic ISSN: 2327-1884. 24–26 Oct. 2016. Tangier, Morocco

Processing Time and Computing Resources Optimization in a Mobile Edge Computing Node

Mohamed El Ghmary, Tarik Chanyour, Youssef Hmimz and Mohammed Ouçamah Cherkaoui Malki

Abstract The deployment of edge computing forms a two-tier mobile computing network where each computation task can be processed locally or at the edge node. In this paper, we consider a single mobile device equipped with a list of heavy off-loadable tasks. Our goal is to jointly optimize the offloading decision and the computing resource allocation to minimize the overall tasks processing time. The formulated optimization problem considers both the dedicated energy capacity and the processing deadlines. Therefore, as the obtained problem is NP-hard and we proposed a simulated annealing-based heuristic solution scheme. In order to evaluate and compare our solution, we carried a set of simulation experiments. Finally, the obtained results in terms of total processing time are very encouraging. In addition, the proposed scheme generates the solution within acceptable and feasible timeframes.

Keywords Mobile edge computing · Computation offloading · Processing time · Optimization · Simulated annealing

1 Introduction

The concept of mobile edge computing (MEC) has emerged and intends to deliver new cloud-based services directly from the network edge. These services generally require good transmission bandwidth, additional data storage, and processing [1].

M. El Ghmary (✉) · T. Chanyour · Y. Hmimz · M. O. Cherkaoui Malki
FSDM, LIIAN Labo, Sidi Mohamed Ben Abdellah University, Fez, Morocco
e-mail: mohamed.elghmary@usmba.ac.ma

T. Chanyour
e-mail: tarik.chanyour@usmba.ac.ma

Y. Hmimz
e-mail: youssef.hmimz@usmba.ac.ma

M. O. Cherkaoui Malki
e-mail: oucamah.cherkaoui@usmba.ac.ma

V. Bhateja et al. (eds.), *Embedded Systems and Artificial Intelligence*,
Advances in Intelligent Systems and Computing 1076,
https://doi.org/10.1007/978-981-15-0947-6_10

Applications execution using this new technique is performed in close proximity to the end users and significantly reduces the end-to-end delay. Moreover, MEC can augment the capabilities of a mobile device by offloading [2] its heavy computation tasks to be executed within a resource-rich edge node. As a result, the execution time of these tasks can be efficiently reduced by offloading heavy tasks to a MEC node.

The first works which dealt with the offloading within MEC environments were interested in studying the multi-user single-task case. In these scenarios, a user only targets to offload only a unique task. They studied resource allocation while they optimize the average task duration [3]. Actually, an application is generally partitioned into multiple tasks, and the offloading decisions must concern each of them. Consequently, even a single user has to handle simultaneously multiple tasks. Therefore, the offloading decision parameters should be selected according to a multi-task scenario.

Recently, the authors of [4, 5] proposed a single-user multi-task offloading by optimizing the communication resources and the local frequency without considering the amount of available local energy. Similarly, we consider the same single-user multi-task offloading environment while we introduce the available user energy as a constraint. Moreover, due to the huge impact of the remote edge server's processing capacity on the consumed energy as well as tasks' processing time, we introduced the server's frequency as a decision variable in the proposed optimization problem.

Our main motivation in this work is the study of an edge node with a single user setting in order to reduce the processing time of its time-constrained tasks while we target extending the battery lifetime of the smart mobile device.

The remainder of this paper is organized as follows. The system model and the optimization problem formulation are presented in Sects. 2 and 3. In Sect. 4, we present the solution to the proposed problem. Evaluation and result are presented in Sect. 5. Finally, Sect. 6 concludes the paper.

2 System Model and Problem Formulation

2.1 System Model

Figure 1 shows a single smart mobile device (SMD) containing a list of off-loadable tasks. In this work, we plan to study the behavior of the offloading process for a multi-task SMD within an edge environment, while we optimize the processing resources available at the edge server as well as at the mobile device. Particularly, the available energy at the SMD for tasks' execution is limited. Besides, in the context of offloading, a computationally intensive application is divided into multiple mutually independent off-loadable tasks [6]. Therefore, according to the available computational and radio resources, some tasks are offloaded to the edge node (EN) for computing. The others are processed locally on the SMD itself. The processing of all

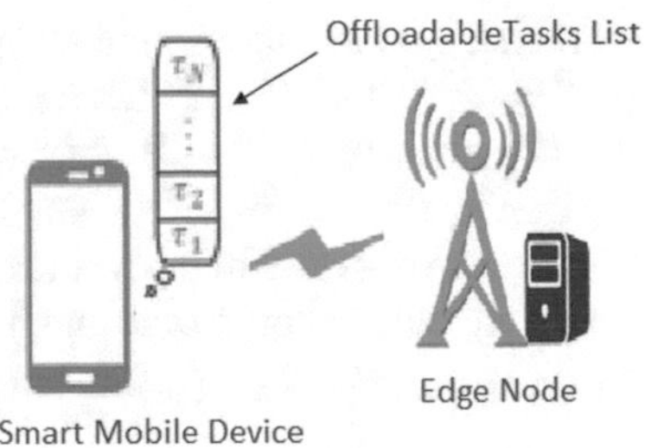

Fig. 1 System model illustration

tasks must happen within the time limit of the application. Additionally, it is assumed that the SMD concurrently performs computation and wireless transmission.

As shown in Fig. 1, a SMD is connected to an edge node (EN) that is equipped with a resource-rich server. This SMD intends to offload a set of independent tasks by the mean of an edge access point (EAP). Moreover, the wireless channel conditions between the SMD and the wireless access point are not considered in this work. Additionally, at the time of the offloading decision and the transmission of tasks, the uplink rate r is assumed to be almost invariable.

Here, the considered SMD holds a list of N independent tasks denoted as $\tau \triangleq \{\tau_1, \tau_2, \ldots, \tau_N\}$. In addition, these tasks are assumed to be computationally intensive, time-constrained and can be processed locally or at the EN. Moreover, the processing time of all tasks cannot exceed a maximum latency $t^{\max}$. Every task represents an atomic input data. Also, it is mainly characterized by two parameters $\langle d_i, \lambda_i \rangle$. The first one denoted d_i [bits] identifies the amount of the input parameters and program codes to transfer from the SMD to the EN. The second one denoted λ_i [cycles] specifies the computation amount needed to accomplish the processing of this task.

The execution nature decision for a task τ_i either locally or by offloading to the edge server is denoted x_i where $x_i \in \{0; 1\}$. $x_i = 1$ indicates that the SMD has to offload τ_i to the edge server, and $x_i = 0$ indicates that τ_i is locally processed.

From this point, all time expressions are given in *Seconds*, and energy consumptions are given in *Joule*. Then, if the SMD locally executes task τ_i, the processing time is $t_i^L = \frac{\lambda_i}{f_L}$. So, for all tasks we have

$$t^L = \sum_{i=1}^{N} (1 - x_i) \frac{\lambda_i}{f_L} \tag{1}$$

Additionally, the τ_i task's local energy consumption is given by: $e_i^L = k_L \cdot f_L^2 \cdot \lambda_i$ [7]. Hence, the total energy consumption while executing all tasks that were decided to be executed locally in the SMD is given by:

$$e^L = k_L \cdot f_L^2 \cdot \sum_{i=1}^{N} \lambda_i (1 - x_i) \tag{2}$$

If task τ_i is offloaded to the edge node, the offloading process completion time is $t_i^O = t_i^{\text{Com}} + t_i^{\text{Exec}} + t_i^{\text{Res}}$, where t_i^{Com} is the time to transmit the task to the EN, and it is given by $t_i^{\text{Com}} = \frac{\mathrm{d_i}}{\mathrm{r}}$. t_i^{Exec} is the time to execute the task τ_i at the EN, and it is given by $t_i^{\text{Exec}} = \frac{\lambda_i}{f_S}$. t_i^{Res} is the time to receive the result out from the edge node. Because the data size of the result is usually ignored compared to the input data size, we ignore this response time and its energy consumption as adopted by [8]. Hence, for the τ_i task $t_i^O = x_i\left(\frac{d_i}{r} + \frac{\lambda_i}{f_S}\right)$, and for all tasks, we have:

$$t^O = \sum_{i=1}^{N} x_i\left(\frac{d_i}{r} + \frac{\lambda_i}{f_S}\right) \tag{3}$$

So, the communication's energy consumption is obtained by multiplying the resulting transmission period by the transmission undertaken power p^T. Thus, this energy is:

$$e^C = \frac{p^T}{r}\sum_{i=1}^{N} x_i d_i \tag{4}$$

Finally, given the offloading decision vector $\mathbb{X}$ for all tasks, the local execution frequency f_L of the SMD, and the server execution frequency f_S at the edge, the total execution time for the SMD is composed of its local execution time, the communication time as well as the execution time at the EN, and it is given by $T(\mathbb{X}, f_L, f_S) = t^L + t^O$. Then, according to Eqs. (1) and (3) and if we note $\Lambda = \sum_{i=1}^{N} \lambda_i$, the total execution time can be formulated as:

$$T(\mathbb{X}, f_L, f_S) = \left\{\frac{\Lambda}{f_L} - \frac{\sum_{i=1}^{N} \lambda_i x_i}{f_L} + \frac{\sum_{i=1}^{N} d_i x_i}{r} + \frac{\sum_{i=1}^{N} \lambda_i x_i}{f_S}\right\} \tag{5}$$

2.2 *Problem Formulation*

In this section, we present our optimization problem formulation that aims to minimize the overall execution time in the offloading process. The obtained problem is formulated as:

$$\mathcal{P}1: \min_{\{x, f_L, f_S\}} \left\{ \frac{\Lambda}{f_L} - \frac{\sum_{i=1}^{N} \lambda_i x_i}{f_L} + \frac{\sum_{i=1}^{N} d_i x_i}{r} + \frac{\sum_{i=1}^{N} \lambda_i x_i}{f_S} \right\}$$

s.t. $(C_{1.1}) x_i \in \{0; 1\}; i \in 1; N$

$(C_{1.2}) F_L^{\min} \leq f_L \leq F_L^{\max}$

$(C_{1.3})\ 0 < f_s \leq F_s;$

$$(C_{1.4}) t^L = \sum_{i=1}^{N} \frac{\lambda_i}{f_L}(1 - x_i) \leq t^{\max}$$

$$(C_{1.5}) t^O = \sum_{i=1}^{N} x_i \left(\frac{d_i}{r} + \frac{\lambda_i}{f_S} \right) \leq t^{\max};$$

$$(C_{1.6}) e^L + e^C = k_L \cdot f_L^2 \cdot \sum_{i=1}^{N} \lambda_i (1 - x_i) + \frac{p^T}{r} \sum_{i=1}^{N} d_i x_i \leq E^{\max}.$$

In this work, each one of the available tasks can be either executed locally or offloaded to the edge node. Thus, every feasible offloading decision solution has to satisfy the above constraints.

The constraint $(C_{1.1})$ refers to the offloading decision variable x_i for task τ_i which equals 0 or 1. The second constraint $(C_{1.2})$ indicates that the allocated variable local frequency f_L belongs to a priori fix interval given by $\left[F_L^{\min}, F_L^{\max}\right]$. Similarly, the allocated variable remote edge server frequency f_S belongs to the interval $\left[0, F_S^{\max}\right]$ in constraint $(C_{1.3})$. The constraint $(C_{1.4})$ shows that the execution time of all decided local tasks must be less than the given latency requirement $t^{\max}$. In the same way, in constraint $(C_{1.5})$, the offloading time of all decided remote tasks must satisfy the same latency requirement $t^{\max}$. The final constraint $(C_{1.6})$ is important especially if the SMD's battery power is critical. It imposes that the total local execution energy must not exceed the tolerated given amount $E^{\max}$.

3 Problem Resolution

In this section, we will introduce how we derive our solution from the obtained optimization problem. In our proposed model, the offloading decision vector for all the tasks is denoted $\mathbb{X}$. Let define the vector that contains the off-loadable tasks' identifiers:

$$\mathbb{X}_1 = \{i \in \mathbb{X} / x_i = 1\} \tag{6}$$

Thus, we can define: $\Lambda_1 = \sum_{i \in \mathbb{X}_1} \lambda_i$ and $D_1 = \sum_{i \in \mathbb{X}_1} d_i$. In addition, given the decision vector $\mathbb{X}_1$, constraint $(C_{1.4})$ can be reformulated as $\frac{\Lambda - \Lambda_1}{t^{\max}} \leq f_L$ and constraint $(C_{1.5})$ can be similarly reformulated as: $\frac{\Lambda_1}{t^{\max} - \frac{D_1}{r}} \leq f_S$. For ease of use, let note:

$$f_L^- = \frac{\Lambda - \Lambda_1}{t^{\max}} \tag{7}$$

$$f_L^+ = \sqrt{\frac{E^{\max} - \frac{p^T D_1}{r}}{k_L(\Lambda - \Lambda_1)}} \tag{8}$$

$$f_S^- = \frac{\Lambda_1}{t^{\max} - \frac{D_1}{r}} \tag{9}$$

Thus, for a given offloading decision vector $\mathbb{X}$, we get the following optimization subproblem:

$$\begin{aligned}
&\mathcal{P}2(\mathbb{X}) : \min_{\{f_L, f_S\}} \left\{ \tfrac{\Lambda - \Lambda_1}{f_L} + \tfrac{D_1}{r} + \tfrac{\Lambda_1}{f_S} \right\} \\
&s.t.\ (C_{2.1})\ F_L^{\min} \le f_L \le F_L^{\max}; \\
&(C_{2.2})\ f_L^- \le f_L; \\
&(C_{2.3}) f_S^- \le f_S \le F_S; \\
&(C_{2.4})\ k_L f_L^2 (\Lambda - \Lambda_1) + \tfrac{p^T D_1}{r} \le E^{\max}.
\end{aligned}$$

For the $\mathcal{P}2$ problem, the objective function $T_1(f_L) = \frac{\Lambda - \Lambda_1}{f_L}$ is a strictly increasing continuous function according to its variable f_L. Hence, by taking into consideration the obtained constraints $(C_{2.1})$, $(C_{2.2})$ and $(C_{2.4})$, we can derive the following function's optimum f_L^* given by:

$$f_L^* = \begin{cases} 0 & \text{if } \mathbb{X} = \mathbb{X}_1 \\ \emptyset & \text{if } E^{\max} \le \frac{p^T D_1}{r} \ \textbf{or}\ f_L^- > F_L^{\max} \ \textbf{or}\ f_L^+ < F_L^{\min} \textbf{or}\ f_L^- > f_L^+ \\ F_L^{\max} & \text{if } f_L^+ > F_L^{\min} \\ f_L^+ & \text{otherwise} \end{cases} \tag{10}$$

For the $\mathcal{P}2$ problem, the objective function $T_2(f_S) = \frac{D_1}{r} + \frac{\Lambda_1}{f_S}$ is strictly decreasing w.r.t. the variable f_S.

So, by taking into consideration the $(C_{2.3})$ constraint, we can derive the following function's optimum f_S^* given by:

$$f_S^* = \begin{cases} \emptyset \ \text{if } f_S^- > F_S \ \textbf{or}\ \frac{D_1}{r} \ge t^{\max} \\ F_S \qquad \text{otherwise} \end{cases} \tag{11}$$

4 Proposed Solutions

Next, the problem relies on determining the optimal offloading decision vector $\mathbb{X}$ that gives the optimal energy consumption. However, to iterate over all possible combinations of a list of N binary variables, the time complexity is exponential (the exhaustive search over all possible solutions requires 2^N iterations). Subsequently, the total time complexity of the whole solution (including Algorithm 1) is $O(2^N)*O(1) = O(2^N)$ that is not practical for large values of N. In the following, we propose a low complexity approximate algorithm to solve this question.

4.1 Brute Force Search Solution

For comparison purpose, we introduce the brute force search method for feasible small values of N. This method explores all cases of offloading decisions and saves the one with the minimum execution time as well as its completion time.

4.2 Simulated Annealing-Based Solution

For our proposed solution, we use a simulated annealing-based method. The simulated annealing technique (SA) is adopted as heuristic solution in the optimization field especially for hard problems. To improve a solution, it employs iterative random solution variation. Interested readers can refer to the following works [9] and [10] for more details about this issue.

In the simulated annealing-based heuristic solution, we start by a random offloading decision state $\mathbb{X}$. Then, at every step, some neighboring state $\mathbb{X}*$ of the current state $\mathbb{X}$ and probabilistically decides between moving the system to state $\mathbb{X}*$ or staying in state $\mathbb{X}$. Practical, a state's variation consists of changing the offloading decision of some tasks among the list. These probabilistic transitions ultimately lead the system to move to states of lower energy. Generally, this step is repeated until getting a good enough energy, or until a given iterations' count has been reached.

The following algorithm presents the pseudo-code of the simulated annealing-based heuristic as described above.

```
Algorithm :Simulated Annealing Pseudo-code
Input: The list τ of N sub-tasks.
Output: the offloading policy 𝕏*.
Initialize: a random policy 𝕏.
Calculate T using τ and 𝕏;
for k = 0 to kmax do
   Temp ← temperature(k/kmax)
   Pick a random neighbour 𝕏_new ← neighbour(𝕏)
   Calculate T_new using τ and 𝕏_new;
   if T_new≠∞ and P(T,T_new,Temp) ≥ random(0,1) then
        𝕏 ← 𝕏_new ;
        T ← T_new ;
   end if
 end for
 𝕏* ← 𝕏 ;
```

In this algorithm, ***random(0,1)*** is a function's call that generates a random number in [0,1]. Neighbor ($\mathbb{X}$) is a function's call that generates a decision vector state near the input $\mathbb{X}$. P($\boldsymbol{T}$, $\boldsymbol{T}_{\mathbf{new}}$, Temp) is an acceptance probability function that depends on the processing times $\boldsymbol{T}$ and $\boldsymbol{T}_{\mathbf{new}}$ of the two states $\mathbb{X}$ and $\mathbb{X}_{\text{new}}$, and on a global parameter Temp called the temperature that varies over iterations.

In our proposed solution, which we denote simulated annealing offloading (SAO), where Temp_0 is the initial temperature constant ΔT is the solutions' processing time variation while changing the task i state.

5 Evaluation

The presented results in this work are averaged for 100 time executions. We implement all the algorithms using C++ language. The transmission bandwidth between the mobile device node and remote edge server is set to $r = 100$ Kb/s. The local CPU frequency f_L of the SMD is optimized between $F_L^{\min} = 1$ MHz and $F_L^{\max} = 60$ MHz. The CPU frequency of the remote edge server node will be optimized under the value $F_S = 6$ GHz. The deadlines t^{max} are uniformly defined from 0.5 s to 2 s. The threshold energy $E^{\max}$ is uniformly chosen in $[0.6, 0.8] * \Lambda.k_L.\left(F_L^{\max}\right)^2$. Additionally, the data size of each one of the N tasks is assumed to be in [30,300] Kb. For the cycle amount of each task, it is assumed to belong to [60,600] MCycles. The transmission power is set to be $p^T = 0.1$ Watt. For the energy efficiency coefficient, we set $k_L = 10^{-26}$. For the initial temperature, we set $\text{Temp}_0 = 100$. We present our results in terms of average decision time and processing time.

We start by studying the average tasks' processing time for each method. Thus, we carried an experiment where we vary the number of tasks parameter between 3 and 26. The experiment's results are depicted in the following tow figures. Figure 2 represents the obtained results for both brute force search-based solution (BFS) and simulated annealing-based solution (SAO). It shows a small distance between the curves representing the realized averaged tasks' processing time. Accordingly, the

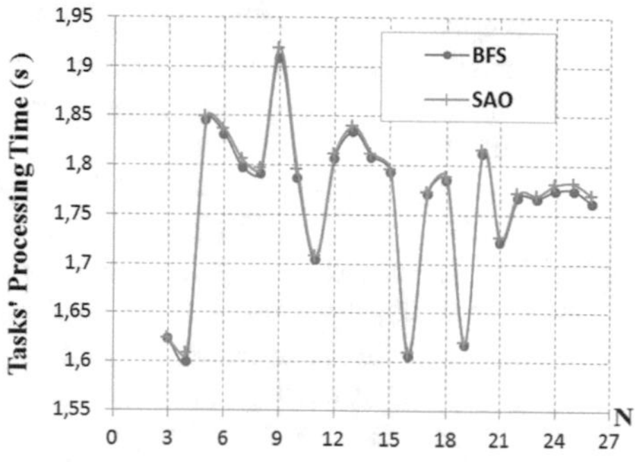

Fig. 2 Tasks' processing time for N between 3 and 26

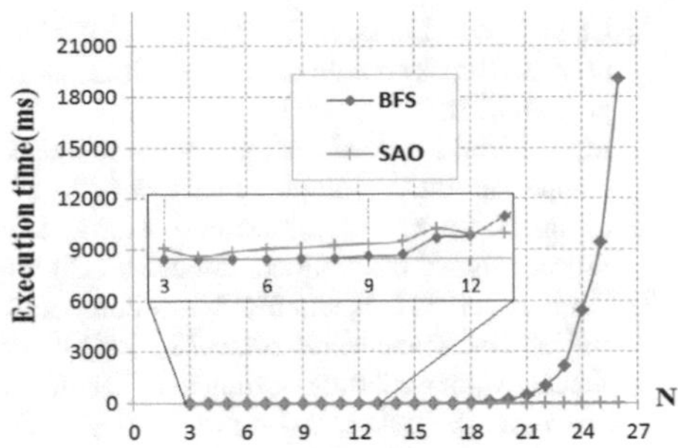

Fig. 3 Execution time for N between 3 and 26

differences between the optimal BFS method and the SAO method vary from 0.00 to 0.82%.

Now, Fig. 3 depicts the average of the execution time in ms to get the offloading decisions for both schemes. While the tasks count N is between 2 and 26, it clearly shows the exponential variation of the BFS execution time w.r.t. N. Additionally, The SAO solution gives a stable execution time that reached only 0.05 ms for $N = 26$.

This experiment shows that our proposed heuristic scheme achieves a good trade-off between the solution's execution time and the accomplished processing delays of the offloaded tasks within the edge node.

6 Conclusion

In this paper, we consider a single mobile device that is energy-constrained and equipped with a list of heavy off-loadable tasks that are delay constrained. We propose a simulated annealing-based heuristic solution to solve a hard decision problem that jointly optimizes the processing time and computing resources in a mobile edge computing node. The obtained results in terms of processing time are very encouraging. In addition, the proposed solution performs the offloading decisions within acceptable and feasible timeframes.

References

1. Mach, P., Becvar, Z.: Mobile edge computing: a survey on architecture and computation off loading. IEEE Commun. Surveys Tutorials **19**(3), 1628–1656 (2017)
2. You, C., et al.: Energy-efficient resource allocation for mobile-edge computation offloading. IEEE Trans. Wireless Commun. **16**(3), 1397–1411 (2017)
3. Chen, M.-H., Liang, B., Dong, M.,: Joint offloading and resource allocation for computation and communication in mobile cloud with computing access point. In: IEEE INFOCOM Conference on Computer Communications, pp. 1–9 (2017)

4. Chen, M.-H., Liang, B., Dong, M.,: Joint offloading decision and resource allocation for multi-user multi-task mobile cloud. In: IEEE International Conference on Communications (ICC), pp. 1–6 (2016)
5. Li, H.: Multi-task offloading and resource allocation for energy-efficiency in mobile edge computing. Int. J. Comput. Techn. **5**(1), 5–13 (2018)
6. Chun, B.-G., et al.: Clonecloud: elastic execution between mobile device and cloud. In: Proceedings of the sixth conference on Computer systems, pp. 301–314 (2011)
7. Chen, X., et al.: Efficient multi-user computation offloading for mobile-edge cloud computing. IEEE/ACM Trans Networking **24**(5), 2795–2808 (2016)
8. Zhang, K., et al.: Energy-efficient offloading for mobile edge computing in 5G heterogeneous networks. IEEE Access **4**, 5896–5907 (2016)
9. Fan, Z., et al.: Simulated-annealing load balancing for resource allocation in cloud environments. In: International Conference on Parallel and Distributed Computing, Applications and Technologies, pp. 1–6 (2013)
10. Chen, L., et al.: ENGINE: Cost Effective Offloading in Mobile Edge Computing with Fog-Cloud Cooperation. arXiv preprint arXiv:1711.01683 (2017)

Viscous Dissipation and Heat Source/Sink Effects on Convection Heat Transfer in a Saturated Porous Medium

Elyazid Flilihi, Mohammed Sriti and Driss Achemlal

Abstract A numerical investigation is performed to analyze the transient laminar free convection over an isothermal inclined plate embedded in a saturated porous medium with the viscous dissipation and heat source/sink effects. The flow in the porous medium is modeled with the Darcy–Brinkman–Forchheimer model. To solve the problem, an explicit procedure of finite difference method with stability and convergence criterion has been used in this work. The effects of different parameters that enter into the problem on the dimensionless streamlines of the velocity field, the isothermal lines distributions, skin friction, and the local Nusselt number are examined. Also, the physical aspects of the problem are discussed in details.

Keywords Unsteady flow · Free convection · Non-Darcian porous medium

1 Introduction

The analysis of convection heat transfer flow in porous media has attracted the attention of many researchers during the past few decades. This type of flow is encountered in geothermal energy systems, extraction of oil and gas through the soil, the storage of radioactive, the biological systems, etc. Many principal past studies concerning these processes can be found in the recent excellent reviews by [1, 2].

Numerous studies of such flows have been focused on the importance of convective heat transport phenomenon in non-Darcian porous media. Islam et al. [3] studied mass transfer flow through an inclined plate with a porous medium. In this numerical investigation, the explicit finite difference method has used to solve the dimensionless system of equations. Also, Mondal et al. [4] have analyzed the unsteady free-convective flow along with a vertical porous plate with variable viscosity and thermal conductivity, with viscous dissipation and heat generation. The results obtained to show that the temperature profile increases for the increase of thermal conductivity

E. Flilihi (✉) · M. Sriti · D. Achemlal
Polydisciplinary Faculty of Taza, Engineering Sciences Laboratory,
Sidi Mohamed Ben Abdellah University, BP.1223 Taza, Morocco

V. Bhateja et al. (eds.), *Embedded Systems and Artificial Intelligence*,
Advances in Intelligent Systems and Computing 1076,
https://doi.org/10.1007/978-981-15-0947-6_11

parameter, Eckert number, and heat generation parameter. Recently, Flilihi et al. [5] investigated the variable heat source and wall radiation effects on boundary layer convection from an inclined plate in non-Darcian porous medium.

This study extended the researches of Flilihi et al. [6, 7] to an unsteady convective boundary layer flow past an inclined plate in a non-Darcian porous medium in the presence of the viscous dissipation and the heat source/sink effects using the Darcy–Brinkman–Forchheimer model, taking into account the convective term. This study finds applications in the fields of petroleum engineering, geothermal energy, etc.

2 Physical Model and Governing Equations

Consider a two-dimensional transient laminar free-convection flow of a fluid over an isothermal inclined plate embedded in a non-Darcian porous medium. Initially, it is assumed that the plate and the fluid are at the temperature T_∞. At $t' > 0$, the temperature of the plate is raised to T_w, which is then maintained constant. The temperature of the fluid away from the plate is T_∞. The physical model and coordinate system are shown in Fig. 1. The Darcy–Brinkman–Forchheimer model is used to describe the flow in the porous medium. Under the Boussinesq and boundary layer approximations, the governing equations are:

$$\begin{cases} \dfrac{\partial u}{\partial x} + \dfrac{\partial v}{\partial y} = 0 \\ \dfrac{\partial u}{\partial t'} + u\dfrac{\partial u}{\partial x} + v\dfrac{\partial u}{\partial y} = \nu\dfrac{\partial^2 u}{\partial y^2} + g\beta(T - T_\infty)\cos\phi - \dfrac{\nu\varepsilon}{K}u - \dfrac{F\varepsilon^2}{K^{1/2}}|u|u \\ \dfrac{\partial T}{\partial t'} + u\dfrac{\partial T}{\partial x} + v\dfrac{\partial T}{\partial y} = \alpha\dfrac{\partial^2 T}{\partial y^2} + \dfrac{\mu}{\rho C_p}\left(\dfrac{\partial u}{\partial y}\right)^2 + \dfrac{Q_0}{\rho C_p}(T - T_\infty) \end{cases} \tag{1}$$

The initial and boundary conditions are:

$$\begin{aligned} &t' = 0 : u = v = 0, \quad T = T_\infty \quad \text{for all} \quad x \quad \text{and} \quad y \\ &t' > 0 : \begin{cases} u = v = 0, T = T_\infty \quad \text{at} \quad x = 0 \\ u = v = 0, T = T_w \quad \text{at} \quad y = 0, x > 0 \\ u = 0, T = T_\infty \quad \text{at} \quad y \to \infty, \quad x > 0 \end{cases} \end{aligned} \tag{2}$$

where x and y are the Cartesian coordinates and t' represents time. u and v are, respectively, the velocity components along the x- and y-axes, T is the fluid temperature. The constants μ, ν, K, α, g, and ρ are, respectively, fluid viscosity, kinematic viscosity, permeability of porous medium, thermal diffusivity, gravitational acceleration, and density. Cp, β, ε, F are, respectively, the specific heat at constant pressure, the coefficient of thermal expansion, the porosity of porous medium, and the empirical constant.

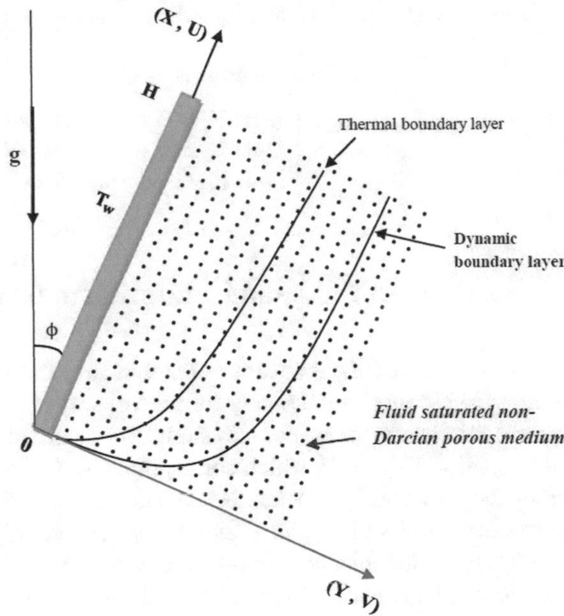

Fig. 1 Physical model and coordinate system

The following dimensionless variables are introduced:

$$\begin{cases} (X,Y) = \dfrac{(x,y)}{H}, \quad (U,V) = \dfrac{(u,v)}{U_r}, \quad \theta = \dfrac{T - T_\infty}{T_w - T_\infty} \\ t = \dfrac{U_r t'}{H} \quad \text{with } U_r = \dfrac{\nu}{H} \end{cases} \tag{3}$$

here U_r is a reference velocity, and H is the height of the plate.

In terms of these variables, and after substitution and development, the system of Eq. (1) becomes:

$$\begin{cases} \frac{\partial U}{\partial X} + \frac{\partial V}{\partial Y} = 0 \\ \frac{\partial U}{\partial t} + U\frac{\partial U}{\partial X} + V\frac{\partial U}{\partial Y} = \frac{\partial^2 U}{\partial Y^2} + \mathrm{Gr}\theta\cos\phi - \frac{1}{\mathrm{Da}}U - \frac{\mathrm{Fr}}{\mathrm{Da}}|U|U \\ \frac{\partial \theta}{\partial t} + U\frac{\partial \theta}{\partial X} + V\frac{\partial \theta}{\partial Y} = \frac{1}{\mathrm{Pr.Re}}\frac{\partial^2 \theta}{\partial Y^2} + \mathrm{Ec}\left(\frac{\partial U}{\partial Y}\right)^2 + Q\theta \end{cases} \tag{4}$$

Here $\mathrm{Gr} = \dfrac{g\beta(T_w - T_\infty)H}{U_r^2}$ is the Grashof number, $\mathrm{Da} = \dfrac{K}{H^2\varepsilon}$ is the Darcy number, $\mathrm{Fr} = \dfrac{F\varepsilon K^{1/2}}{H}$ is the Forchheimer number, $\mathrm{Pr} = \dfrac{\mu}{\rho\alpha}$ is the Prandtl number, $\mathrm{Re} = \dfrac{\rho H U_r}{\mu}$ is the Reynolds number, $\mathrm{Ec} = \dfrac{U_r^2}{C_p(T_w - T_\infty)}$ is the Echert number, and $Q = \dfrac{\nu Q_0}{\rho C_p U_r^2}$ is the heat source/sink parameter.

The transformed initial and boundary conditions for Eq. (3) are now given by:

$$t = 0 : U = V = \theta = 0, \quad \text{for all} \quad X \quad \text{and} \quad Y.$$
$$t > 0 : \begin{cases} U = V = \theta = 0, & \text{at} \quad X = 0 \\ U = V = 0, \theta = 1 & \text{at} \quad Y = 0, X > 0 \\ U = 0, \theta = 0 & \text{at} \quad Y \to \infty, \quad X > 0 \end{cases} \tag{5}$$

3 Solution Method and Non-uniform Grid

The system of nonlinear equations (4) that subject to the initial and boundary conditions (5) is solved numerically for the velocity and temperature using the explicit finite difference method. For this study, we consider a plate of height $X_{\max} = 100$ and regarded $Y_{\max} = 35$ which corresponds to $Y = \infty$, where X-direction is taken along the plate and Y-direction is taken normal to it. Subscripts i and j will be used to represent nodes in X- and Y-directions, respectively.

Physically, the thickness of the boundary layer is much smaller than any characteristic length defined in the streamwise direction. Therefore, the changes in physical properties in the direction parallel to the plate are small compared to the corresponding changes in perpendicular to the plate. Thus, grids in the Y-direction should be much finer close to the plate and spaced elsewhere for a better appreciation of the formation of the boundary layer.

A variable grid size in the Y-axis is calculated with Eq. (7).

$$Y(j) = Y_{\max} \cdot \left[\frac{e^{\alpha_y (j-N)} - A_y}{1 - A_y} \right] \tag{6}$$

$$\Delta Y(j-1) = Y(j) - Y(j-1) \tag{7}$$

with $j = 2, \ldots, N$ and $A_y = e^{\alpha_y (1-N)}$, where α_y is the distribution coefficient in the Y-axis, generally between 0 and 1. N is the number of nodes contained on Y-direction.

4 Numerical Formulation and Solution

The unsteady nonlinear coupled equations (4) subject to the initial and boundary conditions (5) are solved by using an explicit finite difference scheme. The set of approximate finite difference equations corresponding to system of Eq. (4) are:

$$\begin{cases} \frac{U_{i,j}^{k+1}-U_{i-1,j}^{k+1}}{\Delta X} - \frac{V_{i,j}^{k+1}-V_{i-1,j}^{k+1}}{\Delta Y(j)} = 0 \\ \frac{U_{i,j}^{k+1}-U_{i,j}^{k}}{\Delta t} + U_{i,j}^{k}\frac{U_{i,j}^{k}-U_{i-1,j}^{k}}{\Delta X} + V_{i,j}^{k}\frac{U_{i,j}^{k}-U_{i-1,j}^{k}}{\Delta Y(j)} = \mathrm{Gr}\theta_{i,j}^{k+1} \\ \quad + \frac{U_{i,j+1}^{k}-2U_{i,j}^{k}+U_{i,j-1}^{k}}{(\Delta Y(j))^2} - \frac{1}{\mathrm{Da}}U_{i,j}^{k} - \frac{\mathrm{Fr}}{\mathrm{Da}}|U_{i,j}^{k}|U_{i,j}^{k} \\ \frac{\theta_{i,j}^{k+1}-\theta_{i,j}^{k}}{\Delta t} + U_{i,j}^{k}\frac{\theta_{i,j}^{k}-\theta_{i-1,j}^{k}}{\Delta X} + V_{i,j}^{k}\frac{\theta_{i,j+1}^{k}-\theta_{i,j}^{k}}{\Delta Y(j)} = \\ \quad \frac{1}{\mathrm{Pr.Re}}\frac{\theta_{i,j+1}^{k}-2\theta_{i,j}^{k}+\theta_{i,j-1}^{k}}{(\Delta Y(j))^2} + \mathrm{Ec.}\left(\frac{U_{i,j+1}^{k}-U_{i,j}^{k}}{\Delta Y(j)}\right)^2 + Q.\theta_{i,j}^{k} \end{cases} \tag{8}$$

The coefficients $U_{i,j}$ and $V_{i,j}$ are treated as constants, during any one time step. Then, at the end of any time step Δt, the new velocity components, U^{k+1} and V^{k+1}, and the new temperature θ^{k+1} at all interior grid points may be obtained by successive applications of system of equation (8). The region of integration is considered as a rectangle with sides X, $X_{\max} = 100$ and Y, $Y_{\max} = 35$. After performing few tests on sets of mesh sizes to access grid independence taking account a variable grid size in the Y-axis calculated with Eq.(7), the time and spatial step sizes $\Delta t = 0.05$ and $\Delta X = 2$ were found to give accurate results.

5 Skin Friction Coefficient and Local Nusselt Number

The most important physical quantities in this problem are the skin friction coefficient C_f and local Nusselt number Nu, which are given, respectively, by the following equations:

$$C_f = 2\left.\frac{\partial U}{\partial Y}\right|_{Y=0} \quad \text{and} \quad \mathrm{Nu} = -\left.\frac{\partial \theta}{\partial Y}\right|_{Y=0} \tag{9}$$

6 Results and Discussion

A numerical investigation has been made for an unsteady thermal convection flow along with an isothermal inclined plate embedded in non-Darcian porous medium, using the Darcy–Brinkman–Forchheimer model, taking into account the convective term. In this study, we have analyzed the effects of the time t, inclination angle ϕ, Forchheimer number Fr, Echert number Ec, and the heat source/sink parameter Q on the thermal, dynamic skin friction, and heat transfer rate profiles. We notice that the heat source corresponds to $Q > 0$, the heat sink to $Q < 0$, and without heat source/sink to $Q = 0$. Also, we note that the steady state is obtained at $t = 250$, i.e., there is no major change in the velocity profiles after time $t = 250$.

6.1 Streamlines of the Velocity Field

The displayed Figs. 2 and 3 show the effect of Forchheimer number Fr on dimensionless streamlines of the velocity field over an isothermal and impermeable inclined plate ($\phi = 30°$) in a saturated porous medium for Re = 2, Gr = 10, Da = 10, Pr = 0.71 in the presence of viscous dissipation (Ec = 0.01) and internal heat generation ($Q = 0.01$). From the streamlines velocity, we notice that an increasing of Fr values from 0.0 to 2.0 causes a strong increase in Forchheimer drag which decelerates the flow, i.e., reduces velocities.

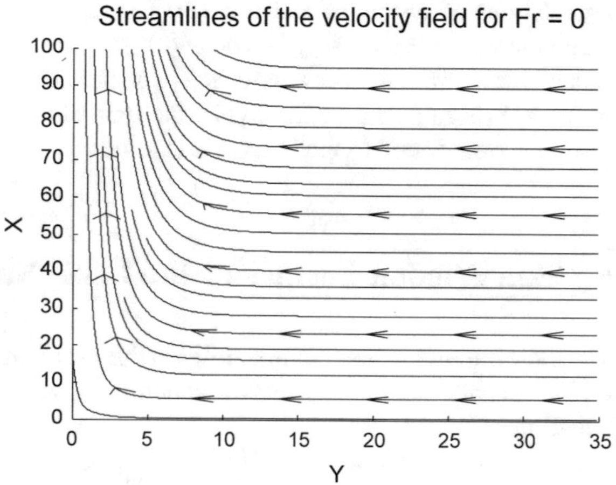

Fig. 2 Streamlines of dimensionless velocity field according to *X* and *Y* for Fr = 0

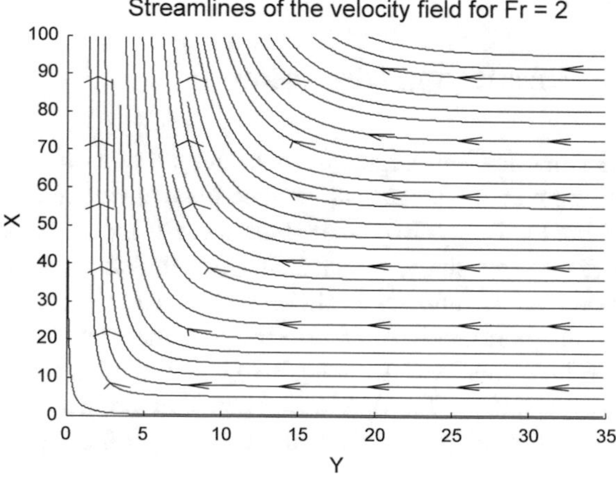

Fig. 3 Streamlines of dimensionless velocity field according to *X* and *Y* for Fr = 2

6.2 Temperature Profiles and Isothermal Lines

Steady-state temperature profiles as a function of Y for different positions X of the plate are plotted in Fig. 4. This figure shows that the thermal boundary layer thickness depends on the position of X. Thus, the thermal boundary layer thickness increases as X increases.

Figures 5 and 6 show for Re = 2, Gr = 10, Da = 10, Pr = 0.71, Fr = 0.5, and Ec = 0.01, and the isothermal line distributions in the boundary layer area of an isothermal and impermeable inclined plate ($\phi = 30°$) placed in a saturated porous medium, for selected values of heat source/sink parameter $Q = -0.03$ and $Q = 0.03$. Here, we notice that the effect of increasing values heat source/sink parameter results in an increase of the thermal boundary layer thickness.

6.3 Skin Friction and Local Heat Transfer Rate Profiles

Figure 7 displays the unsteady local Nusselt number profiles for various values of Forchheimer number Fr. We see that the local Nusselt number decreases quickly with the time t and then remains constant. It is apparent that the lower heat transfer rate occurs as Fr increases. This is clear from the fact that inertia effects tend to slow down the buoyancy-induced flow in the boundary layer and so retard the heat transfer.

In Fig. 8, we present the evolution of the heat transfer rate at the wall for several positions of the plate at $R = 2$, Gr = 20, Da = 10, Pr = 0.71, and Ec = 0.01 with and without heat generation/absorption. From this figure, it is concluded that the presence of an internal heat source in a porous medium makes to reduce the heat transfer rate at the surface for all selected positions of the plate; on the contrary, the

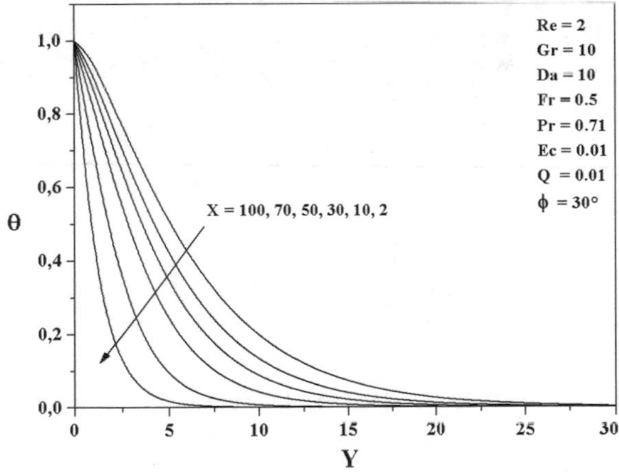

Fig. 4 Temperature profiles at selected position X of the plate

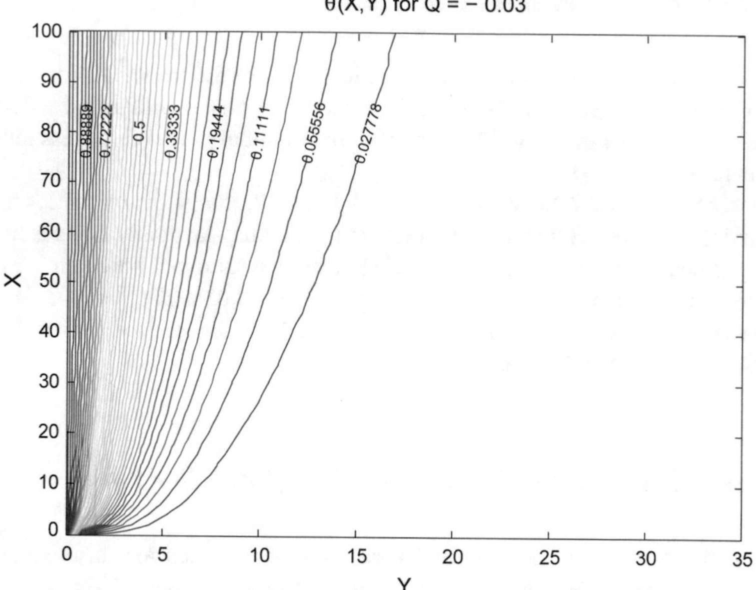

Fig. 5 Dimensionless isothermal lines according to X and Y at $Q = -0.03$

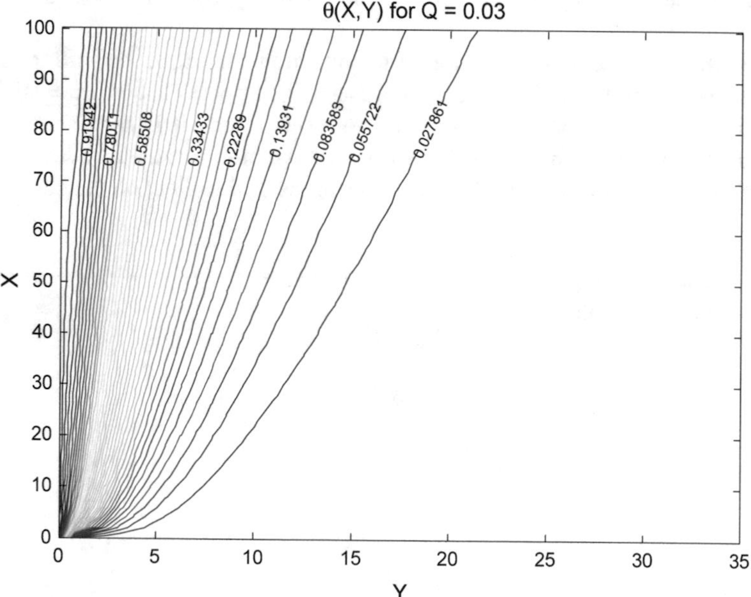

Fig. 6 Dimensionless isothermal lines according to X and Y at $Q = 0.03$

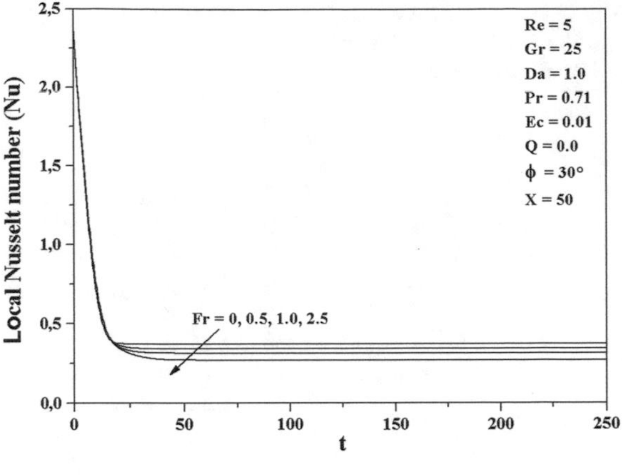

Fig. 7 Unsteady Nu profiles for various values of Fr

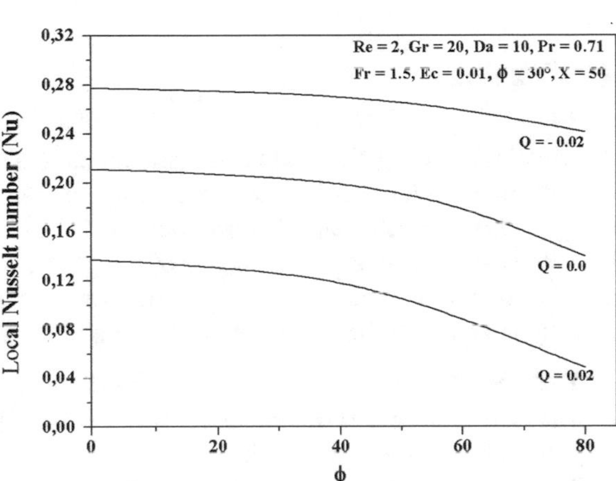

Fig. 8 Steady Nu profiles versus ϕ for selected values of Q

presence of a heat sink in the medium permits to amplify it. This seemed logical because the presence of an absorption of heat in the medium increases the thermal gradients, which leads consequently to a high rate of heat transfer at the surface of the plate.

In Fig. 9, steady-state local skin friction profiles are plotted as a function of X for various values of the Fochheimer number Fr. From this figure, we see that the increase in the Forchheimer number Fr leads to reduce the local skin friction at the wall.

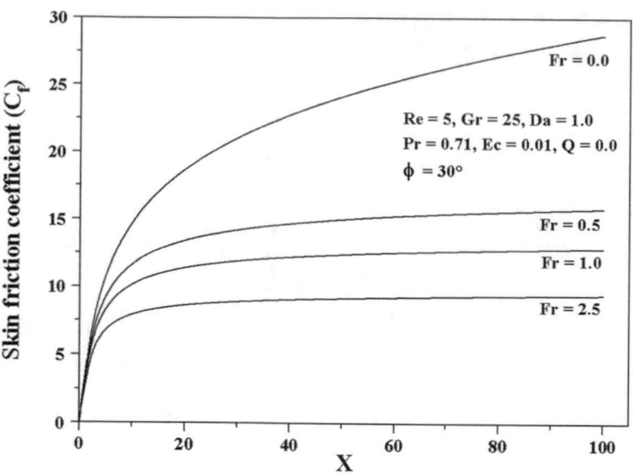

Fig. 9 Steady skin friction profiles as a function of *X* for selected values of Fr

7 Conclusion

A numerical investigation has been made for an unsteady thermal convection flow along with an isothermal inclined plate embedded in non-Darcian porous medium, using the Darcy–Brinkman–Forchheimer model, taking into account the convective term. We have discussed the influence of the physical parameters that enter into the problem on the thermal, dynamic skin friction, and heat transfer rate profiles. The main conclusions of the current analysis are as follows:

1. The thermal boundary layer thickness increases with *X* position of the plate.
2. The increasing in the Forchheimer number Fr leads to reduce the local skin friction and the heat transfer rate at the wall.
3. The heat sink in the medium amplified the heat transfer rate at the wall, while the presence of heat source leads to reduce it.
4. The wall heat transfer rate is optimal for the vertical position of the plate.
5. The inertia effects tend to slow down the buoyancy-induced flow in the boundary layer and so retard the heat transfer.
6. The increasing in values of heat source/sink parameter results in an increase of the thermal boundary layer thickness.

References

1. Nield, D.A., Bejan, A.: Convection in Porous Media, 5th ed. Springer Science, New York (2017). https://doi.org/10.1007/978-3-319-49562-0
2. Nield, D.A., Simmons, C.T.: A Brief Introduction to Convection in Porous Media. Springer Nature (2018)
3. Islam, M., Akter, F., Islam, A.: Mass transfer flow through an inclined plate with porous medium. Am. J. Appl. Math. **3**(5), 215–220 (2015). https://doi.org/10.11648/j.ajam.20150305.12

4. Mondal, R., Azad, M.A.K., Arifuzzaman, S., Hossain, K.E., Ahmmed, S.: Unsteady free convective flow along a vertical porous plate with variable viscosity and thermal conductivity. IOSR J. Math. (IOSR-JM) **12**(1), 64–71 (2016)
5. Flilihi, E., Sriti, M., Achemlal, D., El Haroui, M.,: Variable heat source and wall radiation effects on boundary layer convection from an inclined plate in non-Darcian porous medium. Front. Heat Mass Transf. (FHMT) **9**(23) (2017). https://doi.org/10.5098/hmt.9.23
6. Flilihi, E., Sriti, M., Achemlal, D., El Haroui, M.: Semi-analytical prediction of mixed convection in porous medium using Darcy-Brinkman model. J. Eng. Appl. Sci. **14**(4), 1122–1129 (2019)
7. Flilihi, E., Sriti, M., Achemlal, D.: Numerical solution on non-uniform mesh of Darcy-Brinkman-Forchheimer model for transient convective heat transfer over flat plate in saturated porous medium. Front. Heat Mass Transf. (FHMT) **12**(12) (2019). https://doi.org/10.5098/hmt.12.12

Fast and Efficient Decoding Algorithm Developed from Concatenation Between a Symbol-by-Symbol Decoder and a Decoder Based on Syndrome Computing and Hash Techniques

M. Seddiq El Kasmi Alaoui, Saïd Nouh and Abdelaziz Marzak

Abstract The exchange of data is done through a communication channel that is not entirely reliable and these data are most likely to be altered by errors. In order to detect and correct any errors on the receiver side, error-correcting codes are used; their principle is to add control data on the transmitter side. The Hartmann and Rudolph (HR) algorithm is based on a probabilistic study to determine if a bit of the received sequence is equal to 0 or 1. To do this, it exploits all the codewords of the dual code, which makes its complexity very high. The efficiency and the low complexity of the hard-decision decoder based on hash and syndrome decoding (HSDec) algorithm compared with its competitors encouraged us to use it in a serial concatenation with the HR decoder. In this paper, we propose a mutual cooperation between the HR partially exploited (PHR) and HSDec decoders. The simulation results show the efficiency and the rapidity of the proposed cooperation. The best performance qualities obtained using PHR-HSDec against HR alone prove the robustness of this cooperation.

Keywords Error-correcting codes · PHR · HSDec · Syndrome decoding · Hash techniques

1 Introduction

A digital communication system allows a source to send a message to a recipient using a physical medium such as cable, fiber optic or even spread over a radio channel with the maximum reliably. The system is called digital, which means that the information is represented by a sequence of symbols selected from a finite alphabet (binary, for example).

Recently, communications systems have experienced a high growth in order to satisfy telecommunications markets in a wide variety of fields, namely broadcasting or digital television. Therefore, the quality of the digital transmission evaluated by

---

M. Seddiq El Kasmi Alaoui (✉) · S. Nouh · A. Marzak
TIM Lab, Faculty of Sciences Ben M'sik, Hassan II University, Casablanca, Morocco
e-mail: sadikkasmi@gmail.com

V. Bhateja et al. (eds.), *Embedded Systems and Artificial Intelligence*,
Advances in Intelligent Systems and Computing 1076,
https://doi.org/10.1007/978-981-15-0947-6_12

the bit error ratio (BER) is maintained only by an increase in the transmitted power or by the introduction of a channel coding. Most transmission systems opt for the second solution, channel coding also called "error correcting codes."

Error-correcting codes are tools used to improve the reliability of transmissions over a noisy channel. The method that they use is to send more data on the channel than the amount of information to be transmitted, so redundancy is introduced. If this redundancy is operably structured, it is then possible to correct any errors introduced by the channel. We can then, despite the noise, find all the information originally transmitted.

A large family of error-correcting codes consists of block codes. For these codes, the information is first cut into blocks of constant size and each block is transmitted independently of the others, with a redundancy of its own. The largest subfamily of these codes is what are called linear codes. A linear code of length n and dimension k is a linear subspace C with dimension k of the vector space F_q^n where F_q is the finite field with q elements. Generally, a linear code is denoted by $C(n, k, d)$ where d is minimum distance between distinct vectors of C. The vectors in C are called codewords; they are found using a generator matrix G or generator polynomial g.

The remainder of this paper is structured as follows. In Sect. 2, we present some decoding algorithms as related works. In Sect. 3, we present the proposed cooperation between PHR and HSDec. In Sect. 4, we present the simulation results of the proposed decoder and we make a comparison with other ones. Finally, a conclusion is outlined in Sect. 5.

2 Related Works

Generally, there are two types of decoding algorithms: hard-decision and soft-decision algorithms. Hard-decision algorithms work on the binary form of the received information and use the Hamming distance to decode, in contrast soft-decision algorithms work directly on the received symbols and generally they use the Euclidian distance as a metric to decide the most likely transmitted codeword [1]. Among the works that are interested in the hard-decision algorithms, we find these based on the genetic algorithms (GA) [2–5], the algebraic decoder [6–8] of Berlekamp–Massey based on compute of syndromes and it has an efficient mechanism to localize all corrigible errors. We also find some works based on syndrome decoding and the hash techniques [9, 10], for example in [10] we have presented two fast and efficient decoders that we called HSDec and HWDec.

For soft-decision decoders, we find an algorithm applicable for BCH codes based on test and syndrome computing to localize the error positions [11]. Askali et al. have presented a version Soft In-Hard Out (SIHO) where they have used the MacWilliams's permutation decoding algorithm (McPD) as a hard decoder [12]. To facilitate self-synchronization of the digital communication systems, Shim et al. have proposed forward error-correction codes in communication channels [13]. In [14], a compact Genetic Algorithm (cGA) is used to generate two dual domain soft decision decoders. The low complexity of the HSDec algorithm compared with its competitors encouraged us to combine it with the Chase-2 algorithm [15]. In [16], Fossorier et al. have presented the OSD algorithm which is based on ordered statistics and it applicable on linear block codes. A decoding algorithm whose principle is different to these cited above is presented by Hartmann and Rudolph (HR) in [17]. The main idea behind this decoder is the symbol by symbol decoding of the received sequence. In [18], Nouh et al. have presented a serial concatenation of symbol-by-symbol and word-by-word decoders.

3 The Proposed Cooperation Between PHR and HSDec

The HR algorithm is based on a probabilistic study to determine if the bit r_j, of the received sequence r, equals to 0 or 1. To do this, it exploits all the codewords, of order 2^{n-k}, of the dual code. The use of 2^{n-k} dual codewords makes its complexity very high; to remedy this problem the authors of [18] proposed to use a part of order M of the codewords of the dual code space, then the HR is used only in some symbols of the sequence r. In this case, the HR is called the partial Hartmann–Rudolph decoder (PHR). The formula 1 represents the method proposed by HR to decide if the mth bit of the decoded word "c" is equal to 1 or 0 from the received sequence r.

$$\begin{cases} c'_m = 0 \text{ if } \sum_{j=1}^{2^{n-k}} \prod_{l=1}^{n} \left(\frac{1-\phi_l}{1+\phi_l}\right)^{c^{\perp}_{jl} \oplus \delta_{\mathrm{ml}}} > 0 \\ c'_m = 1 \text{ otherwise} \end{cases} \tag{1}$$

where

$$\delta_{ij} = \begin{cases} 1 \text{ if } i = j \\ 0 \text{ otherwise} \end{cases} \text{ and } \phi_m = \frac{\Pr(r_m|1)}{\Pr(r_m|0)}$$

The bit $c^{\perp}_{jl}$ denotes the lth bit of the jth codeword of the code $C^{\perp}$.

The formula 2 presents the improvement of formula 1 in order to use just the M codewords of the dual code.

$$\begin{cases} c'_m = 0 \text{ if } \sum_{j=1}^{M} \prod_{l=1}^{n} \left(\frac{1-\phi_l}{1+\phi_l}\right)^{c^{\perp}_{jl} \oplus \delta_{\mathrm{ml}}} > 0 \\ c'_m = 1 \text{ otherwise} \end{cases} \tag{2}$$

The hard-decision decoder based on hash and syndrome decoding (HSDec) is a fast and efficient decoder that used the hash techniques, hash table and hash function, to alleviating the temporal complexity of the syndrome decoding algorithm. In HSDec, we have defined a hash function that allowed us to directly reach the error pattern "e" to be used in the decoding phase; therefore, the search time is reduced considerably. For example for the code BCH (63, 39, 9), to look for the vector "*e*" it is enough for us to use the hash function which determines its position in the hash table filled in a structured way, whereas without hashing it will be necessary to do an exhaustive search in a set of dimension $2^{24} = 16{,}777{,}216$.

HSDec works as follows:

```
Inputs:
    - b: the binary word to decode of length n.
    - The parity check matrix H or the generator
      polynomial h(x) of the dual code.
    - TH1: the hash table of 2^(n-k) rows.
    - POW2: the vector of n-k columns contains in each
    cell i the value of the 2^i
Outputs:
    -the corrected word c.
Begin
    Compute the syndrome S of b.
    Position:=hash(S, POW2)
    c :=b⊕TH1[Position]
End
```

The hash function is used to convert the binary vector of length *n*–*k* to decimal, it works as follows:

```
Function hash(S, POW2): integer
Position := Σ_{i=0}^{n-k-1} S[i] * POW2[n - k - 1 - i]
Return Position
End Function
```

$$\text{Position} := \sum_{i=0}^{n-k-1} S[i] * \text{POW2}[n-k-1-i]$$

In this paper, we propose a serial combination between PHR and HSDec; to decode a received sequence, it is initially partially processed with the PHR decoder then with HSDec. We call PHR-HSDec the resulting decoder of this mutual concatenation.

4 Simulation Results of PHR-HSDec

The error-correcting performances will be represented in terms of Bit Error Rate (BER) in each signal-to-noise ratio (SNR $= E_b/N_0$). We note here that without coding/decoding techniques the BER reaches 10^{-5} for a value of SNR equal to 9.6 dB over the Additive White Gaussian Noise (AWGN) channel that can be considered as a binary channel using a Binary Phase Shift Keying (BPSK) modulation. To visualize the efficiency and the rapidity of PHR-HSDec comparing with other competitors, it will be applied for some linear codes.

Figure 1 presents the performances of PHR-HSDec for some BCH codes of length up to 63. For example for the BCH(15, 7, 5) code, the gain of coding is about 2.6 dB, about 4.1 for the BCH (31, 11, 11) code and about 4.6 dB for BCH (63, 45, 7).

In Fig. 2, we make a comparison of the performances of PHR-HSDec and Chase-HSDec decoders for some QR codes of length up to 31. From this figure, we deduce that PHR-HSDec and Chase-HSDec have the same performances for the QR (23,12,7) code; contrariwise for the QR (31, 16, 7) code, the PHR-HSDec guarantees a gain of coding of about 0.4 dB comparing to the competitor Chase-HSDec.

Figure 3 presents a comparison of the performances of PHR-HSDec and Chase-HSDec decoders for BCH (15, 7, 5) and BCH (31, 21, 5). This figure shows that

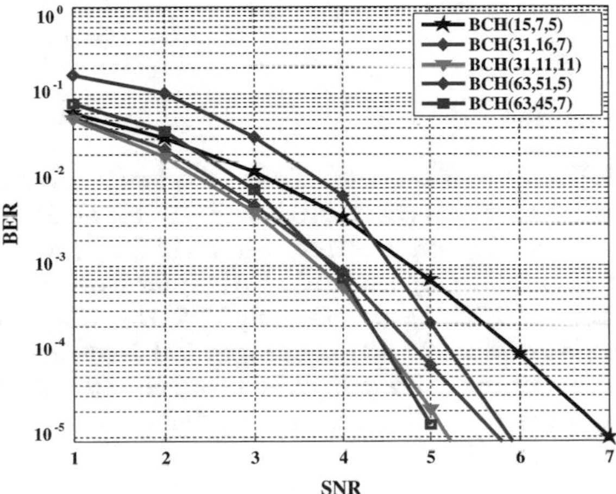

Fig. 1 Performances of PHR-HSDec for some BCH codes of length up to 63

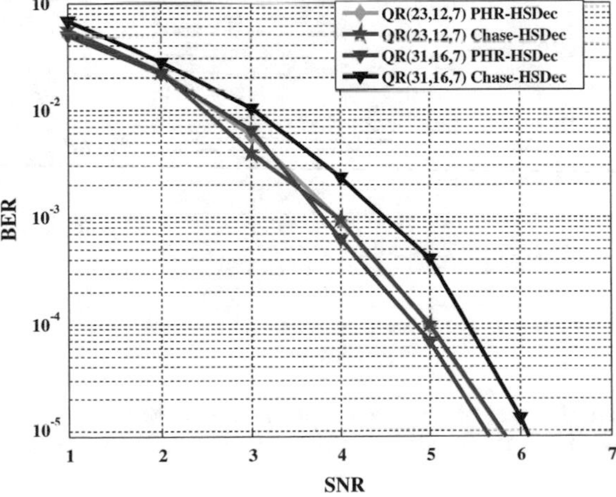

Fig. 2 Comparison of the performances of PHR-HSDec and Chase-HSDec decoders for some QR codes of length up to 31

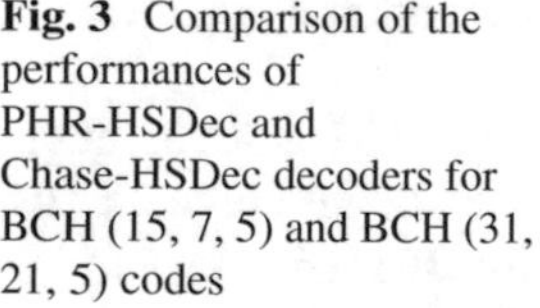

Fig. 3 Comparison of the performances of PHR-HSDec and Chase-HSDec decoders for BCH (15, 7, 5) and BCH (31, 21, 5) codes

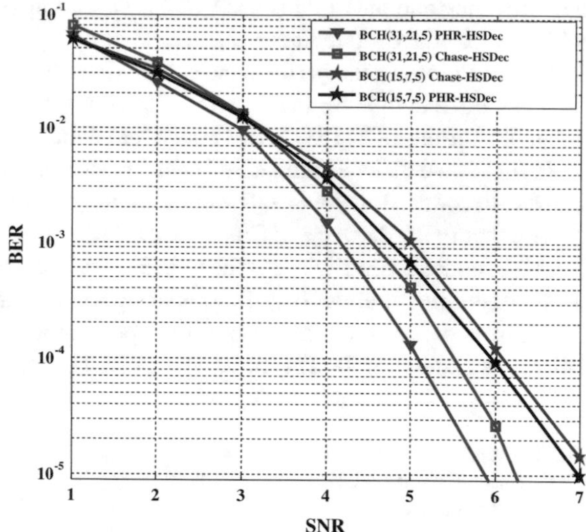

PHR-HSDec passes relatively the concurrent decoder for the BCH (15, 7, 5) code and exceeds it of about 0.4 dB for the BCH (31, 21, 5) code.

Figure 4a represents a comparison of the performances of PHR-HSDec, Chase-HSDec, SD1 [11] and S2W2Dec [18] for BCH (63, 51, 5) code, this figure shows that PHR-HSDec passes relatively S2W2Dec, Chase-HSDec and guarantees a gain of coding of about 0.4 dB comparing to SD1 decoder for this code. Figure 4b represents a comparison of the performances of PHR-HSDec, Chase-HSDec and cGAD [14] for BCH (63, 45, 7) code. From this figure, we deduce that PHR-HSDec exceeds the competitors for this code.

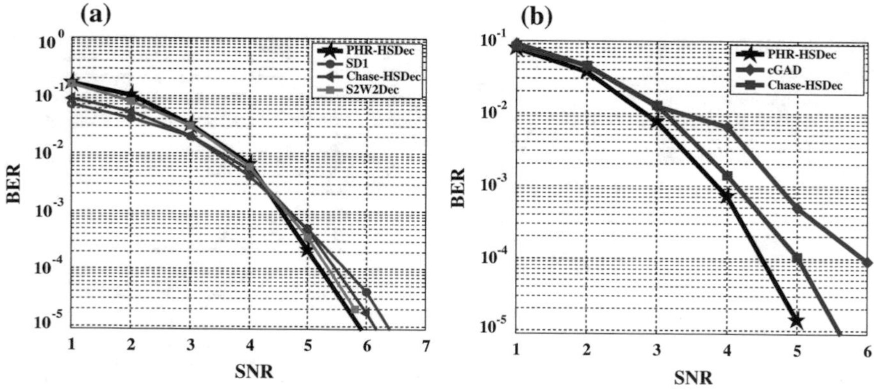

Fig. 4 Comparison of the performances of PHR-HSDec, Chase-HSDec **a** SD1 and S2W2Dec for BCH (63, 51, 5) code; **b** and cGAD for BCH (63, 45, 7) code

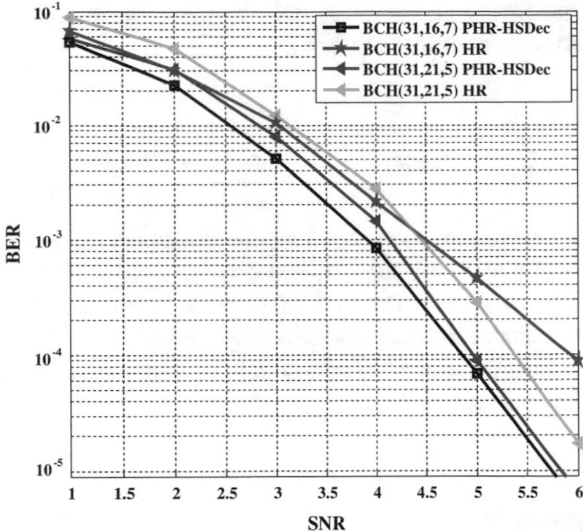

Fig. 5 Comparison of the performances of PHR-HSDec and HR for BCH (31, 16, 7) and BCH (31, 21, 5) codes

To show the efficiency of the proposed combination we present in Fig. 5 a comparison between the proposed decoder PHR-HSDec and the Hartmann and Rudolph (HR) decoder for BCH (31, 16, 7) and BCH (31, 21, 5). From this figure, we notice that the concatenation between HR and HSDec improves the results in a very remarkable way compared to the unique use of HR decoder. Also, we note that the results of PHR-HSDec for the BCH (31, 21, 5) code are obtained with only 40 codewords of the dual code, i.e., just 3.9% of the size of the dual code space, which justifies the decrease of the run time of the proposed concatenation.

To show the rapidity of the proposed PHR-HSDec algorithm, we plot in Fig. 6 the ratio between the execution times of PHR-HSDec and HR algorithms for the BCH (31, 21, 5) code. This figure shows that PHR-HSDec has a very reduced temporal

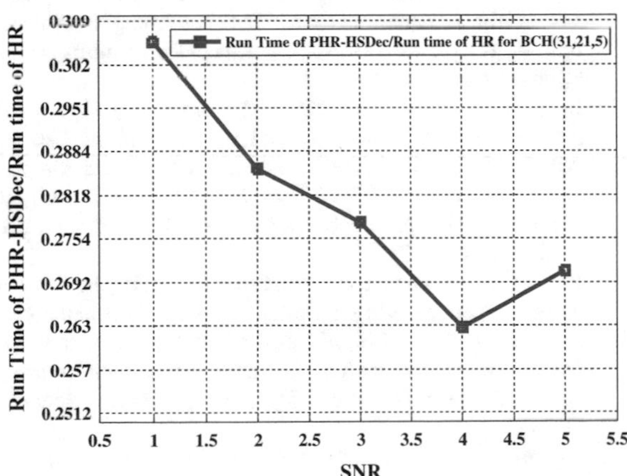

Fig. 6 Representation of the ratio between the run time of PHR-HSDec and HR algorithms for the BCH (31, 21, 5) code

complexity compared to HR, for example, for snr = 5 dB the rate of reduction of run time is more than 70%.

5 Conclusion

In general, iterative decoding consists in decoding a code in successive steps, using several low-cost decoders instead of decoding with a single complex decoder. In this paper, we have presented a fast and efficient decoder that exploits firstly a symbol-by-symbol decoder and secondly another one word-by-word. The quality of the simulation results and the study of the complexity prove the good performance of the concatenation idea.

References

1. Clarck, G.C., Cain, J.B.: Error-Correction Coding for Digital Communication. New York Plenum (1981)
2. Azouaoui, A., Chana, I., Belkasmi, M.: Efficient information set decoding based on genetic algorithms. Int. J. Commun. Netw. Syst. Sci. **5** (2012)
3. Gallager, R.G.: Low-density parity-check codes. IRE Trans. Inf. Theory **8**, 21–28 (1962)
4. Morelos-Zaragoza, R.H.: The Art of Error Correcting Coding, 2nd edn. Wiley (2006)
5. Nouh, S., El Khatabi, A., Belkasmi, M.: Majority voting procedure allowing soft decision decoding of linear block codes on binary channels. Int. J. Commun. Netw. Syst. Sci. **5** (2012)
6. Berlekamp, E.R.: Algebraic Coding Theory, rev. edn. Aegean Park Press (1984)
7. Massey, J.L.: Shift-register synthesis and BCH decoding. IEEE 1969 Trans Information Theory IT-15 **1**, 122–127 (1969)
8. Chen, Y.H., Truong, T.K., Chang, Y., Lee, C.D., Chen, S.H.: Algebraic decoding of quadratic residue codes using Berlekamp-Massey algorithm. J. Information Sci. Eng. **23**, 127–145 (2007)
9. Chen, Y., Huang, C., Chang, J.: Decoding of binary quadratic residue codes with hash table. IET Common. **10**, 122–130 (2016)
10. El Kasmi Alaoui, M.S., Nouh, S., Marzak, A.: Two new fast and efficient hard decision decoders based on hash techniques for real time communication systems. In: Lecture Notes in Real-Time Intelligent Systems, RTIS 2017. Advances in Intelligent Systems and Computing, vol. 756. Springer, Cham (2019)
11. Jung, B., Kim, T., Lee, H.: Low-Complexity Non-Iterative Soft-Decision BCH Decoder Architecture for WBAN Applications. J. Semicond. Technol. Sci. **16** (2016)
12. Askali, M., Nouh, S., Belkasmi, M.: A Soft decision version of the Permutation decoding algorithm. In: NTCCCS 12 Workshop. Oujda, Morocco (2012)
13. Shim, Y.G.: Forward error correction codes in communication channels. Int. J. Control Autom. **10**, 131–144 (2017)
14. Berkani, A., Azouaoui, A., Belkasmi, M., Aylaj, B.: Improved decoding of linear block codes using compact genetic algorithms with larger tournament size. Int. J. Comput. Sci. (14) (2017)
15. El Kasmi Alaoui, M.S., Nouh, S., Marzak, A.: A low complexity soft decision decoder for linear block codes. In: The 1st International Conference On Intelligent Computing in Data Sciences(ICDS). University My Ismail, Superior School of Technology, Meknes Morocco (2017)
16. Fossorier, M.P.C., Lin., S.: Soft decision decoding of linear block codes based on ordered statistics. IEEE Trans. Inf. Theory **184** 1379–1396 (1995)

17. Hartmann, C.R., Rudolph, L.D.: An optimal symbol by symbol decoding rule for linear codes. IEEE Trans. Inf. Theory **22**, 514–517 (1976)
18. Nouh, S., Aylaj, B.: Efficient serial concatenation of symbol by symbol and word by word decoders. Int. J. Innovative Comput. Inf. Control **14** (2018)

SWIMAC: An Ultra-Low Power Consumption MAC Protocol for IR-UWB-Based WSN

Anouar Darif

Abstract Wireless sensor network (WSN) has gained popularity in recent times in residential, commercial and industrial application. In this type of network, the energy efficiency is the key design challenge. The introduction of the IR-UWB technology in the field of WSN was promising for researchers especially for its low power consumption feature. For this reason, MAC layer protocols for IR-UWB-based WSN must be energy efficient to exploit the main features of IR-UWB technology implemented in the physical layer. In this paper, we present and show the good impact in terms of energy consumption for an energy-efficient MAC protocol in WSN based on IR-UWB. This MAC protocol takes advantage of these two key properties by using synchronous periodic beacon transmissions from each network node and its duty-cycling mode. We developed our own class SwiMacLayer under MiXiM platform on OMNet++ to test and evaluate the performance of SWIMAC protocol compared to ALOHA.

Keywords WSN · IR-UWB · SWIMAC · ALOHA · Power consumption · MiXiM · OMNet++

1 Introduction

WSN is one of the most interesting networking technologies since its ability to use no infrastructure communications, it has been used for many applications, including military sensing, data broadcasting [1], environmental monitoring [2], Intelligent Vehicular Systems [3], multimedia [4], patient monitoring [5], agriculture [6], and industrial automation [7]. This kind of network has not yet achieved widespread deployments, though it has been proven able to meet the requirements of many classes of applications. Mobile wireless sensor nodes have some limitations as lower computing capabilities, smaller memory devices, small bandwidth and very lower battery autonomy; these constraints represent the main challenges in the development or deployment of any solution using WSN. Energy consumption is a very important

A. Darif (✉)
LIMATI, University of Sultan Moulay Slimane Faculte Polydisciplinaire, Benimellal, Morocco

V. Bhateja et al. (eds.), *Embedded Systems and Artificial Intelligence*,
Advances in Intelligent Systems and Computing 1076,
https://doi.org/10.1007/978-981-15-0947-6_13

design consideration in WSN, new wireless technologies emerge in the recent few years, providing large opportunities in terms of low power consumption, high and low rate and are promising for environment monitoring applications. IR-UWB technology is one of these new technologies; it is a promising solution for WSN due to its various advantages such as its robustness to severe multipath fading even in indoor environments, its potential to provide accurate localization [8], its low cost and complexity, and low energy consumption [9]. It is necessary to find a very adapt MAC layer protocol to this Technology for keeping his advantages.

The present paper is organized as follows. In Sect. 2, we introduced WSN. In Sect. 3, we presented the IR-UWB technology. Section 4 presents the Synchronized WideMac (SWIMAC) protocol. The simulation and its results are presented in Sect. 5; finally, Sect. 6 concludes the paper.

2 Wireless Sensor Network Overview

A wireless sensor network (WSN) in its simplest form can be defined as a network of (possibly low-size and low complex) devices denoted as nodes that can sense the environment and communicate the information gathered from the monitored field through wireless links; the data is forwarded, possibly via multiple hops relaying, to a sink (Base Station) that can use it locally, or is connected to other networks (e.g., the Internet) through a gateway (see Fig. 1).

The nodes can be stationary or moving.
They can be aware of their location or not.
They can be homogeneous or not.

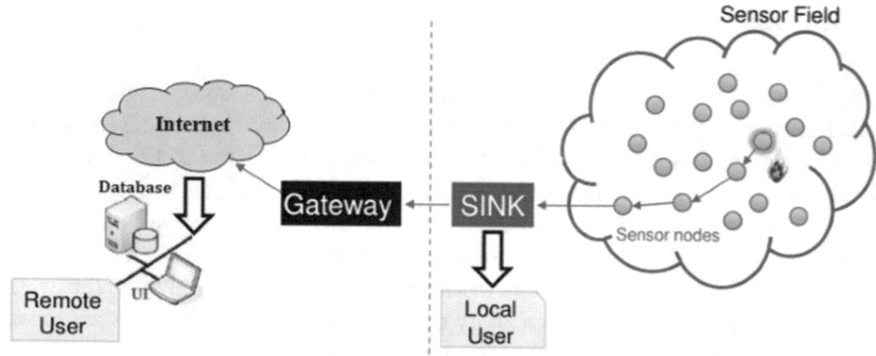

Fig. 1 Sensor network architecture

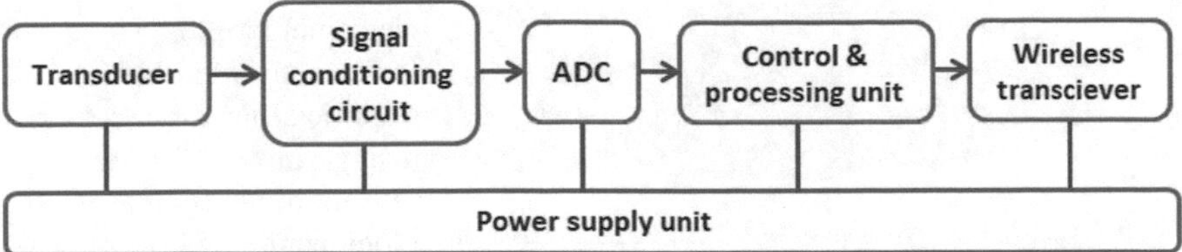

Fig. 2 Architecture of a sensor node

2.1 Sensor Node Architecture

A typical wireless sensor node is a compact-sized hardware unit that acquires desired data from its environment and communicates wirelessly to other nodes in a network so as to relay that data or extracted information to a central station. The architecture and data flow of a typical wireless sensor node [10] are shown in Fig. 2. Depending upon what and how many parameters are to be monitored in an application, it may consist of one or more transducers that measure physical phenomena and produce equivalent electrical outputs, that is, in the form of electrical current or voltage. The output signal from a transducer, which is typically low in amplitude, is amplified with the assistance of signal conditioning circuits so as to ensure that it matches the requirements of the digitization circuits. Following on from the amplification process, a signal is then digitized using an analog-to-digital converter (ADC). In the case of a transducer that produces digital output, along with basic sensing system, both the amplification and the digitization circuits are embedded into the transducer. The digitized signal is then fed to a control and processing unit, typically a micro-controller, which with little or no processing at all, relays the acquired data to its destination using a wireless transceiver. With regard to modules integrated in a wireless sensor node, the control and processing unit plays an important role, as it not only coordinates all the activities within a sensor node but also manages its association with other nodes in a WSN. In addition to these modules, which enable sensing, processing/control and wireless communication capabilities, a sensor node incorporates an energy source such as a battery or some mechanism for energy scavenging [11], through which all of the integrated modules are powered.

2.2 Types of Energy Waste in WSN

All protocols must deal with five types of energy:

- Collisions,
- Overhearing,
- Idle listening,
- Signaling overhead,
- Over-emitting.

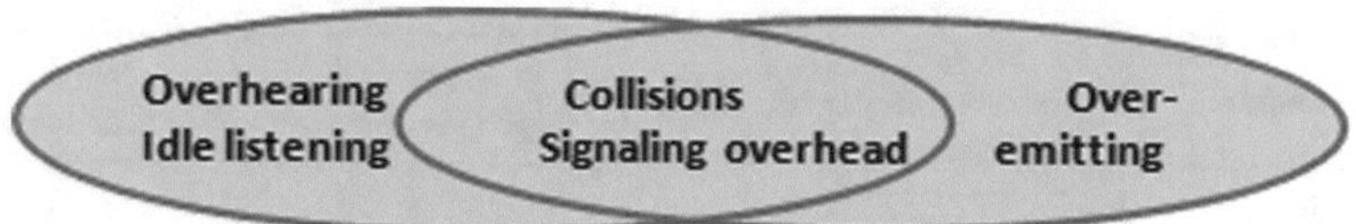

Fig. 3 Types of energy waste in MAC protocols and associated radio states

Collisions happen when transmissions occur simultaneously in such a manner that message reception fails for at least one of the intended recipients. ***Overhearing*** occurs when a transceiver uselessly listens to a message. ***Idle listening*** happens when the radio is in reception mode while no transmission takes place. ***Signaling overhead*** is the energy cost incurred by signaling data such as acknowledgments and synchronization packets. Lastly, ***over-emitting*** is the energy cost associated to transmitting packets when the destination node is not ready to receive them. Figure 3 represents these five cases and the associated radio states [12].

3 IR-UWB

IR-UWB is a promising technology to address MWSN constraints. However, existing network simulation tools do not provide a complete MWSN simulation architecture, with the IR-UWB specificities at the Physical (PHY) and the medium access control (MAC) layers. The IR-UWB signal uses pulses baseband a very short period of time of the order of a few hundred picoseconds. These signals have a frequency response of nearly zero hertz to several GHz. According to [13], there is no standardization; the waveform is not limited, but its features are limited by the FCC mask. There are different modulation schemes baseband for IR-UWB [14]. This paper uses the PPM technique for IR-UWB receiver.

3.1 IR-UWB Pulse Position Modulation

Pulse position modulation can be represented as follows at the transmitter:

$$S(t) = \sqrt{E}\sum_{j} P_0\big(t - jT_{\text{sym}} - \theta_j - b_j T_{\text{sym}}/2\big) \qquad (1)$$

where: E is the pulse energy, $P_0(t)$ is the normalized pulse waveform, T_{sym} is the symbol duration, θ_j is the time-hopping shift for the considered symbol j, b_j is the jth bit value and $T_{\text{sym}}/2$ is the time shift for the modulation.

We are considering an AWGN channel of impulse response:

$$h(t) = \lambda\delta(t - \tau) \tag{2}$$

where λ is the attenuation and τ is the delay, the energy at the receiver is

$$E_u = \lambda^2 E \tag{3}$$

The signal after propagation becomes:

$$r_u(t) = \sqrt{E_u}\sum_j P_0\big(t - jT_{\text{sym}} - \theta_j - b_j T_{\text{sym}}/2 - \tau\big) \tag{4}$$

The received signal can be separated into three components: $r_u(t)$, $r_{\text{mai}}(t)$ and $n(t)$.

Where $r_u(t)$ is the transmitted signal from the source transformed by the channel, $r_{\text{mai}}(t)$ is the multiple access interference caused by simultaneous transmissions and $n(t)$ is the thermal noise.

The thermal noise is a zero-mean Gaussian random variable of standard deviation $N_0/2$ (where N_0 is the thermal noise given by $N_0 = K_B T$, K_B being the Boltzmann constant and T the absolute temperature).

The multiple access interference can be expressed as follows:

$$r_{\text{mai}}(t) = \sum_{n=1}^{N_i} \sqrt{E^{(n)}} \times \sum_j P_0\left(t - jT_{\text{sym}} - \theta_j^{(n)} - \frac{b_j^{(n)} T_{\text{sym}}}{2} - \tau^{(n)}\right) \tag{5}$$

where N_i is the number of interfering signals, $E^{(n)}$ is the received energy, $\tau^{(n)}$ is the channel delay for the considered signal, $\theta_j^{(n)}$ is the time-hopping shift and $b_j^{(n)}$ is the bit value for the jth symbol of the considered interfering signal.

The correlating received signal $S_{Rx}(t)$ with a correlation mask $m(t)$ effect can be expressed as:

$$Z(x) = \int_{\tau}^{\tau+T_s} S_{Rx}(t) m_x(t - \tau)\mathrm{d}t = Z_u + Z_{\text{mai}} + Z_n \tag{6}$$

With:

$$m_x(t) = P_0\big(t - xT_s - \theta_j\big) - P_0\big(t - xT_s - \theta_j - T_s/2\big) \tag{7}$$

The signal contribution (Z_u), the thermal noise contribution (Z_n) and multiple access interference (Z_{mai}) are the decision variable $Z(x)$ component.

With Z_n is Gaussian distributed with zero mean and variance:

$$\sigma_n^2 = N_0(1 - R_0(\varepsilon)) \tag{8}$$

With:

$$R_0(t) = P_0(\varepsilon)P_0(\varepsilon - t)\mathrm{d}\varepsilon \tag{9}$$

Considering Z_{mai} is Gaussian distributed with zero mean and variance:

$$\sigma^2_{\mathrm{mai}} = \frac{1}{T_s}\sigma^2_M \sum_{n=1}^{N_i} E^{(n)} \tag{10}$$

With:

$$\sigma^2_M = \int_{-\infty}^{+\infty}\left(\int_{-\infty}^{+\infty} P_0(t-\tau)(P_0(t) - P_0(t-\varepsilon))\mathrm{d}t\right)^2 \mathrm{d}\tau \tag{11}$$

4 Swimac

4.1 Presentation

SWIMAC is a novel MAC protocol designed for IR-UWB-based WSN. It makes all nodes periodically (period T_W) and synchronously wake up [15], transmit a beacon message announcing their availability and listen for transmission attempts during a brief time T_{Listen}.

Figure 4 illustrates a single-period structure. It starts with a known and detectable synchronization preamble and is followed by a data sequence which announces the node address and potentially other information, such as a neighbor list or routing table information (for instance, cost of its known path to the sink). A small listening time follows T_{Listen}, during which the node stays in reception mode and that allows it to receive a message. ΔT is the time added to T_{Sleep} to make the synchronization between the nodes.

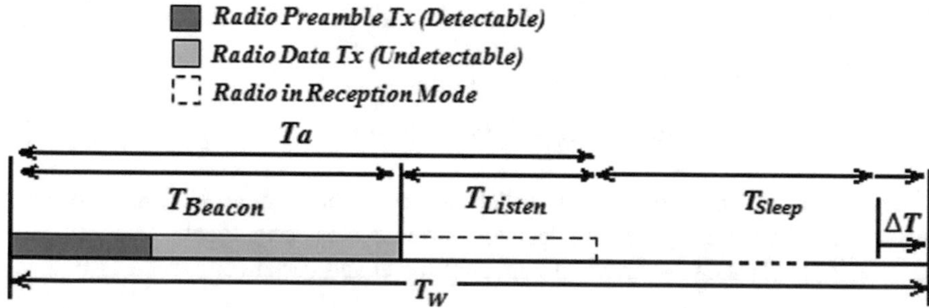

Fig. 4 Detailed view of a SWIMAC period

The whole period composed of T_{beacon} and T_{Listen} is called T_a (time of activity); and its very small compared to the time window T_W. This period is followed by a long sleeping period T_{Sleep} during which nodes save energy by keeping the radio in its sleeping mode.

When a node has a message to transmit, it first listens to the channel until it receives the beacon message of the destination node. This beacon message contains a backoff exponent value that must be used by all nodes when trying to access this destination. If this value is equal to zero, the source node can transmit immediately. Otherwise, it waits a random backoff time, waits for the destination beacon, and transmits its data packet. Because of the unreliability of the wireless channel, packets are acknowledged. If a packet is not acknowledged, or if the destination beacon was not received a retransmission procedure using the backoff algorithm is initiated, until the maximum number of retransmissions maxTx Attempts is reached.

4.2 Power Consumption Models

Each normal T_W interval starts with a beacon frame transmission followed by a packet or a beacon reception attempt, during this start a node must enter transmission mode $\left(E_{\text{Setup}\,Tx}\right)$,transmit its beacon $(T_{\text{Beacon}} P_{Tx})$, switch to reception mode (E_{SwRxTx}) and attempt a packet reception $(T_{\text{Listen}} P_{Rx})$. These costs are regrouped in the beacon energy E_{Beacon}.

$$E_{\text{Beacon}} = E_{\text{Setup}\,Tx} + T_{\text{Beacon}} P_{Tx} + E_{SwTxRx} + T_{\text{Listen}} P_{Rx} \tag{12}$$

In addition, during a time L, a node must sometimes transmit a packet E_{Tx} or receive one E_{Rx}, and sleep the rest of the time E_{Sleep}, resulting to the following average power consumption:

$$\text{P}_{\text{SWIMAC}} = \frac{1}{T_W}\left(E_{\text{Beacon}} + E_{Tx} + E_{Rx} + E_{\text{Sleep}}\right) \tag{13}$$

where

$$E_{Tx} = K \cdot C_{Tx}(P_{\text{out}}) \cdot V_B \cdot T_{Tx} \tag{14}$$

$$E_{Rx} = K \cdot C_{Rx} \cdot V_B \cdot T_{Rx} \tag{15}$$

$$E_{\text{Sleep}} = C_{\text{Sleep}} \cdot V_B \cdot T_{\text{Sleep}} \tag{16}$$

K represents the message length in bytes, P_{out} is the transmission power, C_{Tx}, C_{Rx} and C_{Sleep} represent the current intensities for the three modes, T_{Tx} and T_{Rx} are the time of transmission and reception.

5 Simulations and Results

5.1 OMNet++ and MiXiM Simulation Platform

OMNeT++ is an extensible, modular, component-based C++ simulation library and framework which also includes an integrated development and a graphical runtime environment; it is a discreet events-based simulator and it provides a powerful and clear simulation framework.

MiXiM joins and extends several existing simulation frameworks developed for wireless and mobile simulations in OMNeT++. It provides detailed models of the wireless channel, wireless connectivity, mobility models, models for obstacles and many communication protocols especially at the medium access control (MAC) level. Moreover, it provides a user-friendly graphical representation of wireless and mobile networks in OMNeT++, supporting debugging and defining even complex wireless scenarios [16].

5.2 Simulation Parameters

We performed the simulations in the MiXiM framework with the OMNeT++ network simulator.

To test and evaluate the performance of SWIMAC protocol, we used PhyLayerUWBIR class developed under MiXiM platform on OMNet++ as a physical layer. For the MAC layer, we developed our own class SwiMacLayer.

We used a grid network, where nodes transmit packets to a sink node; also we ran several simulations with different nodes numbers and parameters values to evaluate our new protocol. For the energy consumption, we used the following radio power consumption parameters shown in Table 1. For the radio timing, we used the parameters shown below in Table 2.

Table 1 Energy parameters

Parameter	Value (mW)
P_{Rx}	36.400
P_{Tx}	1.212
P_{Sleep}	0.120
$P_{SetupRx}$	36.400
$P_{steupTx}$	1.212
P_{swTxRx}	36.400
P_{swRxTx}	36.400

Table 2 Timing parameters

Parameter	Value
$T_{SetupRx}$	0.000103 s
$T_{SetupTx}$	0.000203 s
T_{SwTxRx}	0.000120 s
T_{SwRxTx}	0.000210 s
$T_{RxToSleep}$	0.000031 s
$T_{TxToSleep}$	0.000032 s
Bit rate	0.850000 Mbps

5.3 Results

Energy was and is an interesting issue that is still a factor in the development of WSN protocols especially in the MAC and physical layers. This factor affects directly the lifetime of the network. In this section, we present the results obtained using the timing and energy parameters cited in Sect. 5.2.

The energy-efficient of SWIMAC was concretized by the results shown in Figs. 5 and 6. They show that the power consumption of SWIMAC protocol is remarkably less than the ALOHA and Slotted ALOHA MAC protocols. The good result obtained in the case of SWIMAC due to the duty-cycling mode and synchronized method uses by this protocol. It means that, the main advantage to use the duty-cycling mode is to keeping the radio of wireless communication systems in sleep period as much as possible.

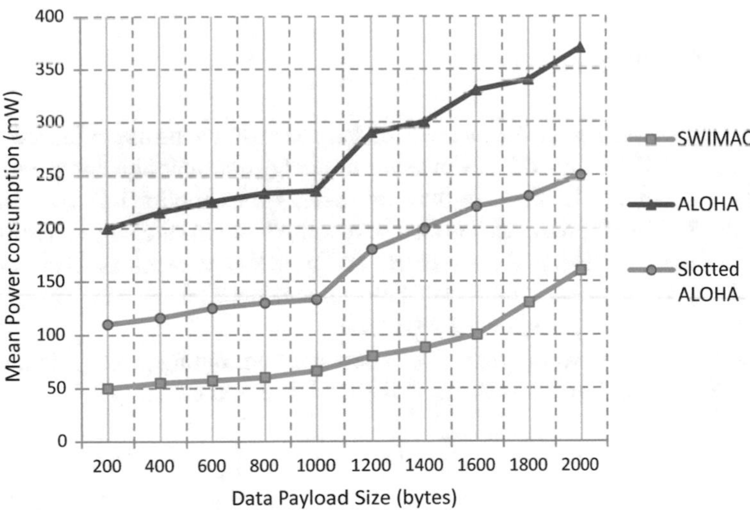

Fig. 5 Nodes power consumption versus data payload size

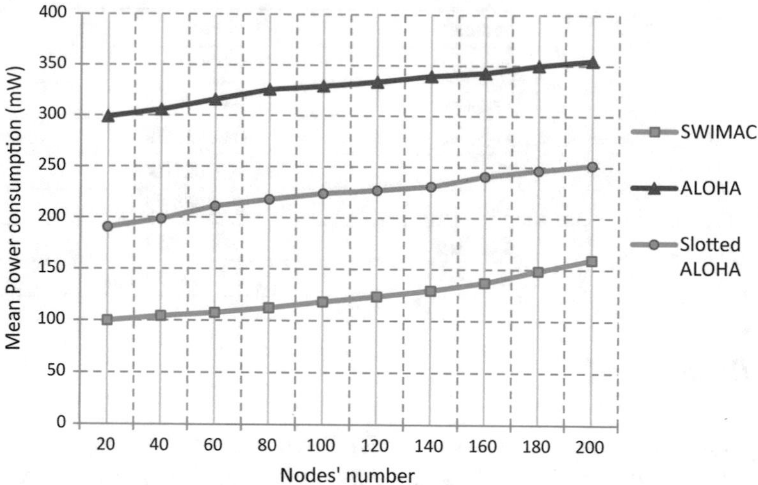

Fig. 6 Nodes power consumption versus nodes' number

Figure 5 shows the result obtained by a nodes' number fixed at 160 nodes and varying the data payload size. It shows clearly that the value of power consumption increase with increasing the data payload size due to the required power for sending all data packet. The result shown in Fig. 6 is obtained by a data payload fixed at 1800 bytes and varying the nodes' number. It shows also the linear dependence between the power consumption and the nodes' number.

6 Conclusion

In this paper, we showed the impact and the gain brought by the use of the SWIMAC protocol for IR-UWB-based WSN in terms of energy consumption compared to the ALOHA and Slotted ALOHA. The ultra-low energy consumption is the main advantage of the SWIMAC protocol; it is also very close to an ideal energy consumption model for the IR-UWB-based transceivers and gave a good result at this level. This result was achieved thanks to the fact that the network nodes sleep in the T_{sleep} periods which occupy a wide range in the T_W periods.

We aim, as a future work, to develop a new adapted routing protocol that will be paired with SWIMAC to largely exploit the IR-UWB features in WSN.

References

1. Sung, W., Wu, T.T., Yang, C.S., Huang, Y.M.: Reliable data broadcast for Zigbee wireless sensor networks. Int. J. Smart Sens. Intell. Syst. **3**(3) (2010, September)
2. Posnicek, T., Kellner, K., Brandl, M.: Wireless sensor network for environmental monitoring with 3G connectivity. **87**, 524–527 (2014)
3. Mouftah, H.T., Khanafer, M., Guennoun, M.: Wireless sensor network architectures for intelligent vehicular systems. In: Symposium International for Telecommunication Techniques (2010)
4. Cevik, T., Gunagwera, A., Cevik, N.: A survey of multimedia streaming in wireless sensor networks: progress, issues and design challenges. Int. J. Comput. Netw. Commun. (IJCNC) **7**(5), 95–114 (2015)
5. Ramesh, T.K., Giriraja, C.V.: Wireless sensor network protocol for patient monitoring system. In: International Conference on Computer Communication and Informatics (ICCCI), pp. 1–4 (2017, January)
6. Kassim, M.R.M., Harun, A.N.: Applications of WSN in agricultural environment monitoring systems. International Conference on Information and Communication Technology Convergence (ICTC), pp. 344–349, (2016, October)
7. Arvind, R.V., Raj, R.R., Raj, R.R., Prakash, N.K.: Industrial automation using wireless sensor networks. Indian J. Sci. Technol. **9**(8), 1–8 (2016, February)
8. Ling, R.W.C., Gupta, A., Vashistha, A., Sharma, M., Law, C.L.: High precision UWB-IR indoor positioning system for IoT applications. In: IEEE 4th World Forum on Internet of Things (WF-IoT), pp. 135–139, (2018, February)
9. Vauche, R., Bourdel, S., Dehaese, N., Gaubert, J., Sparrow, O. R., Muhr, E., Barthelemy, H.: High efficiency UWB pulse generator for ultra-low-power applications. Int. J. Microwave Wireless Technol. **8**(3), 495–503 (2016, May)
10. Akyildiz, I.F., Kasimoglu, I.H.: Wireless sensor and actor networks: research challenges. Ad Hoc Netw. **2**(4), 351–367 (2004)
11. Paradiso, J.A., Starner, T.: Energy scavenging for mobile and wireless electronics. Pervasive Comput. IEEE, **4**(1), 18, 27 (2015 January–March)
12. Rezaei, Z., Mobininejad, S.: Energy saving in wireless sensor networks. Int. J. Comput. Sci. Eng. Survey (IJCSES) **3**(1) (2012,February)
13. Lazaro, A., Girbau, D., Villarino, R.: Analysis of vital signs monitoring using an IR-UWB Radar. In: Progress In Electromagnetics Research, pp. 265–284 (2010, January)
14. Adsul, P.A., Bodhe, K.S.: Performance comparison of BPSK, PPM and PPV modulation based IR-UWB receiver using wide band LNA. Int. J. Comput. Technol. Appl. **3**(4), 1532–1537 (2012, July–August)
15. Darif, A., Saadane, R., Aboutajdine, D.: No communication nodes synchronization for a low power consumption MAC Protocol in WSN based on IR-UWB. J. Commun. Softw. Syst. **10**(2), 90–98 (2014)
16. Köpke, A., Swigulski, M., Wessel, K., Willkomm, D., Haneveld, P.T., Parker, T.E., Visser, O.W., Lichte, H.S., Valentin, S.: Simulating wireless and mobile networks in OMNeT++ The MiXiM Vision. In: Proceedings of International Workshop on OMNeT++ (co-located with SIMUTools '08) (2008, March)

An Optimal Design of a Short-Channel RF Low Noise Amplifier Using a Swarm Intelligence Technique

Soufiane Abi, Hamid Bouyghf, Bachir Benhala and Abdelhadi Raihani

Abstract In this paper, an optimal sizing of a cascode short-channel CMOS low noise amplifier (LNA) with inductive source degeneration by using the ant colony optimization (ACO) and the artificial bee colony (ABC) techniques is presented. The reason for employing this proposed topology is to provide a high conversion gain and improved noise figure (NF) at the operating frequency of 2.3 GHz. In order to validate the obtained results by ACO and ABC techniques, the advanced design system (ADS) simulator is used for this purpose using 90 nm CMOS technology.

Keywords Meta-heuristics · Ant colony optimization · Artificial bee colony · Low noise amplifier · Noise figure · Gain

1 Introduction

Low noise amplifier (LNA) is one among the important components in the analog front end of a radio frequency (RF) receiver. It is usually the first component in a radio receiver used to amplify the weak signals captured by the antenna by adding as little noise and distortion as possible [1]. The most requirements on LNA are low noise and sufficiently high gain. In order to satisfy these requirements, many researchers have tried to optimize parameters of LNA using different optimization techniques.

In this context, meta-heuristic-based techniques were appeared and used [2]. They provide approximate solutions for the optimization problems with a reasonable time contrariwise classical methods. In the literature, some of these meta-heuristics are proposed, such as Tabu search [3], genetic algorithms (GA) [4] and local search (LS)

S. Abi (✉) · B. Benhala
BABA Team, EAB Laboratory, Faculty of Sciences, University of Moulay Ismail, BP 11201 Zitoune, Meknes, Morocco
e-mail: soufianeabi@gmail.com

H. Bouyghf · A. Raihani
SSDIA Laboratory, ENSET Mohammedia, University of Hassan II Casablanca, BP 159 Bd Hassan II, Mohammedia, Morocco

V. Bhateja et al. (eds.), *Embedded Systems and Artificial Intelligence*,
Advances in Intelligent Systems and Computing 1076,
https://doi.org/10.1007/978-981-15-0947-6_14

[5]. However, the meta-heuristics that provide the better results are those of nature inspired; they are inventive, resourceful, efficient, easy to use and are known as swarm intelligence (SI) techniques [6, 7]. The SI techniques concentrate on animal behavior in order to develop some meta-heuristics which can imitate their problem resolution abilities, such as particle swarm optimization (PSO) [8], ant colony optimization (ACO) [9, 10] and artificial bee colony (ABC) [11, 12].

The aim of this work is to employ the ACO and ABC techniques for an optimal sizing of RF CMOS short-channel low noise amplifier (LNA). The remainder of this article is organized as follows: An overview of the ACO and ABC techniques is presented in Sect. 2. Section 3 discusses in detail the application example which is the RF CMOS short-channel LNA. Section 4 presents the results and discussions. Finally, the conclusion is given in Sect. 5.

2 An Overview of the Used Meta-Heuristic Techniques

2.1 Ant Colony Optimization Technique

The ant colony optimization (ACO) is a meta-heuristic-based optimization technique. It has been inspired by the foraging behavior of real ant colonies, and it is based on the indirect communication between the ants by the means of a chemical substance called pheromone. Dorigo et al. [13] introduced this algorithm, and it was successfully applied to solve several problems; see for instance [14, 15] and the traveling salesman problem (TSP) [16]. For solving such problems, ants select the vertex to be visited based on the random probability rule, which is given by (1):

$$p_{ij}^{k} = \begin{cases} \frac{\tau_{ij}^{\alpha} * \eta_{ij}^{\beta}}{\sum_{1 \in N_i^k} \tau_{il}^{\alpha} * \eta_{il}^{\beta}} & j \in N_i^k \\ 0 & \text{otherwise} \end{cases} \tag{1}$$

where N_i^k is the neighbors of vertex i of the kth ant, τ_{ij} is the amount of pheromone trail on edge(i, j), α and β are the weights that control the quantity of pheromone trail, and the visibility value η_{ij} has the following expression:

$$\eta_{ij} = \frac{1}{d_{ij}} \tag{2}$$

d_{ij} is the distance between vertices i and j. The pheromone τ_{ij} on the edge(i, j) is updated as follows:

$$\tau_{ij} = (1 - \rho) * \tau_{ij} + \sum_{k=1}^{m} \Delta\tau_{ij}^{k} \tag{3}$$

where ρ is the pheromone evaporation rate, m is the number of ants, and $\Delta\tau_{ij}^{k}$ is the quantity of pheromone deposited on edge (i, j) by ant k:

$$\Delta\tau_{ij}^{k} = \begin{cases} \frac{Q}{L^k}, & \text{if ant } k \text{ used } (i,\ j) \text{ in its tour} \\ 0, & \text{Otherwise} \end{cases} \tag{4}$$

Q is a constant, and L^k is the length of the tour built by ant k. The pseudo-code of the ACO technique is given as follows:

```
Initialization of the pheromone value randomly
 Do
   For each iteration
     For each ant
       For each variable
         Compute of the probability P according to (1)
         Determine the Pmax
         Deduce the value of Vi
       End
      Compute objective function
     End
     Deduce the best objective function
     Update pheromone values using equations (3) and (4)
   End
    Report the best solution
End
```

Algorithm 1: Pseudo-code of the ACO technique

2.2 *Artificial Bee Colony Technique*

The artificial bee colony (ABC) is a swarm intelligence-based technique proposed by Karaboga [12]. It mimics the intelligent cooperative foraging behavior of honey bees. This ABC technique contains three groups of bees, namely employed, onlooker and scout bees. Employed bees go to find the food sources and get back to the hive and exchange the information with onlooker bees about food sources by waggle dance. Onlooker bees select the food sources according to the dance moves. The food sources that have been abandoned by employed bee become a scout and start looking for new food sources.

The position of the food source corresponds to a feasible solution, and the quantity of nectar of a food source is the fitness of the associated solution. Moreover, the number of onlooker or employed bees is the number of solutions. Initially, the ABC algorithm produces randomly an initial population of *SN* solutions. Each solution x_i (1, 2, …, *SN*) is a vector with *D* elements, where *SN* is the size of employed or onlooker bees and *D* is the number of design variables. The positions of the population are repeated until the requirements are satisfied.

Each employed bee x_i produces a new food source V_i by using (5):

$$v_i^j = x_i^j + \phi_i^j * \left(x_i^j - x_k^j\right) \tag{5}$$

where $k \in \{1, 2, \ldots, SN\}$, $j \in \{1, 2, \ldots, D\}$ are random indexes with $i \neq k$, and ϕ_i^j is a random number chosen between $[-1, 1]$. After the employed bees complete the search mechanism, each onlooker bee selects a food source according to the fitness value given from the employed bees with a probability using (6):

$$P_i = \frac{\text{fit}_i}{\sum_{n=1}^{SN} \text{fit}_n} \tag{6}$$

where fit_i is the fitness value of the solution i. The employed bee changes the position and checks the amount of nectar of the candidate source. The onlooker bee saves the new position and eliminates the old one if the nectar amount is higher to the previous one. The scout's bee produces a new food source using Eq. (7):

$$x_i^j = x_{\min}^j + \text{rand}(0, 1) * \left(x_{\max}^j - x_{\min}^j\right) \tag{7}$$

where $x_{\min}^j$ and $x_{\max}^j$ are the lower bound and the upper bound of the dimension j. The ABC algorithm pseudo-code is given as follows:

```
Produce the initial population xi (i=1, 2... SN), and evaluate the fitness value of the population
   Cycle =1
   Repeat
     For each employed bee
        Produce new solution vi using (5)
        Compute the fitness value fit
     End For
       Compute the probability values Pi by using (6)
     For each onlooker bee
    Select a solution xi according to Pi
    Generate new solution Vi, and Compute the fitness value
     End For
     If an abandoned solution exists for the scout,
    Then generate new solution by using (7)
   Save the best solution
   Cycle = cycle +1
Until the criteria are met
```

Algorithm 2: The ABC algorithm pseudo-code

3 Application Example: Low Noise Amplifier (LNA)

The LNA is an essential component exists at the first stage of a radio receiver circuit, and it is used to amplify very weak signals received by an antenna while adding a small noise as possible [1]. Figure 1 presents a schematic circuit of LNA, which contains a cascoded stage (M1, M2) with inductive source degeneration.

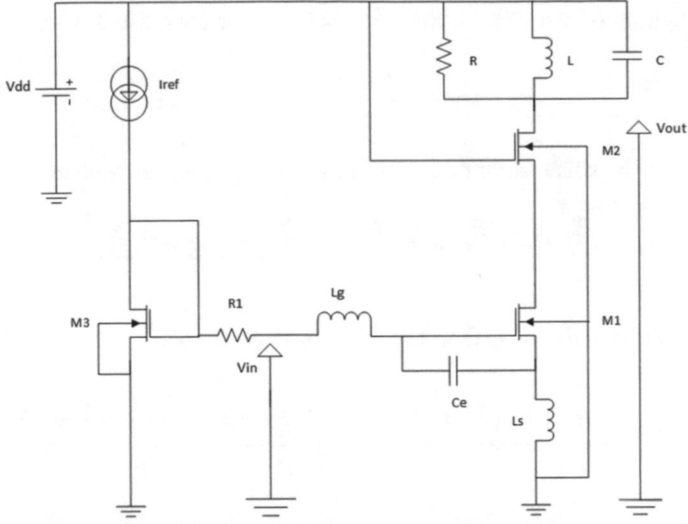

Fig. 1 Schematic of RF CMOS cascode LNA with inductive source degeneration

For the determination of the noise factor expression, the small signal noise analysis is used. However, $\overline{\iota^2}_{n,R_s}$ is the source resistance thermal noise, $\overline{\iota^2}_{n,d}$ is the channel thermal noise, $\overline{\iota^2}_{n,g}$ is the induced gate noise, and $\overline{\iota^2}_{n,\mathrm{R_{out}}}$ is the output resistance thermal noise. These noises can be described as follows [17–19]:

$$\begin{aligned}\overline{\iota^2_{n,R_s}} &= 4\mathrm{KT}\frac{1}{R_s}\Delta f\\ \overline{\iota^2_{n,R_{out}}} &= 4\mathrm{KT}\frac{1}{R_{out}}\Delta f\\ \overline{\iota^2_{n,d}} &= 4\mathrm{KT}\gamma_{\mathrm{short}}g_{d0}\Delta f\\ \overline{\iota^2_{n,g}} &= 4\mathrm{KT}\beta_{\mathrm{short}}g_g\Delta f\end{aligned} \tag{8}$$

The contributions of these four noise sources to the output noise are denoted by $\overline{\iota^2}_{n,o,R_s}$, $\overline{\iota^2}_{n,o,d}$, $\overline{\iota^2}_{n,o,g}$ and $\overline{\iota^2}_{n,o,\mathrm{R_{out}}}$. In addition, they can be expressed as follows:

$$\begin{aligned}\overline{\iota^2}_{n,o,\mathrm{R_s}} &= \frac{g_m}{j2\omega_0 C_{\mathrm{tot}}}\iota^2_{n,R_s}\\ \overline{\iota^2}_{n,o,d} &= -\frac{1}{2}\overline{\iota^2}_{n,d}\\ \overline{\iota^2}_{n,o,g} &= \frac{g_m}{j\omega_0 C_{\mathrm{tot}}}\frac{1-jR_s\omega_0 C_{\mathrm{tot}}}{j2R_s\omega_0 C_{\mathrm{tot}}}\overline{\iota^2}_{n,g}\\ \overline{\iota^2}_{n,o,\mathrm{R_{out}}} &= \overline{\iota^2}_{n,\mathrm{R_{out}}}\end{aligned} \tag{9}$$

The output noise due to the correlation can be represented as follows[19]:

$$\overline{\iota^2}_{n,o,\text{correlation}} = \frac{g_m c}{2\omega_0 C_{\text{tot}}} \sqrt{\overline{\iota^2}_{n,g} . \overline{\iota^2}_{n,d}} \tag{10}$$

Therefore, the noise factor of the LNA can be expressed as follows:

$$F = \frac{\overline{\iota^2}_{n,o,\mathrm{R_s}} + \overline{\iota^2}_{n,o,d} + \overline{\iota^2}_{n,o,g} + \overline{\iota^2}_{n,o,\text{correlation}} + \overline{\iota^2}_{n,o,\mathrm{R_{out}}}}{\overline{\iota^2}_{n,o,\mathrm{R_s}}} \tag{11}$$

Thus, (12) gives the noise factor at resonance as follows:

$$F = 1 + \frac{\frac{1}{4}\gamma g_{d0} + g_m^2 \left(\frac{C_{gs}}{C_{\text{tot}}}\right)^2 (Q^2 + \frac{1}{4}) \beta \frac{1}{5 g_{d0}} + g_m c \left(\frac{C_{gs}}{C_{\text{tot}}}\right) \sqrt{\frac{\gamma\beta}{20}} + \frac{1}{R_{\text{out}}}}{g_m^2 R_\mathrm{s} Q^2} \tag{12}$$

where γ is the white noise factor, β is the gate noise parameter, c is the correlation coefficient, C_{gs} is the intrinsic gate capacitance, C_{tot} is the sum of C_{gs} and additional capacitance Ce, and Q is the quality factor of the input circuit. There is another expression of noise, which is the noise figure that is given by (13) as follows:

$$\text{NF} = 10 \log_{10}(F) \tag{13}$$

The transconductance g_m and output conductance g_{d0} can be expressed by the monomial expressions as follows:

$$\begin{aligned} g_m &= A_0 L^{A1} W^{A2} {I_{ds}}^{A3} \\ g_{d0} &= B_0 L^{B1} W^{B2} {I_{ds}}^{B3} \end{aligned} \tag{14}$$

where L is the channel length (μm), W is the channel width (μm), and I_{ds} is the drain current (A). $A_0 = 0.0423$, $A_1 = -0.4578$, $A_2 = 0.5275$, $A_3 = 0.4725$, $B_0 = 0.0091$, $B_1 = -0.5637$, $B_2 = 0.5305$ and $B_3 = 0.4695$ are constants. The quality factor of the input circuit at the resonant frequency ω_0 is given by

$$Q = \frac{1}{R_{\text{tot}} \omega_0 C_{\text{tot}}} = \frac{1}{2 R_\mathrm{s} \omega_0 C_{\text{tot}}} \tag{15}$$

$$\omega_0 = \frac{1}{\sqrt{L_{\text{tot}} C_{\text{tot}}}} \tag{16}$$

$$R_\mathrm{s} = \frac{g_m L_s}{C_{\text{tot}}} \tag{17}$$

where L_{tot} is the sum of L_s and L_g.

The optimization problem using ACO and ABC techniques can then formulated as follows:

Minimization of the objective function: Noise Factor F (12).
Subject to the following constraints:

$$\begin{aligned}
& L = L_{\text{feature size}} \\
& 1\,\mu\text{m} \le W \le 100\,\mu\text{m} \\
& \frac{C_{gs}}{C_{\text{tot}}} \le 1 \\
& \frac{2}{3}\frac{C_{gs}}{C_{ox}WL} = 1 \\
& \frac{g_m L_s}{C_{\text{tot}}} = 50\,\Omega \\
& I_{ds} \cdot V_{dd} \le 1\,\text{mW} \\
& g_m = A_0 L^{A1} W^{A2} {I_{ds}}^{A3} \\
& g_{d0} = B_0 L^{B1} W^{B2} {I_{ds}}^{B3}
\end{aligned} \tag{18}$$

where $L_{\text{feature size}} = 90\,\text{nm}$ and $V_{dd} = 2\,V$. The setting out $|S_{11}| < -10\,\text{dB}$ and $|S_{21}| \ge 9\,\text{dB}$.

4 Results and Discussion

The design and simulation of RF CMOS LNA for an operating frequency of 2.3 GHz are done using 90 nm CMOS technology. The setting parameters of the ACO and ABC algorithms are shown in Table 1. The number of iteration is 1000, and the number of population is 100. The convergence graphs are illustrated in Fig. 2.

Table 1 Algorithmic parameters

Parameters	ACO	ABC
Evaporation rate (ρ)	0.1	–
Quantity of deposit pheromone (Q)	0.8	–
Heuristics factor (β)	1	–
Pheromone factor (α)	1	–
Number of onlookers bees	–	50% of the swarm
Number of employed bees	–	50% of the swarm
Number of food sources	–	50

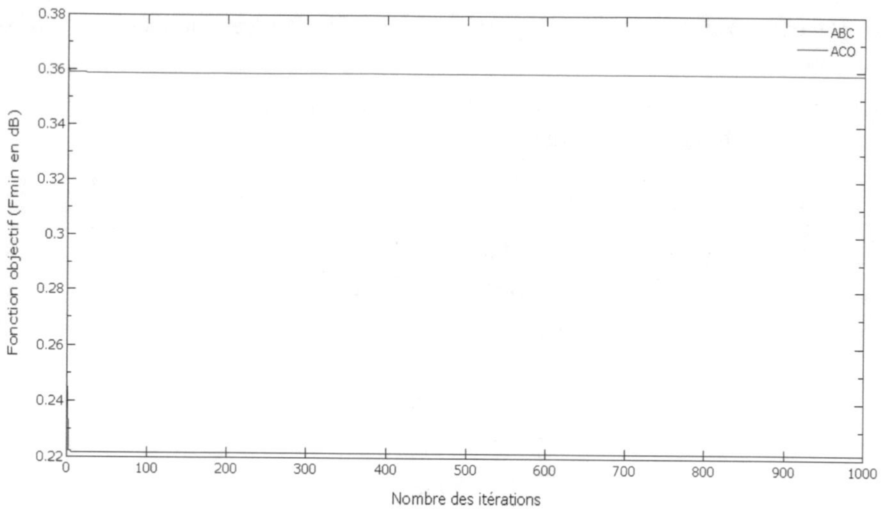

Fig. 2 Convergence graphs of ACO and ABC techniques

The resulting optimal design parameters obtained by ACO and ABC techniques are shown in Table 2.

As seen in Table 2, the ABC technique offers better performance of noise figure than the ACO technique.

The optimization results obtained by using the ACO and ABC techniques are compared with the results obtained by advanced design system (ADS) software. Figures 3 and 4 display the ADS simulation results ($F_{\min}$, S_{21} and S_{11}) using the obtained optimal values by ABC and ACO techniques for RF CMOS LNA, respectively. The comparison between optimization and simulation results of ACO and ABC techniques is summarized in Table 3:

From Figs. 3, 4 and Table 3, we notice that simulation results are in good agreement with those obtained using ABC and ACO techniques.

Table 2 Optimal design results for LNA when input circuit $Q = 4$ and output circuit $Q_{\text{out}} = 5$

Parameters	ACO	ABC
Output conductance (g_{d0})	0.0248 S	0.0080 S
Transconductance (g_m)	0.0210 S	0.0085 S
Gate width (W)	23.635 μm	22.188 μm
Gate length (L)	0.09 μm	0.09 μm
P factor	0.1151	0.1156
Intrinsic capacitance (C_{gs})	19.930 fF	18.516 fF
Source inductor (L_s)	1.648 nH	1.2636 nH
Drain current (I_{ds})	0.5 mA	0.5 mA
Minimum noise figure ($F_{\min}$)	0.3585 dB	0.2215 dB

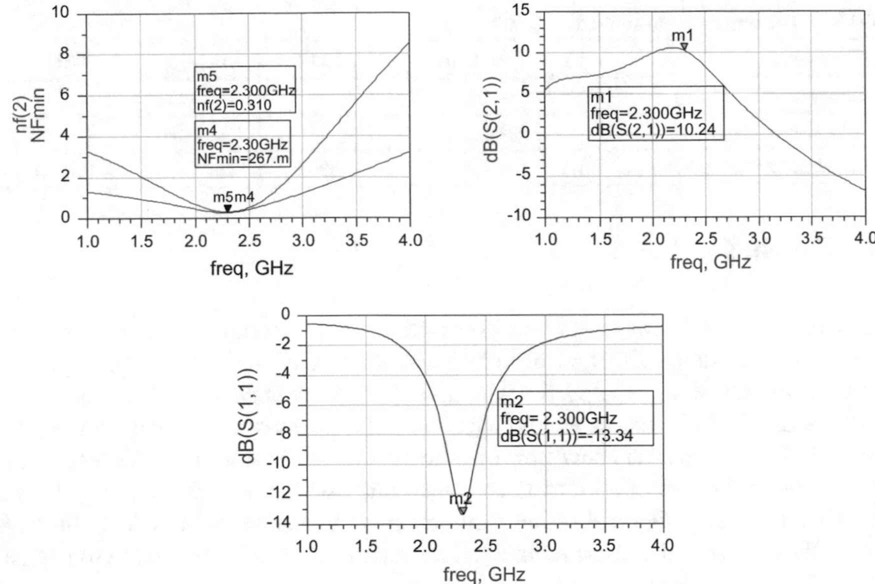

Fig. 3 Simulation results using ABC technique

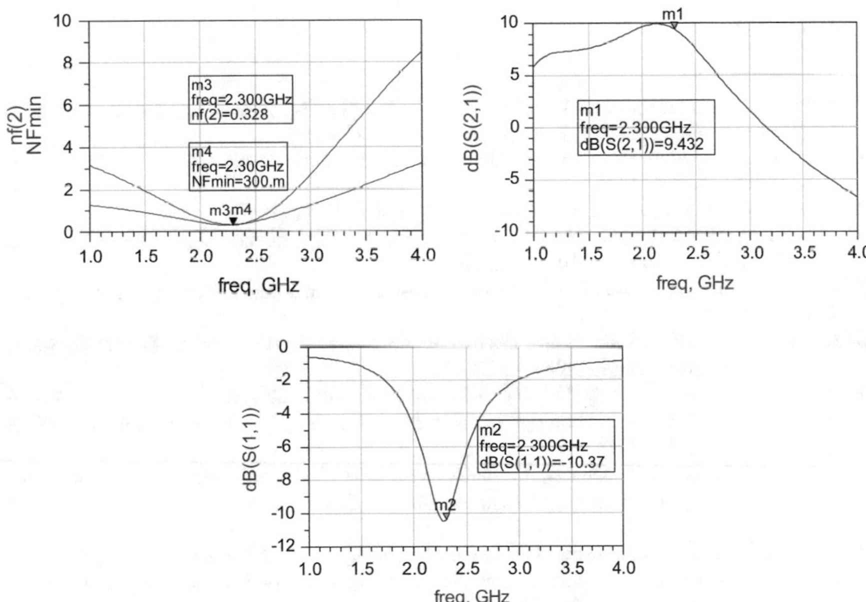

Fig. 4 Simulation results using ACO technique

Table 3 Performance and simulation results

Parameters	ACO_{ADS}	ACO	ABC_{ADS}	ABC
Input return loss (S_{11})	−10.37 dB	–	−13.34 dB	–
voltage gain (S_{21})	9.432 dB	–	10.24 dB	–
Minimum noise figure (F_{min}) (dB)	0.328	0.3585	0.310	0.2215

5 Conclusion

This paper presents two meta-heuristic optimization techniques, namely the ant colony optimization (ACO) and artificial bee colony (ABC) using for getting optimal sizing of a cascode RF CMOS LNA with inductive source degeneration implemented with 90 nm CMOS process technology. The ABC technique provides conversion gain (10.24 dB) and very low noise figure (0.2215 dB), and the ACO technique offers conversion gain (9.432 dB) and very low noise figure (0.3585 dB) at the 2.3 GHz frequency. Optimal sizing parameters are validated through ADS simulations. We can argue that these meta-heuristic techniques can be employed to design the high-performance noise of the LNA circuit.

References

1. Xiaoyu, J.: Optimization of Short-Channel RF CMOS Low Noise Amplifiers by Geometric Programming. Electrical Engineering Theses. Paper 15 (2012)
2. Reeves, C.R.: Modern Heuristic Techniques for Combinatorial Problems. Blackwell Scientific Publications, Oxford (1993)
3. Glover, F.: Tabu search-part II. ORSA J. Comput. **2**(1), 4–32 (1990)
4. Grimbleby, J.B.: Automatic analogue circuit synthesis using genetic algorithms. IEEE Proc.-Circuits, Dev. Syst. **147**(6), 319–323 (2000)
5. Aarts, E., Lenstra, K.: Local Search in Combinatorial Optimization. Princeton University Press, Princeton (2003)
6. Chan, F.T.S., Tiwari, M.K.: Swarm Intelligence: Focus on Ant and Particle Swarm Optimization. I-Tech Education and Publishing (2007)
7. Benhala, B., Pereira, P., Sallem, A. (eds.): Focus on Swarm Intelligence Research and Applications. Computer Science, Technology and Applications series. Nova Science Publishers, ISBN: 978-1-53612-452-1 (2017)
8. Fakhfakh, M., Cooren, Y., Sallem, A., Loulou, M., Siarry, P.: Analog circuit design optimization through the particle swarm optimization technique. J. Anal. Integr. Circuits Sign. Process. (2010)
9. Benhala, B.: Sizing of an inverted current conveyors by an enhanced ant colony optimization technique. In: The International Conference on Design of Circuits and Integrated Systems (DCIS 2016), November 23–25, (2016), Granada, Spain
10. Benhala, B.: An improved ACO algorithm for the analog circuits design optimization. Int. J. Circuits, Syst. Sign. Process. **10**, 128–133 ISSN: 1998-4464 (2016)
11. Benhala, B., Bouyghf, H., Lachhab, A., Bouchikhi, B.: Optimal design of second generation current conveyors by the artificial bee colony technique. In: IEEE International Conference on Intelligent Systems and Computer Vision (ISCV'15), pp. 1–5, March 25–26 (2015), Fez, Morocco

12. Bouyghf, H., Benhala, B., Raihani, A.: Optimal design of RF CMOS circuits by means of an artificial bee colony technique (Chap. 11). In: Benhala, B., Pereira, P., Sallem, A. (eds.) Focus on Swarm Intelligence Research and Applications, pp. 221–246. NOVA Science Publishers, Inc. ISBN: 978-1-53612-452-1 (2017)
13. Dorigo, M., DiCaro, G., Gambardella, L.M.: Ant algorithms for discrete optimization. Artif. Life J. **5**, 137–172 (1999)
14. Dac-Nhuong, L., et al.: Optimizing feature selection in video-based recognition using Max–Min Ant System for the online video contextual advertisement user-oriented system. J. Comput. Sci. **21**, 361–370 (2017)
15. Bhateja, V., Tripathi, A., Sharma, A., Le, B.N., Satapathy, S.C., Nguyen, G.N., Le, D.N.: Ant colony optimization based anisotropic diffusion approach for despeckling of SAR images. In: International Symposium on Integrated Uncertainty in Knowledge Modelling and Decision Making, pp. 389–396. Springer, Cham (2016, November)
16. Jinhui, Y., Xiaohu, S., Maurizio, M., Yanchun, L.: An ant colony optimization method for generalized TSP problem. Prog. Nat. Sci. **e18**, 1417–1422 (2009)
17. Hung, K.K., Hu, P.K., Ko, C., Cheng, Y.C.: A physics-based MOSFET noise model for circuit simulators. IEEE Trans. Electronic Devices **37**, 1323–1333 (1990)
18. Andreani, P., Sjoland, H.: Noise optimization of an inductively degenerated CMOS low noise amplifier. IEEE Trans. Circuits Syst. **48**, 835–841 (2001)
19. Hoe David, H.K., Xiaoyu, J.: The Design of Low Noise Amplifiers in Deep Submicron CMOS Process: A Convex Optimization Approach. Hindawi Publishing Corporation VLSI Design, ID 312639 V (2015)

Study and Design of a See Through Wall Imaging Radar System

M. Abdellaoui, M. Fattah, S. Mazer, M. El bekkali, M. El ghazi, Y. Balboul and M. Mehdi

Abstract Ultra-wideband (UWB) radars are used for several applications such as locating, detecting and preventing collisions. The detection of humans hidden by walls or trapped in burning buildings or avalanche victims is so important. In our research, we aim to design UWB radar to meet those needs. In this paper, we study monostatic and bistatic radars in order to form a general idea about the process of implementation of a radar information processing chain, before adapting it to the UWB technology that will be used in our application later on. The application which consists in detecting and locating targets through walls.

Keywords Ultra-wideband (UWB) technology · Monostatic radar · Bistatic radar · Detection · Localization

1 Introduction

The use of ultra-wideband wireless communications is growing rapidly due to the need to support multiple users and provide more information with higher data rates. Ultra-wideband technology uses very short nanosecond pulses, covering a very large bandwidth in the frequency domain. The big advantages of UWB short pulses are immunity to passive interference (rain, fog, clutter, aerosols, etc.) and the ability to penetrate most building materials such as bricks, wood, drywall and reinforced concrete. The very large bandwidth translates into good radar resolution, which has the ability to differentiate between closely spaced targets [1, 2].

M. Abdellaoui (✉) · S. Mazer · M. El bekkali · M. El ghazi · Y. Balboul
Transmission and Information Processing Laboratory, Sidi Mohamed Ben Abdellah University, Fez, Morocco
e-mail: Abdellaoui.marwa@gmail.com

M. Fattah
ETTI, FPE, My Ismail University, Meknes, Morocco

M. Mehdi
Microwaves Laboratory, Lebanese University, Beirut, Lebanon

V. Bhateja et al. (eds.), *Embedded Systems and Artificial Intelligence*, Advances in Intelligent Systems and Computing 1076,
https://doi.org/10.1007/978-981-15-0947-6_15

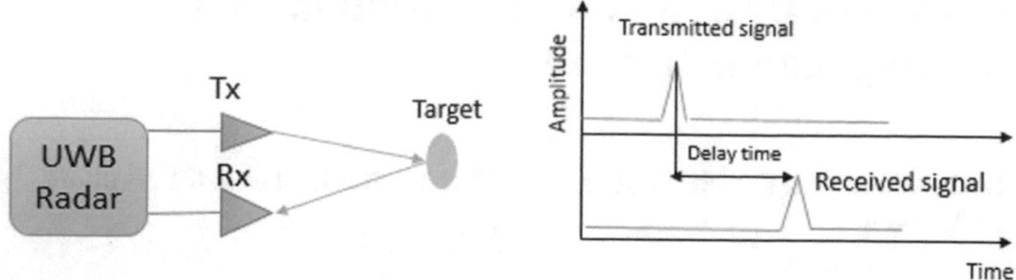

Fig. 1 Radar transmission

The UWB radar generates and transmits a short pulse through TX transmission antenna. The signal propagates in an environment, when it meets the target; a part of the electromagnetic energy is reflected by the object and received by RX, the receiving antenna. The delay between the transmitted signal and the received signal represents the spatial distance between TX—target—RX [3] (Fig. 1).

Radars can be classified according to their structure (monostatic radar, bistatic radar, distributed radar and MIMO radar), their technology (continuous wave (CW), frequency modulated continuous wave (FMCW), stepped frequency continuous wave (SFCW), noise and impulsive) and finally the nature of the measured information (0D system, 1D system, 2D system and 3D system).

This paper deals with monostatic and bistatic radars, defining the basic steps of implementation of a radar information processing chain. The main purpose of this work is to detect and locate targets using monostatic and bistatic radars at first and predict their nature and movements after that using UWB radar.

2 Dimensioning a See Through Wall (STW) Radar

The design of radar must take into account several parameters (cost, portability, accuracy, reliability…). STW radar requires specific features; the wave that it must emit must penetrate the walls with a reasonable attenuation that imposes the choice of a frequency and an adapted power; it must also have a good resolution to discriminate the close targets, which requires the choice of a suitable frequency band and a specific antenna structure.

It must also be taken into account that the propagation of the wave is done in complex indoor environments (propagation through inhomogeneous structures, multiple paths due to multiple reflections in the room) which again obliges us to choose reception stages with good dynamics and adapted signal processing. The key parameters of radar for vision through the walls are resolution, range, pulse repetition frequency (PRF), probability of detection (Pd) and probability of false alarms (Pfa).

The radar design process can be summed up in three main parts, radar design (waveform, transceiver noise characteristics, radiator and collector), system simulation (targets, propagation environment, and synthesis of the signal) and finally

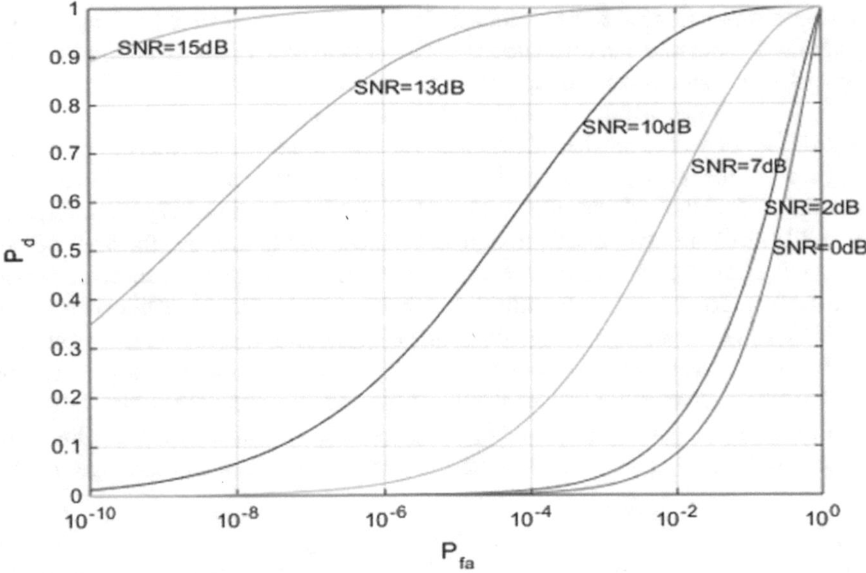

Fig. 2 Receiver operating characteristic curves (ROC); (*SNR must exceed 13 dB in order to satisfy the desired Pfa and Pd values*) [4]

the range detection (detection threshold, adapted filter, non-coherent integration and range detection).

The most critical parameter of a transmitter is the transmission's peak power. It is related to several factors, namely the maximum unambiguous range, the signal-to-noise ratio required at the receiver and the pulse width of the waveform. The SNR required at the receiver is determined using the values of (Pd) and (Pfa); the relationship between these variables can be represented by a characteristic curve of the operation of the ROC receiver (Fig. 2).

To make the radar more practical, we can use pulse integration techniques to reduce the SNR. The peak power at the transmitter can be calculated later through the radar equation [5].

First, we will study monostatic and bistatic radars in order to form a general idea about the process of implementation of a radar information processing chain, before adapting the latter to the UWB technology that will be used in our application, namely the detection and the location of the targets through the walls.

3 Design of Monostatic and Bistatic Radars

The particularity of a bistatic radar is that the transmitter and the receiver are separated, unlike the monostatic radar where these two are the same. In this case, the positions and speeds of the transmitter and the receiver are independent. In what follows a simulation of a basic monostatic radar and a polarimetric bistatic radar system is demonstrated to estimate the range and the speed (velocity) of the targets taking into account the transmitter, the receiver and the targets [5, 6].

Figure 3a shows that the second and third echoes are weaker than the first, as the target is further away from the radar. Noise is an undesirable energy that compromises the ability of the receiver to detect the desired signal. It may be generated within the receiver or enter through the antenna along with the wanted signal [7]. Normally, the received pulses are first passed through an adapted filter to improve the SNR before doing pulse integration, threshold detection, etc. Figure 3b shows the received pulses with the detection threshold after filtering. Adapted filtering (matched filter) improves the SNR, but because signal strength is range dependent, a closer target's return is much stronger than a distant one, which means that the near target will give a noise that has a good chance of crossing the threshold and obscuring the further target.

A time-varying gain can be used to compensate for distance-dependent losses in the received signal so that the threshold is fair for all targets in the range. It will

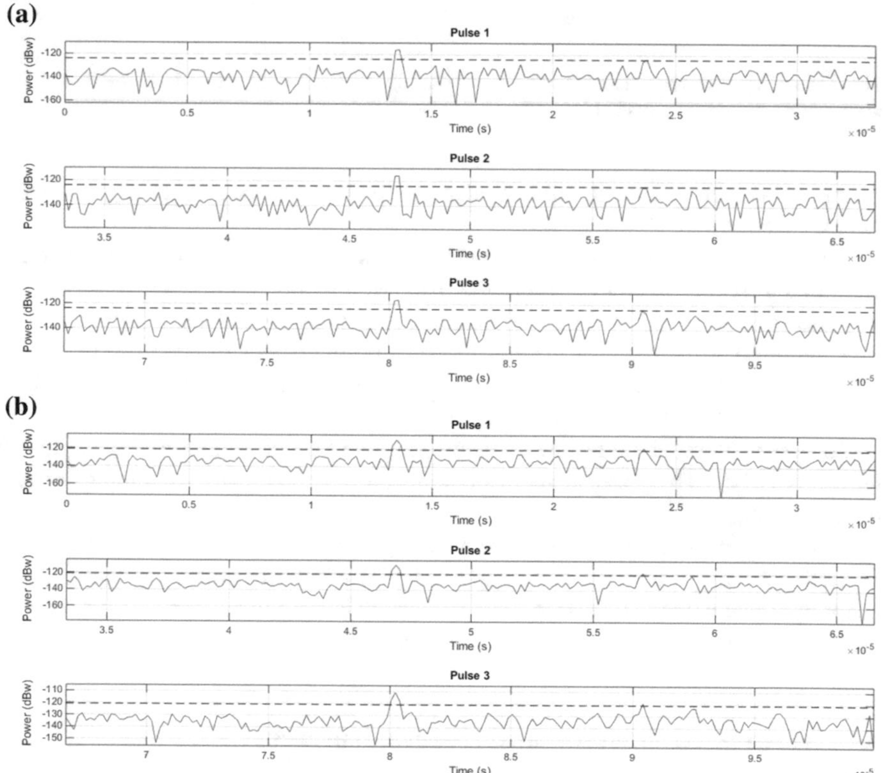

Fig. 3 **a** Representation of the received pulses (here, we show the echoes received from three pulses and the detection threshold before filtering) [4]. **b** Representation of the received pulses (here, we show the echoes received from the same three pulses and the detection threshold after filtering) [4]

be applied to the received pulse so that the returns are as coming from the same reference range (the maximum detectable range) [6].

Figure 4a shows that the threshold is greater than the maximum power level contained in each pulse. Pulse integration will be necessary to make sure that the power of the echoes returned by the targets can exceed the threshold while leaving the noise level below the bar. After integration Fig. 4b, three target echoes can be detected since they are above the threshold. Threshold detection is performed on the integrated pulses. The detection scheme subsequently identifies the peaks and then translates their positions into the target ranges.

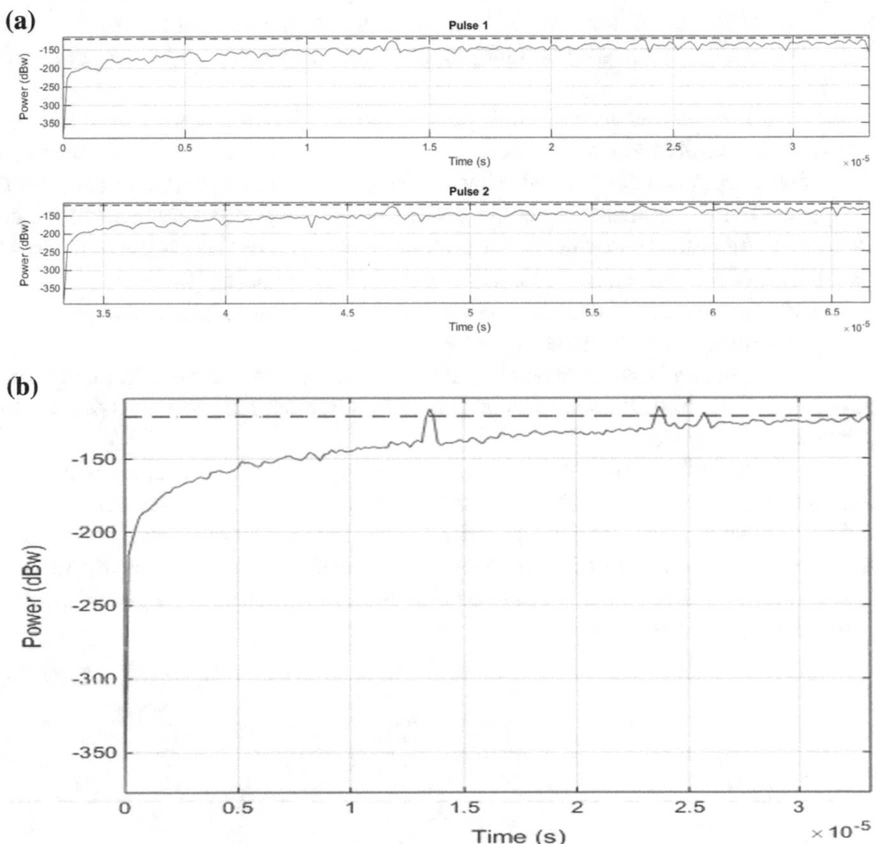

Fig. 4 **a** Representation of two pulses after range normalization (*without integration*) [4]. **b** Representation of the pulse after range normalization (*with integration*)

4 Simulation Results

MATLAB is used to simulate bistatic radar based on monostatic one with the objective to estimate the range and the positions of the targets.

We will assume the existence of two targets in space. The first target is a target point modeled as a sphere; it represents the polarization state of the incident signal. It is located at [15,000–1000–500] m from the emission network and moves at a speed of [100; 100; 0] m/s and the second target is located at [35,000; −1000; 1000] m of the transmission network and approaches at a speed of [−160; 0; −50] m/s. Unlike the first target, the second one returns the state of polarization of the incident signal which means that the *H* horizontal/V vertical polarization components of the input signal become the vertical/horizontal polarization components of the output signal [4, 7].

A single diffusion matrix is a simple polarimetric model for a target. It assumes that whatever the incident and the reflective directions, the power distribution between the *H* and *V* components is fixed. However, even such a simple model can reveal complicated behavior in the simulation as the directions *V* and *H* vary to differentiate the incident directions and reflect. The orientation, defined by the system of the local coordinates of the targets, also affects the polarization mapping [8].

The next section simulates 256 received pulses of a bistatic radar. The reception matrix is formed in a beam toward the two targets.

Figure 5a shows how the receiving array and targets are moving in the system configuration and Fig. 5b shows a range Doppler map generated for all the 64 received pulses at the receive array.

The circles indicate where the targets should appear in the Doppler map. The Doppler range map shows only the return of the first target since the transmission and reception matrix are vertically polarized and the second target maps the vertically polarized wave to a horizontally polarized wave. The signal received from the second target is mainly orthogonal to the polarization of the reception network, which causes a significant loss of polarization.

(a)

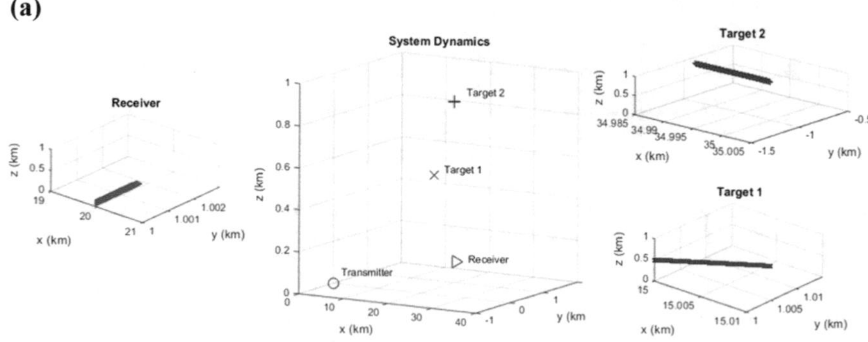

Fig. 5 **a** Receiving target movements. **b** Range Doppler map presentation of the 64 received pulses (*Target 2 not detected*). **c** Range Doppler map presentation of the 64 received pulses (*using circular polarization*)

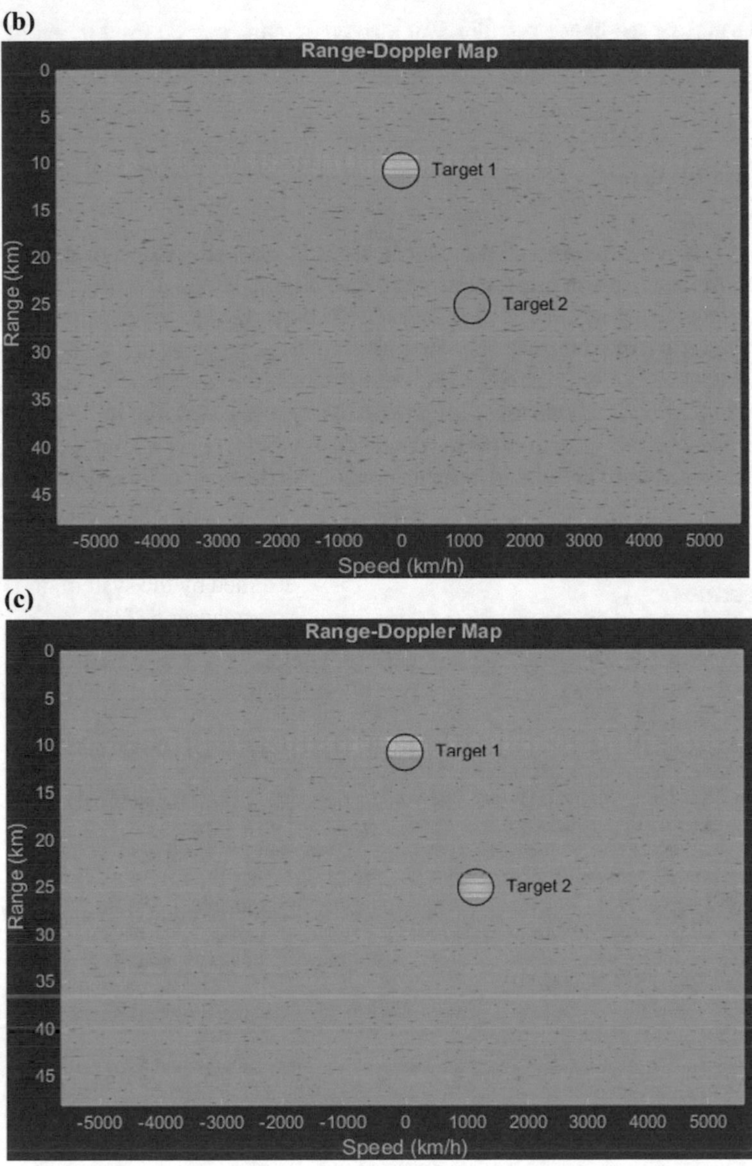

Fig. 5 (continued)

A linear polarized antenna is not often the best choice as a receiving antenna in such a configuration, because no matter how the linear polarization is aligned, there is always an orthogonal polarization. It can be fixed by opting for a circularly polarized antenna at the end of the reception [8, 9].

We can then see how a circularly polarized antenna can be used to avoid losing linear polarized signals due to the polarization diffusion property of a target [9].

5 Conclusion

In this paper, we have presented the basic steps of implementation of a radar information processing chain using monostatic and bistatic radars, in order to detect and locate targets using MATLAB simulations. To improve the performance of a radar system, several parameters must be modified to detect targets such as the pulse repetition frequency (PRF), the SNR required to reach the desired (Pd) and (Pfa) and the maximum required transmission power. We can also add that the type of waveform influences the detection performance of the radar. Several aspects related to waveforms remain to be studied and this constitutes the perspectives of our research.

References

1. Rahayu, Y., Rahman, T.A, Ngah, R., Hall, P.S.: Ultra-wideband technology and its applications. In: 2008 Fifth IFIP International Conference on Wireless and Optical Communications Networks WOCN '08. IEEE
2. Fontana, R.J.: Recent system applications of short-pulse UWB technology. IEEE Trans. Microwave Theory Tech. **52**(9), 2087–2104 (2004)
3. Taylor, J.D.: Ultra-Wideband Radar Technology. CRC Press, Boca Raton (2001)
4. Mathworks.Inc.support/Radar/Examples/Monostaic and bistatic radars
5. Chaliseetal, B.K.: Target localization in a multistatic passive radar system through convex optimization. Signal Process. **102**, 207–215 (2014). Elsevier
6. In: 11th International Conference Interdisciplinarity in Engineering, INTER-ENG 2017, 5–6 October 2017, Tirgu-Mures, Romania (2017)
7. Pardhu, T., Sree, A.K., Tanuja, K.: Design of matched filter for radar applications. Electr. Electron. Eng. Int. J. **3**(4) (2014)
8. Willis, N.J., Griffiths, H.D. (eds.): Advances in bistatic radar. IEEE Aerosp. Electron. Syst. Mag. **23**(7) (2008)
9. Sharma, S., Tripathi, C.C.: A comprehensive study on circularly polarized antenna. In: Second International Innovative Applications of Computational Intelligence on Power, Energy and Controls with their Impact on Humanity (CIPECH)

Analog Active Filter Component Selection Using Genetic Algorithm

Asmae El Beqal, Bachir Benhala and Izeddine Zorkani

Abstract In this paper, we highlight the optimal design of an active fourth-order band-pass filter for radio frequency identification (RFID) system reader to reject all signals outside the band (10–20) kHz and to amplify the low antenna signal with a center frequency of 15 kHz. The filter is designed to have a Butterworth response, and the topology that will be used to implement this filter is Sallen–Key. The values of the passive components are selected from manufactured series; thus, it is very exhaustive to search on all possible combinations of values from those series for an optimized design. The metaheuristics have proved a capacity to deal with such problem effectively. In this work, the metaheuristic genetic algorithm (GA) is applied for the optimal sizing of the fourth-order band-pass filter. SPICE simulations are used to validate the obtained results/performance.

Keywords Metaheuristic · Optimization · Genetic algorithm · Fourth-order band-pass filter · Sallen–Key topology

1 Introduction

Filter circuits are used in a wide variety of applications. In the field of wireless communication system, active band-pass filter is used to reject all signals outside the (10–20) kHz signals and to amplify the low antenna signal for a RFID system which is used to identify tagged objects, people, or animals.

Active fourth-order band-pass filter is made up of many discrete components (resistors and capacitors); the values of those components are calculated by fixing

A. El Beqal (✉) · I. Zorkani
Faculty of Sciences Dhar el Mahraz, University of Sidi Mohamed Ben Abdellah, Fez, Morocco
e-mail: asmae.elbekkal@gmail.com

I. Zorkani
e-mail: izorkani@hotmail.com

B. Benhala
Faculty of Sciences, University of Moulay Ismail, Meknes, Morocco
e-mail: b.benhala@fs-umi.ac.ma

V. Bhateja et al. (eds.), *Embedded Systems and Artificial Intelligence*,
Advances in Intelligent Systems and Computing 1076,
https://doi.org/10.1007/978-981-15-0947-6_16

the values of some components and finding the others [1, 2]; the use of this method not only limits the freedom of the design but also the values of the passive components founded cannot belong to the manufactured series such as E12, E24, E48, E96, or E192. In order to select the discrete components from those industrial series, the use of an exhaustive search on all possible combinations of preferred values to obtain an optimized design is infeasible. Thus, intelligent methods are required.

In the literature, some metaheuristics are used by the analog designers to solve the design problems of integrated circuits and also of discrete component systems, such as simulated annealing (SA) [3], genetic algorithms (GA) [4], tabu search (TS) [5], particle swarm optimization (PSO) [6], ant colony optimization (ACO) [7–9] and artificial bee colony (ABC) [10, 11].

In this work, we focus on the use of the GA method to solve a typical analog circuit-sizing problem: A fourth-order active band-pass filter design mainly composed of discrete elements (capacitors and resistors) which are chosen among industrial series to comply with clearly defined specifications. SPICE software was used for performing simulations in order to check reached performances.

The rest of the paper is arranged as follows. Section 2 briefly discusses about the genetic algorithm (GA). Section 3 describes the application of this algorithm for the optimal design of the analog active band-pass filter. Section 4 shows the simulation results obtained by using GA. Finally, Sect. 5 concludes the paper.

2 Genetic Algorithm: An Overview

The genetic algorithm (GA) was proposed for the first time by John Holland in 1975, it is based on the principles of natural selection and "Survival of the fittest" [12], and the basic components of this method are [13]:

- a fitness function for optimization;
- a population of chromosomes;
- selection of which chromosomes will reproduce;
- crossover to produce the next generation of chromosomes;
- random mutation of chromosomes in the new generation.

The fitness function is the function that the algorithm is trying to optimize, its tests and quantifies how "fit" each potential solution is. The term chromosome refers to a numerical value or values that represent a candidate solution to the problem that the genetic algorithm is trying to solve. Each candidate solution is encoded as an array of parameter values. A genetic algorithm begins with a randomly chosen assortment of chromosomes, which serves as the first generation (initial population). Then, each chromosome in the population is evaluated by the fitness function. The selection operator chooses some of the chromosomes for reproduction based on a probability distribution defined by the user. The fitter the chromosome is, the more likely it is to be selected. The crossover operator consists to swaps a subsequence of two of the chosen chromosomes to create two offspring. The mutation operator has the effect of maintaining diversity in the population and ensures that the GA will not being trapped in a local optimum.

Operations of selection, crossover and mutation continue until the number of offspring is the same as the initial population, so that the second generation is composed entirely of new offspring and the first generation is completely replaced.

Now, the second generation is tested by the fitness function, and the cycle repeats. Thus, the fitness of the entire population will be decreased with the reproduction of the generation.

The general flowchart of a GA is presented in Fig. 1:

The genetic algorithm has been the subject of several studies, and its fields of application are the widest and the most varied. These include, for example, robotics [14], mathematics [15], electric distribution systems [16], traffic light signal parameters optimization, [17] etc. In the next section, we present an application of the GA to the optimal design of the fourth-order band-pass filter.

Fig. 1 Flowchart of a GA

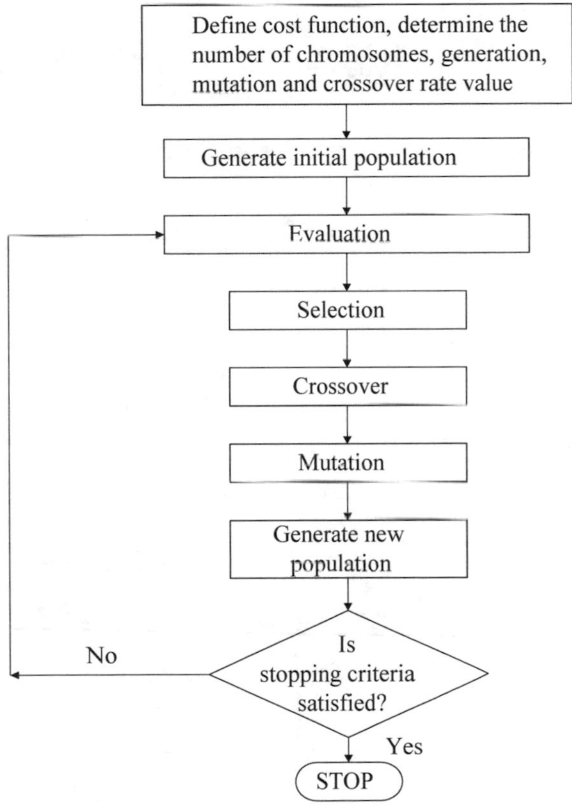

3 Application: Optimal Design of the Fourth-Order Band-Pass Filter

The architecture that has been used to implement the fourth-order band-pass filter is the Sallen–Key topology, and this circuit has the advantage that the quality factor (Q) can be varied via the inner gain (G) without modifying the mid-frequency (f_{m}). A drawback is, however, that (Q) and the gain (A_{m}) at the mid-frequency (f_{m}) cannot be adjusted independently [18]; the second-order Sallen–Key band-pass filter is shown in Fig. 2.

The general transfer function for a second-order band-pass filter is:

$$A(s) = \frac{\frac{A_{\mathrm{m}}}{Q} s}{1 + \frac{1}{Q} s + s^2} \tag{1}$$

The Sallen–Key band-pass circuit in Fig. 2 has the following transfer function:

$$A(s) = \frac{GRC\,\omega_{\mathrm{m}}\,s}{1 + [RC\,\omega_{\mathrm{m}}(3 - G)s] + \left(R^2C^2\omega_{\mathrm{m}}^2 s^2\right)} \tag{2}$$

Through coefficient comparison with Eq. (1), obtain the following equations:

$$f_{\mathrm{m}} = \frac{1}{2\,\pi\,RC} \quad \text{or} \quad \omega_{\mathrm{m}} = \frac{1}{RC} \tag{3}$$

$$G = 1 + \frac{R_2}{R_1} \tag{4}$$

$$A_{\mathrm{m}} = \frac{G}{3 - G} \tag{5}$$

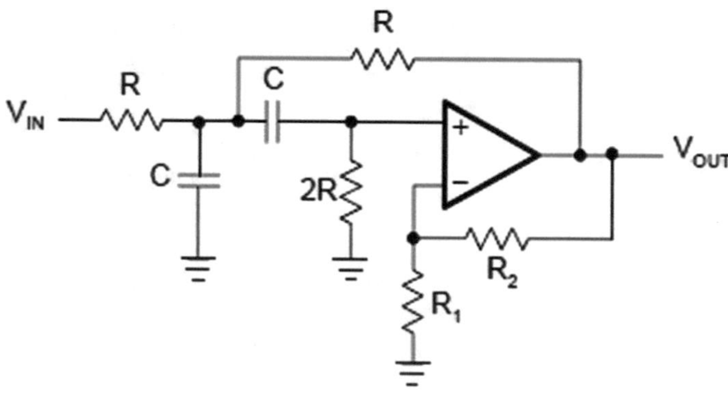

Fig. 2 Schematic diagram of second-order Sallen–Key active band-pass filter

$$Q = \frac{1}{3 - G} \tag{6}$$

A fourth-order Butterworth band-pass filter can be designed by cascading two second-order blocks as the Fig. 3 shows, and the general transfer function of this filter is expressed in (7) [18]:

$$A(s) = \frac{\left(\frac{A_{\text{mi}}}{Q_i}\alpha\, s\right)}{1 + \frac{\alpha\, s}{Q_i} + (\alpha\, s)^2} \times \frac{\left(\frac{A_{\text{mi}}}{Q_i}\frac{s}{\alpha}\right)}{1 + \frac{1}{Q_i} \times \left(\frac{s}{\alpha}\right) + \left(\frac{s}{\alpha}\right)^2} \tag{7}$$

This equation represents the connection of two second-order band-pass filters in series, where:

- A_{mi} is the gain at the mid-frequency, (f_{mi}), of each partial filter.
- Q_i is the pole quality of each partial filter.
- (α) and $(1/\alpha)$ are the factors by which the mid frequencies of the individual filters, (f_{m1}) and (f_{m2}) derive from the mid-frequency (f_{m}) of the overall band-pass filter. Factor (α) is determined by using Eq. (8), [18].

$$\alpha^2 + \left[\frac{\alpha\, \Delta\, \Omega\, a_1}{b_1\left(1 + \alpha^2\right)}\right] + \left(\frac{1}{\alpha^2}\right) - \left[\frac{(\Delta\Omega)^2}{b_1}\right] = 0 \tag{8}$$

where the normalized bandwidth $\Delta\Omega = (1/Q_{\text{BP}})$, (Q_{BP}) is the overall quality factor of the filter with (a_1) and (b_1) being the second-order low-pass coefficients of the desired filter type.

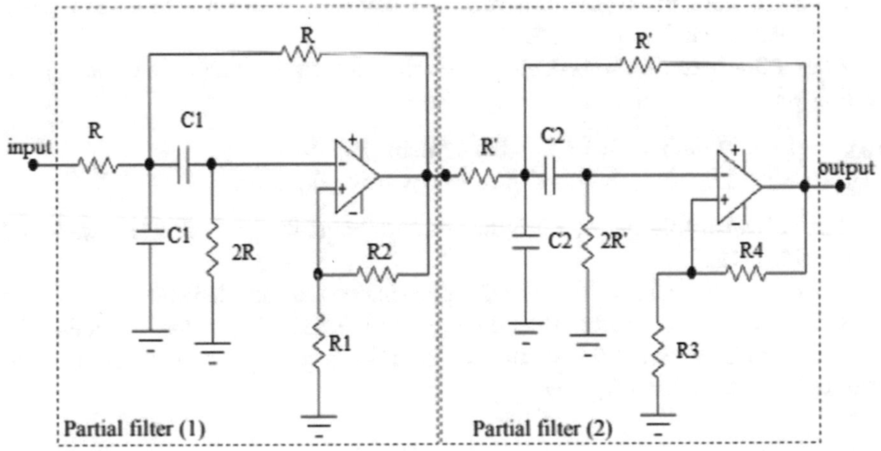

Fig. 3 Schematic diagram of fourth-order Butterworth active band-pass filter circuit using Sallen–Key topology

The mid-frequency (f_{m1}) of partial filter (1) is [18]:

$$f_{m1} = \frac{f_m}{\alpha} \quad (9)$$

The mid-frequency (f_{m2}) of partial filter (2) is [18]:

$$f_{m2} = f_m \cdot \alpha \quad (10)$$

With (f_m) being the mid-frequency of the overall fourth-order band-pass filter. The individual pole quality, (Q_i), is the same for both filters [18]:

$$Q_i = Q_{BP}\left(\frac{(1+\alpha^2)b_1}{\alpha . a_1}\right) \quad (11)$$

With (Q_{BP}) being the quality factor of the overall filter which is defined as the ratio of the mid-frequency of the overall filter (f_m) to the bandwidth of the overall filter (BW) [18]:

$$Q_{BP} = \frac{f_m}{BW} = \frac{f_m}{f_H - f_L} \quad (12)$$

The design specifications of the active band-pass filter are:

- Mid-frequency of the overall filter, $f_m = 15$ kHz ($\omega_m = 94.2$ K rad/s). Q_i is the pole quality of each partial filter.
- Bandwidth of 10 kHz.
- Passband frequencies: $f_L = 10$ kHz; $f_H = 20$ kHz.

For Butterworth filter type, we have $a_1 = 1.4142$ and $b_1 = 1$, and by using Eq. (8), α is calculated and it is equal to 1.2711.

After α has been determined, all quantities of the partial filters can be calculated as follows:

$f_{m1} = 11.8$ kHz ($\omega_{m1} = 74.14$ K rad/s) by using Eq. (9).
$f_{m2} = 19.067$ kHz ($\omega_{m2} = 119.80$ K rad/s) by using Eq. (10).

The individual pole quality, Q_i, is the same for both filters, by using (11) we found Q_i equal to 2.1827.

In order to generate ω_{m1}, ω_{m2} and Q_i approaching the specified values, the values of the resistors and capacitors should be carefully selected. For this, we define the total error (TE) which is the summation of mid-frequency deviation ($\Delta\omega_m$) and quality factor deviation (ΔQ) by:

$$TE = \alpha \, \Delta \, \omega_m + \beta \, \Delta Q \quad (13)$$

where:

$$\Delta\,\omega_{\mathrm{m}} = \frac{|\omega_1 - \omega_{\mathrm{m1}}|}{\omega_{\mathrm{m1}}} + \frac{|\omega_2 - \omega_{\mathrm{m2}}|}{\omega_{\mathrm{m2}}} \tag{14}$$

$$\Delta Q = \frac{|Q_1 - Q_i| + |Q_2 - Q_i|}{Q_i} \tag{15}$$

In terms of the components of the filter, the mid-frequency deviation parameter and the quality factor deviation can be written as:

$$\Delta\,\omega_{\mathrm{m}} = \frac{\left|\frac{1}{R\,C_1} - \omega_{\mathrm{m1}}\right|}{\omega_{\mathrm{m1}}} + \frac{\left|\frac{1}{R\,C_2} - \omega_{\mathrm{m2}}\right|}{\omega_{\mathrm{m2}}} \tag{16}$$

$$\Delta Q = \frac{\left|\frac{1}{3-\left(1+\frac{R_2}{R_1}\right)} - Q_i\right| + \left|\frac{1}{3-\left(1+\frac{R_4}{R_3}\right)} - Q_i\right|}{Q_i} \tag{17}$$

The objective function considered is the total error which is calculated for the different values of α and β, and the decision variables are the resistors and capacitors forming the circuit. Each component must have a value of the standard series (E192), and they present the chromosome for the GA.

4 Result and Discussion

In this section, we applied GA to perform optimization of the analog filter. The optimization technique works on MATLAB codes with the following parameters given in Table 1, and the circuit is simulated in SPICE to obtain the frequency response.

The optimal values of resistors and capacitors forming the active fourth-order band-pass Butterworth filter are selected from the E192 series, the use of those series lie to the fact of having a tolerance less than 1%, and as a consequence, a good accuracy of results is obtained. The optimal values of components and the performance associated with these values for the different values of α and β are shown in Table 2.

From Table 2, it is observed that the design error obtained for $\alpha = 0.01$ and $\beta = 0.99$ is the least as compared to all other values.

Table 1 GA parameters

Population size	900
Generation	1000
Crossover	Two-point crossover
Mutation rate	0.0001
Selection probability	50%

Table 2 Values of components following E192 series and related butterworth band-pass filter performances for the different values of α and β

α	β	R_1 (KΩ)	R_2 (KΩ)	R_3 (KΩ)	R_4 (KΩ)	R (KΩ)	R' (KΩ)	C_1 (nF)	C_2 (nF)	$\Delta\, \omega_m$	ΔQ	TE
0.01	0.99	13.3	20.5	13.3	20.5	1.45	1.02	9.19	8.16	0.0086	0.0022	**0.0023**
0.2	0.8	15.2	23.4	18.4	28.4	1.35	0.988	9.88	8.45	0.0114	0.0087	0.0093
0.5	0.5	11.3	17.40	12.1	18.7	5.36	3.92	2.52	2.13	0.0017	0.0123	0.0070
0.8	0.2	15.0	23.2	15.8	24.3	1.42	1.32	9.53	6.34	0.0059	0.0190	0.0085
0.99	0.01	17.4	26.7	16.4	25.5	1.42	1.38	9.53	6.04	0.0047	0.0451	0.0051

The following figure shows the SPICE simulation of the filter gain for the fourth-order active band-pass Butterworth filter for the smallest design error. The practical mid-frequency using the GA for E192 series is equal to 15.004 kHz. The Fig. 4 shows the SPICE simulation.

Table 3 shows the comparison between the specifications for the desired band-pass filter and the theoretical and practical values obtained by using the optimal values of passive components; we notice that we have a slight difference between specifications and practical values and that proved the validity of the proposed technique.

5 Conclusion

In this study, the use of GA, one of the methods of the heuristic calculation, was proposed for the optimal design of fourth-order band-pass Butterworth filter, in order to make it suitable for RFID application. The selection of passive components has been done from the manufactured series E192. The design of the analog filter with

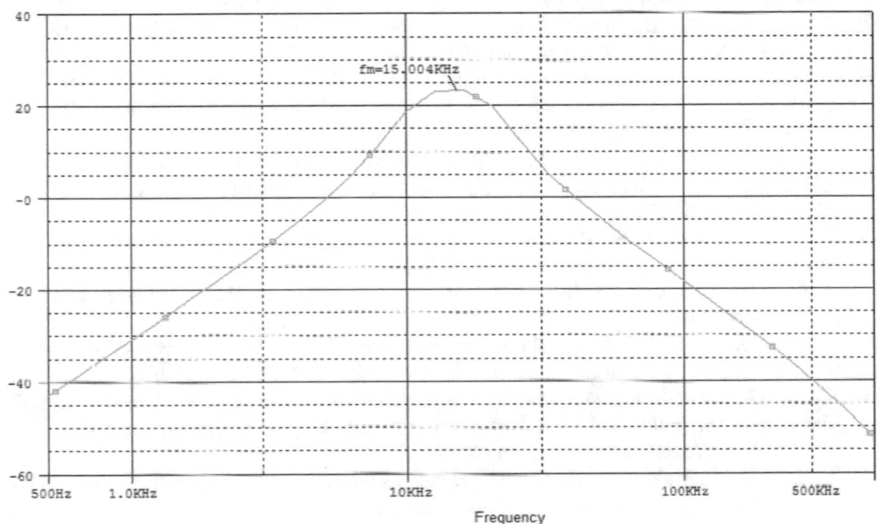

Fig. 4 Frequency responses of fourth-order band-pass butterworth filter by using GA

Table 3 Comparisons between specifications, theoretical, and practices values

	$fm1$ (KHz)	$fm2$ (KHz)	fm (KHz)	Q_{BP}
Specifications	11.80	19.067	15.00	1.5000
Theoretical values	11.94	19.12	15.11	1.4984
Practical values	11.80	19.07	15.004	1.8079

the targeted specifications is successfully achieved by using the GA method; validity of the proposed technique was proved via SPICE simulation.

References

1. Abdul, H.A.Z.A.: Design and simulation of 4th order active bandpass filter using multiple feedback and Sallen key topologies. J. Babylon Univ. Eng. Sci. **22**(2) (2014)
2. Myo, Z.M.M., Aung, Z.M., Naing, Z.M.: Design and implementation of active band-pass filter for low frequency RFID (radio frequency identification) system. In: Proceedings of the International Multi Conference of Engineers and Computer Scientists Hong Kong, vol. 1 (2009)
3. Dreo, J., Pétrowski, A., Siarry, P., Taillard, E.: Metaheuristics for Hard Optimization: Methods and Case Studies. Springer, New York (2006)
4. Benhala, B., Bouattane, O.: GA and ACO techniques for the analog circuits design optimization. J. Theor. Appl. Inf. Technol. (JATIT) **64**(2), 413–419 (2014)
5. Glover, F.: Tabu search-part I. ORSA J. Comput. **1**(3), 190–206 (1989)
6. Chan, F.T.S., Tiwari, M.K.: Swarm Intelligence: focus on ant and particle swarm optimization. I-Tech Education and Publishing (2007)
7. Benhala, B.: An improved aco algorithm for the analog circuits design optimization. Int. J. Circuits, Syst. Signal Process. **10**, 128–133 (2016). ISSN: 1998-4464
8. Kritele, L., Benhala, B., Zorkani, I.: Ant colony optimization for optimal low-pass state variable filter sizing. Int. J. Electr. Comput. Eng. (IJECE) **8**(1), 227–235 (2018)
9. Kritele, L., Benhala, B., Zorkani, I.: Optimal digital IIR filter design using ant colony optimization. In: IEEE 4th International Conference on Optimization and Applications (ICOA'18), pp. 1–5. Mohammedia, Morocco (2018)
10. Bouyghf, H., Benhala, B., Raihani, A.: Optimization of 60-GHZ down-converting CMOS dual-gate mixer using artificial bee colony algorithm. J. Theoretical Appl. Inf. Technol. (JATIT) **95**(4), 890–902 (2017)
11. Bouyghf, H., Benhala, B., Raihani, A.: Optimal design of RF CMOS circuits by means of an artificial bee colony technique. In: Benhala, B., Pereira, P., Sallem, A. (eds.) Chapter 11, Book: Focus on Swarm Intelligence Research and Applications, NOVA Science Publishers, pp. 221–246 (2017)
12. Haupt, R.L., Haupt, S.E.: Practical GeneticAlgorithms. Wiley (2004). ISBN 0-471-45565-2
13. Mitchell, M.: Genetic algorithms: an overview. Complexity **1**(1), 31–39 (1995)
14. Silva, V.G., Tavakoli, M., Marques, L.: Optimization of a three degrees of freedom DELTA Manipulator for well-conditioned workspace with a floating point genetic algorithm. Int. J. Nat. Comput. Res. (IJNCR) **4**(4), 1–14 (2014)
15. Nadimpalli, V.L.V., Wankar, R., Chillarige, R.R.: Innovative genetic algorithmic approach to select potential patches enclosing real and complex zeros of nonlinear equation. Int. J. Nat. Comput. Res. (IJNCR) **6**(2), 18–37 (2017)
16. Taroco, C.G., Carrano, E.G., Neto, O.M.: Robust design of power distribution systems using an enhanced multi-objective genetic algorithm. In: Nature-Inspired Computing Design, Development, and Applications, pp. 179–200. IGI Global (2012)
17. Wijaya, S., Uchimura, K., Koutaki, G.: Traffic light signal parameters optimization using modification of multielement genetic algorithm. Int. J. Electr. Comput. Eng. (IJECE) **8**(1), 246–253 (2018)
18. Mancini, R.: Op Amps for Everyone Design Reference. 2nd ed., Elsevier (2003)

Preliminary Study of Roots by Georadar System

Ahmed Faize and Gamil Alsharahi

Abstract This study has made it possible to draw up a state of the art of the GPR radar domain. This measuring instrument is used in many fields of application such as archeology, space exploration and even in civil engineering. Its non-destructiveness of the probed environment is its main advantage. However, it remains an expensive investment, and the number of models on the market remains low. There are GPR radars operating in different ways in the data collection and in this work were used common offset mode. However, the performance of the radar can be altered. The electromagnetic properties of the soil and the central working frequency can decrease or increase the maximum depth of investigation. Indeed, the higher this frequency, the lower the maximum depth of exploration. This work focuses on the application of the GPR. Our main task is the contribution to the identification, by an adequate data processing and the choice of suitable software, of the objects fleeing underground. In the first place, we realized simulations, by sophisticated software, GPR signals. Then, we went to the application of the GPR at the Faculty of Nador.

Keywords GPR · Cavity · Reflexw

1 Introduction

Ground penetrating radar (GPR) is a similar technique to the seismic imagery reflection [1]. It uses electromagnetic waves that propagate and refract in heterogeneous medium in order to scan, localize and identify quantitative variations within electric and magnetic proprieties of the soil. GPR uses an array of frequencies ranged from 10 MHz to 2.6 GHz. When using lower frequency antenna (between 10 and 100 MHz), investigated depth will be higher (more than 10 m); however, resolution remains lower [2].

A. Faize (✉) · G. Alsharahi
MASI Laboratory, Multidisciplinary Faculty, University Mohammed I, BP. 300, 62700 Nador, Morocco
e-mail: ahmedfaize6@hotmail.com

V. Bhateja et al. (eds.), *Embedded Systems and Artificial Intelligence*,
Advances in Intelligent Systems and Computing 1076,
https://doi.org/10.1007/978-981-15-0947-6_17

The treatment of geophysical data is based on numeric modeling of signals captured by the radar. Within electromagnetic field, numeric modeling appeared at the beginning of the seventies [3] and coincided with the introduction of powerful and accessible microprocessors that allow realization of personal calculation within a short time [4]. It is possible to model the physical proprieties implicated in any geophysical technique by using either forward modeling or inverse modeling. Forward modeling takes as model subsoil that includes all of its appropriate physical properties and uses theoretical equations in order to simulate the response of the receiver. This, in turn, will determine a given technique to measure through the model. However, the modeling uses data collected from the field in order to create a model of subsoil that would feet better with the original data. To obtain reliable data, a number of measures and data processing must be performed. Forward modeling allows us to test and determine how proprieties of subsoil materials will change.

GPR is used for the exploration of the subsoil in several research fields such as the detection of landmines [5], geology [6], civil engineering [7], glaciology [8, 9] and archeology [10]. It was also successfully applied to 2D imaging of cracks and fractures in resistive soils such as gneiss [11, 12].

The study proposed and presented in this work combines, for the first time, simulation studies and in situ detection studies. The advent of economically feasible geophysical methods of detection has significantly altered the landscape to identify and assess potential sites and detect surface gaps. In this work, the equipment used for the study was a GPR GSSI system with a 400 MHz antenna.

This work will focus on the results of the simulations performed by the appropriate software and also the results of many measurements that we made in situ in the garden of the faculty of Nador.

2 Principe of Use of GPR

An example of 2D acquisition is shown schematically in Fig. 1. The radar system, represented by two antennas (T: transmitter and R: receiver), is deployed on the surface along a horizontal profile (Fig. 1a). Figure 1b shows the registration of the propagation time of electromagnetic waves according to the position of the GPR device used. The observed hyperbola in this figure is an index extremely important because it gives, in general, information on the location of buried objects.

2.1 Important Physical Parameters Involved in the Operation of GPR

For homogeneous isotropic materials, the relative propagation velocity v_r can be defined by

$$v_r = \frac{c}{\sqrt{\varepsilon_r}} \quad \text{(m/s)} \tag{1}$$

And the depth can be calculated by

$$d = v_r \frac{t_r}{2} \quad \text{(m)} \tag{2}$$

ε_r is the relative permittivity
t_r is the transit time of the target

In practice, the relative permittivity will be unknown. The propagation velocity must be measured or estimated by direct depth measurement at a physical interface or target. It can also be estimated by calculation means of multiple measurements. From Fig. 1b, we can see that if we have a hyperbolic spreading function, so, we can extract a formula of the relative velocity that is written:

$$v_r = 2\sqrt{\frac{x_{n-1}^2 - x_n^2}{t_{n-1}^2 - t_n^2}} \tag{3}$$

And the depth of the target is written:

$$d_0 = \frac{v_r t_0}{2} \tag{4}$$

The calculation of the depth of a plane reflector is unique. It is done using the so-called common depth point method. If both transmit and receive antennas are moved at equal distances from the common center point, the same apparent reflection position will be maintained.

The depth of the plane reflector can be derived from

$$d = \sqrt{\frac{x_{n-1}^2 t_n^2 - x_n^2 t_{n-1}^2}{t_{n-1}^2 - t_n^2}} \tag{5}$$

The propagation velocity decreases with increasing relative permittivity. The wavelength within the material is related to the speed of propagation by the relation:

$$\lambda_m = \frac{v_r}{f} \quad \text{(m)} \tag{6}$$

f the frequency

The results of these effects are illustrated in Table 1.

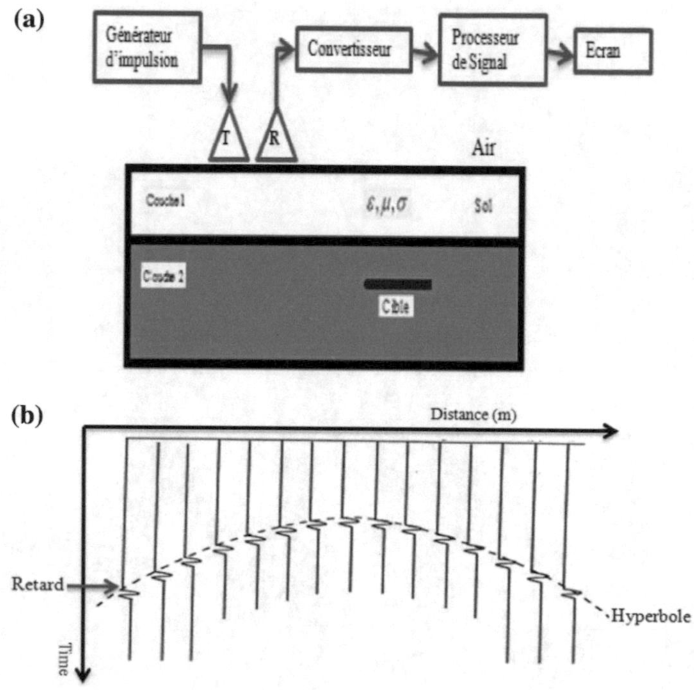

Fig. 1 Principle acquisition of the GPR radar and the B-scan associated

Table 1 Characteristics of propagation materials

			Wavelength λ (cm)	
Material	Relative permittivity ε_r	Propagation speed (cm/ns)	100 MHz	1 GHz
Air	1	30	300	30
Concrete	9	10	100	10
Pure water	80	3.35	33	3

3 Results of Simulations with Reflexw Software

To simulate the GPR signals of the objects by the simulator (Reflexw), one needs a certain number of parameters [13], like the frequency of the antenna used, the geometry of the subsoil and the dielectric permittivity, magnetic permeability and electrical conductivity of the media that are involved in the simulation [14, 15]. In what follows, we give obtained results for the different simulations carried out [16]:

3.1 Simulation 1

To do the simulation by Reflexw, the soil in which the object is buried is simulated by dry sand whose dielectric characteristics are $\varepsilon_r = 3$ and $\sigma = 0.01$ mS/m. The permittivity object $\varepsilon_r = 5$ and the radius $r = 0.01$ m are buried at a depth of 1 m and is located at 4 m along the axis *Ox* (the peak) and a dielectric cylinder of square section of dielectric constant $\varepsilon_r = 9$ and the conductivity $\sigma = 0.0045$ S/m. The cylinder is buried at the same depth, with the ground surface coordinates between (7.5–8.5 m) according to *Ox* and (1–1.5 m) according to *Oy* (Fig. 2a). The obtained results are summarized and represented by the software, as radargram as in Fig. 2b, for example. In this figure, we note the presence of diffraction hyperbolas that indicate

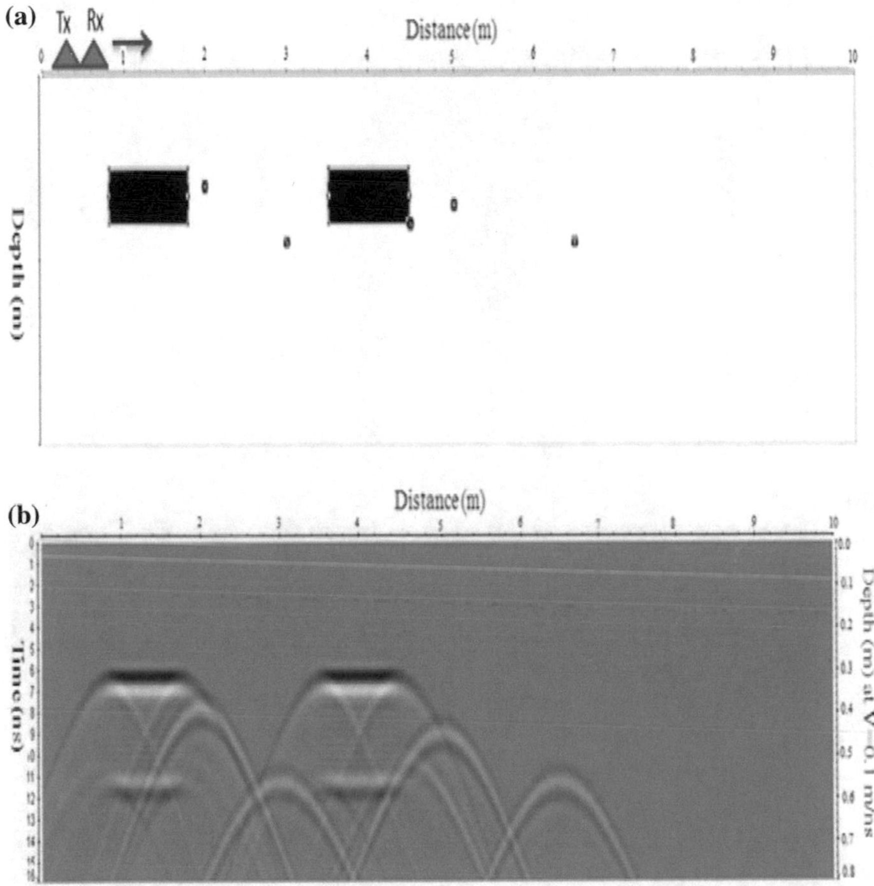

Fig. 2 **a** Diagram showing the position of two square dielectric cylinders and simulated objects (peak) in the ground, **b** radargram of two square dielectric cylinders and simulated objects (peak) in the dry sand for the 400 MHz antenna

the presence of the peak around 2, 4 and 5 m, and these hyperbolas indicate also the presence of the cylinder around 1 and 3 m; it is exactly the depth to which they were supposed to bury these objects.

3.2 Simulation 2

Here, we performed a simulation to see the influence of an empty cavity in the presence of simulated objects (two dielectric square sections). Figure 3b shows the radar system of the simulated system for a frequency of 400 MHz. Note the presence of revealing hyperbolas that indicate the positions of the objects of the system in the

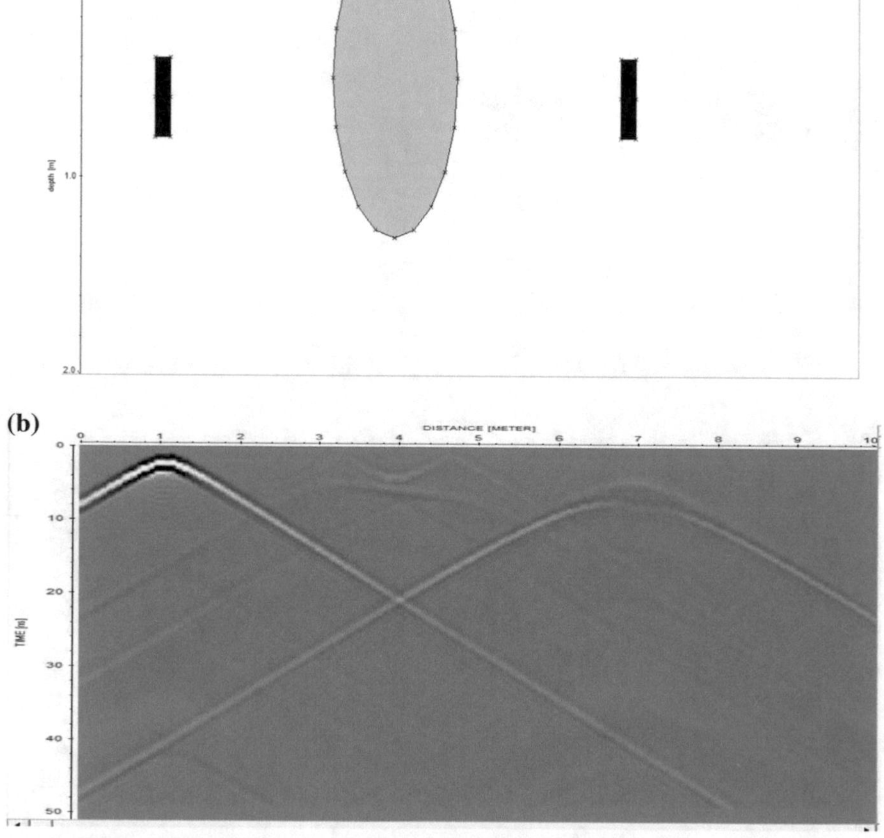

Fig. 3 **a** Diagram showing the position of two square dielectric cylinders and an empty cavity in the dry sad, **b** radargram showing the hyperbolas of two square dielectric cylinders and an empty cavity in the dry sand

expected place. This radargram illustrates the hyperbola due to the total diffraction of the wave inside the cavity and its reflection at the end. We note that the hyperbola in this figure is clearly visible because of the conductivity of the target.

4 Experimental Results

Although the GPR data showed good signal penetration, careful 2D and 3D processing allowed us to obtain better results. The structures studied have helped the civil engineering specialists to plan their next excavation season [17, 18]. In what follows, we describe the most notable results of the geophysical survey carried out in the garden of the faculty in Nador. We propose, in this part, the experimental study, which consists in applying the GPR system to the objects simulated in the previous section.

Treatment of several raw images taken in the garden of the faculty of Nador performed by Reflexw software. In what follows, we describe the most notable results of the geophysical survey carried out in the garden of the faculty polydisciplinary in Nador (Fig. 4).

Figure 5 illustrates the B-Scan achievements with the 400 MHz antenna. Result showed the presence of strong reflectors probably related to bedrock structures or plant roots. The presence of clear anomaly characterized by hyperbolic reflector

Fig. 4 Experimental study in the garden of the faculty of Nador

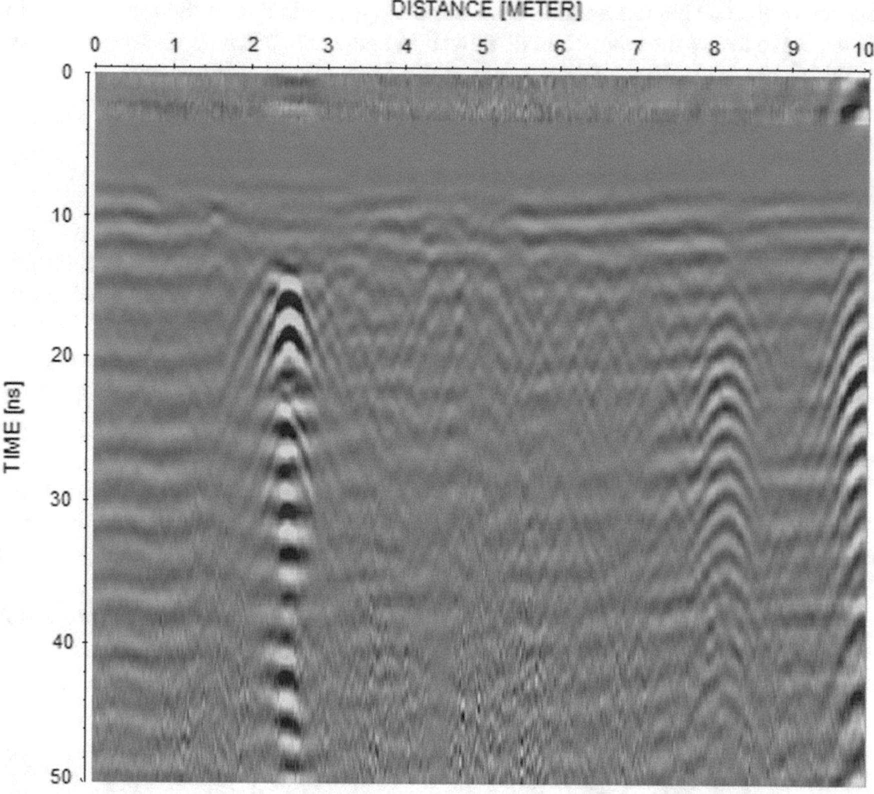

Fig. 5 Radargram after treatment (prospection carried out in the garden of the faculty)

(0.13 m/ns) (red circle), and we note that the soil of the garden of Nador (Fig. 4) was far from ideal for this kind of experiments, because it was full of plant roots, whose percentage content of water is important, which makes them strong reflectors of georadar signals. Therefore, in the radargrams obtained by the GPR system, we saw a large number of hyperboles indicating the positions of these roots (Fig. 6).

5 Conclusion

This work was based essentially on the simulation of the signals taken by the GPR radar, using the Reflexw software. Note that the operation of this software is based on the finite-difference time-domain method (FDTD). The simulations we carried out relate to different dielectric and conductive objects.

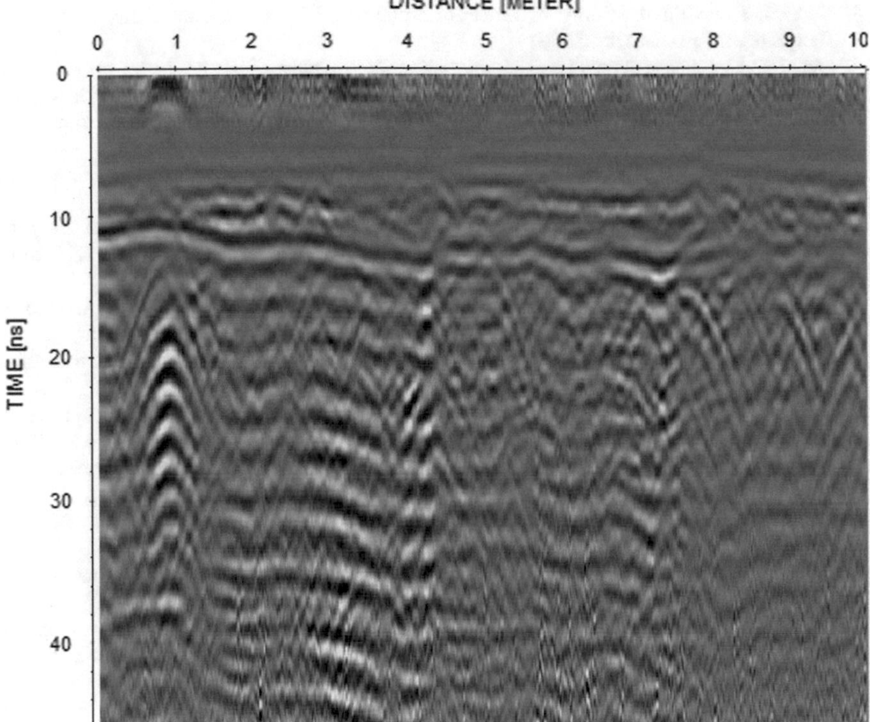

Fig. 6 Radargram after treatment (prospecting carried out in the garden of the faculty of Nador)

To see the value of our results from our simulations with those obtained by the direct application of the GPR device, we conducted our experiments at the multidisciplinary faculty of Nador. The field of our measurements was the garden of the faculty multidisciplinary, note that the soil of this garden was far from ideal for this kind of experiments, because it was full of plant roots, whose percentage content of water is important, which makes them strong reflectors of georadar signals. Therefore, in the radargrams obtained by the GPR system, we saw a large number of hyperboles indicating the positions of these roots.

References

1. Bano, M., Pivot, F., Marthelot, J.M.: Modeling and filtering of surface scattering in ground-penetrating radar waves. First Break **17**, 215–222 (1999)
2. Noon, D.A., Stickley, G.F., Longstaff, D.: A frequency-independent characterization of GPR penetration and resolution performance. J. Appl. Geophys. **40**, 127–137 (1998)
3. Chen, H.W., Tai-Min, H.: Finite difference time domain simulation of GPR data. J. Appl. Geophys. **40**, 139–163 (1998)

4. Ceruzzi, Paule E.: A History of Modern Computing History of Computing. MIT Press, Cambridge, Massachusetts (1998)
5. Hamadi, A.: Analyse et prédiction comportementales du radar GPR polarimétrique de la mission spatiale EXOMARS. Thèse de Doctorat de l'Université De Limoges, France (2010); Gu, Z.W.: The application of ground penetrating radar to geological investigation on ground in cold regions. J. Glaciol. Geocryol. **16**, 283–289 (1994)
6. Ulriksen, C.P.F.: Application of impulse radar to civil engineering. Ph.D. Thesis, Lund University of Technology, Lund, Sweden (1982)
7. Navarro, F.J., Lapazaran, J.J., Machio, F. Martin C., Otero, J.: On the Use of GPR Energetic Reflection coefficients in Glaciological Applications, Geophysical Research, 11 EGU2009-9804, EGU General Assembly (2009)
8. Woodward, J., Burke, M.J.: Applications of ground-penetrating radar to glacial and frozen materials. J. Environ. Eng. Geophys. **12**, 69–85 (2007)
9. Leckebusch, J.: Ground-penetrating radar: a modern three-dimensional prospection method. Archaeolog. Prospect. **10**, 213–240 (2003)
10. Benson, A.K.: Application of ground penetrating radar in assessing some geological hazards: examples of ground water contamination, faults, cavities. J. Appl. Geophys. **33**, 177–193 (1995)
11. Grasmueck, M.: 3-D Ground-penetrating radar applied to fracture imaging in gneiss. Geophysics **61**, 1050–1064 (1996)
12. Faize, A., Driouach, A.: Analytical and experimental study of radar signals GPR Of some objects. Int. J. Curr. Res. **5**(12), 4173–4180 (2013)
13. Driouach, A.: Stouti M Application of the FDTD method to identification of detected objects in archaeology. J. Basic. Appl. Sci. Res. **3**(4), 425–429 (2013)
14. Alsharahi, G., Faize, A., Driouach, A.: 2D FDTD theoretical and simulation to study response of GPR signals in homogeneous and inhomogeneous mediums. Int. J. Commun. Antenna Propag. (IRECAP), **6**(3) June 2016
15. Faize, A., Driouach, A.: The use of radar subsoil GPR for the detection and study of a buried marble and in situ location of possible cracks. Int. J. Eng. Res. Appl. **2**, 1036–1039 (2012)
16. Alsharahi, G., Faize, A., Mint Mohamed Mostapha, A., Driouach, A.: 2D FDTD simulation to study response of GPR signals in homogeneous and inhomogeneous Mediums. Int. J. Communi. Antenna Propag. **6**(3), 153–159 (2016)
17. Alsharahi, G., Mint, M.M., Faize, A., Driouach, A.: Modeling and simulation resolution of ground penetrating radar antennas. J. Electromagn. Eng. Sci. **16**(3), 182–190 (2016)
18. Alsharahi, G., et al.: Detection of cavities and fragile areas by numerical methods and GPR application. J. Appl. Geophys. **164**, 225–236 (2019)

Predictive Agents for the Forecast of CO_2 Emissions Issued from Electrical Energy Production and Gas Consumption

Seif Eddine Bouziane and Mohamed Tarek Khadir

Abstract Energy production is nowadays compulsory for human well-being. However, the current energy systems and increasing demand come with a heavy environmental cost, hence, forecasting the environmental impact of energy production became a crucial task in order to control and reduce pollutant emissions while monitoring energy production. This paper describes an agent-based approach for forecasting CO_2 emissions issued from energy production and consumption in the Algerian city of Annaba using data provided by the Algerian electricity and gas distribution company SONALGAZ. The proposed approach consists of combining artificial neural networks (ANN) forecasting models with an agent-based architecture in order to give the ability to the autonomous agents to forecast the hourly gas consumption and electrical production using dedicated ANNs. Forecasted values will then be used to calculate the equivalent amount of emitted CO_2 for both energy sources.

Keywords Neural network · Agent-based · Short-term forecasting · Carbon dioxide

1 Introduction

Algeria has the tenth-largest proven reserves of natural gas in the world, is the sixth-largest gas exporter, its domestic energy consumption is mainly based on burning natural gas. According to the Annual report from the Algerian Ministry of Energy [1], the production of natural gas in Algeria is reached 96.6 Bm^3 for the sole year 2017, representing 55% of all primary energy production types. The total national energy consumption (including losses) reached 59.6 MTep in 2017. This is equivalent to (760 BTh $\approx$ 76 Bm^3) reflecting an increase of 2.1% when compared to 2016

S. E. Bouziane (✉) · M. T. Khadir
Laboratoire de Gestion Electronique de Documents (LabGED), Department of Computer Science, University Badji Mokhtar, PO-Box 12, 23000 Annaba, Algeria
e-mail: Seifeddine.bouziane@univ-annaba.org

M. T. Khadir
e-mail: Khadir@labged.net

V. Bhateja et al. (eds.), *Embedded Systems and Artificial Intelligence*, Advances in Intelligent Systems and Computing 1076,
https://doi.org/10.1007/978-981-15-0947-6_18

dominated by natural gas 37%, electricity 30% and oil product 27% consumptions percentages. This increase, when compared to 2016, is mainly driven by the increase of:

- Electricity consumption 5.5% increase, knowing that more than 90% of electricity production in Algeria is natural gas driven [2].
- Natural gas consumption with a 1.4% increase.

It is clear from the above statistics that energy consumption is due to the socio-economical growth of the country mainly driven by domestic demand (44%) with a total number of energy subscribers of 9.2 millions against 8.8 millions at the end of 2016. This exponential economic growth comes at a heavy environmental cost, of 145 Million Tons of CO_2 emission [3], mainly due to domestic energy consumption and industry. The trend is alarming, and a better use of production and consumption of energy imposes itself not only for financial reasons but also for environmental and health ones. Therefore, providing energy planners with forecasting tools permitting better resource management and crisis handling imposes itself. This paper describes the basis of a CO_2 forecasting systems based on ANN's agents collaborating in a larger multi-agent system (MAS). The rest of the paper is organized as follows: Sect. 2 focuses on a literature review of the state of the art in terms of energy consumption and CO_2 emission models. The paradigms used in the proposed approach, i.e., ANNs and MAS are described in Sects. 3 and 4 respectively. Section 5 details the proposed system architecture where results and discussions are provided in Sect. 6.

2 Short Literature Review

Various recent researches were conducted to forecast energy production and consumption related to CO_2 emissions. Wang and Zhang [4] Employed a gray forecasting model to estimate the amount of energy-related carbon emission in the Chinese city Baoding, in order to model the carbon emission from electrical production and consumption. An adaptive seasonal model based on the hyperbolic tangent function to define seasonal and daily trends of electricity demand using an Ensemble Kalman filter to estimate the electricity consumption and the carbon emission is developed in [5]. Sheta et al. [6] applied two types of artificial neural networks to forecast the world carbon emission issued from the global energy consumption, the first model is a neural network auto-regressive with exogenous Input while the second model is an Evolutionary Product Unit Neural Network. Wang and Ye [7] employed non-linear gray multivariable models to forecast the CO_2 emissions from fossil energy consumption in China. In the aim of forecasting the carbon emissions in 30 provinces in China, [8] proposed a methodology that consists of a K-means cluster and a Logistic model, K-means was used to divide the carbon emission into five categories, while the Logistic model was built to forecast the carbon emissions in these provinces. In [9], a gray multivariable model is used in order to quantify the carbon emissions from fuel combustion in China from 2014 to 2020. To forecast carbon emission [10] proposed a Gaussian processes regression method based on the particular swarm optimization algorithm, the proposed method was tested with the total CO_2 emissions data from the USA, China, and Japan. In [11], a recurrent neural network using the covariance matrix adaptation evolutionary strategy is trained in order to forecast short-term power demand, wind power generation and carbon dioxide emissions in Ireland. An optimized gene expression programming model using a metaheuristic algorithm to predict the South Korean's carbon emissions in 2030 is proposed in [12]. Huang et al. [13] used gray relational analyses to select the factors that have a strong impact on the CO_2 emissions and the principal component analyses to extract four principal components, finally a long short-term memory recurrent neural network was trained to forecast the carbon emissions. Based on the above short-listed state of the art, ANN emerges as an efficient tool for energy and pollutant predictions and will be combined in this study with multi-agent systems, for that purpose, to benefit from both approaches.

3 ANN for the Forecast of Energy Consumption

Artificial neural networks are a powerful machine learning method inspired by the biological neural networks; they are intended to simulate the biological nervous system and consist of multiple connected nodes called artificial neurons, and each connection is able to transmit a signal from a neuron to many others. The most widely used ANN type for the forecasting problems is the traditional multi-layer perceptron (MLP) [14]. The typical architecture of a MLP consists of three types of layers, an input layer with a number of neurons that equals the number of variables in the input vector, an output layer with a number of neurons corresponding to the number of desired outputs, and finally one or more hidden layers with a number of neurons that should be adjusted of every problem. Training a MLP consists of three main steps, the first is the data extraction, where we chose the most relevant data for the problem, next is the learning stage, where the objective is to find the optimal network configuration in terms of the number of hidden layers and the number of neurons in the hidden layers, the optimization algorithm, etc., there exist several optimization methods such as Levenberg-Marquardt, backpropagation, and quasi-Newton [15].

ANNs were employed in various studies in the field of forecasting the energy consumption and production, to name a few: [16] trained multiple MLPs to forecast the yearly natural gas consumption in Algeria, each MLP was trained specifically to predict the gas consumption in a specific distribution area, and finally summing their results to get the total consumption. Hsu et al. [17] proposed a two-phase ANN-based method for short-term load prediction, the first phase consists of predicting the 24-h load pattern, while the second stage is utilized to predict the minimum and the maximum loads. Jetcheva et al. [18] present a building level neural network-based ensemble model for a day ahead electricity load prediction, the proposed method consists of multiple MLPs where each MLP is trained on a different subset of the data.

4 MAS for the Forecast of Pollution

A multi-agent system is a modeling approach composed of autonomous and interacting agents. MAS has appeared as an encouraging approach for modeling environmental pollution problems [19], diverse recent studies used the agent-based approach to model such problems, to mention a few, [20] proposed an agent-based system to simulate the control of an air pollution crisis, the system consists of a Gaussian plume dispersion model combined with an artificial neural network with both models integrated with a MAS. Dragomir and Opera [21] adopted an agent-based approach to monitor the urban air pollution due to power plant activities; the paper presented a case study of the Romanian town Ploieşti. El Fazziki et al. [22] proposed an agent-based approach for modeling the urban road network infrastructure in order to monitor the on-road air quality, the proposed system used an ANN to predict the air quality and the Dijkstra algorithm to search for the least polluted path in the road network.

5 The Proposed System

Our approach consists of ANN models integrated as agents, to forecast the gas consumption and the electrical energy production. Each ANN is integrated with an Agent to forecast the energy and calculate the equivalent carbon emissions, later the emissions from each agent are then aggregated to calculate the total Carbon emission, as illustrated in Fig. 1.

5.1 The Prediction Models

The first step of the proposed system is to forecast gas consumption and the electrical energy production. For that matter, an ANN for each type of energy is developed. The type of ANNs applied is MLPs with one hidden layer containing 14 neurons

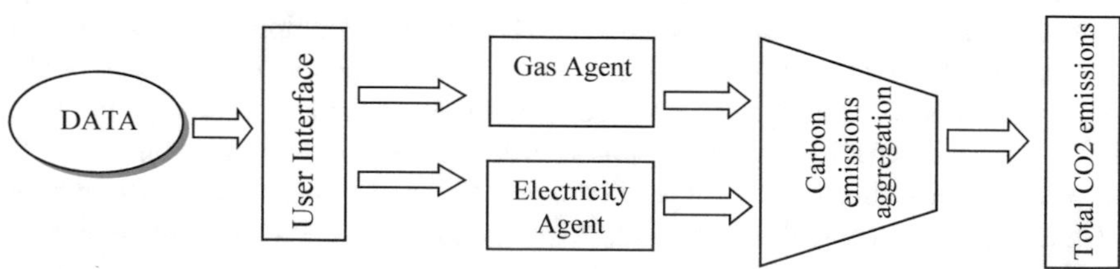

Fig. 1 System's architecture

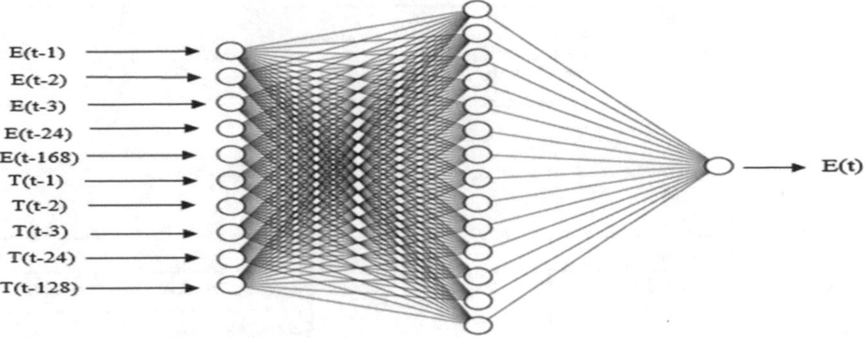

Fig. 2 Architecture of the proposed MLPs

(Fig. 2), and use the backpropagation optimization algorithm. The network topology is fixed after multiple trials selecting each time the best topology in terms of best validation error.

In order to train our models, we divided the available datasets (used in [16] and [23]) into training sets and test sets, where we used the first year for the training to adjust the weights and the biases and the second year for the validation.

After splitting the data, another essential task for the forecasting problems is choosing an appropriate input vector, in our case a 10 dimension vector that consists of the past values of the energy consumption and the temperature is used. Therefore, the energy load E at hour t is explained by the following past values: E_{t-1}, E_{t-2}, E_{t-3}, E_{t-24} and E_{t-168}, which represents respectively the load in the previous three hours, the previous day and the previous week load at the same hour of the day, in addition to the past load values, we used the temperature estimations T of the same selected times, i.e., T_{t-1}, T_{t-2}, T_{t-3}, T_{t-24} and T_{t-168}.

5.2 *The MAS Architecture*

Since there are several existing sources for carbon emissions, an agent-based approach is preferred in order to provide a robust and an autonomous solution for the prediction of the carbon emission. The aim is to develop a MAS architecture that is able to perform an autonomous and simultaneous forecasting of energy consumption and the carbon emissions from different sources, as well as a flexible framework able to be extended to cope with further identified CO_2 emission sources.

The proposed architecture consists of two main autonomous agents, the first agent is the gas agent with the main tasks of forecasting the hourly natural gas consumption using an ANN model and calculating the equivalent carbon emission issued from this consumption, the second agent is the electrical energy agent, its main tasks are forecasting the hourly electric energy production using an ANN, calculating the equivalent quantity of natural gas combusted to produce such load of electricity.

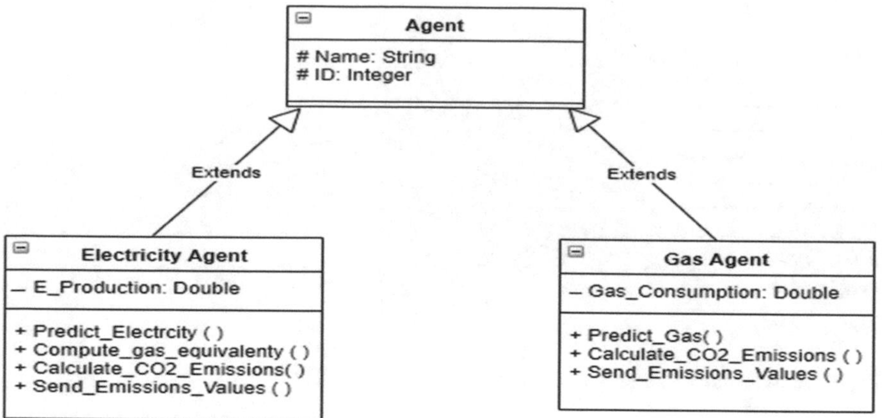

Fig. 3 Class diagram of the system's main Agents

Finally, the system predicts the carbon emission issued from the process of electricity production; furthermore, the two agents communicate autonomously to aggregate their predictions and calculate the total carbon emission from both energy sources. Figure 3 shows a class diagram that describes the main agents of our system.

6 Results and Discussions

To evaluate each model one of the most frequently used evaluation measures, the root-mean-square error (RMSE) given by Eq. (1) is applied. Where a_l is the forecasted value, b_l is the observed value, and L is the number of variables in the input vector. The results obtained by our ANNs gives an RMSE of 0.06 and 0.035 for gas consumption and electricity production models, respectively.

$$\text{RMSE} = \sqrt{\frac{1}{L}\sum_{l}^{L}(a_l - b_l)^2} \tag{1}$$

Figure 4 shows the comparison between the forecasted values and the real values for the two models, using validation data point from the year 2017. After each step of the simulation, each agent calculates the carbon emissions within the forecasted hour based on the emission factors of natural gas mentioned in [24] which indicates that CO_2 equivalency to burning the natural gas is 0.0551 metric tons CO_2/Mcf which is equivalent to 0.0019 metric tons CO_2/m^3. We should point out that in order to calculate the carbon emissions issued from the electrical energy production the agent calculates the equivalent amount of burned natural gas to produce the forecasted amount of electricity knowing that producing 1 kWh of electricity requires, on average, burning 280 m^3 of natural gas.

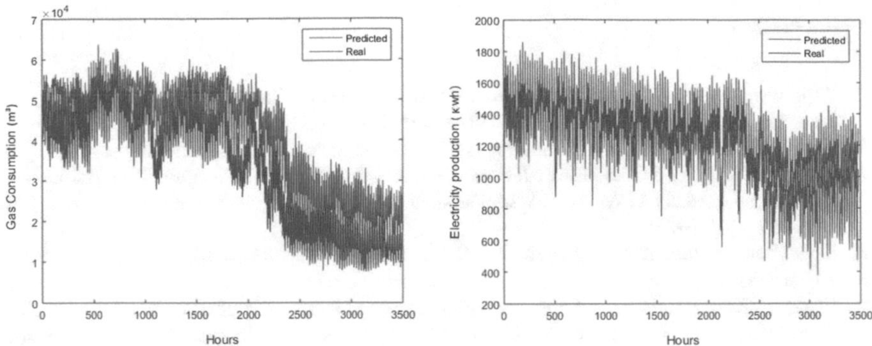

Fig. 4 Comparison of the observed and predicted gas and electricity values

Table 1 Comparison between the obtained CO_2 emissions from the gas consumption and electricity production

Emission source	The maximum hourly emissions (t/m^3)	The minimum hourly emissions (t/m^3)	The average hourly emissions (t/m^3)	The total emissions percentage (%)
Gas consumption	111.12	14.79	61.9	9
Electricity production	950.48	133.20	657.94	91

Table 1 shows a comparison between the CO_2 emissions issued from the electricity generation and the gas consumption that was obtained from the simulation. Based on the obtained results, it indicates that producing electrical energy using natural gas is far more polluting than the community's gas consumption.

7 Conclusion and Perspectives

In this work, we proposed an agent-based approach in the aim of forecasting hourly CO_2 emissions issued from gas consumption and electrical energy generation. Two autonomous agents to forecast both polluting energy sources based on ANN models are developed. The global MAS model then predicts the equivalent carbon emissions. The obtained results are encouraging and prove that the agent-based approach can be a reliable solution to pollution-related problems.

For future work, we aim to improve our simulation tool by improving the accuracy of the ANN models to obtain better predictions, include more pollutant factors such as (NO_x, O_3, etc.) as well as more energy sources in order to construct a more complete and reliable simulator.

References

1. http://www.energy.gov.dz/francais/uploads/MAJ_2018/Stat/Bilan_Energ%C3%A9tique_National_2017_edition_2018.pdf. Last accessed: 2019/01/31
2. http://perspective.usherbrooke.ca/bilan/servlet/BMTendanceStatPays?codeTheme=6&codeStat=EG.ELC.NGAS.ZS&codePays=DZA&optionsPeriodes=Aucune&codeTheme2=6&codeStat2=x&codePays2=DZA&optionsDetPeriodes=avecNomP&langue=fr. Last accessed: 2019/01/31
3. https://data.worldbank.org/indicator/EN.ATM.CO2E.KT?locations=DZ. Last accessed: 2019/01/31
4. Wang, J., Zhang, J.: Research on carbon emissions and its trend forecast of Baoding. In: 2013 Sixth International Conference on Advanced Computational Intelligence (ICACI), Hangzhou, China, pp. 39–44 (2013)
5. Lau, E.T., Yang, Q., Forbes, A.B., Wright, P., Livina, V.N.: Modelling carbon emissions in electric systems. Energy Convers. Manag **80**, 573–581 (2014)
6. Sheta, A.F., Ghatasheh, N., Faris, H.: Forecasting global carbon dioxide emission using auto-regressive with eXogenous input and evolutionary product unit neural network models. In: 2015 6th International Conference on Information and Communication Systems (ICICS), Amman, pp. 182–187 (2015)
7. Wang, Z.X., Ye, D.J.: Forecasting Chinese carbon emissions from fossil energy consumption using non-linear grey multivariable models. J. Clean. Prod. **142**(2), 600–612 (2017)
8. Ma, L., Lin, K., Guan, M., Lin, M.: The prediction of carbon emission in all provinces of China with the K-means cluster based logistic model. In: 2017 International Conference on Service Systems and Service Management, Dalian, pp. 1–6 (2017)
9. Ding, S., Dang, Y.G., Li, X.M., Wang, J.J., Zhao, K.: Forecasting Chinese CO_2 emissions from fuel combustion using a novel grey multivariable model. J. Clean. Prod. **162**, 1527–1538 (2017)
10. Fang, D., Zhang, X., Yu, Q., Jin, T.C., Tian, L.: A novel method for carbon dioxide emission forecasting based on improved Gaussian processes regression. J. Clean. Prod. **173**, 143–150 (2018)
11. Mason, K., Duggan, J., Howley, E.: Forecasting energy demand, wind generation and carbon dioxide emissions in Ireland using evolutionary neural networks. Energy. **155**, 705–720 (2018)
12. Hong, T., Jeong, K., Koo, C.: An optimized gene expression programming model for forecasting the national CO_2 emissions in 2030 using the metaheuristic algorithms. Appl. Energy **228**, 808–820 (2018)
13. Huang, Y., Shen, L., Liu, H.: Grey relational analysis, principal component analysis and forecasting of carbon emissions based on long short-term memory in China. J. Clean. Prod. **209**, 415–423 (2019)
14. Kizilaslan, R., Karlik, B.: Comparison neural networks models for short term forecasting of natural gas consumption in Istanbul. In: 2008 First International Conference on the Applications of Digital Information and Web Technologies (ICADIWT), Ostrava, pp. 448–453 (2008)
15. Suykens, J., Lemmerljng, P.H., Favoreel, W., Crepe, M., Briol, P.: Modelling the Belgian gas consumption using neural networks. Neural Process. Lett. **4**, 157–166 (1996)
16. Laib, O., Khadir, M.T., Chouired, L.: Forecasting yearly natural gas consumption using artificial neural network for the Algerian market. In: 2016 4th International Conference on Control Engineering and Information Technology (CEIT), Hammamet, pp. 1–5 (2016)
17. Hsu, Y.Y., Tung, T.T., Yeh, H.C., Lu, C.N.: Two-stage artificial neural network model for short-term load forecasting. IFAC-PapersOnLine. **51**, 678–683 (2018)
18. Jetcheva, J.G., Majidpour, M., Chen, W.P.: Neural network model ensembles for building-level electricity load forecasts. Energy Build. **84**, 214–223 (2014)
19. Aulinas, M., Turon, C., Sànchez-Marrè, M.: Agents as a decision support tool in environmental processes: the state of the art. In: Whitestein Series in Software Agent Technologies and Autonomic Computing, pp. 5–35 (2009)

20. Ghazi, S., Dugdale, J., Khadir, M.T.: Modelling air pollution crises using multi-agent simulation. In: 2016 49th Hawaii International Conference on System Sciences (HICSS), Koloa, HI, pp. 172–177 (2016)
21. Dragomir, E.G., Oprea, M.: A multi-agent system for power plants air pollution monitoring. IFAC Proc. Vol. **46**, 89–94 (2013)
22. El Fazziki, A., Benslimane, D., Sadiq, A., Ouarzazi, J., Sadgal, M.: An agent based traffic regulation system for the roadside air quality control. IEEE Access **5**, 13192–13201 (2017)
23. Farfar, K.E., Khadir, M.T.: A two-stage short-term load forecasting approach using temperature daily profiles estimation. Neural Comput. Appl. 1–11 (2018)
24. https://www.epa.gov/energy/greenhouse-gases-equivalencies-calculator-calculations-and-references. Last accessed: 2019/01/26

Optimized Type-2 Fuzzy Logic PSS Combined with $H\infty$ Tracking Control for the Multi-machine Power System

Khaddouj Ben Meziane and Ismail Boumhidi

Abstract In this paper, an optimized type-2 fuzzy logic based on power system stabilizer combined with the optimal $H\infty$ tracking control has been developed to design intelligent controllers for improving and enhancing the performance of stability for the multi-machine power system. The type-2 fuzzy logic based on interval value sets is capable for modeling the uncertainty and to overcome the drawbacks of the conventional power system stabilizer. The scaling factors of the type-2 fuzzy logic are optimized with the particle swarm optimization algorithm to obtain a robust controller. The optimal $H\infty$ tracking control guarantees the convergence of the errors to the neighborhood of zero. The simulation results show the damping of the oscillations of the angle and angular speed with reduced overshoots and quick settling time.

Keywords Multi-machine power system · Interval type-2 fuzzy logic · Particle swarm optimization · Power system stabilizer · $H\infty$ tracking control

1 Introduction

With the increasing electric power demand, power systems are bound to stress conditions, which result in undesirable system voltage and frequency. For many years, power system stabilizers (PSSs) have been the choice of control to damp out oscillations and to offset the negative damping of automatic voltage regulators [1].

K. Ben Meziane (✉)
Department of Engineering, Higher Institute of Engineering and Business (ISGA), Fez, Morocco
e-mail: Khaddouj.benmeziane@isga.ma

I. Boumhidi
Department of Physics, LESSI Laboratory, Faculty of Sciences, University Sidi Mohamed Ben Abdellah, Fez 30000, Morocco
e-mail: iboumhidi@hotmail.com

V. Bhateja et al. (eds.), *Embedded Systems and Artificial Intelligence*,
Advances in Intelligent Systems and Computing 1076,
https://doi.org/10.1007/978-981-15-0947-6_19

Nowadays PSS has possibility to be designed and implemented with a new controller based on a modern and sophisticated technology such as artificial intelligent methods, where, the fuzzy logic is really interesting, because it does not need mathematical approach to give control solution [2]. The vast majority of the fuzzy logic controllers (FLC) that have been used to date were based on the traditional type-1 fuzzy logic. However, type-1 fuzzy logic control cannot fully handle or accommodate the linguistic and numerical uncertainties associated with dynamic unstructured environments as they use type-1 fuzzy sets. Where as a possible alternative is using type-2 fuzzy logic controllers (type-2 FLCs) that make use of type-2 fuzzy sets [3]. Type-2 fuzzy logic systems (T2-FLS) are used successfully in many application areas because they offer more advantages in handling uncertainty with respect to type-1 fuzzy systems [4]. Type-2 fuzzy sets, originally presented by Zadeh in 1975 and developed by Mendel and Liang [5, 6] allowing the characterization of a type-2 fuzzy set with a superior membership function and an inferior membership function. Type-2 fuzzy sets can deal with models' uncertainties efficiently because of its three-dimensional membership functions, but selecting suitable parameters of membership functions is not easy. The particle swarm optimization algorithm (PSO) is adopted to find the optimal parameters of the membership functions of the type-2 fuzzy system, as it is easy to understand and simple to implement [7].

In this paper, the optimized type-2 fuzzy logic controller is proposed for designing a power system stabilizer PSS (PSO-T2FPSS) combined with the robust $H\infty$ tracking control to increase the stability of the power system. A speed deviation and its derivative are selected as type-2 fuzzy controller inputs. It is well known that robust control provides an effective approach to deal with uncertainties introduced by variations of system parameters as well as changes of operating conditions [8]. The $H\infty$ control performance for uncertain nonlinear systems is proposed to attenuate the effects caused by modeled dynamics, disturbances and approximate errors. The robust $H\infty$ tracking control has a simplified structure, regulate the output amplitude of the angle rotor and the angular speed deviation to a desired value, and attenuate the oscillations [9, 10]. The main objective of this paper is developing the optimal $H\infty$ tracking control combined with the optimized type-2 fuzzy logic as a power system stabilizer (PSO-T2FPSS) to guarantee the robust performance for damping low-frequency oscillations and to enhance the stability in the multi-machine power system. The effectiveness of the proposed controller (PSO-T2FPSS & $H\infty$) is tested and compared with the interval type-2 fuzzy logic power system stabilizer (T2FPSS) and the (T1FPSS) on the three-machine nine-bus power system. The simulation results show the validity and robustness of the proposed control which improves the oscillation damping.

2 Interval Type-2 Fuzzy Logic System

Membership function in interval type-2 fuzzy logic set as an area called footprint of uncertainty (FOU) which limited by two type-1 membership function those are: upper membership function (UMF) and lower membership function (LMF) [11, 12]. Interval type-2 membership function is shown in Fig. 1 [7].

The output type-2 fuzzy sets are processed by the type-reducer which combines the output sets and then performs a centroid calculation, which leads to type-1 fuzzy sets called the type-reduced set. The defuzzifier can then defuzzify the type-reduced type-1 fuzzy outputs to produce crisp outputs [13, 14].

$$\begin{aligned} Y_{\cos}(X) &= [y_l, y_r] \\ &= \int_{y^l \in [y_l^l, y_r^l]} \cdots \int_{y^M \in [y_l^M, y_r^M]} \int_{f^l \in \left[\underline{f}^l, \overline{f}^l\right]} \cdots \int_{f^M \in \left[\underline{f}^M, \overline{f}^M\right]} 1 \Bigg/ \frac{\sum_{i=1}^{M} f^i y^i}{\sum_{i=1}^{M} f^i} \end{aligned} \tag{1}$$

where:

$$y_l = \frac{\sum_{i=1}^{M} f_l^i y_l^i}{\sum_{i=1}^{M} f_l^i}; \quad y_r = \frac{\sum_{i=1}^{M} f_r^i y_r^i}{\sum_{i=1}^{M} f_r^i} \tag{2}$$

The values of y_l and y_r define the output interval of the type-2 fuzzy system, which can be used to verify if training or testing data are contained in the output of the fuzzy system [6].

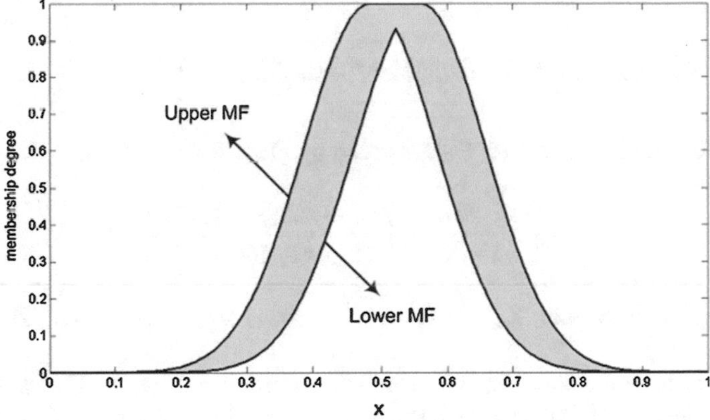

Fig. 1 Membership function interval type-2 fuzzy logic set

3 Proposed Design Procedure and Optimization

3.1 *Particle Swarm Optimization*

The PSO method employs an appropriate model of fitness for particle, to evaluate fitness value of each particle and to record the particle that is the highest fitness value in an iterative procedure [15]. The position corresponding to the best fitness is known as *pbest* and the overall best out of all the particles in the population is called *gbest* [16]. The modified velocity and position of each particle can be calculated using the current velocity and the distance from the $pbest_{j,g}$ to $gbest_g$ as shown in the following formulas [17]:

$$v_{j,g}^{(t+1)} = w.v_{j,g}^{(t)} + c_1 r_1 \left(pbest_{j,g} - x_{j,g}^{(t)} \right) + c_2 r_2 \left(gbest_g - x_{j,g}^{(t)} \right) \quad (3)$$

$$x_{j,g}^{(t+1)} = x_{j,g}^{(t)} + v_{j,g}^{(t+1)}; \quad j = 1, 2, \ldots, n \text{ and } g = 1, 2, \ldots, m$$

W is the inertia weight, which produces a balance between global and local explorations requiring less iteration on average to find a suitably optimal solution. It is determined by the following equation [15]:

$$w = w_{\max} - \frac{w_{\max} - w_{\min}}{\text{iter}_{\max}} \cdot \text{iter} \quad (4)$$

where $w_{\max}$ the initial weight is $w_{\min}$ is the final weight, iter is the current iteration number, is the maximum iteration number.

3.2 *Optimization of Scaling Factors of T2FPSS*

The transfer function of the *i*th PSS is given by [18]:

$$u_{\text{pssi}}(s) = K_{\text{pss}_i} \frac{sT_{Wi}}{1 + sT_{Wi}} \left(\frac{(1 + sT_{1i})(1 + sT_{3i})}{(1 + sT_{2i})(1 + sT_{1i})} \right) \Delta\omega_i(s) \quad (5)$$

where K_{pss_i} = PSS gain, T_{w_i} Washout Time constant, T_{1i}, T_{2i}, T_{3i}, T_{4i} = Time constants.

In this article, the lead-lag PSS controller is replaced by the IT2FL-PSS. The IT2FLC is designed with the help of the interval type-2 fuzzy logic toolbox (IT2FLT) in the MATLAB software [11, 12]. The Gaussian membership function of type-2 is chosen for the two inputs $\Delta\omega$, $\Delta\omega'$, and the output U_{T2FPSS} [19, 20]. A total of 49 rule bases are designed for the optimal performance of the controller which is shown in Table 1 [21].

Table 1 Type-2 fuzzy logic rules

de(t) *e(t)*	NB	NM	NS	Z	PS	PM	PB
NB	NB	NB	NB	NM	NM	NS	Z
NM	NB	NB	NM	NM	NS	Z	PS
NS	NB	NM	NS	NS	Z	PS	PM
Z	NM	NM	NS	Z	PS	PM	PM
PS	NM	NS	Z	PS	PS	PM	PL
PM	NS	Z	PS	PM	PM	PB	PB
PB	Z	PS	PM	PM	PB	PB	PB

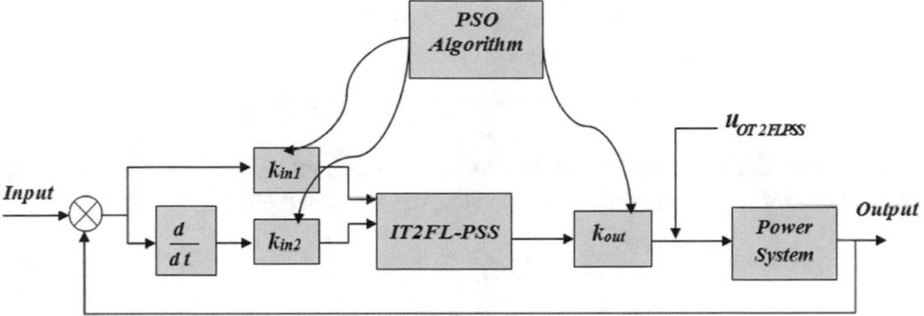

Fig. 2 Tuning for input-output scaling factors of T2FPSS using PSO algorithm

The optimized type-2 fuzzy logic control based on PSS has two Inputs, which are the error and derivative of error and one output as depicted in Fig. 2.

The proposed design procedure is optimizing the input-output scaling factors (k_{in1}, k_{in2} and k_{out}) of the type-2 fuzzy controller. In this case, the membership functions are fixed over a unity universe of discourse of interval $[-1, 1]$.

4 The Proposed Control Design

The state representation of the ith machine of a multi-machine power system can be written in the following form: $x_i = [x_{i1}, x_{i2}, x_{i3}]^{\text{T}} = \left[\delta_i, \omega_i, E'_{qi}\right]$

For $i = 1, 2,\ldots, n$, represents the state vector of ith subsystem, and the control applied is given by:

$$\begin{aligned}\dot{x}_{i1} &= x_{i2}\\ \dot{x}_{i2} &= f_{i1}(X)\\ \dot{x}_{i3} &= f_{i2}(X) + u_i\end{aligned} \tag{6}$$

$$u_i = \frac{1}{T'_{di}} E_{fi} \tag{7}$$

$$f_{i1}(X) = a_i - b_i x_{i2} - c_i x_{i3}^2 - d_i x_{i3} \sum_{j=1, j \neq i}^{n} x_{j3} \begin{Bmatrix} G_{ij} \cos(x_{j1} - x_{i1}) \\ -B_{ij} \sin(x_{j1} - x_{i1}) \end{Bmatrix} \tag{8}$$

$$f_{i2}(X) = -e_i x_{i3} + h_i \sum_{j=1, j \neq i}^{n} x_{j3} \begin{Bmatrix} G_{ij} \sin(x_{j1} - x_{i1}) \\ +B_{ij} \cos(x_{j1} - x_{i1}) \end{Bmatrix} \tag{9}$$

The detail of the parameters (a_i, b_i, c_i, d_i, e_i and h_i) is given in appendix. The tracking error is defined as:

$$\begin{aligned} e_{1i} &= \delta_i - \delta_{ir} = x_{i1} - x_{i1r} \\ e_{2i} &= \dot{e}_{1i} = x_{i2} \\ e_{3i} &= \dot{e}_{2i} = a_i - b_i x_{i2} - c_i x_{i3}^2 - d_i x_{i3} I_{q_i} \end{aligned} \tag{10}$$

Then our design objective is to impose $H\infty$ control so that the following asymptotically stable tracking. In this study, the relative degree is $r = 3$ then

$$e_i^{(3)} + k_{2_i} e_i^{(2)} + k_{1i} \dot{e}_i = 0 \tag{11}$$

For $i = 1, 2 \ldots n$. Where $k = [k_{1i}, k_{2i}, 1]^{\mathrm{T}}$ are the coefficients of the Hurwitz Polynomial:

$$h_i(\lambda) = \lambda^2 + k_{2i}\lambda + k_{1i} \tag{12}$$

From (11), we have:

$$f_i(x) + g_i(x) u_i + k_{2i} \ddot{e}_i + k_{1i} \dot{e}_i = 0 \tag{13}$$

If $f_i(x)$ and $g_i(x)$ are known, we can easily construct the nominal control. From (13), we have:

$$u_{eq_i} = \frac{-1}{g_i(x)} (f_i(x) + k_{1i} \dot{e}_i + k_{2i} \ddot{e}_i) \tag{14}$$

The output tracking error dynamic equation of nonlinear system is described by:

$$\dot{e}_i = A_i e_i + B_i [g_i(x).u_i + u_{hi}] \tag{15}$$

where

$$A_i = \begin{pmatrix} 0 & 1 & 0 & .. & 0 \\ 0 & 0 & 1 & 0. & 0 \\ . & . & . & .. & . \\ -k_{1i} & -k_{2i} & -k_{3i} & .. & -k_{(n-1)i} \end{pmatrix} \text{ and } B_i = \begin{bmatrix} 0 & 0 & 0 & 1 \end{bmatrix}$$

where u_{hi} is a $H\infty$ compensator, defined as:

$$u_{\text{hi}} = -\frac{1}{g_i(x) r_i} e_i^{\text{T}} P_i B_i \tag{16}$$

where r is a positive scalar value and $P = P^{\text{T}} > 0$ is the P solution of the following Riccati-like equation [9]. The control law used in this study is given as:

$$u_i = \frac{-1}{g_i(x)} \left(f_i(x) + k_{1i} \dot{e}_i + k_{2i} \ddot{e}_i + \frac{1}{r_i} e_i^{\text{T}} P_i B_i \right) + u_{(\text{PSO - T2FPSS})_i} \tag{17}$$

5 Simulation of Multi-machine Power System

To evaluate the performance of the proposed control, we performed simulation for the three-machine nine-bus power system as in Fig. 3. Detail of the system data is given in Table 2 [22].

The following equilibrium point: $X_{ir} = (x_{i1r}, x_{i2r}, x_{i3r}) = \begin{bmatrix} \delta_i & \Delta\omega_i & E'_{qi} \end{bmatrix}$. For i = 1, 2, 3 of the three-machine system is considered:

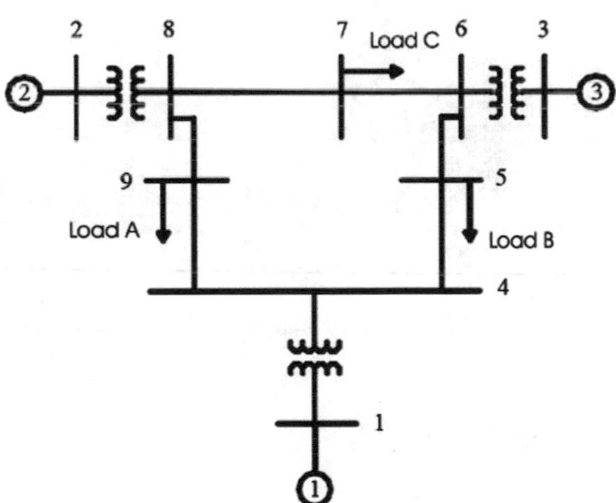

Fig. 3 Three-machine nine-bus power system

Table 2 Nominal parameters values

Parameters	Gen1	Gen2	Gen3
H	23.64	6.4	3.01
X_d	0.146	0.8958	1.3125
X'_d	0.0608	0.7798	0.1813
D	0.3100	0.5350	0.6000
P_m	0.7157	1.6295	0.8502
T'_{do}	8.96	6.0	5.89

$$\text{EP1}: \begin{cases} x_{11r} = 0.0396, \ x_{12r} = 0, \ x_{13r} = 1.0566 \\ x_{21r} = 0.3444, \ x_{22r} = 0, \ x_{23r} = 1.0502 \\ x_{31r} = 0.2300, \ x_{32r} = 0, \ x_{33r} = 1.017 \end{cases}$$

From simulation results, the T1FPSS controller requires more time and more oscillations before the same variables are stabilized compared with T2FPSS. Thus, optimizing the inputs and outputs scaling factors of T2FPSS using PSO algorithm enables to improve the response time and reduce the oscillations. Therefore, the combination between the PSO-T2FPSS and $H\infty$ tracking control in this section has demonstrated a superior performance of the proposed control in terms of the stability and the reduction of the overshoots (Figs. 4, 5, 6, 7, 8 and 9).

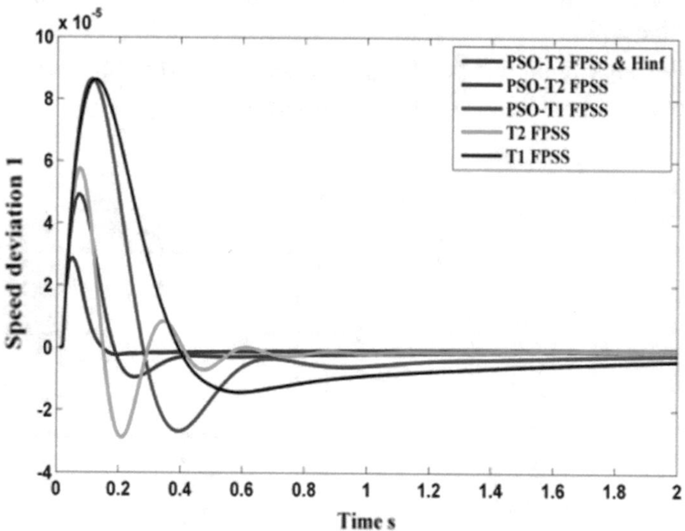

Fig. 4 Speed deviation $\Delta\omega_1$

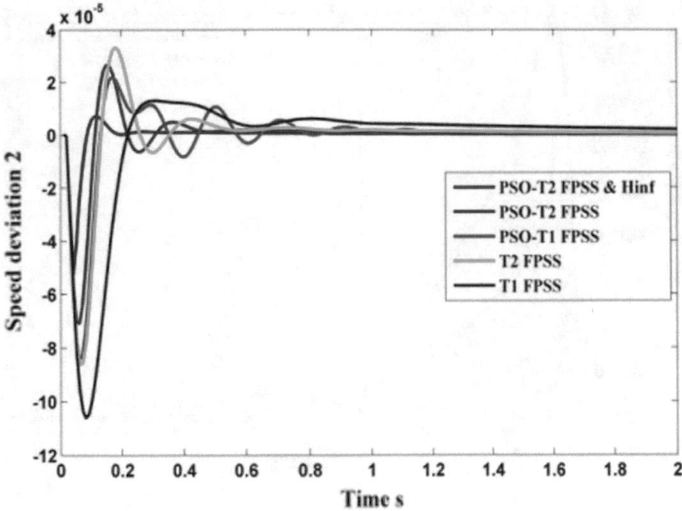

Fig. 5 Speed deviation $\Delta\omega_2$

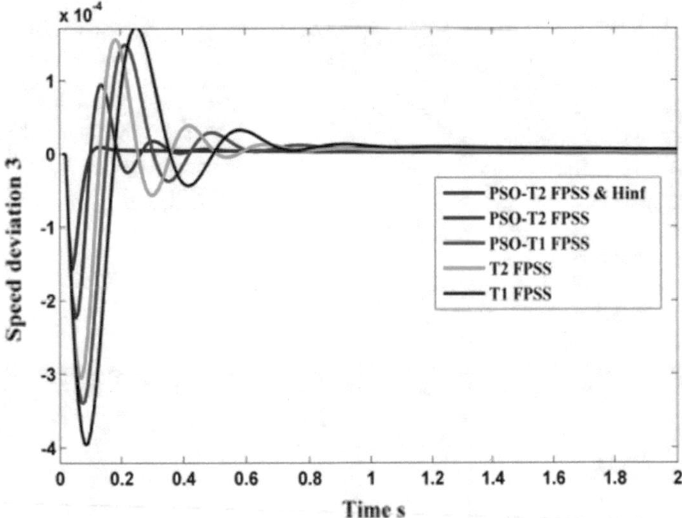

Fig. 6 Speed deviation $\Delta\omega_3$

6 Conclusion

In this paper, the optimized type-2 fuzzy logic based on PSS combined with $H\infty$ tracking control (PSO-T2FPSS & $H\infty$) provides an efficient solution to damp the low-frequency oscillations in the multi-machine power system. In order to find the

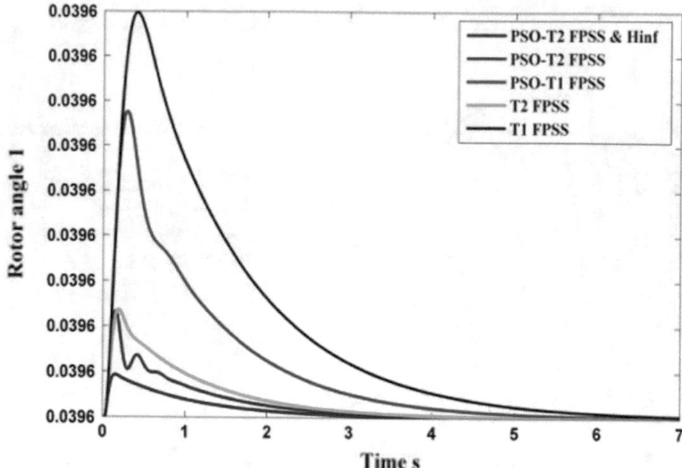

Fig. 7 Rotor angle δ_1

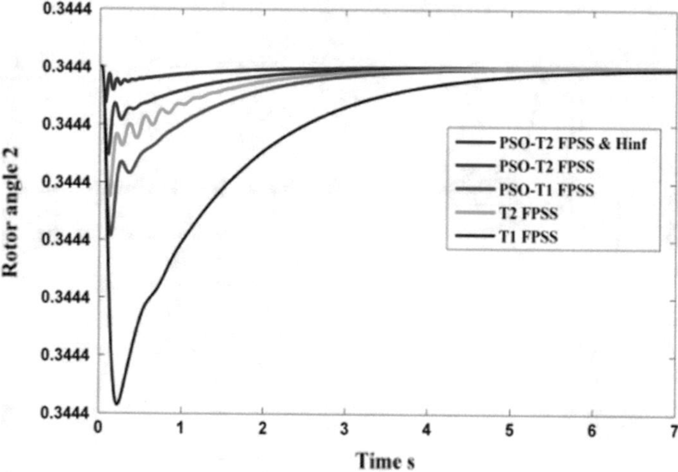

Fig. 8 Rotor angle δ_2

better controller, PSO algorithm has been used to tune the input-output scaling factors of the type-2 fuzzy logic control based on PSS. It was shown that the proposed control can guarantee the overall stability and the robust performance of the multi-machine power systems.

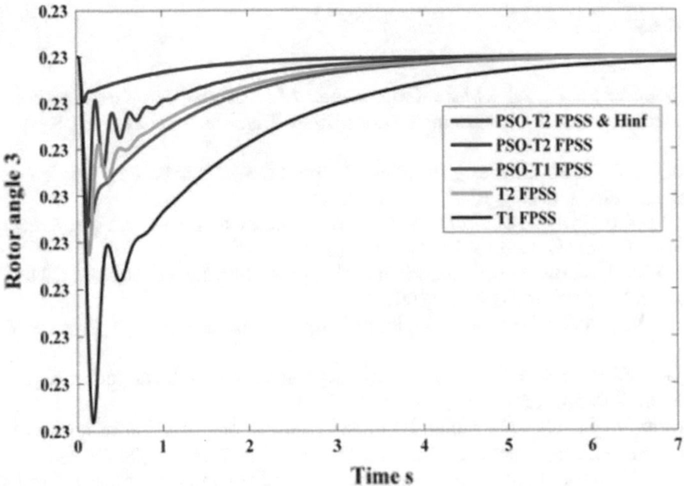

Fig. 9 Rotor angle δ_3

Appendixes

The dynamical model of the *i*th machine is represented by the classical third-order model [17].

$$\begin{cases} \dot{\delta}_i = \omega_i - \omega_s \\ \dot{\omega}_i = \dfrac{\omega_s}{2H_i}\left(Pm_i - D_i(\omega_i - \omega_s) - E'_{qi}I_{qi}\right) \\ \dot{E'}_{qi} = \dfrac{1}{T'_{di}}\left(E_{fi} - E'_{qi} - \left(X_{di} - X'_{di}\right)I_{di}\right) \end{cases}$$

$$\begin{cases} I_{qi} = G_{ii}E'_{qi} + \sum\limits_{j=1, j\neq l}^{n} E'_{qi}\left\{G_{ij}\cos\left(\delta_j - \delta_i\right) - B_{ij}\sin\left(\delta_j - \delta_i\right)\right\} \\ I_{di} = -B_{ii}E'_{qi} - \sum\limits_{j=1, j\neq i}^{n} E'_{qi}\left\{G_{ij}\sin\left(\delta_j - \delta_i\right) + B_{ij}\cos\left(\delta_j - \delta_i\right)\right\} \end{cases}$$

The detailed parameters of the multi-machine power system:

$$a_i = \frac{\omega_s}{2H_i}Pm_i \quad ; \quad b_i = \frac{\omega_s}{2H_i}D_i \quad ; \quad c_i = \frac{\omega_s}{2H_i}G_{ii}$$

$$d_i = \frac{\omega_s}{2H_i} \quad ; \quad e_i = \frac{\left(1 - \left(X_{di} - X'_{di}\right)B_{ii}\right)}{T'_{di}}; \quad h_i = \frac{X_{di} - X'_{di}}{T'_{di}}$$

References

1. Eslam, M., Shareef, H., Taha, M.R., Khajehzadeh, M.: Adaptive particle swarm optimization for simultaneous design of UPFC damping controllers. Electric. Power Energy Syst. **57**, 116–128 (2014)
2. Robandi, I., Kharisma, B.: Design of interval type-2 fuzzy logic based power system stabilizer. Int. J. Electr. Electron. Eng. **3**, 593–600 (2009)
3. Miccio, M., Cosenza, B.: Control of a distillation column by type-2 and type-1 fuzzy logic PID controllers. J. Process. Control **24**, 475–484 (2014)
4. Mendel, J.M.: Uncertainty Rule Based Fuzzy Logic Systems: Introduction and New Directions. Prentice-Hall, Upper-Addle River (2001)
5. Mendel, J.M.: Uncertainty, fuzzy logic, and signal processing. Sign. Process. **80**, 913–933 (2000)
6. Liang, Q.L., Mendel, J.M.: Interval type-2 fuzzy logic systems: theory and design. IEEE Trans. Fuzzy Syst. **8**(5), 535–550 (2000)
7. Maldonado, Y., Castillo, O., Melin, P.: Particle swarm optimization of interval type-2 fuzzy systems for FPGA applications. Appl. Soft Comput. **13**, 496–508 (2013)
8. Saoudi, K., Harmas, M.N.: Enhanced design of an indirect adaptive fuzzy sliding mode power system stabilizer for multi-machine power systems. Electr. Power Energy Syst. **54**, 425–431 (2014)
9. Chen, B.S., Lee, C.H., Chang, Y.C.: $H\infty$ tracking design of uncertain nonlinear SISO systems: adaptative fuzzy approach. IEEE Trans. Fuzzy Syst. **4**, 32–43 (1996)
10. Wenlei, L., Peter, X.L.: Adaptive tracking control of an MEMS gyroscope with H-infinity performance. J. Control Theory Appl. **9**, 237–243 (2011)
11. Mendel, J.M., John, R.I., Liu, F.L.: Interval type-2 fuzzy logic systems made simple. IEEE Trans. Fuzzy Syst. **14**(6), 808–821 (2006)
12. Castillo, O., Martinez, A.I., Martinez, A.C.: Evolutionary computing for topology optimization of type-2 fuzzy systems. Adv. Soft Comput. **41**, 63–75 (2007)
13. Yesil, E.: Interval type-2 fuzzy PID load frequency controller using big bang-big crunch optimization. Appl. Soft Comput. **15**, 100–112 (2014)
14. Khaddouj, B.M., Ismail, B.: An interval type-2 fuzzy logic PSS with the optimal $H\infty$ tracking control for multi-machine power system. Intell. Syst. Comput. Vision (ISCV) IEEE (2015) https://doi.org/10.1109/isacv.2015.7106188
15. Gaing, Z.L.: A particle swarm optimization approach for optimum design of PID controller in AVR system. IEEE Trans. Energy Convers. **19**, 384–391 (2004)
16. Liang, R.H., Tsai, S.R., Chen, Y.T., Tseng, W.T.: Optimal power flow by a fuzzy based hybrid particle swarm optimization approach. Electr. Power Syst. Res. **81**, 1466–1474 (2011)
17. Zineb, L., Khaddouj, B.M., Ismail, B.: Sliding mode controller based on type-2 fuzzy logic PID for a variable speed wind turbine. Int. J. Syst. Assur. Eng. Manag. 1–9 (2019). https://doi.org/10.1007/s13198-019-00767-z
18. Castillo, O., Melin, P.: Optimization of type-2 fuzzy systems based on bio-inspired methods: a concise review. Inf. Sci. **205**, 1–19 (2012)
19. Sanaye-Pasand, M., Malik, O.P.: A fuzzy logic based PSS using a standardized rule table. Electr. Mach. Power Syst. **27**, 295–310 (1999)
20. Braae, M., Rutherford, D.A.: Selection of parameters for a fuzzy logic controller. Fuzzy Sets Syst. **2**, 185–199 (1979)
21. Sambariya, D.K., Prasad, R.: Evaluation of interval type-2 fuzzy membership function and robust design of power system stabilizer for SMIB power system. Sylwan Journal **158**, 289–307 (2014)
22. Colbia-Vegaa, A., Leon-Morales, J., Fridman, L., Salas-Pena, O., Mata-Jimenez, M.T.: Robust excitation control design using sliding-mode technique for multimachine power systems. Electr. Power Syst. Res. **78**, 1627–1634 (2008)

Optimal Sliding Mode Control of a DC Motor Velocity Based on Neural Network and Genetic Algorithm

Mohammed Chakib Sossé Alaoui, Hassan Satori and Khalid Satori

Abstract In this paper, an optimal sliding mode control based on neural network and genetic algorithm are designed for a DC motor velocity. The classical sliding mode control (SMC) can be used for the considered system. However, it presents some drawbacks of chattering, due to the higher needed switching gain in the case of large uncertainties. In order to reduce this gain, the neural network is used for the prediction of model unknown parts and hence enable a lower switching gain to be used. The neural network (NN) is used to improve the nominal model and then reduce the model uncertainties. This enables the sliding mode technique to be used without any chattering problems. The genetic algorithm is used in this study to optimize both, the learning rate of backpropagation algorithm used by the neural network and the variable switching gain of the SMC. The performance of the proposed approach is investigated in simulations by the comparison of the proposed approach with the classical sliding mode control technique.

Keywords DC/DC converter · Pumping system · Sliding mode · Neural network · Genetic algorithm

---

M. C. Sossé Alaoui (✉)
LPESD Laboratory of Pedagogical Engineering and Science Didactics, Regional Center of Education and Training Careers (CRMEF), Kuwait Street, BP 49, 30000 Fez, Morocco

H. Satori · K. Satori
LIIAN Laboratory, Department of Informatics, Faculty of Sciences, Sidi Mohammed Ben Abdellah University, BP 2626, 30000 Fez, Morocco

V. Bhateja et al. (eds.), *Embedded Systems and Artificial Intelligence*, Advances in Intelligent Systems and Computing 1076,
https://doi.org/10.1007/978-981-15-0947-6_20

1 Introduction

Switched-mode DC/DC power converters are used in a wide variety of application, including DC motor drive. Several control strategies have been used for the control of DC/DC converters, especially those of PI and PID types [1, 2]. In this paper, due to the nonlinear of the centrifugal pump, the considered system is described by nonlinear uncertain model. Because it is robust character, the sliding mode approach [3–6] is then preferred. The control action consists of the called equivalent control and robust control component [3, 7], where the equivalent control law is used to make the undisturbed system state to slide in sliding surface and the robust control term is used to force the system state to the so-called sliding surface. However, in the presence of large uncertainties, the duty cycle control of the converter needs a higher switching gain and produces higher amplitude of chattering. In this paper, the NN [8–11] is used for the prediction of model unknown part which is incorporated in the equivalent control and hence enables a lower discontinuous switching gain to be used. To guarantee, an optimal behavior of the processes control, the genetic algorithm is used to optimize both the learning rate of backpropagation algorithm used by the NN and the variable switching gain of the SMC [12, 13].

This paper is organized as follows: Sect. 2 describes the combined DC motor, buck converter with centrifugal pump model and the mathematical model of the system. Section 3 presents the proposed optimal sliding mode controller based on combining neural network with the genetic algorithm. In Sect. 4, we present the simulation results and a concluding remark is given at the end of the paper.

2 System Model

The considered system model as shown in Figs. 1 and 2 can be deduced for the combination of the converter of the “buck” type model, electrical motor model and the centrifugal pump model [14].

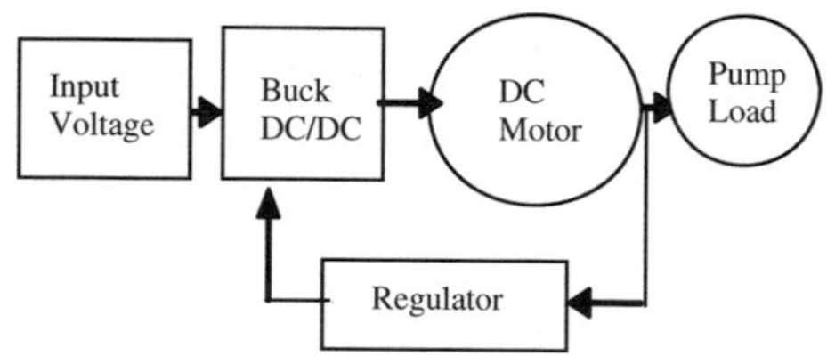

Fig. 1 General diagram

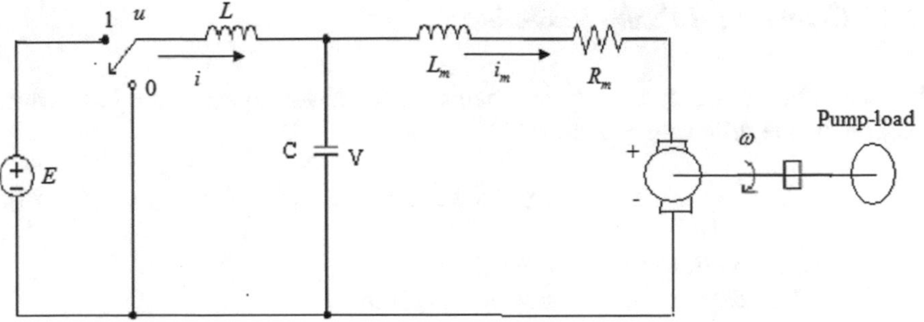

Fig. 2 Combined pump load-DC-motor "buck" converter model

2.1 Power Electronic Converter Model

The power electronic power converter is a "buck" type which is inserted between the generator of input voltage and the motor inductor. Its duty cycle noted u may only take values in the interval [0, 1] of the real line. The state equations of the converter are:

$$L\frac{\mathrm{d}i}{\mathrm{d}t} = -V + uE \quad \text{and} \quad C\frac{\mathrm{d}V}{\mathrm{d}t} = i - i_m \tag{1}$$

2.2 Electrical Motor Modeling

We consider a DC motor with a constant magnetic flux and we neglect the magnetic reaction and the commutation phenomenon. The dynamic model of a DC motor with constant magnetic flux [15] can be expressed as follows:

$$V = R_m i_m + L_m\frac{\mathrm{d}i_m}{\mathrm{d}t} + K_e \cdot \omega \tag{2}$$

The motor torque is:

$$C_m = K_m i_m \tag{3}$$

The parameters of the DC motor are K_e the back emf, K_m the torque constant, L_m the armature inductance, R_m the armature resistance and ω the rotation speed.

2.3 Centrifugal Pump Modeling

The centrifugal pump opposes to the motor a resistant hydrodynamic torque C_r which is given by the following equation [16].

$$C_r = K_r \cdot \omega^2 \tag{4}$$

K_r is the proportionality coefficient.
The mechanical equation of the system is given by:

$$C_m - C_r = J\frac{\mathrm{d}\omega}{\mathrm{d}t} \tag{5}$$

J is the group inertia.

2.4 Mathematical Model of the System

The combination of the different equations describing the system leads to the following model:

$$\begin{gathered} L\tfrac{\mathrm{d}i}{\mathrm{d}t} = -V + uE,\ \boldsymbol{C}\tfrac{\mathbf{d}\boldsymbol{V}}{\mathbf{d}\boldsymbol{t}} = \boldsymbol{i} - \boldsymbol{i_m} \ \text{ and } \ L_m\tfrac{\mathrm{d}i_m}{\mathrm{d}t} = V - R_m i_m - K_e\omega \\ J\tfrac{\mathrm{d}\omega}{\mathrm{d}t} = -B_m\omega + K_m i_m - K_r\omega^2 \end{gathered} \tag{6}$$

Let us consider $x_1 = \omega$. We have:

$$\begin{cases} \dot{x}_1 = x_2 \\ \dot{x}_2 = x_3 \\ \dot{x}_3 = x_4 \\ \dot{x}_4 = f(x_1, x_2, x_3, x_4) + \Delta f + gu \end{cases} \tag{7}$$

The output is: $y = x_1 = \omega$; and Δf is the uncertainties part.
We denote the desired output $y_d(t) = \omega_d$, the tracking error:

$$\begin{aligned} e(t) =& y(t) - y_d(t) = \omega - \omega_d. \\ f(x_1, x_2, x_3, x_4) =& -\,(a_4x_4 + a_3x_3 + a_2x_2 + a_1x_1 + a_{14}x_1x_4 \\ & + a_{23}x_2x_3 + a_{12}x_1x_2 + a_{13}x_1x_3 + a_{22}x_2^2 + a_{11}x_1^2) \end{aligned} \tag{8}$$

$g = b \cdot E,\ \ b = \dfrac{K_m}{LL_mCJ},\ a_4 = \left(\dfrac{B_m}{J} + \dfrac{R_m}{L_m}\right), a_3 = \dfrac{1}{JL_m}(K_eK_m + R_mB_m) + \dfrac{1}{c}\left(\dfrac{1}{L_m} + \dfrac{1}{L}\right),$
$a_2 = \dfrac{B_m}{CJ}\left(\dfrac{1}{L_m} + \dfrac{1}{L}\right) + \dfrac{R_m}{LL_m},\ a_1 = \dfrac{1}{JCLL_m}(K_eK_m + R_mB_m), a_{23} = \dfrac{6K_r}{J},\ a_{14} = \dfrac{2K_r}{J},$
$a_{13} = \dfrac{2K_rR_m}{L_mJ}; a_{12} = \dfrac{2K_r}{L_mJC},\ a_{11} = \dfrac{R_mK_r}{JCLL_m}$

3 Sliding Mode Control Design Based on Neural Network and Genetic Algorithm

3.1 Traditional Sliding Mode Control

The main idea is to find a sliding mode controller for the system defined by (6) to ensure that the system remains in so-called sliding surface in the presence of large uncertainties and with no chattering. The sliding surface is chosen according to the output tracking error $\boldsymbol{e(t)}$ and the relative degree of the system. We define the relative degree $\boldsymbol{r}$ of the system to be the least positive integer $\boldsymbol{i}$ for which the derivative $\boldsymbol{y^{(i)}(t)}$ is an explicit function of the control law $\boldsymbol{u(t)}$, such that:

$$\frac{\partial y^{(r)}(t)}{\partial u} \neq 0 \text{ and } \frac{\partial y^{(i)}(t)}{\partial u} = 0 \text{ for } i = 0, \ldots.., r-1. \quad (9)$$

The switching function can be selected as follows:

$$\sigma(t) = e^{(r-1)}(t) + \beta_{r-2}e^{(r-2)}(t) + \cdots + \beta_1\dot{e}(t) + \beta_0 e(t) \quad (10)$$

where the coefficients $\beta_0 \ldots \beta_{r-2}$ are chosen so that the characteristic polynomial associated with $\sigma(t)$ have its roots strictly in the left half complex plane. Then, the output tracking error $e(t)$ tends asymptotically to zero in a finite time when σ tends to zero in a finite time.

In our application $r = 4$ and the switching function:

$$\sigma = e^{(3)} + 2.2e + 1.55\ddot{e} + \dot{0}.35e \text{ and } \dot{\sigma} = f + gu + 2.2\,x_4 + 1.55x_3 + 0.35x_2$$
$$\dot{\sigma} = F + gu \quad (11)$$

With: $F = f + 2.2x_4 + 1.55x_3 + 0.35x_2$.

To ensure that a sliding mode exists on a switching surface, one has to satisfy the condition given below [3]:

$$\sigma\dot{\sigma} < 0 \quad (12)$$

The duty cycle control law $u(t)$ that satisfies (8) and then forcing approximately the output rotation speed $\omega(t)$ to the desired rotation speed $\omega_d(t)$ is given as:

$$u = u_{\text{eq}} + u_r \quad (13)$$

with: $u_{\text{eq}} = -\frac{F}{g}$ and $u_r = -\frac{m}{g}\text{sat}(\sigma)$

Where $m > 0$ and

$$\text{sat}(\sigma) = \begin{cases} \text{sign}(\sigma) & \text{if}|\sigma| > \varepsilon \\ \frac{\sigma}{\varepsilon} & \text{if}|\sigma| \le \varepsilon \end{cases}$$

With ε is the boundary layer thickness, and m is the positive switching gain to compensate the uncertainties.

3.2 Neural Network Representation

The type of NN used in this study is the multilayer neural network, which is a type of predictive network. The architecture of the proposed network is shown in Fig. 3. The network input is the tracking error e, and the output variable is $\hat{f}$.

$$y_k(X) = W_k^T \sigma_c\left(W_j^T X\right), \quad \text{with } j = 1, \ldots, N \text{ and } k = 1.$$

where $\sigma_c(\cdot)$ represents the hidden-layer activation function considered in as a hyperbolic tangent function,

$$\sigma_c(s) = \tan h(s).$$

$W_k = [w_{k1} w_{k2} \ldots w_{kN}]^{\mathrm{T}}$ are the interconnection weights between the hidden and the output layers, and $W_j = \left[w_{j1} w_{j2} \ldots w_{jn}\right]^{\mathrm{T}}$ are the interconnection weights between the input and the hidden layers. N and n are, respectively, the number of nodes in hidden layer and the number of network inputs. The network weights are adjusted during online implementation by using the gradient descent. The network weights are adjusted during online implementation by using the gradient descent method. The essence of gradient descent consists of iteratively adjusting the weights in the opposite discrepancy according to: direction to the gradient of E, so as to reduce the

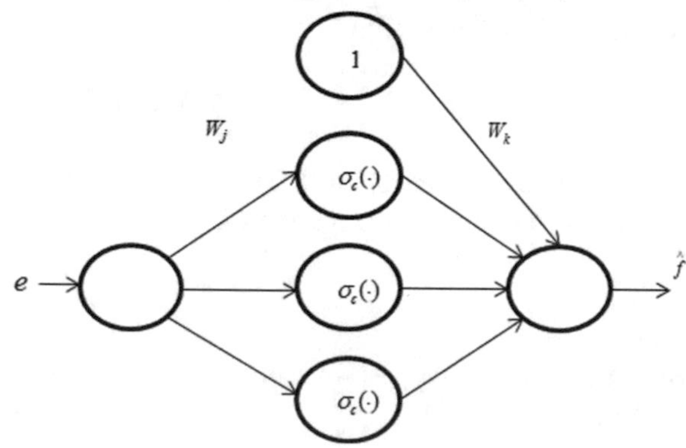

Fig. 3 Used neural network architecture

$$\frac{dw_{kj}}{dt} = -\eta_k \frac{\partial E}{\partial w_{kj}} \quad (14)$$

η_k is the optimal learning rate obtained by the genetic algorithm. The gradient where terms $\frac{\partial E}{\partial w_{kj}}$ can be derived using the backpropagation algorithm [17]. And the cost function E is defined as the error index and the least square error criterion is:

$$E = \frac{1}{2} e_{\text{NN}}^2 \quad (15)$$

Remark: Before incorporating the neural network into the proposed control strategy, the network was trained offline. The objective of offline training is to let the network learn the functional nonlinearities to a certain degree of accuracy before its implementation into the controller, and thus can give faster online adaptation as needed.

3.3 Proposed Optimal Neural Network Sliding Mode Controller Design

This method can be used for systems with small uncertainties. However, in the presence of large uncertainties, this approach produces chattering phenomenon due to the higher needed switching gain. In order to reduce this gain, we propose in this study to use neural network to estimate the unknown dynamics of the system, so that the system uncertainties can be kept small. The neural network prediction error is denoted as:

$$e_{\text{NN}} = f - \hat{f} \quad (16)$$

where $f(x, t) = f_n(x) + \Delta f_n(t)$, and $\hat{f}(x, t)$ is the neural network prediction of unknown nonlinear function f, given in the next section.

With $|e_{\text{NN}}| < e_{\text{NN}}^*$ and e_{NN}^* is the upper bound of the network error prediction assumed known.

Theorem 1 *Consider the system described by (7) in the presence of large uncertainties. If the system control is designed as:*

$$u = \hat{u}_{\text{eq}} - \frac{\alpha_1}{g} \text{sat}(\sigma) \quad (17)$$

where $\hat{u}_{\text{eq}} = -\frac{\hat{F}(x,t)}{g}$, with $e_{\text{NN}}^* < \alpha_1$ and α_1 is the optimal positive switching gain given by the genetic algorithm; the trajectory tracking error will converge to zero.

Proof Consider the candidate Lyapunov function:

$$\dot{V}_1 = \sigma(f_n(x) + \Delta f_n(t) + gu) = \sigma(f(x,t) + gu)$$

By replacing the expression of U given in the theorem we have:

$$\begin{aligned}\dot{V}_1 &= \sigma\Big(f(x,t) - \hat{f}(x,t) - \alpha_1 \mathrm{sat}(\sigma)\Big)\\ &= \sigma e_{\mathrm{NN}} - \alpha_1 \sigma pt \mathrm{sat}(\sigma) \leq |\sigma||e_{\mathrm{NN}}| - \alpha_1 \sigma\, \mathrm{sat}(\sigma)|\sigma| e^*_{\mathrm{NN}} - \alpha_1 \sigma\, \mathrm{sat}(\sigma)\end{aligned} \tag{18}$$

By choosing $e^*_{\mathrm{NN}} < \alpha_1$ we have: For any $L > 0$, if $\sigma > L$, $\mathrm{sat}(\sigma) = \mathrm{sign}(\sigma)$
The function

$$\dot{V}_1 = \big(e^*_{\mathrm{NN}} - \alpha_1\big)|\sigma| < -\eta_1|\sigma|. \tag{19}$$

However, in a small L-vicinity of the origin [4], (boundary layer), $\mathrm{sat}(\sigma) = \frac{\sigma}{L}$ is continuous, the system trajectories are confined to a boundary layer of sliding mode manifold $\sigma = 0$.

3.4 Genetic Algorithms Representation

Genetic algorithms (GAs) are search techniques and stochastic optimization based on genetics and mechanisms of natural selection and evolution [18, 8]. The GA starts with a randomly generated population of individuals, each one made of strings of the design variables, representing a set of points spanning the search space. Each individual is suitably coded into a chromosome made by a string of genes: each gene encodes one of the design parameters, by means of a string of bits, a real number or other alphabets. In order to evaluate and rank chromosomes in a population, a fitness function based on the objective function should be defined. New individuals are then generated by using some genetic operator [13], the classical ones being the crossover, the selection and the mutation [19].

In this study, two GAs have used in order to optimize the switching gain $\boldsymbol{\alpha}_1$ of the additive term and the neural network learning rate η_k. Figure 4 illustrates the architecture of each GA program. The parameters of two AGs are given in Table 1. The objective is to minimize the tracking error of the angular velocity, and the output error of the neural network, and then the fitness functions chosen are the inverse of each tracking error:

$$\mathrm{fitness}_1 = \frac{1}{|e|},\ \mathrm{fitness}_2 = \frac{1}{|e_{\mathrm{NN}}|} \tag{20}$$

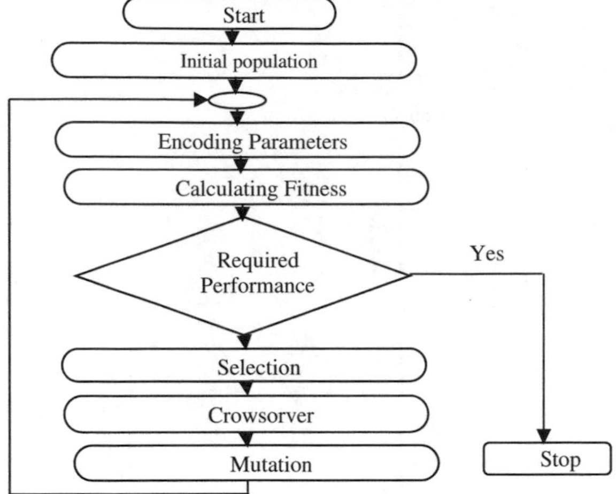

Fig. 4 Genetic algorithm architecture

Table 1 Parameters of two AGs

GA property	Value/method of:	
	Switching gain	Learning rate
Number of generations	10	10
Number of chromosomes in each generation	4	4
Chromosome length	8 bit	16 bit
Selection method	Roulette wheel selection	
Crossover method	Multipoint	

4 Simulation Result

For the shown simulations, we have set the following parameter values:

The considered DC motor is the DC direct-drive brushed motor with a case diameter of 42 mm and 42 W output power in 12 V DC versions. The motor type is: 828,500. The nominal characteristics:

$$V_n = 12\,\text{V}, i_n = 4.25\,\text{A}, L_m = 0.7\,\text{mH}, R_m = 0.81\,\Omega, K_m = K_e = 0.027\,\text{Nm/A}$$
$$K_r = (9.5113)10^{-7}\,\text{W(rad/s)}^{-3}, J = (0.14)10^{-4}\,\text{Kgm}^2, B_m = (4.4992)10^{-5}\,\text{Nm/rad}$$

Centrifugal pump parameters:

$$P_n = 32.5\,\text{W}, \omega_n = \omega_d = 324.4667\,\text{rad/s}.$$

Buck converter type: $L = 0.075$ H, $C = 0.090$ F.

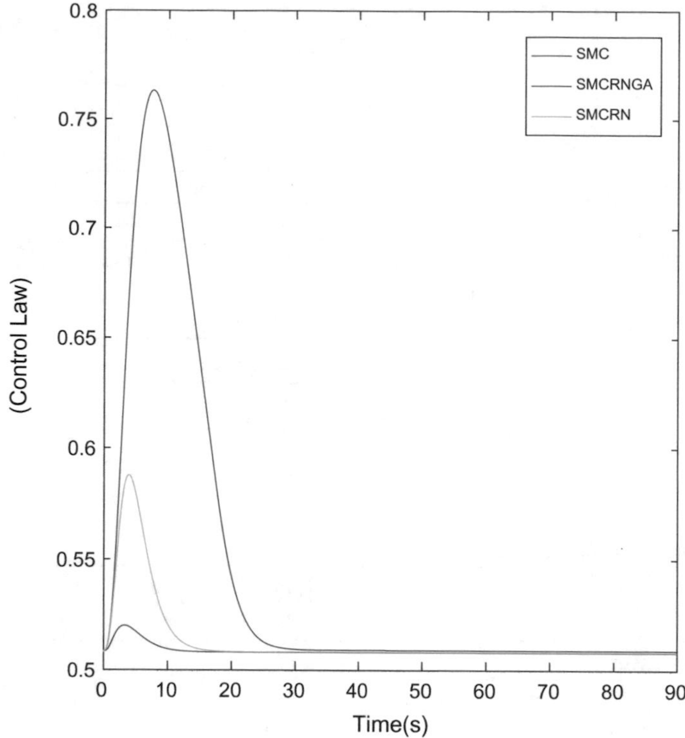

Fig. 5 Control law sliding mode (SMC) combined with neural network (SMCRN) and genetic algorithm (SMCRNGA)

Figure 6 shows the good tracking of the angular velocity to the nominal desired value which is (324.4667 rad/s); however, Fig. 5 shows that the corresponding duty cycle considered as robust control law where its value hold in the interval [0, 1].

5 Conclusion

This paper addressed the robust optimal tracking problem for a DC motor coupling with a centrifugal pump. The designed method is a combination of sliding mode control approach and neural network based on genetic algorithm. The latter has been employed to approximate the unknown nonlinear model function with online adaptation of parameters via BP learning algorithm. This provides a better description of the plant and hence enables a lower switching gain to be used despite the presence of large uncertainties. The GA is used to optimize the learning rate and switching gain.

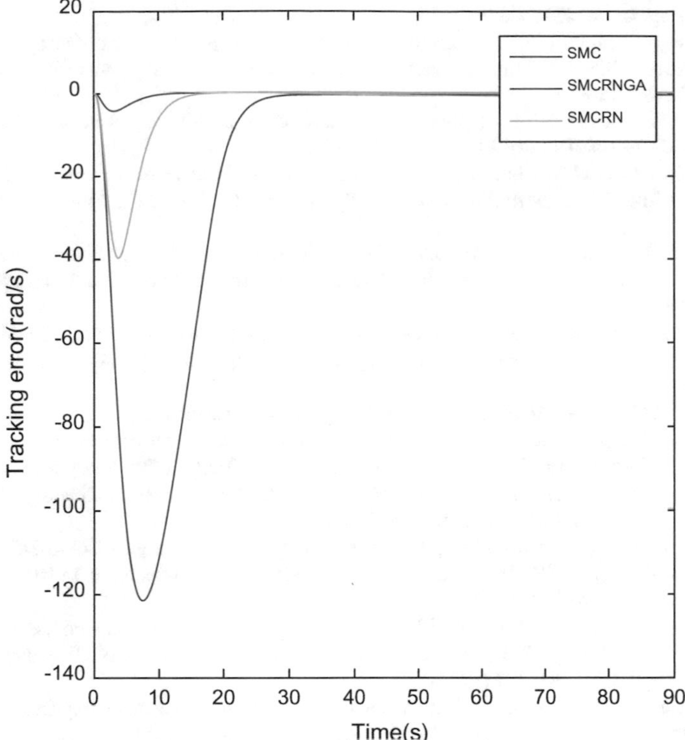

Fig. 6 Tracking error sliding mode (SMC) combined with neural network (SMCRN) and genetic algorithm (SMCRNGA)

The faster convergence of the proposed approach has been obtained. The comparison with the sliding mode control has been realized and simulation results have shown a good performance of the proposed method.

References

1. Jahmeebacus, M.I., Oolun, M.K., Bhurtun, C., Soyjaadah, K.M.S.: Speed sensorless control of a converter+fed Dc mtor. IEEE Trans. Ind. Elect. **43**(4), 492–497 (1996)
2. Millan, A.J., de Guzman, M.I.G., Guzman, V.M., Torres, M.: A close loop DC motor speed control simulation system using spice. In: IEEE International Caracas Conference on Devices, Circuits and Systems, pp. 261–26 (1995)
3. Utkin, V.I.: Sliding Modes in Optimization, and Control Problems. Springer, New York (1992)
4. Slotine, J.J.: Sliding controller design for non linear systems. Int. J. Control **40**(20), 421–434 (1984)
5. Slotine, J.J., Li, W.: Applied nonlinear control. Prentice-Hall, Inc., London (1991)
6. Rao, P.N., Rao, G.V.S.K, Kumar, G.V.N.: A novel technique for controlling speed and position of bearingless switched reluctance motor employing sensorless sliding mode observer. Arab.

J. Sci. Eng. **43**(8), 4327–4346 (2018)

7. Boumhidi, J., Mrabti, M.: Sliding mode controller for robust force control of hydraulic servo-actuator. In: ISEE, IEEE International on Electrical Engineering, Targoviste, Romania, pp. 27–33, 1–2 Nov 2004
8. Hussain, M.A., Ho, P.Y.: Adaptive sliding mode control with neural network based hybrid models. J. Process Control **14**, 157–176 (2004)
9. Seera, M., Lim, C.P., Ishak, D.: Detection and diagnosis of broken rotor bars in induction motors using the fuzzy min-max neural network. Int. J. Natural Comput. Res. **3**(1), 44–55 (2012)
10. Hasanien, H.M., Muyeen, S.M.: Speed control of grid-connected switched reluctance generator driven by variable speed wind turbine using adaptive neural network controller. Electr. Power Syst. Res. **84**, 206–213 (2012)
11. Rodger, J.A.: A fuzzy nearest neighbor neural network statistical model for predicting demand for natural gas and energy cost savings in public buildings. Exp Syst Appl **41**(4), 1813–1829 (2014)
12. Alaoui, M.C.S., Magrez, H.: DC motor velocity neural network sliding mode controller for the combined pumping load-DC motor-buck converter system. J. Mechatron **3**(3), 253–257 (2015)
13. Niu, B., Wang, D., Li H., Xie, X.-J., Alotaibi, N.D., Alsaadi, F.E.: A neural-network-based adaptive control scheme for output-constrained stochastic switched nonlinear systems. IEEE Trans. Syst. Man Cybern. Syst. **49**(2), 418–432 (2019)
14. Linares-Flores, J., Ramirez, H.S.: DC motor velocity control through a DC-to-DC power converter. In: 43rd IEEE Conference on Decision and Control, Atlantis, Paradise Island, Bahamas, pp. 5297–5302, 14–17 Dec 2004
15. Grellet, G., Clerc, G.: Actionneurs Electriques Principes Modèles Commande. Eyrolles (1997)
16. Anis, W.R., Metwally, H.M.: Dynamic performance of a directly coupled PV pumping system. Sol. Energy **53**(4), 369–377 (1994)
17. Rumelhart, D.E., Hinton, G.E., Williams, R.J.: Parallel Distribute Processing. Learning internal representations by error propagation. MIT Press, Cambridge (1985)
18. D' Addona, D.M., Teti, R.: Genetic algorithm-based optimization of cutting parameters in turning processes. In: Forty Sixth CIRP Conference on Manufacturing Systems, vol. 7, pp. 323–328 (2013)
19. Garcia-Martinez, C., Lozano, Herrera, M.F., Molina, D., Sanchez, A.M.: Global and local real-coded genetic algorithms based on parent-centric crossover operators. Eur. J. Oper. Res. **185**, 088–1113 (2008)

Smart Farming System Based on Fog Computing and LoRa Technology

Mohamed Baghrous, Abdellatif Ezzouhairi and Nabil Benamar

Abstract Nowadays, the increased agricultural production at a lower cost is more and more driven by the Internet of Things (IoT) and the cloud computing paradigms. Many researches and projects have been elaborated so far in this context. It aims at reducing human efforts as well as resources and power consumption. Such projects are mainly based on collecting various data pertaining to the agricultural area and sending it to the cloud for further analysis. However, the long distance between sensors/actuators and the cloud leads to a significant increase in latency, which leads to a decrease of performances of the irrigation systems, pesticide monitoring, etc. This paper presents an alternative solution based on Fog Nodes and LoRa technology to optimize the number of nodes deployment in smart farms. Our proposed solution reduces the total latency induced during data transmission toward the cloud for processing.

Keywords Cloud computing · Fog computing · LoRa · Agriculture · IoT · Fog Nodes

1 Introduction

In Morocco, the agriculture is considered as a vital sector, employing about four million people and contributing about 14% of the GDP [1]. Nevertheless, this sector is facing tremendous challenges such as climate change and water scarcity. In these

M. Baghrous (✉)
Renewable Energies and Intelligent Systems Laboratory, Faculty of Science and Technology, Sidi Mohammed Ben Abdellah University, Fez, Morocco
e-mail: mohamed.baghrous@usmba.ac.ma

A. Ezzouhairi
National School of Applied Sciences, Sidi Mohamed Ben Abdellah University, Fez, Morocco
e-mail: abdellatif.ezzouhairi@usmba.ac.ma

N. Benamar
School of Technology, University Moulay Ismaïl, Meknes, Morocco
e-mail: n.benamar@est.umi.ac.ma

V. Bhateja et al. (eds.), *Embedded Systems and Artificial Intelligence*,
Advances in Intelligent Systems and Computing 1076,
https://doi.org/10.1007/978-981-15-0947-6_21

Table 1 LoRa versus other wireless technologies

Heading level	Range	Rate	Power (mA)	Cost
Bluetooth	15 m	3 Mbps	20	low
WiFi	150 m	3 Mbps	75	high
ZigBee	100–200 m	250 Kbps	200	low
LoRa	3 km	2.4 Kbps	110	low

circumstances, the adoption of IoT in the Moroccan agricultural sector can certainly help to face these various challenges as well as improving the agricultural income. On the other hand, it is supposed to reduce considerably farmer's physical effort, water and energy consumption. Actually, smart farming (SF) is widely adopted in USA, China and in many leading countries in this area. SF involves many tools (sensors, robots, etc), which makes farms as an important source of data. In general, SF functionalities are categorized into the following phases: (1) data collection by sensors; (2) processing, store and analyzing this data at cloud and (3) feedback and taking actions by actuators. Transferring data to the cloud for processing takes time and resources, and the delays are not always acceptable in latency-sensitive systems like autonomous drones/robots, fire detection system and livestock control. On the other hand, most of the wireless sensor communication technologies, such as those mentioned in Table 1, cover only a short range which demands additional equipments, especially in the vast farms. The rest of this paper is organized as follows: literature review, background, system architecture based on Fog and LoRa technology, system simulation and conclusion.

2 Literature Review

In the last few years, many researchers have addressed the issues relevant to SF. Indeed, in [2], the authors proposed an intelligent farming (IF) system that uses the concepts of IoT and SF to help farmers to monitor and sense useful information from their farms in order to help in the quality improvement and product quantity. In [3], the proposed model is architecture of IoT sensors that collect data and send it over the WiFi network to the cloud server; their server can take actions depending on the treated information. The work published in [4] proposed a greenhouse monitoring system based on agriculture IoT with a cloud platform; the parameters that are collected by a network of sensors are being logged and stored online based on a cloud computing platform. In [5], authors proposed an animal behavior monitoring platform; based on IoT, it includes an IoT local network to gather data from animals and a cloud platform with processing and storage capabilities. All of these systems and others mentioned in a survey in [6] are mainly based on the cloud which is considered as a promising solution for IoT agricultural applications. However, it still has some limitations. The fundamental limitation is the connectivity between the cloud and the SF devices. The distance between the cloud and the end devices might be an issue for latency-sensitive applications such as autonomous robots used for agricultural purposes [7], fire detection, drones and smart surveillance system. This latency is unavoidable due to the large number of hops that a packet has to cross to reach the cloud. Furthermore, most of the aforementioned proposals are based on short-range communication using ZigBee or WiFi which lacks to ensure coverage of wide agriculture areas. To overcome these issues, we propose Fog Nodes for agricultural purposes based on Fog Computing paradigm and LoRa technology.

3 Background

3.1 Cloud Computing

Cloud computing has been used as an efficient way to process data because of its high computation power and storage capability [8]. However, as cloud computing paradigm is a centralized computing model, most of the computations happen in the cloud. This means that all the data produced in smart farms needs to be transmitted to the cloud. Although the data processing speed has risen rapidly, the network bandwidth has not increased appreciably. This may result in long latency.

3.2 Fog Computing

Fog Computing paradigm was introduced by Cisco Systems in 2012, and in its initial definition, it was considered as an extension of the cloud computing paradigm that provides computation, storage and networking services between sensors and traditional cloud servers [9]. Fog aims to bring networking resources near source of data (the nodes that are generating data). Because of their proximity to the source of the data compared to the cloud data centers, Fog Computing has the potential to offer services that deliver better delay performance.

3.3 Lora Technology

As the name implies, long range (LoRa) is a long-range wireless communications system, promoted by the LoRa Alliance. It is capable of transmitting data over long distances. This system aims at being usable in long-lived battery-powered devices, where the energy consumption is very important [10]. The payload of each transmission can range from 2 to 255 octets, and the data rate can reach up to 50 Kbps when channel aggregation is employed. The LoRa modulation has been patented by

Semtech Corporation [11]. Table 1 compares some parameters including power consumption, transfer rate, transmission range and cost between some popular wireless technologies used in the above systems. Accordingly, LoRa has shown its superiority in many aspects. Its only weakness is the data rate. However, in wireless sensor network applications, this is not an issue.

A. **LoRa parameters**

In LoRa technology, there are four main types of parameters to consider [12]:

- RF Transmission Power (TxP): RF Txp is the maximum power that can be selected to transmit the packets on air. TxP on a LoRa radio can be adjusted from −4 to 20 dBm, but because of limited hardware implementation, the range is often limited to 10 dBm on 433 MHz and 14 dBm in 868 MHz.
- Carrier Frequency (CF): The frequency at which the LoRa module transmits at CF in different channels the LoRa module operates in two bands 863–870 MHz and 433.050–434.790 MHz.
- Spreading Factor: In LoRa modulation, a chirp signal that continuously varies in frequency is used to spread the spectrum of the original data. In this case, the time and frequency offset between the transmitter and receiver are same. As well as the frequency bandwidth of this chirp and the spectral bandwidth of the signal are same.
- Bandwidth (BW): The radio bandwidth of the LoRa module takes any of the three values 125, 250 and 500 kHz.

B. **Time on the air of data packet**

For LoRa, the actual time on the air for a packet can be defined as [12]:

$$T_{\text{packet}} = T_{\text{preamble}} + T_{\text{payload}} \tag{1}$$

where T_{preamble} is transmission time of preamble field and T_{payload} is transmission time of the payload field.

T_{preamble} is defined as:

$$T_{\text{preamble}} = \left(n_{\text{preamble}} + 4.25\right) T_{\text{sym}} \tag{2}$$

The payload transmission time is:

$$T_{\text{payload}} = n_{\text{preamble}} * T_{\text{sym}} \tag{3}$$

where n_{preamble} is:

$$n_{\text{preamble}} = 8 + \max(A * (\text{CR} + 4), 0) \tag{4}$$

And, A is defined as:

$$A = \frac{(8\text{PL} - 4\text{SF} + 28 + 16\text{CRC} - 20\text{IH})}{4(\text{SF} - 2\text{DE})} \tag{5}$$

With the following dependencies:

- PL: number of bytes in payload field.
- IH: 0 when the implicit header mode is used and 1 for vice versa.
- DE: 1 for enabled low data rate optimization and 0 for opposite case.
- CR: code rate.

So, using 1, 2 and 3, the total time on the air

$$T_{\text{packet}} = (n_{\text{preamble}} + n_{\text{payload}} + 4.25)T_{\text{sym}} \tag{6}$$

4 Proposed System Architecture

In this section, we introduce a SF System based on a Fog Nodes with LoRa technology. As shown in Fig. 1, these Fog Nodes are installed in a strategy place allowing it to cover a large number of sensors and actuators. The main role of the proposed nodes is not only to build a bridge between the low-level sensors and the network, but also it is considered as a local server equipped with LoRa module and located near the source of the data. It handles data collected from sensors they are connected to. As mentioned above, a range of time-sensitive tools and technologies (autonomous drones, robots, etc.) has been adopted in smart farms which need data processing in the real time. So, these proposed Fog Nodes are the ideal solution for providing less latency service to SF time-sensitive subsystems. On the other hand, each agricultural area may contain many smart farms and each farm has an area of hundreds of hectares. The proposed Fog Nodes cover a large area because of its LoRa unit, allowing it to handle a large number of sensors and actuators, thus reducing the investment in communications nodes.

The following are some advantages of the proposed system:

- Low latency: In our proposed architecture, Fog Nodes acquire the data generated by sensors, process and store this data locally. It significantly reduces data movement across the Internet to the cloud. Therefore, it enables low latency and meets the demand of real-time processing, especially for latency-sensitive applications such as autonomous drones/robots for agricultural purposes.
- Save bandwidth: Some computation tasks, for example, data preprocessing, data filtering are performed locally at Fog Nodes near the source of the data. Fog Nodes transmitted only the useful data to the cloud. In this way, the proposed system will save the bandwidth effectively and reducing the cost.
- Low energy consumption: The proposed system is based on LoRa technology which combines long-range connectivity with a considerable increase of the battery lifetime at a low-cost using sub-GHz unlicensed spectrum. In addition, there are

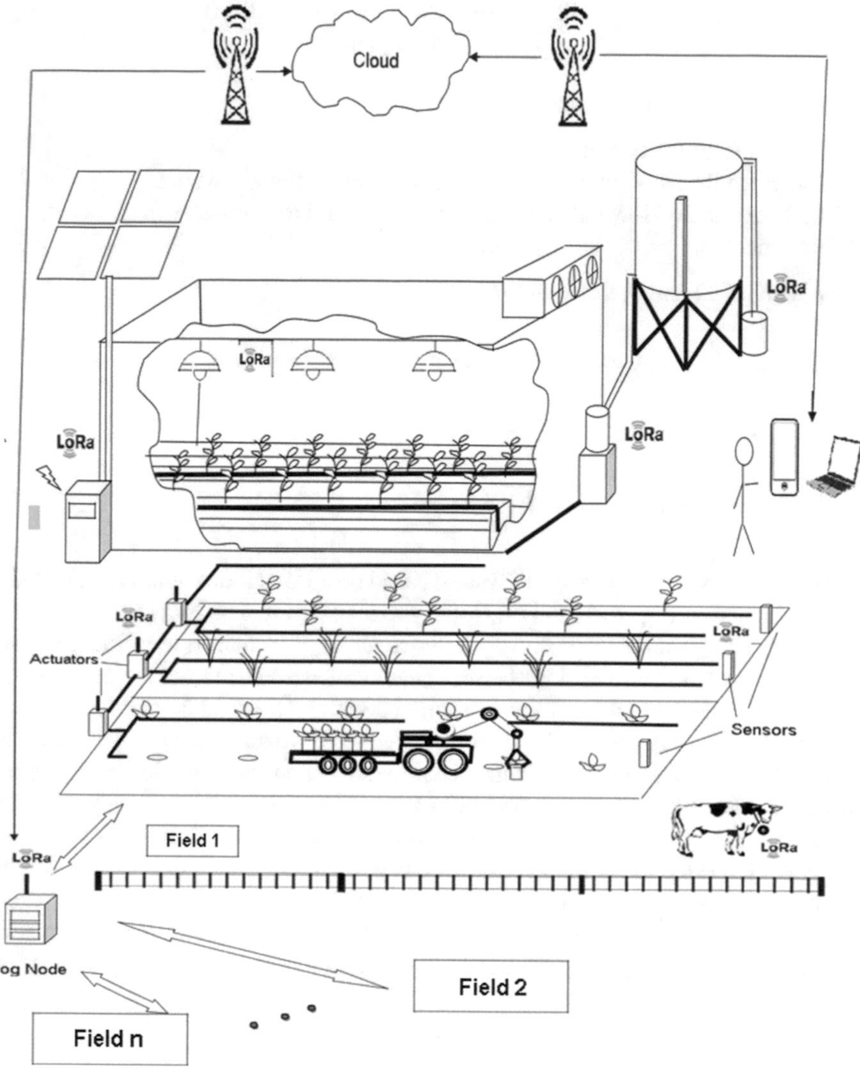

Fig. 1 Proposed system architecture based on fog and LoRa technology

many researches that have proven that energy consumption for Fog Computing architecture is 40.48% less than the traditional cloud computing model [13]. So the combination of Fog Computing and LoRa technology in SF applications will lead to reduce power consumption and decrease the cost.

- Data security: Compared to [2], [3] and [4] which are mainly based on the cloud computing, data in our proposed system is transmitted in a limited range between sensors/actuators and Fog Nodes, it is sent to the cloud over the Internet only when necessary. Therefore, data is protected from attacks that may adversely affect the performance of various systems used in smart farms (Irrigation systems, surveillance systems, etc.).

5 Simulation and Results

We have performed the simulation with the iFogSim simulator based on the network topology that is shown in Fig. 2. It consists of three levels: Cloud level, fog level and sensors/actuators level. In fog level, Fog Nodes (Agri-Fog-Nodes and Agri-Areas) handle with data generated by sensors distributed in smart farms. These nodes configured to be servers close to sensors.

Figure 3 presents the average latency in millisecond as a function of different configurations. The graph shows clearly that latency is much lower when application modules are executed on the fog. This will certainly enhance real-time processing.

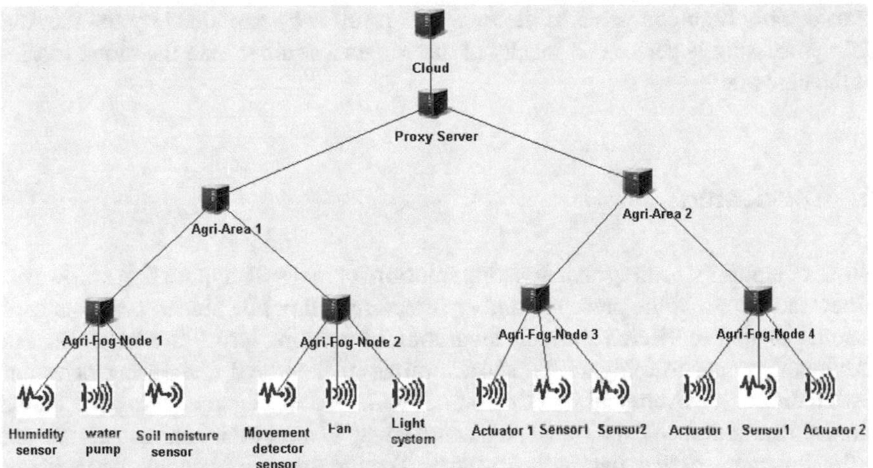

Fig. 2 Simulation network topology

Fig. 3 Average latency of control loop

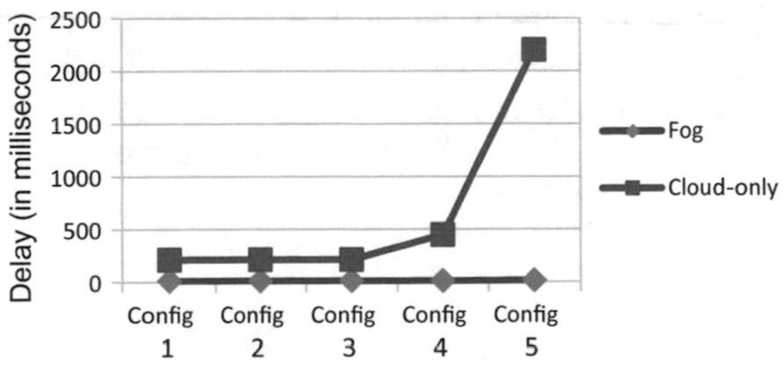

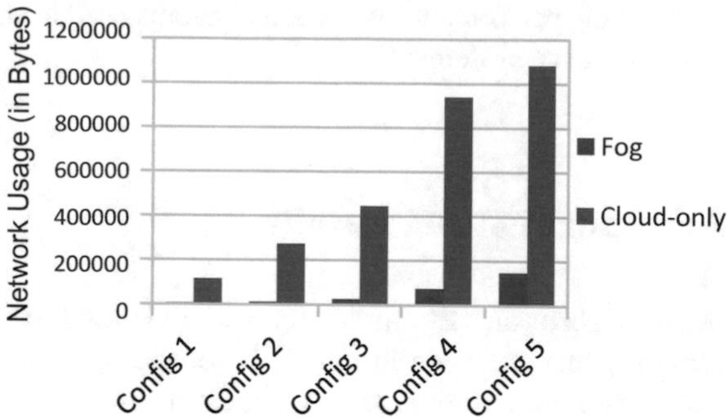

Fig. 4 Network usage

Therefore, it will improve the performance of time-sensitive applications in smart farms.

In any IoT application, more the network is used, more the overhead increases too. Figure 4 illustrates the behavior of network usage in both cloud and fog paradigms. The obtained results clearly show that network overhead in cloud architecture is considerably high compared to the fog. This result is be explained by the fact that data processing is performed locally at the fog nodes rather than the cloud located in the network.

6 Conclusion

Cloud computing is surely an interesting solution for many SF applications. However, it became clear that the present cloud architecture will not be able to handle a great amount of data collected through thousands of sensors across the network. Fog Computing came to mitigate the shortcomings of the cloud computing paradigm within the IoT environment by bringing network, processing and storage resources closer to the source of data. Hence, it paves the way to meet their hard constraints and offloading much of their data to the fog layer. The proposed architecture is considered as an efficient solution to save bandwidth and reduce latency. LoRa technology offers compelling features for SF application including long range, low power consumption and secure data transmission. This paper has presented SF System based on Fog Computing paradigm and LoRa technology. The results showed that the adoption of Fog Computing paradigm in the context of SF will improve the real-time processing, reduce time response and save bandwidth. Moreover, the use of LoRa technology in the communication between nodes will ensure efficient power management and large-scale coverage.

References

1. Department of Agriculture of Morocco. http://www.agriculture.gov.ma/. Accessed 03 May 2018
2. Putjaika, N., Phusae, S., Chen, I.: A control system in an intelligent farming by using arduino. In: IEEE Fifth ICT International Conference, pp. 53–56 (2016)
3. Dagar, R., Som, S., Khatri, S.: Smart farming—IoT in agriculture. In: IEEE International Conference on Inventive Research in Computing Applications, pp. 1052–1056 (2018)
4. Keerthi, V., Kodandaramaiah, N.: Cloud IoT based greenhouse monitoring system. In: Int. J. Eng. Res. Appl. 35–41 (2015)
5. Nóbrega, L., Tavare, A., Cardoso, A., Gonçalves, P.: Animal monitoring based on IoT technologies. In: IEEE IoT Vertical and Topica Summit on Agriculture, pp. 1–5 (2018)
6. Shareef, M., Viswanathan, P.: A Survey: smart agriculture IoT with cloud computing. In: IEEE International conference on Microelectronic Devices (ICMDCS), pp. 1–7 (2017)
7. Ball, D., Ross, P., English, A., Milani, P., Richards, D., Bate, A., Upcroft, B., Wyeth, G., Corke, P.: Farm workers of the future: vision-based robotics for broad-acre agriculture In: IEEE Robotics and Automation Magazine, pp. 97–107 (2017)
8. Mohiddin S., Babu, S., Sharmila, S.: A complete ontological survey of CLOUD forensic in the area of cloud computing. In: Springer-Sixth International Conference on Soft Computing, pp. 38–47 (2017)
9. Linthicim, S.: Connecting fog and cloud computing. IEEE J. Mag. 18–20 (2017)
10. Sasián, F., Gachet, D., Buenaga, M., Aparicio, F.: A dictionary based protocol over LoRa technology for applications in internet of things. In: International Conference on Ubiquitous Computing and Ambient Intelligence, pp. 140–148 (2017)
11. Semtech: LoRa Technology. https://www.semtech.com/. Accessed 15 Jan 2019
12. Semtech-Corporation. https://www.Lorasx1276/77/78/79datasheet/. Accessed 19 Jan 2019
13. Fatima, I., Javaid, N., Iqbal, M., Shafi, I., Anjum, A., Memon, U.: Integration of cloud and fog based environment for effective resource distribution in smart buildings. In: 14th IEEE International Wireless Communications and Mobile Computing Conference, pp. 60–64 (2018)

Behavioral Model of the Superjunction Power MOSFET Devices

Abdelghafour Galadi, Yassine Hadini, Bekkay Hajji and Adil Echchelh

Abstract In this paper, superjunction power metal–oxide–semiconductor field-effect transistor (MOSFET) model is presented. The proposed behavioral model is simple, accurate and uses a reduced number of parameters to describe accurately the static behavioral of the superjunction power MOSFET devices. Unlike the conventional power MOSFET model, this model uses only one standard MOSFET to describe the normal and the quasi-saturation operational regions of the power MOSFET. Consequently, this model gives fast simulation results without any problem of convergence. Using the simulated program with integrated circuit emphasis (SPICE) circuit simulator, simulation results show that the proposed model gives high correlations with the manufacturer datasheet and the measured data.

Keywords Superjunction · Drift layer · SPICE · Simulation · Validation

1 Introduction

The unipolar power metal–oxide–semiconductor field-effect transistor (MOSFET) is widely used in power electronic applications thanks to its low gate drive power, high switching speed and thermal stability. In conventional power MOSFET structure, the breakdown voltage (V_{BR}) is determined by the thickness and the doping level of the extended-drain drift layer (Fig. 1a). To sustain high voltage, the drift layer resistance

A. Galadi (✉)
ENSA de Safi, Cadi Ayyad University, Marrakech, Morocco
e-mail: agaladi@uca.ma

Y. Hadini · A. Echchelh
FSK, Department of Physics, Ibn Tofail University, Kenitra, Morocco
e-mail: Yassine.hadini@gmail.com

A. Echchelh
e-mail: adilechel@gmail.com

B. Hajji
ENSAO, Mohammed Premier University, Oujda, Morocco
e-mail: hajji.bekkay@gmail.com

V. Bhateja et al. (eds.), *Embedded Systems and Artificial Intelligence*,
Advances in Intelligent Systems and Computing 1076,
https://doi.org/10.1007/978-981-15-0947-6_22

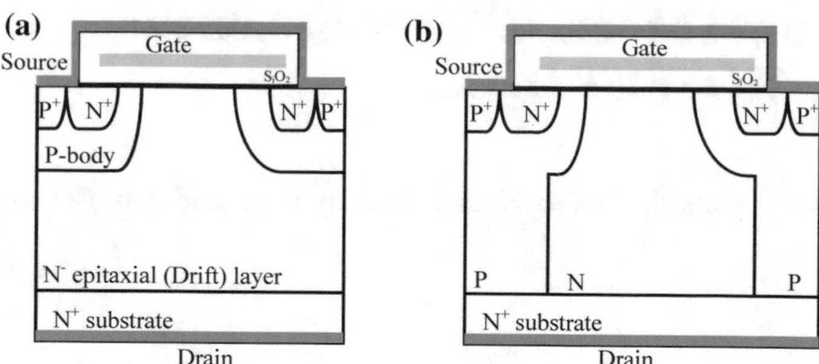

Fig. 1 Power MOSFET structure: **a** conventional structure; **b** superjunction structure

(R_{on}) becomes very large in the on-state of the power MOSFET. In literature, the well-known relationship 'silicon limit' was introduced to give the ideal specific on-resistance (R_{onsp}), which is the product of R_{on} by active area of the device, versus V_{BR} [1]. In conventional power MOSFET structure, R_{onsp} is proportional to the square of V_{BR}. The superjunction concept was introduced in the power MOSFET structure broking the 'silicon limit' and enhancing the R_{on} of the power MOSFET devices [2–4]. In superjunction power MOSFET structure, the drift layer is replaced by alternative P and N layers (Fig. 1b). Consequently, R_{onsp} is linearly proportional to V_{BR}. In superjunction structure, V_{BR} is highly sensitive to the charge imbalance of the P and the N layers [2].

In other words, the static transfer and output characteristics of the superjunction power MOSFET show a pronounced quasi-saturation regime (Figs. 2 and 3) [5–7]. In this regime, the drain current becomes less sensitive to the gate–source voltage. Therefore, the standard MOSFET model, introduced in simulated program with

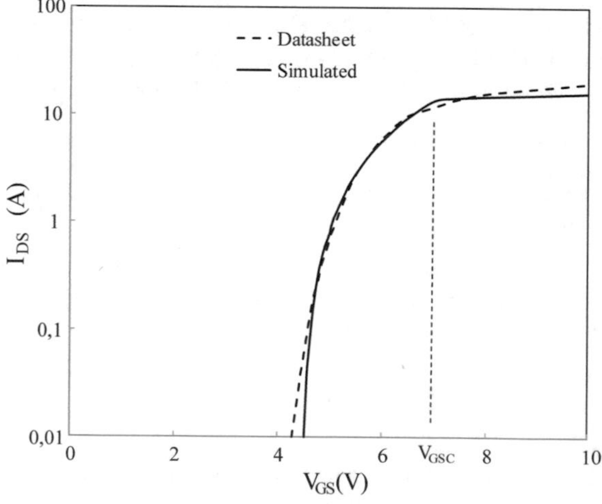

Fig. 2 Device datasheet (dashed) and SPICE simulated (solid) transfer characteristics of the power MOSFET

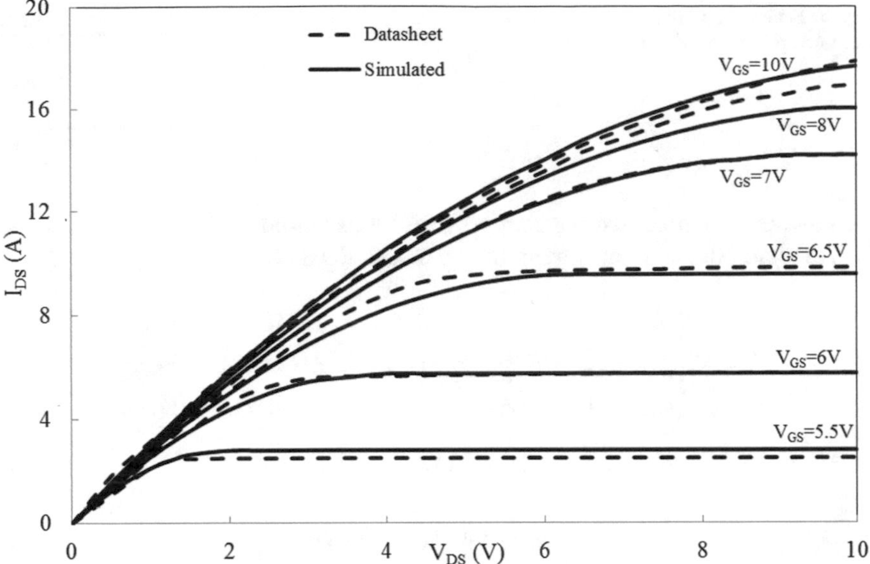

Fig. 3 Device datasheet (dashed) and SPICE simulated (solid) output characteristics of the power MOSFET

integrated circuit emphasis (SPICE) simulator, cannot describe accurately the quasi-saturation region and the degradation of the drain current at high gate voltages in superjunction power MOSFET. In literature, macromodels and compact models were introduced to describe accurately the static behavioral of the power MOSFET [8–12]. These models represent the power MOSFET in circuit simulator by using a subcircuit based on series association of standard MOSFET and resistance or junction field-effect transistor (JFET). Unfortunately, this subcircuit uses an internal drain potential calculated iteratively with large number of parameters using proprietary tools in some cases.

In this paper, the behavioral superjunction power MOSFET model is proposed. The model uses one MOSFET model without an additional effective voltage source or JFET transistor to describe the static characteristics of the superjunction power MOSFET. The drain current behavioral is described accurately by using variable model parameters. In addition, this model uses 10 parameters extracted from measured or manufacturer datasheet data. Finally, datasheet and measured data are presented for the verification of the model.

2 Model Description

The proposed model is shown in Fig. 4. The drain current is governed by the modified standard MOSFET model while the body diode models the breakdown voltage in

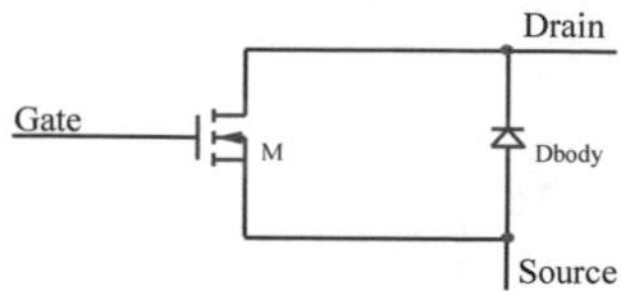

Fig. 4 Schematic of the proposed power MOSFET

the off-state. The modified standard MOSFET uses *n*th-power law MOSFET model [13] to represent the drain current. In this model, the drain current I_{DS} is given by

$$I_{DS} = \begin{cases} 0 & \text{if } V_{GS} \le V_{TM} \\ I_{DSAT}\left(2 - \frac{V_{DS}}{V_{DSAT}}\right)\frac{V_{DS}}{V_{DSAT}}(1 + \lambda_M V_{DS}) & \text{if } V_{DS} < V_{DSAT} \\ I_{DSAT}(1 + \lambda_M V_{DS}) & \text{if } V_{DS} \ge V_{DSAT} \end{cases} \quad (1)$$

$$I_{DSAT} = B(V_{GS} - V_{TM})^n \quad (2)$$

$$V_{DSAT} = K(V_{GS} - V_{TM})^m \quad (3)$$

where V_{GS}, V_{DS} and V_{DSAT} are gate–source, drain–source and drain–source saturation voltage, respectively. V_{TM} is the threshold voltage and I_{DSAT} is the saturation drain current. K and m model parameters control the linear region while B and n control the saturation region of operation in output characteristics.

In superjunction power MOSFET, I_{DS} is sensitive to V_{GS} variations in normal region (low V_{GS} in Fig. 2). On the other hand, I_{DS} becomes less sensitive to the variations of V_{GS} in quasi-saturation region (high V_{GS} in Fig. 2). For low V_{GS}, R_{on} is equivalent to the channel region resistance while R_{on} is equivalent to the drift resistance for high V_{GS} [14]. These two regions are separated by critical gate voltage (V_{GSC}) [14]. In this letter, V_{GSC} is used as a parameter separating the normal and the quasi-saturation operational regions of the power MOSFET. For $V_{GS} < V_{GSC}$, the model parameters will be extracted in normal region ($n = n_1$, $B = B_1$, $K = K_1$ and $m = m_1$) [13, 14]. For $V_{GS} \ge V_{GSC}$, the model parameters will be extracted in quasi-saturation region ($n = n_2$, $B = B_2$, $K = K_2$ and $m = m_2$) [13, 14].

3 Results and Discussion

The proposed model was introduced in a personal SPICE (PSPICE) simulator using netlist description. The static model is verified against datasheet from a 600 V-12 A RJK60S3DPE Renesas superjunction MOSFET [7]. The simulated and datasheet transfer and output characteristics are shown in Figs. 2 and 3, respectively. Obviously, the model gives a good description of the normal and the quasi-saturation regions. In other words, the model is validated by comparing the simulated and measured data [15] (Fig. 5). The model matches closely to the measured data in all

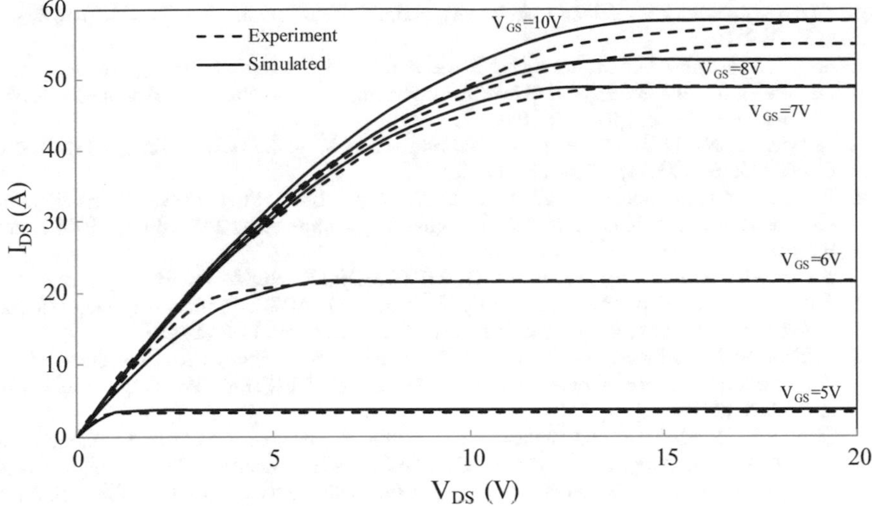

Fig. 5 Simulated (solid) and measured (dashed) output characteristics of superjunction power MOSFET

operational regions. Unlike the conventional power MOSFET models, the proposed model describes more precisely the transition of the drain current from normal to quasi-saturation regime. Finally, the simple model equations and the reduced number of parameters give the model a lower computational complexity enhancing the simulation time.

4 Conclusion

An enhanced superjunction power MOSFET model which accounts for the normal and the quasi-saturation operational regions is presented. Verifications of the model by manufacturer datasheet and experimental data confirm the validity of the model. The ease parameter extraction method, the accuracy and the convergence of the proposed model make it very useful for power electronic circuit design.

References

1. Baliga, B.J.: D-MOSFET structure. In: Advanced Power MOSFET Concepts, pp. 23–61. Springer-Science, New York (2010)
2. Udrea, F., Deboy, G., Fujihira, T.: Superjunction power devices: history, development, and future prospects. IEEE Trans. Electron Devices **64**(3), 713–727 (2017)

3. Fujihira, T.: Theory of semiconductor superjunction devices. Jpn. J. Appl. Phys. **36**(10), 6254–6262 (1997)
4. Saggio, M., Fagone, D. Musumeci, S.: MDmesh: Innovative technology for high voltage power MOSFETs. In: Proceedings of 12th International Symposium on Power Semiconductor Device Ics, Toulouse, France, pp. 65–68 (2000)
5. Darwish, M.N.: Study of the quasi-saturation effect in VDMOS transistors. IEEE Trans. Electron Devices **33**(11), 1710–1716 (1986)
6. Wang, L., Wang, J., Gao, C., Hu, J., Li, P., Li, W., Yang, S.H.Y.: Physical description of quasi-saturation and impact-ionisation effects in high-voltage drain-extented MOSFETs. IEEE Trans. Electron Devices **56**(3), 492–498 (2009)
7. N-channel superjunction power MOSFET. RJK60S3DPE RENESAS datasheet (2012)
8. Scott, R.S. Franz, G.A.: An accurate model for power DMOSFETs including interelectrode capacitances. In: Power Electronics Specialists Conference, pp. 113–119 (1990)
9. Mudholkar, M., Ahmed, S., Ericson, M.N., Frank, S.S., Britton, C.L., Manthoot, H.A.: Datasheet driven silicon carbide power MOSFET model. IEEE Trans. Power Electron. **29**(5), 2220–2228 (2014)
10. Chauhan, Y.S., Anghel, C., Krummenacher, F., Gillon, R., Baguenier, A., Desoete, B., Frere, S., Ionescu, A.M. Declercq, M.: A compact DC and AC model for circuit simulation of high voltage VDMOS transistor. In: Proceedings of the 7th International Symposium Quality Electronic Design (ISQED '06), San Jose, CA, pp. 109–114 (2006)
11. Aarts Annemarie, C.T., Kloosterman, Willy J.: Compact modeling of high-voltage LDMOS devices including quasi-saturation. IEEE Trans. Electron Devices **53**(4), 897–902 (2006)
12. Yao, W., Gildenblat, G., McAndrew, C.C., Cassagnes, A.: SP-HV: a scalable surface-potential-based compact model for LDMOS transistors. IEEE Trans. Electron Devices **59**(3), 542–550 (2012)
13. Sakurai, T., Newton, A.R.: A simple MOSFET model for circuit analysis. IEEE Trans. Electron Devices **38**(4), 887–894 (1991)
14. Galadi A.: Accurate power MOSFET models including quasi-saturation effect. J. Comput. Electron. **15**(2), 619–626 (2016)
15. Oonishi, Y., Ooi, A., Shimatou, T.: Superjunction MOSFET. Fuji Electron. Rev. **56**(2), 65–68 (2010)

Resolution of Brinkman Equations with a New Boundary Condition by Using Mixed Finite Element Method

Omar El Moutea, Hassan El Amri and Abdeslam Elakkad

Abstract This paper considers numerical methods for solving Brinkman equations with a new boundary condition summing Dirichlet and Neumann conditions. We develop here a robust stabilized mixed finite element method (MFEM), and two types of a posteriori error indicator are introduced to give global error estimates; there are equivalent to the true error. We present numerical simulations.

Keywords Brinkman equations · C_{a,μ^*} boundary condition · Mixed finite element methods · Residual error estimator · Numerical simulations

1 Introduction

This work deals with the development of stable numerical methods for the Brinkman equations; these equations are very important in a different domain, for example: in hydrogeology, porous media, and petroleum engineering. By these equations, we can model the flow in complex situations, for example, coupling flow in porous media and surface flow of fluids and we use these equations if we have different domains with variable coefficients. To describe the flow of a viscous fluid, see [1] and in soil mechanics see [2, 3]. Mathematically, this equation is a combination of two partial differential equations; we use a new boundary condition (generalizes the Dirichlet and the Neumann conditions). This boundary condition is used for Stokes

O. El Moutea (✉) · H. El Amri
Laboratory of Mathematics and Applications, ENS - Hassan II University, Casablanca, Morocco
e-mail: mouteaomar@gmail.com

H. El Amri · A. Elakkad
Centre Régional des Métiers d'Education et de Formation (CRMEF), Fes, Morocco
e-mail: elakkadabdeslam@yahoo.fr

V. Bhateja et al. (eds.), *Embedded Systems and Artificial Intelligence*, Advances in Intelligent Systems and Computing 1076,
https://doi.org/10.1007/978-981-15-0947-6_23

problem in [4, 5]. The weak formulation of this equation is a problem of saddle point type which is our case in this study to show the existence, the uniqueness of the solution of this problem see [6, 7]. During the last decades, a posteriori error analysis in problems related to fluid dynamics is a very important subject that has received a lot of attention. For the conforming case, there are different ways to define error estimators by using the residual equation. In particular, for the Stokes problem, Ainsworth and Oden [8], Bank and Welfert [8], and Verfurth [9] introduced several error estimators and show that they are equivalent to the energy norm of the errors.

The plan of the paper is as follows. In Sect. 2, we present the model problem used in this paper. The weak formulation of our problem is presented, and we show the existence and uniqueness of the solution in Sect. 3. The discretization by classical mixed finite elements is described in Sect. 4. In Sect. 5, we perform the same analysis for this introduced two types of a posteriori error bounds of the computed solution. We present a numerical test in Sect. 6.

2 Governing Equations

Let Ω be an open bounded polygonal or polyhedral reservoir in $\mathbb{R}^2$ and $\Gamma = \partial\Omega$ its boundary. The simplest form of Brinkman's equation is to search the unknowns velocity functions and pressure of the fluid satisfying

$$\begin{cases} -\mu^* \nabla^2 \vec{u} + \mu K^{-1}\mathbf{u} + \nabla p = \vec{f} & \textbf{in } \Omega \\ \nabla \cdot \vec{u} = 0 & \textbf{in } \Omega. \end{cases} \tag{1}$$

The function $\vec{f}$ is a momentum source term, μ denotes the fluid viscosity, μ^* the effective viscosity of the fluid, and K is the permeability tensor of the porous media, which may contain multiscale features of the media.

We assume that these functions $\mu,\ \mu^* \in L^\infty(\Omega)$, and the tensor K are symmetric definite positive, which is uniformly elliptic; i.e., there exist two positive constants $\gamma_{\min}, \gamma_{\max}$ such that

$$\gamma_{\min} |\eta|^2 \le \eta^t K \eta \le \gamma_{\max} |\eta|^2 \tag{2}$$

for all $\eta \in L^2(\Omega)$ and $x \in \Omega$.

The problem consists of finding a velocity $\vec{u}$ and a pressure p fields with the C_{a,μ^*} boundary condition defined by

$$a\vec{u} + (\mu^* \nabla \vec{u} - pI)\vec{n} = \vec{t} \ \ \textbf{in } \Gamma = \partial\Omega. \tag{3}$$

We will consider the fluid viscosity and the effective viscosity of the fluid are bounded functions depend on the spaces. In the boundary condition (3), the functions $\vec{t}$, a and μ^* are bounded polynomials such that

$$\alpha_1 \leq \frac{a(x)}{\mu^*(x)} \leq \beta_1 \ \forall x \in \Gamma \tag{4}$$

where the constants $\alpha_1 \in \mathbb{R}^+$ and $\beta_1 \in \mathbb{R}^+$.

Remark 1 Let the functions a and μ^* (two nonzeros defined on $\partial\Omega$ are strictly positive constants), if $a << \mu^*$ then C_{a,μ^*} is the Neumann boundary condition, else if $\mu^* << a$ then C_{a,μ^*} is the Dirichlet boundary condition, for that the boundary condition C_{a,μ^*} generalized Dirichlet–Neumann conditions.

3 Weak Formulation and Existence and Uniqueness of the Solution

Before starting to define the weak formulation of our problem, we define different spaces used in this study see [10]. For more details on the notation or spaces used in this part, see [4, 5].

3.1 The Weak Formulation

The variational formulation of (1)–(3) reads, find $\vec{u} \in H^1(\Omega)$ and $p \in L_0^2(\Omega)$ such that:

$$\begin{cases} \int_\Omega \mu^* \nabla \vec{u} \cdot \nabla \vec{v} + \int_\Gamma a \vec{u} \cdot \vec{v} + \int_\Omega \mu K^{-1} \vec{u} \cdot \vec{v} - \int_\Omega p \nabla \cdot \vec{v} \\ \quad = \int_\Omega \vec{f} \cdot \vec{v} + \int_\Gamma \vec{t} \cdot \vec{v} \\ \int_\Omega q \nabla \cdot \vec{u} = 0 \end{cases} \tag{5}$$

for all $\vec{v} \in H^1(\Omega)$ and $q \in L_0^2(\Omega)$.

To simplify this study, we use these notations

$$a(\vec{u}, \vec{v}) = \int_\Omega \mu^* \nabla \vec{u} \cdot \nabla \vec{v} + \int_\Gamma a \vec{u} \cdot \vec{v} + \int_\Omega \mu K^{-1} \vec{u} \cdot \vec{v},$$

$$L(\vec{v}) = \int_\Omega \vec{f} \cdot \vec{v} + \int_\Gamma \vec{t} \cdot \vec{v}$$

and

$$b(\vec{u}, q) = \int_\Omega q \nabla \cdot \vec{u}.$$

The system is written to find $\overrightarrow{u} \in H^1(\Omega)$ and $p \in L_0^2(\Omega)$ such that:

$$\begin{cases} a(\overrightarrow{u}, \overrightarrow{v}) + b(\overrightarrow{v}, p) = L(\overrightarrow{v}) \\ b(\overrightarrow{u}, q) = 0 \end{cases} \tag{6}$$

for all $\overrightarrow{v} \in H^1(\Omega)$ and $q \in L_0^2(\Omega)$.

3.2 The Existence and Uniqueness of the Solution

In this part, we will study the existence and uniqueness of the variational formulation of our problem (6), for that, we recall important inequalities, which will be used in this analysis. Firstly, we can see that the space $(H^1(\Omega), \|\overrightarrow{v}\|_{J,\Omega})$ is a Hilbert space, which is obliged condition in the existence and uniqueness of the solution, for that we need the following results:

Theorem 1 *There exists two positives constants c_1 and c_2 such that:*

$$c_1\|\overrightarrow{v}\|_{1,\Omega} \leq \|\overrightarrow{v}\|_{J,\Omega} \leq c_2\|\overrightarrow{v}\|_{1,\Omega} \tag{7}$$

for all $\overrightarrow{v} \in H^1(\Omega)$.

Proof See [4, 5].

Theorem 2 *The space $(H^1(\Omega), \|\cdot\|_{J,\Omega})$ is a real Hilbert space.*

Proof $(H^1(\Omega), \|\cdot\|_{1,\Omega})$ is a real Hilbert space, the norms $\|\cdot\|_{1,\Omega}$ and $\|\cdot\|_{J,\Omega}$ are equivalent, then $(H^1(\Omega), \|\cdot\|_{J,\Omega})$ is a real Hilbert space.

We can see, now, the existence and uniqueness of the solution.

Theorem 3 *The function $b\,(\cdot,\cdot)$ is satisfies the velocity–pressure $inf - sup$ condition, there exists a constant $\beta > 0$ such that:*

$$\sup_{\overrightarrow{v} \in H^1(\Omega)} \frac{b(\overrightarrow{v}, q)}{\|\overrightarrow{v}\|_{J,\Omega}} \geq \beta\|q\|_{0,\Omega} \tag{8}$$

for all $q \in L_0^2(\Omega)$.

Proof The same proof of [7] suffices to see that $H_0^1(\Omega) \subset H^1(\Omega)$ and $\|\overrightarrow{v}\|_{J,\Omega} = |\overrightarrow{v}|_{1,\Omega}$ in $H_0^1(\Omega)$.

Using “big” symmetric bilinear form $C[(\overrightarrow{u}, p), (\overrightarrow{v}, q)]$ and the corresponding function $F(\overrightarrow{v}, q)$. The bilinear form $a\,(\cdot,\cdot)$ is positive continuous $H^1(\Omega) - elliptic$ and the bilinear form $b(\cdot,\cdot)$ is continuous satisfies the $inf - sup$ condition. Now, we present a stabilized finite element scheme for the Brinkman problem.

Find $\overrightarrow{u} \in H^1(\Omega)$ and $p \in L_0^2(\Omega)$ such that:

$$C[(\overrightarrow{u}, p), (\overrightarrow{v}, q)] = F(\overrightarrow{v}, q) \tag{9}$$

for all $\overrightarrow{v} \in H^1(\Omega)$ and $q \in L_0^2(\Omega)$. Then, the problem (6) is well-posed, and the form bilinear C satisfies the following propositions.

Proposition 1 *For all* $(\overrightarrow{w}, s) \in H^1(\Omega) \times L_0^2(\Omega)$, *we have*

$$\sup_{(\overrightarrow{v},q)\in H^1\times L_0^2} \frac{C[(\overrightarrow{u}, p), (\overrightarrow{v}, q)]}{\|\overrightarrow{v}\|_{J,\Omega} + \|q\|_{0,\Omega}} \geq \delta(\|\overrightarrow{w}\|_{J,\Omega} + \|s\|_{0,\Omega}). \tag{10}$$

Proof See [10].

4 Mixed Finite Element Approximation

In this section, we use the finite element method to solve this problem see [11]. We consider the family of triangulations T_h, of our domain Ω where $h > 0$. For any triangle $T \in T_h$ and for an element edge E, we define these notations

- ω_T is of triangle sharing at least one edge with element T,
- $\widetilde{\omega}_T$ is the set of triangles sharing at least one vertex with T,
- ∂T is the set of the four edges of T, $\varepsilon(T)$ the set of its edges and N_T vertices.
- ω_E denotes the union of triangles sharing E,
- $\widetilde{\omega}_E$ is the set of triangles sharing at least one vertex whit E.

We let $\varepsilon_h = \bigcup_{T\in T_k} \varepsilon(T)$ denotes the set of all edges split into interior and boundary edges $\varepsilon_h = \varepsilon_{h,\Omega} \bigcup \varepsilon_{h,\Gamma}$ where

$$\varepsilon_{h,\Omega} = \{E \in \varepsilon_h : E \subset \Omega\} \textbf{ and } \varepsilon_{h,\Gamma} = \{E \in \varepsilon_h : E \subset \partial\Omega\}.$$

We denote by h_T the diameter of a simplex, h_E the diameter of a face E of T and $h = \max_{T\in T_k}\{h_T\}$. We define FE spaces $X_h^1 \subset H^1(\Omega)$ and $M^h \subset L_0^2(\Omega)$. The discrete version of (6) is, find $\overrightarrow{u}_h \in X_h^1$ and $p_h \in M^h$ such that:

$$\begin{cases} a(\overrightarrow{u}_h, \overrightarrow{v}_h) + b(\overrightarrow{v_h}, p_h) = L(\overrightarrow{v}_h), \\ b(\overrightarrow{u}_h, q_h) = 0, \end{cases} \tag{11}$$

for all $\overrightarrow{v}_h \in X_h^1$ and $q_h \in M_h$.

Note that, all the results remain valid for these spaces X_h^1 and M^h.

5 A Posteriori Error Estimator

In this section, we use two types of a posteriori error indicator: the first, residual error estimator and, the second, local Poisson problem estimator. These errors give global error estimates where there are equivalent to the true error. For the a posteriori error estimation for stabilized mixed approximations of the Stokes equations see [12].

5.1 A Residual Error Estimator

In this paper, we use MINI element method; they use a function called the "bubble function," which is related to any element of the space meshing. In Ceruse et al. ([13], Lemma 4.1), we established the Clement interpolation estimate.

Our aim is to estimate the velocity error $\vec{u} - \vec{u}_h \in H^1(\Omega)$ and the pressure error $p - p_h \in L_0^2(\Omega)$. The element of contribution $\eta_{R,T}$ is given by

$$\eta_{R,T}^2 = h_T^2 \| \vec{R}_T \|_{0,T}^2 + \| R_T \|_{0,T}^2 + \sum_{E \in \partial T} h_E \| \vec{R}_E \|_{0,E}^2, \tag{12}$$

the components of the residual error estimator $\vec{R}_T$, R_T in (12) are given by

$$\vec{R}_T = \{ \vec{f} + \mu^* \nabla^2 \vec{u}_h - \mu K^{-1} \vec{u}_h - \nabla p_h \}|_T \tag{13}$$

and

$$R_T = \{ \nabla \cdot \vec{u}_h \}|_T. \tag{14}$$

The residual error estimator $\mathbf{R}_E$ is given by

$$\mathbf{R}_E = \begin{cases} \frac{1}{2}[|\mu^* \nabla \mathbf{u}_h - p_h I|] \text{ if } E \in \varepsilon_{h,\Omega} \\ \vec{t} - [a \vec{u}_h + (\mu^* \nabla \vec{u}_h - p_h I) \vec{n}] \text{ if } E \in \varepsilon_{h,\Gamma}. \end{cases} \tag{15}$$

With the key contribution coming from the stress jump associated with an edge E adjoining elements T and S:

$$[[\mu^* \nabla \vec{u}_h - p_h I]] = ((\mu^* \nabla \vec{u_h} - p_h I)|_T - (\mu^* \nabla \vec{u}_h - p_h I)|_S) \vec{n}_{E,T}. \tag{16}$$

The global residual error estimator is given by:

$$\eta_R = \left(\sum_{T \in \tau_h} \eta_{R,T}^2 \right)^{\frac{1}{2}}. \tag{17}$$

Our aim is to bound $\|\overrightarrow{u} - \overrightarrow{u}_h\|_X$ and $\|p - p_h\|_M$ with respect to the norm $\|\cdot\|_J$ for the quotient velocity norm $\|\overrightarrow{v}\|_X = \|\overrightarrow{v}\|_{J,\Omega}$ and the pressure norm $\|p\|_X = \|p\|_{0,\Omega}$. For any $T \in T_h$ and $E \in \partial T$, we define the following functions:

$$\overrightarrow{w}_T = \overrightarrow{R}_T b_T, \ \overrightarrow{w}_E = \overrightarrow{R}_E b_E$$

where this functions verified
• $\overrightarrow{w}_T = \overrightarrow{0}$ on ∂T.
• if $E \in \partial T \cap \varepsilon_{h,\Omega}$
then $\overrightarrow{w}_E = \overrightarrow{0}$ on $\partial \omega_E$,
• if $E \in \partial T \cap \varepsilon_{h,\Gamma}$ *then* $\overrightarrow{w}_E = \overrightarrow{0}$ in the other three edges of triangle T.
• $\overrightarrow{w}_T$ and $\overrightarrow{w}_E$ can be extended to whole of Ω by setting:
$\overrightarrow{w}_T = \overrightarrow{0}$ in $\Omega - \overline{T}$
$\overrightarrow{w}_E = \overrightarrow{0}$ in $\Omega - \overline{\omega_E}$ if $E \in \partial T \cap \varepsilon_{h,\Omega}$.
$\overrightarrow{w}_T = \overrightarrow{0}$ in $\Omega - \overline{T}$ if $E \in \partial T \cap \varepsilon_{h,\Gamma}$.

With these two functions, we have the following lemmas.

Lemma 1 *For any $T \in T_h$ we have:*

$$\int_T \overrightarrow{f} \cdot \overrightarrow{w}_T = \int_T (\mu^* \nabla \overrightarrow{u} - pI) \cdot \nabla \overrightarrow{w}_T + \int_T \mu K^{-1} \overrightarrow{u} \cdot \overrightarrow{w}_T. \tag{18}$$

for all $\overrightarrow{u}$, $\overrightarrow{w}_T \in X_h^1$.

Lemma 2 *(i) if $E \in \partial T \cap \varepsilon_{h,\Omega}$, we have:*

$$\int_{\omega_E} \left(\overrightarrow{f} - \mu K^{-1} \overrightarrow{u}\right) \cdot \overrightarrow{w}_E = \int_{\omega_E} (\mu^* \nabla \overrightarrow{u} - pI) \cdot \nabla \overrightarrow{w}_E. \tag{19}$$

(ii) if $E \in \partial T \cap \varepsilon_{h,\Gamma}$, we have:

$$\int_T \left(\overrightarrow{f} - \mu K^{-1} \overrightarrow{u}\right) \cdot \overrightarrow{w}_E = \int_T (\mu^* \nabla \overrightarrow{u} - pI) \cdot \nabla \overrightarrow{w}_E + \int_{\partial T} (a \overrightarrow{u} \quad \overrightarrow{t}) \cdot \overrightarrow{w}_E. \tag{20}$$

Theorem 4 *For any mixed finite element approximation defined on triangular grids T_h, the residual estimator η_R satisfies:*

$$\|\overrightarrow{e}\|_{J,\Omega} + \|\varepsilon\|_{0,\Omega} \leq C_\Omega \eta_R \tag{21}$$

$$\eta_{R,T} \leq C \left(\sum_{T' \in \omega_T} \left\{ \|\overrightarrow{e}\|_{J,T'}^2 + \|\varepsilon\|_{0,T'}^2 \right\} \right)^{\frac{1}{2}}. \tag{22}$$

Note that, the constant C in the local lower bound is independent of the domain.

5.2 A Local Poisson Problem Estimator

The local Poisson problem estimator:

$$\eta_P = \sqrt{\sum_{T \in T_h} \eta_{P,T}^2} \tag{23}$$

as follows

$$\eta_{P,T}^2 = \|\mathbf{e}_{P,T}\|_{J,T}^2 + \|\varepsilon_{P,T}\|_{0,T}^2 \tag{24}$$

Theorem 5 *The estimator $\eta_{P,T}$ is equivalent to the $\eta_{R,T}$ estimator:*

$$c\eta_{P,T} \leq \eta_{R,T} \leq C\eta_{P,T} \tag{25}$$

Theorem 6 *For any mixed finite element approximation defined on triangular grids T_h the estimator η_P satisfies:*

$$\|\mathbf{e}\|_{J,\Omega} + \|\varepsilon\|_{0,\Omega} \leq C\eta_P \tag{26}$$

and

$$\eta_{P,T} \leq C\left(\sum_{T \in T_h} \left\{\|\mathbf{e}\|_{J,T'}^2 + \|\varepsilon\|_{0,T'}^2\right\}\right)^{\frac{1}{2}} \tag{27}$$

The constant C in the local lower bound is independent of the domain.

6 Numerical Simulation

In this section, we present numerical tests; based on the MFE method presented in this article, we use the simulator Comsol Multiphysics. The results obtained confirm that the errors are reasonable and the numerical computations of our problems have demonstrated that this approach yields a physically realistic flow. In this simulations, we take fluid density = 1000 kg/m^3, permeability = 0.3 m^2, Porosity = 0.4, and the fluid viscosity values ranging from 0.1 Pa $*$ s. We consider the boundary condition C_{a,μ^*}, where $\overrightarrow{t} = (1, 1)$, $a = 10^{-4}$ and $a = 10^3$. For discretization, we use uniform rectangular discretization of our reservoir. Figure 1 presents the velocity for $a = 10^{-4}$ and $a = 10^3$.

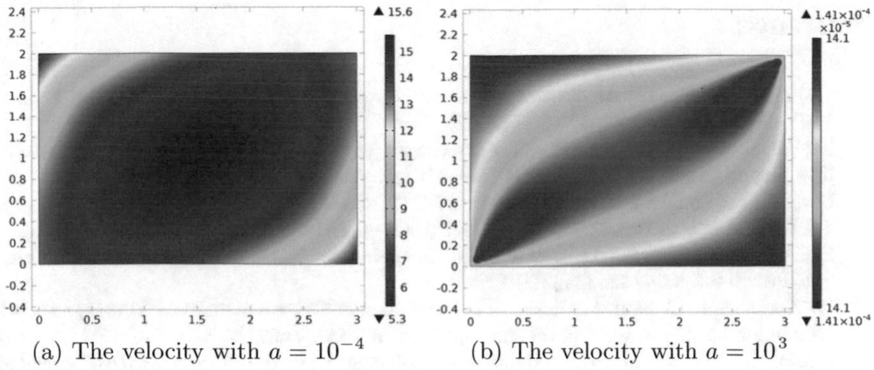

(a) The velocity with $a = 10^{-4}$ (b) The velocity with $a = 10^{3}$

Fig. 1 Velocity of Brinkman equation

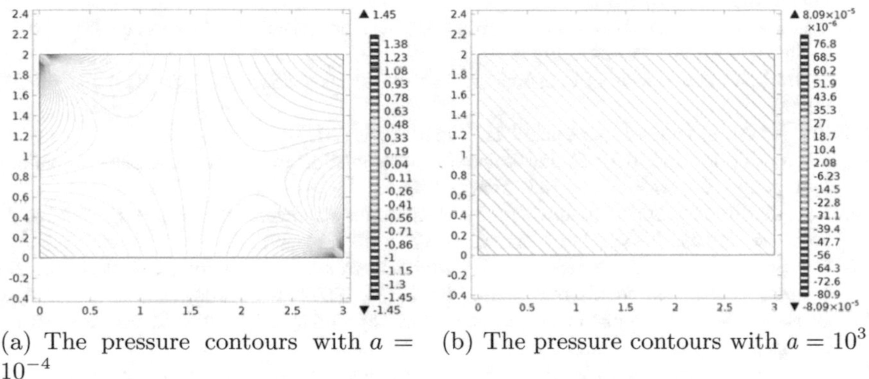

(a) The pressure contours with $a = 10^{-4}$ (b) The pressure contours with $a = 10^{3}$

Fig. 2 Pressure contours of Brinkman equation

Table 1 Errors, the linear error, residual error, and the error of the velocity

h	Number of grids	erru	LinErr	LinRes
0.10	30 × 20	0.25	$1.6e^{-12}$	$8.0e^{-15}$
0.15	20 × 15	0.36	$1.8e^{-12}$	$1.0e^{-14}$
0.20	15 × 10	0.47	$2.4e^{-12}$	$1.9e^{-14}$
0.30	10 × 07	0.68	$3.8e^{-12}$	$2.7e^{-14}$
0.40	08 × 05	0.82	$5.5e^{-12}$	$5.0e^{-14}$

In Fig. 2, we show the pressure contours of Brinkman equation with different parameter in the boundary equation. Now, we present different errors for our problem. In Table 1, LinErr is the linear error, LinRes is the residual error, and erru is the error of the velocity equation. From Table 1, we can see the efficiency of this method, when the mesh is small enough the error approaches zero.

References

1. Wu, D.H., Currie, I.G.: Analysis of a posteriori error indicator in viscous flows. Int. J. Num. Meth. Heat Fluid Flow **12**, 306–327 (2002)
2. Rajagopal, K.R.: On a hierarchy of approximate models for flows of incompressible fluids through porous solids. Math. Models Methods Appl. Sci. **17**(2), 215–252 (2007)
3. Lévy, T.: Loi de Darcy ou loi de Brinkman? C. R. Acad. Sci. Paris Sér. II Méc. Phys. Chim. Sci. Univers Sci. Terre. **292**(12), 871–874, Erratum (17):12–39 (1981)
4. Elakkad, A., Elkhalfi, A.: Analysis of estimators for stokes problem using a mixed approximation. Bol. Soc. Paran. Mat. (3s.) (2018)
5. El-Mekkaoui, J., Elkhalfi, A., Elakkad, A.: Resolution of Stokes equations with the $C_{a;b}$ boundary condition using mixed finite element method. WSEAS Trans. Math. (2015)
6. Brezzi, F., Fortin, M.: Mixed and Hybrid Finite Element Method, Computational Mathematics. Springer, New York (1991)
7. Raviart, P.A., Thomas, J.: Introduction l'analyse numérique des à équations aux dérivées partielles. Masson, Paris (1983)
8. Ainsworth, M., Oden, J.: A posteriori error estimates for Stokes' and Oseen's equations. SIAM J. Numer. Anal **34**, 228–245 (1997)
9. Verfurth, R.: A posteriori error estimators for the Stokes equations. Numer. Math **55**, 309–325 (1989)
10. Ern, A.: Aide-mémoire Eléments Finis. Dunod, Paris (2005)
11. Girault, V., Raviart, P.A.: Finite Element Approximation of the Navier-Stokes Equations. Springer, Berlin, Heiderlberg, New York (1981)
12. Kay, D., Silvester, D.: A posteriori error estimation for stabilized mixed approximations of the Stokes equations. SIAM J. Sci. Comput. **21**, 1321–1336 (1999)
13. Creuse, E., Kunert, G., Nicaise, S.: A posteriori error estimation for the Stokes problem: anisotropic and isotropic discretizations. MMAS **14**, 1297–1341 (2004)
14. Brinkman, H.C.: A calculation of the viscous force exerted by a flowing fluid on a dense swarm of particles. Appl. Sci. Res. **A1**, 27–34 (1948)
15. Elman, H., Silvester, D., Wathen, A.: Finite Elements and Fast Iterative Solvers: With Applications in Incompressible Fluid Dynamics. Oxford University Press, Oxford (2005)
16. Roberts, J., Thomas, J.M.: Mixed and Hybrid Methods, Handbook of Numerical Analysis II, Finite Element Methods 1. P. Ciarlet and J. Lions, Amsterdam (1989)
17. Clement, P.: Approximation by finite element functions using local regularization. RAIRO. Anal. Numer. **2**, 77–84 (1975)
18. Bank, R.E., Weiser, A.: Some a posteriori error estimators for elliptic partial differential equations. Math. Comput. **44**, 283–301 (1985)

Smart Health Monitoring System Using MIT App Inventor 2

Suneetha Merugula, Akhil Chakka and Satish Muppidi

Abstract Even in healthcare industries and organizations are drastically moving toward the automation for patient treatment or monitoring which helps them explore into the other affecting areas in the medical field. Computerization can be done using the Internet of Things (IoT) or Artificial Intelligence or RFID technology. In this paper, we will be discussing regarding the healthcare monitoring using IoT and with the advanced and enhanced wireless technologies like Wi-Fi, Bluetooth, and ZigBee for tracing the patient health using assisted sensors such as DHT11, LM35, HC-05, and heart rate. The sensors connected to the Arduino board will gather those readings and then process the information. In this, all the sensors connected will transmit the data using Bluetooth module to the application and will be notifying the changes observed in the readings continuously to the person connected to the application instead of displaying the sensed data separately one after the other on the display module.

Keywords Sensing · Processing · Monitoring · Internet of Things · Healthcare · Tracing · Data or information

1 Introduction

Arduino is a prototype platform consisting of a circuit board, which can be programmed, and for compiling and dumping the code to the physical board, there is an Arduino integrated development environment (IDE). Before working with these boards, you need to be familiar with C/C++ and the basic idea of the microcontrollers and electronics. The basic Arduino board resides of power USB, voltage regulator, reset button, analog and digital pins, main microcontroller, TX, and RX LEDs.

The system mainly comprises of two phases, the first phase is developing the Android application, and the second phase is the connection of the sensors to the Arduino. For the building of the Android application, we have used the online tool,

S. Merugula (✉) · A. Chakka · S. Muppidi
Department of Information Technology, GMRIT, Rajam, Andhra Pradesh, India
e-mail: suneetha.merugula@gmail.com

V. Bhateja et al. (eds.), *Embedded Systems and Artificial Intelligence*,
Advances in Intelligent Systems and Computing 1076,
https://doi.org/10.1007/978-981-15-0947-6_24

MIT App Inventor 2. There are many paths for building the Android application such as Android Studio, Appery.io, Mobile Roadie, The App Builder, and many others. It is difficult installing such tools which need to support the system specifications, so instead of that we will be using an online tool MIT App Inventor 2 where all the required blocks such as user interface and coding for developing the projects will be available, i.e., all the required magnets and back-end coding are in-built. In this application, we will be having two screens:

Screen-1: In this, we can observe a scan button used for scanning the nearby Bluetooth devices and four types of sensor names and four empty boxes for displaying the values that are received from the Bluetooth module.

Screen-2: In this, it will be asking you to enter your name, contact number, the values that are displayed in the Screen 1, and once you click on the save button, the entered details will be stored in the form of CSV format.

In the second phase, we need to connect the sensors to the Arduino, and the required components or sensors are Arduino board, temperature, and humidity, temperature, HC-05 Bluetooth module, 10 kΩ resistor, breadboard, and jumper wires. After connecting the sensors to the Arduino board, we will connect the board to PC; using the Arduino IDE, we will dump the code into Arduino.

2 Literature Survey

Late years have seen a rising excitement for wearable sensors, and today, a couple of devices are modernly available for individual human administrations, fitness, and development care. In light of current inventive examples, one can instantly imagine a period within the near future when your routine physical examination is gone before by a two multi-day time of steady physiological watching using sensible wearable sensors. Over this break, the sensors would determinedly record signals related to your key parameters and exchange the resulting data to a database associated with your prosperity records [1].

Internet of Things (IoT) can be portrayed as installed gadgets (things) with Internet availability that collaborates with one another, administrations, and individuals on an exhaustive scale. Medicinal service is the protection and improvement of well-being through recognizable proof, finding, treatment, and anticipation of ailments, infection, wounds, and other physical and mental harm in people. Social insurance can add to a significant piece of a nation's economy [2].

Among the variety of utilizations empowered by the Internet of Things (IoT), great and associated human services could be a prominently essential one. In particular, the supply of information at so far mind-blowing scales, and transient longitudes also a substitution age of canny procedure calculations can: (a) encourage a development inside the prescription, from this post-analysis and treat responsive world view, to a proactive system for guess of sicknesses at a beginning time, (b) modify personalization of treatment, and the executives decisions focused on strikingly to precise circumstances and want of the individual, and (c) encourage downsize the estimation of human services while in the meantime rising results [3].

The advanced medicinal services framework empowers therapeutic experts to remotely perform constant observing, early conclusion, and treatment for potential well-being perils. The cutting edge human services are connected for giving progressively productive use of doctors, decreasing the expense for emergency clinic stays, diminishing the expertise level and limiting the recurrence of visits of home-care experts, lessening medical clinic readmission rates, and advancing well-being training at different dimensions. In clinics, where the patient's condition should be always checked, it is typically done by a specialist or other paramedical staff. On account of a crisis, even a little deferral in treatment may represent a danger to the patient's life [4].

Internet of Things (IoT) is such when the web and systems are crushed up. As indicated by the framework, IoT is the blend of implanted gadgets, sensors, programming, and network, and in some cases, it is called the web of everything. In this sense, we have built up a model for gathering well-being information from patients as a human to things and the well-being units got to the information to manage the patient's express whenever remotely. Nearly, individuals of any age are utilizing cell phones, for example, PDAs and tablets for utilizing different applications as a result of being propelled portable advances Accessibility of web it is currently getting to be less demanding to utilize versatile advancements for therapeutic applications [5].

With the advancement of the world, health checking framework is utilized in each field, for example, emergency clinic, home consideration unit, and sports. The normal heartbeat every moment for 25-year old ranges between 140 and 170 bpm while, for a 60-year old, it is around between 115 and 140 bpm and body temperature is 37 °C or 98.6 Fahrenheit. Patients are not knowledgeable with manual treatment which specialists regularly use for following the tally of the heartbeat. Distinctive biomedical sensors like temperature sensor, pulse sensor, and circulatory strain sensor are utilized for observing the well-being condition which is incorporated on a single framework on the chip [6].

Health has prime significance in our everyday life. Sound health is important to do the day by day work legitimately. This task goes for building up a framework which gives body temperature and pulse utilizing LM35 and heartbeat sensor individually. These sensors are interfaced with controller Arduino Uno board. Remote information transmission is done by Arduino through Wi-Fi module. ESP8266 is utilized for remote information transmission on IoT stage, for example, thing talk. With the goal that record of information can be put away over time frame. This information put away on web server so it can see to who logged [7].

The expanded utilization of portable advances and brilliant gadgets in the zone of health has caused an incredible effect on the world. Health specialists are progressively exploiting the advantages these advances bring, in this manner creating a noteworthy improvement in medicinal services in clinical settings and out of them. Moreover, endless normal clients are being served from the upsides of the mobile health (M-Health) applications and medicinal services bolstered by ICT (E-Health) to improve, help and help their health. The Internet of Things is progressively permitting to incorporate gadgets fit for interfacing with the Internet and gives data on the condition of health of patients and gives data progressively to specialists who help [8].

The improvement of a microcontroller-based framework for remote heartbeat and temperature observing utilizing ZigBee. In India, numerous patients are kicking the bucket in view of heart assaults and purpose for that they are not getting opportune and appropriate help. The fixed observing framework can be utilized just when the patient is on bed and this framework is colossal and just accessible in the medical clinics in ICU. The framework is produced for home use by patients that are not in a basic condition but rather should be steady or occasionally observed by clinician or family [9].

3 Approaches

In this system, temperature, pulse, and ECG signals of the patient or person being monitored. There are two different ways to assemble data of the individual from the sensors, one is a microcontroller and other is using Arduino and Raspberry. The main function of this microcontroller is to convert the analog data to digital data and transfer the information using the ZigBee technology which is a big end communication protocol with low power digital signals based on IEEE standard. The sensors connected to the Arduino generate some data and it will be transmitting or storing the data as given in the instructions. For the carrying of the data, we will be using the Bluetooth module or Wi-Fi module. Since there is a chance of establishing a connection with one device at a time to the Bluetooth device, the data can be securely transmitted. The information collected from the sensors is displayed in the LCD module at most two readings at a time. After that, the data is sent to the internet cloud application or server using the Wi-Fi module.

3.1 Disadvantages

- The values from each sensor will be displayed separately one after another or at most two at a time due to which it takes more time to observe each reading when more sensors are connected.
- Sometimes, there might be the possibility of failing to retrieve the collected data and is not possible to check the data if the cloud server fails to respond when needed.
- The data in the existing system will be reserved in the cloud in which the data cannot be shared from person to another automatically.
- If he wants to analyze the present observed data with the past data, it might be difficult to open the cloud server if not connected to the Internet (Figs. 1, 2, and 3).

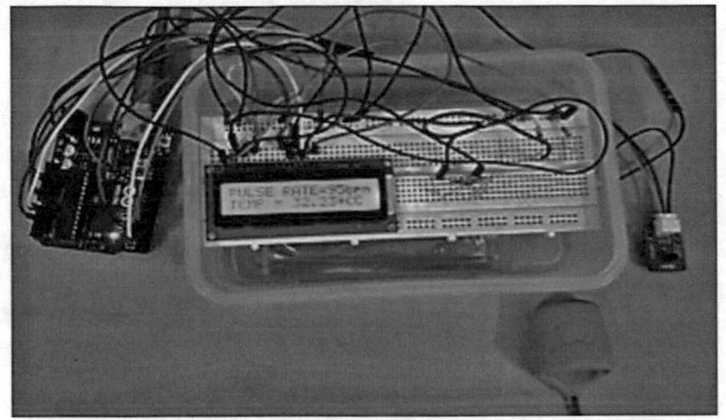

Fig. 1 Interfacing of LCD and sensors

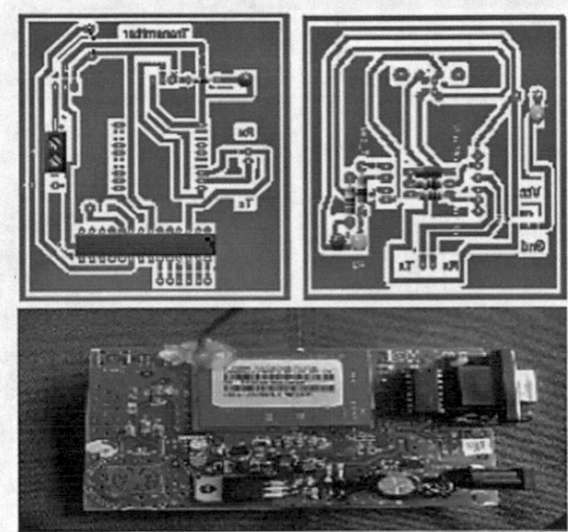

Fig. 2 PCB design of the system

4 Design and Implementation

Before starting the first step, we need to develop an Android application to display the sensed data which is being received through the Bluetooth module. For this purpose, we have used the MIT App Inventor 2 for developing the Android application which consists of two screens, one for displaying the received data and the other for storing and sharing the data.

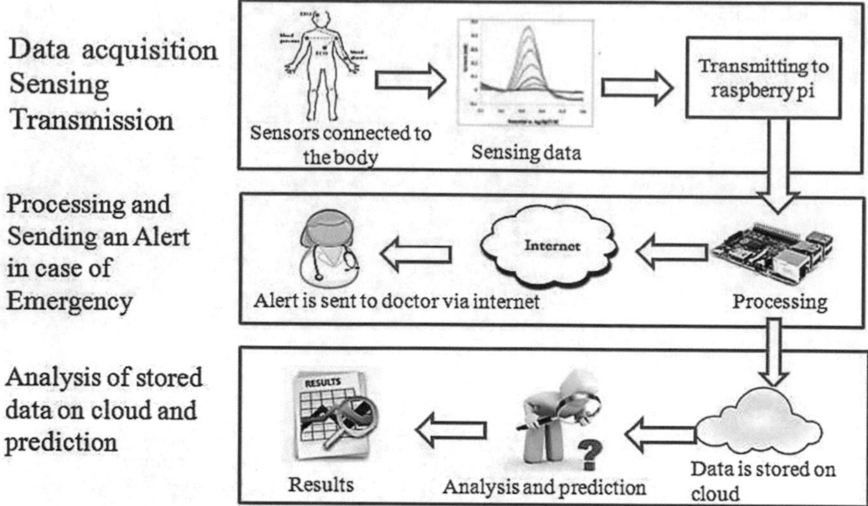

Fig. 3 Connection to the cloud server

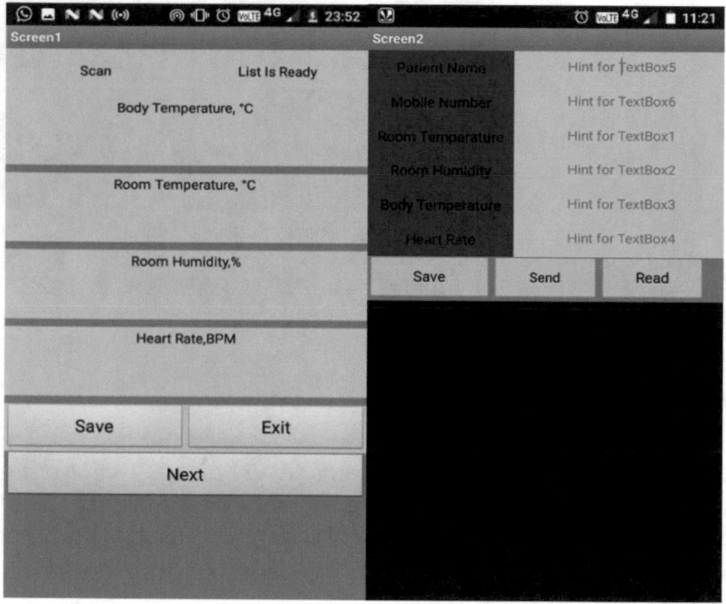

In this system, we will be monitoring the patient or person body temperature and heartbeat. We will also be looking at room temperature and humidity of the person where he will be residing. This system is four-step process sensing, processing, application, alerts. Before starting the first step, we need to gather all the required components and then start the process. The components used to build the system are:

(1) **Arduino UNO or Genuino**: Arduino is a prototype platform consisting of a circuit board, which can be programmed using the C/C++. Arduino board is easy to learn and flexible enough for advanced users. Architects, designers build interactive prototypes, and artists, musicians use it for installations and experiment with new musical instruments.
(2) **Bluetooth module (HC-05)**: This module is a serial port protocol designed for transferring the captured information from Arduino to an application wirelessly. Since only one user or person can connect at a time, the data can be securely transmitted without any loss.
(3) **Temperature and humidity sensor**: The DHT11 sensor is integrated with high-performance, eight-bit microcontroller and ensures high reliability and excellent stability.
(4) **Pulse or heart rate sensor**: The pulse rate sensor is a plug and heart rate sensor is used to calculate the number of heartbeats per minute placing near to patients nerve to capture the information.
(5) **Breadboard**: A breadboard is a modest, simple to utilize bit of equipment for wiring electrical circuits. A breadboard is typically secured with gaps fixed with metal, in which wires and electrical segments, for example, resistors, diodes, and capacitors, can be stopped. The openings are isolated into columns, and gaps inside specific lines are wired together on the underside of the breadboard with the goal that an electric flow can stream down the lines.
(6) **Resistor and jumper wires**: A resistor is a passive two-terminal electrical part that actualizes electrical opposition as a circuit component. In electronic circuits, resistors are utilized to decrease current stream, change flag levels, to separate voltages, inclination dynamic components, and end transmission lines, among different employments.

A jumper wire is an electrical wire, or gathering of them in a link, with a connector or stick at each end, which is typically used to interconnect the parts of a breadboard or other model or test circuit, inside or with other hardware or segments, without welding.

Block Diagram:

See Fig. 4.

In the first step, we will not be connecting the components directly to the Arduino, we will be using a breadboard between the Arduino and the sensors to control the voltage drops so that there will not be any damage for the sensor or module. Then the connections are done open the Arduino IDE in your system or PC and compile the code, and once the code is compiled, dump the code into the Arduino board using the USB cable. As soon as the code is dumped, all the sensors should flash their LEDs if available (Fig. 5).

In step three, we need to first connect application to the Bluetooth module by clicking on the scan button, then click on the list is a ready button to view all the available devices, and connect to the Bluetooth module. Then the collected data will be validated and then it will be transmitted on to the Android application developed by us using the Bluetooth module.

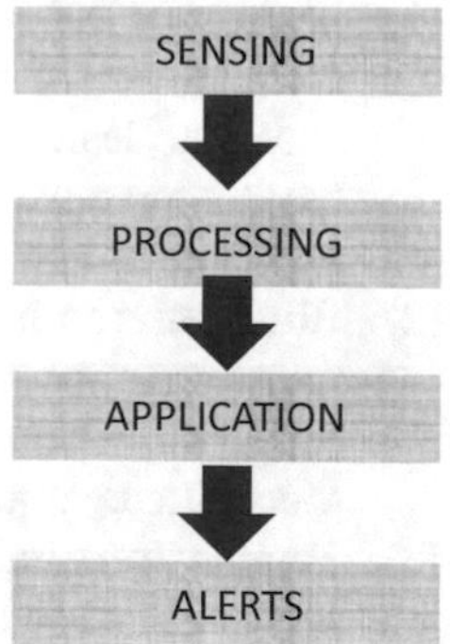

Fig. 4 Block diagram

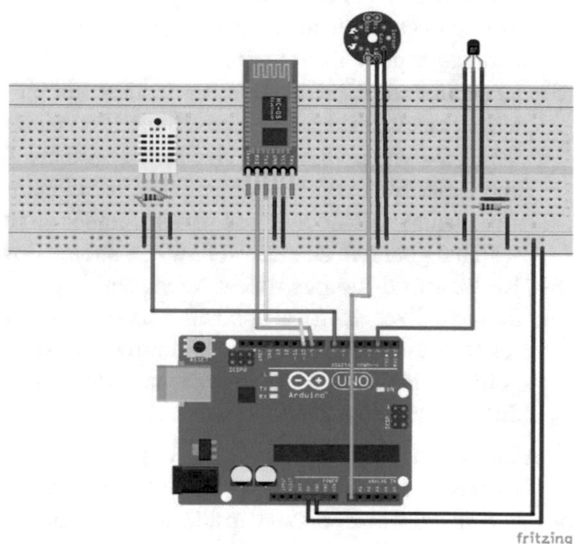

Fig. 5 Interfacing the components

In the last step, once we click on the save button, the data will be stored in the system files which cannot be viewed directly. So in order to view the data in the future purpose, by clicking on the next button, another screen will be opened where you need to manually enter the name, phone number, and the values observed in the previous slide, and if you click on save, the entered data will be stored in CSV format which be shared to anyone else if you click on the send button, the values you have entered in the fields will be sent to the number which you have entered in the phone number field and if click on the read button, then the data you have entered will be displayed below.

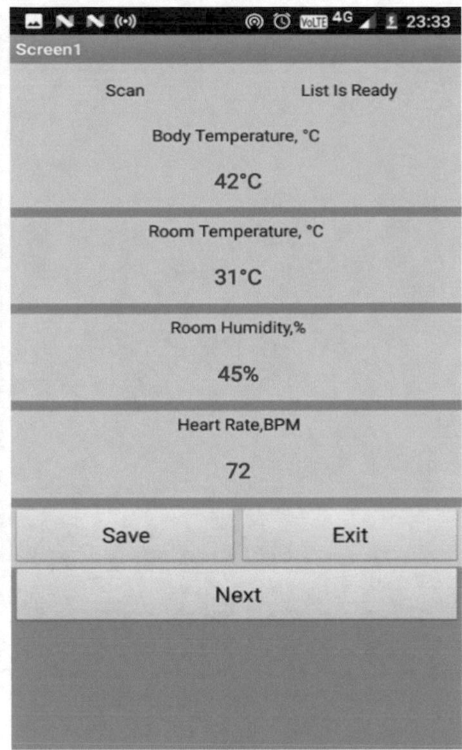

Fig. 6 Displaying data from sensors

5 Results

See Figs. 6, 7, 8 and 9.

6 Conclusion

According to the latest survey of many of the villages and the some of the areas in cities and towns, there is at most one physician, and in some villages, there are not even one also. In the coming futures ahead, even the population of the elderly people will also be increased in which they need continuous monitoring of the patients for the sake of their life. So using this IoT designed health system which does not even need much amount to implement can be used at any places at any time whenever required. No project has been done for integrating and sharing the information automatically through any android application to the patient. If such a system is implemented, then these can be used in the absence of the doctor in the hospital or organization so that the patient can be alerted if any problem occurs.

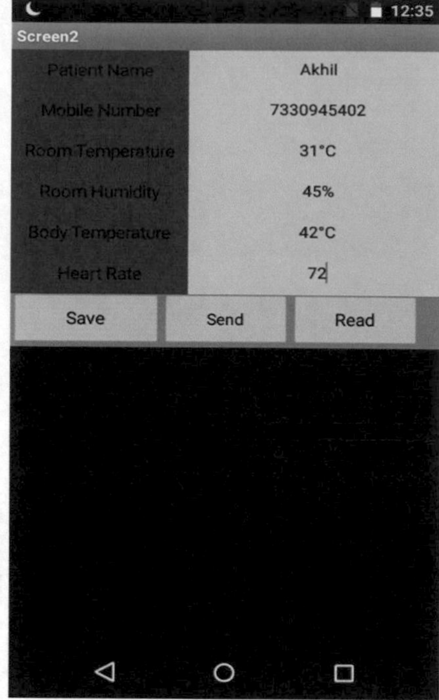

Fig. 7 Storing and sharing

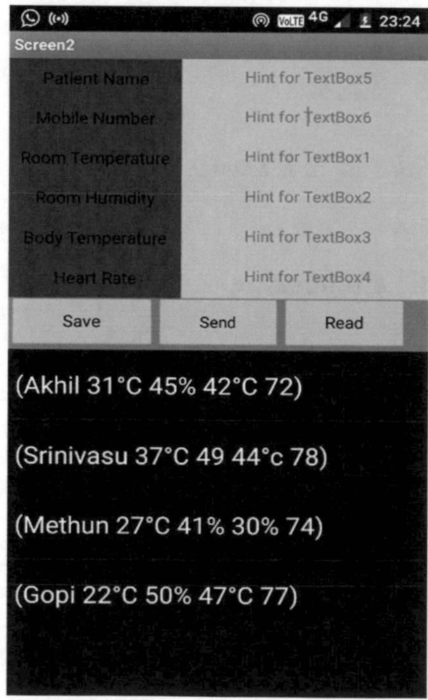

Fig. 8 Storing and message alerts

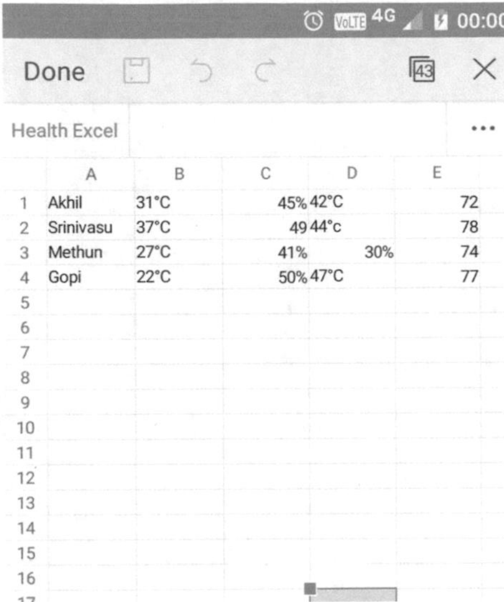

	A	B	C	D	E
1	Akhil	31°C	45%	42°C	72
2	Srinivasu	37°C	49	44°c	78
3	Methun	27°C	41%	30%	74
4	Gopi	22°C	50%	47°C	77

Fig. 9 Reading values from CSV file

References

1. Saha, H.N., Auddy, S., Pal, S., Kumar, S., Pandey, S., Singh, S., Singh, A.K., Sharan, P., Ghosh, D., Saha, S.: Health monitoring using internet of things (IoT). 978-1-5386-2215-5/17/$31.00 © IEEE (2017)
2. Hassanalieragh, M., Page, A., Soyata, T., Sharma, G., Aktas, M., Mateos, G., Kantarci, B., Andreescu, S.: Health monitoring and management using internet-of-things (IoT) sensing with cloud-based processing: opportunities and challenges (2015)
3. Shaikh, S., Chitre, V.: Healthcare monitoring using IoT. In: International Conference on Trends in Electronics and Informatics ICEI (2017)
4. Rathore, D.K., Upamanyu, A., Iulla, D.: Wireless patient health monitoring system. 978-1-4799-1607-8/13/$31.00 © IEEE (2013)
5. Ghosh, A.M., Halder, D., Hossain, S.A.: Remote health monitoring system through IoT. In: 2016 5th International Conference on Informatics, Electronics and Vision (ICIEV)
6. Krishnan, D.S.R., Gupta, S.C., Choudhury, T.: IoT based patient health monitoring system. Int. Res. J. Eng. Technol. (IRJET) **04**(03) (2017)
7. Patil, S., Pardeshi, S.: Health monitoring system using IoT. Int. Res. J. Eng. Technol. (IRJET) **05**(04) (2018)
8. Gomeza, J., Oviedob, B., Zhumab, E.: Patient monitoring system based on internet of things. In: The 7th International Conference on Ambient Systems, Networks and Technologies (ANT 2016)
9. Shelar, M., Singh, J., Tiwari, M.: Wireless patient health monitoring system. Int. J. Comput. Appl. **62**(6) (2013)

Performance Analysis of DYMO and LAR in Grid and Random Environments

K. Narasimha Raju, B. Satish Kumar, G. Hima Bindu and S. K. Sharma

Abstract The network formed randomly with moving wireless links in MANETs. Establishing routes to deliver the packets in this kind of network is really a difficult task due to different deployment patterns and mobility of the nodes. The analysis of the protocols helps to test its suitability due to cost aspect before deploying the network in real time. In this paper, the two popular protocols namely DYMO and LAR are analyzed in grid and random environments. Qualnet simulator is used to conduct the experiments. The result states that each protocol has its significance depending on the situation.

Keywords DYMO · LAR · MANET · Performance

1 Introduction

Ad hoc network provides communication via wireless nodes without any centralized base. In wireless networks (MANET) [1–3], the wireless devices move in their own fashion at each instance. The most advantageous thing in MANET is that it can be deployed with a little interval of time compared to the traditional centralized structures. This flavor attracted so many researchers to create and deploy the MANET environment in different fields. Establishing the routes called routing which is a major challenging issue in MANETs. There many routing protocols were developed in this regard but the two popular protocols among them are DYMO and LAR. Before

K. Narasimha Raju (✉) · B. Satish Kumar
Department of CSE, Lendi Institute of Engineering & Technology, Vizianagaram, India
e-mail: rj.vizagg@gmail.com

B. Satish Kumar
e-mail: vsp.satish@gmail.com

G. Hima Bindu · S. K. Sharma
Department of MCA, Vignan's Institute of Information Technology, Visakhapatnam, India
e-mail: goginenibindu9@gmail.com

S. K. Sharma
e-mail: sharma.santosh83@gmail.com

V. Bhateja et al. (eds.), *Embedded Systems and Artificial Intelligence*,
Advances in Intelligent Systems and Computing 1076,
https://doi.org/10.1007/978-981-15-0947-6_25

operating into the real-time environment due to cost aspect, the protocols should be analyzed in a simulated environment. In this paper, DYMO and LAR routing protocols are evaluated using Qualnet simulator. Section 2 illustrates the description of the protocols (DYMO, LAR) and various deployment situations (placement scenarios). Section 3 describes brief related literature in the connected work. Section 4 presents experimental evaluation methodology considered for simulation. Section 5 shows the results and made a conclusion in section.

2 Routing Protocols and Deployment Models

Routing protocols act as an aid to establish ways for the data packets from an origin to the target. This is a very crucial task in case of MANETs due to random behavior of the mobile nodes and dynamic patterns of the topology. Different routing protocols [4–6] were developed in MANETs to establish the routes. In this paper, DYMO [7] and LAR [8] protocols are evaluated in grid and random environments.

2.1 Dynamic MANET On-demand (DYMO)

The Dynamic MANET On-demand (DYMO) is an on-demand routing protocol in MANETs. In this, source node generates Route Request (RREQ) packets and broadcast them to establish a route. Once the target receives RREQ packet, the target node responds with Route Reply (RREP) packets. Failure in the established routes is intimated through Route Error (RERR) Packets to the source from the nodes.

2.2 Location Aided Routing (LAR) Protocol

In LAR, each mobile node utilizes location information obtained via GPS to establish and maintain the paths. This helps for reducing the incurred overhead in the network. The limitation in applicability of the protocol is that each mobile node should have GPS connection.

2.3 Deployment Models

The deployment strategies or placement patterns illustrate the placement of the nodes to create the communication.

Fig. 1 Random pattern of mobile nodes

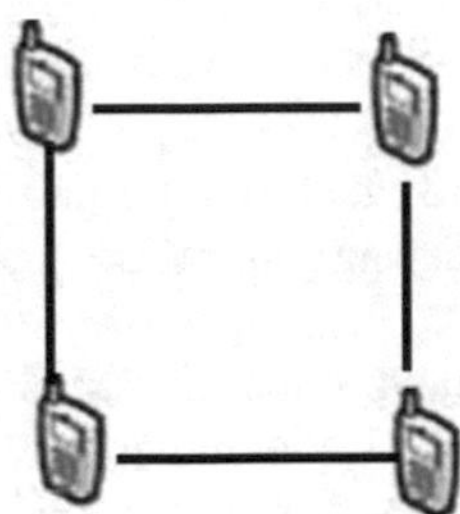

Fig. 2 Grid pattern of mobile nodes

1. Random Placement: The mobile nodes are deployed arbitrarily and placed in a random fashion within an available terrain region. Figure 1 shows one such pattern.
2. Grid Placement: The dynamic nodes are distributed in grid fashion as presented in Fig. 2. In most cases, mobile nodes to be deployed in a grid fashion are suggested to be in square.

3 Related Work

In future generation networks, Chlamtac et al. [9] advised that WSN and MANETs occupy a significant role. They presented various challenges involved in this type of networks.

Different routing protocols are proposed to deliver the data packets over dynamic patterns in MANETs. Srivastava et al. [10] provided a survey and overview on different routing protocols. They compared various protocols in various aspects. Mohseni [11] classified the routing schemes based on their behavior compared with the two routing strategies of routing protocols provided and highlighted various characteristics. Renu et al. [12] presented their work on current protocols and compared the protocols in MANET. Abolhasan et al. [13] provided a description of routing protocols and gave a performance comparison of all routing protocols. They suggested that different protocols may perform well in some situations.

More focus should be kept on performance considerations and evaluation strategies in this type of network. Corson et al. [14] described the characteristics and considerations for performance in MANETs. Kumar et al. [15] studied and analyzed the DSR, AODV, ZRP, and DYMO protocols using random waypoint mobility model under random placement based on CBR. They analyzed the performance using Qualnet 5.0.2. Spaho et al. [16] evaluated the DYMO protocol in various scenarios under VANET environment. They investigated the performance by varying speed and density. They concluded that with increase in number of nodes, packet delivery will be increased.

Sagar et al. [17] evaluated the DYMO and OLSR protocols in both VANET and MANET environments. They analyzed the performance in terms of Packet Delivery ratio and delay. They concluded that DYMO performs well in VANET compared to MANET. Dinesh Singh et al. [18] analysed the performance of LANMAR, LAR1, DYMO and ZRP protocols by varying pause time. They considered jitter, throughput and delay metrics for evaluation.

Garcia-Campos et al. [19] analyzed the LAR, DYMO, DSR, and AODV in urban scenarios. They concluded that LAR yields better results in step mode. Setty et al. [20] analyzed the AODV in uniform, random, and grid patterns. They evaluated the performance in terms of QoS Metrics by conducting simulation studies through speed and network size. Narasimha Raju et al. [21] analyzed the LAR and Fisheye protocols in random and grid environment at 49 nodes.

4 Methodology

The analysis of the protocols helps in a great way to test the suitability before deploying the network in real time. The applicability of each protocol can be studied through a serious of experiments. In this simulation, methodology is considered for evaluating the two protocols, namely DYMO and LAR, which are analyzed in grid and random environments. The Experiments are conducted at 64 nodes with Qualnet tool [22]. The deployment terrain region is of 1500×1500 m^2 area. Table 1 shows various parameters involved in the experimental studies.

5 Results and Discussion

The obtained results are studied in terms of QoS metrics and energy consumption aspects. QoS metrics considered are average end-to-end delay, average jitter, packet delivery ratio, and average throughput. A typical running scenario using Qualnet in grid environment is shown in Fig. 3. From Figs. 4 and 5, it was observed that both

Table 1 Experimental setup for grid and random environments

Simulation environment	
Area	1500 m × 1500 m
Simulation time	600 s
Node deployment	Grid and random
Pause time	0 s
Max speed	10 m/s
Traffic	CBR
Number of items	100
Item size	512 bytes
Protocols	DYMO and LAR
MAC	802.11
Number of nodes	64

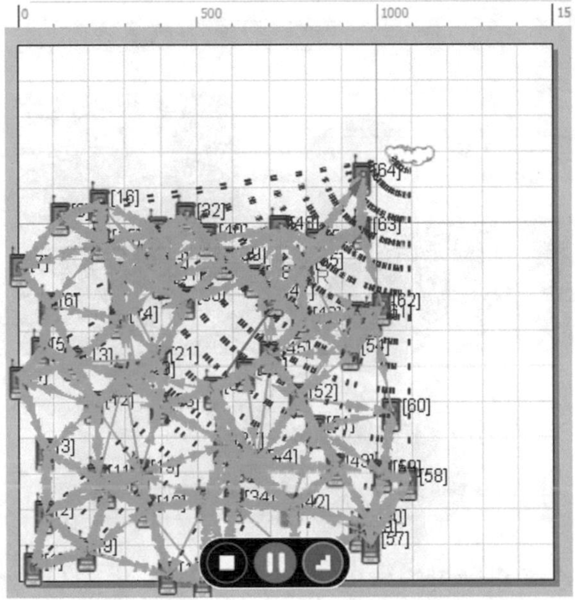

Fig. 3 Typical running scenario in grid environment

protocols perform well in random environment; when compared LAR with DYMO in grid environment, LAR performs well in grid environment in case of packet delivery ratio. From Figs. 6 and 7, it was observed that both protocols perform well in random environment; when compared LAR with DYMO in grid environment, LAR performs well in grid environment in case of average end-to-end delay. From Figs. 8 and 9, it was observed that both protocols perform well in random environment; when compared LAR with DYMO in grid environment, DYMO performs well in grid

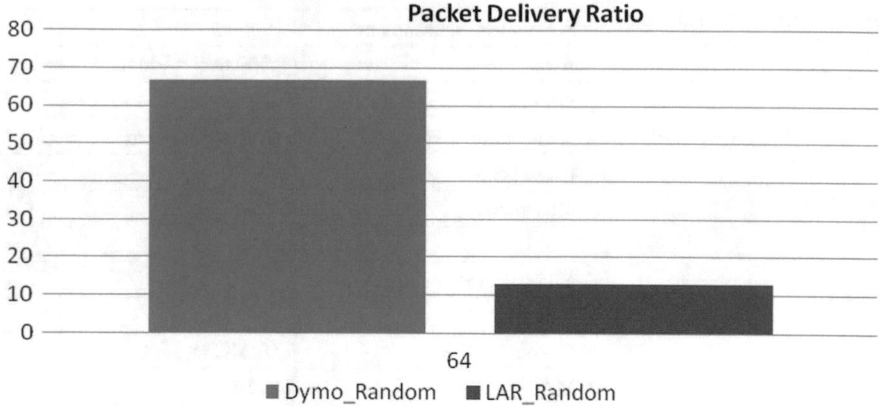

Fig. 4 Packet delivery in percentage for LAR in random and grid

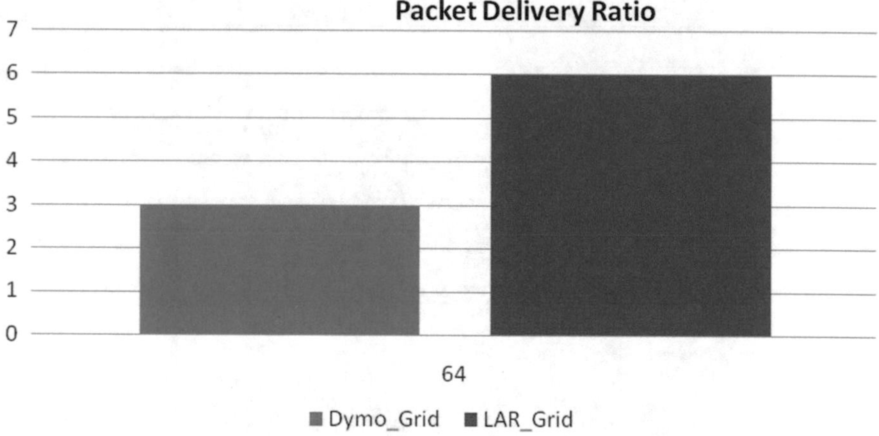

Fig. 5 Packet delivery in percentage for DYMO in random and grid

environment in case of average throughput. From Figs. 10 and 11, it was observed that both protocols perform well in random environment; when compared LAR with DYMO in grid environment, LAR performs well in grid environment in case of average jitter. From Figs. 12 and 13, it was observed that DYMO consumes less energy in transmit mode in grid environment, when compared LAR with DYMO, LAR performs well in both grid and random environments in case of energy transmit mode. From Figs. 14 and 15 it was observed that DYMO consumes less energy in receive mode in random environment; when compared LAR with DYMO, LAR performs well in both grid and random environments in case of energy in receive mode. From Figs. 16 and 17, it was observed that LAR consumes less energy in Idle mode in random environment; when compared LAR with DYMO, DYMO performs well in both grid and random environments in case of energy in idle mode.

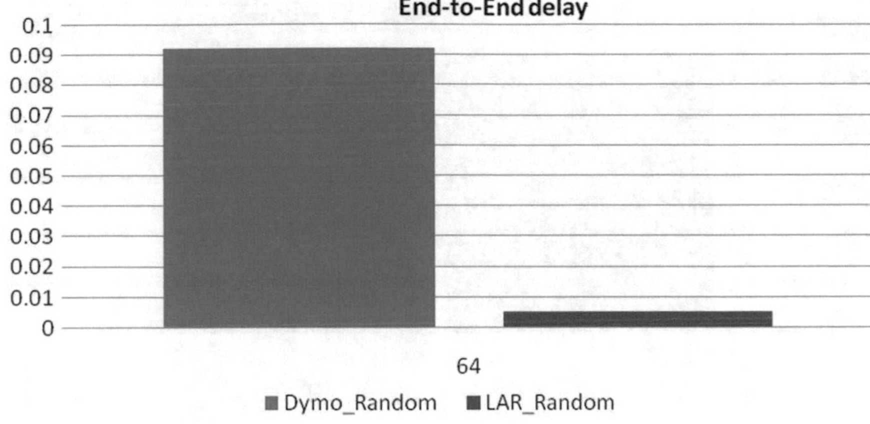

Fig. 6 Average end-to-end delay for DYMO and LAR in random

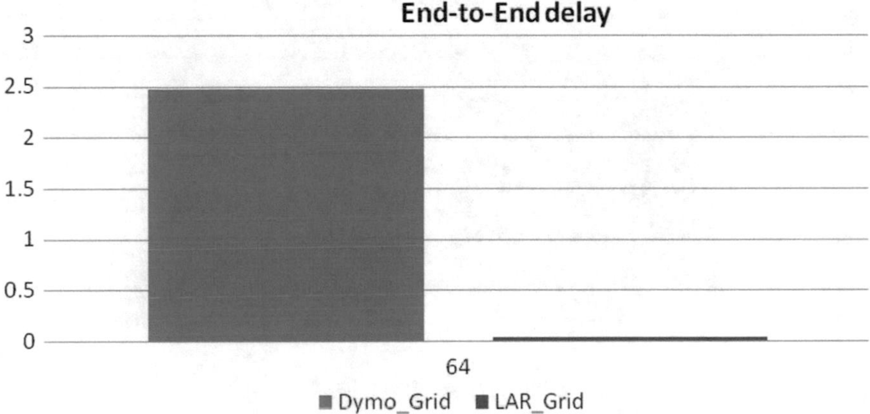

Fig. 7 Average end-to-end delay for DYMO and LAR in grid

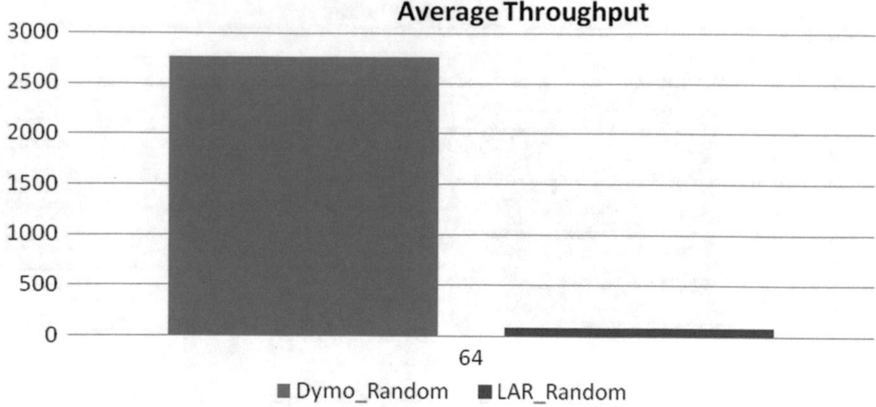

Fig. 8 Average throughput for DYMO and LAR in random

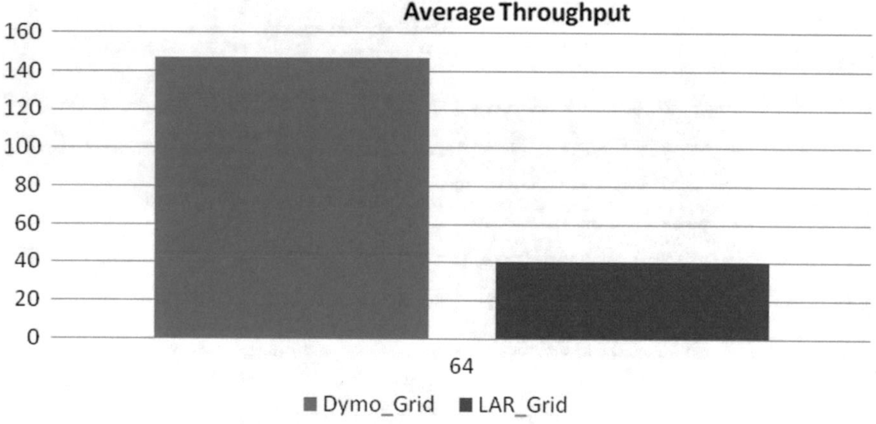

Fig. 9 Average throughput for DYMO and LAR in grid

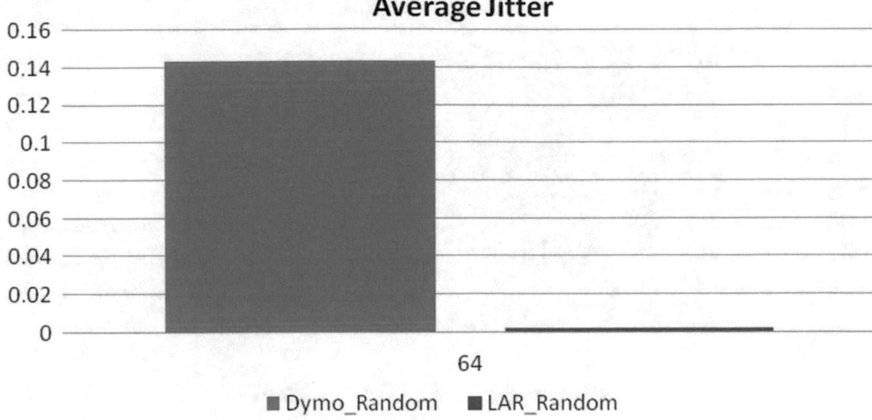

Fig. 10 Average jitter for DYMO and LAR in random

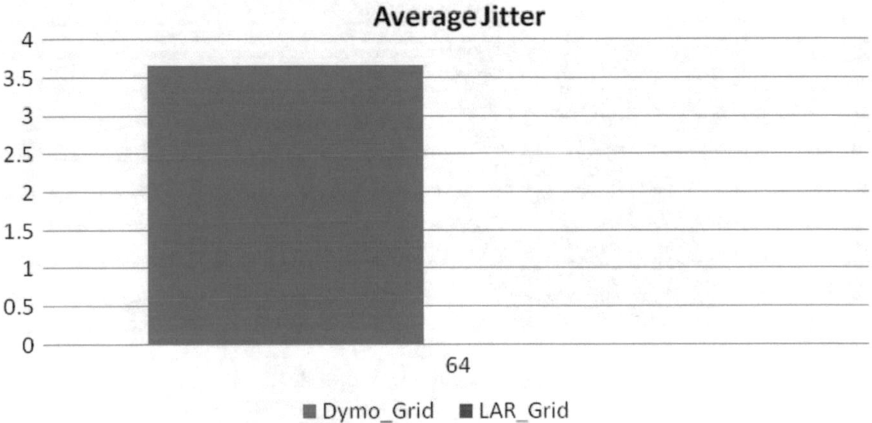

Fig. 11 Average jitter for DYMO and LAR in grid

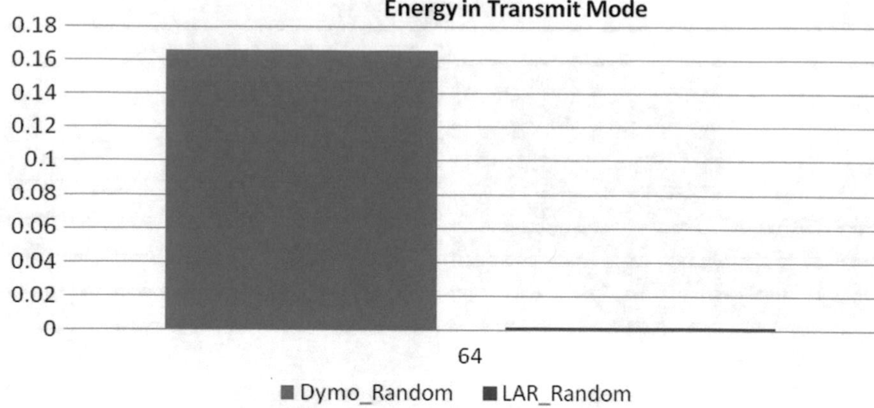

Fig. 12 Energy in transmit mode for DYMO and LAR in random

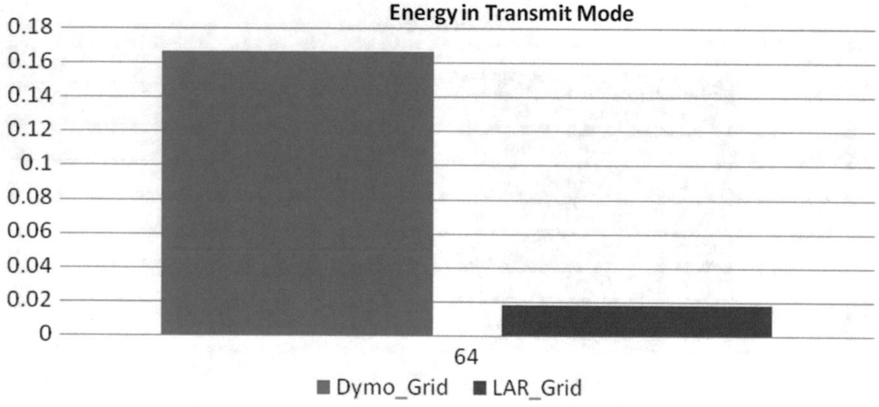

Fig. 13 Energy in transmit mode for DYMO and LAR in grid

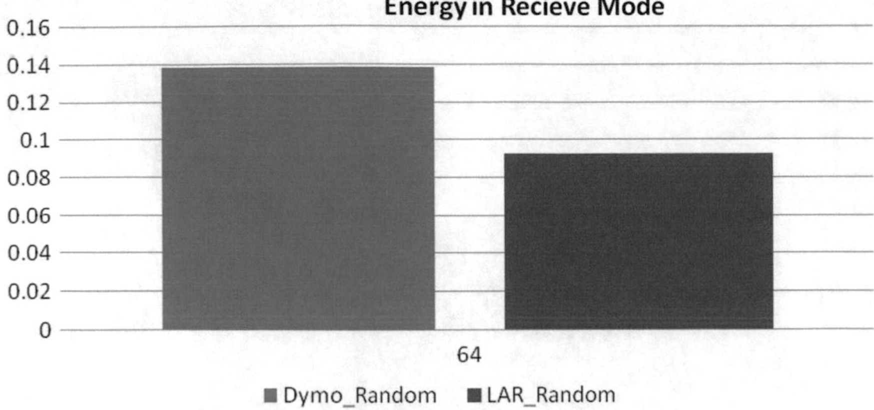

Fig. 14 Energy in receive mode for DYMO and LAR in random

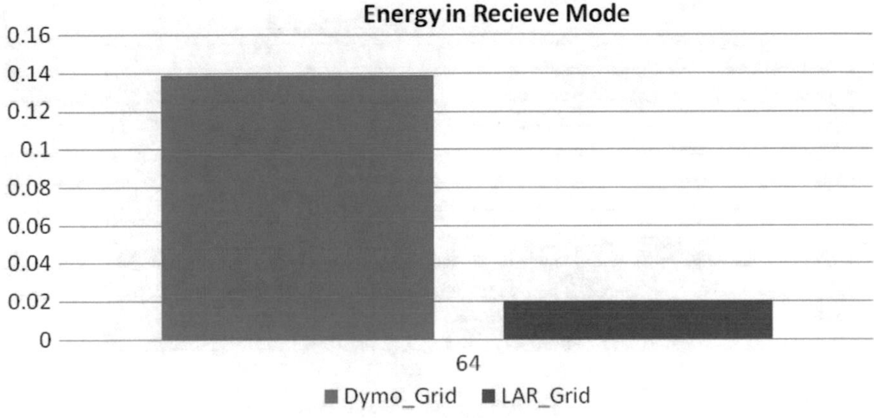

Fig. 15 Energy in receive mode for DYMO and LAR in grid

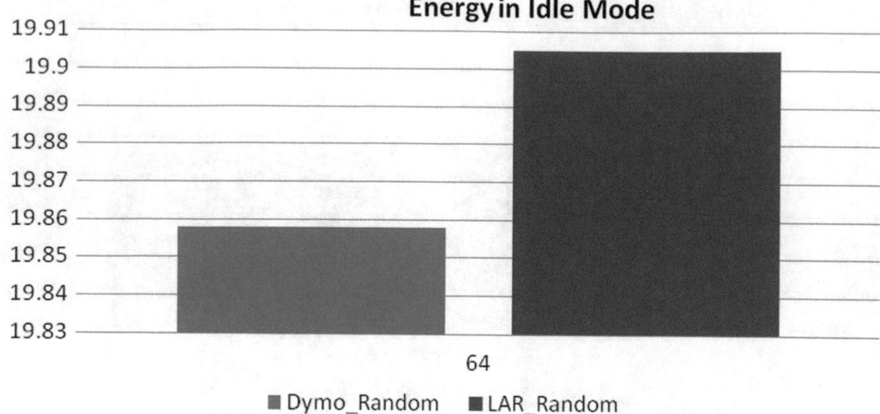

Fig. 16 Energy in idle mode for DYMO and LAR in random

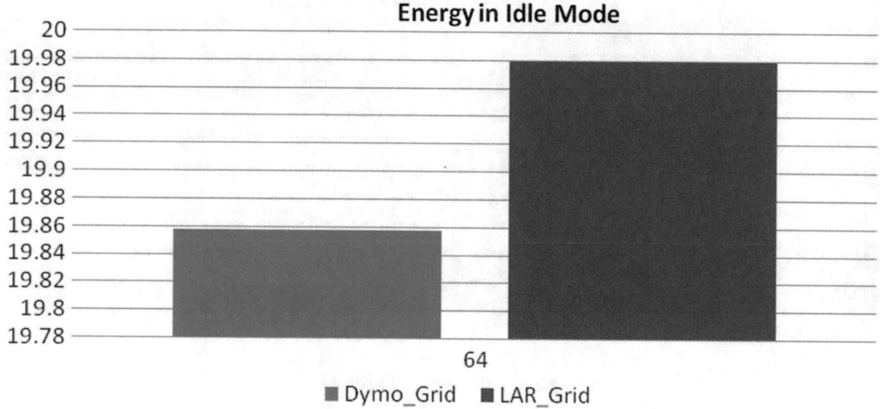

Fig. 17 Energy in idle mode for DYMO and LAR in grid

6 Conclusion and Future Scope

Mobile Ad hoc Network is a temporarily configured network with wireless links. The two popular protocols—DYMO and LAR are analyzed in grid and Random environments. From the experiments it can be concluded that DYMO and LAR protocols exhibit different behavior in Grid and Random environments. In future, the protocols will be evaluated in different mobility models.

References

1. Perkins, C.E.: Ad hoc networking, pp. 53–74. Addison-Wesley (Jan 2001)
2. Tonguz, O.K., Ferrari, G.: Ad hoc wireless networks: a communication—theoretic perspective, p. 330. Wiley (2006)
3. Loo, J., Mauri, J.L., Ortiz, J.H.: Mobile ad hoc networks: current status and future trends, p. 538. CRC Press (2012)
4. Toh, C.K.K.: Ad hoc wireless networks: protocols and systems, pp. 1–324. Prentice Hall PTR Upper Saddle River, NJ, USA (2001)
5. Siva Rammurty, C., Manoj, B.S.: Ad hoc wireless networks architectures and protocols, ISBN 978-81-317- 0688-6 (2011)
6. Boukerche, A.: Algorithms and protocols for wireless, mobile ad hoc networks. Wiley-IEEE Press, p. 500. Online ISBN: 9780470396384 (2009)
7. Chakeres, I., Perkins, C.: Dynamic MANET on-demand (DYMO) routing. Internet draft, draft-ietf-manet-dymo-11.txt, IETF (2007)
8. Ko, Y.B., Vaidya, N.H.: Location aided routing (LAR) in mobile ad hoc networks. In: Proceedings of the Fourth Annual ACM/IEEE International Conference on Mobile Computing and Networking, Dallas, Texas, vol 6, No 4, pp. 307–321 (July 2000)
9. Chlamtac, I., Conti, M., Liu, J.J.: Mobile ad hoc networking: imperatives and challenges. Ad Hoc Netw. **1**(1), 13–64 (2003)
10. Srivastava, A., Mishra, A., Upadhyay, B., Yadav A.K.: Survey and overview of mobile ad-hoc network routing protocols. In: Proceedings of International Conference on Advances in Engineering and Technology Research (ICAETR), IEEE, pp. 1–6 (2014)
11. Mohseni, S., Hassan, R., Patel, A., Razali, R.: Comparative review study of reactive and proactive routing protocols in MANETs. In: Proceedings of 4th International Conference on Digital Ecosystems and Technologies, IEEE, pp. 304–309 (2010)
12. Renu, B., Hardwari Lal, M., Pranavi, T.: Routing protocols in mobile ad-hoc network: a review. In: Quality, Reliability, Security and Robustness in Heterogeneous Networks, Lecture Notes of the Institute for Computer Science, Social Informatics and Telecommunications Engineering, Springer, vol 115, pp. 52–60 (2013)
13. Abolhasan, M., Wysocki, T., Dutkiewicz, E.: A review of routing protocols for mobile ad hoc networks. J. Ad Hoc Netw. **2**(1), 1–2 (2004)
14. Corson, S., Macker, J.: Mobile ad hoc networking (MANET): routing protocol performance issues and evaluation considerations. Network Working Group, RFC 250 l. 1999. http://www.ietf.org/rfc/rfc2501.txt
15. Kumar, J., et al.: Study and performance analysis of routing protocol based on CBR. Procedia Comput. Sci. **85**, 23–30 (2016)
16. Spaho, E., et al.: Performance evaluation of DYMO protocol in different VANET scenarios. In: Proceedings of 15th International Conference on Network-Based Information Systems, IEEE (2012)
17. Sagar, S. et.al.: Evaluating and comparing the performance of DYMO and OLSR in MANETs and in VANETs. In: Proceedings of 14th international Multitopic Conference, IEEE (2011)

18. Singh, D., et al. Comparative performance analysis of LANMAR, LAR1, DYMO and ZRP routing protocols in MANET using random waypoint mobility model. In: Proceedings of 3rd International Conference on Electronics Computer Technology, IEEE (2011)
19. Garcia-Campos, J.M., et al.: Performance evaluation of RRP for VANETs in urban scenarios following good simulation practices. IEEE (2015)
20. Setty, R.: Performance evaluation of AODV in different environments. Int. J. Eng. Sci. Technol. **2**(7), 2976–2981 (2010)
21. Narasimha Raju, K., Anji Reddy, V., Satish Kumar, B.: Performance analysis of LAR and fisheye routing protocols in different environments. Int. J. Comput. Sci. Eng. **9**(11), 650–656 (2017)
22. QualNet Network Simulator. http://www.scalable-networks.com

Intelligent Engineering Informatics

Interactive Voice Application-Based Amazigh Speech Recognition

Mohamed Hamidi, Hassan Satori, Ouissam Zealouk and Khalid Satori

Abstract This paper aims to build an interactive speaker-independent automatic Amazigh speech recognition system. The proposed system offers a methodology to extract data remotely from a distance database using the combined interactive voice response (IVR) and automatic speech recognition (ASR) technologies. We describe our experience to design an interactive speech system based on hidden Markov models (HMMs), Gaussian mixture models (GMMs) and Mel frequency spectral coefficients (MFCCs) based on ten first Amazigh digits and six Amazigh words. The best-obtained performance is 89.64% by using 3 HMMs and 16 GMMs.

Keywords IVR · VOIP · Asterisk · Database server · Amazigh speech recognition

1 Introduction

Interactive voice response (IVR) is a promising technology which automates interactions between callers and phone systems by pressing digits on the telephone or speaking words or short phrases. It enables the user to retrieve/enter information from a database using their voice in real time. The IVR consisted of several technologies working together such as computer telephony, speech recognition, to schedule, receive, enter, and record automated phone calls. This will allow an efficient exchange of information with reducing costs [1]. Shah et al. [2] have studied a VoIP network using the Asterisk server. The system was configured with different security parameters like VPN server, firewall IPtable rules, intrusion detection and intrusion prevention system. Their proposed system architecture was implemented using VMware. The researchers in [3] have examined the various VoIP attacks, and it is present policies according to NIST report. The authors in [4] have considered the DoS attack by categorizing the network into SIP dependent performance matrix and SIP independent matrix in order to evaluate the performance of VoIP.

M. Hamidi (✉) · H. Satori · O. Zealouk · K. Satori
LIIAN Laboratory, FSDM, USMBA, Fez, Morocco
e-mail: mohamed.hamidi.5@gmail.com

V. Bhateja et al. (eds.), *Embedded Systems and Artificial Intelligence*,
Advances in Intelligent Systems and Computing 1076,
https://doi.org/10.1007/978-981-15-0947-6_26

Basu et al. [5] have described the real-time challenges to design telephonic automatic speech recognition system. In their study, they have used the Asterisk server to design a system which poses some queries and the spoken responses of users are stored and transcribed manually for ASR system training. In this work, the speech data are collected from West Bengal.

Aust et al. [6] have created an automatic system that permits users to ask for train traffic information using the telephone. This system connects 1200 German cities. The caller can retrieve information talking fluently with the system which behaves like a human operator. The important components of their system are speech recognition, speech understanding, dialogue control and speech output which is executed sequentially. Bhat et al. [7] created the Speech Enabled Railway Enquiry System (SERES) which is a system that permits users to get the railway information considering the Indian scenario, as a case study to define issues that need to be fixed in order to enable a usable speech-based IVR solution.

Satori et al. [8] have created a system based on hidden Markov models (HMM) using the CMU Sphinx tools. The aim of their work is the creation of automatic Amazigh speech recognition system that includes digits and alphabets of Amazigh language. The system performance achieved was 92.89%. In [9], we present our first experiment to integrate the ten first digits of Amazigh language in an IVR server where the users use speech (ten first Amazigh digits) to interact with the system. This paper describes our experience to design an interactive system allows the users to extract information from a database by using the voice. Our work is based on interactive voice response and automatic speech recognition systems. In our work, the Moroccan Amazigh speech is used to interact with the IVR system.

The rest of this paper is organized as follows: Sect. 2 presents an overview of the VoIP system and related protocols. Section 3 gives an overview on implementation of telephony server. In Sect. 4, Amazigh speech recognition system will be discussed. Finally, Sect. 5 investigates an implementation of our ASRVRS system. We finished by a conclusion.

2 VoIP System and Protocols

2.1 Asterisk

Telephony server Asterisk is an open-source and a development environment for various telecommunication applications programmed in C language. It provides establishment procedures enabling to manipulate communication sessions in progress. Asterisk supports the standard protocols: SIP, H.323 and MGCP and transformations between these protocols. It can use the IAX2 protocol to communicate with other Asterisk servers [10, 11].

2.2 *Session Initiation Protocol*

Session Initiation Protocol (SIP) is a signaling protocol which is responsible for creating media sessions between two or more participants. SIP was defined by Internet Engineering Task Force (IETF) and is a simpler than H.323 and adapted more specifically for session establishment and termination in VOIP [12]. In our word, SIP was used to create user account and to assure inter-network communication.

3 Implementation of Telephony Network

3.1 *VoIP Network*

In a first step, we install the Asterisk server on the host machine, and we create two virtual machines using Oracle VirtualBox [13]. The second step consists of installing Ekiga softphone, on the two virtual machines, to permit the users to communicate with each other. In the sip.conf file, we configure the users' accounts with username, host, password, etc. On the other hand, the extensions.conf file permits to give an identifier number for each user.

3.2 *The Database Server*

The database server (MySQL server) is created as a new virtual host. The ODBC connector which is a database abstraction layer allows Asterisk to communicate with several databases without requiring the developers to create a separate database connector for everyone supported by Asterisk. This helps to save and reduce the maintenance effort and time [11]. The connection between Asterisk and MySQL servers is assured by configuring ODBC files. The database server is integrated in the schematic representation of our network. See Fig. 1.

4 Amazigh Speech Recognition System

4.1 *Speech Recognition*

Speech recognition is the process of decoding the speech signal received by the microphone and converting it into words [14]. The recognized words can be used

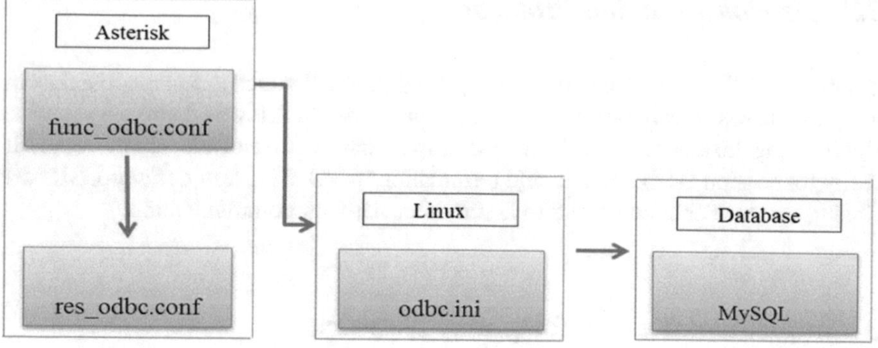

Fig. 1 Database server network

as commands, data entry or application control. Recently, the applications of automatic speech recognition for Moroccan Amazigh language were targeted by our lab researchers [15–20].

4.2 Amazigh Language

Before the implementation of a Speech Recognition Voice Response Server System for any language, it is necessary to have preliminary study of this language. In our case, we choose Amazigh which is a less-resourced Moroccan language. The Amazigh language is widely spoken in a vast geographical area of North Africa. It is spoken by 28% of the Moroccan population [21].

4.3 Hidden Markov Model

The hidden Markov model (HMM) [22] is a popular method in machine learning and statistics for modeling sequences like speech. This model is a finite ensemble of states, where each set is associated with a probability distribution. Transitions among the states are governed by a set of probabilities called transition probabilities. Markov models are excellent ways of abstracting simple concepts into a relatively easily computable form.

Table 1 System parameters

Parameters	Values	Parameters	Values
Sampling rate	16 kHz	Number of token-training	2400
Number of speakers- training	15	Number of token-test	1120
Number of speakers- test	7	MFCC	13

4.4 Training

In order to determine their optimal values for maximum performance, different acoustic models were prepared by varying HMMs (3-5) and GMMs (8-16-32-64). The wave recorded audio data is used in the training phase where the database is partitioned to 70% training and 30% testing in order to ensure the speaker-independent aspect. The system was trained for two different phases. The first one is the identification phase, where the system is trained only by the admin voice. In the second phase, the system was trained by different voices. More technical details about our system are shown in Table 1.

5 Implementation of the Interactive Voice Response (IVR)

5.1 Implementation

The designed system is created using Oracle VirtualBox tool on host machine with 2 GB of RAM and an Intel Core i3 CPU of 1.2 GHz speed. The operating system used in our experience was Ubuntu 14.04 LTS. The system implementation operates in three steps:

The first one is the creation of the VoIP network as described in Sect. 3.1. Secondly, we established a connection with MySQL database which is installed in another virtual machine. This connection is achieved by installing Asterisk add-ons. The port used by MySQL is 3306. This port is opened when the connection is established. The connection of the Asterisk server with the MySQL database is provided in the extensions.conf file while writing the dial plans.

In the last step, the integration of Amazigh automatic speech recognition is assured by Cairo server using Media Resource Control Protocol (MRCP). MRCP is designed to provide a mechanism for a client device requiring audio/video stream processing to control processing resources on the network [23]. Our system architecture is presented in Fig. 2.

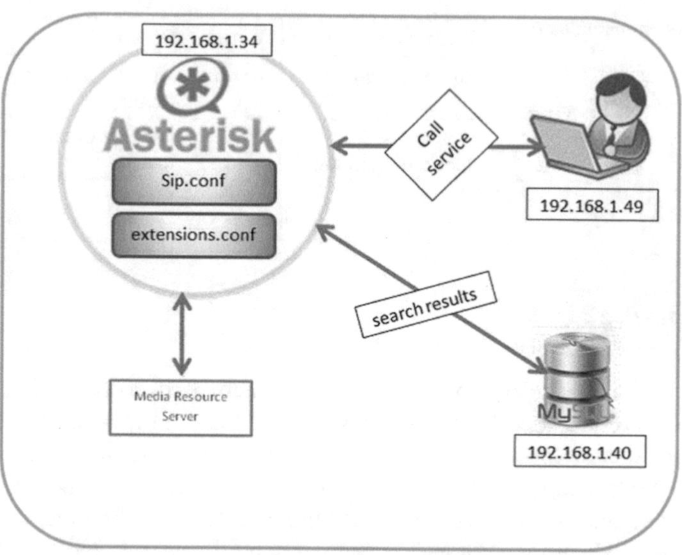

Fig. 2 Proposed architecture

5.2 *Management System Algorithm*

This part describes an algorithm that permits all network's components to react with the system. Before starting the reaction, the system tests the connection between all components. In a first step, it tests the signal with Sphinx than tests the connection with database server. My own voice is configured inside the IVR system to ask the user to enter the ID and password using the Amazigh language. The user authenticates to the server via a login and password using his/her voice. In the next step, the system sends the voice to Sphinx 4 in order to convert it to text, and Sphinx 4 resends the text back to Asterisk. Thus, Asterisk checks up the validity of that information from a MySQL database. Then, the IVR opens the admin manage interface for the data consultation. Table 2 describes the system algorithm.

6 Experimental Results

In order to evaluate the performance of speech recognition system through the IVR server, we perform several tests on different individuals each one of them was asked to command the system by using the Amazigh commands. The recognition rates for each vocal command were recorded to obtain the success of identification and manage for each speaker.

To evaluate the identification of Amazigh speech recognition system performance via IVR server and for exploiting the quality of service to achieve the optimal

Table 2 Algorithm description

Steps	Actions
1	Play welcome file and asking login
2	The user reads his login vocally
3	System sends the request to Sphinx to convert user's voice
4	Sphinx sends login with text to Asterisk
5	Asking password
6	System sends the request to Sphinx to convert user's voice
7	Sphinx converts the password to the txt
8	Sphinx sends password with text to Asterisk
9	If (login = = login && password = = password) go to step 13 if not go to step 2
10	Open the manage interface
11	System requests command
12	The user commands by his voice
13	System gives information to the user
14	Play thanks message and do Hang up

parametrization for placing an interactive speech system of high quality, we have based on different numbers of Gaussian mixtures per model and hidden Markov models. Figure 3 presents the overall test recognition rate for identification phase. Where the best result is 91.43% found with 3 HMMs and 16 GMMs.

The system performances for 3 HMMs are 87.86, 89.64 and 86.87% whereas, the obtained results with 5 HMMs are 85.89, 86.39 and 83.89% were found for using 8, 16 and 32 Gaussian mixture distributions, respectively. By considering the digits recognition analysis, the most frequently recognized Amazigh digits are AMYA and KUZ. The most recognized Amazigh words are AMSSUGUR and ASNMKTA. Table 3 presents the recognition rates of all commands. It is found that 3 HMMs and 16 GMMs parametrization gives the best results.

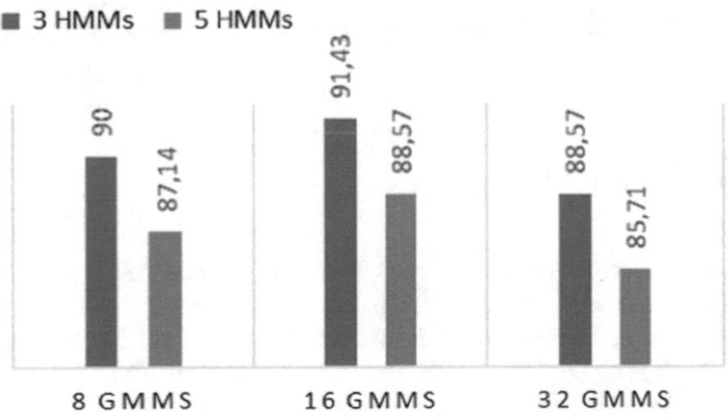

Fig. 3 Test recognition rate for identification phase

Table 3 System recognition rates

	3 HMMs			5 HMMs		
	8	16	32	8	16	32
AMYA	91.43	92.86	88.57	88.57	88.57	88.57
YEN	88.57	91.43	85.71	88.57	85.71	84.29
SIN	85.71	87.14	84.29	84.29	85.71	85.71
KARD	87.14	90.00	90.00	88.57	90.00	82.86
KUZ	92.86	92.86	88.57	85.71	90.00	81.43
SMMUS	90.00	91.43	87.14	88.57	88.57	85.71
SDES	87.14	88.57	88.57	85.71	85.71	84.29
SA	84.29	85.71	84.29	85.71	85.71	78.57
TAM	88.57	90.00	88.57	85.71	85.71	84.29
TZA	85.71	87.14	85.71	85.71	85.71	81.43
AMSSUGUR	88.14	91.43	87.14	86.71	87.14	83.71
ASNMKTA	85.71	91.43	87.14	85.71	82.86	87.14
ASIKI	87.14	88.57	84.29	80.00	87.14	84.29
TILIT	88.57	85.71	88.57	85.71	85.71	81.43
AMATTAL	87.14	90.00	84.29	84.29	82.86	82.86
AZAN	87.71	90.00	87.14	84.71	85.14	85.71
Total	87.86	89.64	86.87	85.89	86.39	83.89

7 Conclusion

In this work, we realize a highly parameterized system combining IVR and ASR technologies. Our designed system allows to extract information stored on a distance database. In our approach, we use the Asterisk server as a backbone an Amazigh speech recognition system and a database server to store the information. The best recognition rate is 89.64% achieved by the parameterization 3 HMMs and 16 GMMs. In our next work, we will try to improve the speech interactive system functionality by extending the tests for other numbers and compound and complicated words.

References

1. Asterisk IVR. http://www.asterisk.org. Accessed Jan 2015
2. Shah, K., Ghrera, S.P., Thaker, A.: A novel approach for security issues in VoIP networks in virtualization with IVR. arXiv preprint arXiv:1206.1748 (2012)
3. Anwar, Z., Yurcik, W., Johnson, R.E., Hafiz, M., Campbell, R.H.: Multiple design patterns for voice over IP (VoIP) security. In: 25th IEEE International Performance, Computing, and Communications Conference, IPCCC 2006, pp. 8-pp (2006, April)

4. Rafique, M.Z., Akbar, M.A., Farooq, M.: Evaluating DoS attacks against SIP-based VoIP systems. In: Global Telecommunications Conference, GLOBECOM 2009, pp. 1–6. IEEE (2009, November)
5. Basu, J., Bepari, M.S., Roy, R., Khan, S.: Real time challenges to handle the telephonic speech recognition system. In: Proceedings of the Fourth International Conference on Signal and Image Processing 2012 (ICSIP 2012), pp. 395–408. Springer, India (2013)
6. Aust, H., Oerder, M., Seide, F., Steinbiss, V.: The Philips automatic train timetable information system. Speech Commun. **17**(3), 249–262 (1995)
7. Bhat, C., Mithun, B.S., Saxena, V., Kulkarni, V., Kopparapu, S.: Deploying usable speech enabled IVR systems for mass use. International Conference on Human Computer Interactions (ICHCI), pp. 1–5 (2013)
8. Satori, H., ElHaoussi, F.: Investigation Amazigh speech recognition using CMU tools. Int. J. Speech Technol. **17**(3), 235–243 (2014)
9. Hamidi, M., Satori, H., Satori, K.: Implementing a voice interface in VOIP network with IVR server using Amazigh digits. Int. J. Multi-disciplinary Sci. **2**(2), 38–43 (2016)
10. Madsen, L., Van Meggelen, J., Bryant, R.: Asterisk: The Definitive Guide. O'Reilly Media, Inc., pp. 121–145, 737–745, 417–478 (2011)
11. Penton, J., Terzoli, A.: Asterisk: A converged TDM and packet-based communications system. In: Proceedings of SATNAC 2003-Next Generation Networks (2003)
12. Handley, M., Schulzrinne, H., Schooler, E., et al.: RFC 2543. SIP: Session Initiation Protocol (1999)
13. Oracle VM VirtualBox. https://www.virtualbox.org/. Accessed Jan 2015
14. Huang, X., Acero, A., Hon, H.W., Foreword By-Reddy, R.: Spoken language processing: a guide to theory, algorithm, and system development. Prentice Hall PTR (2001)
15. Satori, H., Zealouk, O., Satori, K., ElHaoussi, F.: Voice comparison between smokers and non-smokers using HMM speech recognition system. Int. J. Speech Technol. **20**(4), 771–777 (2017)
16. Zealouk, O., Satori, H., Hamidi, M., Satori, K.: Speech recognition for Moroccan dialects: feature extraction and classification methods. J. Adv. Res. Dyn. Control Syst. **11**(2), 1401–1408 (2019)
17. Hamidi, M., Satori, H., Zealouk, O., Satori, K.: Speech coding effect on Amazigh alphabet speech recognition performance. J. Adv. Res. Dyn. Control Syst. **11**(2), 1392–1400 (2019)
18. Zealouk, O., Satori, H., Hamidi, M., Laaidi, N., Satori, K.: Vocal parameters analysis of smoker using Amazigh language. Int. J. Speech Technol. **21**(1), 85–91 (2018)
19. Zealouk, O., Satori, H., Hamidi, M., Satori, K.: Voice pathology assessment based on automatic speech recognition using Amazigh digits. In: Proceedings of the 2nd International Conference on Smart Digital Environment, pp. 100–105. ACM (2018)
20. Mohamed, H., Hassan, S., Ouissam, Z., Khalid, S., Naouar, L.: Interactive voice response server voice network administration using hidden Markov model speech recognition system. In: 2018 Second World Conference on Smart Trends in Systems, Security and Sustainability (WorldS4- IEEE), pp. 16–21 (2018, October)
21. Boukous, A.: Société, langues et cultures au Maroc: Enjeux symboliques (No. 8). Faculté des lettres et des sciences humans-Rabat (1995)
22. Beal, M.J., Ghahramani, Z., Rasmussen, C.E.: The infinite hidden Markov model. In: Advances in Neural Information Processing Systems, pp. 577–584 (2002)
23. Shanmugham, S., Burnett, D.: Media Resource Control Protocol Version 2 (MRCPv2) (2012)

Pathological Detection Using HMM Speech Recognition-Based Amazigh Digits

Ouissam Zealouk, Hassan Satori, Mohamed Hamidi and Khalid Satori

Abstract In the last decade, the automatic speech pathology detection systems based on voice production theory are evolving up to date. Overall, there have not been much speech technology research studies for persons regarding voice disorders which center on Amazigh language. This research project focuses on the build of an automatic speech recognition system based on Sphinx-4 that permits to detect the differences between normal and pathological voices based on the produced speech. The performance in our system was measured using the combinations of different Hidden Markov Models and Gaussian mixture distributions. Results show that the maximum accuracy with the normal voices is greater than the maximum accuracy obtained from the pathological speaker.

Keywords Automatic speech recognition · Pathological voices · Amazigh language · Hidden Markov Model

1 Introduction

Automatic speech recognition (ASR) system is one of the frontiers in human–computer interaction. This process allows converting the acoustic waveform into a sequence of words similar to the data being transferred by the speaker. These systems based on a conventional Hidden Markov Model (HMM) mostly utilize phonemes as basic linguistic units and cepstral features as acoustic observation.

ASR has a vast field of applications, e.g., command recognition, dictation, interactive voice response, learning foreign languages, helping disabled people to interact

---

O. Zealouk (✉) · H. Satori · M. Hamidi · K. Satori
Laboratory Computer Science, Image Processing and Numerical Analysis, Faculty of Sciences Dhar Mahraz, Sidi Mohamed Ben Abdellah University, Fés, Morocco
e-mail: ouissam.zealouk@gmail.com

H. Satori
e-mail: hsatori@gmail.com

M. Hamidi
e-mail: mohamed.hamidi.5@gmail.com

V. Bhateja et al. (eds.), *Embedded Systems and Artificial Intelligence*,
Advances in Intelligent Systems and Computing 1076,
https://doi.org/10.1007/978-981-15-0947-6_27

with society, and medical diagnostic. It is a very promising technology that makes life easier [1, 2].

Voice pathology detection refers to a detection procedure of the pathology in the vocal folds from an input voice. In subjective detection, an experienced physician hears the voice and assesses whether the voice is normal or pathological based on his or her previous knowledge and experience. However, this type of assessment may vary from physician to physician depending on the experience. From a health science point of view, it was proved that the human health condition and the pathological status do affect the human voice [3]. In our previous work, we employed the ASR technology to develop a system which differentiates between smokers and non-smokers voice using the Amazigh digits speech based on the mel-frequency cepstral coefficients (MFCCs) to determine the voices' features [4].

In this paper, a feature based on the voice pathological analysis of a speech signal is proposed. A detection system by means of isolated words is developed, and the HMM technique is used for the classification of normal and pathological speech samples to decide if the speaker has voice pathologies or not.

The rest of this paper is organized as follows. Section 2 presents related works. Section 3 gives an overview of (ASR). Section 4 gives a brief description of the Amazigh language. Section 5 emphasizes the description of Hidden Markov Model. Section 6 shows the technology and method used in this work. The experimental results are presented in Sect. 7. Conclusions are drawn in Sect. 8.

2 Related Works

The detection and classification of voice pathology is a topic which has interested the international voice community. The majority of the work in this domain is centered on automatic diagnostic of pathology by using digital signal processing modes [5, 6]. The first acoustic voice parameters in pathological voice analysis are proposed by Lieberman in 1961 [7]. In [8], a system was developed for short-time jitter and the area under curve amounts to 94.82%. The MFCCs have been utilized also in voice pathology detection. In another study, the objective of Muhammad et al. [9] was detecting voice disorders based on six different cases. The aim of their work is the classification of the type and severity of voice pathologies using Arabic automatic speech recognition (ASR). Godino-Llorente et al. [10] proposed a detection system of pathological voice by means of Gaussian mixture models and short-term Mel cepstral vectors parameters achieved by framing energy together with first derivatives. Wiśniewski et al. [11] also presented automatic speech recognition system using Hidden Markov Model technique to detect the speech disorders. This system achieved a success rate of approximately 70%. For the purpose of voice pathology evaluation, various databases have been generally used by the researchers, and among them Saarbruecken Voice Database and Arabic Voice Pathology Database [12, 13] are commonly used by the scientific community.

3 Automatic Speech Recognition (ASR)

Automatic speech recognition is a derived method of pattern recognition [14]. This recognition operates through two phases which are training and testing where the features extraction process is common in both phases. In the training phase, each reference is learned from spoken examples and stored either in the form of templates obtained by some averaging models that characterize the statistical properties of pattern. In the testing or recognition phase, the feature of test pattern (test speech data) is matched with the trained model of each and every class.

The speech recognition system uses two principal models that are acoustic and language models to obtain the recognition rate of the received speech, that is, the number of correctly recognized words. Acoustic modeling plays a very essential role in developing the accuracy of ASR systems. For the given acoustic observation A, the goal of speech recognition is to find the most probable word sequence $\hat{M}$, that maximizes the posterior probability $P(M/A)$, which is written as:

$$\hat{M} = \underset{M}{\operatorname{argmax}} \; P(A/M)P(M)$$

where A is the acoustic feature of the word sequence M, and $P(M)$ is the language model. The language model includes an ensemble of rules for a language that is utilized as the primary context for recognizing words. This model helps to reduce the search space and resolve acoustic ambiguity [15].

4 Amazigh Language

The Amazigh language or Tamazight is spoken in a vast geographical area of North Africa, from the Canary Islands to the Siwa Oasis in the North, and from the Mediterranean coast to Niger, Mali, and Burkina Faso in the South. Historically, the Amazigh language has been autochthonous and was exclusively reserved for familial and informal domains [16].

In Morocco, the Amazigh language is spoken by some 28% of the population, grouped in three main regional varieties, depending on the area and the communities: Tarifit spoken in northern Morocco, Tamazight in central Morocco and South-East, and Tachelhit spoken in southern Morocco [17, 18].

The allowed syllables in Amazigh language are: V, CV, VC, CVC, C, CC, and CCC where V indicates a vowel while C indicates a consonant [19]. Recently, the Amazigh language is considered among the rich languages targeted by researchers in several domains [20–25].

5 Hidden Markov Model

The Hidden Markov Model (HMM) is a popular statistical tool for modeling a wide range of time series data. It provides efficient algorithms for state and parameter estimation, and it automatically performs dynamic time warping of signals that are locally stretched. Hidden Markov models are based on the well-known chains from probability theory that can be used to model a sequence of events in time. The Markov chain is deterministically an observable event. The most likely word with the largest probability is produced as the result of the given speech waveform. A natural extension of the Markov chain is the Hidden Markov Model, where the internal states are hidden and any state produces observable symbols or observable evidences [26].

6 Technology and Method

6.1 Sphinx-4

The Sphinx-4 is a system of speech recognition designed by Carnegie Mellon University, Sun Microsystems Laboratories, Mitsubishi Electric Research Laboratories, and Hewlett-Packard's Cambridge Research Lab. It developed completely in the Java TM programming language. Sphinx-4 uses newer search strategies and is universal in its acceptance of various kinds of grammars and language models, types of acoustic models, and feature streams [27].

6.2 SphinxTrain

SphinxTrain is CMU tool used in acoustic model development. This is a set of programs and documentation to realize constructing acoustic models for several languages [28].

6.3 Speech Database Preparation

This phase consists of recording the speech signal. Firstly, we used a desktop microphone in clean environment and WaveSurfer tool, while keeping a distance of approximately 5–10 cm between mouth of the speaker and the microphone. The sampling rate used for recording is 16 kHz, with 16 bits resolution for more details on the corpus. Technical parameters are given in Table 1. The database used in our system includes 24 Amazigh speakers aged between 26 and 50 years; this database is divided

Table 1 System parameters

Parameter	Value
Sampling rate	16 kHz
Number of bits	16 bits
Corpus	10 Amazigh digits
Accent	Moroccan Tarifit Berber

into two categories: The first one consists of 22 normal persons and the second contains two speakers having vocal fold disorders (the voice recording with disorder patients has been approved by the ethical committee of Oujda University Hospital). The method followed to record voice is asking those speakers to pronounce the ten first Amazigh digits (ten times for each digit) sequentially. Audio recordings for each speaker were saved in ten ".wav" files; every ".wav" file includes ten repetitions of one number. Then, we divided each file into ten .wav files. Thus, this corpus consists of 2400 tokens.

6.4 Acoustic Model

The acoustic model consists of the sub-words which are named phonemes that collectively form the word. It permits to convert the pronounced words into phonemes and from phonemes to words that are a statistically possible representation of the acoustic image for the speech signal. During the learning and training phase, each acoustic unit or phoneme is represented by a statistical model that describes the data distribution. The speech signal is transformed into a series of features vectors including mel-frequency cepstral coefficients (MFCCs) [29].

The Sphinxbase and Sphinxtrain are used to produce the acoustic model. Every audio file in the training corpus is transformed into a sequence of characteristic vectors. The front end provided by Sphinxtrain computes a set of features files for each file. In this work, the acoustic model was produced by using a speech signal from the Amazigh digits training database.

6.5 Dictionary

In the dictionary file, the correspondence will be specified between the word of the transcription file and the phonemes used in the file extension phone. The dictionary provides pronunciations for each existing word in the language model, and it includes the words we want to train followed by their pronunciation, which divides words into sequences of sub-word units. Our dictionary includes the symbolic representations of the first ten Amazigh digits. The pronunciation dictionary is considered an intermediary between the language model and acoustic model. Examples of presented

```
#JSGF V1.0;
/**
 * JSGF Grammar for amdigits example
*/
grammar amdigits;

    public <greet> = (Amya | Yen | Sin | Krad | Koz | Smmus | Sdes | Sa |
Tam | Tza);
```

Fig. 1 Grammars file of Amazigh digits

digits structure in the dictionary file.

AMYA	A M Y A
KRAD	K R A DD
KUZ	K O Z

6.6 *Language Model*

The language model is defined in three kinds: the simplest that is used for isolated word recognition, the second who is for applications based on command and control, and the last a set of n-gram grammars that are used for free speech form. Each word in the language model should be in the dictionary. In our work, we used a grammar file that includes the ten first Amazigh digits which are shown in Fig. 1.

7 Experimental Results

In order to evaluate the performance of our systems, we carried out two main experiments which were focused on a connected phoneme task constituting isolated ten first Amazigh digits. Each phoneme was modeled by three and five HMMs. The number of mixture in the model of each state was 16. For feature extraction, 13-dimensional MFCCs were used. The first experiment concerns the training and testing of the system with the normal speakers (20 training 2 tests). The second experiment is about testing the system performance with the speakers who have pathological voices (training by using the voice of 20 normal people and testing by two pathological voices). The 70% of wave files in training phase and the rest 30% in testing phase role are not respected to ensure the speaker-independent aspect of our ASR module. Table 2 shows a comparison of the system's performance for the ten first Amazigh digits using 16 GMMs and HMMs.

Table 2 Recognition accuracy (%) of normal and pathological voices

Amazigh digits	Recognition rate (%)			
	Normal voices		Pathological voices	
	16 GMMs		16 GMMs	
	3 HMMs	5 HMMs	3 HMMs	5 HMMs
AMYA	85.00	85.00	30.00	30.00
YEN	80.00	80.00	25.50	25.00
SIN	80.00	80.00	25.00	25.00
KRAD	90.80	90.00	35.00	35.00
KOZ	75.00	80.00	20.00	20.00
SMMUS	90.00	90.00	30.00	35.00
SDES	90.00	90.00	30.00	30.00
SA	80.00	80.00	20.00	20.00
TAM	80.00	85.00	25.00	30.00
TZA	80.00	80.00	25.00	25.00
Total recognition rate (%)	*83.08*	*84.00*	*26.55*	*27.50*

As presented in Table 2, the obtained accuracy by normal speakers was very high than pathological speakers where a significant loss of accuracy on speech recognition for voice disorder samples is observed. The overall system performances are 83.08% and 84.00% which were found for using three and five HMMs, respectively, for normal speakers, while the system correct rates in the pathological case were 26.55% and 27.50% for the both three and five HMMs, respectively. This difference between the recognition rate of normal and speakers suffering from voice disorders is due to the speech signal of a subject with disorders containing lower amplitude than the speech signal of a normal subject [30], in addition to the impairment of mucosal vibration. Based on the results obtained from the experiences, we can see that our system is able to distinguish between the normal and pathological voices. These results are in good agreement with the study [9] which has shown the analysis of voice disorders. That is a very satisfactory result.

8 Conclusion

In this paper, we propose a method to able to show the difference between the normal and pathological speakers by using the automatic speech recognition task. This implemented system was developed with open-source CMU Sphinx-4 depends on the ten first Amazigh digits. In the best of our knowledge, this is the first study that tries to evaluate the accuracy of ASR in Amazigh speech for people with pathological voices. In our future work, we will record a largest number of pathological speakers and investigate the performance of the proposed system in a continuous speech to analyze different kinds of vocal tract disorders.

References

1. Halton, M., Cerisara, C., Fohr, D., Laprie, Y., Smaili, K.: Reconnaissance automatiue de la parole du signal a son interpretation. Monographies and Books, Oxford (2006)
2. Satori, H., Hiyassat, H., Harti, M., Chenfour, N.: Investigation Arabic speech recognition using CMU sphinx system. Int. Arab J. Inf. Technol. **6**(2) (2009)
3. O'Shaughnessy, D.: Automatic speech recognition: history, methods and challenges. Pattern Recogn. **41**(10), 2965–2979 (2008)
4. Satori, H., Zealouk, O., Satori, H., et al.: Voice comparison between smokers and non-smokers using HMM speech recognition system. Int. J. Speech Technol. **20**(4), 771–777 (2017)
5. Dubuisson, T., Dutoit, T., Gosselin, B., Remacle, M.: On the use of the correlation between acoustic descriptors the normal/pathological voices discrimination. EURASIP J. Adv. Signal Process. **2009**(1), 173967 (2009)
6. Michaelis, D., Frohlich, M., Strube, H.W.: Selection and combination of acoustic features for the description of pathologic voices. J. Acoust. Soc. Am. **103**(3), 1628–1639 (1998). https://doi.org/10.1121/1.421305
7. Lieberman, P.: Perturbation in vocal pitch. J. Acoust. Soc. **33**(5), 597–603 (1961)
8. Vasilakis, M., Stylianou, Y.: Sepctral jitter modeling and estimation. Biomed. Signal Process. Control **4**(3), 183–193 (2009)
9. Muhammad, G., Mesallam, T.A., Malki, K.H., Farahat, M., Alsulaiman, M., Bukhari, M.: Formant analysis in dysphonic patients and automatic Arabic digit speech recognition. Biomed. Eng. Online **10**(41), 1–12 (2011)
10. Godino-Llorente, J., Gomez-Vilda, P., Blanco-Velasco, M.: Dimensionally reduction of a pathological voice quality assessment system based on Gaussian mixture models and short-term cepstral parameters. IEEE Trans. Biomed. Eng. **53**(10), 1943–1953 (2006)
11. Wiśniewski, M., Kuniszyk-Jóźkowiak, W., Smołka, E., Suszyński, W.: Automatic detection of disorders in a continuous speech with the Hidden Markov Models approach. In: Computer Recognition Systems 2, vol. 45, pp. 447–453 (2007)
12. Muhammad, G., Alsulaiman, M., Ali, Z., Mesallam, T.A., Farahat, M., Malki, K.H., Al-nasheri, A., Bencherif, M.A.: Voice pathology detection using interlaced derivative pattern on glottal source excitation. Biomed. Signal Process. Control **31**, 156–164 (2017)
13. Woldert-Jokisz, Bogdan. Saarbruecken Voice Database. (2007)
14. Gaikwad, S.K., Gawali, B.W., Yannawar, P.: A review on speech recognition technique. Int. J. Comput. Appl. **10**(3), 16–24 (2010)
15. Huang, X., Acero, A., Hon, H., Foreword, B.: Spoken Language Processing: A Guide to Theory, Algorithm, and System Development. Prentice Hall PTR (2001)
16. Boukous, A.: Société, langues et cultures au Maroc: Enjeux symboliques. Najah El Jadida, Casablanca, Maroc, 1f895
17. Ouakrim, O.: Fonética y fonología del Bereber. Survey: University of Autònoma de Barcelona (1995)
18. Chaker, S.: Textes en linguistique berbère: introduction au domaine berbère. Ed. du C.N.R.S, Paris (1984)
19. Ridouane, R.: Suites de consonnes en berbère: phonétiqueet phonologie. Doctoral Dissertation, Université de la Sorbonne nouvelle-Paris III (2003)
20. Mohamed, H., Hassan, S., Ouissam, Z., Khalid, S., Naouar, L.: Interactive voice response server voice network administration using Hidden Markov Model speech recognition system. In: Second World Conference on Smart Trends in Systems, Security and Sustainability (WorldS4), pp. 16–21. IEEE (2018, October)
21. Zealouk, O., Satori, H., Hamidi, M., Satori, K.: Voice pathology assessment based on automatic speech recognition using Amazigh digits. In: Proceedings of the 2nd International Conference on Smart Digital Environment, pp. 100–105. ACM (2018, October)
22. Hamidi, M., Satori, H., Satori, K.: Implementing a voice interface in VOIP network with IVR server using Amazigh digits. Int. J. Multi. Sci. **2**, 38–43 (2016)

23. Zealouk, O., Satori, H., Hamidi, M., Laaidi, N., Satori, K.: Vocal parameters analysis of smoker using Amazigh language. Int. J. Speech Technol. **21**(1), 85–91 (2018)
24. Hamidi, M., Satori, H., Zealouk, O., Satori, K.: Speech coding effect on Amazigh alphabet speech recognition performance. J. Adv. Res. Dyn. Control Syst. **11**(2), 1392–1400 (2019)
25. Zealouk, O., Satori, H., Hamidi, M., Satori, K.: Speech recognition for Moroccan dialects: feature extraction and classification methods. J. Adv. Res. Dyn. Control Syst. **11**(2), 1401–1408 (2019)
26. Young, S., Evermann, G., Hain, T., Kershaw, D., Moore, G., Odell, J., Ollason, D., Povey, D., Valtchev, V., Woodland, P.: The Book (2002). http://htk.eng.cam.ac.uk
27. Carnegie Mellon University. Sphinx-4. Available http://cmusphinx.sourceforge.net
28. Satori, H., Harti, M., Chenfour, N.: Arabic speech recognition system using cmu-sphinx4. arXiv preprint (2007). arXiv:0704.2201
29. Varela, A., Cuayáhuitl, H., Nolazco-Flores, J.A.: Creating a Mexican Spanish Version of the CMU Sphinx-III Speech Recognition System, vol. 2905. Springer (2003)
30. Zulfiqar, A., Alsulaiman, M., Elmavazuthi, I., et al.: Voice pathology detection based on the modified voice contour and SVM. Biol. Inspired Cogn. Archit. **15**, 10–18 (2016)

Healthcare Social Data Platform Based on Linked Data and Machine Learning

Salma El Hajjami, Mohammed Berrada and Soufiane Fhiyil

Abstract The healthcare system is facing very important challenges in order to improve the whole system performance. Different communities are interested in this subject from different perspectives ranging from technical issues to organizational aspects. An important aspect of this research area is to consider social network data within the system especially because of the rapid and growing development of social networks. It can be general social networks, like Facebook or twitter but also others dedicated as PatientsLikeMe. This social network proliferation generates complex problems and locks when we want to take into account the resulting large amounts of data, created continuously, within the healthcare system. We call these data "social data". The aim of this work is to demonstrate that is possible and feasible to build promising alternatives of the traditional healthcare system to improve the quality of services and reduce cost. In our opinion, taking into account "social data" can provide efficient healthcare decisional support systems to help healthcare operators to make optimal and efficient decisions in dynamic and complex environments. Our approach involves data extraction from multiple social networks, data aggregation, and the development of a semantic model in order to answer high-level users' queries. In addition, we show how an analytical tool can help operators to understand data. Lastly, we present a model of machine learning which aims to detect the Sentiments of users expressed toward a given medication and the "TOP TRENDING" of care and treatments used for a given disease.

Keywords Social network · Social data · Semantic model · Analytical tools · Health information · Machine learning · Sentiment top trending

S. El Hajjami (✉) · M. Berrada
Laboratoire D'Informatique et Physique Interdisciplinaire (LIPI), ENS-Fès, USMBA, Fez, Morocco
e-mail: salma.elhajjami@usmba.ac.ma

S. Fhiyil
Faculté des Sciences Dhar El Mehraz, USMBA, Fez, Morocco

V. Bhateja et al. (eds.), *Embedded Systems and Artificial Intelligence*, Advances in Intelligent Systems and Computing 1076,
https://doi.org/10.1007/978-981-15-0947-6_28

1 Introduction

Today, we are witnessing a veritable deluge of data produced by social media. This data is generated from a large number of Internet applications and Web sites. Facebook is the most popular platform (with more than 1.19 billion active users per month), followed by Twitter (500 million users worldwide). In parallel of general use social networks, we also find platforms dedicated to health, such as Doctissimo (eight million unique visitors each month) [1] and PatientsLikeMe (growing, currently has more than 187,000 members and covers more than 500 patients) [2]. These platforms are examples of medical social networking sites with many user bases, where people with specific diseases can exchange information about their illness, treatment, and experience.

This profusion of digital traces left by users of social media is at the origin of a new phenomenon very popular in recent years, and it is about social data. This phenomenon profoundly transforms many sectors, like health. In this field, the available data represent an unprecedented innovation potential: identification of disease risk factors, assistance with diagnosis, choice and monitoring of treatment effectiveness, pharmacovigilance, epidemiology, etc. In this sense, taking into account "social data" makes it possible to respond to market needs and trends in healthcare institutions [3], and to offer epidemiologists, physicians, and health policy experts an excellent opportunity to formulate evidence-based judgments that will eventually lead to patient care [4].

However, everyone agrees that the treatment of these masses of data, or "social data", is a major issue [5]. Collecting, processing, and analyzing large social media data from unstructured (or semi-structured) sources to extract valuable knowledge are an extremely difficult task that has not been fully resolved. Traditional data management methods, algorithms, frameworks, and tools have become inadequate to handle this large amount of data. This problem has generated a large number of challenges related to different aspects such as knowledge representation, data collection, processing, analysis, and visualization [5]. Some of these challenges include access to very large amounts of unstructured data (management problems), determining the amount of data needed to obtain a large amount of high-quality data (quality vs. quantity), and the processing of the data stream changing dynamically.

However, given the very large number of heterogeneous data from social media, one of the main challenges is to identify and analyze the useful data in order to discover useful knowledge to improve the decision making of individual users and enterprises [6]. As such, the integration and analysis of these data, key elements of patient privacy, and vital tools for healthcare professionals are not a trivial task due to their scalability, complexity, and heterogeneity. Traditional analysis techniques and methods (data analysis) need to be adapted and integrated with the new data paradigms that have emerged for mass data processing.

The main objective of our work is to develop a platform that allows integrating and visualizing knowledge from a large amount of information available on "patientslikeme.com", with the inclusion of "Twitter". We use linked data technologies to aggregate all this data into a semantic model in order to arrive at contextually relevant information and generate knowledge. Thus, we present a model of automatic learning which aims to detect the "TOP TRENDING" drugs used for a given disease. The platform also supports a set of analytical tools that help the user to become aware of socially distributed health information. The paper is organized as follows. Section 2 surveys the related work on healthcare data integration that identifies the research issues involved. In Sect. 3, we present our semantic approach that we used for integrating health knowledge, and in Sect. 4, the visualization results are illustrated, interpreted, and discussed, followed by conclusions and future work in Sect. 5.

2 Related Work

This section discusses the state of the art in the area of integration and analysis of social data available on social networks.

Social media has been integrated into medical practice and has reshaped healthcare services in several ways. It has been proven a viable platform for patients to discuss health-related issues [7] and for researchers to derive health intelligence [8]. However, integrating data from the social web is a challenging task. Several works surveyed approaches to extracting information from the "social web" for health personalization. Research works [9, 10] extract health information from different sources, including the web and social media, sensors, healthcare claims and laboratory images, and physician notes that provide useful health information. The authors of [11] used social media to effectively build novel disease surveillance systems that detect, track, and respond to infectious diseases, such as in the case of the 2009 H1N1 Influenza. Social media has improved healthcare quality with better communication between patients and clinicians, either through generic channels such as Facebook or Twitter [12], or via special sites such as PatientsLikeMe for patients.

Recently, the use of SW technology as a framework for the integration of public data has become popular. Most of the work in this thread follows Linked Open Data (LOD) [13, 14] principles to create links between resources distributed in heterogeneous data sources. In [15, 16], an OWL ontology is used to link the schemas of semi-structured and structured data. In the field of health, a semantic integration model of different health data sources that can help annotate social health blogs is used [17]. The work [18] uses an integrated semantic model to create a machine-readable encoding of the content semantics of various open health data sources, especially social data sources.

Even though there are many sentiment analyses of Tweets in general area [19, 20] and in the health domain [21] using data mining and machine learning approach, most of the works do not apply the results of the sentiment analysis to measure the degree of public concerns or anxiety toward disease. Data mining and machine learning techniques are used to predict disease risks for individuals or to rank diseases by their risks. For instance, in [22, 23], a condition for one patient is predicted using similar patients, based on 13+ million elderly patients' hospital visit records.

3 Semantic Approach for Health Knowledge Integration

In this section, we present our approach that we implemented in Python. We describe in detail the methods and techniques used to extract data from social networks on the Web. We also present a "Matching-Term" algorithm that aims to validate the "tweets" extracted by calculating the similarity between the term used by the user in his "tweet" and the list of medical terminologies extracted from the "UMLS" database [24]. Finally, the "Mapping" of relational model to "RDF" semantic model is described.

3.1 Data Collection

The collection of data from social networks is a big part of our project, and it is the most important phase. In this phase (Fig. 1), we start with the social network "PatientsLikeMe"; this site contains information on patient profiles such as their personal details "Surname, first name, phone …", as well as information on different diseases and their treatments "drugs". We will then extract the "tweets" of users who speak about a given treatment; these treatments are already extracted from "PatientsLikeMe"; and therefore, for each treatment, we must extract a number of "tweets" to find out what are the most popular treatments for a given disease (TOP TRENDING) and also to detect the feelings of users regarding a treatment.

Fig. 1 Data collection

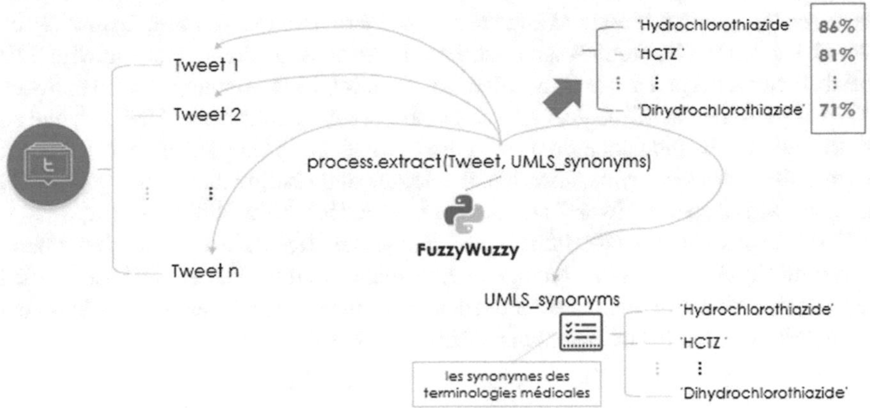

Fig. 2 Cleaning data

3.2 Cleaning Data

After collecting "tweets", we must apply a cleanup before storing them. We use the UMLS [24] Metathesaurus, which provides a common vocabulary and semantics for multiple terms, that refer to the same concept. In this paper, we utilize a "Matching-Term" algorithm (Fig. 3), using the UMLS to recognize identical concepts. This relational database occupies more than "30 GB" of space and contains over 300 million lines. We also use a package called "FuzzyWuzzy" [25] to find the similarity

---

Algorithm 1: Matching Term

Input : tweets, dict_synonyms, threshold

Browse each "tweet" from the "tweets" input dataset ;

Search in the medical dictionary "dict_synonyms" the list of terminologies corresponds to the term (i.e. name of the treatment) used in the "tweet";

Get the different synonyms terms and add the terms in a list "[UMLSTreatement_Terms]";

Use the "FuzzyWuzzy" library which calculates the similarity between a list of entries and a string of characters which is in our list of medical synonyms "[UMLSTreatement_Terms]" and the "tweet";

Initialize the filter rate: (set thresold at 70%);

foreach *ForEach similarity_ratio in FuzzyWuzzy, extract([UMLSTreatement_Terms] , tweet):* **do**

    if *similarity_ratio > threshold* **then**

        store_Tweet_FileCSV ();

        break;

    end

end

Fig. 3 "Matching-Term" algorithm

between the medical synonyms and the medical term that the user used in his "tweet" which has among its features a comparison function of a character string with a list of character strings. However, we give this function in the parameters the collected "Tweet" and the list of medical synonyms that we extracted from "UMLS", and so, if we find that the medical term used in the "Tweet" is similar with at least a medical term in the list of synonyms more than the degree of threading that we set (e.g., 70%), then we will keep the "Tweet" and persist it in an Excel file "CSV" (Fig. 2).

The purpose of this algorithm is to validate the "tweets"; it means that when a surfer on "Twitter" speaking on a given treatment, we must be sure that the medical term (name of treatment) that he used in his "tweet" complies with at least one terminology in the list of synonyms of this treatment.

3.3 Applying Machine Learning to Detect Sentiment

Now that we collected all tweets and cleaned them by "Matching-Term" algorithm, we present the best model constructed on a training dataset, it is called "Sentiment140" [26], and it has been developed by a group of student researchers at University of Stanford, and the way they created this dataset is that any tweet with positive emotions, like :), were positive, and tweets with negative "emotions", like :(, were negative. It contains more than 1.6 million records, and data are annotated in two classes "positive, negative". Therefore, we used this data frame to train our models, and we selected the best model after a set of tuning parameters of every algorithm and also applying dimensionality reduction in order to reduce features, but this last did not give better results, so at the end we did not use dimensionality reduction approach. We tested two approaches in order to present textual data into numerical data "Count Vectorizer" and "TF-IDF Vectorizer" [27], but the one that gave better results in terms of accuracy performance is "TF-IDF Vectorizer". We evaluated our models in a cross-validation of fivefolds, and we used accuracy metric to measure their performance. As we can see in (Fig. 4), we have three best performing models "LinearSVC", "Logistic Regression", "Ridge Classifier" but the best one is logistic regression with 0.82 accuracy score evaluated with cross-validation. Eventually, we used logistic regression model to predict new tweets talking about different treatments in order to detect their sentiments.

3.4 Transformation to RDF

To perform intelligent analyzes across multiple data sources, we use a lightweight ontology to build an integrated knowledge base. Using the entities in the data model, we developed the lightweight ontology by structuring entities and relationships ontologically as a concept hierarchy as shown in Fig. 5.

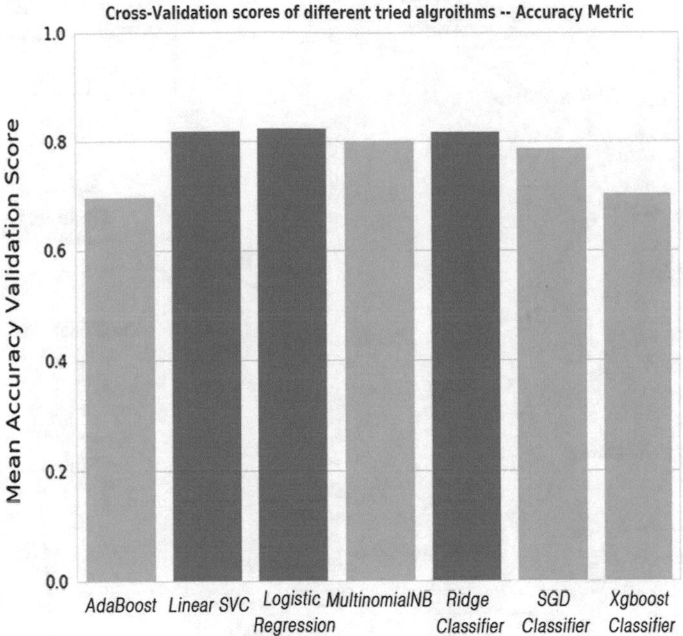

Fig. 4 Cross-validation

An ontology typically consists of classes, properties, relationships, between classes (e.g., IS-A, PART-OF), and instances.

To capture the variety of health information, we developed a conceptual model to capture the relationships among health data entities extracted from different sources. A patient can have a certain disease; this is modeled with the "Patient" class, which has a predicate relation called "hasCondition" with the "Condition" class. A disease can have symptoms and treated with several treatments; this is modeled with the "Condition" class, which has a predicate relation called "show" with the "Symptom" class and another predicate relation called "treatedBy" with "Treatment" class. However, each of the classes "Side-Effects" and "Symptom" have severity levels and their number evaluations, this is modeled through the nodes "EvaluationSeveritySymtpom" and "EvaluationSeverityTreatment" and their properties "hasSeverity", "hasEvaluation". Finally, we model the "tweets" we extracted for each treatment, through the "Treatment" class that has a predicate relation called "hasTweets" with the "Tweets" class. The "Tweets" class has a predicate relation called "hasTweetDate" with the "TweetDate" class which models the fact that a tweet has a own date and also his own feeling.

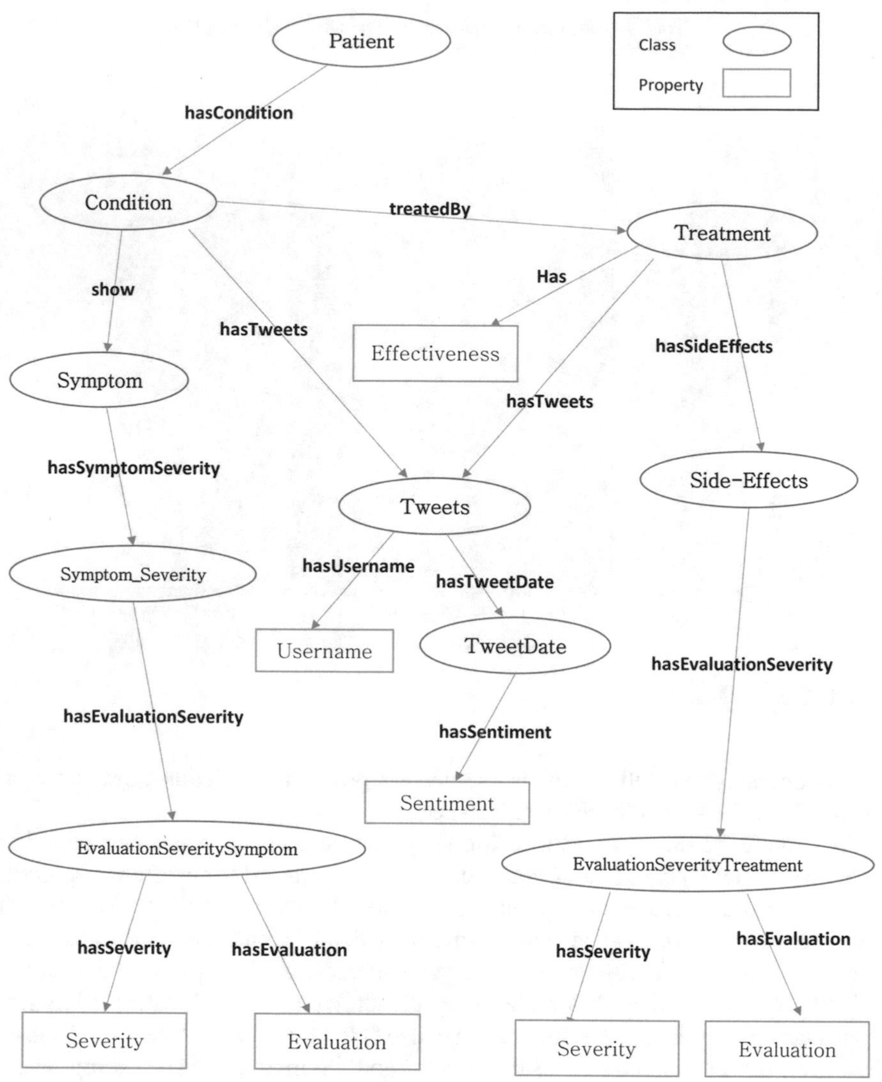

Fig. 5 Semantic model for social health entities

4 Implementation and Results

In this section, we present our platform with various analytical features allowing the extraction of data from social networks "PatientsLikeMe, Twitter", the detection of the most popular treatments TOP TRENDING used by patients and finally the prediction of the sentiments expressed by the patients who have followed these treatments (Fig. 6).

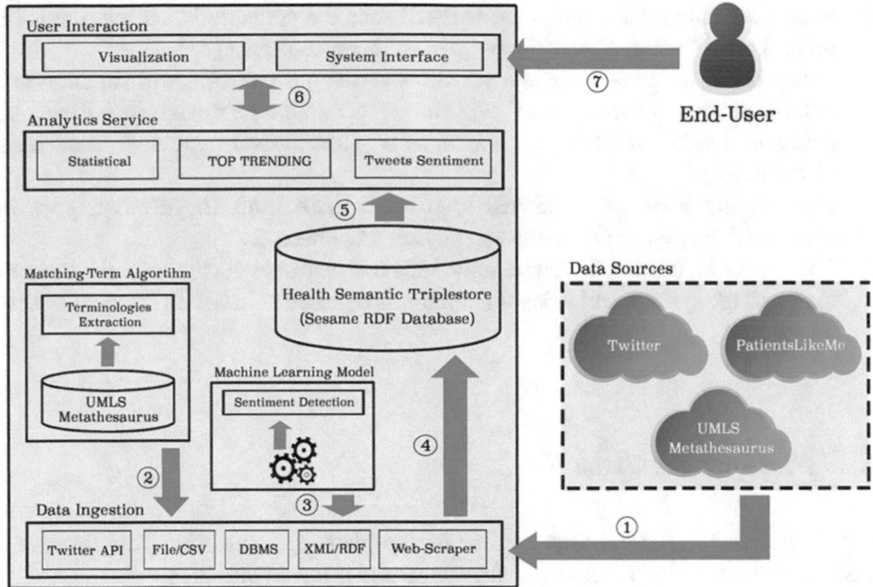

Fig. 6 Platform architecture

4.1 Platform Architecture

The figure above shows the architecture of our platform with all steps taken:

1. On the lower level of the architecture, we have the data ingestion layer. This layer is responsible for the extraction of data from various sources of data on health available to the public. The layer is composed of several connectors, one for each type of data source. Most of the Web sites of health do not provide an API that allows researchers to retrieve data. In addition, we used a Web Scraper called "Scrapy" in order to target Web sites and extract relevant information. We have used data sources like PatientsLikeMe, UMLS, and Twitter (via APIs).
2. After the extraction of data from Twitter's social network, we must also apply our "Matching-Term" algorithm that is designed to validate the extracted "tweets" by comparing the medical term used by the user in his "tweet" and the list of medical synonym terms extracted through the UMLS database, and so, if there is at least one term in the list similar to the term used by the user, then we keep the "tweet" by saving it in an Excel file.
3. In this layer, after having validated all the tweets by our algorithm "Matching-Term", we must now use our machine learning model to predict the sentiment of all tweets and save them into our relational database MySQL.
4. In this layer, we migrate toward the semantic model RDF that we built, we will apply our Python script which allows the mapping from the relational model

to the semantic model, and then, we will save the converted data into the RDF server called "Sesame" that allows you to store RDF data [28].

5. In this layer, we present the analytical tools that our platform will support such as the detection of patient users' sentiments toward the treatments they took, and the identification of the most popular TOP TRENDING treatments used for a given disease.
6. Last but not least, we need just to present these tools in different types of visualizations on our platform such as the "Pie charts …".
7. The last layer, users can interact now with our platform via the visualizations or the interface system, which invokes analysis operations based on the request sent by the user.

4.2 *Platform Analytics Tools*

In this section, we present the interface of the platform, which is in the form of a dashboard containing different visualizations and types of graphics.

4.2.1 Presentation of the Most Popular Treatments

In order to detect the most popular treatments, we must construct a SPARQL query that is designed to query RDF data from the "Sesame" server and then return the results needed to be displayed to the user. Therefore, the patient can select a certain disease; secondly, he should mention the start date and the end date, because popularity is something relative to time. For example, if a treatment is popular in a certain period of time, perhaps it will be more popular in another time, so the date is very important in the ranking of popularity.

Figure 7 shows the most popular treatments according to the disease that the patient has selected; in this case, it is "type 1 Diabetes". The message shows that 922,737 users on Twitter talk around these treatments, and the data are based from 2018-03-06 to 2018-03-28.

4.2.2 Presentation of Patient Data

Figure 8 shows the locations of patients who are diagnosed with the type 1 diabetes; these data came from the profiles of patients that we have extracted from "Patientslikeme".

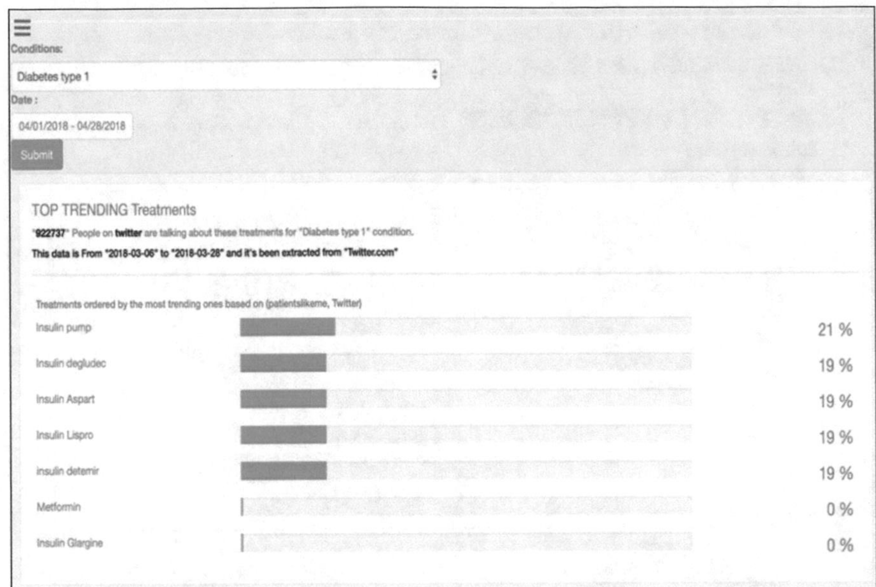

Fig. 7 TOP TRENDING treatments

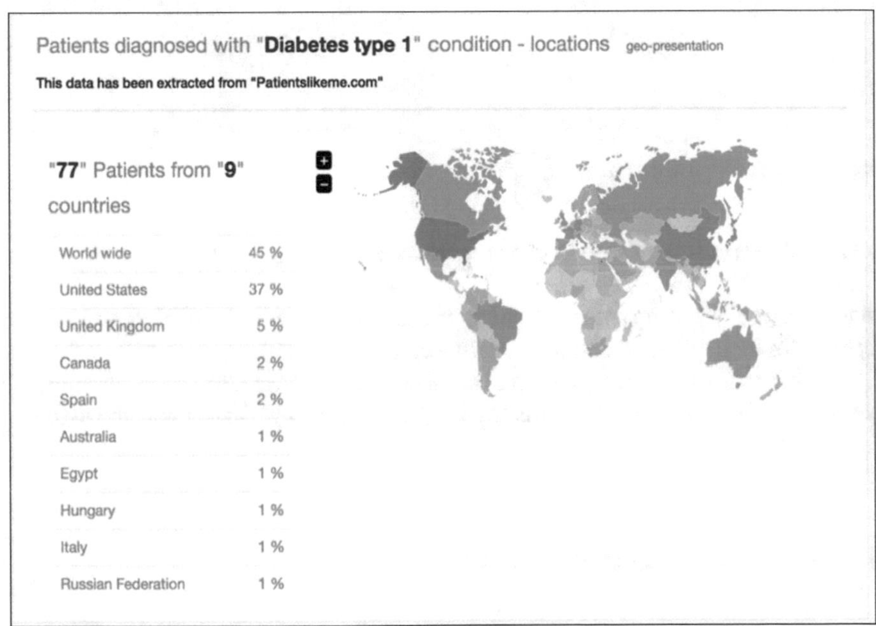

Fig. 8 Patients diagnosed with a given disease

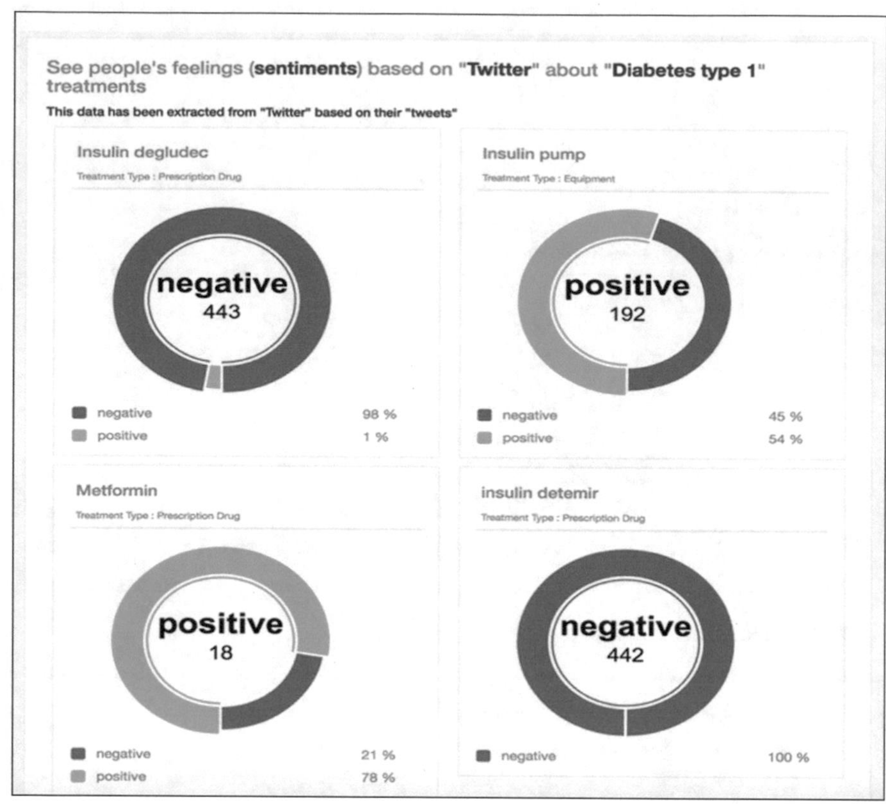

Fig. 9 Feelings of "Twitter" about treatment

4.2.3 Presentation of Sentiments Expressed by Users in Their Tweets

In this section, we also present graphical visualizations about the feelings expressed by users in their "tweets" about a certain treatment (Fig. 9).

The figure above shows the graphical visualizations "Pie Charts" concerning the people's feelings about different treatments. This part was realized using the machine learning model.

5 Conclusion and Perspectives

The field of health is a very complex and dynamic area where decision makers often rely on decision support systems to analyze and compare the actors involved (clinical trials, procedures, etc.). The expansion of the social web in the health field increases considerably the flow of data, where the experiences of patients are shared such as treatments, side effects, etc. These data can be important sources to provide collective intelligence, evaluate, and improve healthcare performance. As a result, the goal of our work was to develop a platform that synthesizes knowledge from a large body of data extracted from "PatientsLikeMe, Twitter" social networks, in order to identify the most used treatments by the patients and also to detect their feelings expressed toward these treatments.

In our future work, we aim to use several sources of medical data from many social networks, such as "Blogs, Forums, Datasets …" which contain, for example, information about patients' discussions, their questions asked in the forums, and the different health data sets that have been published by the government. There are also other sources of data that contain a lot of information such as articles and scientific journals of health. We plan after exploring all these data sources to build a semantic model that will be able to group all of these data together to generate knowledge. We also aim to implement these results in an extended platform that will contain numerous queries and analytical tools to better serve policy makers and patients.

References

1. Doctissimo. http://www.doctissimo.fr/
2. PatientsLikeMe. https://www.patientslikeme.com/
3. Raghupathi, W., Raghupathi, V.: Big data analytics in healthcare: promise and potential. Heal. Inf. Sci. Syst. **2**, 1–10 (2014). https://doi.org/10.1186/2047-2501-2-3
4. Sessler, D.I.: Big data and its contributions to peri-operative medicine. Anaesthesia **69**, 100–105 (2014)
5. Kaisler, S., Armour, F., Espinosa, J.A., Money, W.: Big data: issues and challenges moving forward. In: Proceedings of 46th Hawaii International Conference on System Sciences (HICSS), pp. 995–1004. IEEE (2013)
6. Chen, H., Chiang, R.H., Storey, V.C.: Business intelligence and analytics: from big data to big impact, MISQ **36**(4), 1165–1188 (2012)
7. Househ, M., Borycki, E., Kushniruk, A.: Empowering patients through social media: the benefits and challenges. Health Inf. J. **20**, 50–58 (2014)
8. Ji X, Chun SA, Geller J.: Monitoring public health concerns using Twitter sentiment classifications. In: Proceedings of IEEE International Conference on Healthcare Informatics, pp. 335–344. IEEE, Philadelphia, PA (2013)
9. Raghupathi, W., Raghupathi, V.: Big data analytics in healthcare: promise and potential. Health Inf. Sci. Syst. **2**, 1–10 (2014)
10. Ji, X., Chun, S.A., Geller, J.: Monitoring public health concerns using Twitter sentiment classifications. In: Proceedings of IEEE International Conference on Healthcare Informatics, pp. 335–344. Philadelphia, PA (2013)
11. Brownstein, J.S., Freifeld, C.C., Chan, E.H., Keller, M., Sonricker, A.L., Mekaru, S.R., Buckeridge, D.L.: Information technology and global surveillance of cases of 2009 H1N1 influenza. N. Engl. J. Med. **362**(18), 1731–1735 (2010)
12. Bull, S.S., Breslin, L.T., Wright, E.E., Black, S.R., Levine, D., Santelli, J.S.: Case study: an ethics case study of HIV prevention research on Facebook: the just/us study. J. Pediatr. Psychol. **36**(10), 1082–1092 (2011)
13. Bizer, C.: Evolving the Web into a Global Data Space. In: Fernandes, A.A., Gray, A.G., Belhajjame, K. (eds.). Proceedings of 28th British National Conference on Databases, p. 1. Springer Berlin Heidelberg, Manchester (2011)
14. Bizer, C., Heath, T., Berners-Lee, T.: Linked data—the story so far. Int. J. Semant. Web Inf. Syst. **5**, 1–22 (2009)

15. Skoutas, D., Simitsis, A.: Designing ETL processes using semantic web technologies. In: DOLAP, pp. 67–74 (2006)
16. Skoutas, D., Simitsis, A.: Ontology-based conceptual design of ETL processes for both structured and semi-structured data. IJSWIS **3**(4), 1–24 (2007)
17. Chun, S.A., Mac Kellar, B.: Social health data integration using semantic web. In: Proceedings of the 27th Annual ACM Symposium on Applied Computing, pp. 392–397 (2012)
18. Ji, X., et al.: Linking and using social media data for enhancing public health analytics. J. Inf. Sci. **43.2**, 221–245 (2017)
19. Pang, B., Lee, L.: Opinion mining and sentiment analysis. Found. Trends Inf. Retrieval **2**(1–2), 1–135 (2008) Social Health Records: Gaining Insights into Public Health … 41
20. Zhuang, L., Jing, F., Zhu, X.-Y.: Movie review mining and summarization. In: Proceedings of the 15th ACM International Conference on Information and Knowledge Management, pp. 43–50. Arlington, VAS (2006)
21. Chew, C., Eysenbach, G.: Pandemics in the age of Twitter: content analysis of Tweets during the 2009 H1N1 outbreak. PLoS ONE **5**(11), e14118 (2010)
22. Chawla, N.V., Davis, D.A.: Bringing big data to personalized healthcare: a patient-centered framework. J. Gen. Intern. Med. **28**, 660–665 (2013)
23. Davis, D.A., Chawla, N.V., Christakis, N.A., Barabasi, A.L.: Time to CARE: a collaborative engine for practical disease prediction. Data Min. Knowl. Disc. **20**, 388–415 (2010)
24. Bodenreider, O.: The unified medical language system (UMLS): integrating biomedical terminology. Nucleic Acids Res. **32**(suppl_1), D267–D270 (2004)
25. Gonzalez, J.: fuzzywuzzy Fuzzy String Matching in python. https://github.com/seatgeek/fuzzywuzzy
26. http://help.sentiment140.com/for-students
27. Soucy, P., Mimeau, G.W.: Beyond TF-IDF weighting for text categorization in the vector space model. In: Proceedings of 19th International Joint Conference Artificial Intelligence (IJCAI '05), pp. 1130–1135 (2005)
28. Broekstra, J., Kampman, A., Van Harmelen, F.: Sesame: an architecture for storing and querying RDF data and schema information (2001)

Image Classification Using Legendre–Fourier Moments and Artificial Neural Network

Abderrahmane Machhour, Mostafa El Mallahi, Zakia Lakhliai, Ahmed Tahiri and Driss Chenouni

Abstract The nonlinear structure of the artificial neural network is efficient for the classification; however, the choice of features is a fundamental problem due to their direct impact on the network convergence and performance. In this paper, we present a new method of image classification method based on Legendre–Fourier moments using an artificial neural network. We used LFMs to extract features from images. In result, every image is represented by a descriptor vector; these vectors are inputs of our neural network. We tested this model on Fashion-MNIST database and we got important results; the model's accuracy exceeds 97%. The validity of this proposed method has provided under different transformations.

Keywords Image classification · Legendre–Fourier moments · Artificial neural network · Features · Descriptor vector · Moment invariants

1 Introduction

In recent years, automatic classification has become a major concern in various fields because of the significant role it plays in helping to make decisions and advance artificial intelligence. In the automatic classification of images, we have two main

A. Machhour (✉) · M. El Mallahi · Z. Lakhliai · A. Tahiri · D. Chenouni
Sidi Mohamed Ben Abdellah University, Laboratory of Computer Science and Interdisciplinary Physics LIPI, ENS, Fez, Morocco
e-mail: machhour007@hotmail.com

M. El Mallahi
e-mail: Mostafa.elmallahi@usmba.ac.ma

Z. Lakhliai
e-mail: zakia.lakhliai@usmba.ac.ma

A. Tahiri
e-mail: tahiri_ahmed@hotmail.com

D. Chenouni
e-mail: d_chenouni@yahoo.fr

V. Bhateja et al. (eds.), *Embedded Systems and Artificial Intelligence*, Advances in Intelligent Systems and Computing 1076,
https://doi.org/10.1007/978-981-15-0947-6_29

components, the computer vision and the neural networks. To create models capable of classification and prediction based on data, we need to train our neural network using vector inputs, and each vector is the result of the application of a computer vision technique to extract image features. Among these techniques and methods, we have orthogonal moments [1].

The first use of moments in analysis and instance representation took place in 1961 by Hu et al. [2]. They used geometric moments to extract invariants that are then used in automatic character recognition. This set of moment invariants is translation, scale and rotation independent. But these moments are not orthogonal, and as a consequence, reconstructing the image from it is seemed to be a hard operation. Since then, after more than 50 years of research, a lot of new accomplishments in the theory of moments and moment invariants have been achieved.

The next important step was the introduction of orthogonal moments by Teague et al. in 1980 [3]. These moments are based on continuous orthogonal polynomials such as Zernike and Legendre polynomials [4]. The main advantage of orthogonal moments is that we can use it to reconstruct the image with minimal information redundancy. Since then, many continuous orthogonal moments were consecutively presented, Fourier–Mellin moments [5], Chebyshev–Fourier moments [6], Gaussian–Hermite moments [7] and Gegenbauer moments [8]. The other some methods based on the discrete moments using discrete polynomials are Krawtchouk polynomials [9, 10], Hahn polynomials [11, 12], dual Hahn polynomials [13, 14], Racah polynomials [15], Meixner polynomials [16, 17], Charlier polynomials [18] for image processing. The calculation of discrete moments eliminates the need for numerical approximation and satisfies exactly the orthogonality property in discrete space coordinates of the image.

The calculation of image moments is usually done in a recursive way, and this is expensive for the calculation time and memory. But at the same time, the orthogonal moments are powerful in image representation, since they retain relevant information about the image. In our model (Fig. 1), we built the descriptor vector using Legendre–Fourier moments of different orders and we used these vectors as input of the neural network.

2 Orthogonal Legendre–Fourier Moments

In a unit circle, Legendre–Fourier moments (LFMs) of order p and repetition q are defined as follows:

$$M_{pq} = \frac{(2P+1)}{\pi} \int_0^{2\pi} \int_0^1 f(r,\theta)\left[L_{pq}(r,\theta)\right]^* r\mathrm{d}r\mathrm{d}\theta, \tag{1}$$

The functions of the Legendre–Fourier moments $L_{pq}(r,\theta)$ are defined in polar coordinates as the product of the orthogonal substituted shifted Legendre polynomials

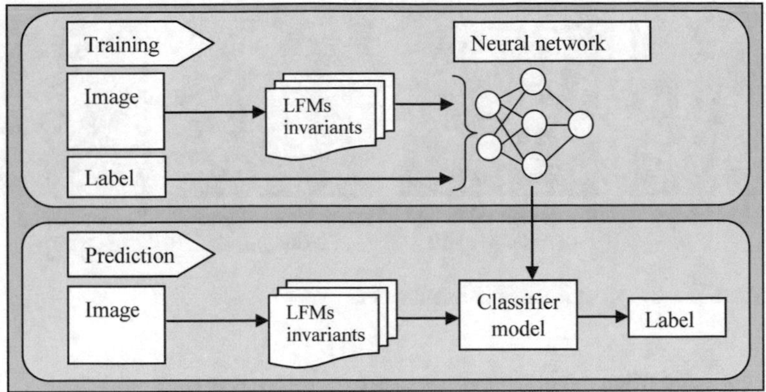

Fig. 1 Proposed classification system

$\overline{P}_p(r)$ (Fig. 2) and the Fourier exponential function $\exp(iq\theta)$. They are two separable functions; the first is the radial function and the second is the angular one.

$$L_{pq}(r,\theta) = \overline{P}_p(r)e^{\hat{i}q\theta}, \tag{2}$$

$\overline{P}_p(r)$ obey the following recurrence relation:

$$\overline{P}_{p+1}(r) = \frac{2p+1}{p+1}\left(2r^2-1\right)\overline{P}_p(r) - \frac{p}{p+1}\overline{P}_{p-1}(r), \tag{3}$$

where $\overline{P}_0(r) = 1$ and $\overline{P}_1(r) = 2r^2 - 1$.

Computation of LFMs in polar coordinates is preferable because the geometric errors which results of a circle-to-square mapping are avoided. Approximation errors

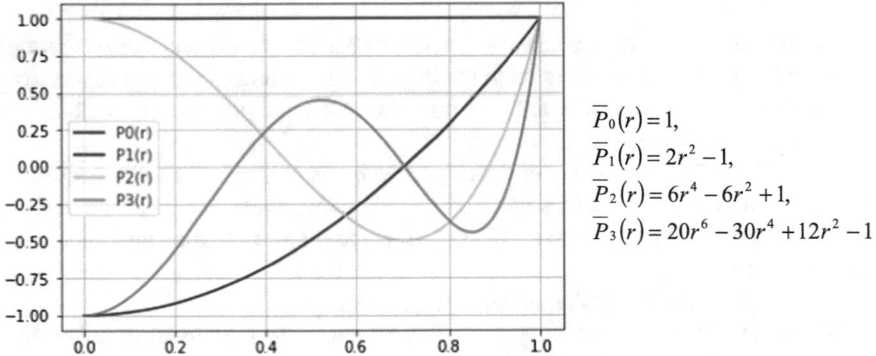

Fig. 2 Plot of substituted radial shifted Legendre polynomial for the first four orders

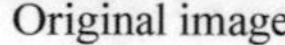
Original image

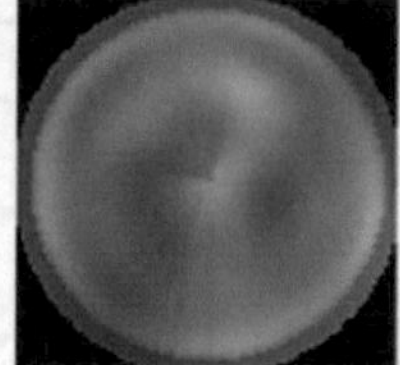
Reconstructed image of order p+q=10

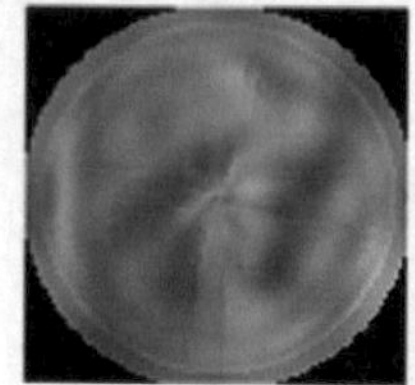
Reconstructed image of order p+q=20

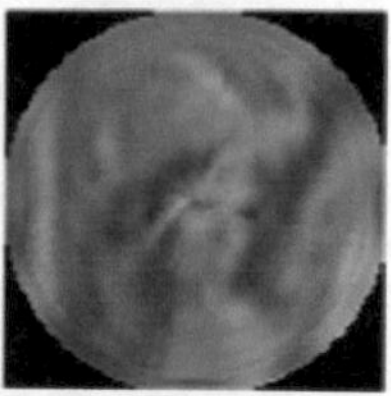
Reconstructed image of order p+q=30

Fig. 3 Lena image reconstruction of different orders

are encountered when double integrations are replaced by double summations. These errors are avoided by exact computation of LFMs in polar coordinates using the following forms [8]:

$$M_{pq} = \frac{2p+1}{\pi} \sum_i \sum_j \hat{f}\big(r_i, \theta_{ij}\big) L_{pq}\big(r_i, \theta_{ij}\big), \tag{4}$$

The function $\hat{f}$ is deduced from the input image function using cubic interpolation.

The orthogonality property allows us to reconstruct an image function (Fig. 3) using the Legendre–Fourier moments as follows:

$$f(r, \theta) \approx \sum_{p=0}^{p_{\max}} \sum_{q=0}^{q_{\max}} M_{pq} \bar{P}_p(r) e^{\hat{i}q\theta} \tag{5}$$

3 Legendre–Fourier Moment Invariants

To classify the images based on their contents, we need to choose and extract features from the images that must distinguish the objects that belong to the same class from the other objects, whatever the size, the inclination and the position of the object on the image. We call these features invariants.

Hosny et al. [8] succeeded in extracting three types of invariants based on LFMs: rotation invariants, scaling invariants and translation invariants. We have revised these invariants and have been able to obtain invariants to rotation, scaling and translation at the same time.

Let f^{RS} be the rotated scaled version of the image f:

We consider $f^{RS}(r, \theta) = f\big(\frac{r}{\lambda}, \theta + \beta\big)$. So, we can express the rotation and scaling Legendre–Fourier moments as following:

$$M_{pq}\left(f^{RS}\right) = \frac{2p+1}{\pi}\int_{0}^{2\pi}\int_{0}^{1} f^{RS}(r,\theta)\left[L_{pq}(r,\theta)\right]^{*} r\mathrm{d}r\mathrm{d}\theta \tag{6}$$

$$M_{pq}\left(f^{RS}\right) = \frac{2p+1}{\pi}\int_{0}^{2\pi}\int_{0}^{1} f^{RS}(r,\theta)\overline{P}_{p}(r)\mathrm{e}^{-\hat{i}q\theta} r\mathrm{d}r\mathrm{d}\theta \tag{7}$$

$$M_{pq}\left(f^{RS}\right) = \frac{2p+1}{\pi}\int_{0}^{2\pi}\int_{0}^{1} f\left(\frac{r}{\lambda},\theta+\beta\right)\overline{P}_{p}(r)\mathrm{e}^{-\hat{i}q\theta} r\mathrm{d}r\mathrm{d}\theta \tag{8}$$

Let $r' = \frac{r}{\lambda}$ then $\mathrm{d}r = \lambda\, \mathrm{d}r'$ and $r = \lambda\, r'$
And let $\theta' = \theta + \beta$ so $\mathrm{d}\theta = \mathrm{d}\theta'$ and $\theta = \theta' - \beta$
The equation becomes:

$$M_{pq}\left(f^{RS}\right) = \lambda^{2}\frac{2p+1}{\pi}\int_{0}^{2\pi}\int_{0}^{1} f\left(r',\theta'\right)\overline{P}_{p}\left(\lambda\, r'\right)\mathrm{e}^{-\hat{i}q(\theta'-\beta)} r'\mathrm{d}r'\mathrm{d}\theta' \tag{9}$$

$$M_{pq}\left(f^{RS}\right) = \lambda^{2}\frac{2p+1}{\pi}\int_{0}^{2\pi}\int_{0}^{1} f\left(r',\theta'\right)\overline{P}_{p}\left(\lambda\, r'\right)\mathrm{e}^{-\hat{i}q\theta'}\mathrm{e}^{iq\beta} r'\mathrm{d}r'\mathrm{d}\theta' \tag{10}$$

$$M_{pq}\left(f^{RS}\right) = \mathrm{e}^{iq\beta}\lambda^{2}\frac{2p+1}{\pi}\int_{0}^{2\pi}\int_{0}^{1} f\left(r',\theta'\right)\overline{P}_{p}\left(\lambda\, r'\right)\mathrm{e}^{-\hat{i}q\theta'} r'\mathrm{d}r'\mathrm{d}\theta' \tag{11}$$

where $\overline{p}_{p}\left(\lambda\, r'\right)$ is the scaled form of $p_{p}\left(r'\right)$. We can write the scaled substituted shifted Legendre polynomials $\overline{p}_{p}\left(\lambda\, r'\right)$ in terms of the substituted shifted Legendre polynomials as following [8]:

$$\overline{P}_{p}\left(\lambda\, r'\right) = \sum_{k=0}^{p}\left(\sum_{i=k}^{p}\left(\lambda^{2i}S_{pi}T_{ik}\right)\right)\overline{P}_{k}\left(r'\right) \tag{12}$$

where:

$$S_{pi} = (-1)^{p-1}\frac{(p+i)!}{(p-i)!(i\,!)^{2}} \tag{13}$$

$$T_{ik} = \frac{(2k+1)(i\,!)^{2}}{(i+k+1)!(i-k)!} \tag{14}$$

After the substitution, the expression of $M_{pq}\left(f^{RS}\right)$ becomes:

$$M_{pq}\left(f^{RS}\right) = \mathrm{e}^{\hat{i}q\beta}\frac{2p+1}{\pi}\sum_{k=0}^{p}\left(\sum_{i=k}^{p}\lambda^{2i+2}S_{pi}T_{ik}\times\int_{0}^{2\pi}\int_{0}^{1}f\left(r',\theta'\right)\overline{P}_{k}\left(r'\right)\mathrm{e}^{-\hat{i}q\theta'}r'\mathrm{d}r'\mathrm{d}\theta'\right)$$

And, we can simplify its form as follows:

$$M_{pq}\left(f^{RS}\right) = \mathrm{e}^{\hat{i}q\beta}\frac{2p+1}{\pi}\sum_{k=0}^{p}\left(\sum_{i=k}^{p}\lambda^{2i+2}S_{pi}T_{ik}M_{kq}(f)\right) \tag{15}$$

Replacing p and q with 0, we find:

$$M_{00}\left(f^{RS}\right) = \lambda^{2}M_{00}(f)$$

From Eq. (15), we can extract a scaling invariant as follows:

$$\omega_{pq} = \mathrm{e}^{\hat{i}q\beta}\sum_{k=0}^{p}\frac{2p+1}{2k+1}\left(\sum_{i=k}^{p}(M_{00}(f))^{-(i+1)}S_{pi}T_{ik}\right)M_{kq}(f) \tag{16}$$

Applying the magnitude to the equation yields:

$$\left|\omega_{pq}\right| = \left|\mathrm{e}^{\hat{i}q\beta}\right|\times\left|\sum_{k=0}^{p}\frac{2p+1}{2k+1}\left(\sum_{i=k}^{p}(M_{00}(f))^{-(i+1)}S_{pi}T_{ik}\right)M_{kq}(f)\right| \tag{17}$$

And we know that $\left|\mathrm{e}^{\hat{i}q\beta}\right| = 1$ then:

$$\left|\omega_{pq}\right| = \left|\sum_{k=0}^{p}\frac{2p+1}{2k+1}\left(\sum_{i=k}^{p}(M_{00}(f))^{-(i+1)}S_{pi}T_{ik}\right)M_{kq}(f)\right| \tag{18}$$

We can conclude from this last equation that $\left|\omega_{pq}\right|$ is invariant to rotation and scaling transformations.

4 Experimentation

Initially, to test the invariance of the extracted characteristic $\left|\omega_{pq}\right|$ to the rotation and the scaling transformations, we applied to the famous image of Lena three types of transformations, first we applied a rotation of angle $\beta = \pi/2$ and a scaling of ratio $\lambda = 2$ and a transformation consisting of a rotation $\beta = \pi/2$ and a scaling $\lambda = 2$ at the same time. As a result, four different images were obtained (Fig. 4) and $\left|\omega_{pq}\right|$ was calculated from these images. The results were acceptable; $\left|\omega_{pq}\right|$ with its different orders remains invariant to different transformations (Table 1).

Fig. 4 Lena image and some transformations

Table 1 LFM invariants of Lena image before and after rotation and scale transformations

Image	$\lvert\omega_{1,1}\rvert$	$\lvert\omega_{1,2}\rvert$	$\lvert\omega_{2,1}\rvert$	$\lvert\omega_{3,1}\rvert$	$\lvert\omega_{3,4}\rvert$	$\lvert\omega_{4,1}\rvert$	$\lvert\omega_{4,3}\rvert$	$\lvert\omega_{4,4}\rvert$
Original	0.2268	0.1526	0.3731	0.5121	0.2094	0.6410	0.3616	0.2576
$\beta = \pi/2$	0.2277	0.1405	0.3749	0.5150	0.2002	0.6456	0.3272	0.2465
$\lambda = 2$	0.2230	0.1581	0.3667	0.5033	0.2289	0.6300	0.3613	0.2824
$\beta = \pi/2$, $\lambda = 2$	0.2160	0.1385	0.3558	0.4892	0.2034	0.6139	0.3052	0.2510

During the features extraction phase, we chose to work on the Fashion-MNIST dataset, which is a challenging dataset because it contains 70,000 grayscale images in 10 classes. The images display different articles of clothing at resolution 28 by 28 pixels, with pixel values ranging between 0 and 255. Each image is assigned to a single label. The labels are an array of integers, ranging from 0 to 9. These correspond to the class of clothing the image represents (Table 2).

Image moments of first orders are ideal for the classification of objects based on their shapes. Then to work on the shapes of the clothes, we carried out a binarization of the images using a threshold function before the calculation of LFM invariants (Fig. 5).

Table 2 Fashion MNIST image classes

Label	Class	Samples
0	T-shirt/top	
1	Trouser	
2	Pullover	
3	Dress	
4	Coat	
5	Sandal	
6	Shirt	
7	Sneaker	
8	Bag	
9	Ankle boot	

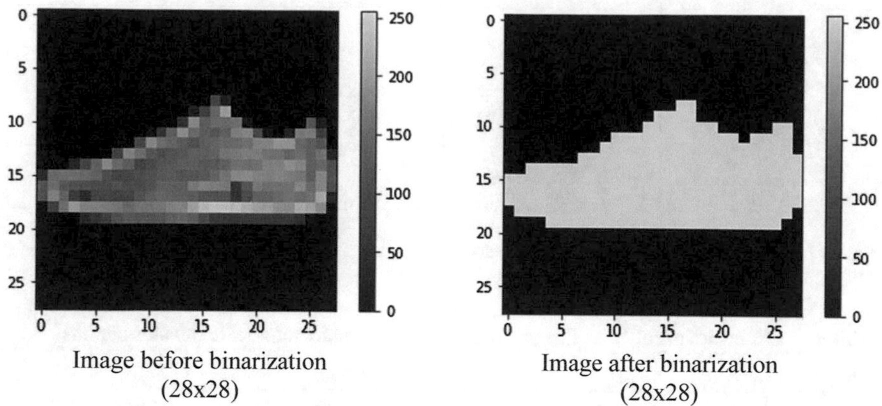

Image before binarization
(28x28)
Image after binarization
(28x28)

Fig. 5 A sample before and after binarization

The calculation of LFM invariants is carried out in the polar coordinates to avoid the geometric errors [19]. And, it is done in a recursive way, and this is a little expensive for the calculation time and memory. We computed 16 LFM invariants $\left\{|\omega_{p,q}|/(p,q) \in \{1,2,3,4\}^2\right\}$ for every sample of the first 1000 images of the dataset without any fast or accelerated computation method (Table 3).

All our numerical experiments were performed with 2.60-GHz Core i5-3320 M laptop with six gigabytes of RAM. The code is designed using Python 3.6. And, it took almost 20 s to calculate 16 invariants of one sample, which is equivalent to five and a half hours for 1000 samples.

To build our artificial neural network we used Keras, it is a model-level library for Python that provides an appropriate way to define and train deep learning models, and it is used for fast prototyping, advanced research and production.

Keras allowed us to build an artificial neural network in a sequential way, i.e., to add the layers one by one, specifying for each layer the number of neurons as well as the chosen activation function and other configurations.

The input data of our model is vectors with 16 dimensions (16 LFM invariants), and the labels (Classes) are scalars {0;1;2;3;…;9}. A type of network that performs well on such a problem is a simple stack of fully connected (dense) layers with ReLu activations.

First, we trained the ANN to classify 80% of the items that belong to two categories (Trouser, Ankle boot) and we tested the model on the remaining 20%, the model's test accuracy exceeds 97%. Then we made several classifications by increasing the number of categories each time (Table 4).

We have noticed that the accuracy has decreased in the case of 10 classes and this is due to the confusion that the neural network has encountered when classifying the elements that belong to the different classes but have the same shape. This cause appeared when we trained the ANN to classify only the elements of the three categories: pullover, shirt and coat. The train accuracy was about 69% and the test accuracy was about 54%.

Table 3 LFM invariants of the first four dataset images after binarization

Label	$\lvert\omega_{1,1}\rvert$	$\lvert\omega_{1,2}\rvert$	$\lvert\omega_{1,3}\rvert$	$\lvert\omega_{1,4}\rvert$	…	$\lvert\omega_{4,1}\rvert$	$\lvert\omega_{4,2}\rvert$	$\lvert\omega_{4,3}\rvert$	$\lvert\omega_{4,4}\rvert$
2	0.300	0.178	0.112	0.031	…	0.841	0.496	0.314	0.088
9	0.750	0.833	0.655	0.126	…	2.115	2.312	1.802	0.352
6	0.200	0.430	0.093	0.113	…	0.566	1.209	0.261	0.318
0	0.264	0.833	0.257	0.228	…	0.737	2.351	0.718	0.651

Table 4 Classification accuracy according to number of classes

Number of classes	2	3	4	5	6	10
Labels	Trouser, Ankle boot	Trouser, Ankle boot, Pullover	Trouser, Ankle boot, Pullover, T-shirt	Trouser, Ankle boot, Pullover, T-shirt, Sneaker	Trouser, Ankle boot, Pullover, T-shirt, Sneaker, Bag	All labels
Training samples	144	223	309	391	468	800
Test samples	36	55	77	97	116	200
Train accuracy (%)	98	98	96	92	89	73
Test accuracy	97	92	83	86	79	61

5 Conclusion

In this paper, a new method for grayscale image classification has been presented based on the invariants extracted from the Legendre–Fourier moments. The calculation of the moments was carried out in the polar coordinates to avoid the geometric errors. Consequently, the classification has achieved a good level of accuracy by using only moments of low orders. Our next goal is to increase our model's accuracy and to use an accelerated algorithm of LFM invariants calculation in order to work on LFM invariants of higher orders and classify color images of higher resolution.

References

1. Amakdouf, H., El Mallahi, M., Zouhri, A., Qjidaa, H.: Classification and recognition of 3D image of Charlier moments using multilayer perceptron architecture. Procedia Comput. Sci. **127**(2018), 226–235 (2018)
2. Hu, M.K.: Visual pattern recognition by moment invariants. IRE Trans. Inf. Theory **8**(2), 179–187 (1962)
3. Teague, M.R.: Image analysis via the general theory of moments. J. Opt. Soc. Am. **70**(8), 920–930 (1980)
4. El Mallahi, M., Zouhri, A., Amakdouf, H., Qjidaa, H.: Rotation scaling and translation invariants of 3D radial shifted Legendre moments. Int. J. Autom Comput. **15**(2), 169–180 (2018)
5. Sheng, Y., Shen, L.: Orthogonal Fourier-Mellin moments for invariant pattern recognition. JOSA A **11**(6), 1748–1757 (1994)

6. Ping, Z., Wu, R., Sheng, Y.: Image description with Chebyshev-Fourier moments. J. Opt. Soc. Am., A Opt. Image Sci. Vis. **19**(9), 1748–54 (2002)
7. Yang, B., Dai, M.: Image analysis by Gaussian-Hermite moments. Signal Process **91**(10), 2290–2303 (2011)
8. Hosny, K.M.: Image representation using accurate orthogonal Gegenbauer moments. Pattern Recognit. Lett. **32**(6), 795–804 (2011)
9. El Mallahi, M., Mesbah, A., El Fadili, H., Zenkouar, K., Qjidaa, H.: Compact computation of Krawtchouk moments for 3D object representation. WSEAS Trans. Circ. Syst. **13**. E-ISSN: 2224-266X (2014)
10. El Mallahi, M., Zouhri, A., El-Mekkaoui, J., Qjidaa, H.: Three dimensional radial Krawtchouk moment invariants for volumetric image recognition. Pattern Recogn. Image Anal. **27**(4), 810–824 (2017)
11. El Mallahi, M., Mesbah, A., Qjidaa, H.: Fast algorithm for 3D local feature extraction using Hahn and Charlier moments. Advances in Ubiquitous Networking 2. Lecture Notes in Electrical Engineering, vol. 397. Springer, Singapore (2016)
12. Mesbah, A., El Mallahi, M., Qjidaa, H.: Fast and efficient computation of three-dimensional Hahn moments. SPIE, J. Electron. Imaging. **25**(6), 061621 (2016)
13. El Mallahi, M., Mesbah, A., Qjidaa, H.: 3D radial invariant of dual Hahn moments. Neural Comput. Appl. **30**(7), 2283–2294 (2018)
14. Flusser, J., Suk, T., Zitová, B.: Moments and Moment Invariants in Pattern Recognition. Wiley, Chichester (2009)
15. El Mallahi, M., Zouhri, A., Mesbah, A., El Affar, I., Qjidaa, H.: Radial invariant of 2D and 3D Racah moments. Multimedia Tools Appl. **77**(6), pp. 6583–6604 (2018)
16. El Mallahi, M., Zouhri, A., Mekkaoui, J., Qjidaa, H.: Radial Meixner moments for rotational invariant pattern recognition. IEEE 2017 Intelligent Systems and Computer Vision, ISCV (2017)
17. El Mallahi, M., Zouhri, A., Qjidaa, H.: Radial Meixner moment invariants for 2D and 3D image recognition. Pattern Recogn. Image Anal. **28**(2), 207–216 (2018)
18. El Mallahi, M., Mesbah, A., Qjidaa, H.: Radial Charlier moment invariants for 2D object/image recognition. IEEE, International Conference on Multimedia Computing and Systems -Proceedings, Procedia Computer Science, vol. 127, pp. 226–235 (2018)
19. Xin, Y., Pawlak, M., Liao, S.: Accurate computation of Zernike moments in polar coordinates. IEEE Trans. Image Process. **16**(2), 581–587 (2007)

Use of New Communication Technologies in the Behavioral and Cognitive Awareness of Road Users: Toward Safe Driving

Halima Ettahiri and Taoufiq Fechtali

Abstract Driver fatigue is a serious problem resulting in many thousands of road accidents each year. It is not possible to calculate the exact number of sleep-related accidents, but research shows that driver fatigue may be a contributory factor in up to 20% of road accidents, and up to one-quarter of fatal and serious accidents. These types of crashes are about 50% more likely to result in death or serious injury as they tend to be high-speed impacts because a driver who has fallen asleep cannot brake or swerve to avoid or reduce the impact. Drivers are aware when they are feeling sleepy, and so make a conscious decision about whether to continue driving or to stop for a rest. It may be that those who persist in driving underestimate the risk of falling asleep while driving. Or it may be that some drivers choose to ignore the risks (in the way that drivers drink too). For this reason, we include in this paper the analysis of records of EEG signals that we collect from 20 volunteers, 10 of them are deprived of sleep for one night. The interpretation of this data is very important to provide insight into the problem; it also helps us understand the mental fatigue of the driver so we could know if the driver is able to drive or not.

Keywords Driver · Fatigue · Sleep · EEG signals · Mental fatigue

1 Introduction

Recently, a lot of research has been performed to prevent road accidents. Thus, in this paper, we would like to understand the causes of the accident. Electroencephalogram (EEG) signals help us to understand the mean of human–computer interaction, which requires very little in terms of physical abilities. When training the computer to recognize and classify EEG signals, users could manipulate the machine by simply thinking about what they want to do within a limited set of choices.

H. Ettahiri (✉) · T. Fechtali
Faculty of Sciences and Techniques Mohammedia, University of Hassan 2, Mohammedia 28806, Maroc

V. Bhateja et al. (eds.), *Embedded Systems and Artificial Intelligence*, Advances in Intelligent Systems and Computing 1076,
https://doi.org/10.1007/978-981-15-0947-6_30

For many years, neurologists have speculated that electroencephalography activity on other electrophysiological measures of brain function might sustain a protocol that allows us to send messages and commands to external world such as Brain Computing Intelligence (BCI) [1].

In our paper, the goal is to understand the state of the brain when the user is tired and deprived of sleep. Therefore, we preferred to define the notion of sleep before we learn how to control it in the laboratory. In addition, even if we observe the sleep of others, so much of the change they experience in the functions of their brains and bodies is not easily noticed from the outside. Sleep scientists have explored these changes in depth, and their definition of sleep is tied to characteristic patterns of brain waves and other physiological functions.

In this study, we assessed the impact of one night of sleep deprivation on mental fatigue. In fact, we selected twenty volunteers to participate in this study; none of the subjects had a history of medical, neurological, or psychiatric disorders, neither medication nor drug intake. The participants were randomly assigned to one of the two groups: the normal group and the deprivation sleep group. In fact, we had collected the EEG signals using the Open VIBE interface developed by INRIA, and the equipment is ENOBIO 8 which allowed us to perform tests in three sessions: each session had duration of 7 min with 40 trials. In each trial, the user must imagine the movement of the right and left hand. Altogether in this paper, we examine the application of SVM and LDA, TREE to EEG signals to choose the best classification for this type of dataset.

1.1 Background and Related Works

Since the power of the modern computers expands over our understanding of the human brain, such as transmitting signals directly to someone's brain that would allow them to hear, feel, or see different sensory inputs, considering the potential to manipulate machines and computers with nothing more than a thought, for people who are having a physical or mental condition that limits their movements, senses, or activities. Development of a brain–computer interface could be the most important technological breakthrough in decades. Brain–machine interfaces (BMI) are systems based on EEG signals. In fact, these systems allow direct communication between an individual and a machine that does not rely on standard human communication channels. For many years, neurologists have speculated that electroencephalography activity on other electrophysiological measures of brain functions might sustain a protocol that allows us to send messages and commands to external world such as BCI [2]. In the last years, productive BCI research programs have been carried out, supported by the new consideration of brain function. In the meantime, the diversity and low cost of equipment have encouraged scientists to explore more brain functions [3].

The main objective of the study of BCI technology is to develop a new communication approach and a technological system to detect people that suffer from neuromuscular disorders. In our paper, the goal is to understand the state of the brain when the user is tired and deprived of sleep. This is why we preferred to understand the notion of sleep before we learn how to control it in the laboratory [4]. We all have at least a vague notion of what sleep is, but that does not mean that defining this mysterious part of our lives is simple. After all, a detailed analysis of our own sleep is not really an option, given that we rarely know that we are sleeping when we are asleep. In addition, even if we observe the sleep of others, so much of what they experience changes in the functions of their brains and bodies is not easily seen from the outside. Sleep scientists have explored these changes in depth, and their definition of sleep is tied to characteristic patterns of brain waves and other physiological functions [5]. Sleep is essential to proper brain functioning and the lack of it results in performance impairment during everyday cognitive tasks. Sleep deprivation seems to exert specific local effects on the brain, rather than global effects. Recent electroencephalographic and neuroimaging studies revealed that the prefrontal cortex (PFC) is more responsive to sleep deprivation than other brain areas, which is expected since the frontal region has a greater restorative need than other areas of the brain, thus the electroencephalographic changes after one night of sleep deprivation.

1.2 Materials

Enobio® is a wearable, wireless electrophysiology sensor system for the recording of EEG. Using the superb Neuroelectrics Cap, Enobio 8 is ideal for out-of-the-lab applications [6, 7]. It comes integrated with an intuitive, powerful user interface for easy configuration, recording, and visualization of 24-bit EEG data at 500 S/s, including spectrogram and 3D visualization in real time of spectral features (Fig. 1).

- Design a BCI experiment with two different groups: One of the groups must be under drowsiness effects.
- Extract features with standard methods such as Power Spectral Density, Hurth Parameters, Adaptive Autoregressive Coefficients, or Common Spatial Patterns.
- Classify the features extracted with linear classifiers. A standard linear classifier will be tested.
- Select and apply one state-of-the-art classifier.
- Evaluate results and test with other classifiers or features in order to improve the obtained results.

(awake one night), so I will analyze them per the standards of neurosciences and the already existing classifiers.

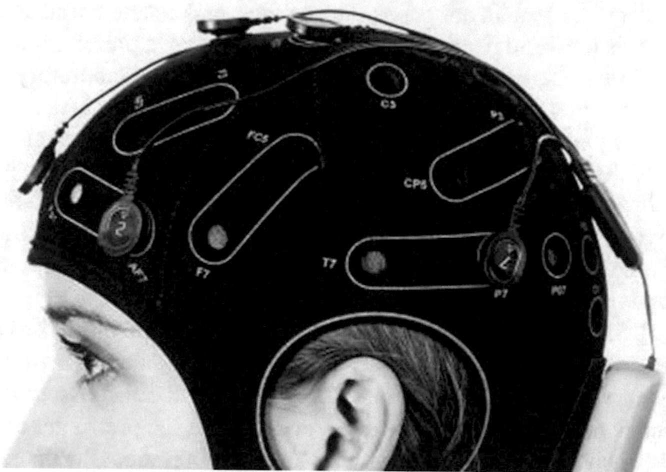

Fig. 1 Image of the cap ENOBIO

2 Methods

After collecting the EEGs of the 20 volunteers, using the ENOBIO 8 equipment, indeed the material allows us to perform tests in three sessions, each session has for duration of 7 min with 40 trials, and in each trial, the user must imagine the movement of the right and left hand. Therefore, we will begin the analysis of the data in order to make a comparison between the two different groups of persons.

The first method we choose to analyze our dataset by representing the power spectrum of each subject (open vibe) allows us to represent the different spectral components of a signal and performing harmonic analysis. It is used in physics, engineering, and signal processing.

Thus, the interpretation of the results, the EEG signals will enable us to define mental fatigue and its impact; we examine the application of these classifiers to the problem of EEG classification and compare the result to those obtained using SVM and linear discriminant analysis.

2.1 Data Preprocessing

Brain workload is characterized by brain activities evolution in the frequency range of [1–30 Hz] [8]. In our experiment, the first step of artifacts rejection, using ICA, was necessary before any data processing. This filtering step aims at reducing the effects of artifacts during the computation and estimation of spectral powers in θ and α bands [9].

We decided to focus our estimating approach on θ and α since according to many previous works these two rhythms reflect changes in the cognitive and memory performance [10]. The introduced classifier is based on the short-time Fourier transform (STFT) in computing the EEG power spectrum. To get preprocessing data, we did apply two functions to calculate the STFT, first [9] the Power Spectral Density (PSD) is computed in order to represent the frequency distribution of the signal power according to the frequencies that compose it; the mathematical formula used to calculate PSD is expressed as:

$$\Gamma_x = |X|^2 * T$$

where Γ_x is the integration time and X is Fourier transformation which represents the frequency spectrum of X which the original signal and can modelized in 2D and 3D figure to visualize the state of the brain across the experiment.

The second selection from the sleepy group (Fig. 2: deprivation group) was characterized by slower frequencies <4 Hz, those waves are deep relaxation, in full awakening, reached especially by the experienced meditators, and they are called θ waves. The activity of the brain (measured by the electroencephalogram) shows that its functioning is similar to that of sleep deprivation and meditating. Moreover, meditators often find that during a retreat, during which they practice intensively, they have less need to sleep. When one sleeps for one night, the next night is richer in dreams: one is in paradoxical sleep less recuperative, whereas when one meditates regularly, this effect does not take place, and one can resume a rhythm of 'normal' sleep more quickly.

The most famous tested rhythm of brain waves is alpha wave, that rhythm can be observed clearly on the occipital and posterior zones (Fig. 3), and the frequency will vary between 8 and 12 Hz. Due to the results we got with four volunteers (selection

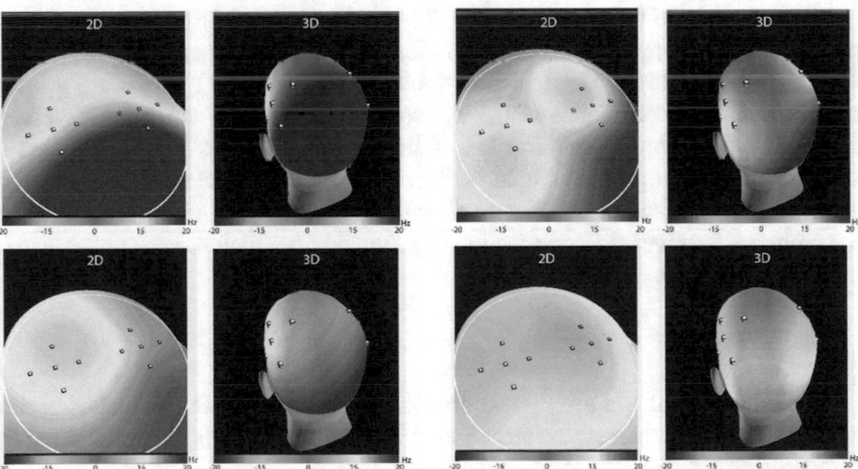

Fig. 2 Mean powers comparison of the θ activity (from the second group)

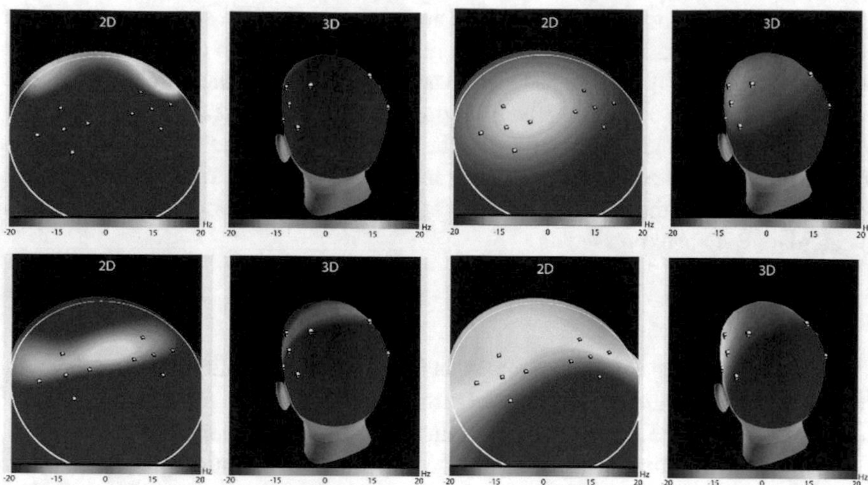

Fig. 3 Mean powers comparison of the α activity (from the first group)

from the normal group), the alpha rhythm was important in the central and posterior regions (Fig. 3). Alpha rhythm characterizes a state of appeased conscience and is mainly emitted when the subject has eyes closed, the mean powers show us the apparition of alpha waves for the normal group during the experiment, and those rhythms characterize light relaxation and calm awakening (Table 1).

2.2 *Workload Classification*

Many methods have been proposed for the classification of EEG signals in binary (normal versus sleepy) classification problems; we proposed to classify those type of dataset by classifiers like a support-vector machine (SVM) which are a set of supervised learning techniques designed to solve discrimination note 1 and regression problems. SVMs are a generalization of linear classifiers; SVMs can be used to solve discrimination problems, that is, decide which class a sample belongs to, or regression, that is, predict the numerical value of a variable. The resolution of these two problems involves the construction of a function h which has an input vector x matches an output y:

$$y = h(x)$$

In addition to that, we used linear discriminant analysis (LDA) which is part of predictive discriminant analysis techniques. It is about explaining and predicting the membership of an individual to a predefined class (group) based on its characteristics measured using predictive variables.

Table 1 Accuracies of (Subject 1, Subject 2) from normal versus (Subject 3 and Subject 4) from deprivation group using classifiers LDA and SVM and KNN and TREE

[Subject 1]	'Tree'	'LDA'	'SVM'	'KNN'
'BP'	0.7875	0.8750	0.8875	0.7875
'CSP'	0.9125	0.9375	0.9250	0.8375
'BP + CSP'	0.9500	0.9375	0.9250	0.8625
[Subject 2]	'Tree'	'LDA'	'SVM'	'KNN'
'BP'	0.7275	0.7750	0.7275	0.7345
'CSP'	0.8225	0.8375	0.8115	0.8375
'BP + CSP'	0.9500	0.9375	0.9250	0.8625
[Subject 3]	'Tree'	'LDA'	'SVM'	'KNN'
'BP'	0.6833	0.7466	0.6749	0.4330
'CSP'	0.5499	0.6416	0.6249	0.4749
'BP + CSP'	0.5916	0.5750	0.6666	0.4666
[Subject 4]	'Tree'	'LDA'	'SVM'	'KNN'
'BP'	0.6686	0.7543	0.7344	0.4330
'CSP'	0.5659	0.6234	0.6249	0.4749
'BP + CSP'	0.5999	0.3833	0.4833	0.4333

For the two classifiers, *k*-nearest neighbors (KNN) and decision tree (TREE), a set of data is used for which the value of the target variable is known in order to build the tree (so-called labeled data), and then the results are extrapolated to all the test data.

With our dataset, we got a high accuracy for the SVM for the first group and LDA for the second group.

3 Discussions

Mental fatigue is a neurological disorder. For its detection, encephalography (EEG) is commonly used as a clinical approach. The manual detection by doctors of EEG brain signals is a time-consuming and laborious process, which could be a burden on neurologists and affect their efficiency. Several automatic techniques have been proposed using traditional approaches to assist neurologists in detecting mental fatigue tiredness versus normal. The recognition of mental fatigue from EEG signals is a classification problem. It involves extraction of the discriminatory features from EEG signals and then performing classification. In the article below, we give an overview of the related state-of-the-art techniques, which use different feature extraction and classification methods for classification of mental fatigue from EEG signals extracted hand-crafted features. In this paper, we employed the SVM classifier to classify these features and achieved maximum accuracy. However, the maximum accuracy for the subjects is 88.7%. With another classifier: 'TREE,' we achieved the accuracy of 75.77% for the same dataset. In most of the work, the common classifier used to distinguish between fatigue and normal events is a support-vector machine (SVM).

4 Conclusion

In this paper, we proposed some framework systems for sleep detection for drivers; the recognition of mental fatigue was classified by SVM TREE, and LDA. Therefore, the maximum accuracy was detected in our subjects with the SVM classifier. In the present study, we assessed the impact of one night of sleep deprivation on the mental fatigue. In fact, twenty university students participate in this study. None of the subjects had a history of medical, neurological, or psychiatric disorders, neither medication nor drug intake, so the participants were randomly assigned to one of the two groups: the sleep group and the deprivation sleep group. Each group contains 10 volunteers. Thus, we got at the previous selection of subjects a good classification performance in those experiments.

References

1. Battapady, H., Lin, P., Holroyd, T., Hallett, M., Chen, X., Fei, D.-Y., et al.: Spatial detection of multiple movement intentions from SAM-filtered single-trial MEG signals. Clin. Neurophysiol. **120**, 1978–1987 (2009). https://doi.org/10.1016/J.CLINPH.2009.08.017
2. García-Laencina, P.J., Rodríguez-Bermudez, G., Roca-Dorda, J.: Exploring dimensionality reduction of EEG features in motor imagery task classification. Expert Syst. Appl. **41**, 5285–5295 (2014). https://doi.org/10.1016/J.ESWA.2014.02.043
3. Nunez, P.L., Srinivasan, R.: Electric Fields of the Brain: the Neurophysics of EEG. Oxford University Press (2006)
4. Page, A., Shea, C., Mohsenin, T.: Wearable seizure detection using convolutional neural networks with transfer learning. In: 2016 IEEE International Symposium of Circuits and Systems, pp. 1086–1089. IEEE. https://doi.org/10.1109/iscas.2016.7527433 (2016)
5. Pfurtscheller, G.: Central beta rhythm during sensorimotor activities in man. Electroencephalogr. Clin. Neurophysiol. **51**, 253–264 (1981). https://doi.org/10.1016/0013-4694(81)90139-5
6. Pfurtscheller, G., Aranibar, A.: Event-related cortical desynchronization detected by power measurements of scalp EEG. Electroencephalogr. Clin. Neurophysiol. **42**, 817–826 (1977). https://doi.org/10.1016/0013-4694(77)90235-8
7. Pfurtscheller, G., Aranibar, A.: Occipital rhythmic activity within the alpha band during conditioned externally paced movement. Electroencephalogr. Clin. Neurophysiol. **45**, 226–235 (1978). https://doi.org/10.1016/0013-4694(78)90006-8
8. Zammouri, A., Ait Moussa, A., Mebrouk, Y.: Brain-computer interface for workload estimation: assessment of mental efforts in learning processes. Expert Syst. Appl. **112**, 138–147 (2018). https://doi.org/10.1016/J.ESWA.2018.06.027

9. Dasari, D., Shou, G., Ding, L.: ICA-derived EEG correlates to mental fatigue, effort, and workload in a realistically simulated air traffic control task. Front. Neurosci. **11**, 297 (2017). https://doi.org/10.3389/fnins.2017.00297
10. Ko, L.-W., Komarov, O., Hairston, W.D., Jung, T.-P., Lin, C.-T.: Sustained attention in real classroom settings: an EEG study. Front. Hum. Neurosci. **11**, 388 (2017). https://doi.org/10.3389/fnhum.2017.00388

Explainable AI for Healthcare: From Black Box to Interpretable Models

Amina Adadi and Mohammed Berrada

Abstract As artificial intelligence penetrates deeper into work and personal life, it raises questions about trust and transparency. These questions are of greater consequence in healthcare where decisions are literally a matter of life and death. In this paper, we reflect on recent investigations about the interpretability and explainability of artificial intelligence methods and discuss their impact on medicine and healthcare.

Keywords Explainable AI · Machine learning · Healthcare

1 Introduction

Artificial intelligence (AI) in general and machine learning (ML) in particular have significant potential for disruptive socioeconomic impact. In healthcare, these technologies hold the potential of intelligentization of clinical practices. Unfortunately, the acceptance of AI in clinical settings is still limited because of trust and transparency issues.

The development of techniques for making systems "intelligible" or "explainable" is a growing focus of research within AI. This work introduces an emergent field of study called Explainable Artificial Intelligence (XAI) [1] and discusses the opportunities that this field can potentially present for the use of AI in healthcare.

The remainder of the paper is organized as follows: Sect. 2 presents the applications of AI in healthcare and the resulting limitations. Section 3 introduces the field of XAI. Section 4 discusses the opportunities of XAI for healthcare. Section 5 gives our concluding remarks.

A. Adadi (✉) · M. Berrada
Computer and Interdisciplinary Physics Laboratory (LIPI), ENS Fez Sidi Mohammed Ben Abdellah University, Fez, Morocco
e-mail: amina.adadi@gmail.com

M. Berrada
e-mail: mohammed-berrada@usmba.ac.ma

V. Bhateja et al. (eds.), *Embedded Systems and Artificial Intelligence*, Advances in Intelligent Systems and Computing 1076,
https://doi.org/10.1007/978-981-15-0947-6_31

2 AI in Healthcare

Healthcare and medicine are at the avant-garde of a radical and swift transition toward data dependency. This process has been enabled by simultaneous advances in data acquisition and the development of networked system technologies, and it is becoming the main driver behind the development of whole novel techniques for biological data management and analysis. This situation should be seen as an unprecedented opportunity for AI in general, and ML and related techniques for knowledge extraction from data in particular. In fact, the link between AI and healthcare is not new, AI and healthcare share a well-established past. Healthcare was one of the first practical applications for early AI systems like Dendral. Today, AI is considered a game changer in this industry. It is employed in a myriad of settings including hospitals, clinical laboratories, and research facilities. Healthcare-based AI systems are some of the most well-funded initiatives in the technology sector and major technology companies—including Google, Microsoft, and IBM—are investing in the development of AI for healthcare and related research. Herein, we emphasize the transformative aspect of this technology by underlying its applications in healthcare along with noted limitations.

2.1 Applications of AI in Healthcare

In order to draw up the "big picture" of AI applications in healthcare, we have gathered much of the existing base of evidence for the use of AI in healthcare, laying out the opportunities and later the limitations. Table 1 classifies AI applications in healthcare by area of application and level of impact.

As depicted in Table 1, AI penetration in healthcare is beginning to have an impact at three levels: for clinicians (every type of clinician, ranging from specialty doctor to paramedic), predominantly via rapid, accurate image interpretation; for health systems, by improving workflow and the potential for reducing medical errors; and for patients, by enabling them to process their own data to promote health. As modern healthcare systems are mostly patient centered, ultimately the impact of using AI concerns patients and the care quality provided to them. However, the direct impact

Table 1 Areas of application of AI in healthcare

Area of application	Level of impact	References
Diagnosis and prognosis	Clinicians	[2]
Drug development	Clinicians	[3]
Population health	Clinicians	[4]
Healthcare organization	Health systems	[5]
Patient facing applications	Patients	[6]

is mostly at the level of clinicians, since most current AI-based medical systems are used as tools to assist clinicians in their practices.

Many of today's AI applications in healthcare appear to fall under five key healthcare areas, from our standpoint, each of these areas is spanning its own spectrum, but together they constitute a proper representation of the healthcare landscape:

(a) *AI for Diagnosis and Prognosis*

Timely, accurate detection and/or prediction of a disease are one of the most interesting and challenging tasks for physicians, especially in the case of rare or difficult to diagnose illnesses. Unfortunately, medical practitioners can only analyze a limited number of images and samples, and their diagnoses are subject to human error. AI, on the other hand, can process millions of samples in quick time, consistently and reliably, every time. AI is also holding out the hope of detecting diseases even before the symptoms manifest by correlating a multitude of variables.

(b) *AI for Drug Development*

Drug development is a time-consuming and cost-intensive process encompassing the early stages of research, preclinical testing, clinical trials, and review and approval. Speed up one of these steps in this long process would have big implications down the entire chain; for example, predicting the likelihood of toxicity in the earliest stages before undergoing the clinical trials would considerably save in terms of time and cost. Given the complexity and time involved, scientists are forced to restrict the number of test combinations. AI and ML algorithms, however, have no such limitations and are capable of learning to make predictions based on previously learned data and even prioritize experimentation.

(c) *AI for Population Health*

AI has the potential to be used to aid the early detection of infectious disease outbreaks and sources of epidemics, especially in developing countries where this issue is particularly pressing considering the general lack of medical infrastructure and the limited access to treatments. By receiving training from different data sources such as satellites, previous outbreaks, weather, rainfall, and the total number of positive cases, AI-based systems can make predictions about the likelihood of an infectious disease outbreak like AIDS, malaria, influenza, or BSE. They can forecast the spread behaviors of these epidemic diseases well before they occur and hence helping in controlling its impact and reducing casualties by implementing adequate countermeasures (quarantine, vaccination, medical treatment).

(d) *AI and Patient Facing Applications*

AI holds the hope to raise patient engagement to the next level through Intelligent Virtual Assistant and Medical Virtual Assistant. Now, medical assistance has gone beyond wearables by urging patients to not just manage their goals, but also to actually help them look after their health as a real assistant would. Which opens up the possibility to telemedicine, this way of delivering health care is particularly useful for chronic diseases and thanks to AI, it is becoming a mainstream healthcare model.

(e) *AI for Healthcare Organization*

Aside from the aforementioned applications, AI can also help in scenarios in which current IT solutions may not be optimal such as: scheduling patients and staffing optimization, billing, and collections. By doing so, AI is adding great value to the clinical process by taking over a big chunk of clinical and outpatient services, leaving physicians free to focus on more critical activities.

2.2 Limitations of AI in Healthcare

Arguably and despite the existence of plenty of evidence supporting their usefulness, AI methods are likely not to be adopted in routine medical practice beyond a limited number of niche applications unless some core limitations are addressed. Indeed, opportunities of using AI come with associated challenges. We explicitly outline the core limitations as follows:

- The potential for AI to make erroneous decisions.
- Decision-making and liability, who is responsible when AI is used to support decision-making.
- Difficulties in validating the outputs of AI systems.
- The risk of inherent bias in the data used to train AI systems.
- Ensuring the security and privacy of potentially medical sensitive data.
- Securing public trust in the development and use of AI technology in healthcare.
- Effects on the roles and skill requirements of healthcare professionals.

We believe that techniques provided by the emergent field of Explainable Artificial Intelligence are the solutions of choice to overcome these trust and transparency issues and to enhance AI acceptance in clinical settings. Hence, we propose to explore XAI initiatives in order to identify potential projections in healthcare.

3 Explainable AI

3.1 Emergence of XAI

XAI is a research field that aims to make AI systems results more understandable to humans. In ML community, this goal is known as interpretable machine learning. While this field is relatively new, the problem of explainability has existed since the mid-1970s when researchers studied explanation for expert systems [7]. Recently,

XAI has received a clear renewed attention from academia and practitioners. The re-emergence of this research topic is the direct result of the unstoppable penetration of AI/ML across industries and its crucial impact in critical decision-making processes, without being able to provide detailed information about the chain of reasoning that leads to certain decisions made by it. Therefore, the social, ethical, and legal pressure calls for new AI techniques that are capable of making decisions explainable and understandable.

For commercial benefits, for ethics concerns, or for regulatory considerations, XAI is essential if users are to understand, appropriately trust, and effectively manage AI results. Generally, explanations are needed to justify results, particularly when unexpected decisions are made, to control and improve, a model that can be explained and understood is one that can be more easily improved, and ultimately to discover and learn new facts.

Despite its importance, bringing interpretability to AI systems is a very challenging technical issue. Explainability of intelligent systems has run the gamut from traditional expert systems, which are totally explainable but inflexible and hard to use, to deep neural networks (DNN), which are effective but virtually impossible to see inside. Arguing that AI/ML interpretability is a challenging issue, does not mean that all AI/ML techniques have the same level of opacity. Indeed, there are algorithms that are more interpretable than others are, and there is often a trade-off between accuracy and interpretability.

3.2 *Methods for Explainability*

In the quest to make AI system explainable, several explanation methods and strategies have been proposed in relatively short period, especially for ML algorithms. Explainability methods are classified according to three criteria [1]: (i) the complexity of interpretability, (ii) the scoop of interpretability, and (iii) the level of dependency from the used ML model. We note that, since explainability in AI is still an emerging field, the classes of methods belonging to the proposed taxonomy are neither mutually exclusive nor exhaustive. However, this can be a good yardstick to compare and contrast across multiple methods.

(i) *Complexity Related Methods*

The complexity of an AI model is directly related to its interpretability. Generally, the more complex the model, the more difficult it is to interpret and explain. Thus, the most straightforward way to get to interpretable AI/ML would be to design an algorithm that is inherently and intrinsically interpretable. A common challenge, which hinders the application of this approach, is the trade-off between interpretability and accuracy. An alternative approach to interpretability in AI is to construct a high complex uninterpretable black-box model with high accuracy and subsequently use a separate set of techniques to perform, what we could define as a reverse engineering process to provide the needed explanations without altering or even knowing the inner works of the original model. This class of methods offers then a post hoc explanation. Though it could be significantly complex and costly, most recent works done in XAI field belong to post hoc class, it includes natural language explanations [8], visualizations of learned models [9], and explanations by example [10].

(ii) *Scoop Related Methods*

Interpretability implies understanding an automated model; this supports two variations according to the scoop of interpretability: understanding the entire model behavior or understanding a single prediction. Accordingly, we distinguish between two subclasses: (i) global interpretability and (ii) local interpretability

- Global interpretability facilitates the understanding of the whole logic of a model and follows the entire reasoning leading to all the different possible outcomes. This class of methods is helpful for population-level decisions, such as an epidemic outbreak.
- Local interpretability is explaining the reasons for a specific decision prediction. This scoop of interpretability is used to generate an individual explanation, generally, to justify why the model made a specific decision for an instance. Popular methods belonging to this class include LIME [11] and LOCO [12].

(iii) *Model Related Methods*

Another important way to classify model interpretability techniques is whether they are model agnostic, meaning they can be applied to any types of AI/ML algorithms, or model-specific, meaning techniques that are applicable only for a single type or class of algorithm.

- Model-specific interpretability: this type of methods is limited to specific model classes. Intrinsic methods are by definition model-specific. The drawback of this practice is that when we require a particular type of interpretation, we are limited in terms of choice to models that provide it, potentially at the expense of using a more predictive and representative model.
- Model-agnostic interpretability: this class of methods is not tied to a particular type of ML model. In other words, this class of methods separates prediction from explanation. Model-agnostic interpretations are usually post hoc; they are generally used to interpret DNNs and could be local or global interpretable models. In the interest of improving interpretability AI models, a large amount of model-agnostic methods have been developed recently using range techniques from statistics, machine learning, and data science. These broadly fall into four technique types [1] illustrated in Table 2.

Figure 1 describes the XAI's methods taxonomy that illustrates the state of current thinking in terms of explaining AI-based systems.

Table 2 Techniques of model-agnostic method

Technique	Sub- techniques
Visualization	Surrogate models
	Partial dependence plot (PDP)
	Individual conditional expectation (ICE)
Knowledge extraction	Rule extraction
	Model distillation
Influence methods	Sensitivity analysis
	Layer-wise relevance propagation (LRP)
	Feature importance
Example-based explanation	Prototypes and criticisms
	Counterfactuals explanations

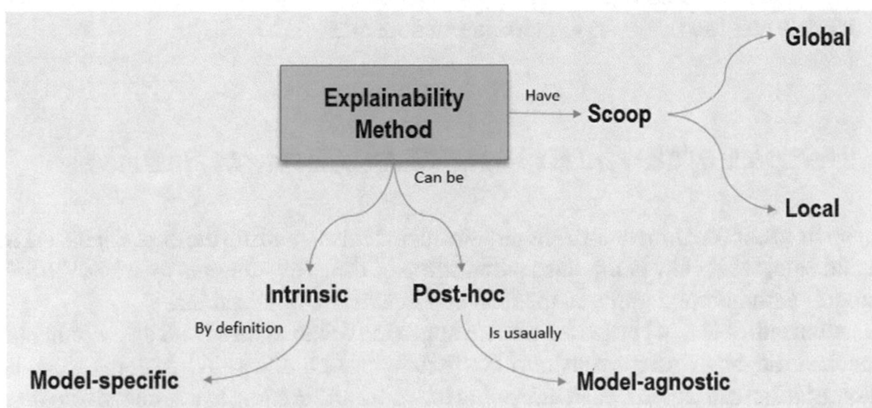

Fig. 1 XAI methods taxonomy [1]

4 Explainable AI for Healthcare

4.1 Characteristics of Explainable Healthcare AI

We argue that research in XAI would greatly advance the healthcare ecosystem. XAI calls for confidence, safety, security, privacy, ethics, fairness, and trust. All these aspects are crucial for healthcare where we deal with life-changing decisions. However, explainable AI/ML for healthcare has very specific characteristics that, we believe, have to be taking into consideration while designing and implementing explainable solution for this sector [13]:

- *Adaptability*: transparency may mean different things to different people, the output of the intelligent system has to be understandable/adaptable to the knowledge of all kind of users (doctors, nurses, patients, administrators…).
- *Context-awareness*: the explanation should make sense in the domain of application (diagnosis, surgery, drug development…) and to the user of the system.
- *Consistency*: the explanation should be consistent across different models and across different runs of the model. Having different explanations for the same instance (same patient) can be confusing for users, hence the need for consistency.
- *Generalizability*: where need it the explanation should be generalizable. Sometimes, many of the instances have similar explanations. However, the explanations could not be generalizable to the whole population.
- *Fidelity*: The expectation that the explanation and the predictive model align well with one another. To reach fidelity, an explanation should be sound and complete. An explanation is sound if it adheres to how the model actually works, and it is complete if it encompasses the complete extent of the model. Intrinsically, interpretable methods are perfectly sound by definition comparing to post hoc models that can have varying leveling of soundness.

4.2 Review of Current Explainable Approaches in Healthcare

In order to analyze the state-of-the-art advancements regarding the explainable AI in healthcare, we review in this subsection some of the early attempts to imbue AI/ML models with interpretability in medical and healthcare applications.

Monteath et al. [14] proposed a novel approach for an assisted and an incremental medical diagnosis using decision tree; their approach allows AI systems to work alongside human experts, each informing the other and coming to a decision together.

Kocbek et al. [15] built an interpretable model for predicting polypharmacy from drug prescription records for newly diagnosed chronic patients. This work focuses on increasing explainability without significant loss of predictive performance.

Zhenga et al. [16] proposed a novel and explainable method to classify a subset of cardiac pathologies using deep learning of cardiac motion and shape.

Hicks et al. [17] introduced, Mimir, an interpretative method that directly adds explainability to DNN models in medical problems by producing structured and semantically correct reports, composed of text and images.

Wu et al. [18] proposed a human-in-the-loop method to interpret internal representations of DNN models for diagnostic classification of mammograms, by labeling the behavior of internal units. Network Dissection (NetDissect) has been used for quantifying interpretability as a measure of how well individual DNN units align with sets of human-interpretable concepts.

The list of described work does not aim to be a systematic review, although it intends to cover the widest possibly palette of early attempts to address the explainability issue in healthcare and medicine. From this light scan of literature, it is clear that the problem of explainability is just starting to gain attention in the healthcare community. Most of the current works deal with the performance–interpretability trade-off and make use of post hoc methods. However, there is no report about the quality evaluation of the generated explanation or its compliance with the healthcare domain requirements. The lack of explanation evaluation is obvious in the current literature. Moreover, a notable commonality is that all methods somehow replicate human interpretation procedures. This emphasizes the need to treat interpretability at the human cognition level, beyond technical detail.

4.3 Future of Explainability in Healthcare

The motivation for model explanations in healthcare is clear. Initial works in this area open up a number of future directions. In what follows, we discuss some future research opportunities and expected trends in this research field.

- *Human Computing Interaction*

Merely, providing an explanation for an algorithm's prediction is insufficient. Indeed, the user's understanding and trust of the system partly depend on the way he interacts with the machine. The manner in which interpretations are shared with the end users, incorporated into user workflows, and utilized must be carefully considered. Given HCI's [61] core interest in technology that entails understanding and better empowering users, techniques from this research field must be adopted in developing transparent systems for healthcare.

- *Human in the Loop*

It is important to consider the human factor when attempting to enhance model interpretability in general and the importance of integrating the medical expert in the process of developing strategies to guarantee the interpretability and explainability of medical data models.

- *Explanation Evaluation*

To further progress in this field, it is also important to develop formalized rigorous evaluation metrics and methods for interpretability. Consider the scenario in which multiple models offer different explanations; for instance, the challenge is to figure out which explanations are the best. As observed before, in literature, there is no clear way to quantify explainability, the related line of work is just in its infancy. In this paper, an initial reflection about the quality criteria of explanation in the respect of healthcare characteristics has been discussed.

- *Others Explainable Intelligent Systems*

Most of the existing works in literature focus on explainability in ML, which is just one type of AI. However, the same issues also confront other intelligent systems.

Particularly, (i) explainable AI planning and (ii) explainable agent is beginning to gain recognition as a promising derived field of XAI. This represents a potential opportunity for healthcare, especially for medical robots. Robots are increasingly involved in critical tasks such as surgery, an agent capable of explaining the reasons behind their actions will surely impact this line of research.

5 Conclusion

Acknowledged as a key limitation of AI application in healthcare; explainability has been discussed in this paper through an emerging field called explainable AI. We argue that research in this field would increase the acceptance of AI in healthcare sector. In this sense, we proposed key requirements to develop explainable medical approaches, we reviewed early attempts to integrate explainability in AI-based medical systems, and we discussed potential future research directions.

References

1. Adadi, A., Berrada, B.: Peeking inside the black-box: a survey on explainable artificial intelligence (XAI). IEEE Access **6**, 52138–52160 (2018)
2. Liu, X., Chen, K., Wu, T., Weidman, D., Lure, F., Li, J.: Use of multimodality imaging and artificial intelligence for diagnosis and prognosis of early stages of Alzheimer's disease. Transl. Res. **194**, 56–67 (2018)
3. Kit-Kay, M., Mallikarjuna, R.P.: Artificial intelligence in drug development: present status and future prospects. Drug Discov. Today (2018)
4. Kumar, R.: Epidemic outbreak prediction using artificial intelligence. Int. J. Inf. Technol. Comput. Sci. **10**, 49–64 (2018)
5. Baldwin, J.L., Singh, H., Sittig, D.F., Giardina, T.D.: Healthcare, patient portals and health apps: pitfalls, promises, and what one might learn from the other. Healthcare **5**, 81–85 (2017)
6. Hsieh, F.S., Lin, J.B., Scheduling patients in hospitals based on multi-agent systems. Modern Advances in Applied Intelligence, pp. 32–42 (2014)
7. Swartout, W.R., Moore, J.D.: Explanation in Expert Systems: A Survey. University of Southern California (1988)
8. Krening, S., Harrison, B., Feigh, K., Isbell, C., Riedl, M., Thomaz, A.: Learning from explanations using sentiment and advice in RL. IEEE Trans. Cogn. Dev. Syst. (2016)
9. Mahendran, A., Vedaldi, A.: Understanding deep image representations by inverting them. In: IEEE Conference on Computer Vision and Pattern Recognition (CVPR) (2015)
10. Mikolov, T., Sutskever, I., Chen, K., Corrado, G., Dean, J.: Distributed representations of words and phrases and their compositionality. In: Advances in Neural Information Processing Systems (NIPS), pp. 3111–3119 (2013)
11. Ribeiro, M.T., Singh, S., Guestrin, C.: Why should I trust you?: explaining the predictions of any classifier. In: Proceedings of the 22nd ACM SIGKDD International Conference on Knowledge Discovery and Data Mining, pp. 1135–1144 (2016)
12. Lei, J., G'Sell, M., Rinaldo, A., Tibshirani, R.J., Wasserman, L.: Distribution-free predictive inference for regression. J. Am. Stat. Ass. pp. 1–18 (2018)
13. Ahmad, M.A., Eckert, C., Teredesai A., Kumar, V.: Explainable AI in Healthcare. Available on line at https://learning.acm.org/webinars/healthcareai (2018)

14. Monteath, I., Sheh, R.: Assisted and incremental medical diagnosis using explainable artificial intelligence. In: Proceedings of the 2nd Workshop on Explainable Artificial Intelligence, pp. 104–108 (2018)
15. Kocbek, S., Kocbek, P., Stozer, A., Zupanic, T., Groza, T., Stiglic, G.: Building interpretable models for polypharmacy prediction in older chronic patients based on drug prescription records. PeerJ Life Environ. Sci. (2018)
16. Zhenga, Q., Delingettea, H., Ayache, N.: Explainable cardiac pathology classification on cine MRI with motion characterization by semi-supervised learning of apparent flow. Available online at https://arxiv.org/pdf/1811.03433.pdf (2018)
17. Hicks, S.A., Eskeland, S., Lux, M., de Lange, T., Randel, K.R., Jeppsson. M., Pogorelov. K., Halvorsen. P., Riegler. M.: Mimir: an automatic reporting and reasoning system for deep learning based analysis in the medical domain. In: Proceedings of the 9th ACM Multimedia Systems Conference (MMSys), pp. 369–374 (2018)
18. Wu, J., Peck, D., Hsieh, S., Dialani, V., Lehman, C.D., Zhou, B., Syrgkanis, V., Mackey, L., Patterson, G.: Expert identification of visual primitives used by CNNs during mammogram classification. In: SPIE Medical Imaging 2018: Computer-Aided Diagnosis (2018)

An Evolutionary Algorithm Approach to Solve the Hybrid Berth Allocation Problem

Issam El Hammouti, Azza Lajjam, Mohamed El Merouani and Yassine Tabaa

Abstract The berth allocation problem (BAP) is classified among the biggest problems confronted in a container terminal. In this article, we addressed the problem in the dynamic hybrid case and we developed an evolutionary algorithm based on a hybrid genetic algorithm (HGA) as a resolution approach. Finally, computational experiments and comparisons are realized to demonstrate the quality of our results.

Keywords Evolutionary algorithm · Hybrid genetic algorithm · Container terminal · Planning and scheduling tasks

1 Introduction

Recent shipping statistics marked a significant increase in freight traffic, particularly in container flows. Consequently, the port authorities were obliged by this increase to search effective solutions to operate in the most efficient way, that is, minimizing the costs of vessel's stay at the port and maximize the quality of services.

The optimal management of port operations is considered as the basis of a container terminal efficiency. Indeed, to serve the arrived ships at the terminal, port operators must plan a series of operations which include: the assignment of vessels to berths, the quay cranes assignment which are in charge of loading/unloading containers, the allocation of yard trucks which transfer the containers from the quay area to the yard area. Finally, the storage area where containers are stored see Meisel [1].

I. El Hammouti (✉) · M. El Merouani · Y. Tabaa
Faculty of Science, UAE, Tetuan, Morocco
e-mail: Issam.elhammouti@gmail.com

M. El Merouani
e-mail: m_merouani@yahoo.fr

Y. Tabaa
e-mail: Yassine.tabba@gmai.com

A. Lajjam
College of Sciences, UEA, Tetuan, Morocco
e-mail: Azza.lajjam@gmail.com

V. Bhateja et al. (eds.), *Embedded Systems and Artificial Intelligence*, Advances in Intelligent Systems and Computing 1076,
https://doi.org/10.1007/978-981-15-0947-6_32

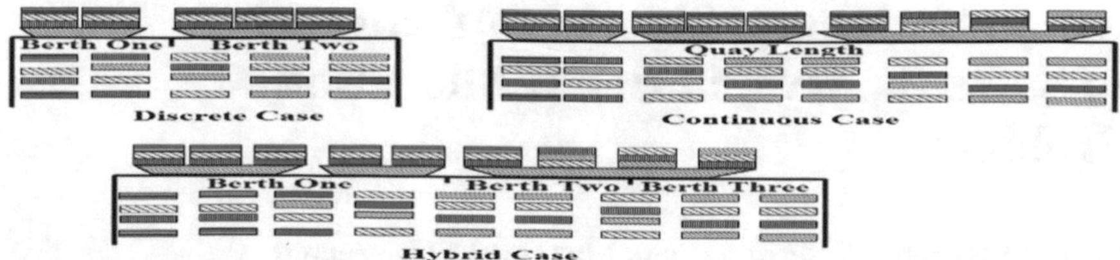

Fig. 1 Type of berth space

In this paper, we focused on the BAP, which is considered as one of the most important problems that have a direct impact on the port efficiency. It consists of planning an optimal assignment of vessels to berths for their berthing in order to minimize the costs of vessel's stay at the port.

According to the previous publications, the BAP depends on several factors: type of the quay (discrete, continuous or hybrid) (Fig. 1), the type of ship arrival (dynamic or static) and the type of handling service (static or dynamic). In the discrete problems, the quay is divided into a number of the positions fixed in advance; in contrast, in the continuous case, the quay is not divided it is used according to the needs of the ships in terms of their length. Finally, in the hybrid case like in the discrete case, the quay is partitioned into berths, but large vessels may occupy more than one berth while small vessels may share a berth.

In the dynamic problem, the ships which have not arrived yet at the port before the planning starts are taken into account, whereas all the ships are already in the port before the planning starts in the static one. Finally, in the static handling problems, the duration of ships services provided is a known parameter. However, in the dynamic case, the work duration of each berth depends, in particular, on the number of containers to be loaded/unloaded or on cranes available.

The primary objective of this paper is to develop an evolutionary algorithm based on HGA as an approach for solving the berth allocation problem in the dynamic and hybrid case (DHBAP) with the objective of minimizing the staying cost for all vessels at the port.

The rest of the article is organized as follows. The next section presents the literature review concerning the BAP. Then a description of our problem is proposed in Sect. 3. Next, in Sect. 4, an evolutionary algorithm based on HGA is adapted as a method of resolution. While Sect. 5 presents comparative experiments to demonstrate the quality of our results. Finally, a conclusion is presented in Sect. 6.

2 Literature Review

In the literature, the BAP has been approached by several scientific articles. However, the difference between these publications consists often in the type of the quay considered or in the type of the ship's arrival. In the following, we present a large list of the main works on the dynamic berth allocation problem (DBAP).

Imai et al. [2] were the first who introduce the dynamic discrete BAP (DDBAP); they solved the problem using a heuristic based on Lagrangian relaxation method. Two years later, Imai et al. [3] improved their model considering different service priorities between ships; they resolved the problem using a genetic algorithm (GA). Kim and Moon [4] proposed a meta-heuristic based on simulated annealing (SA) method for solving the dynamic continuous BAP (DCBAP). In [5], the authors implemented a heuristic based on a tabu search for the DDBAP and DCBAP; they assessed solution quality found with their heuristic by making a comparison with the exact solution found by CPLEX.

Buhrkal et al. [6] presented three different mathematical programming models of the DDBAP and proposed a formulation of the problem as a generalized set partition problem (GSPP). For testing their formulation, they used the instances from Cordeau et al. [5] and obtained the optimal solutions using CPLEX. In [7], a meta-heuristic based on clustering search (CS-SA) with simulated annealing is presented as an alternative for solving the DDBAP which improves the results presented by Buhrkal et al. [6] and Cordeau et al. [5].

Recently, Lin et al. [8] developed two simulated annealing methods each one of them is based on different strategies to assign the incoming vessels to available berth along the quay. In this paper, the authors aim to minimize the total weighted service time and the deviation cost from vessels' preferred berth in a continuous terminal. To minimize the total service time for each vessel in a dynamic and discrete container terminal, a dynamic programming-based meta-heuristic is proposed in Nishi et al. [9]. In this work, different comparisons with other methods of the literature have been realized by the authors to show the performance of the proposed meta-heuristic.

3 Problem Description

In this section, we define the main objective of our problem which is found in the best position for each calling vessel at the port terminal with minimization of the staying cost of each one of them. Figure 2 shows a representation of an optimal solution to DHBAP with 3 berth and 20 vessels.

Evidently, for solving the DHBAP problem, we should respect a set of restrictions both regarding the terminal and the vessels. In the following, we present a list of important restrictions which we should take into account:

- The main objective is to minimize the total stay cost of vessels at the port.
- At the same berth, each vessel must be served at an order of service different, note that although two vessels are simultaneously at a berth, their service orders must be different.
- The starting time of the handling of a vessel j at a berth i is the maximum between the arrival time of vessel j and time when berth i becomes idle for the first time.

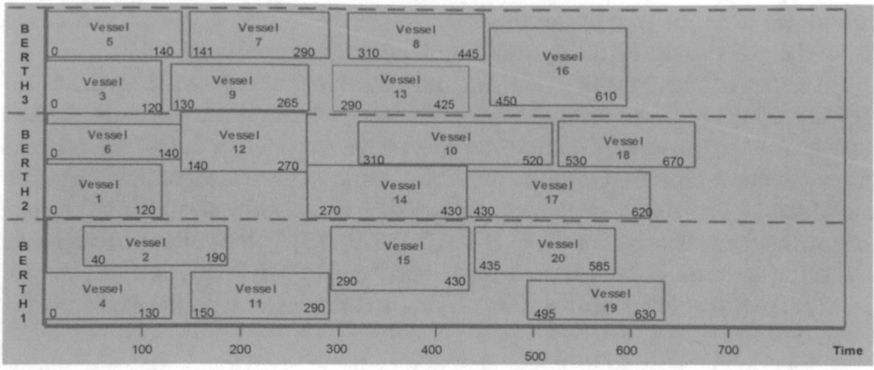

Fig. 2 The Gantt representation of a solution with 3 berth and 20 vessels in DHBAP

- The planning horizon.
- The vessel j departure time is the addition of their starting time of the handling and their handling time.
- If the service order of the vessel j' is greater than that of vessel j, then the latter is served earlier than the vessel j'.
- If two vessels are served simultaneously, their services coincide in time.
- Two vessels can be served at the same berth if their total length is equal or less than the berth length.
- The waiting time for each vessel j is the subtraction of their starting time of the handling and their arrival time.
- The number of vessels served simultaneously to a specific berth does not exceed two.

In Issam et al. [10], we can find a detail description and a mathematical formulation for DHBAP which is the basis of our method of solution.

4 The Proposed Evolutionary Algorithm

4.1 *Principle of the Algorithm*

Given that the BAP treated in this paper is an NP-complete problem to which we cannot find an optimal solution in a reasonable time. In this section, we developed a genetic algorithm based on the population of individuals evolution during a number maximal of iterations combined with a local search algorithm. Each individual in the population represents a solution to the problem (Fig. 3). The goal of our algorithm is to find an optimal solution or a closest to the optimality. So, this is done first, through the selection operation that allows us to select only the best individuals of a population and eliminate the worst according to the fitness value of each one.

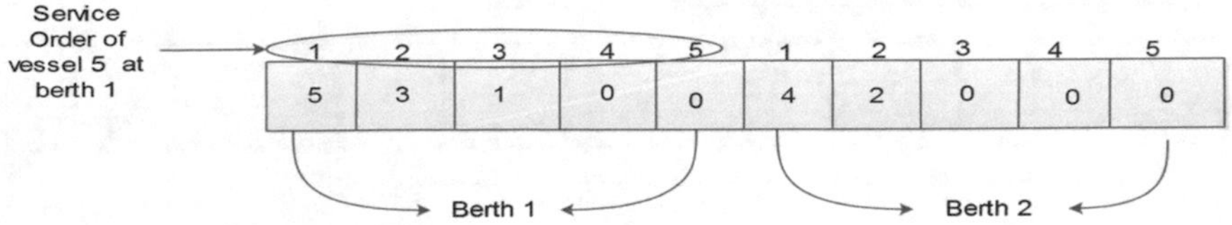

Fig. 3 Solution representation

Therefore, we obtain a new population that contains only the best-selected individuals of the previous population. The second way to find optimal solution is by the crossover operation (Fig. 4) that allows us to create new individuals (solutions) called children through other individuals called parents and the last way is by the mutation operation (Fig. 5) allows us to explore as much as possible the solutions space. Finally, at each improvement of the best current solution, we applied a local search algorithm (Fig. 6), thus increasing the probability of finding the solution optimal global. Both the crossover and mutation operations are based on the probability of crossover Pc and mutation Pm consecutively. The general procedure of our HGA is summarized in (Fig. 7).

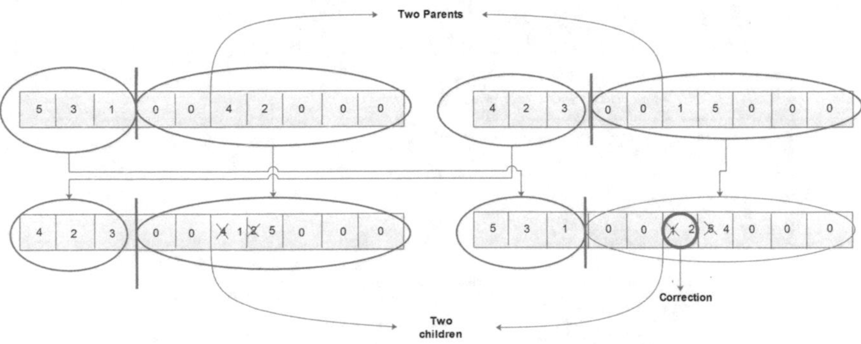

Fig. 4 Crossover operation

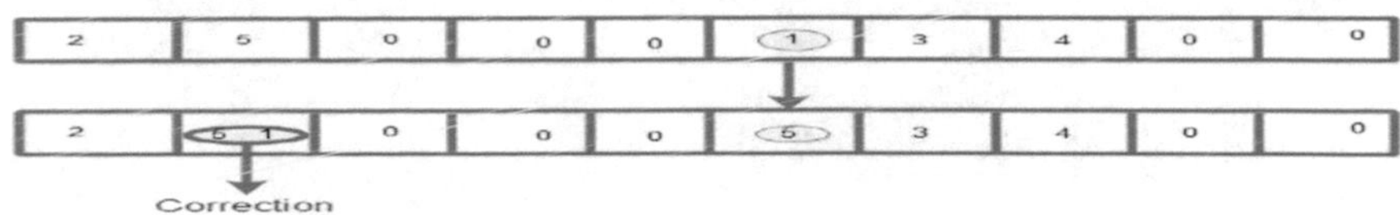

Fig. 5 Mutation operation

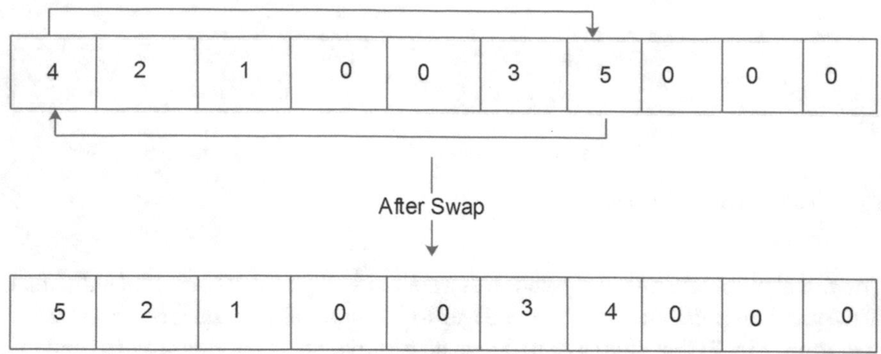

Fig. 6 Local search

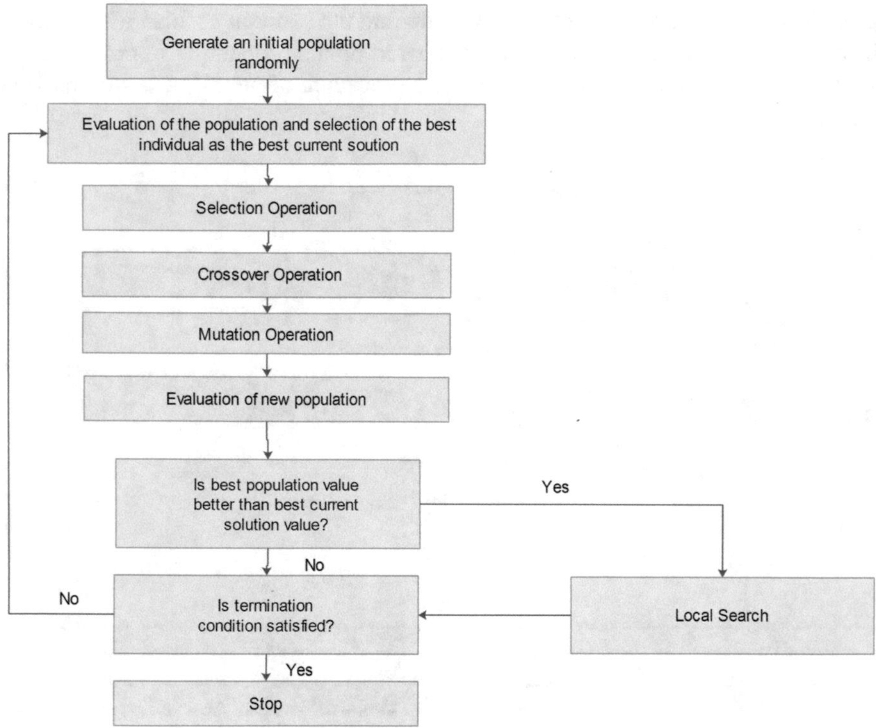

Fig. 7 Flowchart of hybrid genetic algorithm

4.2 Solution Representation

A solution is represented as an array with $n*m$ cell, where m represents the number of berths and n represents the number of vessels. Each vessel is represented by a number ranges from 1 to n, while each berth is represented by a number ranges from 1 to m. For example, Fig. 3 represents a solution to a problem with five vessels and two berths, the first five cells represent the vessels assigned to the berth one, while the latest five cells represent the vessels assigned to the berth two. The position of each cell represents the service order for vessel that is in this cell.

4.3 Local Search Algorithm

After each improvement in the best solution, a local search algorithm is applied to improve the exploitation of our algorithm. The local search is based on a swap between two elements randomly selected of the best current individual. See Fig. 4.

5 Computational Results

The HGA was implemented with the programming language C on a core Intel (R) TM i3 CPU @ 2.30 GHz PC computer (Acer); the testing instance of the proposed HGA was compared with that provided by a commercial solver (CPLEX 12.12) with a running time maximal of 2 h for each experiment. In addition, the HGA uses four main parameters to start, namely population initial size (P_{in}), number of maximal iterations ($T_{\max}$), crossover probability (P_{c}) and mutation probability (P_{m}). In our case, after varieties of experiences, we have configured the algorithm according to the following values: ($P_{\text{in}} = 100$; $T_{\max} = 50$; $P_{\text{m}} = 0.5$; $P_{\text{c}} = 0.8$).

The instances used in this study were generated randomly on horizon of seven days and classified into three categories (M_1: small instances, M_2: medium instance and M_3: large instances); the problem sizes for each category are: 10 ships with 2 berths for M_1, 30 ships with 7 berth for M_2 and 50 ships with 12 berth for M_3. For each problem size, a set of five instances are generated.

5.1 Data Generation

Real problems instances were generated based on the traffic observed in the Tangier container terminal where the number of loaded and unloaded containers ranges from 16,000 to 10,000, while the number of cranes for each berth was generated between 2 and 5. The time when a berth i becomes idle on the planning horizon was generated according to Imai et al. [1] and Cordeau et al. [5] and considered the same for all berths. In order to simplify the calculations, the waiting and the handling cost per hour for each vessel are supposed to be equal to 2 and 1, respectively. Finally, arrival times of vessels within the week period are randomly generated.

In Table 1, column 1 presents the size of the problem, while columns 2, 3 and 4 expose the objective function value, running time and the Gap, respectively, found by CPLEX. Finally, columns 5 and 6 contain the objective function value and running time, respectively, obtained by HGA. So, the results specified in the table show that the function objective value found by two methods as well as the running time is admissible for the instances of category M_1. However, for M_2 and M_3, it is obvious that the results obtained with HGA are more reliable in comparison with those obtained with CPLEX because when we move from small size instances from larger size instances which are more realistic, the CPLEX becomes very limited either in terms of memory and execution time.

Table 1 Computational results for three categories of problems (small, medium and large instances)

Vessel*Berth	CPLEX			Hybrid genetic algorithm	
	Objective value (hours)	Time (sec)	Gap[a] (%)	Objective value (hours)	Time (sec)
10*2 (1)	1226	164.88	0.00	1226	0
10*2 (2)	1587	120.5	0.00	1587	0
10*2 (3)	1500	150.80	0.00	1500	1
10*2 (4)	1890	155.6	0.00	1890	1
10*2 (5)	1221	110.78	0.00	1221	0
10*2 (6)	1770	158.8	0.00	1770	1
10*2 (7)	1325	120.70	0.00	1325	0
30*7 (1)	30,236	3600	86.51	3056	5
30*7 (2)	25,232	3600	61.29	7997	4
30*7 (3)	21,256	3600	49.23	8013	5
30*7 (4)	21,965	3600	47.52	8106	5
30*7 (5)	20,236	3600	50.69	6801	4
30*7 (6)	20,256	3600	60.25	8203	6
30*7 (7)	21,256	3600	45.25	8062	5
50*12(1)	31,589	3600	70.65	10,890	7
50*12(2)	25,485	3600	56.32	12,469	6
50*12(3)	26,056	3600	50.36	12,341	7
50*12(4)	28,036	3600	69.34	14,457	6
50*12(5)	28,896	3600	65.98	11,521	6
50*12(6)	22,125	3600	72.38	15,892	6
50*12(7)	29,256	3600	55.20	13,569	6

[a]Gap is calculated as (feasible solution value − lower bound) * 100/lower bound

6 Conclusion

In this work, an evolutionary algorithm based on hybrid genetic algorithm is developed as an alternative to solve the hybrid dynamic berth allocation problem. The comparison realized between our results, found by the hybrid genetic algorithm, and those found by commercial solver (CPLEX) showed that our method is an interesting alternative for solving the DHBAP. In the future works on DHBAP, we will take into account the uncertainty of vessel's arrival and handling time. Furthermore, we will deal with the HBAP in other types of terminals as bulk terminals.

References

1. Meisel, F.: Seaside Operations Planning in Container Terminals. Physica-Verlag, Berlin (2009)
2. Imai, A., Nishimura, E., Papadimitriou, S.: The dynamic berth allocation problem for a container port. Transp. Res. Part B: Methodological, 401–417 (2001)
3. Imai, A., Nishimura, E., Papadimitriou, S.: Berth allocation with service priority. Transp. Res. Part B: Methodological, 437–457 (2003)
4. Kim, K.H., Moon, K.C.: Berth scheduling by simulated annealing. Transp. Res. Part B: Methodological, 541–560 (2003)
5. Cordeau, J.F., Laporte, G., Legato, P., Moccia, L.: Models and tabu search heuristics for the berth-allocation problem. Transp Sci. 526–538 (2005)
6. Buhrkal, K., Zuglian, S., Ropke, S., Larsen, J., Lusby, R.: Models for the discrete berth allocation problem: a computational comparison. Transp. Res. Part E: Logistics Transp. Rev. 461–473 (2011)
7. de Oliveira, R.M., Mauri, G.R., Lorena, L.A.N.: Clustering search for the berth allocation problem. Expert Syst. Appl. 5499–5505 (2012)
8. Lin, S.W., Ting, C.J., Wu K.C.: Simulated annealing with different vessel assignment strategies for the continuous berth allocation problem. Flex. Serv. Manuf. J. 1–24 (2017)
9. Nishi, T., Okura, T., Lalla-Ruiz, E., Voß, S.: A dynamic programming-based matheuristic for the dynamic berth allocation problem. Ann. Oper. Res. 1–20 (2016)
10. Issam, E.H., Azza, L., Mohamed, E.M., Yassine, T.: A multi-objective model for discrete and dynamic berth allocation problem. In: Proceedings of the 2nd International Conference on Big Data, Cloud and Applications, p. 107. ACM (2017)

Association Models to Select the Best Rules for Fuzzy Inference System

Mariem Bounabi, Karim El Moutaouakil and Khalid Satori

Abstract In the past decade, the fuzzy inference system (FIS) has seen explosive growth in popularity among the researchers and industrialists. The fuzzy classification system is based on the most knowing process of the FIS in order to map features as inputs to outputs classes. Generally, to make a fuzzy classifier decision it is necessary to select manually the best rules that require the intervention of experts. In the current paper, we propose an automatic method to select the best rules using associations models: Apriority and Filter Associations. The proposed method process in five steps: preprocessing (feature reduction), determining membership function for every input and output, nominal representation of the database basing on the membership functions, call for the association model to select the most important rules, and finally post-processing of the obtained rules. In this work, we lead to appropriate rules set tested on the iris data for the classification task. In addition, the proposed method is highly simple to be implemented to control the even complex system. Our system achieved promising results, which demonstrate the effectiveness of the proposed approach.

Keywords Fuzzy logic · Fuzzy inference system · Inference rules · Association models · Fuzzy classifier

M. Bounabi (✉) · K. Satori
Computer Sciences, Imaging and Numerical Analysis Laboratory (LIIAN), 3000 Fes, Morocco
e-mail: bnb-meriem@hotmail.com

K. Satori
e-mail: khalidsatori@gmail.com

K. El Moutaouakil
Engineering Sciences and Applications Laboratory/Artificial Intelligence Team for New Technologies, 93040 Tetouan, Morocco
e-mail: karimmoutaouakil@yahoo.fr

V. Bhateja et al. (eds.), *Embedded Systems and Artificial Intelligence*,
Advances in Intelligent Systems and Computing 1076,
https://doi.org/10.1007/978-981-15-0947-6_33

1 Introduction

The classical membership functions were valued as Boolean logic, varying in the range {True, False} or {0, 1}, therefore, their applications in imprecise and uncertain domains are very limited [1]. Hence, the fuzzy logic (FL), invented by Zadeh [2], is the extension of the classical logic whose membership functions have values in the range [0, 1]. Based on this conversion, lately, FL has been attracting significant attention of the researchers and industrialists; in other words, Fuzzy logic is a mathematical theory namely the fuzzy set theory, where the main aim is the manipulation of natural language' uncertain notions. In this context, a new membership function has been defined: μ_A: X -> [0, 1] implies that x belongs to the fuzzy set A, with a degree of truth equal to $\mu_A(x)$.

Generally, the fuzzy set theory can be used in a wide range of domains where the information is incomplete or imprecise, such as classification, bioinformatics, operational research, decision support, pattern recognition, data mining, information systems, and many other areas of artificial intelligence. In this work, we deal with the fuzzy classification applications where the membership function $\mu_c(x)$ determines the belonging degree for each linguistic variable x to the class c.

To make a fuzzy classification decision, it is necessary to follow the three principal phases for fuzzy inference system (FIS). The first process is fuzzification methods [3], which consist of characterizing the linguistic variables used by the system. Next, the inferences phase [4], where the inference engine is applied for condensing information of a system based on a set of rules, expressed by experts, well define the problem. To make a decision, several values of linguistic variables are suitably defined by membership functions that are linked together by rules. Generally, when we talk about fuzzy deductions or inferences, there are two types of inference rules [5]:

- Inference with a single rule: This rule is applied in non-technical fields when choosing an element (which represents the optimum) from a set.
- Inference with several rules: In this case, when one or more variables require different decision-making according to the values reached by these variables; the latter are linguistic variables and therefore fuzzy variables. This problem is essential for adjustment and control problems, and decision-making leads to the execution of a certain operation.

Each rule delivers a partial conclusion, which is aggregated to the other rules to provide a conclusion (aggregation). In general, to determine the rules that govern an FLS, associated with a given problem, it is necessary to contact an expert. Importantly, this task becomes more difficult if the number of input and/or output is very large or the universe of discourse is very large too. For example, in the retrieval information field, we deal with the selected attributes methods [6], and in the OCR field, some statistical method leads to the vector of more than 80 components [7]. To overcome this problem, we propose a new automatic method to select the best rules basing on association models. The proposed method process in five steps: pre-treatment (feature reduction), determining membership function for every input and output, nominal representation of the data set basing on the membership functions, call for the association model to select the most important rules, and finally post-processing of the obtained rules. Several algorithms, for extracting frequent items (association rules), are proposed in the literature such as Apriori algorithm [8], Close [9], OCD [10], Partition [11], dynamic itemset counting (DIC) [12], Tertius algorithm [13]. In this work, we use the Apriori and Filter Associations because of their simplicity. The set of processes used in this approach will be detailed in the rest of the paper as follows: We start with the main methods of associations to select the best rules basing on the confidence level. Secondly, we delineate all the necessary preprocessing steps for our new approach. The description of the fuzzy classification for the iris data is given in the third part illustrated by some experimental results given in the last part.

2 Methodology

Once the features are represented as input to FIS, it is necessary to call for some association models (AM) in order to select automatically the best inferences rules. In this approach, we use as AM the Apriori Filter Association. In the rest of this section, we describe the principal of the AM and the process of the used algorithms.

2.1 Associations Models

The association models topic is considered as part of unsupervised symbolic learning approaches, used in the field of data mining [14]. An association is, usually, composed of a set of items I $\{i_1, i_2, i_3, \text{etc.}\}$. A set of transactions T $\{t_1, t_2, t_3, t_4, \text{etc.}\}$ corresponds to a learning set which will be used to determine the association's rules in the following step of the model. The volume of the transaction is the number of items contained in the transaction. An important notion for a set of items is its support σ, which refers to the number of observed transactions that contain it.

$$\sigma(X) = \text{Card}\left\{t_i / X \subseteq t_i, t_i \in T\right\} \quad (1)$$

The support (S) and confidence (C) are the performance measures of an association rule, defined as follows:

$$S(A \text{->} B) = \sigma(A \cup B)/N \quad (2)$$

It must be mentioned that a rule with low support can be observed only by chance. For a set of a transaction, we can generate rules by founding all association rules with support $\geq$ minsup and confidence $\geq$ minconf, where minsup and minconf are thresholds for support and confidence.

2.2 Apriori Model

Apriori algorithm is the most well-known association rule algorithm [15] and the first which including pruning steps to account for the growth in the number of candidate itemsets. The algorithm process is summarized in three steps:

- Generating sets of items;
- Calculating the frequency of item sets;
- Keeping the sets of items with minimal support, which represent the sets of frequent items.

The Apriori algorithm has some limitations, as the course of the initial data is recurrent. The computation of the supports and the generations of the rules are very expensive in term of time-consuming.

2.3 Filter Association

To run the Filter Association, an arbitrary association, such as Apriori, is applied on data that has been proceeded through an arbitrary filter.

Like the association, the structure of the filter is based exclusively on the training data, and the filter will process test instances without changing their structure [13]. Filter Association is an efficient algorithm than Apriori algorithm based on above two factors (Number of cycles performed, large item sets) because the Apriori algorithm generates the number of cycles performed and generate large items set which degrade the performance of the algorithm.

3 The Automatic Selection Approach of Rules

The designers of the system define the set of rules basing on their knowledge, once the linguistic variables are defined, as input to the fuzzy classifier system. In our approach, we choose automatically the appropriate rules, and we can summarize our approach in five steps:

Step 1: Preprocessing;
Step 2: Membership functions for every input and output;
Step 3: Nominal representation of the data set basing on the membership functions;
Step 4: Association models to select the best rules basing on the confidence level;
Step 5: Post-treatment.

Pre-treatment. Consider the set of data $D = \{x^1 \ldots x^k \ldots x^N\} \subset \mathrm{IR}^n$ by decomposing

$\forall i \in [1, n]$ the interval $[\min_k x_i^k, \max_k x_i^k]$ we define suitable membership functions. The mainly used membership functions are triangular forms, trapezoidal

Table 1 Some nominal representation of the iris data set

The nominal representation of:	Is:
[d = [3.3,1.0], Iris-versicolor]	[low, low, Iris-versicolor]
[d = [5.1,1.8], Iris-virginica]	[mean AND great, mean AND great, Iris-virginica]
[d = [5.6,2.4], Iris-virginica]	[mean AND great, great, Iris-virginica]
[d = [6.7,2.0], Iris-virginica]	[great, great, Iris-virginica]
[d = [5.0,1.7], Iris-versicolor]	[mean AND great, mean AND great, Iris-Versicolor]

forms, Gaussian forms, sigmoid forms, and polynomial forms. The choice of the membership function is not straight ward and the only experience can help the designer.

Transformation of numerical data to nominal data. The second process deals with the nominal representation of the data set, where for each simple attributed, we determine which fuzzy membership functions that fall within and the component of D is substituted with these fuzzy sets using the logical operators AND and OR. Given the numerical data $x^k = \left[x_1^k, \ldots, x_i^k, \ldots, x_R^k\right]$.

If $\forall k \in [1, N]$ $\forall i \in [1, R]$ x_i^k is A_i^k and B_i^k, The nominal representation of x^k is $\bar{x}^k = \left[A_1^k, B_1^k, \ldots, A_i^k, B_i^k, \ldots, A_R^k, B_R^k\right]$. In this sense $D = \left\{\bar{x}^1, \ldots, \bar{x}^k, \ldots, \bar{x}^N\right\}$ $\subset \mathrm{IR}^{2n}$ Where R is the dimension of the data after reduction (step 1) (Table 1).

Association models to select the best rules basing on the confidence level. In the previous part of the current paper, we use the association models such as Apriori and Filter Association to select best rules, which are discussed in the second part of this paper.

Post-treatment of selected rules. The association models produce three kinds of rules:

- Rejected Rules (RR), those don't give any information about classes noted RR;
- Explicit Rules (ER), those give only information about classes as consequence noted ER;
- Implicit Rules (IR), those give information about classes but not as consequence noted IR.

The first type of rules is suppressed from the selected file of selected rules. The second kind of rules is maintained without any modification. To transform the third type of rules into explicit ones, we use the following results:

$$A \vee (B \wedge C) \cong (A \vee B) \wedge (A \vee C) \tag{4}$$

$$A \wedge (B \vee C) \cong (A \wedge B) \vee (A \wedge C) \tag{5}$$

$$\overline{(B \vee C)} \cong \bar{B} \wedge \bar{C} \tag{6}$$

$$(\overline{B \vee C}) \cong \bar{B} \wedge \bar{C} \tag{7}$$

$$(A \vee B \to C) \cong (A \to C) \vee (B \to C) \tag{8}$$

$$(A \to C \vee B) \cong (A \wedge \bar{B} \to C) \tag{9}$$

$$A \to C \vee B \cong \bar{B} \wedge \bar{C} \to \bar{A} \tag{10}$$

Example
Using iris data set, we notice some examples for the three kinds of rules:

- For a RR: petal width IS low -> petal length IS low.
- As an IR, we found: class IS Iris-versicolor -> petal length IS low AND petal width IS low. This latter is transformed into ER using algebraic results.
- The third example is the ER: Petal length IS low AND petal width IS low -> class IS Iris-versicolor, where the selection of the final class is reasonable.

4 Experimentation and Results

4.1 Data Set

The iris flower data, as most known data in the classification field [16], is used in the experimentation part, to prove the effectiveness of our approach.

4.2 Results and Discussion

In this part, we use the proposed method to classify the iris data with two dimensions [petallength and petalwidth] and two classes [Iris-versicolor and Iris-virginica]. We use trapezoidal forms as membership functions.

The output membership function is a binary function 0 for Iris-versicolor class and 1 for Iris-virginica class. Table 2 gives the set of rules produced by Filter Association and Apriori models.

After transforming IR into the ER, we obtain the set of rules presented in Table 3. In this context, associations filter and Apriori models lead to some set of rules. It should be noted that the selection of the best rules by one of the association models is based on confidence superior to 0.9.

Concerning the inference task, we have tested three methods: minimum, product and the best system is the one based on minimum. Table 4 gives the confusion matrix

Table 2 Description of the selected rules by types and confidence value

Selected rules		Type	Confidence
IF	THEN		
Petallength IS low AND petalwidth IS low	Class IS Iris-versicolor	ER	1
Petalwidth IS low AND class IS Iris-versicolor	Petallength IS low	IR	0.98
Petallength IS low AND class IS Iris-versicolor	Petalwidth IS low	IR	0.98
Class IS Iris-versicolor	Petallength IS low	IR	0.96
Petalwidth IS great	Class IS Iris-virginica	ER	0.96
Class IS Iris-versicolor	Petallength IS low AND petalwidth IS low	IR	0.94
Class IS Iris-versicolor	Petallength IS low AND petalwidth IS low	IR	0.92
Petalwidth IS low	Class IS Iris-versicolor	ER	0.92
Class IS Iris-versicolor	Petalwidth IS low	IR	0.96
Petalwidth IS low	Petallength IS low	RR	0.90

Table 3 The rules after transforming IR into the ER

Rule	Active	If petal length	And petalwidth	Then class
1	+	Low	Low	0
2	+	Mean	Low	1
3	+	Great	Low	1
4	+		Great	1
5	+		Low	0
6	+	Low	Great	1

Table 4 Confusion matrix for the fuzzy logic basing on the selected set of the rules association model

	Iris-versicolor	Iris-virginica
Iris-versicolor	45	3
Iris-virginica	0	50

associated with the proposed system. In this regard, two data were rejected because of the lack of a suitable rule. To improve the quality of our classifier, we can select other rules while decreasing the confidence value to 0.8.

Table 5 gives the performance of the proposed system used for classification that gives recognition rate of 96.00%.

The recall of the proposed fuzzy classifier system is of 96%, it is possible to play on a good choice of input and output membership functions to improve the recognition rate.

Table 5 Performance of the system based on Filter Associations and Apriori models

	TP	FP	Precision	Recall	F-measure	Class
	0.90	0.00	1.00	0.90	0.95	Iris-versicolor
	1.00	0.06	0.94	1.00	0.97	Iris-virginica
Mean	0.95	0.03	0.97	0.95	0.96	

5 Conclusion

Differently to the traditional technique, to select inferences rules for the fuzzy classifier that based on the intervention of the experts, the new technique focuses on the automation choice of rules. Therefore, our approach uses the most popular associate models, which are the Apriori and Filter Association. Furthermore, the new fuzzy classifier system gives the satisfied results where the recognition rate is 96%. Even if we test on the average dataset, our approach encourages practitioners in the field of control of intelligent systems to use it for even complex problems.

References

1. Ross, T.J.: Fuzzy Logic with Engineering Applications. Wiley (2009)
2. Zadeh, L.A.: Information and control. Fuzzy Sets **8**(3), 338–353 (1965)
3. Shaocheng, T., Jiantao, T., Tao, W.: Fuzzy adaptive control of multivariable nonlinear systems 1. Fuzzy Sets Syst. **111**(2), 153–167 (2000)
4. Jang, J.S.: ANFIS: adaptive-network-based fuzzy inference system. IEEE Trans. Syst. Man Cybern. **23**(3), 665–685 (1993)
5. Zadeh, L.A.: Is there a need for fuzzy logic. Inf. Sci. **178**(13), 2751–2779 (2008)
6. Bounabi, M., El Moutaouakil, K., Satori, K.: A probabilistic vector representation and neural network for text classification. In: Springer Conference (2018)
7. Aharrane, N., Dahmouni, A., El Moutaouakil, K., Satori, K.: A robust statistical set of features for Amazigh handwritten characters. Pattern Recogn. Image Anal. **27**(1), 41–52 (2017)
8. Blanchard, J., Kuntz, P., Guillet, F., Gras, R.: Mesure de qualité des règles d'association par l'intensité d'implication entropique. Revue Nationale des Technologies de l'Information (1), 33–44 (2004)
9. Pasquier, N.: Extraction de Bases pour les Règles d'Association à partir des Item sets Fermés Fréquents. In: INFORSID'2000 Congress, pp. 56–77 (2000, May)
10. Mannila, H., Toivonen, H., Verkamo, A. I.: Efficient algorithms for discovering association rules. In: KDD-94: AAAI Workshop on Knowledge Discovery in Databases, pp. 181–192 (1994, July)
11. Agrawal, R., Shafer, J.C.: Parallel mining of association rules. IEEE Trans. Knowl. Data Eng. **8**(6), 962–969 (1996)
12. Brin, S., Motwani, R., Ullman, J.D., Tsur, S.: Dynamic itemset counting and implication rules for market basket data. Acm Sigmod Record **26**(2), 255–264 (1997)
13. Agarwal, R., Srikant, R.: Fast algorithms for mining association rules. In: Proceedings of the 20th VLDB Conference, pp. 487–499 (1994, September)
14. Agrawal, R., Srikant, R.: Fast algorithms for mining association rules. In: Proceedings of the 20th International Conference on Very Large Databases, VLDB, vol. 1215, pp. 487–499 (1994, September)

15. Bathla, H., Kathuria, K.: Apriori algorithm and filtered association in association rule mining. Int. J. Comput. Sci. Mob. Comput. **4**, 299–306 (2015)
16. Fisher, R.A., Marshall, M.: Iris data set. RA Fisher, UC Irvine Machine Learning Repository, 440 (1936)

Robust Loss Function for Deep Learning Regression with Outliers

Lamyaa Sadouk, Taoufiq Gadi and El Hassan Essoufi

Abstract In regression analysis, the presence of outliers in the data set can strongly distort the classical least squares (known as "L2") estimator and lead to unreliable results (due to the large abnormal error registered by outliers compared to the error of the majority of the training samples). To deal with this, several robust-to-outliers methods in robust statistics have been proposed in the statistical literature. However, in the context of deep regression networks, very few efforts have been carried out to deal with outliers and most deep regression algorithms make use of the traditional L2 loss function. In this paper, we consider the issue of training deep neural networks in the context of robust regression. As such, we introduce a robust deep regression model which is based on a novel robust loss function. With this latter, our model is able to adapt to an outlier distribution, without requiring any hard threshold on the proportion of outliers in the training set. Experimental evaluations on a head pose estimation dataset show that our model generalizes well to noisy datasets, compared to other state-of-the-art techniques.

Keywords Robust regression · Deep learning · Neural networks · Outlier detection

1 Introduction

Statistical estimation is one of the fundamental tools in numerous fields in engineering and science. Most often, statistical estimators are derived based on the assumption of a Gaussian-distributed noise and yield poor results when there are even small deviations from that assumption. To overcome those problems, robust estimation theory has been introduced by considering a large family of statistical models as well as possible outliers that deviate from the general model [4]. An estimator is called robust

L. Sadouk (✉) · T. Gadi · E. H. Essoufi
Faculty of Science and Technology Settat, University Hassan Ist,
Casablanca, Morocco
e-mail: lamyaa.sadouk@gmail.com

V. Bhateja et al. (eds.), *Embedded Systems and Artificial Intelligence*,
Advances in Intelligent Systems and Computing 1076,
https://doi.org/10.1007/978-981-15-0947-6_34

if a large deviation from the assumed statistical model (recorded from an outlier) has a low impact on the overall performance.

In the last years, deep neural networks revolutionized related fields of research. As such, the goal of this paper is to apply these robust regression methods in the context of deep learning. In this paper, we introduce a novel loss function within a convolutional neural network (ConvNet) regressor that addresses datasets with outliers by reducing the impact of outliers in the model fitting process. Indeed, in the context of feed-forward neural networks, we can employ a robust estimator (instead of the regular LS or L2 loss function) for minimizing the cost/error function, whereby corrupted training samples (e.g., outliers) with unusually large errors are downweighted such that they minimally influence the training backpropagation process.

1.1 Robust Regression

Robust Regression has long been studied in statistics [4, 9, 12] and in computer vision [2, 10]. The most common robust statistical techniques are: the *M-estimators*, *sampling methods*, *trimming methods* and *robust clustering*. *M-estimators* [4] minimize the sum of a positive-definite function of the residuals and attempt to reduce the influence of large residual values. The minimization is carried out with weighted least squares techniques, with no proof of convergence for most M-estimators. *Sampling methods* [10] such as least-median-of-squares or random sample consensus (RANSAC), estimate the model parameters by solving a system of equations defined for a randomly chosen data subset. The main drawback of such methods is that they require complex data-sampling procedures and it is tedious to use them for estimating a large number of parameters. *Trimming methods* [12], such as least trimmed squares (LTS), rank the residuals and down-weight or discard data points associated with large residuals. They are typically cast into a (nonlinear) weighted least squares optimization problem, where the weights are modified at each iteration, leading to iteratively re-weighted least squares problems.

1.2 Deep Learning for Regression

For the last few years, the deep learning research has been rapidly developing. In the context of regression where continuous values are to be estimated, several deep learning techniques have been recently introduced. For the human pose estimation task, studies [8, 11, 15] attempted to estimate positions of the body joints on the image plane. As for facial landmark detection task, the work of [13] predicts image locations of facial points. In the scheme of object and text detection, the goal is to predict regressed values which represent a bounding box for localization [5, 14].

However, most of the research community still keeps one element nearly completely fixed: When it comes to regression, most of the studies train their deep learning models based on the standard least squares loss function, which is sensitive to outliers. Despite the large literature on regression-based deep learning, only three works attempted to provide a robust deep regressor which tackles the issue of outliers [1, 3, 7]. In [1], authors achieve robustness by choosing the Tukey's biweight M-estimator as the minimizing loss function. In [3], a robust deep regressor is derived by unfolding a gradient descent method for a generalized least trimmed squares objective. This regressor is trained on triplets of unknown parameters, linear models and noisy observations with outliers. Another attempt toward a robust deep regressor was made by [7] which makes use of an optimization algorithm that alternates between the unsupervised detection of outliers using expectation-maximization, and the supervised training with cleaned samples using stochastic gradient descent.

Our paper comes as an extension of works [1, 3, 7]. The main contribution of this paper is to improve the performance of deep regressors in the presence of outliers with a robust regression approach based on training a ConvNet with novel loss function. In Sect. 2, we describe the basic components of our approach. We choose to demonstrate our approach on a pose estimation dataset. Accordingly, we describe experimental details (Sect. 3), then we discuss experiments and results (Sect. 4). Finally, Sect. 5 concludes our work.

2 Methodology

The neural network is fed by an input $\boldsymbol{x} : \Omega \rightarrow \mathbb{R}$ and its corresponding label (target) which is a vector $\boldsymbol{y} = (y_1, \ldots, y_J)$ such that $y_j \in \mathbb{R}$. Having a set of training data with outliers $\left\{(\boldsymbol{x}^{(s)}, \boldsymbol{y}^{(s)})\right\}_{s=1}^{S}$ composed of S samples, we wish to train a regression model represented by a function $\omega(\cdot)$ mapping between $\boldsymbol{x}^{(s)}$ and $\boldsymbol{y}^{(s)}$ as follows: $\hat{\boldsymbol{y}}^{(s)} = \omega\left(\boldsymbol{x}^{(s)}, \boldsymbol{\theta}\right)$, with $\hat{\boldsymbol{y}}^{(s)}$ being the network output and $\boldsymbol{\theta}$ the parameters to be tuned. The training process is accomplished through the minimization of the cost function $L(\cdot)$ defined as,

$$L(\boldsymbol{\theta}) = \frac{1}{N} \sum_{i=1}^{N} \sum_{j=1}^{J} \ell(r_j^{(i)}) \tag{1}$$

where N is the number of instances per batch, J the number of elements within the target vector, $\ell(\cdot)$ is the loss function, and $r_j^{(i)}$ the residual of the jth value of the target vector at an instance i which is given by, $r_j^{(i)} = y_j^{(i)} - \hat{y}_j^{(i)}$, with $y_j^{(i)}$ and $\hat{y}_j^{(i)}$ being j^{th} value of the target and output vectors, respectively, at i.

In this chapter, we start by giving an overview of popular loss functions $\ell(\cdot)$ used for regression including the standard one and the robust ones (Sect. 2.1). Next, Sect. 2.2 defines our novel loss function.

2.1 Background on Popular Loss Functions for Regression

L-estimates. The two most common and standard estimators are the least absolute deviations (LAD) and ordinary least squares (OLS). The LAD regression estimator (also known as L1-estimator) minimizes the sum of the absolute values of the residuals and is defined as, $\ell(r_j^{(i)}) = \left|r_j^{(i)}\right|$. It is quite robust and is generally not affected by outliers, but its derivatives are not continuous which makes it impractical to use as a loss function for training neural networks. On the other hand, the ordinary least square (OLS) estimate (also denoted as L2-estimate) which minimizes the sum of squared residuals, $\ell(r_j^{(i)}) = (r_j^{(i)})^2$, gives a more stable and closed form solution. Nonetheless, the L2-estimate gives outliers excessive weight by squaring the value of the residual and tries to adjust the model according to these outlier values, even on the expense of other samples.

M-estimates. Compared to the L-estimates which are not robust with respect to bad leverage points, the M-estimates are one of the robust regression estimation methods. They are a generalization of the maximum likelihood estimator proposed in [4] and are defined as, $\ell(r_j^{(i)}) = \rho(r_j^{(i)})$ such that ρ is a symmetric function with unique minimum at zero. An example of such estimate is the *Huber loss* (Fig. 1a), a function which is quadratic in small values of $r_j^{(i)}$ and grows linearly for large values of $r_j^{(i)}$ as follows,

$$\ell(r_j^{(i)}) = \begin{cases} \frac{1}{2}(r_j^{(i)})^2 & \text{if } \left|r_j^{(i)}\right| \leq c \\ c\left|r_j^{(i)}\right| - \frac{1}{2}c^2 & \text{otherwise} \end{cases} \tag{2}$$

where c is the hyper-parameter that controls how small the loss should be to go from the linear to the quadratic forms (c usually being set to 1.345). However, despite the fact that the Huber loss is convex, differentiable and robust to outliers, setting its parameter c is not an easy task.

Another robust M-estimate which is more robust than Huber loss is the Tukey's biweight function (plotted in Fig. 1a) which is defined as,

$$\ell(r_j^{(i)}) = \begin{cases} \frac{c^2}{6}(1 - (1 - (\frac{r_j^{(i)}}{c^2})^2)^3) & \text{if } \left|r_j^{(i)}\right| \leq c \\ \frac{c^2}{6} & \text{otherwise} \end{cases} \tag{3}$$

where c is a tuning constant (usually set to 4.6851). This function has the property of suppressing the influence of outliers during backpropagation by reducing the magnitude of their gradient close to zero, as shown in Fig. 1b. However, even though it is robust to outliers, it is non-convex and non-differentiable.

2.2 Our Proposed Loss Function

The idea is to come up with a robust loss function that has advantages over existent robust loss functions (mentioned above) and that generalizes well on deep learning models. Our loss function has a partial derivative w.r.t. The residual which is inspired from the sigmoid function. Instead of having a partial derivative that looks like step function, as it is the case for the L1 loss partial derivative, we want a smoother version of it that is similar to the smoothness of the sigmoid activation function. To this matter, we start by defining our loss function partial derivative before defining the loss function itself.

Partial derivative of our loss function. Let's consider the sigmoid function as a function of $r_j^{(i)}$ which is given by, $\sigma(r_j^{(i)}) = 1/(1+e^{-r_j^{(i)}})$. Add steepness or smoothness to this latter gives us $\sigma_w(r_j^{(i)})$ which is given by,

$$\sigma_w(r_j^{(i)}) = \frac{1}{1+e^{-wr_j^{(i)}}} \tag{4}$$

where w is a weighting factor. The goal is to come up with a loss function whose partial derivative w.r.t $r_j^{(i)}$ acts like a smooth step function being around -1 for large negative residuals and around +1 for large positive residuals (just like the gradient of Huber and L1). And, knowing that the function σ_w is in the range [0, 1], we rescale this latter to become within the range $[-1, 1]$ and to obtain the following partial derivative of $\ell(\cdot)$,

$$\frac{\partial \ell(r_j^{(i)})}{\partial r_j^{(i)}} = -1 + 2\frac{1}{1+e^{-wr_j^{(i)}}} \tag{5}$$

where $\lim_{r_j^{(i)} \to -\infty} \frac{\partial \ell(r_j^{(i)})}{\partial r_j^{(i)}} = -1$, $\lim_{r_j^{(i)} \to +\infty} \frac{\partial \ell(r_j^{(i)})}{\partial r_j^{(i)}} = +1$, and $\frac{\partial \ell(r_j^{(i)})}{\partial r_j^{(i)}} = 0$ at $r_j^{(i)} = 0$. Figure 1b illustrates this partial derivative as well as partial derivatives of other previously mentioned loss functions (L1, L2, Huber and Tukey).

Our loss function. The resulting loss function can be expressed as,

$$\ell(r_j^{(i)}) = (r_j^{(i)}) + \frac{2}{w}\log(1+e^{-w*r_j^{(i)}}) - \frac{2}{w}\log 2 \tag{6}$$

where the term $\frac{2}{w}\log 2$ is added in order to have the loss function equal to 0 at $r_j^{(i)} = 0$. This loss function as well as loss functions mentioned in the previous section is depicted in Fig. 2b.

Varying the weighting factor w. Through Fig. 2a, we can visualize multiple variants of our loss function with varying weighting factor ($w = \{1, 2, 3, 5, 20\}$). We notice that, as w increases, the loss function gets steeper and approaches the L1 loss function, thereby attributing relatively high errors to good estimated outputs (i.e., for

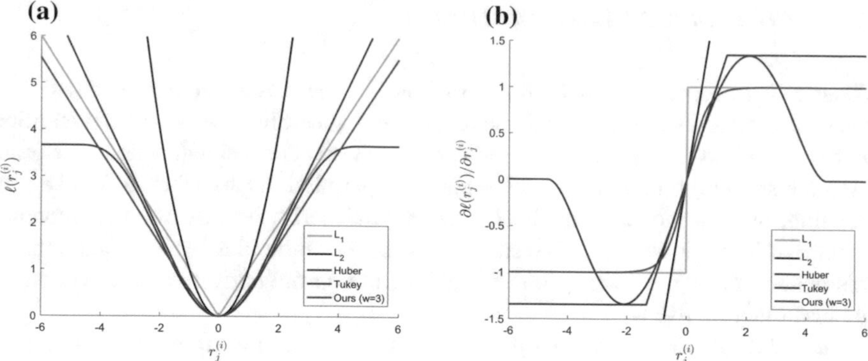

Fig. 1 Visualization of: **a** commonly used loss functions: L1, L2, Huber ($c = 1.345$) and Tukey ($c = 4.685$) as well as ours ($w = 3$), **b** their partial derivatives

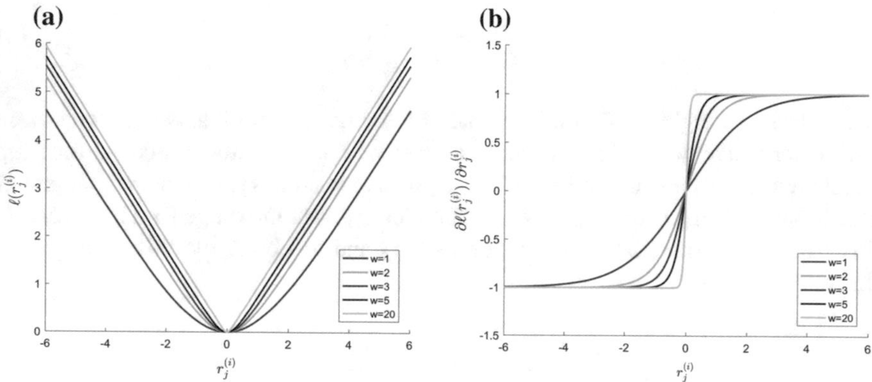

Fig. 2 Visualization of: **a** multiple variants of our loss function with varying weighting factor $w = \{1, 2, 3, 5, 20\}$, **b** their corresponding partial derivatives

$r_j^{(i)} \approx 0$). On the other hand, as w decreases, the loss function flattens and gives less penalty to outliers (to bad estimated output for which $r_j^{(i)}$ is very far from 0). This is confirmed by plots of partial derivatives of those variants (see Fig. 2b).

2.3 *Comparing Our Loss Function to Other Loss Functions*

By looking at Table 1 which compares between existent loss functions and ours, this latter seems to have the following advantages: (i) unlike Huber and Tukey which are composed of two functions each, our loss function is composed of a single function, (ii) unlike the Huber loss which has a hard threshold C imposed, our loss function offers a smoother transition at $r_j^{(i)} = C$ with no threshold required, and

Table 1 Comparison with existent loss functions for regression

Properties	L2	L1	Huber	Tukey	Ours
Convex	Yes	Yes	Yes	No	Yes
Differentiable	Yes	No	Yes	Yes	Yes
Robust to outliers	No	Yes	Yes	Yes	Yes
Bounded	No	No	No	Yes	No

(iii) as opposed to the Tukey loss, our loss function offers a convex optimization that guarantees one optimal solution (globally optimal), whereas the Tukey loss has a non-convex optimization which may result in multiple locally optimal points and may take a lot of time to identify whether the solution is global. Hence, the efficiency in time of the convex optimization problem is much better when training a neural network.

3 Experimental Setup

Dataset. We evaluate our loss function on the 2D human pose estimation task. To this matter, we conduct our experiments on the LSP dataset [6], a publicly available dataset containing 2000 pose annotated images of mostly sports people gathered from Flickr. Each image has been annotated with 14 joint locations (illustrated by yellow dots in the output image of Fig. 3) which are represented by pixel coordinates (x and y coordinates). The purpose is to train our model to estimate the 2D body skeleton as a set of joints. To do so, we assume that each individual is localized within a bounding box with normalized body pose coordinates.

Data processing. This step is similar to [1]. Indeed, after rescaling input images to 120×80, we normalize them by subtracting the mean image (taken from the training images) from them. Moreover, in order to prevent overfitting of the ConvNet, data augmentation is conducted by performing random rotations and flipping and by

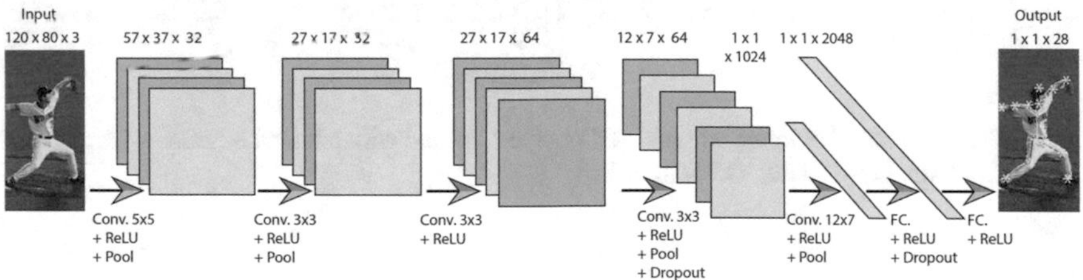

Fig. 3 Architecture of our convolutional neural network

adding a small Gaussian noise to the label vector $\boldsymbol{y}$ of the augmented instances. As for the label vector, it is rescaled to the range [0, 1].

Training details. The ConvNet model hyper-parameters are set according to the architecture shown in Fig. 3. Other hyper-parameters are set as follows: 0.01 for the learning rate, 0.9 for the momentum, 0.5 for the dropout and 230 for the batch size. Training of the ConvNet is performed using a five-fold cross validation.

Performance measure. The evaluation metric used in our study is the mean pixel error (MPE) which consists of averaging over errors of pixel coordinates of all 14 joints locations.

4 Experiments and Results

4.1 Baseline Evaluation

Using the LSP dataset, we train our ConvNet with our loss function based on different values of w (weighting factor) {2.25, 3, 3.5, 5}. Figure 4 depicts the convergence of each ConvNet and shows that the best performance is obtained when $w = 3.5$. This confirms our earliest assumption that: (i) a large w makes our loss function steeper at $r_i = 0$ and look like the L1 loss, which therefore over-penalizes small residuals and also over-penalizes large residuals (as seen in Fig. 2b), and (ii) a small w flattens out our loss function. This results in under-penalizing large residuals (which is a good property for getting rid of outliers) but also in under-penalizing small residuals (which is not a good property for inliers since the network does not learn properly from these latters).

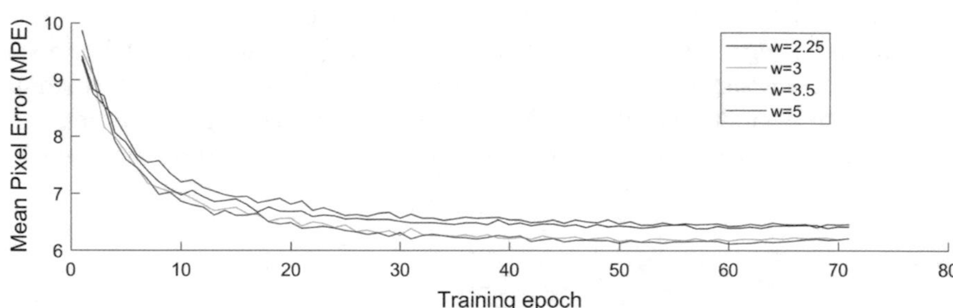

Fig. 4 Convergence of different variants of ConvNets trained with our loss function with different values of w (weighting factor) {2.25, 3, 3.5, 5}

Table 2 Comparative results between the ConvNets trained on commonly used loss functions for regression as well as ours, in terms of mean pixel error (MPE)

L2	Huber	Tukey [1]	Ours ($w = 3.5$)
6.2720	6.3376	5.9587	6.1104

4.2 Comparison with Other Techniques

In this section, the performance of our approach based on our loss function (using $w = 3.5$) is compared to previously mentioned loss functions: L2, Huber and Tukey. After training a ConvNet for each of these loss functions, we obtain results of Table 2. Through obtained results, we show that our approach yields a better performance than the classical approach (ConvNet with L2 loss function) and has comparable results to state-of-the-art robust deep regression techniques.

5 Conclusion

In this paper, we propose a robust deep regression model that addresses the issue of outliers thanks to the use of a novel robust loss function. After defining our loss function, we showed that this latter has advantages over common robust loss functions. Experimental validation conducted on the human pose estimation task showed that our robust deep regression ConvNet: (i) results in a better generalization and a faster convergence than the standard deep regression, (ii) it has higher or comparable performance results to ConvNets trained on robust loss functions. Finally, our robust deep regression model is simple and could be easily used for handle regression given datasets with outliers.

References

1. Belagiannis, V., Rupprecht, C., Carneiro, G., Navab, N.: Robust optimization for deep regression. In: Proceedings of the IEEE International Conference on Computer Vision, vol. 2015 Inter, pp. 2830–2838. https://doi.org/10.1109/ICCV.2015.324, 1505.06606 (2015)
2. Black, M.J., Rangarajan, A.: On the unification of line processes, outlier rejection, and robust statistics with applications in early vision. Int. J. Comput. Vision **19**(1), 57–91 (1996). https://doi.org/10.1007/BF00131148
3. Diskin, T., Draskovic, G., Pascal, F., Wiesel, A.: Deep robust regression. In: IEEE 7th International Workshop on Computational Advances in Multi-Sensor Adaptive Processing (CAMSAP), pp. 1–5 (2017)
4. Huber, P.: Robust Statistics. Springer, Berlin, Heidelberg (2011)
5. Jaderberg, M., Simonyan, K., Vedaldi, A., Zisserman, A.: Reading text in the wild with convolutional neural networks. Int. J. Comput. Vision **116**(1), 1–20 (2016). https://doi.org/10.1007/s11263-015-0823-z

6. Johnson, J., Everingham, M.: Clustered pose and nonlinear appearance models for human pose estimation. In: BMVC (2010)
7. Lathuilière, S., Mesejo, P., Alameda-Pineda, X., Horaud, R.: DeepGUM: learning deep robust regression with a Gaussian-uniform mixture model. In: Proceedings of the European Conference on Computer Vision (ECCV), vol. 11209 LNCS, pp. 202–217. https://doi.org/10.1007/978-3-030-01228-1_13 (2018)
8. Li, S., Chan, A.B.: 3D human pose estimation from monocular images with deep convolutional neural network. In: Asian Conference on Computer Vision, vol. 9004, pp. 332–347. https://doi.org/10.1007/978-3-319-16808-1_23 (2014)
9. Maronna, R., Martin, R., Yohai, V., Salibián-Barrera, M.: Robust Statistics: Theory and Methods (with R). Wiley, Hoboken (2018)
10. Meer, P., Mintz, D., Rosenfeld, A., Kim, D.Y.: Robust regression methods for computer vision: a review. https://doi.org/10.1007/BF00127126 (1991)
11. Pfister, T., Simonyan, K., Charles, J., Zisserman, A.: Deep convolutional neural networks for efficient pose estimation in gesture videos. In: Deep Convolutional Neural Networks for Efficient Pose Estimation in Gesture Videos, vol. 9003, pp. 538–552. https://doi.org/10.1007/978-3-319-16865-4_35 (2014)
12. Rousseeuw, P.J., Leroy, A.M.: Robust Statistics for Outlier Detection, vol. 589. Wiley, Hoboken. https://doi.org/10.1002/widm.2 (2005)
13. Sun, Y., Wang, X., Tang, X.: Deep convolutional network cascade for facial point detection. In: IEEE Conference on Computer Vision and Pattern Recognition, pp. 3476–3483. https://doi.org/10.1109/CVPR.2013.446 (2013)
14. Szegedy, C., Toshev, A., Erhan, D.: Deep neural networks for object detection. In: Advances in Neural Information Processing Systems, vol. 26 (NIPS 2013), pp. 2553–2561. https://doi.org/10.1109/CVPR.2014.276, arXiv:1312.2249v1 (2013)
15. Toshev, A., Szegedy, C.: DeepPose: human pose estimation via deep neural networks. In: Proceedings of the IEEE Conference on Computer Vision and Pattern Recognition, pp. 1653–1660. https://doi.org/10.1109/CVPR.2014.214, 1312.4659 (2014)

Fast and Stable Bio-Signals Reconstruction Using Krawtchouk Moments

A. Daoui, M. Yamni, H. Karmouni, O. El Ogri, M. Sayyouri and H. Qjidaa

Abstract Discrete orthogonal moments such as Tchebichef, Krawtchouk, Hahn, Meixner and Charlier are powerful tools for signal and image reconstruction. The analysis of large size signals by discrete orthogonal moments is limited by the very high computation time and by the numerical instability of these values especially for high orders. In order to accelerate time and guarantee the numerical stability of Krawtchouk moments, we propose in this article a fast and stable method based on the symmetry properties of Krawtchouk polynomials. This method has enabled us to considerably reduce the calculation time and maintain the stability of Krawtchouk moment. The results of the simulations carried out clearly show the effectiveness of the proposed method compared to conventional methods and compared to other types of moments.

Keywords 1D signal reconstruction · Bio-signals reconstruction · Krawtchouk moments · Discrete orthogonal moments · Discrete orthogonal polynomials

---

A. Daoui (✉) · M. Sayyouri
Laboratory of Engineering, Systems and Applications (LISA), National School of Applied Sciences, Sidi Mohamed Ben Abdellah-Fez University, Fez, Morocco
e-mail: achraf.daoui@usmba.ac.ma

M. Sayyouri
e-mail: mhamed.sayyouri@usmba.ac.ma

M. Yamni · H. Karmouni · O. El Ogri · H. Qjidaa
CED-ST, STIC, Laboratory of Electronic Signals and Systems of Information LESSI, Dhar El Mahrez Faculty of Science, Sidi Mohamed Ben Abdellah-Fez University, Fez, Morocco
e-mail: mohamed.yamni@usmba.ac.ma

H. Karmouni
e-mail: hicham.karmouni@usmba.ac.ma

O. El Ogri
e-mail: omar.elogri@usmba.ac.ma

H. Qjidaa
e-mail: qjidah@yahoo.fr

V. Bhateja et al. (eds.), *Embedded Systems and Artificial Intelligence*, Advances in Intelligent Systems and Computing 1076,
https://doi.org/10.1007/978-981-15-0947-6_35

1 Introduction

Applications of discrete orthogonal moments are numerous in several domains for 2D and 3D images such as classification [1], compression [2] and reconstruction [3]. It is also possible to apply these for the analysis and processing of one-dimensional signals. Bio-signals (1D and 2D) are widely studied in several scientific studies [4–6] because of their importance in providing physicians with the necessary information on human health (brain, muscles, heart, blood pressure, etc.). Thus, the diagnosis of many diseases becomes possible on the basis of the analysis of the latter. Since biological signal recordings are generally large in size, it is necessary to use fast and stable methods to analyze them. Discrete orthogonal moments such as Tchebichef [7, 8], Krawtchouk [9, 10], Hahn [11], Meixner [12] and Charlier [13] are considered powerful tools in signal analysis (1D, 2D and 3D) because of their robustness to different types of noise. Most discrete orthogonal moments are limited to high orders of moments. Also, the calculation of the coefficients of their base polynomials is recursively performed and rearranged in a 2D matrix, and this operation generally requires a significant computation time and causes the calculated coefficients to fluctuate numerically, especially for high polynomial orders. Motivated by the minimization of the computation time of the polynomial coefficient matrix which will be used for the computation of moments and also for the reconstruction of a 1D signal, and motivated also by the maintenance of the numerical stability of the computed polynomial coefficients, we use in this article the discrete orthogonal moments of Krawtchouk because of their important symmetry properties. These will be well exploited to ensure digital stability and significantly reduce signal reconstruction time. Several simulations and comparisons are performed to validate the speed and stability of the proposed method for the reconstruction of a bio-signal.

2 Computation of Krawtchouk Polynomials

The normalized Krawtchouk polynomials are defined by the following relation [2]:

$$\tilde{K}_n^p(x, N) = K_n^p(x, N)\sqrt{\frac{\omega(x; p, N)}{\rho(n; p, N)}}. \tag{1}$$

with

$$\omega(x; p, N) = \binom{N}{x} p^x (1-p)^{N-x} \text{ and } \rho(n; p, N) = (-1)^n \left(\frac{1-p}{p}\right)^n \frac{n!}{(-N)_n}. \tag{2}$$

The calculation of Krawtchouk polynomial values from Eq. (1) is very time-consuming and complex due to the terms factorial, power and Pochhamer. To overcome this problem, we will introduce the recursive calculation methods.

2.1 Recurrence Relationship with Respect to Polynomial Order n

The normalized Krawtchouk polynomials with respect to the order n are computed by the following three-term recursive relation [14]:

$$\tilde{K}^p_{n+1}(x, N) = \sqrt{\frac{1-p}{p} \cdot \frac{n+1}{N-n}} \cdot \frac{Np - 2np + n - x}{(N-n)p} \tilde{K}^p_n(x, N) - \sqrt{\left(\frac{1-p}{p}\right)^2 \cdot \frac{n(n+1)}{(N-n)(N-n+1)}} \cdot \frac{n(1-p)}{(N-n)p} \tilde{K}^p_{n-1}(x, N). \quad (3)$$

the first two orders ($\tilde{K}^p_0(x, N)$ and $\tilde{K}^p_1(x, N)$) of the Krawtchouk polynomials are given by the following relationships:

$$\tilde{K}^p_0(x, N) = \sqrt{\omega(x; p, N)} \text{ and } \tilde{K}^p_1(x, N) = 1 - \frac{x}{Np}\sqrt{\omega(x; p, N)} \quad (4)$$

The symmetry property is applied for the particular case ($p = 0.5$) of local parameter using the following formula:

$$\tilde{K}^p_n(x) = (-1)^n \tilde{K}^p_{N-n-1}(N - x - 1). \quad (5)$$

The calculation of Krawtchouk polynomials using the recursive relationship to the order n leads to the fluctuation of the numerical values of the coefficients, especially for high orders ($n \geq 100$) [2]. To reduce this problem, the researchers propose the use of the recurrence relationship with respect to variable x instead of the order n.

2.2 Recursive Computation with Respect to the Polynomial Variable x

The three-term recurrence relationship of the Krawtchouk polynomials with respect to variable x is given by [10]:

$$\tilde{K}^p_n(x+1) = \frac{-n + p(N-x-1) + x(1-p)}{\sqrt{p(N-x-1)(1-p)(x+1)}} \tilde{K}^p_n(x) + \frac{\sqrt{x(1-p)p(N-x)}}{\sqrt{p(N-x-1)(1-p)(x+1)}} \tilde{K}^p_n(x-1). \quad (6)$$

The initial conditions are given by:

$$\tilde{K}_n^p(0) = \sqrt{\frac{(N-n)p}{n(1-p)}}\tilde{K}_{n-1}^p(0) \text{ with } \tilde{K}_0^p(0) = \sqrt{(1-p)^{N-1}}$$

$$\tilde{K}_n^p(1) = \frac{-n+p(N-1)}{p(N-1)}\sqrt{\frac{(N-1)}{(1-p)}}\tilde{K}_n^p(0). \tag{7}$$

Note that in Eq. (7), the calculation of the Krawtchouk polynomial coefficients can only be done for the interval $x = [0, N/2 - 1]$, and the rest of the coefficients are calculated using the following symmetry relationship with respect to the variable x:

$$\tilde{K}_n^p(x) = (-1)^{n+x-1}\tilde{K}_{N-n-1}^p(N-x-1). \tag{8}$$

The computation of the coefficients of the polynomial Krawtchouk matrix by using either the three-term recursion relation with respect to the order $n$or with respect to the variable x leads to the propagation of the numerical errors especially for the higher orders of the polynomials ($n = 1000$ for example). To overcome this problem and also to reduce the computation time of the polynomial coefficient matrix, we will use, in the following section, a method based on the combination of the following elements: the symmetry properties given by equations Eqs. (5) and (8) with other diagonal symmetry relationships of the Krawtchouk polynomials.

3 Fast and Stable Calculation of Krawtchouk Polynomials

In this section, we present the essential steps of the proposed method based on the symmetry properties of Krawtchouk polynomials. These properties are relative to the first diagonal ($n = x$) and the secondary diagonal ($n = N - x - 1$). The symmetry with respect to the first diagonal is given by the following relations [10]:

$$\tilde{K}_n^p(x) = \tilde{K}_x^p(n) \text{ with } x = 0, 1, \ldots, N-1;\ n = 0, 1, \ldots, x-1 \tag{9}$$

For the secondary diagonal, we have:

$$\tilde{K}_{N-x-1}^p(N-n-1) = (-1)^{N-n-x-1}\tilde{K}_n^p(x);$$
$$x = 0, 1, \ldots, N-1; n = 0, 1, \ldots, N-x-1 \tag{10}$$

The proposed algorithm consists of three essential steps: The first is the division of the matrix plan into three zones (Area 1, Area 2 and Area 3). The second step is the calculation of the coefficients of the Krawtchouk polynomials only for Area 1 using Eq. (6), and then, we deduce the coefficients of the Krawtchouk polynomials for Area 2 using Eq. (10). For Area 3, we use equation Eq. (9) to calculate the coefficients in this region. Note that for the particular case $p = 0.5$ of the local parameter of

Krawtchouk, the polynomial coefficients are calculated only in the Area I-(1), and the rest of the coefficients in this region are deduced using Eq. (8). Figure 1 shows the distribution of the matrix plane of the Krawtchouk polynomials.

Figure 2a represents the curves of the Krawtchouk polynomials for the first five orders for $p = 0.5$, while Fig. 2b represents the curves of the Krawtchouk polynomials for all polynomial orders from 0 to 1000 using the proposed method.

It can be noted from Fig. 2b that the Krawtchouk polynomials are calculated up to the last order ($n = 1000$) which shows that the use of the proposed method is stable since it preserves the numerical stability of the polynomial coefficients. After calculating the Krawtchouk polynomials using the proposed algorithm, we will define the Krawtchouk moments for a one-dimensional signal in the next section.

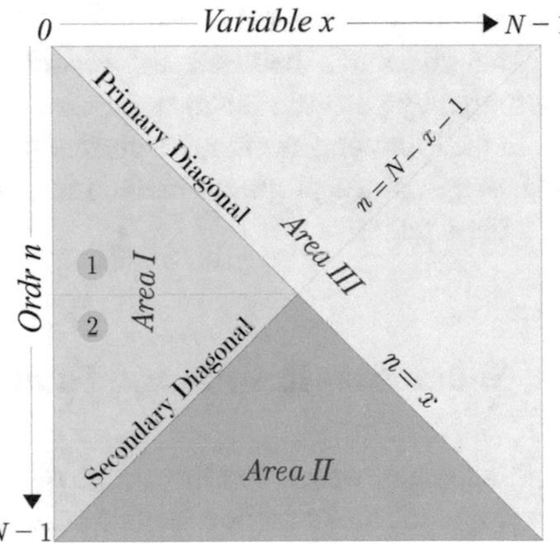

Fig. 1 The distribution of the matrix plane $n - x$ into three zones (Area 1, Area 2 and Area 3)

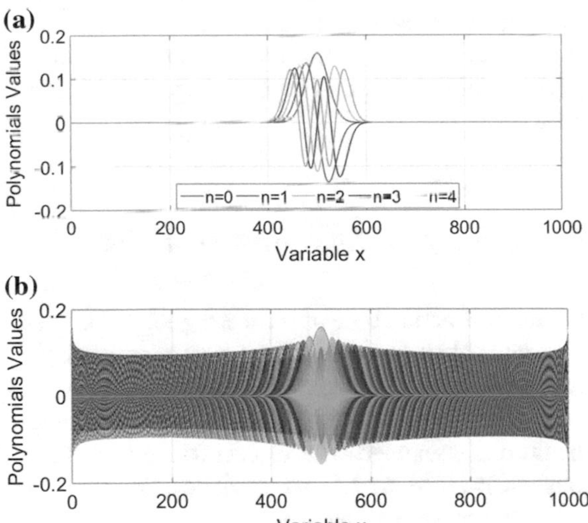

Fig. 2 **a** Plot of the first five orders ($n = 0, 1, 2, 3, 4$) of Krawtchouk polynomials with the parameter $p = 0.5$ and **b** plots of Krawtchouk polynomials for the orders ($n = 0, 1, \ldots, 1000$) with $x = 1, 2, \ldots, 1000$

4 Krawtchouk Moments for a One-Dimensional Signal

The Krawtchouk moments for a one-dimensional signal are defined by the following matrix formulation [2]:

$$KM_n = K_n^T f \text{ with } K_n = \left[\widetilde{K}_0^p(x), \widetilde{K}_1^p(x), \ldots, \widetilde{K}_{N-1}^p(x)\right]^T \tag{11}$$

where f denotes signal vector length N.

The reconstructed signal from the moments of Krawtchouk is calculated by:

$$\hat{f} = KM_n K_n \tag{12}$$

$\hat{f}(x)$ is the reconstructed signal.

The difference between the reconstructed signal and the original signal is measured by percent root-mean-square-difference PRD (%) defined in [15].

In the following section, we present the results of the simulations which show the advantages of the proposed method in terms of speed and quality of reconstruction of a bio-signal.

5 Simulation Results and Discussions

In this section, we present the results of the simulations that show the advantages of the proposed reconstruction method based on the symmetry properties that present the Krawtchouk moments, on the one hand to accelerate the reconstruction time of large bio-signals ($N = 1000$), and on the other hand, the proposed method allows to have a better quality of reconstructed signal. For this purpose, an electrocardiogram (ECG) type signal is selected from among the recordings [7], which also contains other types of biological signals such as electroencephalogram (EEG) and electromyogram (EMG).

5.1 Bio-Signal Reconstruction Time

The reconstruction time of the test signal "ECG" will be performed by Krawtchouk moments calculated using three methods: the first is the one described in Sect. 3, this method will be named later in this paper by (proposed method), the second is calculated according to the order n (Sect. 2.1) that will be named (method 1), and the third is computed with respect to the variable x (Sect. 2.2) that will be named (method 2). The signal reconstruction time and the quality of the reconstruction by the proposed method will also be compared to other types of moments such as Tchebichef moments [7] and Hahn moments [2]. The reconstruction of the test signal

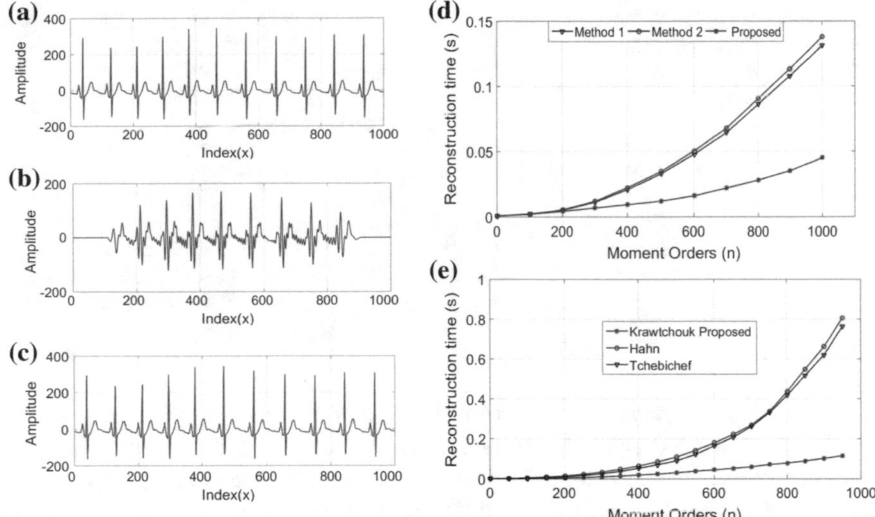

Fig. 3 **a** The ECG test signal, **b** and **c** the signals constructed by the proposed method for the orders $n = 200$ and $n = 1000$ successively, **d** the average reconstruction time using method 1, 2 and the proposed method, and **e** represents the average reconstruction time using the proposed method compared to Hahn moments [2] and Tchebichef moments [7]

"ECG" will be carried out by the different methods for the orders of the moments $n = 0, 50, 100, \ldots, 1000$ successively. The reconstruction time at each order of moments is calculated 100 times, and the average reconstruction time is plotted in Fig. 3b.

Figure 3d clearly shows that the reconstruction time of the test signal by Krawtchouk moments is considerably reduced when using the proposed method compared with conventional methods (method 1 and method 2). The improvement of signal reconstruction time is calculated by the criterion of execution-time improvement ratio (ETIR) which is defined by: $\text{ETIR} = (1 - \text{Time1}/\text{Time2}) \times 100$, with $T1$ represents the reconstruction time of the proposed method and $T2$ represents the reconstruction time of method 2. Indeed, we find that $\text{ETIR} = 72.41\%$ at the moments order ($n = 1000$). The result of the comparison (Fig. 3e) of reconstruction time between Krawtchouk moments by the proposed method with Hahn moments [2] and Tchebichef moments [7] shows that the proposed method is too fast compared to the latter methods.

5.2 *Quality of Reconstructed Bio-Signal*

In this subsection, we present a comparison in terms of reconstruction quality of bio-signals using the PRD (%) criteria.

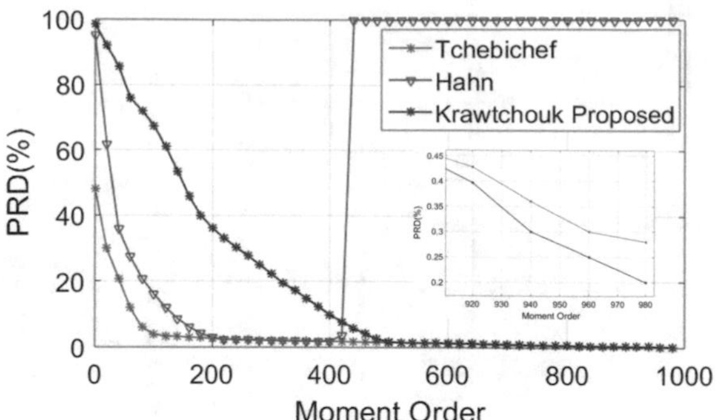

Fig. 4 PRD (%) of "ECG" signal using Tchebichef [7], Hahn [2] and Krawtchouk proposed moments

The result of the reconstruction error of the "ECG" signal (Fig. 4) clearly shows that the reconstructed signal becomes very similar to the original signal when the order of moments becomes high for Krawtchouk moments via the proposed method and for Tchebichef moments [7], and on the other hand, we notice that for Hahn moments, PRD (%) becomes 100%, i.e., the reconstructed signal does not resemble the original signal because of the numerical instability of Hahn moments for high orders. It can also be noted that the Krawtchouk moments via the proposed method offer a better quality of reconstruction compared to that of Tchebichef moment for higher orders.

6 Conclusion

In this paper, we have proposed a fast and stable method for the reconstruction of a large size biological signal using discrete Krawtchouk orthogonal moments. The advantages of this method are obtained by using the symmetry properties of Krawtchouk polynomials. The results of the simulation obtained clearly show the efficiency (fast and stable) of the proposed method compared to existing methods for the reconstruction of large size biological signals.

References

1. Zitová, B., Flusser, J., Invariants to convolution and rotation. In: Rodrigues, M.A. (ed.) Invariants for Pattern Recognition and Classification, pp. 23–46. World Scientific (2000)
2. Zhu, H., Liu, M., Shu, H., Zhang, H., Luo, L.: General form for obtaining discrete orthogonal moments. IET Image Process. **4**, 335–352 (2010)
3. Teague, M.R.: Image analysis via the general theory of moments. J. Optical Soc. Am. **70**(8), 920–930 (1980)
4. Siddharth, Gupta, R., Bhateja, V.: A log-ratio based unsharp masking (UM) approach for enhancement of digital mammograms. In: CUBE (2012)

5. Pandey, A., Yadav, A., Bhateja, V.: Contrast improvement of mammographic masses using adaptive Volterra filter. In: Proceedings of the Fourth International Conference on Signal and Image Processing 2012 (ICSIP 2012), pp. 583–593. Springer, India (2013)
6. Nilanjan, D., Bhateja, V., Hassanien, A.: Medical Imaging in Clinical Applications. Springer International Publishing (2016). https://doi.org/10.1007/978-3-319-33793-7
7. Abdulhussain, S.H., Ramli, A.R., Al-Haddad, S.A.R., Mahmmod, B.M., Jassim, W.A.: On computational aspects of Tchebichef polynomials for higher polynomial order. IEEE Access **5**, 2470–2478 (2017)
8. Camacho-Bello, C., Rivera-Lopez, J.S.: Some computational aspects of Tchebichef moments for higher orders. Pattern Recogn. Lett. **112**, 332–339 (2018)
9. Honarvar, B., Flusser, J.: Fast computation of Krawtchouk moments. Inf. Sci. **288**, 73–86 (2014)
10. Abdulhussain, S.H., Ramli, A.R., Al-Haddad, S.A.R., Mahmmod, B.M., Jassim, W.A.: Fast recursive computation of Krawtchouk polynomials. J. Math. Imaging Vision, 60(3), 285–303, 2018
11. Sayyouri, M., Hmimid, A., Qjidaa, H.: Improving the performance of image classification by Hahn moment invariants. JOSA **A30**(11), 2381–2394 (2013)
12. Sayyouri, M., Hmimid, A., Qjidaa, H.: A fast computation of novel set of Meixner invariant moments for image analysis. Circuits Syst. Signal Process. **34**(3), 875–900 (2015)
13. Karmouni, H., Hmimid, A., Jahid, T., Sayyouri, M., Qjidaa, H., Rezzouk, A.: Fast and stable computation of the Charlier moments and their inverses using digital filters and image block representation. Circuits Syst. Signal Process., 1–19 (2018)
14. Flusser, J., Suk, T., Zitová, B.: 2D and 3D image analysis by moments. Wiley (2016)
15. Pooyan, M., Taheri, A., Moazami-Goudarzi, M., Saboori, I.: Wavelet compression of ECG signals using SPIHT algorithm. Int. J. Signal Process. **1**(3), 4 (2004)
16. https://www.physionet.org/physiobank/database/

Improving Skin Cancer Classification Based on Features Fusion and Selection

Youssef Filali, My Abdelouahed Sabri and Abdellah Aarab

Abstract Recently, skin cancer has been rapidly increasing in terms of the number of melanoma cases due to the skin exposure to the sun. Melanoma is the deadliest skin cancer in the world. It is necessary to use a computer-aided diagnostic to help and facilitate the early detection of the skin cancer. In this paper, the proposed approach uses a fusion of shape, texture and color features that contains the lesion to classify the skin cancer. An image decomposition using the multi-scale is used, which gives two components: object and texture components. The object component will be used in the segmentation to identify the region of interest. The features are then extracted from the texture component, the shape of the lesion and color that contain the lesion. After combining all the features, a feature selection is moderate to keep only the best one. The classification is performed using the support vector machine classifier to classify skin cancer. The accuracy of our proposed approach is 92%, showing the effectiveness of our system.

Keywords Image processing · Skin cancer · PDE multi-scale decomposition · Texture, shape and color analysis · Features selection

1 Introduction

Nowadays, the most important task in medicine and health is cancer. Skin cancer is one of the most common forms of cancer and the number of deaths in all the word is rapidly increasing. There are two types of skin cancer: malignant melanoma and non-malignant melanoma (see Fig. 1). The skin cancer usually begins in the basal cells or the squamous cells, which are the last layer of the skin. The varying amounts

Y. Filali (✉) · M. A. Sabri
Department of Computer Science, Dhar Mahraz, Sidi Mohmed Ben Abdellah University, Fez, Morocco
e-mail: youssef.filali1@usmba.ac.ma

A. Aarab
Department of Physics, Dhar Mahraz, Sidi Mohmed Ben Abdellah University, Fez, Morocco

V. Bhateja et al. (eds.), *Embedded Systems and Artificial Intelligence*,
Advances in Intelligent Systems and Computing 1076,
https://doi.org/10.1007/978-981-15-0947-6_36

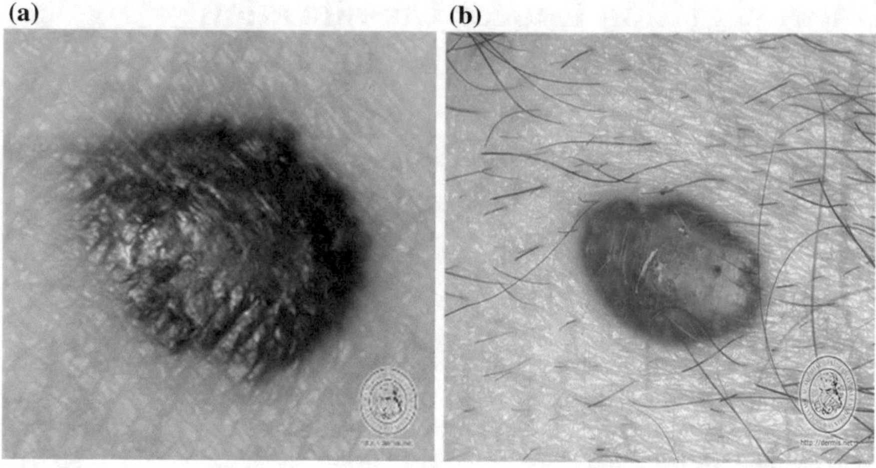

Fig. 1 Example of the database: "melanoma skin cancer" (**a**) and non-melanoma skin cancer (**b**)

of skin exposure to the sun are one of the main factors of the skin cancer. The only way to stop this is an early diagnostic, which increases the rate of a cured patient.

Images acquisitions, segmentation, features extraction and classification are the main steps in computational systems for an early diagnosis of skin lesions. The skin lesion can be captured using two techniques: Dermoscopic and Macroscopic, but some problems occur with these images, such as artifacts like hair, illumination and resolution. This makes the analysis of the skin cancer very difficult.

One of the most important steps in image processing is the pre-processing step which can remove the artifacts, smooth and facilitate the detection of the lesion with keeping pertinent information that contains the border of the lesion. The second step is the segmentation, which has a crucial role in the identification of the type of the lesion. Many segmentation algorithms can be applied in skin cancer to get only the region of interest (ROI). The Otsu algorithm, the active contour, the region-based algorithms are used in this field [1, 2]. Clustering has been also utilized in the segmentation of the skin cancer such as K-mean and Fuzzy c-means algorithms [3]. Recently, deep learning has been also applied to segment the skin cancer [4, 5].

Many features extractions are used to classify skin cancer. The most common features' extraction is similar to what dermatologists use in their diagnoses. All these features relay on the ABCD (asymmetry, border, color and diameter) rules. A lot of works in the literature try to use these rules in the diagnosis of the images or the use of texture, shape and color parameters [6, 7]. The last step is the classification that consists on the use of the feature extracted to interpret the information about the skin lesion. Many algorithms are utilized in the classification step such as support vector machine, decision tree, k-nearest neighbors and logistic regression [1, 8–10]. The performance of the classifier depends firstly on quality of the precedent step (segmentation and features extracted).

From our previous work, we have noticed that the texture and color which the lesion contain inside and outside are very different and can play a good descriptor to decide on the type of skin cancer. The aim of this paper is to extract more features from the inside and the outside of the lesion to better characterize the skin cancer. In our approach, a total of 110 features will be extracted from shape of the lesion, texture and color. These features are not all significant for building a good machine learning model. That is why we need to choose the best and relevant characteristics using a features selection. In this paper, a pre-processing using PDE decomposition is applied that gives two components object and texture; the object component will be segmented using the Otsu algorithm, and the result of the segmentation will give the features shape. The texture component will be used to extract textural features. For the color features, it will be extracted from the input images. Then, these features will be merged then selected using the ReliefF algorithms. Finally, we will employ these features as an input of a machine learning algorithms to classify the skin lesion. The rest of the paper is organized as follows; Sect. 2 presents description of the proposed approach for the detection and classification of the skin lesion. The metrics evaluation and the given results will be discussed in Sect. 3, finally a conclusion.

2 The Proposed Approach

2.1 Global Architecture for Our Approach

From methods and techniques in the introduction, Fig. 2 explains our proposed approach to classify the skin cancer, which involves into four major steps:

- Image pre-processing: the input image will be decomposed using the multi-scale decomposition into object and texture components.
- Image segmentation: The segmentation is made on the result of the object component by the decomposition.
- Features engineering: The features are extracted separately from texture, shape and color. The textural features are extracted from texture component, shape features are extracted from the segmentation component, and color features are extracted from the original image. A fusion of all these features flowed by a features selection.
- Classification: Once the features are prepared, we call some statistical model to characterize our data as the final step in our approach.

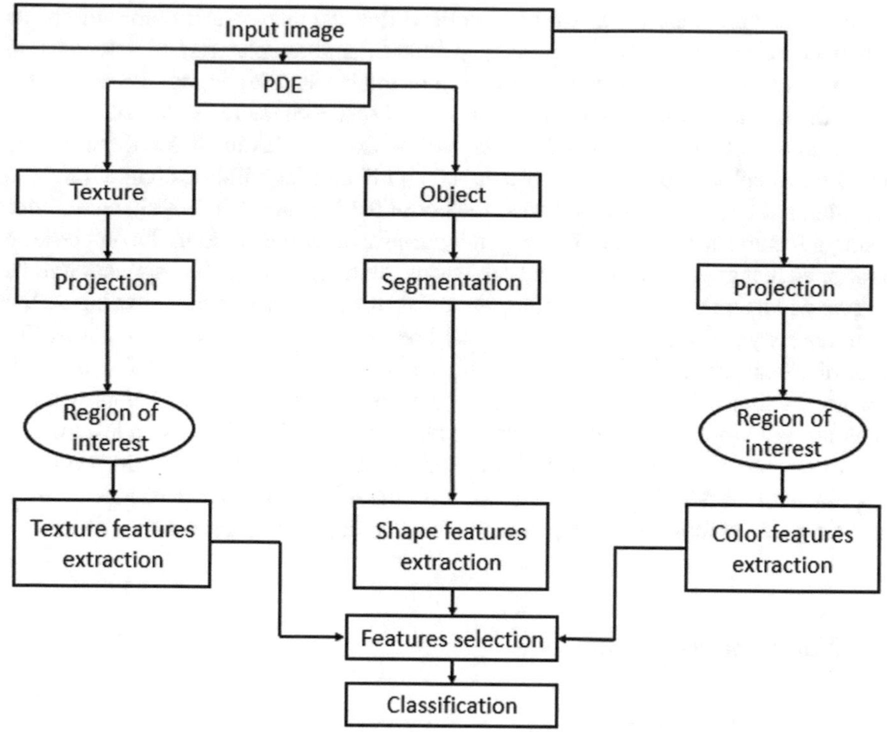

Fig. 2 Workflow of the proposed approach

2.2 Baselines of Our Approach

Pre-processing

The first step is the pre-processing which facilitates the detection of the lesion and removes the artifact that contains the lesion. The method is to use a model of multi-scale decomposition to extract the texture from the object. Many studies are proposed to use a partial differential equation (PDE). The aim of this is to decompose the image into two components. The first component will contain the geometric information of the image; the second component will contain the texture and noise. In our approach, we use the decomposition using Aujol model based on our previous work. The effectiveness of our approach is seen in our articles [6, 7].

Segmentation

The segmentation is most important step in image processing. A bad result of segmentation will lead certainly to a bad classification result, however, the use of a powerful classifier that exists on the literature. So, the segmentation had a major impact on the result of classification. In our approach, we segmented only the object component obtained from the decomposition using the partial differential equation

(PDE) without any noise. For that, we do not need a complex algorithm for the segmentation, and only the Otsu gives a good segmentation result.

Features engineering

Here, we discuss all about the features that can characterize the lesion. The features used are texture, shape and color. We will have a vector of features that contains 110 features but not all of them are relevant. For that, we will proceed to a features selection step to identify the best features needed to represent efficiently the data.

Shape features extraction: As in the diagnosis of dermatologists, the shape of the lesion plays an important role in deciding the type of the lesion. The shape of the lesion was calculated using different parameters: area; greatest diameter; smallest diameter; perimeter; eccentricity and extent from (FS1 to FS6). We add also two important features in shape that are as follows:

- Equivalent diameter:

$$\text{FS7} = 3^{*} A/L^{*}\pi \tag{1}$$

where L is the length of the major axis passing through lesion center.

- Circularity:

$$\text{FS8} = 4^{*}\pi^{*} A/L^{*} p \tag{2}$$

where p is the lesion perimeter

Color features extraction: The goal is to quantify the color variation in the lesion. The features extracted are maximum, minimum, mean and variance of the pixels intensities inside the lesion of the R, G and B plan. After converting the RGB image to HSV color system, the maximum, minimum, mean and variance of the pixels intensities inside the lesion of the H, S and V plan are calculated. We define these features from FC1 to FC24.

We calculate the same features but in the outside of the lesion from FC25 to FC48. Then, we create a vector features that contain the difference between the inside of the lesion and the outside of the lesion. This allows enhancing the differentiation of the malignant and benign lesions from FC49 to FC72.

Texture features extraction: In order to quantify the texture present in a lesion, a set of statistical texture descriptors were employed. These statistics features are contrast; correlation; energy; homogeneity; entropy; inverse difference moment; smoothness; standard derivation; kurtosis; root mean square.

We calculate the texture inside the lesion from (FT1 to FT10) and outside of the lesion from (FT11 to FT20) that can create a new vector with the difference of texture between the inside and outside of the lesion from (FT21 to FT30).

Features selection: Before the selection step, we should normalize the feature, to simplify the features selection. For that, we will use the Zscore transformation [9]. The step of feature selection is to select the number of features that can represent the data exactly and improve the classification score. We will focus only on the ReliefF algorithm, which tries to determine the nearest neighbors from several samples that are selected randomly from the data set.

Classification

The last step after the features selection is to use these features to classify the skin cancer. The dataset is divided into two sets: training sets that consist in building a model that learns from sample and test set to evaluate the performance of the model. Then, we will use a cross validation to maintain the fairness of performance. A lot of classifier is used to classify skin cancer. The choice of the classifier is taken from our previous work [8]. That used a comparison between different classifier with different kernel and distance. The best score is given by the support vector machine (SVM) using the quadratic kernel. For that, we opt also for the same classifier.

3 Experimentation and Result

3.1 Dataset

The dataset used to evaluate the proposed approach contains 180 images of the pigmented skin lesion take from "atlas dermoscopic" [11]. The database contains also the ground of truth of the images made by expert of dermatology.

3.2 Metrics Measure

For the evaluation of the classification results, we will use the sensitivity (TP rate), specificity (TN rate) and accuracy measures.

- $\text{Sensitivity} = \text{TP}/(\text{TP} + \text{FN})$ (3)
- $\text{Specificity} = \text{TN}/(\text{FP} + \text{TN})$ (4)
- $\text{Accuracy} = (\text{TP} + \text{TN})/(\text{TP} + \text{TN} + \text{FP} + \text{FN})$ (5)

With TP is number of true positives; TN is number of true negatives; FP is number of false positives; FN is number of false negatives.

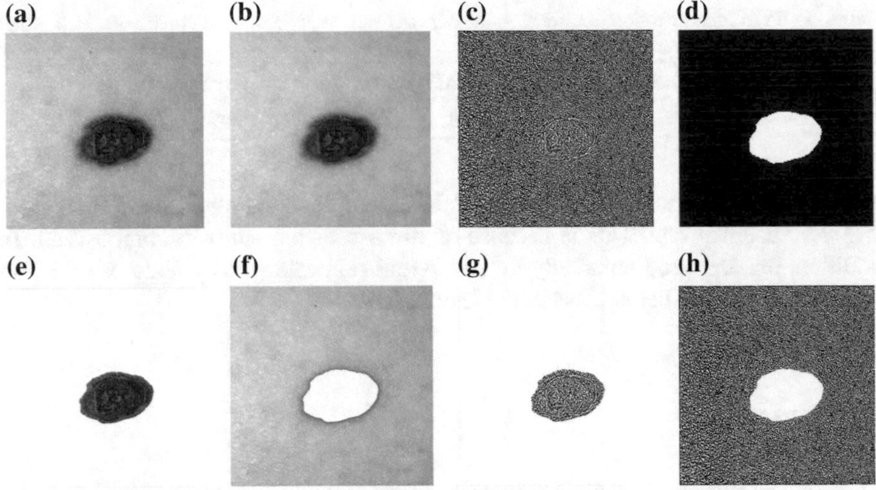

Fig. 3 Example of the image result: **a** original image; **b** object component; **c** texture component; **d** segmentation result using Otsu; **e** color of feature extraction inside the lesion; **f** color of feature extraction outside the lesion; **g** texture of feature extraction inside the lesion; **h** texture of feature extraction outside the lesion

3.3 Result and Discussion

In this part, the given results of features extraction, features section and classification are presented and discussed.

In order to remove the artifacts that the lesion contains, the input image was decomposed into texture and object component using the Aujol model decomposition (see Fig. 3b, c). The decomposition leads us to have a good recognition of the object that makes a better segmentation result. The result of the segmentation Fig. 3d will help to extract the features of the shape of the lesion.

In Fig. 3e, f, we calculate the color existing inside and outside of the lesion by projecting the segmented result on the original image. Finally, in Fig. 3g, h, the features are extracted inside and outside of the lesion using the textural features by projecting in the segmented result on the texture component.

A fusion of all these features, which are not all relevant, will be selected using the ReliefF algorithm.

The features selected are **three** features that are **the greatest diameter** (FS2), **the mean of red** (FC3) and **the root mean square of difference texture between inside and outside** (FT30).

Table 1 presents a comparison between our proposed approach with other approaches from the literature. Alin M. Solomon [12] uses the color and gray-level co-occurrence matrix (GLCM) as a features extraction and got 81.17% accuracy. Almansour [13] uses texture features based on gray-level co-occurrence matrix (GLCM) in addition to local binary pattern (LBP) and color features and got 87.32%

Table 1 SVM classification accuracy of the proposed approach in comparison with other approaches from the literature

	Solomon [12]	Almansour [13]	Proposed approach
Accuracy (%)	81.17	87.32	92.00

accuracy rate. In contrast, our proposed is higher than the other approaches with 92.00% accuracy rate. This is because of the use of the Aujol decomposition. In addition, features used separately for lesion and skin followed by features selection are very efficient to better classify the lesion.

4 Conclusion

To help dermatologists in their diagnosis, many solutions in image processing and analysis have been proposed. The major work that we performed in this paper is to use many features extracted that can describe the lesion, and this helps us improve the classification process. This approach is based on the model decomposition to get object and texture identification, Otsu for segmentation, shape, texture and color for features extraction, ReliefF as a feature selection and support vector machine (SVM) with quadratic kernel as a classifier. The proposed approach achieved good results and good classification rate.

References

1. Victor, A., Ghalib, M.R.: Automatic detection and classification of skin cancer. Int. J. Intell. Eng. Syst. **10**(3), 444–451 (2017)
2. Fan, H., Xie, F., Li, Y., Jiang, Z., Liu, J.: Automatic segmentation of dermoscopy images using saliency combined with Otsu threshold. Comput. Biol. Med. **85**, 75–85 (2017)
3. Sabri,M.A., Ennouni, A., Aarab, A.: Automatic estimation of clusters number for K-means. In: 4th ICIST, pp. 450–454 (2016)
4. Garnavi, R.: Skin lesion segmentation using deep convolution networks guided by local unsupervised learning. IBM J. Res. Dev. **61**(4), 1–8 (2017)
5. Codella, N.C.F., et al.: Deep learning ensembles for melanoma recognition in dermoscopy images. IBM J. Res. Dev. **61**(4), 1–15 (2017)
6. Filali, Y., Ennouni, A., Sabri, M.A.: Multiscale approach for skin lesion analysis and classification. In: ATSIP (2017). https://doi.org/10.1109/atsip.2017.8075545. ISBN: 978-1-5386-0551-6
7. Filali, Y., Sabri, M.A.: An improved approach for skin lesion analysis based on multiscale decomposition. In: ICEIT, pp. 1–6 (2017)
8. Filali, Y., Ennouni, A., Sabri, M.A., Aarab, A.: A study of lesion skin segmentation, features selection and classification approaches. In: ISCV, pp. 1–4 (2018)
9. Deepa, S.N., Aruna Devi, B.: A survey on artificial intelligence approaches for medical image classification. Indian J. Sci. Technol. **4**(11), 1583–1595 (2011)
10. Bhateja, V., Gautam, A., Tiwari, A., Satapathy, S.C., Nhu, N.G., Le, D.N.: Haralick features-based classification of mammograms using SVM. In: Information Systems Design and Intelligent Applications, pp. 787–795 (2018)

11. http://www.dermoscopy.org/atlas/base.htm
12. Solomon, A.M., Murali, A., Sruthi, R.B., Sreekavya, M.K., Sasidharan, S., Thomas, L.: Identification of skin cancer based on colour subregion and texture. Int. J. Eng. Sci. Comput. **6**(7), 8331–8334 (2016)
13. Almansour, E., Jaffar, M.A.: Classification of Dermoscopic Skin Cancer Images Using Color and Hybrid Texture Features. Int. J. Comput. Sci. Netw. Secur. **16**(4), 135–139 (2016)

A Serious Game for Teaching Python Programming Language

Alaeeddine Yassine, Mohammed Berrada, Ahmed Tahiri and Driss Chenouni

Abstract Computer programming is a multidisciplinary course that promotes critical thinking and problem-solving skills. By learning how to program, students learn how to transform abstract problems into logical steps of instructions as a solution. Still, learning a programming language is challenging especially if students lack the prerequisites. This paper discusses the potential of using gamified challenges to teach and learn fundamental concepts of Python programming language. To this end, a recently developed serious game is presented and discussed. The proposed game offers a complete course with different tests ranging from simple quizzes to the most complex puzzles. The game effectiveness is validated by comparing to what extent it respects some learning principles of games for learning.

Keywords Serious games · Mobile learning · Mobile application · Programming language · Gameplay · Quiz game

1 Introduction

Technology has become an integral part of the teaching learning process because it has a great potential of activating student's interest. The twenty-first-century learners are digital natives, most of them spend countless hours glued to different devices such as mobile, laptops, video game consoles, and this generation is less fearful of using technology. For instance, one of the current trends in the educational landscape is

A. Yassine (✉) · M. Berrada · A. Tahiri · D. Chenouni
Laboratory of Computer Science and Interdisciplinary Physics (LIPI), Ecole Normale Supérieure (ENS), Sidi Mohamed Ben Abdellah University, Fes, Morocco
e-mail: alaeyassine@gmail.com

M. Berrada
e-mail: mohammed.berrada@gmail.com

A. Tahiri
e-mail: tahiri_ahmed@hotmail.com

D. Chenouni
e-mail: d_chenouni@yahoo.fr

V. Bhateja et al. (eds.), *Embedded Systems and Artificial Intelligence*, Advances in Intelligent Systems and Computing 1076,
https://doi.org/10.1007/978-981-15-0947-6_37

serious games. Games for learning use the entertaining side to promote learning, and the purpose is to boost student engagement and to make learning enjoyable and fun. One of the application areas of serious games is computer sciences especially learning programming languages. Introductory computer science is part of most engineering curriculums; it is based on the assumption that students need to acquire knowledge and skills related to computer sciences. Among the skills that students should acquire at early stages is critical thinking and problem-solving skills [1]. One of the main benefits of learning how to program is being able to understand the functioning of computers, promoting self-learning, critical thinking, problem-solving and social skills. It is a multidisciplinary course in which students learn how to solve problems by writing lines of codes. However, learning programming is perceived by most of students as a difficult task with multiple barriers like misguided perception. This can be explained by the fact that most of the students think that learning a programming language is difficult and requires a complex cognitive effort. Furthermore, such courses require perquisites and a deep understanding of abstract concepts at early stages of learning.

We posit that the focus should be on learner-centered educational approach allowing students to take control of their learning. We believe that serious games put student at the heart of the scenario and give them the opportunity to work actively in a rich and interactive environment. Using a serious game to teach programming language syntax is a challenging research area and has the potential to reduce the high dropout rate since it could facilitate the comprehension of difficult concepts. Such a goal can be reached by presenting learning concepts in a thoughtful way. For this purpose, we developed a serious game to teach Python programming language. The game is about exploring an environment with interactive elements.

Where most of serious games lack a real scenario and just use graphic elements to attract the learner's attention, our application goes further by teaching Python language through a well-thought scenario, involving colorful characters with their own expressions and roles, various environments and a storyline. Moreover, the game has a pedagogical background. The learning concepts are organized using Solo Taxonomy

This paper presents an overview of a serious game that we have designed and implemented for teaching and learning Python programming language. The reminder of this article is organized as follows: Sect. 2 focuses on related works; we discuss and compare serious games used in computer sciences. In Sect. 3, we expose our solution according to several learning principles that video games feature to motivate and engage player.

2 Related Works

An analysis of related works has revealed that games for learning programming are divided into three categories: the first one is devoted to teach and learn a programming language as a whole (C, Java, JavaScript, Ruby, etc.). The second category is focused on a particular concept considered as the most difficult concept to teach and learn. Finally, the third category aims at teaching fundamental concepts of programming such as variables, loops and conditions regardless of any programming language in particular.

If used in classrooms, serious games should be based on an educational policy. However, most of serious games developed so far do not state any educational approach or instructional model in the design step. This could be explained by the fact that most games are focused on making programming fun rather teaching students how to program. We also posit that most of the games are intended to introduce novice learners to computer science and foster their self-confidence.

Although there exists many gameplay techniques such as writing codes, drag and drop, point and click, we postulate that drag and drop is the most used technique to best learn programming. Drag and drop is more suitable for novice learner (even kids) than advanced ones. In this case, learners are provided with pieces of code or objects and should place them correctly so to create interaction with the game (Table 1).

To sum up, serious games used for teaching programming are various. The target audience are usually kids but could be used at any level to foster learning. The purpose of this particular category of games is to engage students in the process of learning how to program. Some games are focusing on programming concepts, critical thinking or logic. Others are more technical and aim to teach syntax to use in order to construct a working program. Table below reports the most popular serious games related to computer sciences subject field with different gameplay techniques and focusing on various programming languages.

Table 1 Serious games related to computer sciences subject field

Name	Description	Level	Target platform	Language	Gameplay
Light-Bot [1]	Teaching programming concepts such as loops and conditions	Novice	Multicross platform	– [a]	Drag and drop
Ruby warrior [2]	Teach open-source programming language Ruby. The learner must write lines of code to make the character evolve in various courses	Novice	Web-based	Ruby	Writing codes
Robot ON! [3]	A puzzle-like serious game to teach the basics of C++ programming language. Players should demonstrate their understanding of a program's behavior. To progress in the game, the player is asked to complete a set of comprehension tasks.	Novice	Desktop application	C++	Puzzle

(continued)

Table 1 (continued)

Name	Description	Level	Target platform	Language	Gameplay
Code Spells [4, 5]	It is a 3D game designed to teach Java programming language. The game uses the metaphor of wizardry: the learner is a magician who needs to learn how to remember spells (Java programs) and write and execute new ones	Novice	Windows	Java	Role-play game
Move the turtle [6]	Move the turtle is a serious game that teaches how to plan complex operations based on simple steps. The players use predefined commands to build a program and help the turtle get diamonds	Novice	IOS	–	Point and click
Code combat [7]	The player must write code directly in a dedicated area to make the hero move and advance through a maze. Each maze in the game has its own traps and rewards	Novice	Web-based	JavaScript and Python language	Writing codes
Hour of code [8]	Make programming accessible to a larger audience. Building solutions in the form of algorithms that can be translated into code. The player must guide a bird toward an enemy, represented by a green pig	Novice	Web-based	JavaScript	Drag and drop
Checkio.org [9]	Coding games for beginners and advanced programmers. Uncoding challenges and various tools for solving codes	Novice	Web-based	Python and Javascript	Drag and drop

[a]Not related to a particular programming language

3 Evaluation of the Game According to the Principles of Learning in Video Games

We created a prototype mobile application we named *The Stronglobes.* It is a serious game including a complete course on Python language with various quizzes and riddles to solve. For the time being, the application is only available in English. It is a cross-platform application, it is built with unity, and it can be deployed in various platforms such as Windows, Android, iOs and Mac. To date, the size of the application is around 50 Mb, making it lightweight and fast to download. The scenario is simple and involves protagonist characters in constant battle against ruthless viruses. The virus menace wants to steal knowledge about Python to use its power and conquer the universe. The learner must gather and learn enough knowledge through courses and pass the tests that await him on the way.

For managing and storing knowledge, we used an ontology. We build the instructional model using Protégé Editor. The main benefits of using ontology are to provide communication between systems, reuse and organization of knowledge. Domain ontology is a standard-based way for knowledge sharing.

In this section, we will present our application according to several learning principles that video games feature to motivate and engage players. According to Gee [10], video games motivate and engage players physically and mentally by giving clear goals and rich environments, creating a feeling of complete immersion in the game. These principles are very popular in the video game industry, proposed by James Paul Gee, and they were cited more than 10,937 time in the literature to this day. We will limit ourselves to the evocation of seven principles among those that exist.

3.1 Identity

Learning requires personal investment and can only be effective when the learner is convinced of the benefits of his learning. The player acquires an identity through the game and feels invested with the mission that is proposed to him. For example, a game whose subject is surgery would put the player in the role of a doctor in a hospital to take care of patients. In our context, the player embodies the role of a funny character called a *Stronglobe*. What makes identifying with this character easy is the fact that the player can customize their appearance and create a custom hero. The player thus creates a link with the game through a hero with a unique visual.

3.2 Interaction

The immersion is a principal characteristic of a game. A good game should put the focus on maintaining a relationship with a player by providing feedbacks or dialog while interacting with different objects and characters of the game. In our context, the game communicates regularly with the player through a set of textual and graphic messages. For example, each course unit in our application has a set of interactive buttons allowing him to consult the information in different ways. The learner is not a spectator but is the central actor of his training.

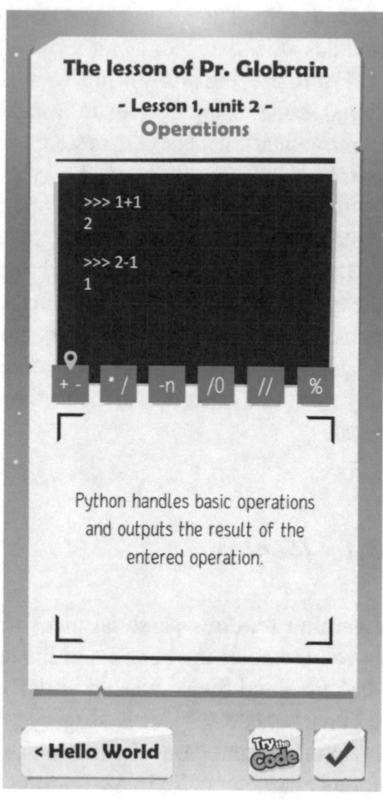

Fig. 1 Second unit of the first lesson in *The Stronglobes*

3.3 Agency

Through environments, characters, actions and decisions, games make players feel like their own bodies are inside the game. The players then deeply invest their bodies and minds in the game. Our game offers various game environments filled with interactive elements such as courses, different kinds of tests and colorful characters. There are even activities not related to learning programming that we included in the application to push the learner to explore and reinforce his feeling of belonging to the game environment (Fig. 1).

3.4 Well-Ordered Problems

Games provide students with increasing levels of difficulty. First levels usually provide players with basic concepts that students should remember. Still, what is particular about games is that, while progressing into the game, the player is also developing problem-solving logic and analytical skills. These skills can be used to solve a variety of problems. As we aim to teach Python language, we start with basic concepts that all students need to master and smoothly move forward to help learner solve complex programming problems and even complete his own courses.

3.5 Just-in-Time and on Demand

Information in our game is delivered using two ways: Information is shown to player under just-in-time (JIT) concept to make immediate use of it in the right context of the game or on-demand access when the user asks explicitly for it. For instance, the help icon, which is available in all scenes, could be used on demand when the player gets stuck. It provides him/her with hints to complete the current goal.

3.6 Pleasantly Frustrating

According to Gee [10], a problem leads to good learning when the learner feels a challenge and feels that an accomplishment has been made by solving it. The learner must also feel that despite the difficulty, he can succeed, get the results and overcome the difficulties. In our serious game, we tried hard to establish a series of tests that allow the player to evaluate his knowledge and advance in the game. The risk of failure being present, it is not necessarily negative because the error automatically generates a correction that is an extension of course that completes the information contained in the game.

3.7 Smart Tools and Distributed Knowledge

Games use mechanisms that players know well. They help students gain knowledge and promote sharing with a community. The game techniques used in our context are drag and drop and point and click as well as code writing as depicted in Fig. 3. Code writing also features in the app through a series of tests where the learner must write code to interact with the hero in order to achieve certain goals. These techniques are currently widely exploited thanks to the advent of tactile technologies. Finding these mechanisms in our application is a crucial advantage since the player discovers a new playing environment without necessarily asking him to adopt a different way of playing (Fig. 2).

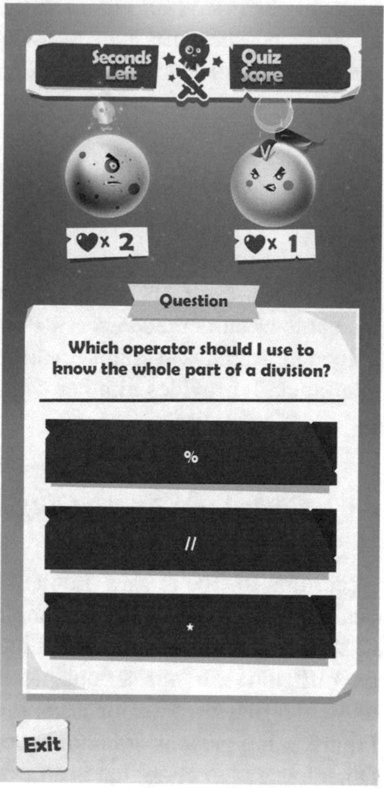

Fig. 2 Comprehension quiz in *The Stronglobes*

4 Conclusion

In this paper, we have presented a serious game for teaching Python programming language. We have integrated simple and intuitive techniques, allowing easy handling of the game. Once the game is available, we will conduct a preliminary test on a brunch of learners to collect their impressions and opinions in order to improve the game and meet their expectations. The question to be asked now is the future of the game and its evolution. Knowing that the learning methods evolve constantly, we must then evolve our game in a way that would make it adapted to the expectations of future learners. Indeed, the evolution could take place on several levels: a multi-player dimension for a network gaming experience, the integration of other notions of different languages to create wealth in content and more vivid interest from a wider audience.

Fig. 3 Code writing in *The Stronglobes*

References

1. Gouws, L.A., Bradshaw, K., Wentworth, P.: Computational thinking in educational activities. In: Proceedings of the 18th ACM Conference on Innovation and Technology in Computer Science Education—ITiCSE '13, p. 10 (2013)
2. Ruby Warrior—Popular Free Ruby Programming Tutorial Game [Online]. Available: https://www.bloc.io/ruby-warrior#/. Accessed 5 Mar 2019
3. Miljanovic, M.A., Bradbury, J.S.: Robot on! In: Proceedings of the 5th International Workshop on Games and Software Engineering—GAS '16, pp. 33–36 (2016)
4. Esper, S., Foster, S.R., Griswold, W.G., Herrera, C., Snyder, W.: CodeSpells. In: Proceedings of the 14th Koli Calling International Conference on Computing Education Research—Koli Calling '14, pp. 05–14 (2014)
5. CodeSpells, CodeSpells | Craft Custom Spells [Online]. Available: https://codespells.org/. Accessed 10 Sept 2017
6. Movetheturtle, Move the turtle—programming for kids on the iPhone and iPad [Online]. Available: http://movetheturtle.com/. Accessed 12 Sept 2017
7. Code Combat, CodeCombat—learn how to code by playing a game [Online]. Available: https://codecombat.com/. Accessed 5 Sept 2017
8. Wilson, C., Cameron, : Hour of code. ACM Inroads **5**(4), 22 (2014)
9. https://checkio.org
10. Gee, J.P.: What Video Games Have to Teach Us About Learning and Literacy. Palgrave Macmillan (2007)

Lip Movement Modeling Based on DCT and HMM for Visual Speech Recognition System

Ilham Addarrazi, Hassan Satori and Khalid Satori

Abstract This paper presents a system that recognizes the lip movement for lip-reading system. Four lip gestures are recognized: rounded open, wide open, small open and closed. These gestures are used to describe visually the speech. Firstly, we detect the mouth region from frame using Viola–Jones algorithm. Then, we use DCT to extract the mouth features. The recognition is performed by a HMM which achieves a high performance of 84.99%.

Keywords Audiovisual speech recognition · Automatic speech recognition · Lip reading · Viseme · Features extraction

1 Introduction

The performance of automatic speech recognition (ASR) systems degrades in noisy environment. The limitations in ASR systems can be reduced by adding other sources of information provided by the visual information such as speaker's lip movement. The audiovisual speech recognition (AVSR) system is made by the addition of the visual speech information (lip-reading) to the auditory information in order to improve the accuracy of the speech when the noise is present [1].

However, lip-reading, called also visual speech recognition (VSR), is used to interpret speech using the movement of speaker's lips. It plays an essential role in AVSR system realization. In view of the fact that the VSR is not influenced by the noise, it is considered as one of sources for improving the speech.

Considering the important interaction between the speech and visual lip movement, the information of mouth shape can be very useful for understanding the speech

I. Addarrazi (✉) · H. Satori · K. Satori
Department of Mathematics and Computer Science FSDM, USMBA, Fez, Morocco
e-mail: ilham.adrz@gmail.com

H. Satori
e-mail: hssatori@gmail.com

K. Satori
e-mail: khalidsatori@gmail.com

V. Bhateja et al. (eds.), *Embedded Systems and Artificial Intelligence*,
Advances in Intelligent Systems and Computing 1076,
https://doi.org/10.1007/978-981-15-0947-6_38

contents. However, visemes are determined as a lip shape used during visual speech processing. It can be considered as the basic units of visual information in form of articulatory mouth shapes of the speech. It is simply represented as movement of lip shape (e.g., the viseme corresponds to phoneme /m/ is represented as a closed mouth). To study all viseme categories, the movement of lip shape is recognized first and then is mapped to phoneme.

The aim of this work is building a system for recognizing the four famous movement of mouth shape which a human eye can notice during speech: rounded open, wide open, small open and closed. These four categories will be used as the characteristic of speech in lip-reading recognition system. Our proposed system is based on DCT extracted mouth features and modeled using HMM to distinguish the four mouth states.

The paper is organized as follows. Section 2 describes the related works. Section 3 presents the face and mouth detection task. Section 4 gives the extracted features from mouth using DCT. Section 5 describes the modeling of mouth shape using HMM. The experimental results and data database description are discussed in Sect. 6. The last section refers to conclusions and future works.

2 Related Works

Several VSR systems have been proposed in the literature. The first system was developed by Petajan et al. [2]. The authors exploited the geometrical feature of the mouth for constructing a lip-reading system to enhance the speech recognition system.

The lip-reading system proposed by Puviarasan et al. [3] uses discrete cosine transform (DCT) and discrete wavelet transform (DWT) to extract the visual features. Thereafter, in order to recognize the visual speech, the extracted features are used as an input to the hidden Markov model (HMM). The experimental results of this system show that the use of DWT gives a high performance compared to the DCT (91.0% for the DCT and 97.0% for DWT).

Shin et al. [4] propose a real-time lip-reading system to be integrated with an ASR system in order to recognize isolated Korean words. This system uses active appearance model (AAM) to extract the geometrical features which are recognized using three classifiers: HMM, ANN and K-NN.

Petridis et al. [5] propose an end-to-end VSR system based on long short-memory (LSTM) networks. This model composes of two streams: extract features from the mouth pixels and difference images. The temporal dynamics in each stream are created by an LSTM, and the fusion of the two streams is done using bidirectional LSTM (BLSTM) approach.

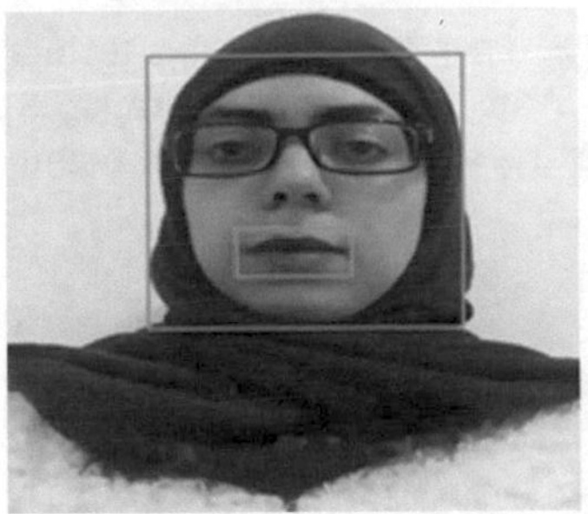

Fig. 1 Results of detection of face and mouth using Viola–Jones method

3 Face and Mouth Detection

The first task on the lip motion recognition is to identify a region of interest (ROI), which is a rectangle around the mouth region of the speaker. This part includes the most visual speech information in lip-reading systems. To detect the ROI, the face is detected in face in advance.

In order to detect the face of speakers, Viola–Jones approach is used considering its high detection accuracy and its capacity to execute in real time [6]. This algorithm finds the speaker's face in the image by generating a new representation named integral image. This representation aims to compute quickly Haar features. Thus, these features are selected and trained using Adaboost. The weak classifiers trained are cascaded to a strong classifier which is used to attain the detection at the end. As shown in Fig. 1, the Viola–Jones algorithm returns a rectangle centered on the detected face region. The same procedure is applied to detect mouth from the detected face [7].

The percentage of the detection using Viola–Jones algorithm reaches 99% and 96.6% for face and mouth, respectively.

4 Region of Interest Features Extraction

Feature extraction is the most important part in VSR systems. Ideally, the visual features must be more informative and robust, to be extracted. A large variety of techniques have been proposed for lip features extraction task. Generally, these techniques are classified into three general groups: (a) the appearance-based approaches where features are directly captured from the region of interest (ROI) pixels; (b) the geometric-based where the features take the form of the mouth shape, height, width and area; (c) hybrid approaches that the combination of (a) and (b). The use of discrete cosine transform (DCT) as an appearance-based features technique in lip-reading or AVSR systems gives a high performance. Thus, the proposed lip movement recognition system applied DCT to extract features from the mouth region of speakers.

4.1 DCT

Feature extraction by the DCT technique consists of two steps: (a) the extraction of DCT coefficients from the entire ROI image; (b) the selection of the DCT coefficient to construct the lip feature vectors.

DCT coefficients

DCT is a method that transforms the pixel values of an image into its elementary frequency components. The application of DCT on each ROI image produces DCT coefficient.

For an $M * N$ image, the DCT coefficients are computed as follows:

$$F(u, v) = a_u a_v \sum_{x=0}^{M-1} \sum_{y=0}^{N-1} f(x, y) \cos\frac{(2x+1)u\pi}{2N} . \cos\frac{(2X+1)v\pi}{2N} \quad (1)$$

where

- $F(u, v)$: are the DCT coefficients.
- $f(x, y)$: is the intensity of the pixel in row x and column y.
- And:

$$a_u = \begin{cases} \frac{1}{\sqrt{M}}, & u = 0 \\ \sqrt{\frac{2}{M}}, & 1 \le u \le M-1 \end{cases} \quad (2)$$

$$a_v = \begin{cases} \frac{1}{\sqrt{N}}, & v = 0 \\ \sqrt{\frac{2}{N}}, & 1 \le v \le N-1 \end{cases} \quad (3)$$

DCT generates a coefficients matrix which has the same dimension of the input ROI image. The high coefficients value is located in the upper left corner and low-value coefficients in the bottom right of the matrix (see Fig. 2).

DCT zigzag coefficients

The DCT coefficient is extracted in a zigzag way and saved in a vector sequence as shown in Fig. 3. The aim of the zigzag scanning is to arrange low frequency coefficients in top of vector. By applying DCT, an image of a mouth is represented by the vector of features.

Fig. 2 Applying DCT on ROI image

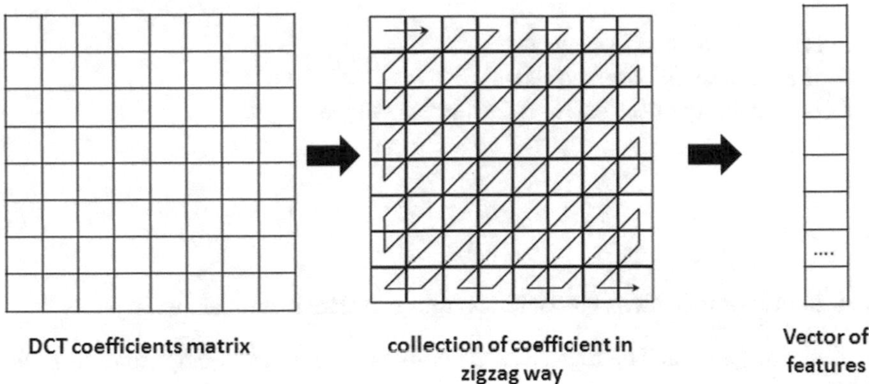

Fig. 3 Procedure of zigzag scanning

To reduce the necessary calculations for the recognition, only 100 coefficients are used to represent the mouth image. This vector will be provided to a classifier for classification.

4.2 Clustering Using K-Means

K-means is a clustering algorithm that separated observations into k clusters. The aim of this technique is to determine groups of associated features. By applying k-means, '*K*' values, named means, are computed. These values are used to classify each item to its corresponding cluster. Thus, '*k*' represents the number of groups or clusters that are used for classification.

The processes mentioned above convert each frame of the input lip image into a DCT vector of 100 dimensions. These features are clustered into four clusters to represent the four states of mouth.

5 Hidden Markov Models (HMM)

Hidden Markov models (HMM) have been widely used for many classification and modeling issues such as speech recognition [8], voice interface in VOIP network [9], audiovisual speech recognition [1] and voice recognition system [10].

Hidden Markov models can be characterized with the parameter set $\lambda = (N, M, \Pi, A, B)$:
where:

- N is a finite set of hidden states.
- M is a finite set of observed states.
- Π is probability of starting in hidden states, where:

$$\sum_{i=1}^{N} \Pi_i = 1 \tag{4}$$

- A: Matrix of transitions probabilities from one state to another, with:

A ➔ $\{a_{ij}\}_{N*N}$ and a_{ij} represents the probability of transitioning from state i to j where

$$a_{ij} = p(X_t = j | X_{t-1} = i) \tag{5}$$

- B: Matrix of probabilities of observations, with $B = \{b_{ij}\}_{N*M}$ where b_{ij} represents the probability of state i in N emitting observation j in M.

For training, the HMM is created for every movement of mouth shape using Baum–Welch algorithm. In this work, four HMMs are trained in order to recognize four movements of the mouth using observation sequence O obtained from the features vectors extracted. Regarding testing step, an unknown movement of mouth data is recognized. Four probability values are computed, and the HMM of movement which has the high probability value is accepted as the recognition result.

6 Experimental Results

6.1 AmDigit_AVSR Database Description

To validate the performance of our proposed shape movement recognition method, we select the samples from the AmDigit_AVSR database [11]. The fact that the first ten Amazigh digits contain the all proposed movements of the mouth and all these categories of mouth movement will be mapped into Amazigh viseme, encourage us to use this database: AMYA, YEN, SIN, KRAD, KOZ, SMMUS, SDES, SA, TAM

Table 1 AmDigit_AVSR database description

Parameter	Value
Sampling rate	16 kHz
Audio data file format	.wav
Video data file format	.avi
Resolution	1280 * 720
Images per second	25
Speakers	20 females, 20 males
Number of repetitions per word	10
Token number	4000

and TZA. Also, the AmDigit_AVSR database is the first audiovisual speech database for Amazigh language. This database consists of 4000 video and audio files for ten first digit uttered by 40 speakers. All parameters of AmDigit_AVSR are described in Table 1.

The images contained in the database facilitate the evaluation of the lip movement representations, which is the main goal of this work.

6.2 Experiment Result

In this work, we recognize the movement of lip shape to be used as the characteristic of some digit in lip-reading recognition system (viseme). The visual shape of the lip is represented by series of the visual vectors extracted from the ROI of speaker in AmDigit_AVSR database grouped into four clusters. Left-to-right HMM models with three states are used to model each category of shape movement. Four mouth movement shapes such as closed, wide open, small open and rounded open were tried to recognize in this paper. In this experiment, there are four HMM models trained to recognize all four categories of mouth shape movement. Figure 4 displays the four different shape of the mouth used in this proposed system.

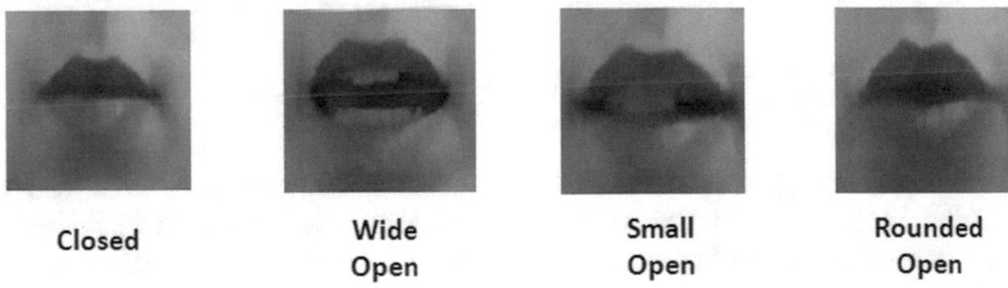

Fig. 4 Four shapes of the mouth used

Table 2 Confusion matrix showing the movement of mouth recognition results

		Predicted mouth shape				Omitted
		Closed	Wide open	Small open	Rounded open	
Actual mouth shape	Closed	**28**	0	2	0	0
	Wide open	0	**26**	2	1	1
	Small open	2	1	**23**	2	2
	Rounded open	0	2	3	**25**	0

Table 3 Recognition rate of mouth shape movement

Mouth shape movement	Recognition rate (%)
Closed	93.33
Wide open	86.66
Small open	76.66
Rounded open	83.33

For training, 70 image sequences for each mouth shape movement are collected from AmDigit_AVSR corpus. For testing, we collect 30 mouth image sequences from difference speaker who pronounced the Amazigh digits. Hence, there are 120 image sequences in total for test. The performance of the proposed system using HMM is presented using the confusion matrix in Table 2.

The total recognition rate for each movement can then be described as in Table 3.

Based on the results, the proposed approach is highly accurate for most movements. The low recognition rate for mouth shape movement 'small open' (76.66%) is due to the confusion to distinguish this shape movement with others. Hence, it is necessary to consider this constraint in future work.

The comparison of the proposed approach results is inadvisable by cause of the diverse databases used and recognition steps. According to [12], three lip movements are obtained: mouth opening, sticking out the tongue and forming puckered lips, by using an artificial neural network and different mouth region.

7 Conclusion

In this paper, a simple approach to recognize mouth shape movement is to be used to describe the characteristic of the speech in VSR system. The proposed work is able to detect faces and mouth of speakers using Viola–Jones approach. The DCT features of the mouth are extracted for classifying the shape of mouth. The recognition was based on HMM achieving a high performance of 84.99%.

As future work, the designed system will be used in VSR system to recognize speech visually using the information from mouth shape movement. Additionally, we will try to study other form of lips using other audiovisual databases.

References

1. Makhlouf, A., Lazli, L., Bensaker, B.: Evolutionary structure of hidden Markov models for audio-visual Arabic speech recognition. Int. J. Signal Imaging Syst. Eng. **9**, 55–66 (2016)
2. Petajan, E.D.: Automatic lipreading to enhance speech recognition (speech reading) (1984)
3. Puviarasan, N., Palanivel, S.: Lip reading of hearing impaired persons using HMM. Expert Syst. Appl. **38**, 4477–4481 (2011)
4. Shin, J., Lee, J., Kim, D.: Real-time lip reading system for isolated Korean word recognition. Pattern Recogn. **44**, 559–571 (2011)
5. Petridis, S., Li, Z., Pantic, M.: End-to-end visual speech recognition with LSTMs. In: 2017 IEEE International Conference on Acoustics, Speech and Signal Processing (ICASSP), pp. 2592–2596. IEEE (2017)
6. Wang, Y.Q.: An analysis of the Viola-Jones face detection algorithm. Image Process. Line **4**, 128–148 (2014)
7. Addarrazi, I., Satori, H., Satori, K.: Amazigh audiovisual speech recognition system design. In: 2017 Intelligent Systems and Computer Vision (ISCV), pp. 1–5. IEEE (2017)
8. Satori, H., ElHaoussi, F.: Investigation Amazigh speech recognition using CMU tools. Int. J. Speech Technol. **17**, 235–243 (2014)
9. Mohamed, H., Hassan, S., Ouissam, Z., Khalid, S., Naouar, L.: Interactive voice response server voice network administration using hidden Markov model speech recognition system. In: 2018 Second World Conference on Smart Trends in Systems, Security and Sustainability (WorldS4), 16–21. IEEE (2018)
10. Zealouk, O., Satori, H., Hamidi, M., Laaidi, N., Satori, K.: Vocal parameters analysis of smoker using Amazigh language. Int. J. Speech Technol. **21**, 85–91 (2018)
11. Ilham, A., Hassan, S., Khalid, S.: Building a first Amazigh database for automatic audiovisual speech recognition system. In: Proceedings of the 2nd International Conference on Smart Digital Environment ACM, pp. 94–99 (2018)
12. Dalka, P., Czyzewski, A.: Human-computer interface based on visual lip movement and gesture recognition. IJCSA **7**, 124–139 (2010)

Fast and Stable Computation of Charlier-Meixner's Bivariate Moments Using a Digital Filter

O. El Ogri, H. Karmouni, M. Yamni, A. Daoui, M. Sayyouri and H. Qjidaa

Abstract In this paper, we propose a new method for the fast and stable computation of Charlier-Meixner's bivariable moments by using the digital filters based on the Z transformation and the image block representation. To guarantee speed and numerical stability of the Charlier-Meixner's moments, we treat the images as a set of blocks where each block will be processed independently, and the Charlier-Meixner's moments are calculated from each block in each slice. The application of this method allowed us a significant reduction of the processed information and the space of the image by using the moments of Charlier-Meixner for very small orders. The performances of the proposed method are demonstrated through several simulations on different image bases.

Keywords Charlier-Meixner's moments · Image reconstruction · Image block representation

O. El Ogri (✉) · H. Karmouni · M. Yamni · A. Daoui · H. Qjidaa
CED-ST, STIC, Laboratory of Electronic Signals and Systems of Information LESSI, Dhar El Mahrez Faculty of Science, Sidi Mohamed Ben Abdellah-Fez University, Fez, Morocco
e-mail: omar.elogri@usmba.ac.ma

H. Karmouni
e-mail: hicham.karmouni@usmba.ac.ma

M. Yamni
e-mail: mohamed.yamni@usmba.ac.ma

A. Daoui
e-mail: achraf.daoui@usmba.ac.ma

H. Qjidaa
e-mail: hassan.qjidaa@usmba.ac.ma

M. Sayyouri
Engineering, Systems and Applications Laboratory, National School of Applied Sciences, Sidi Mohamed Ben Abdellah University, BP 72, My Abdallah Avenue Km. 5 Imouzzer Road, Fez, Morocco
e-mail: mhamed.sayyouri@usmba.ac.ma

V. Bhateja et al. (eds.), *Embedded Systems and Artificial Intelligence*, Advances in Intelligent Systems and Computing 1076,
https://doi.org/10.1007/978-981-15-0947-6_39

1 Introduction

The use of image moments is one of the most debated topics in image analysis and pattern recognition. The basic idea of the theory of moments is the projection of the data space on a complete orthogonal basis to extract often useful information. These are extracted as the projection coefficients called moments. All of continuous and discrete orthogonal moments have separable basic functions that can be expressed as two separate terms by producing the two same classical orthogonal polynomials with one variable. Recently, a novel set of discrete and continuous orthogonal moments based on the bivariate orthogonal polynomials have been introduced into the field of image analysis and pattern recognition [1, 2]. These series of bivariate polynomials are solutions of the second-order partial differential equations [1]. Most of the work has focused on discrete moments based on discrete bivariate orthogonal polynomials of Tchebichef–Krawtchouk [3, 4], Tchebichef–Hahn, Hahn–Krawtchouk [4], Charlier–Tchebichef, Charlier–Krawtchouk and Charlier–Hahn [5], these moments are better compared to the moments based on the same two classical orthogonal polynomials with a variable in terms of image representation capacity and robustness to noise, and no attention was paid to the study of discrete orthogonal moments based on the bivariate orthogonal polynomials of Charlier-Meixner. Although these works have accelerated the computation of moments and their inverse, these methods have shown limits for large images. To reduce the computational time cost of moments, several algorithms are introduced in literature [6–15], and no attention was paid to accelerate discrete orthogonal moments based on the bivariate orthogonal polynomials of Charlier-Meixner. These difficulties have led us to adopt, in this paper, a new precise and stable method to calculate the Charlier-Meixner's moments of an image by using digital filters [16–18] to reconstruct images. In this proposed method, we used the block image analysis method [6, 7]. This method is purely local, where each block is processed independently. Thus, the reduction of the image space allows the use of discrete orthogonal moments of weak orders during the calculation process. Indeed, the calculation of discrete orthogonal moments of the image by the blocks offers a quality of representation of the image. Thus, several simulations were performed to demonstrate the effectiveness of the proposed algorithms in terms of numerical stability, computation time and quality of the reconstruction and robustness vis-à-vis the noise.

This document is organized in several sections. The first defines the bivariate polynomials of Charlier-Meixner. The second introduces the digital filter to accelerate the calculation of the moments of Charlier-Meixner and its inverses. The third section proposes an image block representation associated with an IBR-ISR intensity slice representation in the representation of the image to calculate precise moments on each defined image block and an inverse transform; these blocks are finally assembled to reconstruct the original image. In the last section, we perform tests on gray-scale images using the proposed method, to demonstrate the effectiveness of our methods in terms of time and reconstruction quality.

2 Bivariate Discrete Orthogonal Polynomials of Charlier-Meixner

The bivariate discrete orthogonal of Charlier-Meixner is defined as the product of Charlier and Meixner discrete orthogonal polynomials with one variable as follows [8–10]:

$$\begin{gathered} CM_{n,m}(x, y, a_1, \gamma, \mu) = C_n^{a_1}(x) \times M_m^{(\gamma,\mu)}(y) \\ 0 \le n \le N-1; 0 \le m \le M-1 \end{gathered} \tag{1}$$

where $a_1 \succ 0$, $\gamma \succ 0$ and $0 \prec \mu \prec 1$ are strictly positive real.
whereas $C_n^{a_1}(x)$ and $M_m^{(\gamma,\mu)}(y)$ are the discrete orthogonal polynomial with one variable of Charlier and Meixner. The bivariate discrete orthogonal of Charlier-Meixner is orthogonal on the set $D = \{(n, m) : 0 \le n \le N-1;\ 0 \le m \le M-1\}$ with respect to the weight function of the Charlier-Meixner discrete orthogonal polynomials is defined as:

$$\omega_{CM}(x, y) = \omega_C(x) \times \omega_M(y) = \frac{e^{-a_1} a_1^x}{x!} \times \frac{\mu^x \Gamma(\gamma + y)}{\Gamma(1 + y)\Gamma(\gamma)} \tag{2}$$

In the next subsections, we will present the discrete orthogonal polynomials with one variable of Charlier and Meixner [8–10].

2.1 Charlier Polynomials

The nth Charlier polynomials are represented by using hypergeometric function as [6, 8, 9]:

$$C_n^{a_1}(x) = {}_2F_0\left(-n, -x; -\frac{1}{a_1}\right) \tag{3}$$

Equation (3) can be expressed in terms of binomial $\binom{x}{k}$ as:

$$C_n^{a_1}(x) = \sum_{k=0}^{n} \Gamma_k(n, a_1) \binom{x}{k} \tag{4}$$

where $\Gamma_k(n, a_1)$ is a lower triangle matrix:

$$\Gamma_k(n, a_1) = \frac{(-n)_k}{(a_1)^k} \tag{5}$$

2.2 Meixner's Polynomials

The nth Meixner polynomials are represented by using hypergeometric function as [7, 10]:

$$M_n^{(\gamma,\mu)}(x) = (\gamma)_n\, {}_2F_1(-n, -x; \gamma; 1 - 1/\mu) \tag{6}$$

Equation (6) can be expressed in terms of binomial $\binom{x}{k}$ as:

$$M_n^{(\gamma,\mu)}(x) = \sum_{k=0}^{n} \Gamma_k(n, \gamma, \mu)\binom{x}{k} \tag{7}$$

where $\Gamma_k(n, \gamma, \mu)$ is a lower triangle matrix:

2.3 Charlier-Meixner's Moments

For a digital image $f(x, y)$ with size $N \times M$, that is $x \in [0, N-1]$ and $y \in [0, M-1]$, the $(n+m)$th order Charlier-Meixner's moments is defined as follows:

$$\mathrm{CMM}_{nm} = \sum_{x=0}^{N-1}\sum_{y=0}^{M-1} \widetilde{C}_n^{a_1}(x)\widetilde{M}_m^{(\gamma,\mu)}(y) f(x, y) \tag{8}$$

We can replace it by for later purpose:

$$\mathrm{CMM}_{nm} = A_{nm}\sum_{x=0}^{N-1}\sum_{y=0}^{M-1} C_n^{a_1}(x) M_m^{(\gamma,\mu)}(y) \tilde{f}(x, y) \tag{9}$$

where

$$A_{nm} = \frac{1}{\sqrt{\rho_{CM}(n, m)}} = \frac{1}{\sqrt{\rho_C(n)\rho_M(m)}} \tag{10}$$

with $\rho_C(n)$ and $\rho_M(m)$ is the squared norm of Charlier and Meixner, respectively, as [8–10]

$$\tilde{f}(x, y) = \sqrt{\omega_C(x)\omega_M(y)}\, f(x, y) = \sqrt{\omega_{CM}(x, y)}\, f(x, y) \tag{11}$$

3 Fast Computation of Charlier-Meixner's Moments Based on Digital Filters

In this section, it is shown that a cascaded feed-back digital filter outputs can be sampled at the earlier time intervals. The second subsection is meant to show how the Charlier-Meixner's moments can be computed efficiently in terms of the cascaded digital filter outputs. Finally, a matrix representation of *2D* Charlier-Meixner's moments expresses the direct calculation of these moments based on the filter outputs.

3.1 Digital Filter Outputs

The transfer function of a cascaded filter $H_n(z) = \frac{1}{(1-z^{-1})^{n+1}}$ can be created by a series of n cascaded accumulators. Such a filter as [6, 7], given an input the image $f(x)$, produces a sampled output at earlier points of $N - n - 1$ as

$$y_n(N-n-1) = \sum_{x=0}^{N-1} f(x)\binom{x}{n} \tag{12}$$

This formulation shows that the digital filter outputs are sampled at earlier points, $N-1, N-2, N-3 \ldots N-n-1$. Meanwhile, this set of output values starts to decrease after $n/2$ moments orders. The *2D* accumulator grid as [6, 7] and consists of one row and $n+1$ columns based on single-pole single-zero filters. The filter output for an input scaled image $\tilde{f}(x, y)$ with size $N \times M$ is given by the following expression:

$$y_{nm}(N-n-1, M-m-1) = \sum_{x=0}^{N-1}\sum_{y=0}^{M-1} \tilde{f}(x, y)\binom{x}{n}\binom{y}{m} \tag{13}$$

3.2 Computation of Charlier-Meixner's Moments Using a Digital Filter

In this subsection, a direct relationship between Charlier-Meixner's moments and digital filter outputs is derived.

Then, by using the *2D* Charlier-Meixner's moments definition in Eq. (9), the binomials form of Charlier and Meixner polynomials in Eqs. (4) and (7) we have

$$\mathrm{CMM}_{nm} = A_{nm} \sum_{k=0}^{n} \sum_{l=0}^{m} \Gamma_k(n, a_1)\Gamma_l(m, \gamma, \mu) y_{kl} \tag{14}$$

In theory, an original image function $f(x, y)$ can be represented from an infinite series of Charlier-Meixner's moments.

$$f(x, y) = \sum_{n=0}^{\infty} \sum_{m=0}^{\infty} \mathrm{CMM}_{nm} \widetilde{C}_n^{a_1}(x) \widetilde{M}_m^{(\gamma,\mu)}(y) \tag{15}$$

4 Computation of Meixner's Moments and Its Inverse by IBR and ISR

In this subsection, we will first present a brief introduction of the image block representation (*IBR*) for binary images. In the second subsection, we will portray a detailed description of the image slices representation (*ISR*) for gray-scale images. In the last subsections, we will use the *IBR* algorithm and *ISR* algorithm for the computation of Charlier-Meixner's moments and its inverse.

4.1 Binary Images

By applying the *IBR* algorithm [7, 19], the binary image is described as the following relation:

$$f(x, y) = \{b_i, i = 1, 2, \ldots\ldots, K\} \tag{16}$$

where b_i is the ith block and K is the total number of blocks as [19]. Each block is described by (x_{1,b_i}, x_{2,b_i}) and (y_{1,b_i}, y_{2,b_i}). These latter are the coordination of the upper left and down right corner in vertical and horizontal axes, respectively.

4.2 Gray-Scale Images

In this connection, the image is decomposed into a number of binary slices, and each slice is represented by a set of blocks [7]. These blocks are defined as a homogenous rectangular region. By applying the *ISR* approach, each gray-scale image is represented as a set of binary slices $f_i(x, y)$, and the image is described as the following relation:

$$f(x, y) = \sum_{i=1}^{L} f_i(x, y) \tag{17}$$

where $\boldsymbol{L}$ is the number of slices and f_i is the intensity function of the ith slice. In the case of a binary image, L is 1, and thus, $f(x, y) = f_1(x, y)$.

After decomposing gray-scale image into several slices into two levels, we can apply the algorithm *IBR*. The gray-scale image $f(x, y)$ can be redefined in terms of blocks of different intensities as follows:

$$\begin{aligned} f(x, y) &= \{f_i(x, y), i = 1, 2, \ldots\ldots, L\} = \{b_{ij}, j = 1, 2, \ldots\ldots, K_i\} \\ f_i(x, y) &= \{b_{ij}, j = 1, 2, \ldots\ldots, K_i\} \end{aligned} \tag{18}$$

where b_{ij} is the block of the edge i and K_i is a number of image blocks with intensity. Each block is described by $(x_{1,b_{ij}}, x_{2,b_{ij}})$ and $(y_{1,b_{ij}}, y_{2,b_{ij}})$. These latter are the coordination of the upper left and down right corner in vertical and horizontal axes, respectively.

4.3 Algorithm to Compute the Charlier-Meixner's Moments and Its Inverse via IBR and ISR

In order to improve the quality of the image reconstruction processes, we can reconstruct the whole image with a finite number of Charlier-Meixner's moments using cascaded digital filters. This fact can be accomplished by reconstructing each separate block as sub-image of size $N_i \times M_i$ using the image block representation for binary images (*IBR*) and image slice representation for gray-scale images (*IBR-ISR*). Finally, we all collect the sub-image to get the reconstructed image, as it is shown in Fig. 1.

Figure 1 summarizes the block diagram of the Charlier-Meixner moment and its inverse by *IBR-ISR* via digital filters.

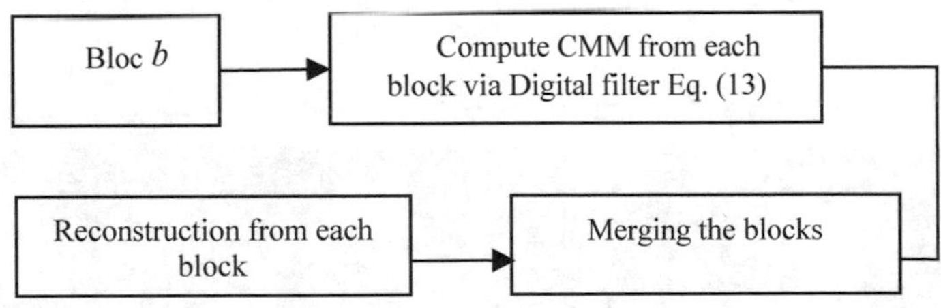

Fig. 1 Extraction of blocks using ISR algorithm

5 Experimental Results and Discussions

In this section, we give experimental results to validate the theoretical results developed in the previous sections. This section is divided into two subsections. In the first subsection, we will compare the time computation of Charlier-Meixner's moments by the direct, *LBBRM* [6] and the proposed method for gray-scale images. In the second part, we will test the ability of Charlier-Meixner's moments for the reconstruction of gray-scale images, and finally, we will compare the ability of image reconstruction by Charlier-Meixner's, Charlier and Meixner moments.

5.1 The Computational Time of Charlier-Meixner's Moments

In this subsection, we will compare the computational time of Charlier-Meixner's moments by three methods: the direct method based on Eq. (8), the proposed method based on *LBBRM* [6] and the application of algorithms *IBR-ISR*, as defined previously, by Eqs. (9) and (18).

In the example, a set of four gray-scale images with a size of 256×256 pixels, shown in (Fig. 2), are used. The number of blocks and slices of these images are $NB = 53701$, $NS = 216$ for Lena, $NB = 55308$; $NS = 239$ for House, $NB = 61610$; $NS = 222$ for Pirate and $NB = 53410$; $NS = 238$ for Lake.

The process of calculating discrete orthogonal moments Charlier-Meixner is carried out using:

- The slice-block (*IBR-ISR*) method: Knowing that each block is a rectangular region, the moments are calculated for orders ranging from 0 to the size of each block for each of the four images.
- The method of partitioning the image to blocks (*LBBRM*) [6]: each of the four images contains 4096 blocks of size $[4 \times 4]$ and 1024 blocks of size $[8 \times 8]$. The moments are calculated for orders ranging from 0 to 4 for blocks of size $[4 \times 4]$ and from 0 to 8 for blocks of size $[8 \times 8]$.

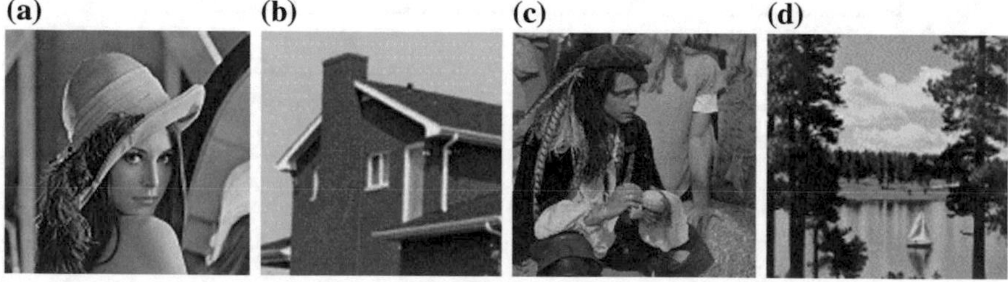

Fig. 2 Set of test gray-scale images **a** Lena, **b** House, **c** Pirate and **d** Lake

Table 1 Average computation time in second for gray-scale images using three methods

Method	Direct method	LBBRM size [8 × 8]	LBBRM size [4 × 4]	IBR-ISR
Average time (s)	0.0377	4.2109e − 004	2.2873e − 004	8.0874e − 005

Table 1 represents the average time of calculation of Charlier-Meixner discrete orthogonal moments for the four gray-scale images using the three methods located earlier. Note that the algorithm was implemented on a PC Dual Core 2.10 GHz, 2 GB of RAM.

The table shows that the proposed *LBBRM* method is faster than the direct one because the computation of Charlier-Meixner's moments by two methods depends only on the number of blocks instead of the image's size.

5.2 *Image Reconstruction Using Charlier-Meixner's Moments*

In this subsection, we will try to illustrate the performances of the proposed approaches through our algorithm. This later is tested on two gray-scale images of size 256 × 256 "Lena" and "Livingroom." The reconstruction test is performed through using the moments of Charlier-Meixner with the *LBBRM* and proposed method. We will calculate the mean squared error (*MSE*) as [6, 7] between the original and the reconstructed image. The method of reconstruction by *IBR-ISR*: The reconstruction order is equal to the block size. The advantage of this method is that only the intensity of the first and last pixels of each block is calculated. The reconstruction results of the test images using Charlier-Meixner's moments are shown in Table 2. Moreover, Charlier-Meixner polynomials depend on parameters μ and β which give a wide choice in the reconstruction.

Table 2 Mean squared error (MSE) and the reconstructed images

Method	Reconstruction by IBR-ISR	Reconstruction by LBBRM of size [4 × 4]	Reconstruction by LBBRM size [8 × 8]
Lena	MSE=0.0826	MSE= 0.5185	MSE= 0.7553
Livingroom	MSE=0.0176	MSE=0.4523	MSE= 0.7429

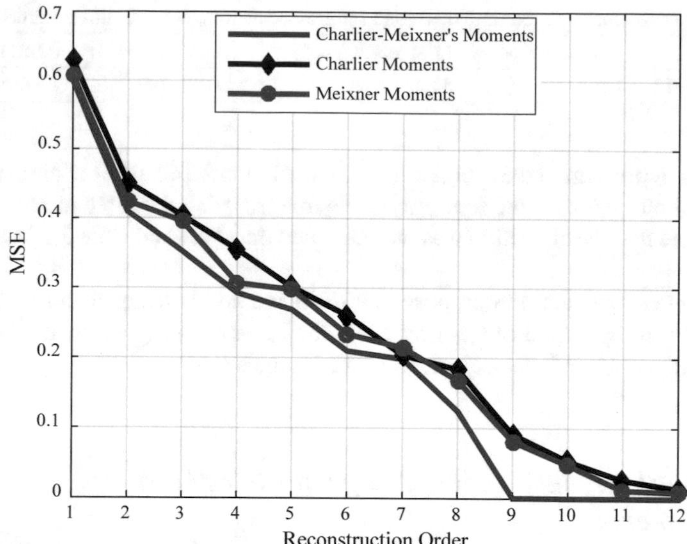

Fig. 3 The MSE using Charlier-Meixner, Charlier and Meixner's moments for gray-scale image "Lena" by proposed method

Table 2 shows that the reconstruction errors by the proposed method are smaller by using the Charlier-Meixner's moments. This also shows that the proposed method is effective in terms of the reconstruction quality for gray-scale images.

Finally, we will compare the reconstruction ability of Charlier-Meixner's moments with Charlier and Meixner's moments using the proposed method. For this, we compare the *MSE* for the two images: the gray-scale image "Lena" by Charlier-Meixner's, Charlier and Meixner moments. Figure 3 shows the plotted curve of *MSE* for the gray-scale image. For easier comparison, the three curves of *MSE* are plotted in the same figure. The results of these experiments show that Charlier-Meixner's moments have better reconstruction ability Charlier and Meixner moments.

6 Conclusion

In this paper, we have proposed a new method for computing Charlier-Meixner's moments that combines two main concepts, digital filters and block image representation. In this method, we adopted two algorithms based on partitioning the image of small regions. Thus, the Charlier-Meixner's moments are calculated from each block of the image. The significant reduction in the calculation time of the Charlier-Meixner's moments has allowed us to reduce the image reconstruction time and to increase the reconstruction quality for orders of very small moments, which shows the effectiveness of the proposed method. In terms of computing time and in terms of reconstruction quality.

References

1. Koornwinder, T.: Two-variable analogues of the classical orthogonal polynomials. In: Theory and Application of Special Functions, Proceedings of the Advanced Seminar Madison, pp. 435–495. University of Wiscons in Press, Academic Press (1975)
2. Dunkl, C.F., Xu, F.: Orthogonal Polynomials of Several Variables, vol. 81. Encyclopedia of Mathematics and its Applications. Cambridge University Press, Cambridge (2001)
3. Karmouni, H., Jahid, T., El Affar, I., Sayyouri, M., Hmimid, A., Qjidaa, H., Rezzouk, A.: Image analysis using separable Krawtchouk-Tchebichef's moments. In: International Conference on Advanced Technologies for Signal & Image Processing (ATSIP'2017), 22–24 May 2017, Fez, Morocco
4. Zhu, H.: Image representation using separable two-dimensional continuous and discrete orthogonal moments. Pattern Recogn. **45**(4), 1540–1558 (2012)
5. Hmimid, A., Sayyouri, M., Qjidaa, H.: Image classification using a new set of separable two-dimensional discrete orthogonal invariant moments. J. Electron. Imaging **23**(1), 013026, 18 (2014)
6. Karmouni, H., Hmimid, A., Jahid, T., Sayyouri, M., Qjidaa, H., Rezzouk, A.: Fast and stable computation of the Charlier moments and their inverses using digital filters and image block representation. Circuits Syst. Signal Process. (2018)
7. Jahid, T., Hmimid, A., Karmouni, H., Sayyouri, M., Qjidaa, H., Rezzouk, A.: Image analysis by Meixner moments and a digital filter. Multimed Tools Appl. (2017)
8. Karmouni, H., Jahid, T., Sayyouri, M., et al.: Fast reconstruction of 3D images using Charlier discrete orthogonal moments. Circuits Syst. Signal Process., 1–28 (2019)
9. Karmouni, H., Jahid, T., Sayyouri, M., et al.: Fast 3D image reconstruction by cuboids and 3D Charlier's moments. J. Real-Time Image Process., 1–17 (2019)
10. Jahid, T., Karmouni, H., Sayyouri, M., et al.: Fast algorithm of 3D discrete image orthogonal moments computation based on 3D cuboid. J. Math. Imaging Vision, 1–21 (2018)
11. Karmouni, H., Jahid, T., Lakhili, Z., Hmimid, A., Sayyouri, M., Qjidaa, H., Rezzouk, A.: Image reconstruction by Krawtchouk moments via digital filter. In: 2017 Intelligent Systems and Computer Vision (ISCV), pp. 1–7. IEEE (2017, April)
12. Jahid, T., Karmouni, H., Hmimid, A., Sayyouri, M., Qjidaa, H.: Image moments and reconstruction by Krawtchouk via Clenshaw's recurrence formula. In: 2017 International Conference on Electrical and Information Technologies (ICEIT), pp. 1–7. IEEE (2017, November)
13. Jahid, T., Karmouni, H., Hmimid, A., Sayyouri, M., Qjidaa, H.: Fast computation of Charlier moments and its inverses using Clenshaw's recurrence formula for image analysis. Multimedia Tools Appl., 1–19 (2018)
14. Karmouni, H., Jahid, T., Sayyouri, M., Hmimid, A., El-Affar, A., Qjidaa, H.: Fast and stable computation of the Tchebichef's moments using image Block representation and Clenshaw's formula. In: International Conference on Advanced Intelligent Systems for Sustainable Development, pp. 261–273. Springer, Cham (2018, July)
15. Karmouni, H., Jahid, T., Sayyouri, M., Hmimid, A., El-Affar, A., Qjidaa, H.: Image analysis by Hahn moments and a digital filter. In: International Conference on Advanced Intelligent Systems for Sustainable Development, pp. 707–718. Springer, Cham (2018, July)
16. Soundrapandiyan, R., Mouli, P.C.: An approach to adaptive pedestrian detection and classification in infrared images based on human visual mechanism and support vector machine. Arab. J. Sci. Eng., 1–13 (2017)
17. Bhateja, V., Urooj, S., Misra, M., Pandey, A., Lay-Ekuakille, A.: A polynomial filtering model for enhancement of mammogram lesions. In: 2013 IEEE International Symposium on Medical Measurements and Applications (MeMeA). IEEE (2013)
18. Bhateja, V., Misra, M., Urooj, S.: Non-linear polynomial filters for edge enhancement of mammogram lesions. Comput. Methods Programs Biomed. **129**, 125–134 (2016)
19. Hmimid, A., Sayyouri, M., Qjidaa, H.: Fast computation of separable two-dimensional discrete invariant moments for image classification. Pattern Recogn. **48**(2), 509–521 (2015)

Amazigh Digits Speech Recognition System Under Noise Car Environment

Ouissam Zealouk, Mohamed Hamidi, Hassan Satori and Khalid Satori

Abstract Automatic speech recognition (ASR) in Amazigh speech, particularly Moroccan Tarifit accented speech, is a less researched area. Some efforts to develop ASR on Moroccan accented Amazigh speech in the clean environment have been studied in our previous works. In this paper, we analyze the effect of noise car at different signal-to-noise ratio (SNR) on Amazigh accented on the first ten digits with different decibel values. Various techniques are used for isolated speech recognition like hidden Markov model (HMMs) and Mel-frequency cepstral coefficients (MFCC). The experimental result presents that recognition rate of 88.22% in a clean environment, and 59.26% and 33.83% rates in noisy condition at SNR 10 dB and 20 dB, respectively.

Keywords Speech recognition · Hidden Markov models · Noisy environment · Amazigh digits

1 Introduction

In the last few years, automatic speech recognition (ASR) systems become more popular as an input mechanism in several applications especially in word processors where dictation software is being introduced. Most of the research has been done in Arabic, English, Japanese, and other languages, but very few researches can be found in the Amazigh language. Some efforts to develop ASR on Amazigh speech in the clean environment have been reported in the literature [1, 2].

In most of the practical applications of automatic speech recognition (ASR), the input speech is contaminated by background noise where background noise has always posed a serious problem in the speech recognition system. For this reason, robust speech recognition has become an important focus area of speech research.

O. Zealouk (✉) · M. Hamidi · H. Satori · K. Satori
Laboratory Computer Science, Image Processing and Numerical Analysis, Faculty of Sciences Dhar Mahraz, USMBA, Fez, Morocco
e-mail: ouissam.zealouk@gmail.com

V. Bhateja et al. (eds.), *Embedded Systems and Artificial Intelligence*,
Advances in Intelligent Systems and Computing 1076,
https://doi.org/10.1007/978-981-15-0947-6_40

In this paper, we analyze the effect of noise on Amazigh digits. In the best of our knowledge, it is the first attempt to develop a noisy Amazigh speech recognition system-based noisy car environment. In the best of our knowledge, it is the first attempt toward developing a noise-robust Amazigh speech recognition system.

Apart from introduction in Sect. 1, the paper is organized as follows. Some of the related works are presented in Sect. 2. Section 3 gives an overview of Amazigh language. The speech recognition system is introduced in Sect. 4. Section 5 presents the system implementation. Finally, Sect. 6 investigates the experimental results. We finished by a conclusion.

2 Related Works

The authors in [3, 4] present that the approaches to robust noise fall into one of the two approaches. Feature enhancement ways try to eliminate the corrupting noise from the observations prior to recognition. There is a great algorithms number that falls into this category. In [5], different signals processing and acoustic feature extraction process are compared to find the most robust front-end configuration suitable for recognition in environments with various noise types and several signal-to-noise ratios (SNRs). As reported by [6], they aim to recognize the alphabets and digits of the Arabic language in noisy condition. Their work focused on analysis and investigation of Arabic alphadigits in the noisy environment from an ASR perspective. As a noisy speech simulation, they added white Gaussian noise to the clean speech at different signal-to-noise ratio (SNR) level. In another study [7], the researchers describe their speech experiments with a small vocabulary for an automatic continuous speech recognition system for Polish. They work in the computer game voice interface under clean and noisy conditions to analyze the noise impact on speech recognition performance. In [8], it was studied the noise robustness performance of deep neural network (DNN)-based acoustic models. To improve the accuracy, they propose three methods which are DNN analogs to feature-space, model-space noise adaptive training, and dropout training. On the other hand [9], authors describe the in-route navigation system. The achieved text from the speech recognition system can be utilized for hands-free applications by using smartphones. Li et al. in [10] provide a global overview of modern noise-robust methods for ASR as well as the pros and cons using noise-robust ASR in applications and future research in this domain. As ASR is applied to several applications, the authors give a view of ASR kinds and phases. The speech recognition process along with the hidden neural networks is demonstrated [11]. In [12], the Empirical Mode Decomposition (EMD) and Variational Mode Decomposition (VMD) algorithms are used for the speech enhancement in the noisy environment. They have been investigated that VMD outperforms EMD due to its self-optimization methods as well as adaptively using Wiener filter.

Table 1 The ten first digits with different transcriptions

Digits	English transcription	Arabic transcription	Tifinagh transcription	Syllables
0	AMYA	أميا	ⴰⵎⵢⴰ	VC-CV
1	YEN	يان	ⵢⴰⵏ	CVC
2	SIN	سين	ⵙⵉⵏ	CVC
3	KRAD	كراض	ⴽⵔⴰⴹ	VC-CVC
4	KOZ	كوز	ⴽⴽⵓⵣ	CVC
5	SMMUS	سموس	ⵙⵎⵎⵓⵙ	CC-CVC
6	SDES	سضيس	ⵙⴹⵉⵚ	CCVC
7	SA	سا	ⵙⴰ	CV
8	TAM	تام	ⵜⴰⵎ	CVC
9	TZA	تزا	ⵜⵣⴰ	CC-CV

3 Amazigh Language

The Amazigh language or Tamazight which is a branch of the Afro-Asiatic (Hamito-Semitic) languages nowadays covers the northern part of Africa which is extended from the Red Sea to the Canary Isles and from the Niger in the Sahara to the Mediterranean Sea.

In Morocco, this language is severed, due to historical, geographical, and sociolinguistic factors, into three principal regional sets which are Tarifite in North, Tamazight in Central Morocco and South-East, and Tachelhite in the South-West and the High Atlas [13–15]. Table 1 shows a description of the 10 first Amazigh digits used in our system.

4 Automatic Speech Recognition

An automatic speech recognition system works by pattern matching that digitizes pronounced words with computer models of speech patterns to produce a text transcription. ASR is the procedure which interprets human speech on a computer. An ideal automatic speech recognition system would operate speech signal into an error-free, word-for-word transcription of that speech [16]. However, due to a number of

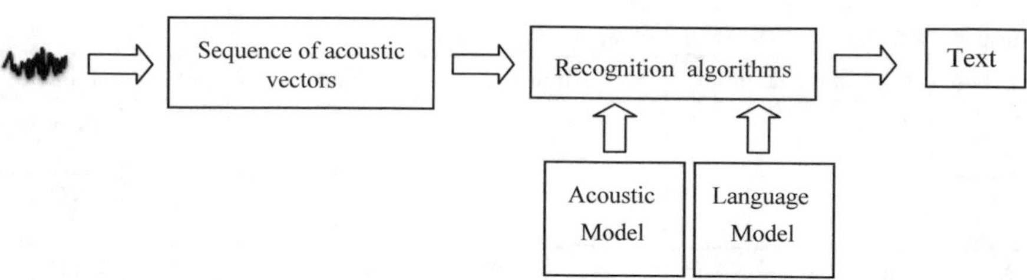

Fig. 1 ASR system components

factors, including the huge variations in normal human speech, perfect, verbatim transcription is not currently feasible. Figure 1 shows a block diagram of ASR system. Several projects have carried out on ASR systems in a car environment. It has been vastly applied to the vehicle operation command, speech-controlled navigation, and speech-controlled cabin systems [9].

4.1 Preprocessing

The speech signals analogic captured by a transducer such as a microphone or telephone must be digitized according to the Nyquist theorem. Form this last, the signal must be sampled more than twice the rate of the highest frequency intended in the analysis. Generally, a sampling rate among 8 and 20 kHz is utilized for speech recognition application. For normal microphones, 16 kHz sampling rate is used while 8 kHz is recommended for the telephonic channel [17].

4.2 Feature Extraction

Feature extraction is a method that allows finding a set of utterance properties which have an acoustic connexion with the speech signal [18]. In this process, the feature extractor keeps useful information and it discards the irrelevant one. To do this, successively some speech signal portion is used for processing, called window size. Data acquired in a window is named a frame. Generally, frame size ranges between 10 and 25 ms with an overlap of about 50–70% among consecutive frames. The data in this analysis interval is multiplied with a windowing function [19]. Several windows kinds like Rectangular, Hamming, Bartlett, Blackman, or Gaussian can be utilized. Then the features are extracted on the frame by frame basis. There are various techniques to extract features like LPCC, MFCC, and PLP [20].

4.3 Acoustic Model Generation

To recognize the unknown speech, some reference models are required to compare with them. These references called acoustic models. These last are divided into two kinds which are words and phoneme models [21], where the words are modeled as a whole. This model is utilized for small vocabulary system. The phoneme model is used for large vocabulary systems. In the phoneme model-based system, adding a new word to the vocabulary is manageable as sounds corresponding to the phone sequence of the newly added word may be already known to the system.

4.4 Language Model

Language model defines the utilized word in a speech application where each word must be included in the dictionary file. A language model refers to a set of limitations on the words sequence admitted in a given language [22]. These restrictions can be performed for example by the grammar rules via statistics on each word estimated on a training speech data. In fact, there are some words which have comparable sound phonemes humans normally do not find obstacles in the recognition phase. Provided this context to an ASR system is the goal of a language model.

5 System Implementation

This section describes the implementation of the speech system developed based upon the system architecture presented in the previous section.

5.1 System Description

In this section, a speaker-independent automatic Amazigh speech recognition system has been developed based on Mel-frequency cepstral coefficients (MFCCs) and HMMs in the car noise environment. The system has been implemented in the Ubuntu 14.04 operating environment which is a Linux platform. Also, it has been designed to recognize the ten first Amazigh digits. Initially, HMM learning has been performed using Sphinx tools. Then the unknown speech was transcribed using Sphinx 4.

Table 2 System details

Parameter	Value
Input file format	wave
Sampling rate	16,000 Hz
Bit rate (bits per sample)	16
Type of channel used	mono
Numbers of MFCC coefficients	13
Kind noisy	car
Decibels (dB)	10 and 20
Training token	1800
Clean testing token	600
Noisy testing token	600

5.2 Corpus Preparation

To train and test the speech recognition system, the data recorded from certain numbers of speakers. We use a microphone as an input device keeping a 5–10 cm distance between the speaker mouth and the microphone.

The system was trained using the voices of 18 Moroccan Amazigh people mixed among male and female of age group ranges between 20 and 40 years. The sounds were recorded at a sampling rate of 16 kHz on the monochannel using the Wavesurfer tool. Sixteen bits per sample were used which divides the element position of a sample into 65,536 (216) possible values. Speech files were saved in .wav format, where each speaker was asked to pronounce each digit of the vocabulary ten times, thus giving a total of 1800 (18 * 10 * 10) speech files. Table 2 shows more speech corpus technical details. Data was recorded in the room environment. Our testing data is divided into two sets. The test data is collected with the same specification used for training.

- The first data set includes the recorded voices in a clean environment (six speakers).
- The second set contains voices recorded in a noisy car environment (six speakers).

6 Results

After executing the Sphinx 4 with the generated acoustic models and language models, word utterances from each speaker were used to evaluate the recognizer. The system was tested on Baum–Welch algorithm and Viterbi algorithm. The performance was measured using combinations of HMM three states and 16 number of Gaussian mixture distribution. Table 3 presents all system recognition rates.

In the case of clean environment test, the system recognition rate for all digits is 88.22%. The best-found rate is 91.66% with "KRAD" digits and the lower one is "SA" with 85%.

Table 3 The total system performances

	Clean	10 dB	20 dB
AMYA	88.33	58.33	33.33
YEN	86.66	55.00	28.33
SIN	86.66	56.67	25.00
KRAD	91.66	80.00	63.33
KUZ	87.80	61.66	38.33
SEMMUS	90.00	51.67	31.67
SEDISS	90.00	46.67	23.33
SA	85.00	66.67	35.00
TAM	88.33	58.33	30.00
TZA	87.80	60.00	30.00
Total (%)	88.22	59.26	33.83

In the case of a noisy environment, depending on the testing corpus subset 600 tokens for the first ten digits, the system performances for 10 dB and 20 dB are 59.26% and 33.83%, respectively. By considering the digits recognition analysis, the most frequently recognized digit is KRAD for the both 10 and 20 dB. The more affected digits are SIN and SEDISS perhaps due to the included of some alphabets which not support this kind of noise.

7 Conclusion

In this paper, we have described a task-oriented speech recognition system for Amazigh language under clean and noisy conditions. The implementation system process is based on CMU Sphinx 4. We presented the steps to create the acoustic model and the language model, using both the grammar and the statistic N-gram model. The speech recognition in noisy conditions presents that the accuracy was hardly affected for 10 and 20 dB. The degradation of performance was observed if recognition was tested in-car environment.

References

1. Satori, H., ElHaoussi, F.: Investigation Amazigh speech recognition using CMU tools. Int. J. Speech Technol. **17**(3), 235–243 (2014)
2. Telmem, M., Ghanou, Y.: Estimation of the optimal HMM para recognition system usin. Procedia Comput. Sci. **127**, 92–101 (2018)
3. Macho, D., Mauuary, L., NoÈ, B., Cheng, Y.M., Ealey, D., Jouvet, D., Kelleher, H., Pearce, D., Saadoun, F.: Evaluation of a noise-robust DSR front-end on Aurora databases. In: Proceedings of the ICSLP, Denver, Colorado (2002)

4. Yu, D., Deng, L., Droppo, J., Wu, J., Gong, Y., Acero, A.: A minimum-mean-square-error noise reduction algorithm on Mel-frequency cepstra for robust speech recognition. In: International Conference on Acoustics, Speech and Signal Processing, pp. 4041–4044. IEEE (2008)
5. Kim, D.S., Lee, S.Y., Kil, R.M.: Auditory processing of speech signals for robust speech recognition in real-world noisy environments. IEEE Trans. Speech Audio Process. **7**(1), 55–69 (1999)
6. Alotaibi, Y., Mamun, K., Ghulam, M.: Noise effect on Arabic alphadigits in automatic speech recognition. In: IPCV, pp. 679–682 (2009)
7. Janicki, A., Wawer, D.: Voice-driven computer game in noisy environments. IJCSA **10**(1), 31–45 (2013)
8. Seltzer, M.L., Yu, D., Wang, Y.: An investigation of deep neural networks for noise. An investigation of deep neural networks for noise robust speech recognition. In: 2013 IEEE International Conference on Acoustics, Speech and Signal Processing, pp. 7398–7402. IEEE (2013)
9. Levitt, H.: Automatic speech recognition and its applications. In: Issues Unresolved: New Perspectives on Language and Deaf Education, p. 133 (1998)
10. Li, J., Deng, L., Gong, Y., Haeb-Umbach, R.: An overview of noise-robust automatic speech recognition. IEEE/ACM Trans. Audio Speech Lang. Process. **22**(4), 745–777 (2014)
11. Katyal, A., Kaur, A., Gill, J.: Automatic speech recognition: a review. Int. J. Eng. Adv. Technol. (IJEAT), 2249–8958 (2014)
12. Ram, R., Mohanty, M.N.: Comparative analysis of EMD and VMD algorithm in speech enhancement. Int. J. Nat. Comput. Res. (IJNCR) **6**(1), 17–35 (2017)
13. Fadoua, A.A., Siham, B.: Natural language processing for Amazigh language: challenges and future directions. In: Language Technology for Normalisation of Less-Resourced Languages, vol. 19 (2012)
14. Hamidi, M., Satori, H., Satori, K.: Implementing a voice interface in VOIP network with IVR server using Amazigh digits. Int. J. Multidisciplinary Sci. **2**, 38–43 (2016)
15. Mohamed, H., Hassan, S., Ouissam, Z., Khalid, S., Naouar, L.: Voice response server voice network administration using hidden Markov model speech recognition system. In: 2018 Second World Conference on Smart Trends in Systems, Security and Sustainability (WorldS4), pp. 16–21. IEEE, 2018
16. Zealouk, O., Satori, H., Satori, K.: Voice comparison between smokers and non-smokers using Amazigh digits. Int. J. Multi-disciplinary Sci. **1**, 106–110 (2016)
17. Kumar, K., Aggarwal, R.K., Jain, A.: A Hindi speech recognition system for connected words using HTK. Int. J. Comput. Syst. Eng. **1**(1), 25–32 (2012)
18. Hamidi, M., Satori, H., Zealouk, O., Satori, K.: Speech coding effect on Amazigh alphabet speech recognition performance. J. Adv. Res. Dyn. Control Syst. **11**(2), 1392–1400 (2019)
19. Gaikwad, S.K., Gawali, B.W., Yannawar, P.: A review on speech recognition technique. Int. J. Comput. Appl. **10**(3), 16–24 (2010)
20. Zealouk, O., Satori, H., Hamidi, M., Satori, K.: Speech recognition for Moroccan dialects: feature extraction and classification methods. J. Adv. Res. Dyn. Control Syst. **11**(2), 1401–1408 (2019)
21. Satori, H., Zealouk, O., Satori, K., ElHaoussi, F.: Voice comparison between smokers and non-smokers using HMM speech recognition system. Int. J. Speech Technol. **20**(4), 771–777 (2017)
22. Zealouk, O., Satori, H., Hamidi, M., Satori, K.: Voice pathology assessment based on automatic speech recognition using Amazigh digits. In: Proceedings of the 2nd International Conference on Smart Digital Environment, pp. 100–105. ACM (2018)

Detection of SMS Spam Using Machine-Learning Algorithms

Fatima Zohra El Hlouli, Jamal Riffi, Mohamed Adnane Mahraz, Ali El Yahyaouy and Hamid Tairi

Abstract Short message service (SMS) is considered as one of the most popular means of communication, it allows to the mobile phone users to exchange a short text message with a low cost. Its growing popularity and its dependence on mobile phone has increased the number of attacks, caused by sending an unsolicited message like SMS spam. In this paper, we address a comparative study, between multilayer perceptron (MLP), support vector machine (SVM), random forest and k-nearest neighbors (KNN). For extracting the feature vectors the bag-of-words (BOW) and the TF-IDF methods are applied. These feature vectors are used as input for training and testing the different machine-learning classifiers mentioned above. The results of different machine-learning classifiers, based on their accuracy, precision, recall, F-measure, and ROC (receiver operating characteristic) curve have shown that the MLP outperforms SVM, random forest and KNN in SMS spam detection. Although the MLP has achieved the highest accuracy by using the BOW, than by using the TF-IDF method.

Keywords SMS spam detection · Bag-of-words · TF-IDF · MLP · SVM · Random forest and KNN

F. Z. El Hlouli (✉) · J. Riffi · M. A. Mahraz · A. El Yahyaouy · H. Tairi
Laboratory Computer, Imaging and Numerical Analysis (LIIAN), Department of Computer Science, Faculty of Sciences, University Sidi Mohammed Ben Abdellah, 30050 Atlas, Fez, Morocco
e-mail: fzelhlouli@gmail.com

J. Riffi
e-mail: riffi.jamal@gmail.com

M. A. Mahraz
e-mail: adnane_1@yahoo.fr

A. El Yahyaouy
e-mail: ali.yahyaouy@usmba.ac.ma

H. Tairi
e-mail: htairi@yahoo.fr

V. Bhateja et al. (eds.), *Embedded Systems and Artificial Intelligence*, Advances in Intelligent Systems and Computing 1076,
https://doi.org/10.1007/978-981-15-0947-6_41

1 Introduction

SMS is a text communication platform, one SMS contains less than 160 characters. It is simple and inexpensive. It can be sent simultaneously to one or more mobiles, it no need Internet connexion like e-mail. SMS spam is generally any unwanted message sent to the mobile phone users. SMS spam is classified as scam or fraud, it can take form of text or a link to a number to call, a link to a website for more information (like card details, username, password), or a link to a website to download an application. It can include advertisement, promotions mostly on the end of the year holidays. The SMS spam is the best way to the attackers to steal the private information from the mobile phone users. So to ensure a high level of security and protection, different approaches are proposed.

Gupta et al. [1] aimed to compare eight different algorithm classifiers (SVM, NB, decision tree, logistic regression, random forest, AdaBoost, ANN, CNN) using two datasets collected from previous research [2, 3], and evaluate them basing to their accuracy, precision, recall, and CAP Curve. The experimental results show that convolutional neural network (CNN) method has achieved the highest accuracy of 99.10% and 98.25% for the two datasets.

Choudhary and Jain [4] have used ten features for classification (Presence of mathematical symbols, presence of URLs, presence of dos, presence of special symbols, presence of emotions, lowercased words, uppercased words, presence of mobile number, keyword specific, and message length), and five machine-learning algorithms namely: Naïve Bayes, logistic regression, J48, decision table and random forest. The results obtained are, that random forest classification algorithm has given a best results with 96.1% true positive rate and 1.02% false positive rate.

Uysal et al. [5] presented the impact of feature extraction on SMS spam classification combining bag-of-words model and six structural features (Message length, number of terms, uppercase character ratio, non-alphanumeric character ratio, numeric character ratio, and presence of URL), particularly for Turkish and English languages. Experimental work using Naïve Bayes and SVM classifiers indicated that the combinations of BOW and structural features has given better performance rather than bag-of-words features alone.

Popovac et al. [6] proposed CNN model composed on two convolutional layers, using TF-IDF method for extracting features, evaluated on Tiago's dataset in order to classify spam from not-spam message. The experimental results has showed that the CNN has provided better results on the same dataset with AUC score of 95.5% and accuracy of 98.4%, it has shown that CNN can be useful for SMS spam detection and similar classification.

In this paper, our aim at first is to pre-process the collection of SMS spam, and to extract the matrix of features using the BOW and the TF-IDF methods. In our work, the goal is to evaluate the performance of the different machine-learning algorithms, to provide the best algorithm for detecting SMS spam, also the best method for extracting features. The rest of this paper is organized as follows: Sect. 2 presents methods and materials those are used in this work. Section 3 presents the results and discussions. Section 4 contains the conclusion and the future work.

2 Methods and Materials

The main objective of our comparative study is to classify the SMS messages as soon as it received on the mobile phone. In this work, after pre-processing the dataset, we extracted the features from the messages (ham and spam) using Bow and TF-IDF methods. These feature vectors are used for training and testing purposes. Figure 1 shows the system architecture of detection SMS spam using machine-learning algorithms. In the testing phase, the classifier defines whether a new message is a spam or not.

2.1 *Data Pre-Processing*

Our dataset is a large text file, each line corresponds to a text message. The first manipulation of cleaning the text message is to make it lowercase, removing stop-words as 'and,' 'or,' 'in' and punctuations. The tokenization is required to transform a text message to a list of words (tokens), also stemmatization is necessary for converting the different forms for the same word in their root (goes, going, gone => go).

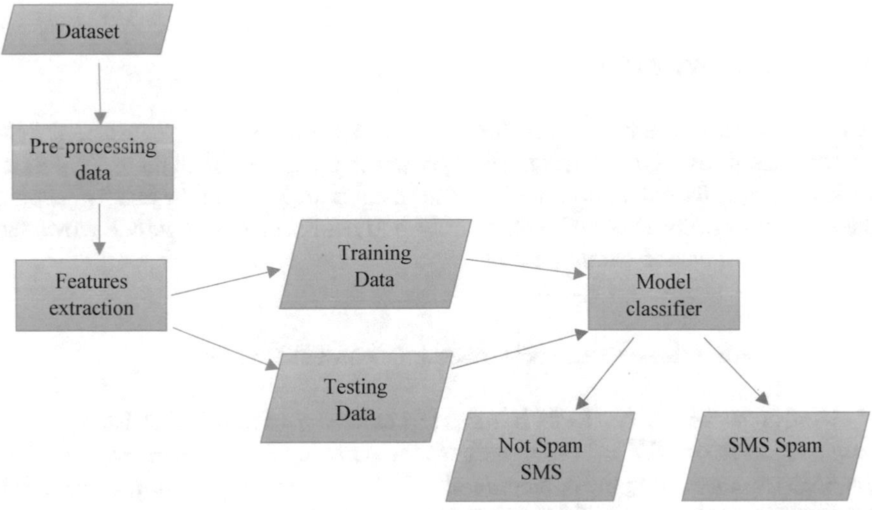

Fig. 1 Architecture of detection SMS spam using ML algorithms

```
def bag_of_words(tab_SMS):
  Initialize dic=[]
    for token in tab_SMS.tokenize():
      if(token not in dic):
       dic=dic+token
  numrows = len(tab_SMS)
  numcols = len(dic)
  bow_features = initialize the matrix with (numrows, numcols)
  for token in tab_SMS.tokenize():
    for word in dic :
      if(word in token):
        idx =getindex from word in token
        add the length of idx to bow_features in a row i
      else:
        add 0 to bow_features in a row i
    i=i+1
return bow_features
```

Fig. 2 Bag-of-words pseudo code

2.2 *Features Extraction*

After pre-processing the dataset, each SMS must be converted to a vector of features which the classifier can train with it, for this, bag-of-words and TF-IDF methods have been applied to transform each SMS of dataset to a vector of features. Before applying either of these methods, the first column of the dataset was changed. Ham and spam labels were converted to values 0 and 1. The two methods mentioned above are described as:

2.2.1 Bag-of-Words (BOW)

A bag-of-words representation is the one in which each SMS is represented by a vector of the size of the vocabulary |V|. The matrix composed of all these 5774 SMS which are input for our algorithms. It creates a unigram model of the text by keeping the number of occurrences of each word in a given document. Figure 2 shows the pseudo code of bag-of-words method.

2.2.2 TF-IDF (Term Frequency-Inverse Document Frequency)

If a word appears in several SMS, it is less representative of the SMS than a word that only appears in one SMS, TF-IDF reduces the number of occurrences of a word in the SMS by considering the number of all SMS in the dataset containing this word. Mathematical equations of TF * IDF are as follows [5]:

```
def  TF_IDF (bow_features):
  numrows = len(bow_features)
  numcols = len(bow_features [0])
  IDF= initialize the matrix with (numcls)
  TFIDF_Features= initialize the matrix with (numrows,numcols)
  N=Nbre of SMS in the dataset
  k=0
 for j in numcols:
    for i in numrows :
      if(bow_features [i][j]!=0):
        k+=1
    IDF[i][j]=log₂(N divide k)
 for i in numrows :
    for j in numcols:
      if(bow_features [i][j]!=0):
        TF= bow_features [i][j] divide the sum of bow_features in a row i
        add the TF*IDF[i] to TFIDF_Features in (i,j)
      else:
        add 0 to TFIDF_Features in (i,j)
return TFIDF_Features
```

Fig. 3 TF-IDF pseudo code

$$\mathrm{TF}(i, j) = \frac{(\text{Term } i \text{ frequency in document } j)}{(\text{Total terms in document } j)} \quad (1)$$

$$\mathrm{DF}(i) = \log_2\left(\frac{\text{Total documents}}{\text{documents with term } i}\right) \quad (2)$$

Figure 3 shows the pseudo code of TF-IDF method.

2.3 *Machine-Learning Classifiers*

After extracting features, accuracy is being evaluated using four supervised machine-learning algorithms: Multilayer perceptron, support vector machines, random forest, and k-nearest neighbors. These algorithms are described as follows.

2.3.1 Multilayer Perceptron MLP

A multilayer perceptron is a deep learning technique, so it is a feed-forward network connecting at least three layers of nodes in an oriented graph, an input layer, a hidden layer, and an output layer. An MLP uses backpropagation as a supervised learning technique that must a set of training data with the relatively desired outputs to adjust the weights iteratively using the back propagated errors function of the errors returned. The input layer contains the inputs features of the SMS. The first hidden layer receives the weighted inputs from the input layer and sends data from the previous layer to the next one. Finally, the output layer contains the classification result where the output ‘0’ indicates non-spam and the output ‘1’ indicates spam [7, 8].

The best value of parameters is {‘Hidden_Layer_Size’: 200, ‘alpha’: 0.001}.

Alpha is a parameter for regularization term.

The activation function for hidden layer is Relu, it returns

$$f(x) = \max(0, x) \tag{3}$$

The solver function by default is adaptive moment estimation (Adam) [8]:

$$m_t = \beta_1 m_{t-1} + (1 - \beta_1) g_t \tag{4}$$

$$v_t = \beta_2 v_t - 1 + (1 - \beta_2) g_t^2 \tag{5}$$

where g_t denotes the gradient at time step t of objective function. m_t and v_t updates biased first moment estimate and second raw moment estimate of the gradients, respectively.

$\beta_1, \beta_2 \in [0, 1]$ presents exponential decay rates for the moment estimates.

Good default settings for the tested machine-learning problems are $\eta = 0.001$, $\beta_1 = 0.9$, $\beta_2 = 0.999$ and $\epsilon = 10^{-8}$.

So for Computing bias-corrected first moment estimate and second raw moment estimate:

$$\widehat{m}_t = \frac{m_t}{1 - \beta_1} \quad \text{and} \quad \hat{v}_t = \frac{v_t}{1 - \beta_2} \tag{6}$$

To update parameters of function objective:

$$\theta_{t+1} = \theta_t - \frac{\eta}{\sqrt{\widehat{V}_t + \in}} \cdot \widehat{m}_t \tag{7}$$

2.3.2 Support Vector Machines SVM

Support vector machine is a supervised machine-learning approach in our work used for data classification. The objective of the support vector machine algorithm is to find the optimal hyperplane as the decision surface should maximize the distance between positive and negative data points. The radial basis function (RBF) kernel was preferred in this study due to its proven performance in text classification research before [8]. It is a reasonable first choice. This kernel nonlinearly maps samples into a higher dimensional space so it unlike the linear kernel case when the relation between class labels and attributes is nonlinear. SVC classifier using an RBF kernel has two parameters [9, 10]: gamma and C.

The best value of parameters of RBF are {'C': 10, 'gamma': 0.1}

While C is a penalty parameter of the error term, it controls the trade off between classifying the training data correctly and maximizing the margin of the decision function.

Gamma is a parameter for non linear hyperplanes. The higher the gamma value it tries to exactly fit the training data point, the 'curve' of the decision boundary is high, which creates islands of decision-boundaries around data points.

The principal goal of SVM, is to produce a best model based on the training data, and can be able to predict a class label of the test data.

Radial Basis Function (RBF):

$$K\left(x_i, x_j\right) = \exp\left(-\gamma \|x_i - y_i\|^2\right), \quad \gamma \text{ is a Kernel parameter } \gamma > 0 \tag{8}$$

$$x_i, y_i \in R^m \quad i = 1, 2, \ldots, m$$

$||x_i - y_i||^2$ is the squared Euclidean distance between two feature vectors x_i, y_i.

2.3.3 Random Forest

Random forests is supervised learning algorithm, it is a combination of decision trees. Random forest builds several decision trees and merges them to get a more accurate and stable prediction, to classify a new object which is performed by each tree, the trees marks their votes for that class. The class having most number of votes decides the classification label 'Spam or Ham.'

The best value of random forest parameters: criterion = 'gini' [11], n_estimators = 31 n_estimators is the number of trees in the forest.

Criterion is the function to measure the quality of a split, this parameter is tree-specific.

$$I_G(f) = \left(1 - \sum_{i=1}^{m} f_i^2\right) \tag{9}$$

where m is the number of classes and f_i is the probability that a tuple in training dataset D belongs to class Ci.

2.3.4 K-Nearest Neighbors KNN

KNN is a simple supervised machine-learning algorithm, it used for classification and regression problems, it is a special type of algorithm that does not use a statistical model. It is 'non-parametric,' and it is based only on training data (feature vectors and labels). This type of algorithm is called memory-based algorithm. More specifically, the principle of this model consists to classify data point based on the labels of the *k*-nearest neighbor by the majority vote. The distance function for KNN is a Euclidean distance [8]:

$$\mathrm{d}(x, y) = \|x - y\| = \sqrt{(x - y) \cdot (x - y)} = \left(\sum_{i}^{m} (x_i - y_i)^2 \right)^{1/2} \tag{10}$$

$$x, y \in R^m \quad i = 1, 2, \ldots.., m$$

where $||x - y||$ is the Euclidean distance between two data vectors *x*, *y*.

In our work, the best value of *K* is 13.

2.3.5 Evaluation Metrics

To evaluate a spam detection system, we considered the standard metrics, true positive rate, true negative rate, false negative rate, false positive rate, precison, recall, F-measure, ROC curve, as described as follows (Fig. 4):

- True positive rate (TPR): The percentage which the machine-learning classifier predicts correctly the SMS as spam.
- True negative rate (TNR): The percentage which the machine-learning classifier predicts correctly the SMS as not spam.
- False negative rate (FNR): The percentage which the machine-learning classifier predicts the SMS spam as not spam.
- False positive rate (FPR): The percentage which the machine-learning classifier predicts the not-spam SMS as SMS spam.
- Precision: It means the correctness, the percentage of messages that the classification algorithm classified as spam and it were spam. It is given as:

$$\text{Precision} = \frac{\text{TP}}{\text{TP} + \text{FP}} \tag{11}$$

Fig. 4 Confusion matrix

		SMS Spam	Not-Spam SMS
Predict	SMS as Spam	TP	FP
	SMS as not Spam	FN	TN

– Recall: It means the completeness, it calculates as:

$$\text{Recall} = \frac{\text{TP}}{\text{TP} + \text{FN}} \tag{12}$$

– F-measure, it is the harmonic mean of precision and recall.

$$\text{F} - \text{measure} = \frac{2 * \text{Precision} * \text{Recall}}{(\text{Precision} + \text{Recall}).} \tag{13}$$

– Receiver operating characteristics (ROC) curve—is plotted between true positive rate and false positive rate for different threshold values.

3 Results and Discussions

In this paper, we have used python language, Google Colaboratory tensorflow-gpu as runtime environment.

Before evaluating our machine-learning algorithms, our dataset is a collection of SMS spam collected from mobile phone research. It is defined as follows.

The dataset of SMS spam has been created by Tiago A. Almeida and José María Gómez Hidalgo. It contains 5574 messages in English [2], extracted from free sources for research.

425 SMS from Grumbletext website, 3375 SMS from the NUS SMS Corpus (NSC), 450 SMS from Caroline Tag's PhD Thesis and 1324 SMS from the SMS Spam Corpus v.0.1 Big. It contains a collection of 747 SMS spam and 4827 legetimate messages (ham). The dataset is a large text file contains one message per line. Each line is composed by two columns: column 1 contains the label (ham or spam) and column 2 contains the raw text.

After pre-processing and extracting the features, our aim is to search the best classifier for detecting the SMS spam. To choose the best parameters for each classifier, the GridsearchCV() function from scikit-learn library is implemented. We have used cross validation of tenfold in which 75% of data is used for training and 25% of data is used for testing the different model of machine-learning algorithms.

Figure 5 shows the accuracy scores using the BOW and the TF-IDF methods.

Figure 5 shows that MLP and SVM classifiers have the higher accuracy with the bag-of-words method comparing to random forest and KNN.

Figures 6 and 7 shows the receiver operating characteristics (ROC) curve for our proposed algorithms.

AUC—ROC curve is the area under the curve.

MLP and SVM have the same value of AUC-ROC in each figure. The higher values of AUC-ROC are, respectively, 94% in Fig. 6 and 93% in Fig. 7. These model classifiers are almost perfectly capable to distinguish between SMS spam and legitimate SMS with value of 94% and 93% (Table 1).

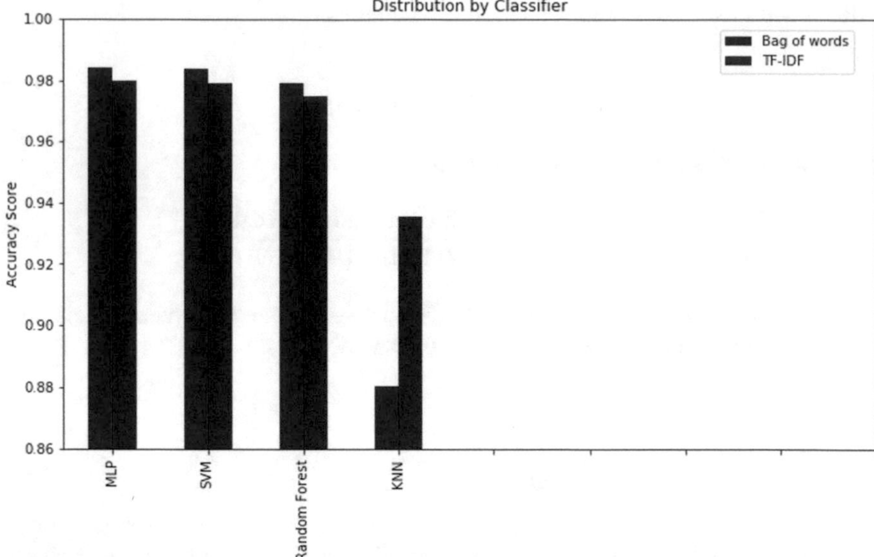

Fig. 5 Machine-learning classifiers performance

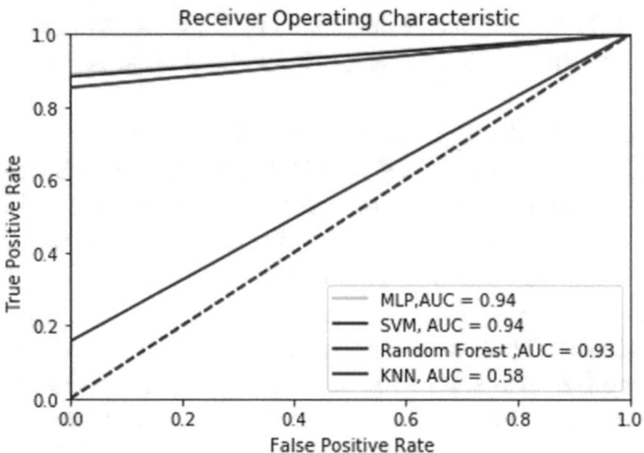

Fig. 6 ROC curve of ML. Classifiers using bow model

For bag-of-words method, KNN classifier shows the least precision rate 96.48%, the least recall rate 15.65%, and the least f-measure 56.06%, also a lower accuracy with value 88.01%. On the other hand, the MLP shows the best classification results with a higher accuracy of 98.42%, recall rate of 88.88%, precision rate of 99.37%, and f-measure of 94.11%.

For TF-IDF method, KNN algorithm improves their results of accuracy value 93.53% instead of 88.01%, recall rate of 56.21%, precision rate of 98.26% and f-measure of 71.51%; however, it shows usually the less results. While the MLP has

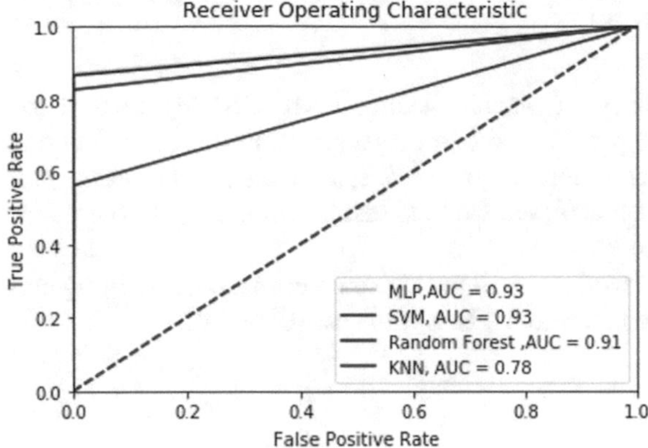

Fig. 7 ROC curve of ML. Classifiers using TF-IDF model

Table 1 Performance metrics of classifiers using bow and TF-IDF models

	Accuracy		Recall		Precision		F-measure	
	Bow (%)	TF-IDF (%)	Bow (%)	TF-IDF (%)	Bow (%)	TF-IDF (%)	Bow (%)	TF-IDF (%)
MLP	98.42	97.98	88.88	87.06	99.37	98.87	94.11	92.59
SVM	98.34	97.91	88.38	86.56	99.1	98.86	93.83	92.3
R. forest	97.91	97.48	85.35	82.58	99.31	98.7	90.09	90.46
KNN	88.01	93.53	15.65	56.21	96.48	98.26	56.06	71.51

the best accuracy of 97.98%, often the best recall, precision, and f-measure of values, respectively, 87.06%, 98.87% and 92.59%.

The MLP has the best results using the two methods of extracting features comparing to the other classifiers, but using the bag-of-words method, the MLP achieves the best accuracy (98.42% instead of 97.98%), recall (88.88% instead of 87.06%), precision (99.37% instead of 98.87%), and the best f-measure (94.11% instead of 92.59%).

Taking into consideration the results obtained in our work and comparing it with the paper [4], listed in introduction, MLP has produce better precison using the bag-of-words method on the same dataset. It shows that MLP can be useful for SMS spam detection and similar classification tasks.

4 Conclusion

Various algorithms of machine-learning MLP, SVM, random forest, KNN, was implemented to provide the best classifier using the bag-of-word and the TF-IDF methods, applied to the Tiago's SMS spam dataset. The results show that the multilayer perceptron achieves the best results comparing to the other classifiers with accuracy of 98.42%.

In our future work we will try to use other machine-learning classifiers, in order to give a best accuracy and a best value of AUC-ROC.

References

1. Gupta, M., Bakliwal, A., Agarwal, S., Mehndiratta, P.: A comparative study of spam SMS detection using machine learning classifiers. In: 2018 Eleventh International Conference on Contemporary Computing (IC3), pp. 287–293
2. SMS Spam Dataset 'Collection V.1'. Available online at http://www.dt.fee.unicamp.br/~tiago/smsspamcollection/
3. Spam SMS Dataset 2011–12. Available online on request at http://precog.iiitd.edu.in/requester.php?dataset=smsspam
4. Choudhary, N., Jain, A.K.: Towards filtering of SMS spam messages using machine learning based technique. In: Advanced Informatics for Computing Research: First International Conference, ICAICR 2017, Jalandhar, pp. 18–30, 17–18 Mar 2017 (Revised selected papers)
5. Uysal, A.K. Gunal, S. Ergin, S. Gunal, E.S.: The impact of feature extraction and selection on SMS spam filtering. 2013 Elektronika Ir Elektrotechnika, 67–72
6. Popovac, M., Karanovic, M., Sladojevic, S., Arsenovic, M., Anderla, A.: Convolutional neural network based SMS spam detection. In: 26th Telecommunications Forum TELFOR 2018, pp. 807–810
7. Goh, K.L., Lim, K.H., Singh, A.K.: Multilayer perceptrons neural network based Weh spam detection application. In:2013 IEEE China Summit and International Conference on Signal and Information Processing (ChinaSIP), Beijing, China, pp. 636–640
8. Kingma, D.P., Ba, J.L.: Adam: a method for stochastic optimization. In: ICLR 2015 Conference, pp 34–48
9. Kim, J., Kim, B., Savarese, S.: Comparing image classification methods: K-nearest-neighbor and support-vector-machines. In: American-Math'12/CEA'12 Proceedings of the 6th WSEAS International Conference on Computer Engineering and Applications, and Proceedings of the 2012 American conference on Applied Mathematics, pp. 133–138
10. Hsu, C.-W., et al.: A practical guide to support vector classification. Available at https://www.csie.ntu.edu.tw/~cjlin (2016)
11. Sedhai, S., Sun, A.: Semi-supervised spam detection in twitter stream. IEEE Trans. Comput. Soc. Syst., 169–175 (2018)

State of the Art of Deep Learning Applications in Sentiment Analysis: Psychological Behavior Prediction

Naji Maryame, Daoudi Najima, Rahimi Hasnae and Ajhoun Rachida

Abstract With the explosion of web 2.0, we are witnessing a sharp increase in Internet users such as a vertiginous evolution of social media. These media constitute a source of rich and varied information for researchers in sentiment analyses. This paper describes a tool for sifting through and synthesizing reviews by identifying the main deep learning techniques applied for sentiment analysis especially when it comes to psychological behavior prediction.

Keywords Sentiment analysis · Deep learning · Psychological behavior · Depression prediction

1 Introduction

Due to the development of electronic and information technologies, the volume of electronic text files has become too large for people to process manually. It has brought challenges and opportunities for the development of natural language processing techniques. As a sub-domain of data classification, text classification is the task of assigning predefined categories to free text documents. Among areas of its application, we note sentiment analysis [1] which is the focus of our research work.

Sentiment analysis has become very popular research area that has appealed huge burst of research activity for several years [2–8] because it has wide applications in the real world [1]. A common purpose of these studies is "to determine the attitude of a speaker or a writer with respect to some specific topic" [9]. Its techniques can be roughly divided into machine learning approach, lexicon-based approach, hybrid approach [10] and recently, we talk about deep learning.

Lexicon-based approach often uses predefined dictionaries of terms annotated with "positive" or "negative" scores [11]. So it is limited when it comes to texts with

N. Maryame (✉) · D. Najima · A. Rachida
Smart System Laboratory (SSL), ENSIAS, Rabat, Morocco
e-mail: maryame.naji@gmail.com

D. Najima · R. Hasnae
Lyrica, ESI, Rabat, Morocco

V. Bhateja et al. (eds.), *Embedded Systems and Artificial Intelligence*, Advances in Intelligent Systems and Computing 1076,
https://doi.org/10.1007/978-981-15-0947-6_42

small sizes, abbreviations, different contexts, empty words or ironic sentences. The strength of machine learning approach "lies in their ability to analyze the text of any domain and produce classification models that are tailored to the problem at hand" [11]. However, it suffers from some issues [12]: Dependency and consistency since it requires a lot of training data and it is domain-dependent [12].

To overcome these issues "Representation learning [-as substitute solution] is a set of methods that allow a machine to be fed with raw data and to automatically discover the representations needed for detection or classification." [13].

This paper describes a tool for sifting through and synthesizing reviews by identifying the main deep learning techniques used for sentiment analysis, especially, those most propitious in recent years. Indeed, it is a crucial step in fulfilling our research objective which is to develop an intelligent recommendation system. This last will benefit persons, through a positive accompaniment and the early alert, in case of complex situations (depression, suicide or feeling of revenge).

The rest of the paper is organized as follows: Sect. 2 introduces sentiment analysis and the main NLP problems to resolve to aim the best performance. Section 3 presents an overview of deep learning techniques for sentiment analysis, especially when it comes to psychological behavior prediction. Finally, Sect. 4 synthesizes our research work and shows the best DL model applied to sentiment analysis and responding to our research goal with high accuracy.

2 Sentiment Analysis Facing NLP Problems

"Sentiment analysis, also called opinion mining, is the field of study that analyzes people's opinions, sentiments, evaluations, appraisals, attitudes, and emotions toward entities such as products, services, organizations, individuals, issues, events, topics, and their attributes" [2]. In other words, the goal of this technique is mainly to achieve human-like performance. For that, [1] argued that there are (at least) 15 NLP problems that need to be solved. The idea was to represent sentiment analysis as suitcase composed of these problems and organized into three layers: syntactic, semantic, and pragmatic [8]. Figure 1 illustrates these problems and their level of relationship in the context of sentiment analysis.

With respect to syntactic layer, it aims to preprocess informal text by transforming it to text with basic sentences structure and plain language. This operation is divided into five mini processes: microtext normalization, sentence boundary disambiguation, part-of-speech tagging, text chunking, and lemmatization.

Besides, semantics layer aims to deconstruct the text obtained from the syntactic layer into concepts which is a key for semantic-aware analysis of the text. Thus we obtain the bag of concepts unlike the bag of words for the syntactic layer. This stage is divided on its part to fives mini processes which are word sense disambiguation, concept extraction, named entity recognition, anaphora resolution, and subjectivity detection.

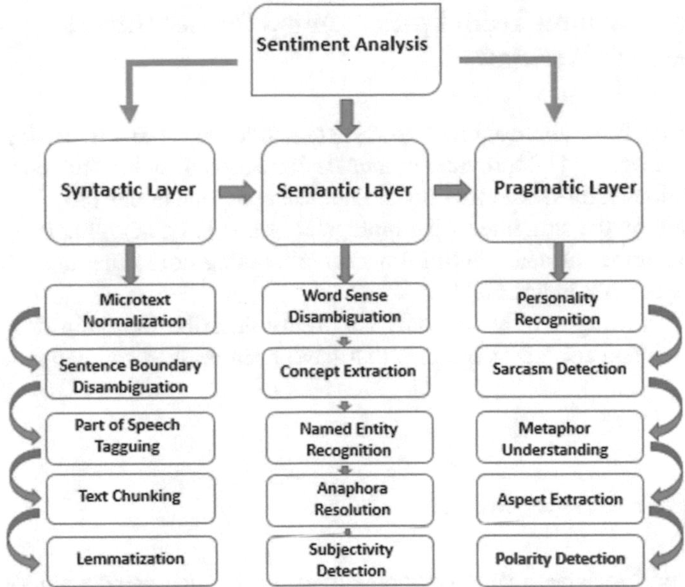

Fig. 1 The 15 NLP problems for SA

While, the goal of pragmatic layer lies on extracting meaning from both sentence structure and semantics obtained from syntactic and semantic layers, respectively. The idea here is to perform information about a user and interpret metaphors to extract the opinion targets and the polarity of text. Thus, for this stage there are also five mini processes: personality recognition, sarcasm detection, metaphor understanding, aspect extraction, and finally polarity detection.

Polarity detection is the most popular sentiment analysis task. In fact, many research works even use the terms "polarity detection" and "sentiment analysis" interchangeably. "This, however, is just one of the many NLP problems that need to be solved to achieve human like performance in sentiment analysis."

In our context, we focus on analyzing texts from social media and especially from medical social media. So according to our previous article [7], we have listed many problems that researchers would face when conducting sentiment analysis on those spaces. Indeed, we are interested in our research team, in the development of intelligent systems that analyze sentiments in health forums. Our goal is communicating with those spaces users and recommending them services that reply to their expected needs.

In the rest of this paper, we attempt to introduce deep learning, which has known a great success on sentiment analysis, by sifting through and synthesizing some researcher's reviews.

3 Deep Learning Techniques Applied for Sentiment Analysis: Overview

Deep learning is the application of neural networks which learn tasks using networks of multiple layers [14]. It provides automatic extraction of richer and more powerful representation features than traditional feature-based techniques [15].

"Inspired by the structure of the biological brain" [14], neural networks consist of a large number of neurons (information processing units) that are organized in layers, which work in unison [14].

In the following several sections, we briefly describe the main deep learning architectures and related techniques that have been applied to sentiment analysis tasks.

3.1 Deep Learning Models

Deep learning includes many networks architectures such as convolutional neural networks (CNNs), recurrent neural networks (RNNs), recursive neural networks (RecNNs) and many more [14].

CNN is a special type of feed-forward neural network originally employed in the field of computer vision. It plays the role of feature extractor [14] and includes three types of layers: convolution, pooling, and fully connected [16]. Unlike this last, RNN can use its internal "memory" to process a sequence of inputs, which makes it popular for processing sequential information. While RecNN, is a generalization of RNN.

3.2 Deep Learning for Sentiment Analysis

Researchers have mainly studied sentiment analysis at three levels of granularity: document level, sentence level, and aspect level [14]. Otherwise, "since the word is the basic computational unit of natural language" [17], we present word embedding which results are needed as input to subsequent sentiment analysis tasks [17].

Figure 2 gives some NLP challenges faced in sentiment analysis, distributing them into their groups based on the parsing level, at which these issues occur.

Word embedding is a representation that aims at representing aspects of word meaning and then capturing the rich relational structure of the lexicon.

About **Document-level sentiment classification**, many research and works used deep learning techniques. For example, [6] aimed to learn document representation by considering sentences relationships. The idea was to use a CNN or LSTM to represent sentences and then encode their semantics and relations in the document. In contrast, [18] focused on information about both user and product by using a neural network

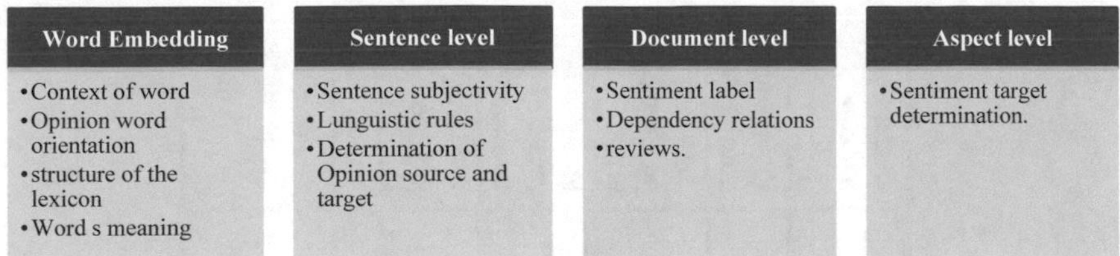

Fig. 2 SA and granularity level

model. This last can be divided into two sub-models: an LSTM to learn document representation and a deep memory network consisting of multiple layers to predict sentiment polarity of a document. On their side, [19] have implemented two different approaches for sentiment analysis of Arabic hotels' reviews: A supervised machine learning (SVM) approach and a deep learning approach (RNN). Results show that SVM performs RNN but this last is faster.

Sentence-level sentiment analysis focuses on classifying the sentiment polarities of a given sentence. As an example, [20] proposed an extension of the basic CNN model for better sentence representation by enlarging the number of CNN layers. Indeed, deeper neural networks structures enhance features encoding. From theme, [21] applied the same principle, but this time by investigating RecNN. In fact, RecNN is applied over the input word embedding and the framework proposed use three-layer RecNN that make the results of their study performing.

Different from predefined levels, **aspect-level sentiment classification** considers both the sentiment and the target information [14], then, this level becomes more challenging. Many researchers focused on this item and used deep learning techniques to overcome those challenges. Irsoy and Cardie [22], for example, proposed a neural network model for aspect phrase grouping that use simple multilayer neural networks to learn aspect representation. On their side, [23] proposed to extend LSTM by considering the target while LSTM can capture relations between the target and its context words. Then the given target is regarded as a feature which is concatenated with context features for aspect sentiment classification.

3.3 Comparison Studies

Based on the previous detailed study on different deep learning techniques applied to sentiment analysis, a comparison table for the predefined models is presented. Table 1 provides different evaluation for each technique according to different categories: date of article publication, field of application, data source, features used, DL models, scales (classification levels), and evaluation metrics.

About features, we discussed in our last research [7] that the improvement of sentiment analysis accuracy requires its personalization and then the consideration of specific factors: emotion object, user context, users influence, and their history, etc.

Table 1 Comparison studies about deep learning application for SA

Article	Field of app and data src	DL model	Comparison with	Scales (pt)	Features				**Layers**	**Evaluation metrics**
					Profile	Interest	Relations	Syntax		
[24]	Hotel reviews	RNN	SVM, Baseline	2	Gender, age.	Yes	Friends' expression	Yes	7	F1, accuracy, execution-time
[25]	Twitter	CNN	Baseline algorithms+LSTM	2	Behavioral information	No	Users' exchanges	Yes	5	Accuracy, precision, F1, recall
[26]	Spanish tweets	CNN	SVM NB	2	No	No	No	Yes	4	Precision, recall, F1 measure
[15]	Movie reviews from twitter	MG	Ensemble methods	2	No	No	No	No	–	Accuracy, precision, F1, recall
[27]	Arabic and English tweets	CNN and LSTM	SVM, NB, MaxEnt, random forest	5	Age, gender, conversations	No	Structure of social network	No	–	AvgRecall, F1, accuracy

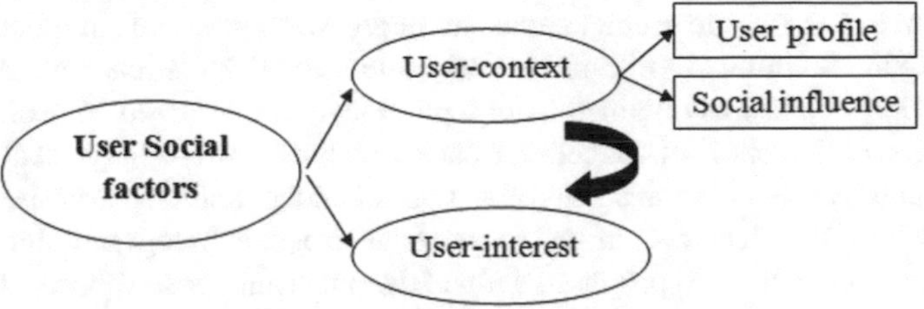

Fig. 3 User social factors

Figure 3 illustrates this idea. In our case, profile (age, sex, etc.), interest, user relations (relations deduced from exchanges), syntax (grammatical rules and expression relations) are given importance.

According to the Table 1, we note that CNN as deep learning model is more used and shows efficient performance and high accuracy in many studies directed toward sentiment analysis especially when it comes to accuracy as an evaluation metric.

Otherwise, it is clear that taking into consideration many features are not applied in all studies using deep learning to analyze sentiments; [15] as an example did not consider any feature or social factors introduced in our previous article [7].

Also, about analysis scales, most of the times researchers classify under two scales and attend the best results in terms of accuracy compared to the use of other machine learning basis algorithms. About the field of application and data source, deep learning is more applied to analyze reviews or tweeter user's expression.

4 Psychological Behavior Prediction: Depression

"Understanding emotions in textual conversations can be a challenging Problem in the absence of facial expressions and voice modulations" [28]. Depression as a behavior figure "is a mental illness often correlated with loss of interest, guilt feeling, low self-esteem, poor concentration, and in the worst case, having suicidal thoughts" [16]. About this illness, the health minister of Morocco (our country) has revealed that 26% of the population suffers from depression [29]. Besides, according to the conclusions of Public Health, France, nearly 9.8% is depressed with wide disparities depending on age, gender, or professional situation [30]. In America, major depression rose by a third between 2013 and 2016 [31].

According to those statistics, psychological behavior is a crucial step to fulfill our research objective which is to develop an intelligent recommendation system to benefit patients with chronic diseases in case of complex situations as depression.

Given its importance, depression prediction has been treated by many studies taking advantages of the increase of computing power and machine learning algorithms [16, 28, 32, 33].

The aim of [33] was to predict persistent depressive symptoms in older adults by using machine learning algorithms instead of traditional statistical approach (logistic regression). Data was obtained from a previously randomized controlled trial to evaluate the effectiveness of cooperative care for depression. Acharya et al. [28] proposed an application of the deep neural network concept and CNN for the diagnosis of depression. The data was in the form of electroencephalograph signals. From them, Chatterjee et al. [16] proposed a novel deep learning based approach to detect emotions—happy, sad, and angry in textual dialogues. The essence of their approach lies in combining both the sentiment and semantic representations of user expression by using LSTM. The objective of researchers in [32] was to compare performances of different machine learning algorithms for the screening of anxiety and depression among 470 seafarers. The data was obtained from interviews taking on consideration several sociodemographic and occupational factors.

Table 2 gathers those studies and shows, as a conclusion, that several articles have treated the subject of depression but not all in the same way. In fact, according to the literature review, few who interested in this topic by taking advantage of textual expressions while using deep learning as analyzing techniques. Nevertheless, we are observing that the real challenge in these studies is not the detection itself but rather the combination of feeling (depression in our case) with other features.

Table 2 Depression prediction research

Article	Data source	Features	Research subject	Analyzing techniques
[16]	Textual dialog from tweeter	Semantic presentation of expression	Detecting emotions (happiness, sadness, and anger)	Deep learning: LSTM
[28]	Electroencephalograph signals	–	Depression diagnosis	Deep learning: CNN
[32]	Interviews with seafarers	Sociodemographic factors (age, marital status, etc.)	Depression diagnosis	Machine learning (SVM, NB)
[33]	Previous randomized controlled trial (paper and pencil' questionnaire)	Baseline demographic data (age, sex, etc.) and psychometric data	Depression prediction	Machine learning (extreme gradient boosting)

5 Synthesis

Sentiment analysis has become a very popular research area that has appealed a huge burst of research activity for several years. Each study takes into consideration a particular component: factors, algorithm, and field of application.

As a result of our research, and to aim to our research objective predefined above we conclude the following:

- We are interested in analyzing patients' expressions and concluding psychological behavior symptoms especially those of depression;
- We are interested to use deep learning algorithms to analyze the textual expression and by integrating social factors to attend better analysis accuracy.

Indeed, according to the state of the art, psychological behaviors prediction is not treated yet the way we aim to do, which shows that the next steps will aim to achieve our interest.

6 Conclusion

This paper presents a state of the art related to sentiment analysis in general and psychological behavior prediction in particular (depression as a figure) which requires the use of sophisticated and new techniques. Thus, through this article, we introduced the main deep learning techniques used for sentiment analysis, especially, those most propitious in recent years.

Certainly, to develop an intelligent system able to benefit patients with chronic diseases, we need to analyze their sentiments to better recommend them solutions.

Obviously, whatever this field of research is evolving rapidly, it is not saturated yet. Indeed, psychological behavior detection by taking advantage of textual expressions and while using deep learning as analyzing techniques has not been studied previously. Thus, we attempt in our following researches to conceive a deep learning architecture for depression prediction from textual expression. Otherwise, we would not avoid integrating social factors that may improve analysis accuracy.

References

1. Wan, Y., Gao, Q:. An ensemble sentiment classification system of twitter data for airline services analysis. In: 2015 IEEE International Conference on Data Mining Workshop (ICDMW), IEEE, pp. 1318–1325 (2015)
2. Liu, B.: Sentiment analysis and opinion mining. Synthesis Lectures on Human Language Technologies, vol. 5, No. 1, pp. 1–167 (2012)
3. Pang, B., Lee, L., et al.: Opinion mining and sentiment analysis. Foundations and Trends® in Information Retrieval, vol. 2, No. 1–2, pp. 1–135 (2008)

4. Bhadane, C., Dalal, H., Doshi, H.: Sentiment analysis: measuring opinions. Proc. Comput. Sci. **45**, 808–814 (2015)
5. Chen, T., Xu, R., He, Y., et al.: Improving sentiment analysis via sentence type classification using BiLSTM-CRF and CNN. Exp. Syst. Appl. **72**, 221–230 (2017)
6. Pak, A., Paroubek, P.: Twitter as a corpus for sentiment analysis and opinion mining. In: LREc (2010)
7. Maryame, N.A.J.I., Najima, D., Rachida, A.: Spatio-temporal context for improving sentiment analysis accuracy. In: 2018 6th International Conference on Multimedia Computing and Systems (ICMCS), IEEE, pp. 1–6 (2018)
8. Tang, D., Qin, B., Feng, X., Liu, T.: Effective LSTMs for target-dependent sentiment classification. In: Proceedings of the International Conference on Computational Linguistics (COLING 2016) (2016)
9. Li, N., Wu, D.D.: Using text mining and sentiment analysis for online forums hotspot detection and forecast. Decis. Support Syst. **48**(2), pp. 354–368 (2010)
10. Medhat, W., Hassan, A., Korashy, H.: Sentiment analysis algorithms and applications: a survey. Ain Shams Eng. J. **5**(4), 1093–1113 (2014)
11. Katz, G., Ofek, N., Shapira, B:. ConSent: context-based sentiment analysis. Knowl.-Based Syst. **84**, 162–178 (2015)
12. Cambria, E., Poria, S., Gelbukh, A.: Sentiment analysis is a big suitcase. IEEE Intell. Syst. **32**(6), 74–80 (2017)
13. Lecun, Y., Bengio, Y., Hinton, G.: Deep learning. Nature **521**(7553), 436 (2015)
14. Zhang, L., Wang, S., Liu, B.: Deep learning for sentiment analysis: a survey. Wiley Interdiscipl. Rev. Data Min. Knowl. Discov. **8**(4), e1253 (2018)
15. Araque, O., Corcuera-Platas, I., Sanchez-Rada, J.F., et al.: Enhancing deep learning sentiment analysis with ensemble techniques in social applications. Exp. Syst. Appl. **77**, 236–246 (2017)
16. Chatterjee, A., Gupta, U., Chinnakotla, M.K., et al.: Understanding emotions in text using deep learning and big data. Comput. Hum. Behav. **93**, 309–317 (2019)
17. Tang, D., Zhang, M.: Deep learning in sentiment analysis. In: Deep Learning in Natural Language Processing. Springer, Singapore, pp. 219–253 (2018)
18. Tang D., Qin B., Liu T.: Document modelling with gated recurrent neural network for sentiment classification. In: Proceedings of the Conference on Empirical Methods in Natural Language Processing (EMNLP 2015) (2015)
19. Rosenthal, S., Farra, N., Nakov, P.: SemEval-2017 task4: Sentiment analysis in twitter. In: Proceedings of the 11th International Workshop on Semantic Evaluation (SemEval-2017), pp. 502–518
20. Chen, H., Sun, M., Tu, C., Lin, Y., Liu, Z.: Neural sentiment classification with user and product attention. In: Proceedings of the Conference on Empirical Methods in Natural Language Processing (EMNLP 2016), (2016)
21. Denil, M., Demiraj, A., Kalchbrenner, N., Blunsom, P., de Freitas, N.: Modelling, visualizing and summarizing documents with a single convolutional neural network (2014)
22. Irsoy, O., Cardie, C.: Deep recursive neural networks for compositionality in language. In: Advances in Neural Information Processing Systems, pp. 2096–2104 (2014)
23. Xiong, S., Zhang, Y., Ji, D., Lou, Y.: Distance metric learning for aspect phrase grouping. In: Proceedings of COLING, pp. 2492–2502 (2016)
24. Cambria, E., Poria, S., Gelbukh, A., et al.: Sentiment analysis is a big suitcase. IEEE Intell. Syst. **32**(6), 74–80 (2017)
25. Al-Smadi, M., Qawasmeh, O., Al-Ayyoub, M., et al: Deep recurrent neural network versus support vector machine for aspect-based sentiment analysis of Arabic hotels' reviews. J. Comput. Sci. **27**, 386–393 (2018)
26. Alharbi, A.S.M., et al.: Twitter sentiment analysis with a deep neural network: an enhanced approach using user behavioral information. Cognit. Syst. Res. (2018)
27. Paredes-Valverde, M.A., Colomo-Palacios, R., Salas-Zárate., M.D.P., et al.: Sentiment analysis in Spanish for improvement of products and services: a deep learning approach. Sci. Progr. **2017**, (2017)

28. Acharya, U.R., Oh, S., Hagiwara, Y., et al.: Automated EEG-based screening of depression using deep convolutional neural network. Comput. Methods Progr. Biomed., **161**, 103–113 (2018)
29. Rosenthal, S., Farra, N., Nakov, P., SemEval-2017 task 4: Sentiment analysis in Twitter. In: Proceedings of the 11th International Workshop on Semantic Evaluation (SemEval-2017), pp. 502–518 (2017)
30. El Ouilani, Z.: 16 Oct 2018. Santé Mentale: Un Marocain Sur Quatre Est Dépressif. On the BLOG 360. Consulté le 01 Mar 2019. http://fr.le360.ma/societe/sante-mentale-un-marocain-sur-quatre-est-depressif-176610
31. Guenne, L.: 16 Oct 2018. Près d'un français sur dix victime de dépression en 2017. On website France Inter. Consulted the 25 Feb 2019. https://www.franceinter.fr/societe/pres-d-un-francais-sur-dix-victime-de-depression-en-2017
32. Sau, A., Bhakta, I.: Screening of anxiety and depression among the seafarers using machine learning technology. Informatics in Medicine Unlocked (2018)
33. Hatton, C.M., Paton, L.W.: Mcmillan, D., et al.: Predicting persistent depressive symptoms in older adults: a machine learning approach to personalised mental healthcare. J. Affect. Disord. **246**, 857–860 (2019)
34. Repport, 10 May 2018. Dépression majeure: impact sur la santé en général. On the website The Heath of America BCBS. Consulted the 01 May 2019. https://www.bcbs.com/the-health-of-america/reports/major-depression-the-impact-overall-health

Automatic Synthesis Approach for Unconstrained Face Images Based on Generic 3D Shape Model

Hamid Ouanan, Mohammed Ouanan and Brahim Aksasse

Abstract Faces captured in unconstrained conditions represent a myriad of challenges like occlusions, non-frontal poses, etc. Compared to the other variations, pose variation is one of the most difficult problems in face recognition. To this end, we propose a new automatic synthesis approach of frontal facing view of unconstrained face images which are based on the use of a generic 3D shape model and accurate localization of facial feature points. On another hand, we have used the same approach as data augmentation technique for enriching a small dataset with important facial appearance variations by manipulating the faces that it contains so as to improve the convolutional neural network (CNN) performances and avoid overfitting. This is done by using a camera model to generate multiple views covering possible poses synthesized from the 3D face model used.

Keywords Unconstrained conditions · 3D face model · Pose estimation · Overfitting · Deep learning

1 Introduction

In computer vision, the pose of an object refers to its relative orientation and position with respect to a camera. The problem of how to perform a pose-invariant face recognition on uncooperative subjects has been a key question in face recognition

The original version of this chapter was revised: The affiliation of the authors has been amended. The correction to this chapter is available at https://doi.org/10.1007/978-981-15-0947-6_86

H. Ouanan (✉)
National School of Applied Sciences, Sultan Moulay Slimane University, Beni Mellal, Morocco
e-mail: ham.ouanan@gmail.com

M. Ouanan · B. Aksasse
Dept. of Computer Science, M2I Laboratory, Faculty of Science and Techniques, Moulay Ismail University, BP 509 Boutalamine, 52000 Errachidia, Morocco
e-mail: ouanan_mohammed@yahoo.fr

B. Aksasse
e-mail: baksasse@yahoo.com

V. Bhateja et al. (eds.), *Embedded Systems and Artificial Intelligence*,
Advances in Intelligent Systems and Computing 1076,
https://doi.org/10.1007/978-981-15-0947-6_43

applications [1]. Pose-invariant face recognition [2–4] has been one of the most challenging tasks in the field of computer vision. The innate features of the faces that distinguish one face from another do not greatly vary from one individual to another. Moreover, the magnitudes of the image variations caused by different poses are often larger than magnitudes of the variations of the innate characteristics. Therefore, the extraction of the innate pose variations-free characteristics becomes a very difficult task. Head pose estimation is most commonly seen as the process of deducting the orientation of a person's head relative to the view of a camera.

In our proposed approach, we have adopted a 3D generic face model, which has the capacity to roughly mimic the out-of-plane rotations the face can undergo. The accurate 3D model of a face will be the 3D data of the individual produced by a laser range finder scanning. This process requires the use of sophisticated equipment, which is not readily available. Besides, we cannot have a 3D model for each individual that we will photograph for our face recognition system because this process is non-automated and time-consuming. For this reason, we have used the same textured 3D face generic model as a reference to align all query images. The goal is to produce a better frontal view of the individual by treating the head as a rigid or a non-rigid object, rotating, and translating it with 6 degrees of freedom. This is in opposition to the state-of-the-art methods that adjust the 3D facial shape to fit facial appearances of the query image [5].

2 Face Pose Estimation

This section is meant to explain, from a geometric point of view, the 3D pose estimation problem formulation using a pinhole camera model. Figure 1 shows some examples of these types of cameras.

We begin by rendering this reference model S in a frontal view using a suitable projection matrix $C_M(3 \times 4)$ (Eq. (1)). To this end, we have selected a suitable rotation matrix $R_M(3 \times 3)$ and translation vector t_M to

Fig. 1 Example of two standard pinhole cameras. A simple webcam creative

produce a frontal view I_R of the generic 3D face model used in our proposed method. The reference frontal view I_R, serves as our reference coordinate system. While producing the reference view I_R, we store for each of its pixels p' (Eq. (2)) the 3D point coordinates of the point located on the surface of the 3D model for which:

$$C_M = A_M[R_M t_M] \tag{1}$$

$$p' \approx C_M P \tag{2}$$

where

A_M is a 3×3 matrix, which contains the camera calibration parameters such as the focal length, the scale factor, and the optical center point coordinates.
R_M is a 3×4 matrix, which corresponds to the Euclidean transformation from a world coordinate system to the camera coordinate system.
And t_M is the translation vector.

The head pose estimation problem is often referred to as Perspective-n-Point (PnP) problem in computer vision jargon. In this problem, the goal is to estimate the pose of a calibrated camera given a set of n 3D points in the object and their corresponding 2D image projections along with the intrinsic camera parameters (focal length, principal point, aspect ratio, and skew) and determine the 6 degrees-of-freedom (DOF) pose of the camera in the form of its rotation (roll, pitch, and yaw) and translation with respect to the world. This follows the perspective projection model of cameras (Fig. 2).

2.1 *The Intrinsic Parameters*

In the previous section, we defined A_Q as a matrix with the internal camera calibration parameters usually referred to as camera calibration matrix. This can be expressed by the following equation (Eq. (3)):

$$A_Q = \begin{bmatrix} \alpha_x & s & x_0 \\ 0 & \alpha_y & y_0 \\ 0 & 0 & 1 \end{bmatrix} \tag{3}$$

where

- α_x and α_y are the scale factor defined in each coordinate direction x and y.
- $c = [x_0; y_0]^{\mathrm{T}}$ represents the principal point coordinates, which is the intersection of the optical axis and the image plane.
- S is referred to the skew angle. In modern cameras, this value is usually zero.

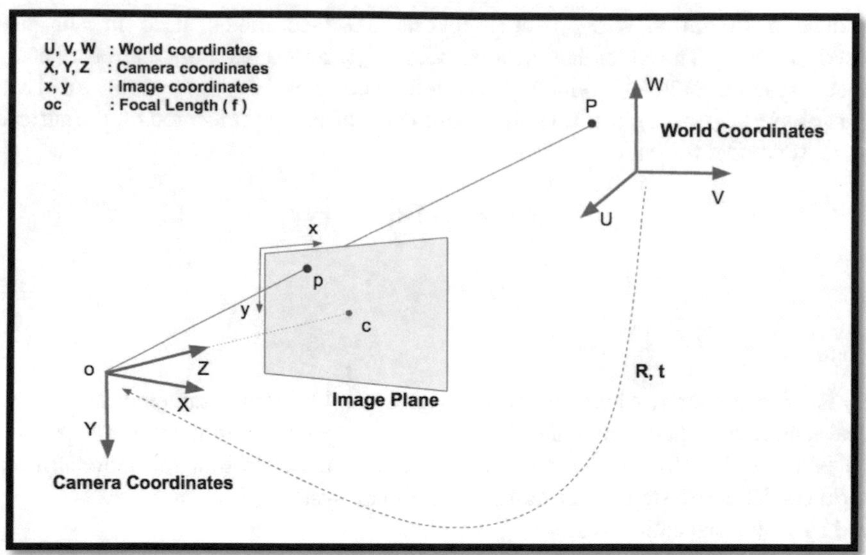

Fig. 2 Problem formulation

In order to simplify the problem, a common solution is to often set the principal point c at the image center. Furthermore, we assume that the pixels have a squared shape, which leads us then to take α_x and α_y with equal values.

2.2 *The Extrinsic Parameters*

We previously defined $[R_Q t_Q]$ (Eq. (4)) as a 3×4 matrix which corresponds to the Euclidean transformation from a world coordinate system to the camera coordinate system. In fact, this matrix is the horizontal concatenation of the rotation matrix and the translation vector which is often referred to as the camera pose.

$$[R_Q t_Q] = \begin{bmatrix} R_{11} & R_{12} & R_{13} & t_1 \\ R_{21} & R_{22} & R_{23} & t_2 \\ R_{31} & R_{32} & R_{33} & t_3 \end{bmatrix} \quad (4)$$

There are a few preliminary aspects of the problem that is common to all sorts of solutions of PnP. The assumption made in most solutions is that the camera is already calibrated. Thus, its intrinsic parameters are already known, e.g., the focal length, principal image point, skew parameter, and other parameters. Some methods, such as the unified Perspective-n-Point (UPnP) [6] or the direct linear transform (DLT) [7] applied to the projection model, are exceptions to this assumption as they estimate these intrinsic parameters as well as the extrinsic parameters that make up the pose of the camera that the original PnP problem is trying to find. For each solution to PnP, the chosen point correspondences cannot be coplanar. In addition, PnP can have multiple solutions, and choosing a particular solution would require post-processing of the solution set. Furthermore, using more point correspondences can reduce the impact of noisy data when solving PnP. PnP is prone to errors if there are outliers in the set of point correspondences. Thus, RANdom SAmple Consensus (RANSAC) [8] can be used in conjunction with existing solutions to make the final solution for the camera pose more robust to outliers.

Therefore, estimating the 3D head pose in a face image means finding the extrinsic parameters in the 3D model's coordinate, which in other words are the orientation and position of the input image with respect to the camera. Accordingly, we need the 2D–3D correspondences between the points in the query image and points on the surface of the 3D generic face model used in our proposed method. This can be attained by matching query points to points on a rendered frontal view of the model. However, estimating correspondences between a real and a synthetic image can be tremendously hard. Instead, we use a robust landmark detection algorithm [6] which finds the same facial key points in both images. Therefore, 68 facial landmarks like the tip of the nose, the chin, the left corner of the left eye, the right corner of the right eye, the left corner of the mouth, and the right corner of the mouth referred to as $p_i = (x_i, y_i)^{\mathrm{T}}$ are detected in this image using the method invented by Zhu and Ramanan [9]. This method is chosen because of its accuracy in real-world face images.

We again use the same approach to detect the same facial key points in the frontal view of the 3D model I_R, giving us points $p_i' = \left(x_i', y_i'\right)^{\mathrm{T}}$; for each point detected we have associated the 3D coordinates $P_i = (U_i, V_i, W_i)^{\mathrm{T}}$ correspond in the surface of the 3D model. Using these detected points, we form the correspondences (p_i, P_i), from 2D pixels in the query photograph to 3D points on the model, in order to compute the 3D pose.

We then compute the camera matrix that approximates the one used to capture the query image (Eq. (5)) (Fig. 3, step 4).

$$C_Q = A_Q[R_Q t_Q] \tag{5}$$

where

C_Q is the estimated camera matrix for the input view.
A_Q is a 3×3 matrix, which contains the camera calibration parameters.
R_Q is a 3×3 matrix, which corresponds to the rotation transformation from a world coordinate system to the camera coordinate system. And t_Q is the translation vector.

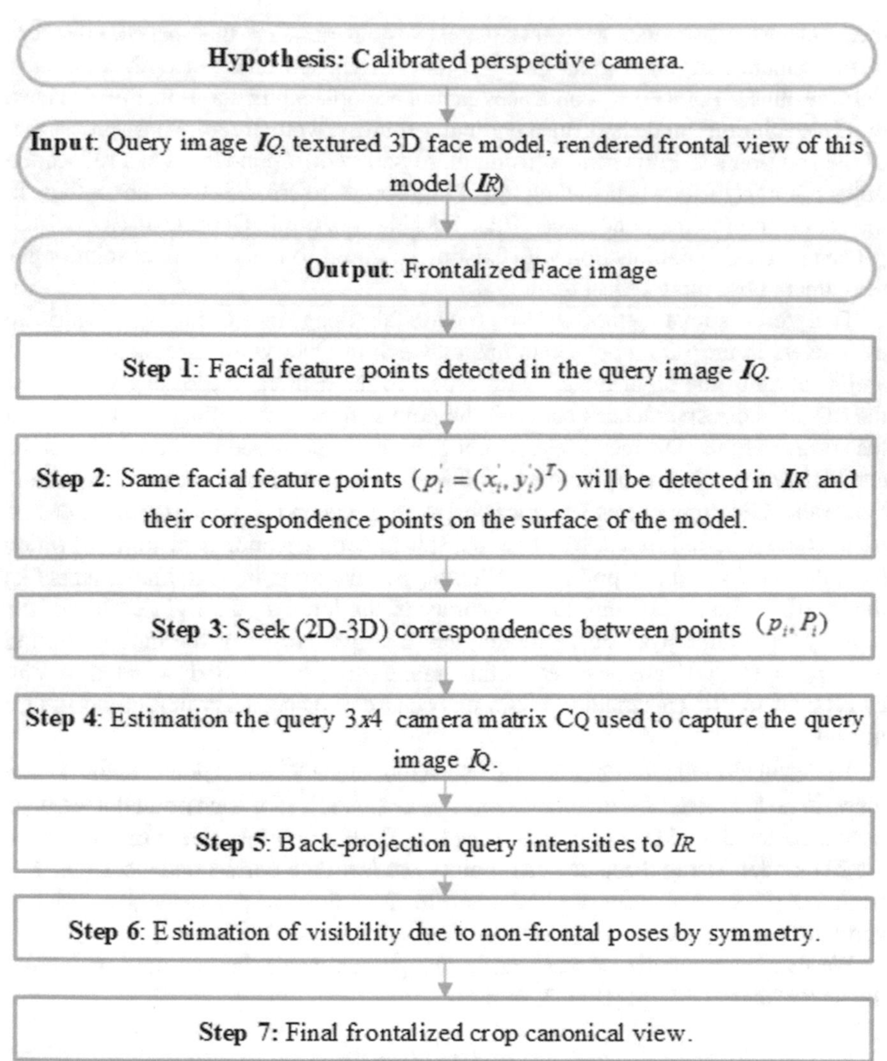

Fig. 3 An overview of the face alignment proposed method

3 Face Pose Normalization: (Frontal Pose Synthesis)

Once the camera pose is known, we thus obtain a perspective camera model mapping the 3D face generic model used to the query image. Hence, we use this matrix C_Q to map any point on the 3D surface onto the input image, thereby providing the desired texture map, e.g., for every pixel coordinate $p' = (x', y')^{\mathrm{T}}$ (Eq. (6)) in the reference view. From Eq. 6, we have the 3D location $P = (X, Y, Z)^{\mathrm{T}}$ on the surface of the reference which was projected onto p' by the estimated matrix C_Q:

$$p' \approx C_Q P \tag{6}$$

Finally, we project the query facial features back onto the reference coordinate system using the geometry of the 3D model to produce a frontal view of the face. Figure 3 illustrates the main stages of the proposed strategy to produce the frontal view of a facial image in unconstrained scenarios.

Out-of-plane rotation of the head causes some facial features to be invisible than others. Therefore, in order to correct the pose, we have transferred the appearances by symmetry from one side of the face to another. However, this operation may introduce problems whenever one side of the face is occluded by anything other than the face itself leaving the result recognizable.

4 Data Augmentation Strategy

Face recognition systems have recently known significant breakthroughs. Much of this progress is attributed to the success of deep learning technology [10, 11] which learns discriminative face representations from large training sets. This technology works on the basis of the assumption that is by collecting massive training sets. Therefore, in order to train face recognition systems based on deep CNNs, very large training sets are needed with millions of labeled face images. To this end, we have proposed a multi-stage strategy to build large face image datasets. The different stages of the proposed strategy are summarized in Fig. 4.

We call this synthetic dataset the Puball-dataset. The statistics of the proposed image dataset are shown in Table 1.

For comparison, we performed a comparative test of the base of images created with large datasets recognized in this area. The table above illustrates this benchmark in terms of the number of identities and the total number of images. The scores shown in this table show that the size of the dataset constructed mainly exceeds the size of the dataset used by Facebook to train their facial recognition system.

5 Conclusion

In this chapter, we have tackled one of the most difficult problems in face recognition, namely pose variations. For this reason, we have proposed a new approach to map the face image to a frontal view. After that, we described the strategy to build a new large-scale dataset with more than five million faces labeled for identity based on a smart synthesis augmented strategy that encompasses rendering pipeline to increase the pose and lighting variability.

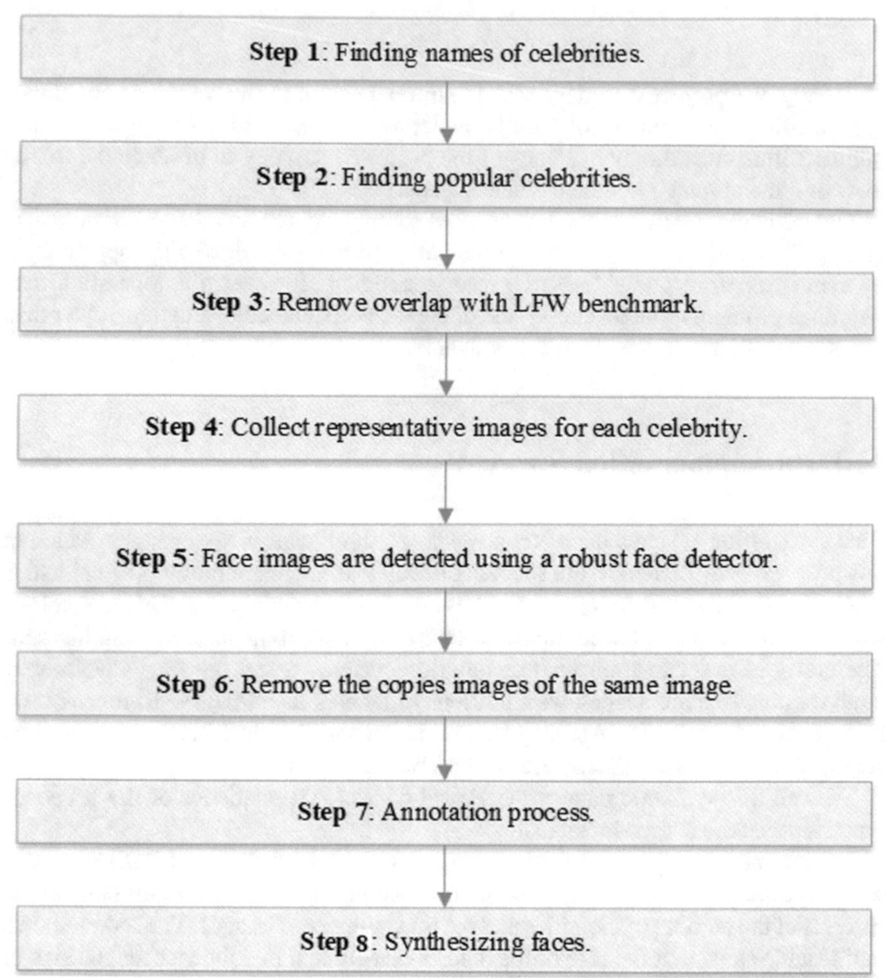

Fig. 4 Main stages of the dataset building process

Table 1 Dataset comparisons

No.	Dataset	#Persons	Total #images
1	SFC [12]	4,030	4.4 M
2	Google [13]	8 M	200 M
3	LFW [14]	5,749	13,233
4	CelebFaces [15]	10,177	202,599
5	Chen et al. [16]	2,995	99,773
8	Puball-dataset (ours)	4,000	5 M

References

1. Ouanan, H., Ouanan, M., Aksasse, B.: Implementation and optimization of face detection framework based on openCv library on mobile platforms using Davinci's technology. Int. J. Imaging Robot. **16**(2), 118–131 (2016)
2. Ding, C., Tao, D.: A comprehensive survey on pose-invariant face recognition. ACM Trans. Intell. Syst. Technol. **7**(3), 01–42 (2016)
3. Ouanan, H., Ouanan, M., Aksasse, B.: Novel approach to pose invariant face recognition. Proc. Comput. Sci. (2017)
4. Ouanan, H., Ouanan, M., Aksasse, B.: Myface: unconstrained face recognition. Lecture Notes in Networks and Systems (2017)
5. Jiang, L., Zhang, J., Deng, B., Li, H., Liu, L.: 3D face reconstruction with geometry details from a single image. arXiv:1702.05619 (2017)
6. Penate-Sanchez, A., Andrade-Cetto, J., Moreno-Noguer, F.: Exhaustive linearization for robust camera pose and focal length estimation. IEEE Trans. Pattern Anal. Mach. Intell. **35**(10), 2387–2400 (2013)
7. Abdel-Aziz, Y.I., Karara, H.M., Hauck, M.: Direct linear transformation from comparator coordinates into object space coordinates in close-range photogrammetry. Photogramm. Eng. Remote Sens. **81**(2), 103–107 (2015)
8. Zuliani, M.: Ransac toolbox for matlab. Available at http://www.mathworks.com/matlabcentral/fileexchange/18555 (2008)
9. Zhu, X., Ramanan, D.: Face detection, pose estimation, and landmark localization in the wild. In: IEEE Computer Vision and Pattern Recognition (CVPR), pp. 2879–2886 (2012)
10. Ouanan, H., Ouanan, M., Aksasse, B.: Face recognition using deep features. In: International Conference on Advanced Information Technology, Services and Systems. Springer, Cham, pp. 78–85 (2017)
11. Ouanan, H., Ouanan, M., Aksasse, B.: Non-linear dictionary representation of deep features for face recognition from a single sample per person. Proc. Comput. Sci. **127**, 114–122 (2018)
12. Taigman, Y., Yang, M., Ranzato, M.A., Wolf, L.: DeepFace: closing the gap to human-level performance in face verification. In: Proceedings of IEEE Conference on IEEE Computer Vision and Pattern Recognition, pp. 1701–1708 (2014)
13. Schroff, F., Kalenichenko, D., Philbin, J.: Facenet: a unified embedding for face recognition and clustering. In: Proceedings of the IEEE Conference on Computer Vision and Pattern Recognition (2015)
14. Huang, G.B., Ramesh, M., Berg, T., Learned-Miller, E.: Labeled faces in the wild: a database for studying face recognition in unconstrained environments, vol. 1, no. 2, pp. 07-49. University of Massachusetts, Amherst, Technical Report (2007)
15. Sun, Y., Wang, X., Tang, X.: Deep learning face representation from predicting 10,000 classes. In: Proceedings of the IEEE Conference on Computer Vision and Pattern Recognition (2014)
16. Chen, D., Cao, X., Wang, L., Wen, F., Sun, J.: Bayesian face revisited: a joint formulation. In: Proceedings of the European Conference on Computer Vision, pp. 566–579 (2012)

Approach Based on Artificial Neural Network to Improve Personalization in Adaptive E-Learning Systems

Ibtissam Azzi, Adil Jeghal, Abdelhay Radouane, Ali Yahyaouy and Hamid Tairi

Abstract Exploring the role of artificial intelligence in improving learner's performance in E-learning systems is an important motivating research area that tries to combine E-learning and traditional tutoring opportunities. Presenting a personalized learning is one of the opportunities that is important in order to increase the effectiveness of an individual learning. In this paper, we investigate the use of soft computing technique to handle the personalization problems in E-learning systems, particularly the problem of the course design regarding the learner's background (prerequisites). In this mind, we present an approach based on artificial neural network which is able to provide learner with the most suitable learning materials. To have information about the learner's knowledge background, the Web data are used as the input of the neural network. The architecture of the artificial neural network is described, and the performances of our approach are illustrated by simulation test.

Keywords Adaptive E-learning systems · Personalization · Learning materials · Knowledge background · Artificial neural networks · Web mining

I. Azzi (✉) · A. Yahyaouy · H. Tairi
LIIAN, Department of Informatics Faculty of Science Dhar-Mahraz, University of Sidi Mohamed Ben Abdellah, P.B 1796, Atlas-Fez, Morocco
e-mail: ibtissam.azzi@usmba.ac.ma

A. Yahyaouy
e-mail: ayahyaouy@yahoo.fr

H. Tairi
e-mail: htairi@yahoo.fr

A. Jeghal
Groupe Sup'Management, Fez, Morocco
e-mail: adil.jeghal@usmba.ac.ma

A. Radouane
Département d'informatique, Centre Régional Des Métiers de l'éducation et de Formation, C. R. M. E. F., Fez, Morocco
e-mail: radouaneabdelhay@yahoo.fr

V. Bhateja et al. (eds.), *Embedded Systems and Artificial Intelligence*,
Advances in Intelligent Systems and Computing 1076,
https://doi.org/10.1007/978-981-15-0947-6_44

1 Introduction

When developing adaptive E-learning systems, the individual needs of learners for providing an effective learning process remain as the most problem because the learning process is a variable that depends on each learner's needs and motivation [1]. Many efforts have been made in recent decades to personalize E-learning leading the increasing effectiveness of learning and providing learners with the appropriate content corresponding to the desired preferences. In the adaptation content-based case where resources and activities dynamically change their actual content, the goal of personalized E-learning systems must focus on how to choose the suitable learning materials to design an appropriate course. Indeed, by analyzing the existing studies, it is pointed out that the most of the computer-aided educational system provides the same leaning materials to all learners without dealing of his level of knowledge [2]. Therefore, in contrast to a traditional computer-aided instructions, course design in E-learning systems aims to take into account the use of the knowledge about the learner concepts learning acquisition rate to support flexible individualized learning [3]. To achieve this goal, many approaches have been proposed where the way to use the knowledge about the learner background is a pretest evaluation [4]. However, this way creates problem that is mainly related to the automatic diagnosing of the learner background. Thus, research focusing on automatic treatment to analysis data relating to pedagogical objects or those relating to learners becomes particularly important in E-learning systems that adapt learning materials to learner preferences. The most of these researches concern the introduction of data mining techniques to analyze data and then propose a personalization strategy following this analysis [5], [6] and [7]. Also, the artificial intelligence techniques have been exploited to investigate ways to automatically analysis data relating to learners to achieve personalization strategies; namely, the learner style identification problem was considered in this investigation [8], [9] and [10]. Our work is interested with the automatic diagnosing of the learner background problem, and in this article, we will use this automatic diagnosing to provide learning personalization in the sense to give to the learner the suitable learning materials. So, the approach we will propose will lead the E-learning system to imitate instructor's role in designing the course. Then, our automation idea comes from the direct analysis of the extracted Web data by using the artificial neural network where self-organizing map is used to cluster the learning objects (LO) according to the learning acquisition rate of the topic concepts contained in these learning objects. The time spent and the visit frequency are considered as parameters for estimating the learning acquisition rate of the topic concepts. Then, the backpropagation algorithm was employed to classify the learning objects in which the learner has assigned a learning acquisition rate related to the topic concepts containing in these learning objects. Thus, the system will be able to choose the suitable contents to improve the learner performance. The remainder of the paper is organized as follows: In Sect. 2, we will present some overview on the artificial intelligence techniques used in adaptive E-Learning systems. Section 3 will describe and explain our approach. The conducted experiments will be presented in Sect. 4. Section 5 will conclude the paper.

2 Artificial Intelligence Techniques in E-learning Systems

In the adaptive E-learning area, the most tasks involving artificial intelligence techniques are prediction, recommendation, filtering and classification with the common objective to adapt learning to learner's preferences and needs. Among the techniques used are fuzzy logic, Bayesian networks, decision trees, neural networks and genetic algorithm. These different techniques have different capabilities in analyzing the data related to learner, and none of the techniques is suitable for all tasks and context. For the classification task, the artificial intelligence techniques are particularly used for identifying learner's learning style [9], [11] and [12]. In this context, these different techniques have been used by considering different learning style models and by processing the data gathered from educational systems. By doing so, the artificial intelligence techniques lead to automatically detect learners' learning styles in automatic way that has the potential to be more accurate and less error-prone. The reasons to use these artificial intelligence techniques are, for a Bayesian network case, its natural representation of probabilistic information, its efficiency and its support to encode uncertain expert knowledge. To build a Bayesian network, the structure of the network must be defined, and then, the network's parameters must be set [13]. The structure of the network can be elicited from data or can be defined by an expert. For decision trees case, this algorithm is employed because of its simplicity, and the rules of the classification are visible and easy to understand, and it is appropriate when many attributes are relevant [14]. Two phases are needed to build this algorithm, namely building phase and pruning phase. In the building phase, the training dataset is recursively partitioned until all the instances in a partition belong to the same class. In the pruning phase, the nodes are pruned to prevent overfitting and to get a tree with higher accuracy. The decision tree algorithm is structured by a root node which represents the attribute that is selected as the base to build the tree, the internal nodes which represent attributes that reside in the inner part of the tree and leaves which represent the classes to infer. The possible values for the attribute the branch initiates are represented by branches between nodes. For neural networks case, these computational models are used because of its speed of execution and its ability to be updated quickly with extra parameters and learn from specific examples. A neural network is defined as a set of connected input–output units, where each connection has a weight associated with it. The weights are adjusted during the learning phase so as to be able to predict the correct class label of the input tuples [15].

The attempt of this paper is to use an artificial neural network for learning objects classification so as to provide personalized design course. The used artificial neural network is trained by employing a data-driven. The following section describes the data that the structure of the network can be elicited from.

3 Approach for Improving Personalization

In E-learning systems, the most important task to be considered in personalization approaches is the selecting of learning materials to be made available to learner so as to satisfy their learning preferences. However, this task when it is treated in a traditional setting, namely the designing of the adaptation rules by means of a content selection rules and concept selection rules, requires tedious work that is closely linked to the instructor's competencies. The automatic way using artificial intelligence techniques represents an alternative to this instructor's tedious work. Our approach is staged within this automatic way and consists of presenting an automatic method of selection of learning materials. An artificial neural network model is used to select a set of suitable learning objects for a learner. The proposed approach is based on the analysis of the learner's background traces of every concept appertaining to a topic's learning objects that the learner has followed. These traces are represented by the data resulting from the interaction of the learner with the system. To extract the useful information or pattern from the Web data for providing input data of algorithms, Web mining techniques are used, namely the web usage mining which consists of capturing the identity of Web users along with their browsing behavior at a Web site. This type of mining includes data collection, data pretreatment, knowledge discovery and pattern analysis [16, 17]. The methodology of our approach is described in the following subsection.

The methodology of our approach consists of the two following steps:

First step:

This step is performed with the objective to cluster the collections of LOs into classes based on concepts similarity of the LOs.

The learners' behaviors which correspond to a specific topic are first captured and have been converted into data to be processed.

To represent the learner's concepts knowledge acquisition rate for every concept of the specified topic, the extracted sequences are defined as per time spent and as per number of visits of each learning object in each session because these two parameters reflect the degree of focus on learning and the attention of learners to the concept's learning objects.

We can calculate the percentage represented by this time spent in relation to the learning time duration assigned by the instructors.

Then, according to this percentage of the time spent and number of visits, we can attribute the knowledge acquisition rate of each concept, and then, we can learn about this concept background for each learner. Table 1 illustrates this attribution. Thus, for example, if the percentage of the time spent is comprised between 70 and 100, and the number of visits is comprised between 0 and 1, then the concept knowledge acquisition rate is equal to 1, and then, the value of the concept background level is set to 1.

Then, the data to be considered in the learning object clustering process are defined by:

Table 1 Attribution of the concept's knowledge acquisition rate

% time spent/allocated learning time	Number of visits	Concept knowledge acquisition rate	Concept background level
[1, 30]	=1	0	0
[1, 30]	>1	0.25	0
[30, 50]	=1	0.25	0
[30, 50]	>1	0.5	0.5
[50, 70]	=1	0.5	0.5
[50, 70]	>1	0.75	1
[70, 100]	=1	0.75	1
[70, 100]	>1	1	1

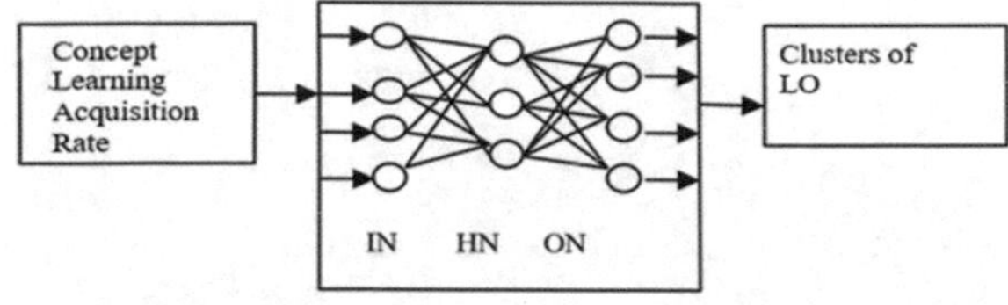

Fig. 1 ANN architecture for LO clustering

Learner's id, all of the id's of the LO's, the set of id's of the concepts and the set of the basic level of the concept for all the concepts and for a specified learner.

The learning objects are clustered into classes based on concepts similarity of the LO by using the self-organizing map neural network (SOM).

Figure 1 illustrates the architecture of the artificial neural network (ANN) created.

3.1 Self-Organizing Map Neural Network (SOM) Algorithm

The SOM is an unsupervised neural network that the training is data-driven, without a target condition that would have to be satisfied (as in a supervised neural network). The SOM algorithm is summarized as shown in Algorithm 1.

Algorithm 1. SOM Neural Network Algorithm

– *Initialization*

Choose random values for the initial weight vectors wj.

– *Sampling*

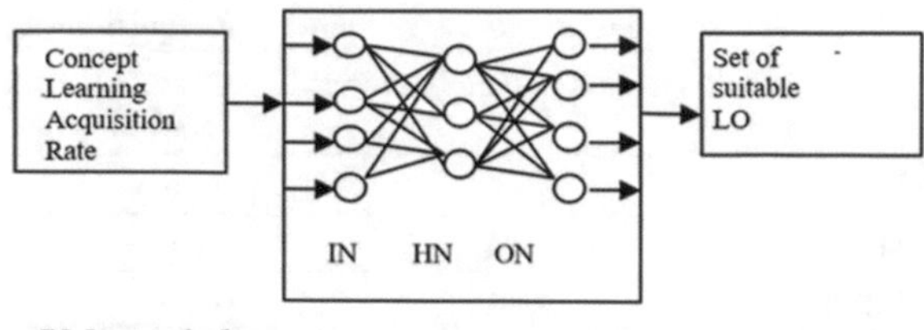

Fig. 2 ANN architecture for LO classification

Draw a sample training input vector x from the input space.

- *Matching*

  Find the winning neuron I(x) that has weight vector closest to the input vector, i.e., the minimum value of $d_j(X) = \sum_{i=1}^{D}\left(x_i - w_{ji}\right)^2$.
- *Updating*

  Apply the weight update equation $\Delta w_{ji} = \eta(t) T_{j,I(x)}(t)\left(x_i - w_{ji}\right)$ *where* $T_{j,I(x)}(t)$ *is a Gaussian neighborhood and* $\eta(t)$ *is the learning rate.*
- *Continuation*

  Keep returning to sampling step until the feature map stops changing.

Second step:

The second step consists of training the neural network in order to classify each learner to any classes of the LO.

To perform this step, three phases have been taken.

Phase 1:

A neural network to train LO data classification is created. The architecture of the network is illustrated by Fig. 2.

Phase 2:

In this phase, the network is trained using a classic backpropagation (BP) algorithm. The BPNN algorithm is summarized as in Algorithm 2.

3.2 Backpropagation Neural Network (BPNN) Algorithm

The backpropagation neural network (BPNN) learns through the algorithm called backpropagation. With backpropagation, the input data is repeatedly presented to the neural network until producing the desired output. So, the neural networks consist of large number of neurons, connected with each other to solve the specific problem. Neural networks are organized in layers, where layer consists of many interconnected nodes. Neural networks work on the learning process which involves the adjustments of weights which are assigned to the connections between the neurons. By using the conjugate gradient technique, the optimum output weights is calculated, where error at each node is calculated and is back propagated to the hidden layer where the hidden weights are calculated so as to minimize mapping error. The obtained weight values decide how the input data are related to the output data. Weight values are determined at the time of training for each iteration through the neural network. Thus, the weights produced during the training phase will be used for testing the network to classify each learner to a suitable LO class.

Algorithm 2. Backpropagation Neural Network (BPNN) Algorithm

- *Initialize two nodes for the input layer, three nodes for the hidden layer and p nodes for the output layer*

  where p is the number of output classes

- *Activate unit$_i$ in the input layer $X_i = a_i\ (m)$ where i = 1, 2 input parameters and $a_i(m)$ is the input vector*
- *Activate unit$_j$ in the hidden layer $Y_j^h = \sum W_{ij}(x) + b_i$ where j = 1 to 3 (hidden layer)*
- *Output signal from unit$_j$ in the hidden layer*

$$\theta_j^h = \frac{1}{1 + \exp(-X_j^h)}$$

- *Activate unit$_k$ in the output layer*

  $Z_k^o = \sum W_k^j \theta_j^h$ *where k = 1 to p output classes*

- *Mean square error for the kth output $E = \frac{1}{n}\sum(d_k - o_k)^2$*

  where d_k is the actual output, o_k is the experimental output and n is the number of patterns in the examples

Phase 3:

This phase consists of determining which class of LO that matched the learner's background level of each concept.

So, after the training procedure, the neural network could be applied for the classification of new data. Thus, the classification is performed by the trained backpropagation neural network according to the following steps:

- Activation of the network input nodes by using data sources that match the data sources used to train the network.
- Forward the data through the neural network, i.e., the interconnected nodes.
- Activation of the network output nodes.
- Activation of network output nodes decides the outcome of the classification.

4 Experiments

Due to unavailability of real data, we generated synthetic data simulating 65 learners and 120 LOs over 25 concepts.

The concepts are considered related to a given course. The number of concepts that are contained in any LO is random.

A part of the simulated data that reflects the learner's performance corresponding to all the concepts contained in the learning objects LO1, LO2,… LO5 is illustrated in the Table 2. In particular, we consider the percentage of time spent and the frequency of visits. The set of percentage of time spent is randomly generated, and each value belongs to [1, 100], and the set of number of visits is randomly generated and each value belongs to [1, 4].

For the clustering experiment, from the collected data, the weight value of each concept is attributed, and then SOM technique is employed to cluster the collections of LO into classes based on concepts similarity of the LO. A network has been created with 25 input neurons and has been trained with different dimension sizes of the map in order to get the best clustering result. Using our data, the best result was obtained when the size of the map dimension is 5×5.

The network managed to cluster the LO into 11 classes. The output of this experiment is the LOs with their class id which later will be trained by the neural network in order to classify each learner to any classes of the LO in the second experiment. Table 3 shows the results of the clustering.

In the classification experiment, the neural network is created where the input layer of the neural network represents the concepts of the course and where the input vector is a set of values belonging to the set $\{0, 0.5, 1\}$. The output layer is assigned to the classes of learning object that have been identified previously. Then, the neural network is trained using a backpropagation algorithm.

Error at each node is calculated and is backpropagated to the hidden layer where the hidden weights are calculated so as to minimize mapping error.

The neural network has been tested toward 65 learners collected data of performance based on the level of concepts knowledge background.

In our experiment, the error has been minimized to 0.0005 after 100 iterations when the hidden layer node is set to 25 neurons.

The result shows that the classifier had selected the most matching LOs with the learner background level. Table 4 presents the results of the classification.

Table 2 Part of the data of learner's performance on every Los

LO 1	LO 2	LO 3	LO 4	LO 5
(01,00;1)	(5,00;1)	(5,00;1)	(1,00;1)	(1,00;1)
(01,00;2)	(17,50;2)	(5,50;1)	(5,00;1)	(5,00;1)
(03,75;0)	(25,00;1)	(8,75;1)	(5,00;1)	(8,00;1)
(10,00;1)	(30,00;1)	(10,00;0)	(8,00;1)	(17,50;0)
(10,00;3)	(31,25;1)	(10,00;1)	(10,00;1)	(20,00;2)
(15,00;1)	(37,50;2)	(15,00;2)	(20,00;2)	(22,50;1)
(15,50;0)	(50,00;1)	(16,25;1)	(21,25;1)	(27,50;1)
(15,50;1)	(52,50;1)	(17,50;0)	(36,25;1)	(35,00;1)
(16,25;1)	(55,00;2)	(17,50;1)	(40,00;1)	(50,00;1)
(20,00;2)	(56,25;2)	(20,00;1)	(50,00;1)	(50,00;2)
(20,00;3)	(57,50;0)	(22,50;2)	(50,00;2)	(55,00;1)
(21,75;2)	(60,00;1)	(23,75;1)	(52,50;2)	(55,00;1)
(22,00;0)	(62,50;0)	(25,00;1)	(56,25;1)	(57,50;2)
(22,00;1)	(62,50;2)	(25,00;2)	(60,00;1)	(60,00;1)
(22,00;1)	(63,75;2)	(27,50;0)	(62,50;2)	(62,50;1)
(22,00;2)	(65,00;2)	(27,50;1)	(65,00;2)	(65,00;2)
(22,00;2)	(65,00;2)	(30,00;1)	(65,00;3)	(65,00;3)
(22,50;1)	(65,00;4)	(30,00;1)	(65,00;3)	(65,00;4)
(22,50;1)	(66,25;3)	(31,25;2)	(66,25;3)	(67,50;2)
(22,50;2)	(66,25;5)	(37,50;1)	(66,25;3)	(67,50;3)
(27,50;1)	(68,75;1)	(37,50;1)	(68,75;1)	(67,50;3)
(27,50;2)	(68,75;3)	(37,50;1)	(68,75;4)	(67,50;3)
(30,00;0)	(68,75;5)	(38,75;1)	(68,75;4)	(67,50;3)
(30,00;1)	(70,00;1)	(40,00;0)	(70,00;1)	(67,50;3)
(31,25;1)	(70,00;2)	(42,50;2)	(70,00;3)	(67,50;4)
(35,50;1)	(70,00;2)	(45,00;1)	(70,00;3)	(70,00;2)
(35,75;1)	(70,00;3)	(46,25;1)	(70,00;3)	(70,00;2)
(37,50;1)	(70,00;3)	(47,50;2)	(70,00;3)	(70,00;2)
(37,50;2)	(70,00;3)	(50,00;1)	(70,00;3)	(70,00;2)
(40,00;1)	(70,00;3)	(50,00;1)	(70,00;4)	(70,00;3)
(42,50;2)	(70,00;4)	(50,00;1)	(70,00;4)	(70,00;3)
(45,00;0)	(70,00;4)	(50,00;1)	(70,00;4)	(70,00;3)
(46,25;1)	(70,00;4)	(50,00;2)	(70,00;4)	(70,00;3)
(47,50;1)	(70,00;4)	(52,50;2)	(70,00;5)	(70,00;3)
(52,50;0)	(71,25;5)	(52,50;2)	(71,25;3)	(70,00;3)
(52,50;1)	(72,50;2)	(52,50;2)	(72,50;2)	(70,00;3)

(continued)

Table 2 (continued)

LO 1	LO 2	LO 3	LO 4	LO 5
(52,50;1)	(72,50;2)	(52,50;3)	(72,50;2)	(70,00;4)
(52,50;3)	(72,50;3)	(52,50;3)	(72,50;3)	(70,00;4)
(55,00;1)	(72,50;4)	(55,00;2)	(72,50;3)	(70,00;4)
(56,25;3)	(73,75;5)	(56,25;3)	(73,75;4)	(70,00;4)
(58,75;3)	(75,00;1)	(57,50;2)	(75,00;3)	(70,00;4)
(59,50;2)	(75,00;3)	(57,50;2)	(75,00;4)	(72,50;3)
(59,50;3)	(76,25;2)	(57,50;4)	(76,25;2)	(72,50;3)
(59,50;3)	(76,25;3)	(58,75;3)	(76,25;3)	(72,50;3)
(60,00;4)	(76,25;3)	(60,00;1)	(76,25;3)	(75,00;2)
(60,00;4)	(76,25;4)	(60,00;2)	(76,25;3)	(75,00;2)
(61,00;3)	(76,25;4)	(60,00;3)	(76,25;3)	(75,00;3)
(61,50;3)	(77,50;1)	(60,00;4)	(77,50;3)	(75,00;3)
(62,50;3)	(77,50;2)	(62,50;1)	(77,50;4)	(75,00;3)
(63,75;3)	(77,50;4)	(63,75;2)	(77,50;4)	(75,00;4)
(65,00;2)	(78,75;3)	(63,75;4)	(78,75;4)	(75,00;4)
(65,00;3)	(80,00;2)	(65,00;3)	(80,00;2)	(77,50;2)
(65,00;4)	(80,00;3)	(65,00;3)	(80,00;3)	(77,50;3)
(70,00;2)	(80,00;4)	(65,00;3)	(80,00;3)	(77,50;3)
(87,00;3)	(80,00;4)	(67,50;2)	(80,00;3)	(77,50;3)
(87,00;4)	(80,00;4)	(67,50;2)	(80,00;4)	(77,50;4)
(87,00;5)	(81,25;2)	(67,50;3)	(81,25;2)	(77,50;4)
(87,25;4)	(81,25;3)	(67,50;3)	(81,25;4)	(80,00;1)
(87,50;2)	(82,50;3)	(67,50;3)	(82,50;3)	(80,00;1)
(87,50;2)	(83,75;3)	(67,50;4)	(83,75;2)	(80,00;2)
(87,50;2)	(83,75;4)	(70,00;1)	(83,75;3)	(85,00;1)
(87,50;3)	(85,00;2)	(71,25;1)	(85,00;3)	(85,00;1)
(87,50;3)	(85,00;3)	(72,50;2)	(85,00;3)	(95,00;0)
(87,50;3)	(85,00;3)	(95,00;0)	(85,00;4)	(95,00;1)
(87,50;4)	(95,00;0)	(95,00;1)	(95,00;0)	(95,00;1)

According to the result of the testing of the network illustrated in Table 4, we can observe that goodness of the classification is assured, and the most of the good classifications were the samples with bigger size of hidden nodes. Consequently, the most relevant LOs at the learner's basic level have been selected while achieving our goals that are to avoid pretest evaluation, to automatically learn about prerequisites and to imitate instructor's role in designing the course.

Table 3 Results of the clustering

Clusters of LOs	Number of LOs
Cluster1	12
Cluster2	7
Cluster3	11
Cluster4	10
Cluster5	15
Cluster6	13
Cluster7	8
Cluster8	14
Cluster9	11
Cluster10	7
Cluster11	12

Table 4 Classification results

Hidden nodes size	Mean squared error
13	2.9296
20	0.0779
25	0.0005

5 Conclusion

The work of this paper is interested with the automatic diagnosing of the learner background problem. This automatic diagnosing provides learning personalization in the sense that it gives to the learner the suitable learning materials. So, the proposed approach leads the E-learning system to imitate instructor's role in designing the course. Then, our automation idea comes from the direct analysis of the extracted Web data by using the artificial neural network where self-organizing map is used to cluster the learning objects according to the learning acquisition rate of the topic concepts contained in these learning objects. The time spent and the visit frequency are considered as parameters for estimating the learning acquisition rate of the topic concepts. Then, the backpropagation algorithm was employed to classify the learning objects that match the learner' background related to topic concepts. Thus, the system is able to choose the suitable contents to improve the learner's performance by considering his concepts background knowledge without the need to use pretest evaluation since the topic concepts which the learner has assigned a learning acquisition rate has been identified automatically.

References

1. Stankov, S., Rosić, M., Žitko, V, Grubišić, A.: TEx-Sys model for building intelligent tutoring systems. Comput. Educ. **51**, 1017–1036 (2008)
2. Brusilovsky, P.: Methods and techniques of adaptive hypermedia. In: User Modeling and User-Adapted Interaction, vol. 6, pp. 87–129. Kluwer Academic Publishers (1996)
3. Brusilovsky, P., Peylo, C.: Adaptive and intelligent web-based educational systems. Int. J. Artif. Intell. Educ. **13**, 156–169 (2003)
4. Norsham, I., Norazah, Y., Puteh, S.: Adaptive course sequencing for personalization of learning path using neural network. Int. J. Adv. Softw. Comput. Appl. **1**(1), 2009
5. Sun, P.C., Tsai, R.J., Finger, G., Chen, Y.Y., Yeh, D.: What drives a successful E learning? an empirical investigation of the critical factors influencing learner satisfaction. Comput. Educ. **50**(4), 1183–1202 (2008)
6. Johnson, S.D., Aragon, S.R., Shaik, N.: Comparative analysis of learner satisfaction and learning outcomes in online and face-to-face learning environments. J. Interact. Learn. Res. **11**(1), 29–49 (2000)
7. Gunawardena, C.N., Zittle, F.J.: Social presence as a predictor of satisfaction within a computer-mediated conferencing environment. Am. J. Distance Educ. **11**(3), 8–26 (1997)
8. Deborah, L.J., Sathiyaseelan, R., Audithan, S., Vijayakumar, P.: Fuzzy-logic based learning style prediction in e-learning using web interface information. Sadhana **40**(2), 379–394 (2015)
9. Crockett, K., Latham, A., Mclean, D., O'Shea, J.: A fuzzy model for predicting learning styles using behavioral cues in an conventional intelligent tutoring system. In: Proceedings of the 2003 IEEE International Conference on Fuzzy Systems (FUZZ), IEEE, Chester, England, pp. 1–8, 7–10 July 2013. ISBN:978-1-4799-0021-3
10. Abrahamian, E., Weinberg, J., Grady, M., Stanton, C.M.: The effect of personality-aware-computer-human interfaces on learning. J. Univer. Comput. Sci. **10**, 27–37 (2004)
11. Zatarain, C.R., Estrada, L.M.B., Angulo, V.P., Garcia, A.J., Garcia, C.A.R.: A learning social network with recognition of learning styles using neural networks. In: Proceeding of the 2nd Mexican Conference on Pattern Recognition (MCPR 2010), Springer, Puebla, Mexico, pp. 199–209, 27–29 Sept 2010
12. Kolekar, S.V., Pai, R.M., Pai, M.M.M.: Prediction of learner's profile based on learning styles in adaptive E-learning system. Int. J. Emerg. Technol. Learn. **12**(6), 31–51 (2017)
13. Brusilovsky, P., Millán, E.: User models for adaptive hypermedia and adaptive educational systems. In: Brusilovsky, P., Kobsa, A., Nejdl, W. (eds.) The Adaptive Web. Lecture Notes in Computer Science, vol. 4321, pp. 3–53. Springer, Berlin, Heidelberg (2007)
14. Crockett, K., Latham, A., Mclean, D., Bandar, Z., O'Shea, J.: On predicting learning styles in conversational intelligent tutoring systems using fuzzy classification trees. In: IEEE International Conference on Fuzzy Systems, pp. 2481–2488 (2011)
15. Han, J., Kamber, M., Pei, J.: Data mining: concepts and techniques. Morgan Kaufmann (2006)
16. Mobasher, B.: Web usage mining. In: Web Data Mining: Exploring Hyperlinks, Contents, and Usage Data pp. 1216–1220. https://doi.org/10.1007/978-3-642-19460-3 (2006)
17. Umadevi, Uma Maheswari, B., Nithya, P.: Design of E learning application through web mining. Int. J. Innov. Res. Comput. Commun. Eng. **2**(8) (2014). (An ISO 3297: 2007 Certified Organization)

Image Retrieval System Based on Color and Texture Features

El Mehdi El Aroussi and Silkan Hassan

Abstract The local binary pattern (LBP) operator and its variants extract the textural information of an image by considering the neighboring pixel values. A single or joined histogram can be derived from the LBP code which can be used as an image feature descriptor in some applications. However, the LBP-based feature is not a good candidate in capturing the color information of an image, making it less suitable for measuring the similarity of color images with rich color information. To overcome this problem, we propose a fast and efficient indexing and image search system based on color and texture features. The color features are represented by combining 2D histogram and statistical moments, and texture features are represented by the local binary pattern (LBP). To assess and validate our results, many experiments were held in color space HSV. Detailed experimental analysis is carried out using precision and recall on colored Brodatz, KTH TIPS, Stex, and USPTex image databases. Experimental results show that the presented retrieval method yields about 8% better performance in precision versus recall and about 0.2 in average normalized modified retrieval rank (ANMRR) than the method using wavelet moments.

Keywords LBP · 2D histogram · Statistical moments · CBIR · HSV

1 Introduction

The number of digital images in the form of personalized and enterprise collections have grown immensely. Hence there is a growing demand for powerful image indexing and retrieval in an automatic way. However, with such widespread use and availability of images, the goal of an image retrieval system is to retrieve a set of images from a collection of images such that this set meets the user's requirements. An image

E. M. El Aroussi (✉) · S. Hassan
Laboratory of Research in Optimization, Emerging Systems, Networks and Imaging, Faculty of Sciences of El Jadida, University Chouaïb Doukkali, El Jadida, Morocco
e-mail: Elaroussi.e@ucd.ac.ma

S. Hassan
e-mail: silkan.h@ucd.ac.ma

V. Bhateja et al. (eds.), *Embedded Systems and Artificial Intelligence*, Advances in Intelligent Systems and Computing 1076,
https://doi.org/10.1007/978-981-15-0947-6_45

retrieval system provides the user with a way to access, browse, and retrieve images efficiently from databases. The techniques of image retrieval system are divided into two categories: text-based and content-based. In text-based image retrieval system, database images are represented by adding text strings [1]. This system gives relevant set of images which are matched to the annotation of query image. This is a simple approach. However, this system has many limitations like tedious manual annotation which is subjective. In content-based image retrieval (CBIR) system, the images are indexed according to image intensity contents such as color, texture, and shape. The image contents are converted into numerical values called features [2]. CBIR algorithms are used to extract features of images. Based on these features, it is possible to retrieve images from databases which are similar to a chosen query image. Many of the recently developed descriptors, which represent aggregation of local features, are very effective in image retrieval. The most widely used such methods include the bag-of-visual words (BoVW), convolution neural network (CNN), and local binary pattern (LBP). This method applied the LBP algorithm under a specific filter bank which can yield a promising result on abnormal detection accuracy. The method [3] utilized LBP feature on the spoof fingerprint detection under multi-scale condition, while the method [4] incorporated the LBP feature on the eye detection system. As reported in [3] and [4], the LBP feature offers promising results on the spoof fingerprint detection and eye detection system. The method in [5] proposed contrast enhancement which is performed using a sigmoidal transformation mechanism followed by extracting a set of 14 Haralick features. For classification purposes, a support vector machine (SVM) classifier is used which sorts the input mammogram into either normal or abnormal subclasses. The method [6] proposed a new technique for edge detection in color images, employing the concept of Hilbert transform. The color image at the input is initially transformed using a RGB color triangle, followed by the application of the proposed edge detection technique. The method in [7] presents an efficient image indexing and search system based on color and texture features. The color features are represented by combines 2D histogram and statistical moments, and texture features are represented by a gray-level co-occurrence matrix (GLCM). The method in [8] presented a new improvement of LBP-based feature by adding the scale-adaptive texton and subuniform-based circular shift for handing the scale-invariant texture image classification. In [9] the authors propose a segmentation approach based on mathematical morphology, quadtree decomposition for mask generation, thresholding, and snake models. The feature extraction stage is steered by a shape model based on principal component analysis (PCA). The method in [10] presented a new improvement of LBP-based feature by adding the scale-adaptive texton and subuniform-based circular shift for handing the scale-invariant texture image classification. The method in [11] extended the superiority of the LBP feature on the face recognition system. The method [12] proposed the rotation-invariant co-occurrence the LBP adjacency information for the HEp-2 cell classification system. It employed the support vector machine (SVM) on performing the image classification which shows a better performance compared to the former schemes. The method in [13] extended the superiority of the LBP feature on the face recognition system. This method investigated the derivative of LBP on handling illumination problem, rotation and angle variations, as well as the facial expression variations on face recognition system with a promising result. An application of LBP feature on abnormal image detection has been reported in [14] under endoscopy video. This approach is very useful on detecting mucosal abnormalities for the medicine and health application. This method applied the LBP algorithm under a specific filter bank which can yield a promising result on abnormal detection accuracy. The method [15] utilized LBP feature on the spoof fingerprint detection under multi-scale condition, while the method [16] incorporated the LBP feature on the eye detection system. As reported in [15] and [16], the LBP feature offers promising results on the spoof fingerprint detection and eye detection system. The method in [17] proposed a computer vision approach applied to video sequences extracting global features of human motion. From the skeleton, the authors extract the information about human joints. From the silhouette, the authors get the boundary features of the human body. The binary and gray-level images contain different aspects about the human motion.

The method in [18] proposed a new technique on the object detection with the shape-based non-redundant LBP which gives a better performance compared to the other schemes. Most of methods employed the LBP scheme under the multi-resolution scenario to obtain more representative feature descriptors. Most of them applied the LBP operator on the filter banks, steerable filter, image decomposition, etc. The method in [12, 19–22] generated an image feature from the Gabor filter banks with the LBP operator. The fusion strategies between the Gabor filter banks and LBP operator have been demonstrated to give a successful results on the image classification system [12, 19, 20], face recognition [21, 22], object tracking [20], and image retrieval system [23].

The rest of this paper is organized as follows: The brief color histogram, color moments, and local binary pattern feature are introduced in Sect. 2. The proposed method for image retrieval and classification is presented in Sect. 3. Extensive experimental results are reported in Sect. 4. Finally, conclusions are drawn in Sect. 5.

2 Color and Texture Features

2.1 Color Histogram

Color histogram (CH) is the probability of occurrence of colors in a predefined range of colors in an image [24]. The color histogram can be built for any kind of color space, although the term is more often used for three-dimensional spaces such as RGB, HSV, L*a*b*, and YCbCr.

Let M and N be the number of rows and columns in an image. We divide each component of color space in bin_r; bin_g, and bin_b bins. Also let $R(i;j)$, $G(i;j)$, and $B(i;j)$ be the intensity values in each component of color space; index_r, index_g, and index_b be the bin index values of intensity in each component; and $H\left(\text{index}_r, \text{index}_g, \text{index}_b\right)$

be the histogram of color image using three components. The normalized histogram is obtained by dividing $H(\text{index}_r, \text{index}_g, \text{index}_b)$ by the size of image. The pseudocode of color histogram in the RGB color space is given as follows:
Initialization:

for $index_r = 0 to bin_r - 1$
for $index_g = 0 to bin_g - 1$
for $index_b = 0 to bin_b - 1$

$H(index_r, index_g, index_b) = 0$

end for
end for
end for

Histogram updation:

for i = 0 to M 1
for j = 0 to N 1

$index_r = R(i, j) * bin_r/256$
$index_g = R(i, j) * bin_g/256$
$index_b = R(i, j) * bin_b/256$
$H(index_r, index_g, index_b) = H(index_r, index_g, index_b + 1)$

end for
end for.

x is a floor function which returns largest integer smaller than x.

2.2 Color Moments

Color moments have been proved efficient in representing color distributions of images. This approach involves calculating the mean, the variance, and the third moment for each color channel, for providing a unique number used to index.

They are defined as:

$$\text{Mean} : \mu_i = \frac{1}{N}\sum_{j=1}^{N} P_{ij}\pi r^2$$

$$\text{Standard Deviation} : \sigma_i = \frac{1}{n}\sqrt{\sum_{j=1}^{N}\left(P_{ij} - \mu_i\right)^2}$$

$$\text{Skewness} : s_i = \left(\frac{1}{N} \sum_{j=1}^{N} (P_{ij} - u_i)^3 \right)^{\frac{1}{3}}$$

2.2.1 Combination of Color Descriptor

We normalize the distances using Iqbal's method [25]; this method ensures that the distance is normalized between 0 and 1. After the normalized distances, we use a linear combination of distances. $D(Q, I) = W_{\text{HC}} * D_{\text{HC}}(Q, I) + W_{\text{Mon}} * D_{\text{Mon}}(Q, I)$

where W_{HC} and W_{Mon} take a value between 0 and 1. This value can be a new program argument. $D_{\text{HC}} - D_{\text{Mon}}$ is [24] used in [26].

2.3 *Local Binary Pattern and Its Variants*

LBP is a texture descriptor that characterizes the texture image using two steps: thresholding and encoding. Given a pixel of image, LBP computes the local grayscale difference by comparing it with each of its neighbor pixel(s) in a selected radius of R. A zero is assigned when the difference is less than zero, and in remaining values, it is taken as 1. A decimal number is computed using these thresholds in the encoding step. In the case of a color image, it is firstly converted into grayscale before LBP computation. The LBP considers the information of the neighboring pixels to form the LBP code by comparison of the center pixel (currently processed pixel) value with its surrounding neighbors. For computing the LBP code, an input image F of size $M \times N$ in RGB color space is first converted into the inter-band-average representation as

$$g(x, y) = \frac{1}{3}[f_R(x, y) + f_G(x, y) + f_B(x, y)]$$

For $x = 1, 2, \ldots, M$ and $y = 1, 2, \ldots, N$, the symbol (x, y) denotes the pixel position of an image, while R, G, and B represents the red, green, and blue color space. Given a center pixel g_c and its neighboring pixel values g_p, the LBP code of image pixel (x, y) can be computed as [27]:

$$\text{LBP}_{PR}(x, y) = \sum_{p=0}^{p-1} S(g_p - g_c) 2^p \text{ where } S_x = \begin{cases} 1 & x \geq 0; \\ 0 & x < 0. \end{cases}$$

P and R denote the number of neighborhood pixels and radius of neighborhood, respectively.

The LBP histogram is derived as:

$$H_{\mathrm{LBP}}(k) = \sum_{x=1}^{M} \sum_{y=1}^{N} h(\mathrm{LBP}_{PR}(x, y), k); k \in [0.k]$$

$$\text{where } h(x, y) = \begin{cases} 1, \text{ if} x = y \\ 0 \ \text{ else.} \end{cases}$$

K denotes the maximum value of the LBP code (LBP pattern value). The pattern with uniformity constraint can be further applied to further reduce the LBP histogram bin. The circular binary representation is restricted into a limited transition or discontinuities, i.e., U (LBP $P, R \leq 2$). This additional restriction can be implemented using the lookup table to further speed up the computation time. The uniformity constraint is defined as:

$$U(\mathrm{LBP}_{PR}) = \left|S\left(g_{p-1} - g_c\right) - S(g_0 - g_c)\right| + \sum_{p=1}^{p-1}\left|S\left(g_p - g_c\right) - S\left(g_{p-1} - g_c\right)\right|$$

In the case of the uniformity constraint, the dimensionality of the LBP feature is $P \times (P - 1) + 3$.

In our implementation, the number of the neighboring pixel P is set at 8, implying the feature dimensionality is 59.

3 Proposed Method

The proposed image retrieval and classification systems employ two different image features, i.e., color and texture features. Herein, the color moments and color histogram represent the color feature, and LBP-based feature captures the textural information of an image. The two feature descriptors are fused together to represent an image. These two features can be employed to measure the similarity degree between two images. Figure 1 presents the way to create the final descriptor of the proposed method. The distance metric measures the similarity between two images by utilizing the similarity weighting constants. A good choice on similarity weighting constants yields better performance on image retrieval and classification systems.

There is no exact method to select and adjust the similarity weighting constants. Thus, selecting the suitable scaling factors can be experimentally determined which leads to better performance.

In this work, the particle swarm optimization (PSO) is exploited to determine the optimum similarity weighting constants to achieve better image retrieval and classification performances. PSO is population-based optimization approach which can search the optimum value on a given fitness function under predefined searching spaces. In the image retrieval and classification systems, the search space of PSO

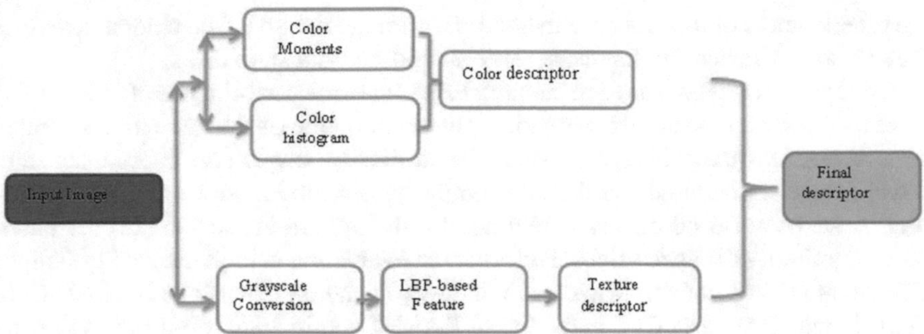

Fig. 1 Architecture of the proposed method

algorithm can be set as [0.1], indicating the minimum and maximum allowable similarity weighting constants. Thus, the similarity weighting constants can be initialized with any value in range [0.1] which is regarded as a single swarm in the PSO computation. In the optimization process, PSO may contain several swarm particles in which each particle owns one combination of similarity weighting constants, i.e., $\overleftarrow{x} = \{\omega_1, \omega_2, \omega_3\}$. Each swarm particle is iteratively updated based on the fitness function.

Herein, the fitness function is the image retrieval or classification performance. The PSO consists of two important steps, i.e., the similarity distance calculation and fitness function computation. The similarity distance computation involves the similarity weighting constants recorded by each swarm particle, whereas the fitness function computation employs the similarity distance obtained for each swarm particle.

The PSO fitness function for determining the optimum similarity weighting constants can be formally defined as follows.

$$\max\left\{\omega_1^*, \omega_2^*, \omega_3^*\right\} P\{q|\omega_1, \omega_2, \omega_3\}$$

where $\left\{\omega_1^*, \omega_2^*, \omega_3^*\right\}$ denotes the optimum similarity weighting constants. The symbol $q = \left(q_1, q_2, \ldots .q_{N_T}\right)$ is a set of training images in the PSO optimization process. The symbol $P\{*\}$ denotes the fitness function, i.e., image retrieval or classification performance. The PSO is executed until the maximum number of iterations is achieved or the maximum fitness function is met.

4 Experimental Results

For the evaluation of our contributions, extensive experiment results are reported in this section to demonstrate the effectiveness of the proposed system. Five textural image databases are utilized in this experiment to have an in-depth investigation of the effectiveness of the proposed method. Two image features, i.e., color and texture feature, are extracted for the image retrieval and classification tasks.

Created data sets are stored in memory to evaluate the performance of index structures using an average of several queries. The set of queries is the same for each index structure. In the image retrieval system, the similarity between the query image and target image is measured based on the similarity distance score from their descriptors. A set of retrieved images is returned by the system in ascending order based on the similarity distance values. Performance evaluation is implemented in Eclipse JEE Mars (JAVA) under a Microsoft Windows 7 environment on an Intel (R) Core (TM) i3 with 2.50 GHz CPU and 4 GB of RAM. Experimental result shows that the proposed method can yield processing speed at approximately 0.6 mini seconds per test picture. The performance accuracy of the image retrieval system is investigated using the average precision and recall rate, and average retrieval rate. For the image classification task, the nearest neighbor is employed to assign a class label of a given (query) image using a set of training images. The performance accuracy of the image classification is investigated using the percentage correct prediction of testing set.

4.1 Image Retrieval

The aim the CBIR is to have relevant images that fit the search request following the need of the user. The more the system provides answers that correspond to the user's need, the more efficient the system. To check the retrieval performance and robustness of the proposed method, in this experiment four texture image databases are employed, such as colored Brodatz, KTH TIPS, Stex, and USPTex texture image databases.

The colored Brodatz, KTH TIPS, Stex, and USPTex image databases consist of 2800, 810, 7616, and 2292 images, respectively. Each class of color Brodatz, KTH TIPS, Stex, and USPTex consists of 25, 81, 16, and 12 similar images, respectively. All images in these databases are of size 128×128, except for KTH TIPS which is of size 200×200. We display the best 10 similar images retrieved to prove the performance of our proposed algorithm. The results are presented separately (Figs. 2, 3, 4, 5). The images found to correspond to the 10 images showing similarities with the query. They are sorted out and posted taking into account distance between the descriptor of the query and that of the images already safeguarded in the database.

4.2 Retrieval Precision/Recall Evaluation

The performance of image recovery by the proposed method is evaluated using the average precision value (P (L)), average recall value (R (k)), and average retrieval rate (ARR). These values indicate the successfulness of the image retrieval scheme to retrieve a set of similar images. These values represent the percentage of relevant

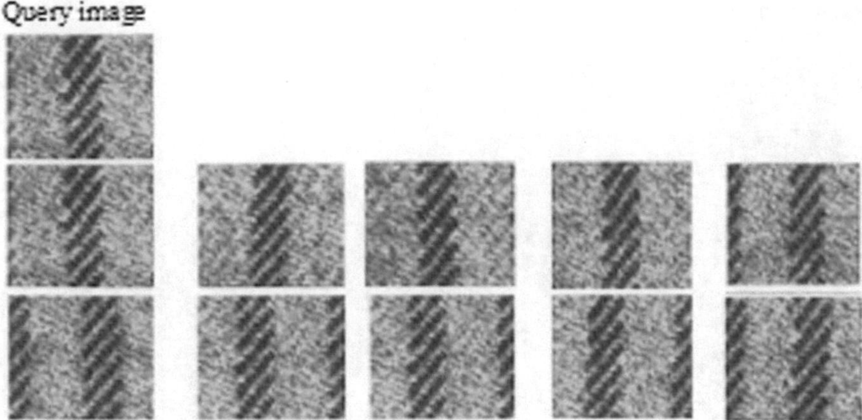

Fig. 2 Example of a query image and similarity search results in the colored Brodatz

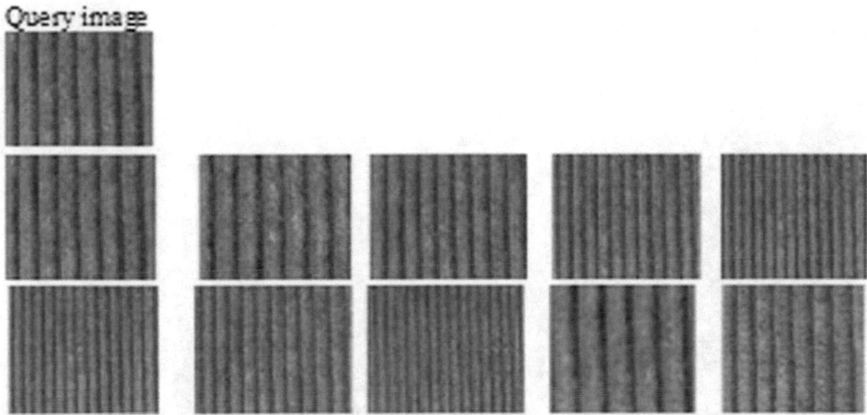

Fig. 3 Example of a query image and similarity search results in the KTH TIPS

images returned by image retrieval system. The performance of the image classification is measured with the percentage of correct prediction over all testing sets. A training set is needed for the prediction of the class label of testing set. Figure 6 shows the precision and recall rates over various image databases for the proposed method.

In we did a comparison between the proposed method and the some methods the most competitive [28–30]. The performance of the image retrieval is investigated using ARR score, and the performance of image classification is examined using percentage of correct classification result. The image feature dimensionality of the proposed scheme is controlled lower or identical to the most competitive former schemes. The experimental setting of the proposed method is set as identical to the

Fig. 4 Example of a query image and similarity search results in the Stex

Fig. 5 Example of a query image and similarity search results in the USPTex

former schemes over various image databases. The performance of the proposed method is delivered in form A/B. The symbol *A* indicates the performance with the similarity weighting constant, where symbol *B* denotes the performance of the proposed method with the optimum similarity weighting constants obtained from PSO algorithm. Table 1 summarizes the image retrieval performance between the

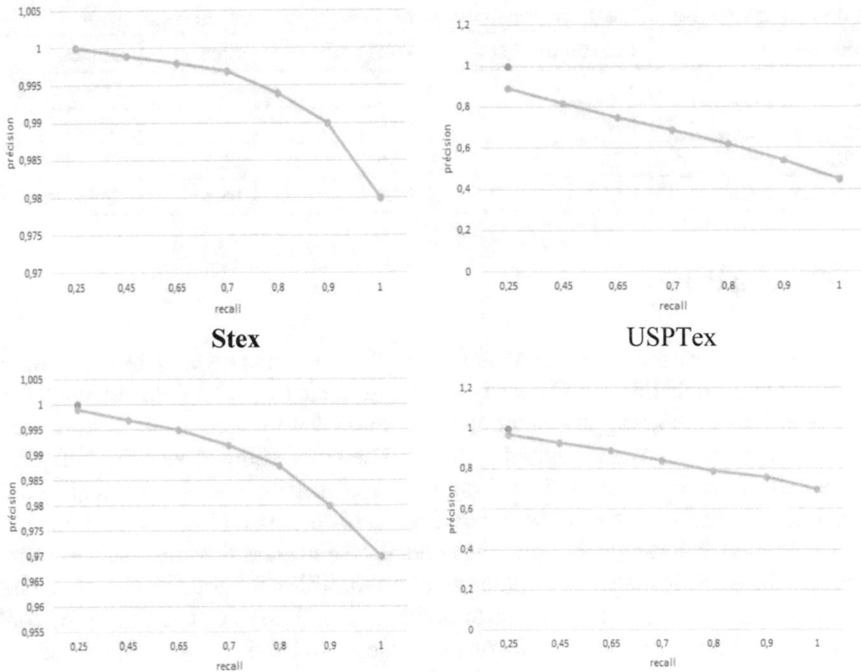

Fig. 6 Precision and recall rates over various image databases

Table 1 Image retrieval performance comparisons between the proposed method and the some methods, the most competitive

Method	Colored Brodatz	KTH TIPS	Stex	USPTex
[28]	71.18	70.23	75.78	72.58
[29]	73.38	71.34	83.2	82.38
[30]	84.11	72.60	95	94.16
Proposed method	86	74.16	96.4	95.5

proposed method and the existing methods under colored Brodatz, KTH TIPS, Stex, and USPTex image databases. Table 2 reports the image classification results between the proposed method and competitive methods in the Brodatz, KTH TIPS, Stex, and USPTex image databases.

As it can be seen from Tables 1 and 2, the proposed method outperforms the existing schemes in terms of the image retrieval and classification tasks. Thus, the fusion of the color and texture features is proved as an effective candidate in the practical application of image retrieval and classification systems.

Table 2 Classification accuracy between the proposed method and competitive methods

Method	Colored Brodatz	KTH TIPS	Stex	USPTex
[28]	82.5	79.8	86.3	83.8
[29]	84.64	80.4	91.5	92.5
[30]	92.3	82.6	98.3	97
Proposed method	94.22	85.4	98.5	98.45

5 Conclusions

This paper has presented a new method for indexing and image search by combining color and texture feature for similarity search in databases. The color features are represented by combining 2D histogram and statistical moments, and texture features are represented by LBP-based operator. The color compensates the difficulty of the LBP-based operator on describing the color distribution. As reported in the experimental results, the fusion of combining 2D histogram and statistical moments and LBP-based features yields a promising retrieval and classification result and outperforms the existing methods. For the future possibilities, other color spaces can be explored to capture the better color information computation to yield a better performance. The additional image feature, fusion strategy, and various similarity distances can be further explored as well. Moreover, the proposed image feature can also be extended to other research domains such as image annotation, shape matching, and face recognition.

References

1. Liu, Y., Zhang, D., Lu, G., Ma, W.Y.: A survey of content-based image retrieval with high-level semantics. Pattern Recogn. **40**, 262–282 (2007)
2. Ma, W., Zhang, H.: Content-based image indexing and retrieval. In: Borko, F. (ed.) Handbook of Multimedia Computing, pp. 227–251. CRC Press LLC, Florida (1999)
3. Jia, X., Yang, X., Cao, K., Zang, Y., Zhang, N., Dai, R., Zhu, X., Tian, J.: Multi-scale local binary pattern with filters for spoof fingerprint detection. Inform. Sci. **268**, 91–102 (2014)
4. Gu, J., Liu, C.: Feature local binary patterns with application to eye detection. Neurocomput. **113**, 138–152, (2013)
5. Bhateja, V., Gautam, A., Tiwari, A., Satapathy, S.C., Nhu, N.G., Le, D.N.: Haralick features-based classification of mammograms using SVM. In: Information Systems Design and Intelligent Applications, pp. 787–795. Springer, Singapore (2018)
6. Gupta, A., Ganguly, A., Bhateja, V.: A novel color edge detection technique using hilbert transform. In: Proceedings of the International Conference on Frontiers of Intelligent Computing: Theory and Applications (FICTA), pp. 725–732. Springer, Berlin, Heidelberg (2013)
7. El Houssif, N., Silkan, H.: Content-based image retrieval approach using color and texture applied to two databases (Coil-100 and Wang). In: International Conference on Soft Computing and Pattern Recognition, pp. 49–59. Springer, Cham (2017)
8. Kwitt, R., Uhl, A.: Lightweight probabilistic texture retrieval. IEEE Trans. Image Process. **19**(1), 241–253 (2010)

9. Lira, P.H., Giraldi, G.A., Neves, L.A.: Segmentation and feature extraction of panoramic dental X-ray images. In: Oral Healthcare and Technologies: Breakthroughs in Research and Practice, pp. 470–485. IGI Global (2017)
10. Li, Z., Liu, G., Yang, Y., You, J.: Scale- and rotation-invariant local binary pattern using scale-adaptive texton and subuniform-based circular shift. IEEE Trans. Image Process. **21**(4), 2130–2140 (2012)
11. Tan, X., Triggs, B.: Enhanced local texture feature sets for face recognition under difficult lighting onditions. IEEE Trans. Image Process. **19**(6), 1635–1650 (2010)
12. Nosaka, R., Fukui, K.: HEp-2 cell classification using rotation invariant co-occurrence among local binary patterns. Pattern Recogn. **47**(7), 2428–2436 (2014)
13. Subrahmanyam, M., Maheswari, R.P., Balasubramanian, R.: Local maximum edge binary patterns: a new descriptor for image retrieval and object tracking. Sig. Process. **92**(6), 1467–1479 (2012)
14. Manjunath, B.S., Ma, W.Y.: Texture feature for browsing and retrieval of image data, IEEE Trans. Pattern Anal. Mach. Intell. **18**(8), 837–842 (1996)
15. Hoang, M.A., Geusebroek, J.M.: Measurement of color texture. In: Proc. Workshop Texture Analysis and Machine Vision, pp. 73–76 (2002)
16. Davarzani, R., Yaghmaie, K., Mozaffari, S., Tapak, M.: Copy-move forgery detection using multiresolution local binary patterns. Forensic Sci. Int. **231**(1–3), 61–72 (2013)
17. Arantes, M., Gonzaga, A.: Recognition of human silhouette based on global features. Int. J. Natural Comput. Res. (IJNCR) **1**(4), 47–55 (2010)
18. Murala, S., Maheshwari, R.P., Balasubramanian, R.: Local tetra patterns: a new feature descriptor for content- based image retrieval. IEEE Trans. Image Process. **21**(5), 2874–2886 (2012)
19. Guo, J.M., Prasetyo, H., Su, H.S.: Image indexing using the color and bit pattern feature fusion. J. Vis. Commun. Image Represent. **24**(8), 1360–1379 (2013)
20. Satpathy, A., Jiang, X., Eng, H.L.: LBP-based edge-texture features for object recognition. IEEE Trans. Image Process. **23**(5), 1953–1964 (2014)
21. Subrahmanyam, M., Wu, Q.M.J., Maheshwari, R.P., Balasubramanian, R.: Modified color motif co-occurrence matrix for image indexing. Comput. Electric. Eng. **39**(3), 762–774 (2013)
22. Suruliandi, A., Meena, K., Rose, R.R.: Local binary pattern and its derivatives for face recognition. IET Comput. Vis. **6**(5), 480–488 (2012)
23. Liu, L., Long, Y., Fieguth, P.W., Lao, S., Zhao, G.: BRINT: Binary rotation invariant and noise tolerant texture classification. IEEE Trans. Image Process. **23**(7), 3071–3084 (2014)
24. Swain, M., Ballard, D.: Color indexing. Int. J. Comput. Vis. **7**(1), 11–32 (1991)
25. Iqbal, Q., Aggarwal, J.K.: Combining structure, color, and texture for image retrieval: a performance evaluation. In: Proceedings of 16th International Conference on Pattern Recognition, vol. 2, pp. 438–443. IEEE.gov (2002)
26. El Asnaoui, K., Aksasse, B., Ouanan, M.: Content-based color image retrieval based on the 2-D histogram and statistical moments. In: 2014 Second World Conference on Complex Systems (WCCS), pp. 653–656. IEEE (2014)
27. Ojala, T., Pietikainen, M., Maenpaa, T.: Multiresolution gray-scale and rotation invariant texture classification with local binary patterns. IEEE Trans. Pattern Anal. Mach. Intell. **24**(7), 971–987 (2002)
28. Kokare, M., Biswas, P.K., Chatterji, B.N.: Texture image retrieval using new rotated complex wavelet filters. IEEE Trans. Syst. Man Cybern. **35**(6), 1168–1178 (2005)
29. Kwitt, R., Uhl, A.: Image similarity measurement by Kullback-Leibler divergences between complex wavelet subband statistics for texture retrieval. In: Proceedings of 15th IEEE International Conference on Image Processing (ICIP 2008), pp. 933–936 (2008)
30. Liu, P., Guo, J.M., Chamnongthai, K., Prasetyo, H.: Fusion of color histogram and LBP-based features for texture image retrieval and classification. Inf. Sci. **390**, 95–111 (2017)

Image Segmentation Based on K-means and Genetic Algorithms

Lahbib Khrissi, Nabil El Akkad, Hassan Satori and Khalid Satori

Abstract In this paper, we studied image segmentation to which we applied a combination of the genetic algorithm and the cooperation between unsupervised classification by K-means and contour detection by the Sobel filter to improve image segmentation results by the K-means method alone. First, we will apply the segmentation process by combining two methods in the following way: We have hybridized the K-means method which is used to classify pixels into classes (regions), with the Sobel gradient filter which will then detect the edges of these regions, and then we will apply the genetic algorithm and by scanning further in the response space, try to find better quality class centers. This process is withdrawn until they are unable to find two sufficiently similar neighboring regions. The effectiveness of the proposed method was studied on a number of images. It is also compared by the K-means algorithm.

Keywords Segmentation · Genetic algorithm · Clustering · K-means

1 Introduction

Segmentation is a fundamental and important step in many different applications of computer vision like telecommunications (TV, video, advertising ...), medicine (radiography, ultrasound ...), biology, astronomy, geology, the industry (robotics, security), meteorology, architecture, printing, and weapons (military application). Segmenting an image consists in creating a subset of the image in subsets, called regions, as the set of pixels belongs to a region that checks for homogeneity criteria. A region is a set of related pixels with common properties that differentiate them from pixels in neighboring regions. Once the image is segmented, it is possible to make measurements on the regions obtained and thus to characterize each of them. The purpose of segmentation is to extract entities from an image to apply specific

L. Khrissi (✉) · N. El Akkad · H. Satori · K. Satori
LIIAN, Department of Computer Science, FSDM, Sidi Mohamed Ben Abdellah University, Fez, Morocco
e-mail: lahbib.khrissi@usmba.ac.ma

V. Bhateja et al. (eds.), *Embedded Systems and Artificial Intelligence*,
Advances in Intelligent Systems and Computing 1076,
https://doi.org/10.1007/978-981-15-0947-6_46

processing and interpret the image content. Regarding the segmentation of images, there are always difficulties because of the complexity of the natural images and the definition of the level of precision on the result. To date, there are many image segmentation methods that can be grouped into four main classes of algorithms:

- Segmentation based on regions [1, 2]. For example, there is region growth, decomposition/fusion.
- Segmentation based on contours [3, 4].
- Segmentation based on a global approach of the image, for example: thresholding, histogram, and approaches based on the color cloud.
- Segmentation based on classification methods, for example (KNN [5], SVM [6], K-means [7, 8]).
- Segmentation based on cooperation between the first segmentations [9].

In this article, we will focus on a segmentation approach by combining unsupervised segmentation that aims to automatically separate the image into clusters using the K-means algorithm and by genetic algorithms to optimize the parameters of the K-means algorithm and to search for an optimal configuration in the space of possible segmentations.

We will first present a brief description of the functioning of these algorithms, as well as the stakes of our approach.

The rest of this paper is organized as follows: We will discuss in the second part the approach we have proposed. The third part will be devoted to the presentation of the experiments and the discussion of the results obtained, and the conclusion will be addressed in the last part.

2 Presentation of the Proposed Approach

In this part, we will discuss our method which is based on the combination of the genetic extraction algorithm and the class extraction algorithm by hybridization between K-means and Sobel. The K-means clustering algorithm is based on optimizing the similarity scale between each cluster with the lowest and highest value for values in clusters. In other words, K-means tries to reduce similarity between regions and increase similarity within the region. Our objective is therefore to propose a classification technique based on the genetic algorithm, which involves the optimization of the final classes. The flowchart of this method is presented in Fig. 1.

2.1 K-means Algorithm

K-means defined by McQueen [10] is one of the simplest algorithms for automatic classification of data. The main idea is to choose randomly a set of centers fixed a priori and iteratively seek the optimal partition.

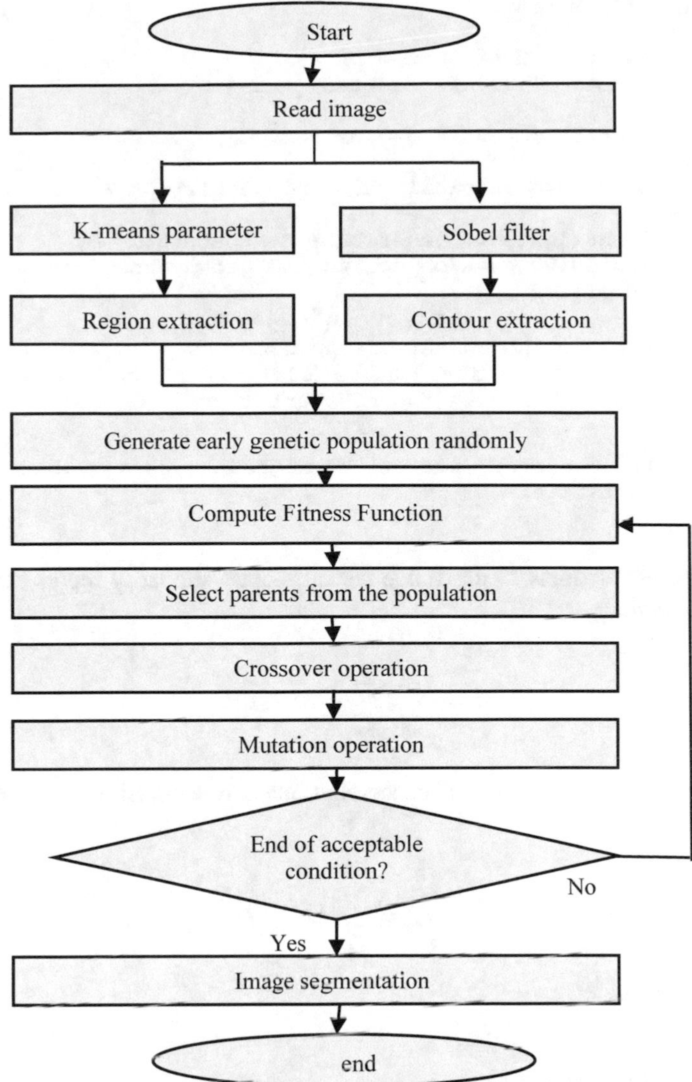

Fig. 1 Flow chart of the proposed method

Each individual is assigned to the nearest center. After the allocation of all the data, the average of each group is calculated; it constitutes the new representatives of the groups. When they have reached a stationary state (no data changes group), the algorithm is stopped.

```
Algorithm: K-means
Input
      Set of N data, noted by x
      Number of groups desired, noted by k
Output
      A partition of K groups {C1, C2, ... Ck}
Begin
      1) Random initialization of centers Ck ;
Repeat
      2) Assignment: generate a new partition by
      assigning each object to the group whose center
      is closest;
```

$$x_i \in C_k \text{ si } \forall j \left|x_i - \mu_k\right| = \min_j \left|x_i - \mu_k\right| \quad (1)$$

```
      With µk the center of the class K;
      3) Representation: Calculate the centers
      associated with the new partition;
```

$$\mu_k = \frac{1}{N}\sum_{x_j \in C_k} x_i \quad (2)$$

```
      Until convergence of the algorithm to a stable
      partition;
End.
```

This process attempts to maximize the intra-class similarity represented as an objective function:

$$J = \sum_{x=1}^{k} \sum_{x_j \in C_i} d(x_j - C_i) \quad (3)$$

In the case of the Euclidean distance, this function is called the quadratic error function.

$$J = \sum_{x=1}^{k} \sum_{x_j \in C_i} \left\|x_j - C_i\right\|^2 \quad (4)$$

2.2 Genetic Algorithms

Genetic algorithms [11] are a programming technique inspired by natural evolution. They are particularly adapted to problem optimization, introduced by C. Darwin in 1895 and based on the principle that the most suitable individuals have a better longevity as well as a better offspring. They are used in several fields, such as: segmentation [12] and 3D reconstruction [13, 14].

In fact, classical genetic algorithms rely heavily on universal coding in the form of fixed-length 0/1 chains and a set of genetic operators: mutation and crossing. An individual under this coding, called a chromosome, represents a configuration of the problem. "Genetic" operators are defined to operate randomly on one or two individuals without any knowledge of the problem.

(1) The individual: In AGs, the individual is a solution of the problem that will be coded, usually represented by a binary string of 0 and 1.
(2) Population: It consists of several individuals representing potential solutions (configuration) of the given problem.
(3) Coding of individuals: associate with each solution a genetic fingerprint (code)
(4) The evaluation function (fitness): It assigns a cost to each individual and can judge it fit or not to reproduce.
(5) Genetic operators:

(a) **Selection**

The evaluation of the performance of each individual is carried out by the objective function. This function is calculated from the parameter values of each individual in the population. This makes it possible to obtain a value of the evaluation function for each individual. In our case, we assume that the size of the population is q, so we have q values in total for the entire population.

After evaluating the performance of each individual, a classification step is made from assigning a probability of reproduction to each individual. Here, we used the Davis procedure [15]: The individual with the best performance receives a relative weight (VF_l) of q^a (usually a $\in [1, 1.5]$). The second-best individual receives a VF_l from $(q-1)^a$ to the last individual who receives a $VF_l = (q-(q-1))^a = 1$. Next, the reproduction PR_l of each individual is calculated by the following formula:

$$PR_l = \frac{VF_l}{\frac{1}{q}\sum_{l=1}^{q} VF_l} \tag{5}$$

To calculate the selection frequency of individuals, we use a stochastic remainder selection procedure developed by Goldberg [16]. The selection frequency of each individual is determined from the entire part of PR_l.

We, then, come to determine the set of individuals who are selected for selection and who are the best solutions obtained.

(b) **Crossing**

This phase represents the choice of pairs of individuals who will cross to create offspring. For this, we randomly choose a pair of individuals from the selection set. The idea is to take two individuals p_l, $p_s(l, s\, \varepsilon[1, q], l \neq s)$ and to check if there is crossing, according to a probability of crossing p_c. That is, if the crossing probability satisfies the condition of Michalewicz [17] ($\text{rand}_1 < p_c$, with rand_1 is a random number ε [0, 1]), we use the approach of Wright [18] to train offspring either:

$$\begin{cases} O_1 = 0.5p_l + 0.5p_s \\ O_2 = 1.5p_l - 0.5p_s \\ O_3 = -0.5p_l + 1.5p_s \end{cases} \quad (6)$$

We keep the two best individuals among the three obtained, so that the size of the population remains constant.

(c) **Mutation**

In this paper, we use a non-uniform mutation developed by Michalewicz [17]. It is an adaptive mutation, and it offers a good balance between the exploration of the research space and the refinement of chromosomes. The older the generations, the less this operator spreads the parameters of the convergence zone. If there is a mutation, that is to say, if the mutation probability p_m satisfies the condition of Michalewicz ($\text{rand}_2 < p_m$, with rand_2 is a random number ε [0, 1]), the change of a parameter p_{lj} is carried out as follows:

$$p'_{lj} = \begin{cases} p_{lj} + \varphi\left[\max\left(p_{lj}\right) - p_{lj}\right]\text{si}\sigma' < 0.5 \\ p_{lj} - \varphi\left[p_{lj} - \min\left(p_{lj}\right)\right]\text{si}\sigma' \geq 0.5 \end{cases} \quad (7)$$

Or φ is the function expressed as follows:

$$\varphi(h) = \text{h.r.}\left(1 - \frac{g'}{G'}\right)^b \quad (8)$$

Or r and σ' are two random numbers belonging to [0, 1], g' is the current generation, G' is the maximum generation and b is the non-uniform mutation coefficient.

2.3 *Use of Genetic Algorithms in Image Segmentation*

Genetic algorithms have been proven to solve some complex optimization problems. Large classes of applications are envisaged to date by the use of AGs. Note that AGs have been applied to image segmentation either to optimize the parameters of an existing segmentation algorithm or to search for optimal configuration in a space of possible segmentations. However, they face certain difficulties.

To program the AGs, we first start by coding the population of channels; there are many ways of coding them. Then, we program the three operators who can all be in the form of relatively simple code segments.

The programming of AGs can be summarized as follows:

(1) Initiation of the population.
(2) Evaluate each individual of this population.
(3) As long as the criterion is not satisfied do

(3.1) generate a new population
(3.2) apply the selection operator
(3.3) apply the transfer operator
(3.4) to evaluate the individuals of this population.

(4) Show the best state encountered during the search.

3 Experimentation and Discussion

The application we have designed is a method combining segmentation and genetic algorithms. We build mergeable regions as follows: A chromosome is represented by a pixel which is coded on 8 bits. The stage of generation of the initial population is done in a completely random way. The selection operator makes it possible to choose a set of individuals from the initial population whose fitness function is as minimal as possible. The crossing operator can generate two children from both parents selected by the selection operator. The mutation operator will make it possible to have the new generation obtained by the crossing operator of the parents selected among the individuals of the population evaluated. The chromosomes that satisfy the criterion of homogeneity specify a fusion of the two regions. The number of generations is chosen since it is necessary to adapt it to the size of the image. The process stops until the convergence.

The K-means algorithm is based on optimizing the similarity scale between each cluster with the lowest value and the highest value for the values in the clusters. In other words, K-means tries to reduce the similarity between the clusters and to increase the similarity within the cluster. Our proposed method is a classification technique based on the genetic algorithm, which involves the optimization of the final classes. This is a simple criterion that intuitively develops clusters. As in the K-means algorithm, we must minimize the distance for a good grouping. However, unlike the K-means algorithm, it can be trapped in values that are not optimal. In this spirit, the simplicity of the K-means algorithm and the ability of the genetic algorithm to avoid trapping in the optimal local environment are combined to provide our GA-based clustering approach. Experiments and simulations show the performance and quality of this work.

In view of the images in Fig. 2, we can see that the combination approach between K-means and the proposed genetic algorithms leads to the best image segmentation results compared to those obtained from K-means alone.

The dice similarity index is coefficient which is region bound, and it calculates overlay and segmentation results:

$$\mathrm{SDC} = (2 * \mathrm{TP})/(2 * \mathrm{TP} + \mathrm{FP} + \mathrm{FN}) \tag{9}$$

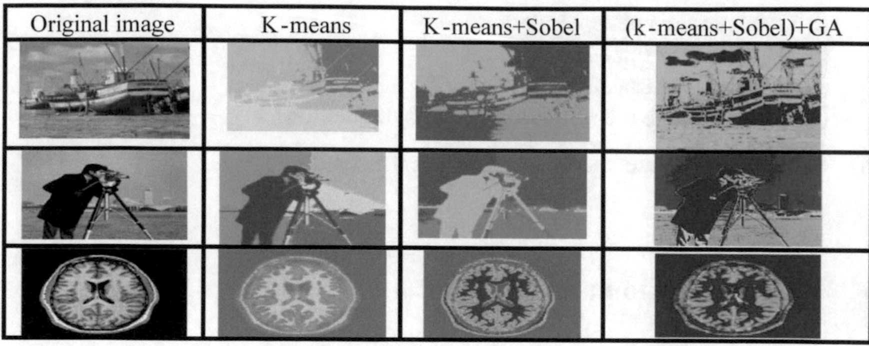

Fig. 2 Results obtained by the different methods

K-means (%)	K-means + Sobel (%)	(K-means + Sobel) + GA (%)
69.31	76.21	93.45

4 Conclusion

In this article, we have presented the most commonly used methods for image segmentation. Currently, the trend in image segmentation is to combine two or more methods in order to take advantage of each to best address the variability of images and the wealth of information they contain. In this respect, we proposed a segmentation method by combining genetic algorithms and hybridization of the K-means algorithm and the Sobel filter. We tested our method on several images. Based on the results obtained, we have found that the results of our approach are better than those obtained by the K-means method alone.

References

1. Muñoz, X., Freixenet, J., Cufi, X., Marti, J.: Strategies for image segmentation combining region and boundary information. Pattern Recogn. Lett. **24**, 375–392 (2003)
2. Fan, J. Yau, D.K.Y., Elmagarmid, A.K., Aref, W.G.: Automatic image segmentation by integrating color-edge extraction and seeded region growing. IEEE Trans. Image Process. **10**, 1454–1466 (2001)
3. Canny, J.: A computational approach to edge detection. IEEE Trans. Pattern Anal. Mach. Intell. **8**, 679–698 (1986)
4. Deriche, R.: Fast algorithms for low-level vision. IEEE Trans. Pattern Anal. Mach. Intel. **12**, 78–87 (1990)
5. Seema, W., Keshavamurthy, B.N., Hussain, A.: Region-based segmentation of social images using soft KNN algorithm. Proc. Comput. Sci. **125**, 93–98 (2018)
6. Yang, H.Y., Wang, X.Y., Wang, Q.Y., Zhang, X.J.: LS-SVM based image segmentation using color and texture information. J. Vis. Commun. Image Represent. **23**, 1095–1112 (2012)

7. Rachid, S., et al.: Comparison between K mean and fuzzy c-mean methods for segmentation of near infrared fluorescent image for diagnosing prostate cancer, IEEE (2015)
8. Khan, S.S., Amir, A.: Cluster centre initialization algorithm for k-means cluster. Pattern Recogn. Lett. **25**, 1293–1302 (2004)
9. El-merabet, Y., Meurie, C., Ruichek, Y., Sbihi, A., Touahni, R.: Orthophotoplan segmentation based on region merging for roof detection. In: IS&T/SPIE Electronic Imaging-Image Processing: Machine Vision Applications V, vol. 8300. Burlingame, USA (2012)
10. Celeux, G., Diday, E., Govaert, G., Lechevallier, Y., Ralam-Bondrainy, H.: Classification Automatique des Données. Bordas, Paris (1989)
11. Maulik, U., Bandyopadhyay, S.: Genetic algorithm-based clustering technique. Pattern Recogn. **33**(9), 1455–1465 (1999)
12. Anubha, K., Himanshu, Y., Anurag, J.: Image segmentation using genetic algorithm. Int. J. Sci. Eng. Res. **5**(2) (2014)
13. El akkad, N., El hazzat, S., Saaidi A., Satori, K.: Reconstruction of 3D scenes by camera self-calibration and using genetic algorithms. 3D Res. (Springer) **7**(6), 1–17 (2016)
14. Mostafa, M., Elakkad, N., Saaidi, A., Gadhi Nazih, A., Satori, K.: Robust method for camera calibration with varying parameters using hybrid modified genetic simplex algorithm. J. Theoret. Appl. Inf. Technol. **51**(3), 363–373 (2013)
15. Davis, L.: Handbook of Genetic Algorithms. Van Nostrand Reinhold, New York (1991)
16. Goldberg, D.E.: Genetic Algorithms in Search. Optimization and Machine Learning, pp. 1–432. Addison-Wesley (1989)
17. Michalewicz, Z.: Genetic Algorithms + Data Structures = Evolution Programs. Springer, Berlin (1996)
18. Wright, A.: Genetic Algorithms for Real Parameter Optimization, pp. 205–218. Morgan Kaufmann, San Mateo, CA (1991)

Silhouettes Based-3D Object Reconstruction Using Hybrid Sparse 3D Reconstruction and Volumetric Methods

Soulaiman El Hazzat, Mostafa Merras, Nabil El Akkad, Abderrahim Saaidi and Khalid Satori

Abstract This paper presents a hybrid approach for 3D object reconstruction from multiple images taken from different viewpoints. The proposed method allows to obtain a complete and automatic reconstruction from limited number of images. It begins with a sparse 3D reconstruction based on camera self-calibration and interest point matching between images. The integration of sparse approach allows us to automatically estimate the projection matrices without using a turn-table (controlled environment) often used in the Shape from Silhouette approach. In addition, it offers the possibility of an accurate estimation of the initial bounding box of the object. This bounding box is discretized into voxels afterward. Then, the reconstruction process consists in using the image Silhouettes and the photo-consistency test to finally have a volumetric textured model that can be transformed into surface model by applying the marching cube algorithm. The experiments on real data are performed to validate the proposed approach; the results indicate that our method can achieve a very satisfactory reconstruction quality.

Keywords Sparse 3D reconstruction · Self-calibration · Shape from Silhouettes · Bounding box · Photo-consistency

1 Introduction

3D reconstruction from images is an important topic in the field of computer vision. It is to perceive the 3D environment from a set of 2D images taken from different

---

S. El Hazzat (✉) · M. Merras · N. El Akkad · A. Saaidi · K. Satori
LIIAN, Department of Computer Science, Faculty of Sciences Dhar-Mahraz, Sidi Mohamed Ben Abdellah University, Fes, Morocco

M. Merras
Department of Computer Science, High School of Technology, Moulay Ismaïl University, Meknes, Morocco

A. Saaidi
LSI, Department of Mathematics, Physics and Informatics, Polydisciplinary Faculty of Taza, Sidi Mohamed Ben Abdellah University, Taza, Morocco

V. Bhateja et al. (eds.), *Embedded Systems and Artificial Intelligence*,
Advances in Intelligent Systems and Computing 1076,
https://doi.org/10.1007/978-981-15-0947-6_47

viewpoints. Its importance is reflected by the diversity of application area: robotics, quality control, monitoring, medicine, pattern recognition, etc.

Several methods have been proposed to solve this problem. Stereo methods allow to make 3D reconstruction from calibrated stereo images. The simplest stereovision system is the binocular stereovision [1], it uses only two images. But for a complete 3D reconstruction, we must use more than two images. It is called multi-view stereo (MVS) [2–7]. There are stereo methods that start from an initial volume discretized into voxels to find the final form [8]. Others are based on the research of matches between the images. These matches can be sparse, quasi dense or dense [1, 2]. The structure from motion approach (SFM) [5, 9–11] allows to automatically find both the 3D geometry of the scene and the camera positions, these calculations are based on interest points detection and matching between different images. The Shape from Silhouette approach [12, 13] is based on the use of Silhouette images to estimate an approximate volumetric shape of the object from a set of calibrated images. First, it consists to estimate a bounding box that is discretized into voxels, then to remove all projected voxels outside Silhouette images and keep the other ones. The result of the reconstruction is a bounding shape called Visual Hull. The quality of the reconstruction depends strongly on the number of images used and the selected resolution. There are other approaches based on the photo-consistency test [14, 15] to improve the reconstruction result obtained by the Shape from Silhouette approach and obtaining textured results close to reality.

In this work, we propose a hybrid approach that uses at first the structure from motion approach with camera self-calibration for the automatic estimation of projection matrices and obtaining a sparse 3D point cloud. This 3D point cloud will be used for an exact estimation of bounding box. Then, Silhouette images and the photo-consistency test allow to estimate the final 3D model from the bounding box discretized into voxels.

The remainder of this paper is organized as follows. In Sect. 2, we present the 3D reconstruction background used in this paper. Section 3 describes our approach to 3D objects reconstruction. We present experimental results in Sect. 4. The conclusion is presented in Sect. 5.

2 Background

2.1 Self-calibration and Sparse 3D Reconstruction Procedure

From a set of images taken from different viewpoints, we want to make the self-calibration of the camera and the sparse 3D reconstruction using matching points. The paragraphs below show the necessary steps.

A. **Matching and fundamental matrix estimation**

The treatment consists in extracting the interest points in each image. In this work, we use the Harris detector [16]. Then, for the matching of these points we use ZNCC measure [17]. For the estimation of the fundamental matrix and elimination of false matches, the RANSAC Algorithm [18] was used. This estimation is based on the equation: $m_2^T F m_1 = 0$

With (m_1, m_2) is a pair of corresponding points.

B. **Camera self-calibration**

The calibration or self-calibration of the cameras [19–24] is a very interesting step to solve the problem of multi-view 3D reconstruction. So, without knowing the camera parameters, it is possible to make only a projective 3D reconstruction.

In this work, we used the method [19] for camera self-calibration. This method is based on the use of fundamental matrix and the projection of two points of the scene in the images for the estimation of intrinsic camera parameters.

C. **Sparse 3D reconstruction**

In this step, structure from motion approach [5, 9–11] was used to estimate the camera motion and a sparse 3D reconstruction of matching points.

2.2 *Shape from Silhouettes*

We are interested in the volumetric approach [13] that consists to reconstruct a volume containing the object called "Visual Hull," which can be calculated from the Silhouettes of the object observed in different views. The major drawback of the Shape from Silhouette approach is that it does not allow reconstructing the concavities of the object. This problem can be solved by the use of Silhouette images and color information (color consistency between images). The volumetric approach begins first with the estimation and discretization of a bounding box into voxels. The reconstruction process consists in projecting the voxels in the image plans (see Fig. 1). If the projections are within the Silhouette images, then the voxels belong to the reconstruction (they are preserved). Otherwise, the voxels do not belong to the reconstruction (they are removed). The result of the reconstruction is composed of the set of voxels remaining.

3 Proposed Approach

3.1 *Description*

The methods based on the Shape from Silhouettes approach [12, 13] often use a turn-table (controlled system) to easily obtain the movements of the camera around the object and to facilitate the camera calibration.

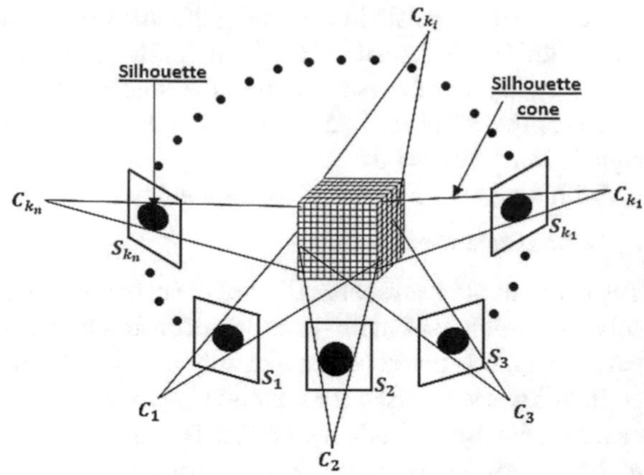

Fig. 1 Shape from Silhouette principle

In this work, we did not impose any constraint on the reconstruction system. Then, for the estimation of the projection matrices and the bounding box we use the structure from motion approach with camera self-calibration. Sparse 3D reconstruction is used for the reliable estimation of the bounding box. This bounding box is discretized into voxels. After that, the reconstruction process consists in projecting the voxels in the images plans. The projection voxels inside the Silhouettes that verify the test of photo-consistency are maintained whereas the others are eliminated. The final result is a volumetric model of the object composed of a set of voxels (voxels remaining). Figure 2 shows the flowchart of our approach.

3.2 Steps

The proposed approach is comprised of several steps:

1. Using the structure from motion approach with self-calibration for the automatic estimation of camera parameters and having a sparse 3D reconstruction.
2. The result of the 3D reconstruction (3D point cloud) is used to estimate the bounding box.
3. Discretization of the bounding box into voxels.
4. Extraction of Silhouette images [25, 26].
5. Projection of the voxels in the image plans.
6. The projected voxels inside the Silhouettes that verify the photo-consistency test are kept and the other voxels are eliminated.
7. The result of the reconstruction is a volumetric model. It can be converted to a surface model by applying the marching cube algorithm.

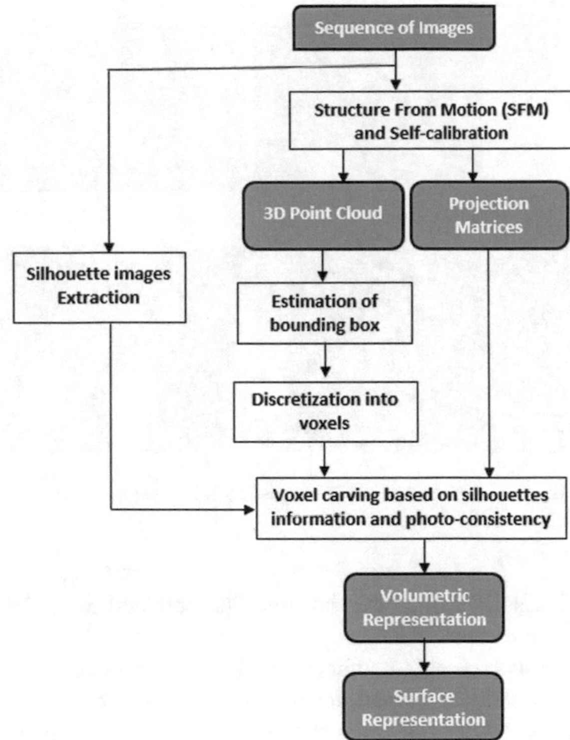

Fig. 2 Scheme of the proposed approach

4 Experiments Results

The presented approach was implemented in Java, using JAMA library and the Java 3D API, and all experiments are executed on a machine HP 650 Intel Core i3, 2.30 GHz CPU, 4 GB of RAM. We present the results of two data sets with different resolutions.

4.1 First Sequence

A digital camera was used to take 14 images with resolution of 640 × 480 pixels. Three images are presented in Fig. 3a. First, the structure from motion approach and the camera self-calibration are applied to estimate the camera parameters as well as to recover a sparse 3D point cloud. This approach is based on the detection and matching of interest points between the different images. In this work, we used Harris detector [16] to detect the interest points and the ZNCC correlation measure for the matching of these points. Figure 3b shows a result of interest point matching

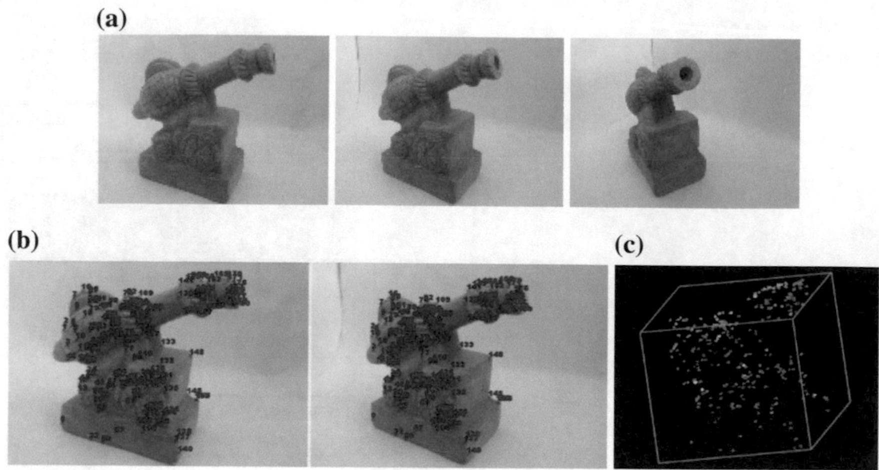

Fig. 3 **a** Three images of the sequence, **b** interest points detection, **c** sparse 3D reconstruction and bounding box

obtained between two images. The obtained 3D point cloud and the bounding box are presented in Fig. 3c.

First, the obtained bounding box is discretized into voxels. After that, we use the Silhouettes information and photo-consistency test to obtain a result of volumetric reconstruction. Results of Silhouettes extraction are presented in Fig. 4a. Figure 4b shows three views of volumetric reconstruction.

4.2 Second Sequence

Here, 12 images with resolution of 615 × 500 pixels were used for 3D reconstruction. Figure 5 shows the obtained results. Two images are presented in Fig. 5a. Silhouettes of object are presented in Fig. 5b. The obtained 3D model is shown in Fig. 5c.

From the obtained 3D models shown in Figs. 4 and 5, we can conclude that the results of our approach are very satisfactory for three-dimensional reconstruction.

5 Conclusion

Our work was motivated by the observation that there is a wider demand for virtual 3D models in real engineering applications. Toward that end, we have proposed a hybrid approach for 3D reconstruction able to have a volumetric, complete and automatic 3D reconstruction of objects. Our method is based on the use of the structure from motion approach with the camera self-calibration for an automatic estimation of

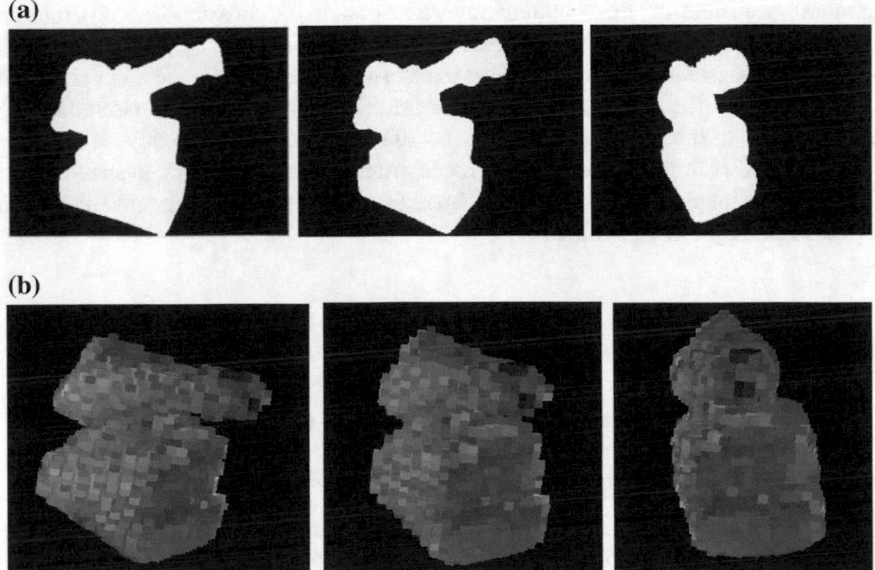

Fig. 4 **a** Extraction of Silhouette images, **b** different views of 3D reconstruction result (resolution 30 × 30 × 30)

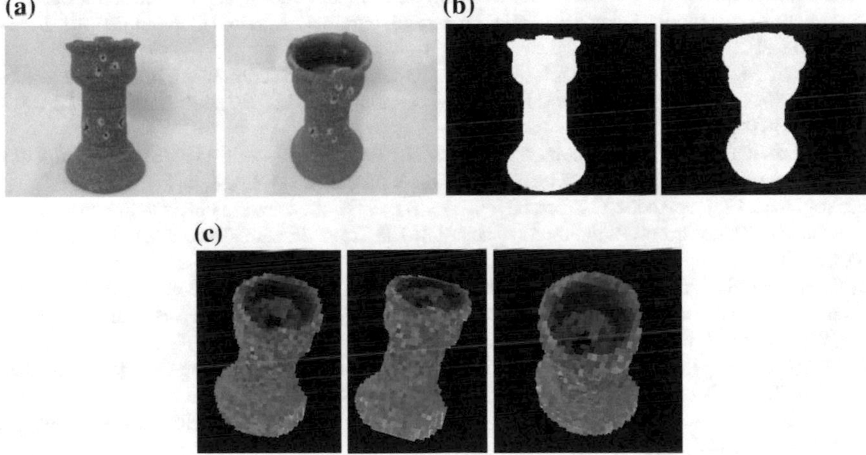

Fig. 5 **a** Two images of the sequence. **b** Extraction of Silhouette images. **c** Different views of 3D reconstruction result (resolution 30 × 30 × 50)

camera parameters and the bounding box. After that, to obtain a dense reconstruction we used Shape from Silhouette approach based on Silhouette images combined with the photo-consistency test that allows us to have a textured volumetric reconstruction that can be transformed into surface reconstruction by applying the marching cube algorithm. Our future work will focus on 3D reconstruction directly of real-life videos. Another improvement, the 3D reconstruction process will be used to collect data from a filmed scene. This data can be used as to apply an intelligent strategy in a particular engineering problem.

References

1. Da, F., Sui, Y.: 3D reconstruction of human face based on an improved seeds-growing algorithm. Mach. Vis. Appl. **22**, 879–887 (2011)
2. Furukawa, Y., Ponce, J.: Accurate, dense, and robust multiview stereopsis. IEEE Trans. Pattern Anal. Mach. Intell. **32**(8), 1362–1376 (2010)
3. El Hazzat, S., Merras, M., El Akkad, N., et al.: Enhancement of sparse 3D reconstruction using a modified match propagation based on particle swarm optimization. Multimed. Tools Appl. (2018). https://doi.org/10.1007/s11042-018-6828-1
4. Merras, M., Saaidi, A., El Akkad, N., et al.: Multi-view 3D reconstruction and modeling of the unknown 3D scenes using genetic algorithms. Soft Comput. **22**, 6271 (2018). https://doi.org/10.1007/s00500-017-2966-z
5. El Hazzat, S., Merras, M., El Akkad, N., et al.: 3D reconstruction system based on incremental structure from motion using a camera with varying parameters. Visual Comput. **34**(10), 1443–1460 (2018)
6. Merras, M., El Hazzat, S., Saaidi, A., et al.: 3D face reconstruction using images from cameras with varying parameters. Int. J. Autom. Comput. **14**, 661 (2017). https://doi.org/10.1007/s11633-016-0999-x
7. El Hazzat, S., Saaidi, A., Karam, A., Satori, K.: Incremental multi-view 3D reconstruction starting from two images taken by a stereo pair of cameras. 3D Res. **6**(1), 1–18 (2015)
8. Vogiatzis, G., Hernández, C., Torr, P.H.S., Cipolla, R.: Multi-view stereo via volumetric graph-cuts and occlusion robust photo-consistency. IEEE Trans. Pattern Anal. Mach. Intell. **29**(12) (2007)
9. El Hazzat, S., Saaidi, A., Satori, K.: Structure from motion for 3D object reconstruction based on local and global bundle adjustment. In: 2015 Third World Conference on Complex Systems (WCCS), pp. 1–6 (2015)
10. Wang, G., Wu, J.: Perspective 3-D Euclidean reconstruction with varying camera parameters. IEEE Trans. Circuits Syst. Video Technol. **19**, 1793–1803 (2009)
11. Del Bue, A.: A factorization approach to structure from motion with shape priors. In: Proceedings CVPR (2008)
12. Laurentini, A.: The visual hull concept for Silhouette based image understanding. IEEE PAMI **16**(2), 150–162 (1994)
13. Mulayim, A.Y., Yilmaz, U., Atalay, V.: Silhouette-based 3-D model reconstruction from multiple images. IEEE Trans. Syst. Man Cybern. B Cybern. **33**(4), 582–591 (2003)
14. Kutulakos, K.N., Seitz, S.M.: A theory of shape by space carving. Int. J. Comput. Vis. **38**(3), 199–218 (2000)
15. Montenegro, A.A., Carvalho, P.C.P., Gattass, M., Velho, L.: Adaptive space carving. In: 3DPVT 2004, 6–9 Sept 2004, Greece
16. Harris, C., Stephens, M.: A combined corner and edge detector. In: Alvey Vision Conference, pp. 147–151 (1988)

17. Di Stefano, L., Mattoccia, S., Tombari, F.: ZNCC-based template matching using bounded partial correlation. Pattern Recogn. Lett. **26**, 2129–2134 (2005)
18. Fischler, M.A., Bolles, R.C.: Random sample consensus: a paradigm for model fitting with applications to image analysis and automated cartography. Commun. ACM **24**(6), 381–395 (1981)
19. El Akkad, N., Merras, M., Saaidi, A., Satori, K.: Camera self-calibration with varying intrinsic parameters by an unknown three-dimensional scene. Visual Comput. **30**(5), 519–530 (2014)
20. El Akkad, N., Merras, M., Saaidi, A., Satori, K.: Camera self-calibration with varying parameters from two views. WSEAS Trans. Inf. Sci. Appl. **10**(11), 356 (2013)
21. Merras, M., et al.: Camera self-calibration with varying parameters by an unknown three dimensional scene using the improved genetic algorithm. 3D Res. **6**(1), 1–14 (2015)
22. Merras, M., El Akkad, N., Saaidi, A., et al.: Camera calibration with varying parameters based on improved genetic algorithm. WSEAS Trans. Comput. **13**, 129–137 (2014)
23. El Akkad, N., Merras, M., Saaidi, A., Satori, K.: Robust method for self-calibration of cameras having the varying intrinsic parameters. J. Theoret. Appl. Inf. Technol. **50**(1) (2013)
24. El Akkad, N., Saaidi, A., Satori, K.: Self-calibration based on a circle of the cameras having the varying intrinsic parameters. ICMCS, 161–166 (2012)
25. Lee, W., Woo, W., Boyer, E.: Silhouette segmentation in multiple views. IEEE TPAMI **33**(7), 1429–1441 (2011)
26. Zeng, G., Quan, L.: Silhouette extraction from multiple images of an unknown background. In: Proceedings of ACCV, pp. 628–633 (2004)

2D Brain Tumor Segmentation Based on Thermal Analysis Model Using U-Net on GPUs

Abdelmajid Bousselham, Omar Bouattane, Mohamed Youssfi and Abdelhadi Raihani

Abstract Brain tumor segmentation allows separating normal and abnormal pixels. In clinical practice, stills a challenging task, due to the complicated structure of the tumors. This paper aims to improve the process of segmentation based on brain tumor thermal profile. Brain tumors are a fast proliferation of abnormal cells, which thermally represent a heat source. In this work, we segment brain tumors using U-Net fully convolutional neural network based on the change on the temperature in the tumor zone. The temperature distributions of the brain including the tumor were generated using the Pennes bioheat transfer equation and converted to grayscale thermal images. Next, U-Net was applied to segment tumors from thermal images. A dataset containing 276 thermal images was created to train the model. As the process of training the model is time-consuming, we used massively parallel architecture based on graphical processing unit (GPU). We tested the model in 25 thermal images, and we obtained a precise segmentation with Accuracy = 0.9965, Prccision = 0.9817, Recall = 0.9513, and F1 score = 0.9338. The training time was 20 h in NVIDIA GTX 1060 GPU. The obtained results prove the effectiveness of deep learning and thermal analysis of brain tumors to reinforce segmentation using magnetic resonance imaging (MRI) to increase the accuracy of diagnosis.

Keywords MRI · Bioheat transfer · CNN · U-Net · GPU

A. Bousselham (✉) · O. Bouattane · M. Youssfi · A. Raihani
Laboratory SSDIA, ENSET Mohammedia, University Hassan 2 Casablanca, Casablanca, Morocco
e-mail: abdelmajid.bousselham@gmail.com

O. Bouattane
e-mail: o.bouattane@gmail.com

M. Youssfi
e-mail: med@youssfi.net

A. Raihani
e-mail: abraihani@yahoo.fr

V. Bhateja et al. (eds.), *Embedded Systems and Artificial Intelligence*, Advances in Intelligent Systems and Computing 1076,
https://doi.org/10.1007/978-981-15-0947-6_48

1 Introduction

A brain tumor is an uncontrollable proliferation of abnormal cells in the brain. We distinguish two categories of brain tumors, primary and secondary. Primary tumors begin in the brain and still in the brain region, whereas secondary tumors start in other organs in the body and spread to the brain region. MRI is the standard medical technique for brain tumor diagnosis, as it provides rich information about its structure. Several methods were developed in the literature for brain tumor segmentation [1–3], such as threshold-based methods [4, 5], classification and clustering methods [6, 7], deformable model methods [8, 9], and deep learning methods [10, 11].

Precise and accurate brain tumor segmentation from MRI is a critical task in clinical practice due to their complicated structures and overlapping intensities with healthy tissues. In this paper, we used the thermal profile of brain tumors to increase the accuracy of segmentation. Tumors have a high heat generation; the temperature in the tumorous region is high compared to surrounding healthy tissues [12]. Toward enhancing brain tumor accuracy, in our previous paper [12], we introduced a new approach by using thermal information of brain tumor to reinforce segmentation from MRI. The temperature was calculated based on Pennes bioheat equation [13], and the temperature profile presented a high variation in tumor borders, is the reason we used Canny edge detector to estimate tumor contours. However, the Canny edge detector is still not a powerful method to segment brain tumors, as they have complicated structures, and finds some difficulties to detect contours when there is a low gradient of temperature in tumor borders. In the present paper, we used fully convolutional neural networks, as they showed impressive results in recent years in medical image analysis.

Convolutional neural networks (CNNs) were extensively used in recent years for brain tumor segmentation. Some of CNN-based models have been developed for brain tumor segmentation [11, 15, 16]. U-Net architecture [14] is the most well known in biomedical image analysis [15], it is a fully convolutional neural network developed using encoder–decoder architecture applied to image pixel-level labeling prediction. Training the U-Net model in CPU is time-consuming. Therefore, we used NVIDIA GPUs as a massively parallel platform to speed up the process of training. GPUs are extensively used in recent years to accelerate segmentation in medical image analysis [17, 18]. In this paper, we used U-Net architecture applied in thermal images and trained on the graphical processing unit (GPU) toward more effective brain tumor segmentation using MRI.

2 Methods

2.1 Brain Temperature Calculation

To calculate the temperature in the brain with tumor, we used Pennes bioheat equation [13, 19], it is a partial differential equation described in the following formula:

$$\rho C_P \frac{\partial T}{\partial t} = \mathrm{K} \cdot \left(\frac{\partial^2 T}{\partial x^2} + \frac{\partial^2 T}{\partial y^2} \right) + \omega_b \rho_b C_{pb} (T_a - T) + Q_m, \tag{1}$$

Discretization, initial, and boundary conditions are described in our recent work [12]. The discretized form of Pennes equation in a 2D Cartesian grid is presented as follows:

$$T_{i,j}^{n+1} = T_{i,j}^{n} + \frac{\Delta t \mathrm{K}_{i,j}}{\rho_{i,j} C_{i,j} \Delta x^2} \cdot \begin{bmatrix} T_{i-1,j}^{n} + T_{i+1,j}^{n} + T_{i,j-1}^{n} \\ + T_{i,j+1}^{n} - 4T_{i,j}^{n} \end{bmatrix} + \frac{\Delta t}{\rho_{i,j} C_{i,j}} \left[(\omega_b)_{i,j} (\rho_b)_{i,j} (C_{Pb})_{i,j} \left(T_a^n - T_{i,j}^n \right) + Q_{i,j} \right] \tag{2}$$

Normal brain tissues' and tumors' thermal properties are mentioned in [12].

2.2 U-Net-Based Fully Convolutional Network

U-Net is a CNN which has been created for biomedical image segmentation [14], and in recent works it was used to segment brain tumors [21–23]. The idea of the network is based on the fully convolutional network (FCN) [20], and its architecture was extended to be trained with fewer images and segmentation that is more accurate. The architecture of the network is shown in Fig. 1; it introduces symmetry into the FCN network by increasing the size of the decoder to fit the encoder and replacing the sum operation in the jumped connections by a concatenation.

The output results of U-Net architecture are highly affected by its parameters. We adopted stochastic gradient-based optimization to minimize the cost function. Adaptive moment estimator (Adam) [24] was used to estimate the parameters with a learning rate of 0.0001, a batch size of 8, and a number of epochs of 100. The network weights were initialized by a normal distribution with a mean of 0 and a standard deviation of 0.01, and network biases were initialized as 0, and binary cross-entropy was used as a loss function.

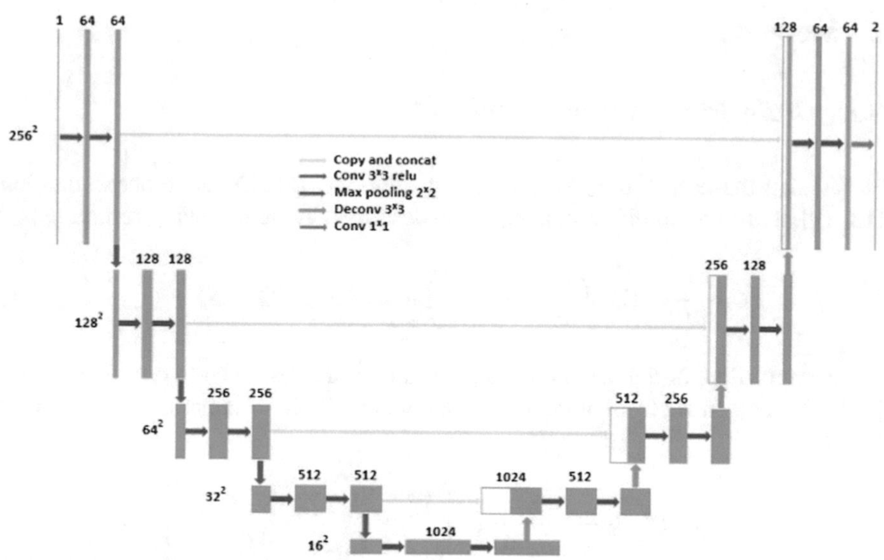

Fig. 1 The used U-Net architecture

2.3 *Datasets*

To evaluate the proposed approach, we used synthetic MRI data, as it has the ground truth of normal tissues and tumors, as they are needed to calculate the temperature of thermal images using Pennes equation. The U-Net network was trained in thermal images generated from BRATS 2013 [1, 25]. It is a public database that contains 50 synthetics datasets; we have taken 276 2D images from 25 patients to generate thermal images for training; all the images contain a tumor inside. An additional 25 thermal images were generated from the rest of 25 patients to test the model.

2.4 *Implementation Platform*

U-Net architecture is implemented using Deeplearning4j (DL4J) [26]. It is an open-source, distributed, deep learning library for Java and Java virtual machine (JVM) that allows us to build, train, and test a wide variety of deep learning methods. DL4J supports parallel training of neural networks using GPU and multi-GPU architectures and also works with a cluster of CPUs or GPUs using Apache Spark and distributed GPUs.

To evaluate the network in thermal images, we used NVIDIA GeForce GTX 1060 graphics card (compute capability 6.1). This GPU is based on Pascal architecture with 1280 cores and 6 GB GDDR5 of memory. Windows 7 (64 bits) with a CPU Intel i7-4770k with four cores of 3.50 GHz of clock speed, eight threads, and 16 GB of memory was used as a platform to train and test the network.

Four segmentation evaluation metrics were used in this work, namely Accuracy, Precision, Recall, and F1 score, which are defined in the following equations:

$$\text{Accuracy} = \frac{(\text{TP} + \text{TN})}{(\text{TP} + \text{FN} + \text{TN} + \text{FP})} \tag{3}$$

$$\text{Precision} = \frac{\text{TP}}{\text{TP} + \text{FP}} \tag{4}$$

$$\text{Recall} = \frac{\text{TP}}{\text{TP} + \text{FN}} \tag{5}$$

$$\text{F1 score} = \frac{2 * (\text{Precision} * \text{Recall})}{(\text{Precision} + \text{Recall})} \tag{6}$$

where TP is the number of true positives, FP is the number of false positives, TN is the number of true negatives, and FN is the number of false negatives.

3 Results and Discussion

In this section, the segmentation results using U-Net architecture are presented. Figure 2 shows the ground truth of normal tissues and tumors, flair images, thermal images in grayscale level, and thermal images in color format of five slices taken from the first five patients, each slice from one patient. As demonstrated in our recent study [12], the temperature is high in the tumor zone compared to healthy brain tissues, the maximum temperature is always in the tumor center, and as we move to tumor contours, the temperature is reduced.

The U-Net architecture was trained in 276 grayscale-level thermal images, which are obtained by converting temperature calculated using Pennes equation to a grayscale image. The segmentation results evaluation metrics are provided in Table 1. It can be observed that U-Net architecture yields a precise segmentation of brain tumors, which can be improved further in the future works by considering more training datasets.

In our recent study, we proved the importance of thermal behavior of brain tumors to reinforce segmentation. The segmentation from temperature distribution was performed using Canny edge detector which is not a robust method when we deal with complex geometries of tumors and noisy images. In this work, we used CNN as it represents a powerful method widely used in recent years for brain tumor segmentation. The results can be improved further by considering more training datasets and used to correct the segmentation in conventional MRI images.

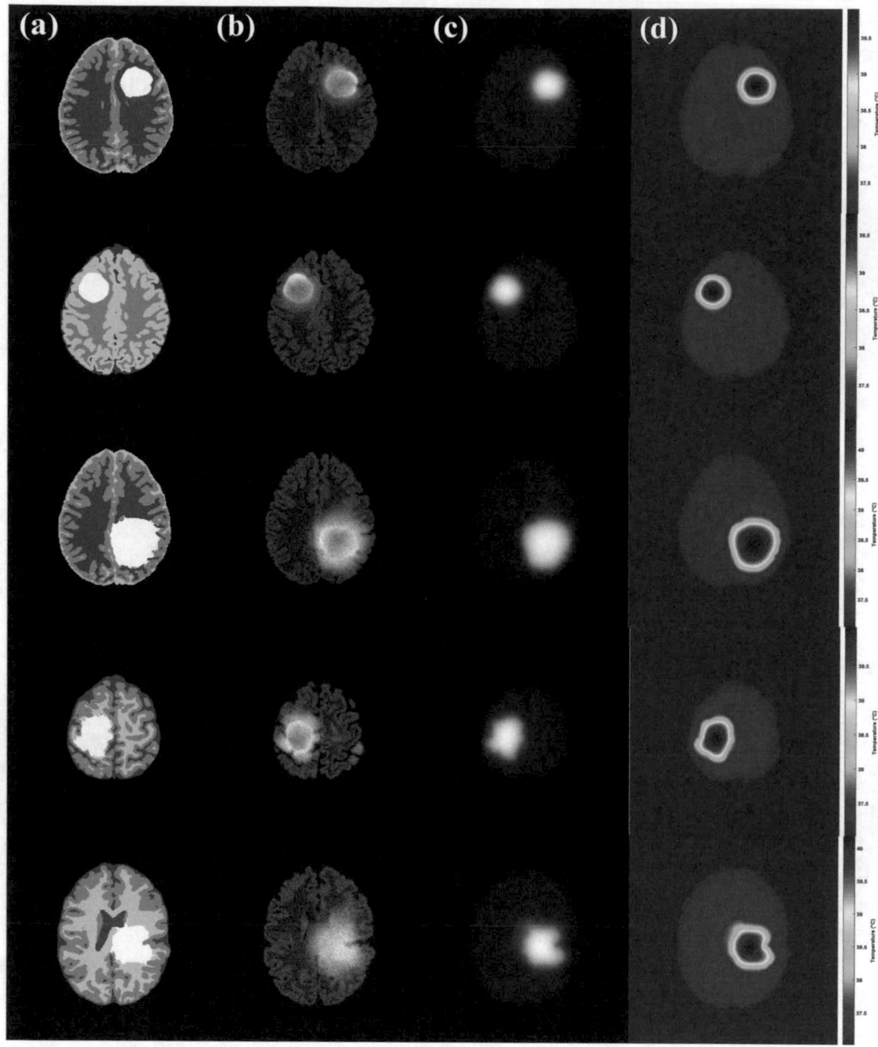

Fig. 2 Synthetic flair and thermal images of five patients with tumors of different volumes in different locations. **a** Ground truth of brain tissues and tumors. **b** Flair images. **c** Thermal images in grayscale level. **d** Thermal images

Table 1 The segmentation evaluation metrics for U-Net segmentation in thermal images

Accuracy	Precision	Recall	F1 score
0.9965	0.9817	0.9513	0.9338

4 Conclusion

In this work, we presented an approach to segment brain tumors from thermal images using a U-Net architecture based on the changing temperature in the tumor zone. The thermal images were obtained using Pennes bioheat equation and converted to grayscale images. Next, U-Net was applied to segment tumors. The training process takes significant time, which is the reason why we used a massively parallel implementation based on NVIDIA GPU and Deeplearning4j. The obtained results will be exploited in the future works to reinforce the segmentation in standard MRI images such as T1, T1c, T2, and flair sequences. Next, we will use an anisotropic Pennes equation for more precise temperature calculation, as the brain tissues are highly anisotropic. Also, we will extend the approach for 3D brain tumor segmentation with parallel and distributed architecture based on GPUs and Apache Spark for more speed in training.

Acknowledgements This work is supported by the grant of the National Center for Scientific and Technical Research (CNRST—Morocco) (No. 13UH22016).

References

1. Menze, B.H., et al.: The multimodal brain tumor image segmentation benchmark (BRATS). IEEE Trans. Med. Imaging **34**, 1993–2024 (2015)
2. Gordillo, N., Montseny, E., Sobrevilla, P.: State of the art survey on MRI braintumor segmentation. Magn. Reson. Imaging **31**(8), 1426–1438 (2013)
3. Angulakshmi, M., Lakshmi Priya, G.G.: Automated brain tumor segmentation techniques—a review. Int. J. Imaging Syst. Technol. **27**, 66–77 (2017)
4. Singh, J.F., Magudeeswaran, V.: Thresholding based method for segmentation of MRI brain images. In: International Conference on I-SMAC (IoT in Social, Mobile, Analytics and Cloud) (I-SMAC), pp. 280–283 (2017)
5. Ilhan, U., Ilhan, A.: Brain tumor segmentation based on a new threshold approach. Proc. Comput. Sci. **120**, 580–587 (2017)
6. Wu, M.N., Lin, C.C., Chang, C.C.: Brain tumor detection using color-based K-means clustering segmentation. In: Third International Conference on Intelligent Information Hiding and Multimedia Signal Processing (IIH-MSP 2007), vol. 2, pp. 245–250 (2007)
7. Dhanalakshemi, P., Kanimozhi, T.: Automatic segmentation of brain tumor using k means clustering and its area calculation. Int. J. Adv. Electr. Electron. Eng. **2**(2), 130–134 (2013)
8. Ibrahim, R.W., Hasan, A.M., Jalab, H.A.: A new deformable model based on fractional Wright energy function for tumor segmentation of volumetric brain MRI scans. Comput. Methods Programs Biomed. **163**, 21–28 (2018)
9. Thapaliya, K., Pyun, J.-Y., Park, C.-S., Kwon, G.-R.: Level set method with automatic selective local statistics for brain tumor segmentation in MR images. Comput. Med. Imaging Graph. **37**(7), 522–537 (2013)
10. Havaei, M., Davy, A., Warde-Farley, D., Biard, A., Courville, A., Bengio, Y., Pal, C., Jodoin, P.M., Larochelle, H.: Brain tumor segmentation with deep neural networks. Med. Image Anal. **35**, 18–31 (2017)
11. Pereira, S., Pinto, A., Alves, V., Silva, C.A.: Brain tumor segmentation using convolutional neural networks in MRI images. IEEE Trans. Med. Imaging **35**(5), 1240–1251 (2016)

12. Bousselham, A., Bouattane, O., Youssfi, M., Raihani, A.: Towards reinforced brain tumor segmentation on MRI images based on temperature changes on pathologic area. Int. J. Biomed. Imaging, Article ID 1758948 (2019)
13. Pennes, H.H.: Analysis on tissue arterial blood temperature in the resting human forearm. Appl. Physiol. **1**(2), 93–122 (1948)
14. Ronneberger, O., Fischer, P., Brox, T.: U-net: convolutional networks for biomedical image segmentation. In: Navab, N., Hornegger, J., Wells, W.M., Frangi, A.F. (eds.) MICCAI 2015. LNCS, vol. 9351, pp. 234–241. Springer, Cham (2015)
15. Litjens, G., et al.: A survey on deep learning in medical image analysis. Med. Image Anal. **42**, 60–88 (2017)
16. Kamnitsas, K., Ledig, C., Newcombe, V., Simpson, J., Kane, A., Menon, D., Rueckert, D., Glocker, B.: Efficient multi-scale 3D CNN with fully connected CRF for accurate brain lesion segmentation. Med. Image Anal. **36**, 61–78 (2017)
17. Ait Ali, N., Cherradi, B., El Abbassi, A., Bouattane, O., Youssfi, M.: GPU fuzzy c-means algorithm implementations: performance analysis on medical image segmentation. Multimed. Tools Appl. **77**(16), 21221–21243 (2018)
18. Sriramakrishnan, P., Kalaiselvi, T., Rajeswaran, R.: Modified local ternary patterns technique for brain tumour segmentation and volume estimation from MRI multi-sequence scans with GPU CUDA machine. Biocybern. Biomed. Eng. **39**(2), 470–487 (2019)
19. Wissler, E.H.: Pennes' 1948 paper revisited. J. Appl. Physiol. **85**(1), 35–41 (1998)
20. Shelhamer, E., Long, J., Darrell, T.: Fully convolutional networks for semantic segmentation. IEEE Trans. Pattern Anal. Mach. Intell. **39**(4), 640–651 (2017)
21. Kermi, A., Mahmoudi, I., Khadir, M.T.: Deep convolutional neural networks using U-Net for automatic brain tumor segmentation in multimodal MRI volumes. Brain Lesion Glioma Mult. Scler. Stroke Trauma. Brain Injuries **11384**, 37–48 (2019)
22. Marcinkiewicz, M., Nalepa, J., Lorenzo, P.R., Dudzik, W., Mrukwa, G.: Segmenting brain tumors from MRI using cascaded multi-modal U-Nets. Brain Lesion Glioma Mult. Scler. Stroke Trauma. Brain Injuries **11384**, 13–24 (2019)
23. Luna, M., Park, S.H.: 3D patchwise U-Net with transition layers for MR brain segmentation. Brain Lesion Glioma Mult. Scler. Stroke Trauma. Brain Injuries **11383**, 394–403 (2019)
24. Kingma, D.P., Ba, J.: Adam: a method for stochastic optimization. arXiv:1412.6980 (2014)
25. Kistler, M., Bonaretti, S., Pfahrer, M., Niklaus, R., B¨uchler, P.: The virtual skeleton database: an open access repository for biomedical research and collaboration. J. Med. Internet Res. **15**(11), e245 (2013)
26. DL4J website: https://deeplearning4j.org/. Accessed 25 Mar 2019

Evaluate the Performance of Port Container Using an Hybrid Framework

Mouhsene Fri, Kaoutar Douaioui, Nabil Lamii, Charif Mabrouki and El Alami Semma

Abstract This work intends to integrate feedforward neural network (FNN) and data envelopment analysis (DEA) in a single framework to evaluate the performance of operations in the port container terminal. The proposed framework is based on three steps. In the first step, we identify the performance measures objectives and the indicators affecting our system. In the second step, a DEA-based oriented inputs model (DEA-CCR) is used to compute the efficiency scores of the system, based on the obtained scores, the data is divided into training and testing datasets. In the last step, an improved crow search algorithm (ICSA) is employed as a new method for training FNNs to determine the efficiency scores. In ICSA, the so-called Levy flights are used to enhance the convergence rate of CSA and prevent it from getting stuck in local optima. To demonstrate the efficacy of the proposed framework, it is utilized to evaluate the performance of two ports container terminal mainly: Tangier and Casablanca. The results are compared with a standard BBO, GA and PSO-based learning algorithm. The new trainer ICSA is also investigated and evaluated using four different classification datasets selected from the UCI machine learning repository and on three approximation functions datasets. The experimental results

M. Fri (✉) · K. Douaioui · N. Lamii · C. Mabrouki · E. A. Semma
LMII-Faculty of Sciences and Technology, Hassan 1st University, PO Box 577, Settat, Morocco
e-mail: frimouhsene@gmail.com

K. Douaioui
e-mail: kaoutar.douaioui@gmail.com

N. Lamii
e-mail: n.lamii@uhp.ac.ma

C. Mabrouki
e-mail: charif.mabrouki@uhp.ac.ma

E. A. Semma
e-mail: Semmaalam@yahoo.fr

M. Fri
CELOG-ESITH, ULFA, PO Box 7731, Casablanca, Morocco

V. Bhateja et al. (eds.), *Embedded Systems and Artificial Intelligence*, Advances in Intelligent Systems and Computing 1076,
https://doi.org/10.1007/978-981-15-0947-6_49

show that ICSA outperforms both BBO, GA and PSO for training FNNs in terms of converging speed and avoiding local minima.

Keywords Port container terminal (PCT) · Performance measurement system (PMS) · Data envelopment analysis (DEA) · Feedforward neural network (FNN) · Levy flights · Crow search algorithm

1 Introduction

Port container terminals have a great importance in the international logistics chain. Their role has continuously evolved from the first generation of port terminals to the present. United Nations Conference on Trade and Development [1] defines the fourth generation of port terminals as geographically separated areas, which have common or administratively centralized operators.

Generally, port terminals are complex environments with many aspects: social, economic, political and cultural where different organizations, institutions and functions interact at different levels. One of the challenging task is to managing separate spaces in the ports terminal which is responsible for 90% of global trade goods, which represent a complex situation. Due to this complexity, many performance measurement systems at port terminals or the whole dry port-sea port system have been developed. In a paper by Bentaleb et al. [2], they presented the state of the art on the different models developing in the whole dry port-sea port system. In 2016, a MACBETH multi-criteria approach was used by Bentaleb et al. [3] to evaluate the performance indicators of the whole dry port-sea port system. Despite the merits of the abovementioned works, the challenges are there are no efficient performance measurement systems in ports, so we looked to develop a performance measurement system on the maritime side.

In recent years, the data envelopment analysis (DEA) and artificial neural network (ANN) have been widely used as a nonparametric tools for evaluating the performance of operations in port container terminal. The first combination between the neural networks (NNs) and DEA was first proposed by Athanassopoulos and Curram [4] for forecasting the number of employees in the healthcare industry. In the proposed model, they considered the DEA as preprocessing methodology for training step, while the ANNs are then trained on selecting samples as a nonlinear forecasting model. In a paper by Costa and Markellos [5], they compared ANNs with corrected ordinary least squares (COLS) and DEA. The proposed approach is applied to the London underground efficiency analysis. Their results indicate that the ANNs find and reveal results than CLOS and DEA in regard to the decision-making about the impact of constant versus variable returns to scale or congestion areas. As two non-parametric models, there are many similarities between ANNs and DEA models [4] such as:

1. Neither DEA nor ANNs make assumptions about the functional form that links its inputs to outputs.

2. DEA seeks a set of weights to maximize the technical efficiency, whereas ANNs seek a set of weights to derive the best possible fit through observations of the training dataset.

The main scope of DEA is to provide an estimation of efficiency surfaces by solving mathematical programming model. However, a major problem faced by DEA is that the derived frontier may be warped if the data is affected by statistical noises [6]. The artificial neural network (ANN) has been widely used as good alternative tools to estimate the efficiency frontiers for decision makers [7]. One of the most important problems faced by neural networks is the training process. In this process, the goal is to find the best combination of connection weights and biases that minimize the mean squared error (MSE). In general, training algorithms can be classified into two groups: gradient-based algorithms versus stochastic search algorithms. The most widely applied gradient-based training algorithms are backpropagation (BP) algorithm [8]. However, this method has some drawbacks, such as slow speed of convergence and get trapped on local minima. On the other hand, the nature-inspired metaheuristic algorithms, as an alternative trainer, are proved more efficient in escaping from local minima for optimization problems.

In the literature, several metaheuristic methods have been used as a trainer for FNNs. In a paper by Mirjalili et al. [9], the biogeography-based optimizer is used to train FNNs. The reported numerical results show that the accuracy of BBO in terms of converging speed, avoiding local optima training process is better than compared algorithms. To the best of our knowledge, there are no previous works attempts to dealing with the performance of operations in port container terminal using FNNs and DEA. This paper presents a hybrid framework to evaluate the performance of operations in port container terminal. The proposed framework is based on three steps. In the first step, we identify the performance measures objectives and the indicators affecting our system. In the second step, we compute the efficiency scores by the CCR model (oriented inputs). In the last step, an improved crow search algorithm (ICSA) is employed as a new method for training FNNs to determine the efficiency scores. In ICSA, the so-called Levy flights are used to enhance the convergence rate of CSA and prevent it from getting stuck in local optima.

The rest of the paper is organized as follows. Section 2 provides the performance measurement system. Section 3 provides the methodologies utilized in this paper. Section 4 reports the numerical results and discussion. Finally, our conclusions and future work are presented in Sect. 5.

2 Performance Measurement System

Today, evaluating the performance of operations in port container terminal is one of the important tasks for the managers of ports. In a paper by Zhu [10], they assume that it is difficult to assess the performance of an organization when there are several performance measures related to a system or operation, including several organizations in the case of port terminals. However, the growing competitiveness in the port needs a higher level of performance. Over the past decades, many researchers have been studying the evaluation of the efficiency and performance of port terminals, particularly those of container ports and terminals. Some researchers have addressed the theory and methodologies of port and terminal assessment and measurement of performance including classification research [11]. The performance measurement system mainly depends on the overall aims of the company [12], which leads us to determine the aims of the port terminals. This work presents the objectives to be achieved to have a global performance of the container terminal as well as the performance indicators involved in achieving these objectives. Based on identifying the performance indicators mentioned in the literature, the organizations (ESPO, UNCTAD …) and the reports of the port terminals, we have selected about 312 indicators. In the first step, a selection is made based on the five criteria provided by the work of the literature illustrated in Table 1. In the second step, the critical study on each indicator is done by:

1. Eliminating redundant indicators
2. Grouping indicators of the same type
3. Keeping the indicators influencing operational performance.

Table 1 List of criteria

Criteria	Description
The relevance of policy	Monitor the strategy and progress of the activity
Informative	Provide relevant information to the activity
Measurable	Measure following a reliable procedure
Representative	Gives information simple to interpret
Pratical	Simple to monitor

(a)

Operational and logistic performance	Administrative management (AM)
	Management of ship services (MSS)
	Management of quay operations (MQO)
	Yard operations (YO)
	Employee performance (EP)
Financial performance	Financial wealth (FW)
	Financial health (FH)
	Investement (I)
Physical performance	Port sizes (PS)
	Port equipement (PE)
	Exploitation of technology (ET)
Commercial performance	Services and organizations (SO)
	Competitive position (CP)
	Safety and security (SS)

(b)

On the last stage, we have limited our list of performance indicators to 14 indicators. Table 2 presents the overall performance measurement system containing all system performance indicators.

Table 2 Overall data and results

Terminal port	I1	I2	I3	I4	I5	I6	I7	I8	I9	I10	I11	I12	I13	I14	Target performance (*100)
E1	0,66	0,72	0,29	0,8	0,03	0,21	0,3	0,07	0,84	0,68	0,4	0,82	0,08	0,09	0,90
E2	0,15	0,3	0,94	0,16	0,54	0,75	0,69	0,8	0,63	0,3	0,18	0,44	0,92	0,83	1,12
E3	0,03	0,55	0,52	0,53	0,56	0,04	0,93	0,04	0,61	0,85	0,04	0,32	0,96	0,27	0,77
E4	0,77	0,88	0,8	0,55	0,02	0,77	0,9	0,09	0,52	0,43	0,54	0,01	0,45	0,21	0,99
E5	0,03	0,65	0,86	0,92	0,72	0,58	0,96	0,45	0,98	0,68	0,35	0,86	0,86	0,62	1,12
E6	0,21	0,35	0,74	0,99	0,39	0,2	0,98	0,98	0,69	0,14	0,71	0,89	0,53	0,31	0,62
E7	0,81	0,3	0,77	0,47	0,17	0,4	0,32	0,4	0,06	0,14	0,7	0,78	0,16	0,47	0,92
E8	0,21	0,05	0,11	0,9	0,76	0,77	0,48	0,24	0,6	0,15	0,85	0,79	0,1	0,57	0,54
E9	0,92	0,77	0,5	0,21	0,27	0,16	0,53	0,24	0,61	0,91	0,99	0,28	0,41	0,83	1,04
E10	0,26	0,27	0,8	0,54	0,35	0,02	0,99	0,98	0,11	0,44	0,85	0,19	0,96	0,62	1,14

(a) Ports datasets

Algorithm	Parameter
Maximum of generation = 250	
Population size = 50,200	
BBO	Habitat modification probability (=1)
	Immigration probability ([0, 1])
	Step size for numerical integration of probabilities (=1)
	Max immigration ($I = 1$) and Max emigration ($E = 1$)
	Mutation probability (=0.005)
PSO	Topology (full connected)
	Cognitive constant ($C1 = 1$)
	Social constant ($C2 = 1$)
	Inertia constant ($w = 0.3$)
GA	Selection (Roulette wheel)
	Crossover: single point (probability = 1)
	Mutation: uniform (probability = 0.01)
ICSA	Index beta = 1.5
	$A_p = 0.4$
	$fl = 2$
	$beta = 1.5$

(b) Parameters of algorithms

Algorithm	AVG+STD	Performance-error (*100%)
BBO	3.1126 ± 0.04898	12.49
PSO	2.0034 ± 0.19391	23.58
GA	2.0397 ± 0.20059	23.58
ICSA	**3.3304 ± 1.6851e–07**	**5.74**

(c) Ports datasets results

(continued)

Table 2 (continued)

Port	Target (%)	Obtained-performance				Rank
		BBO (%)	PSO (%)	GA (%)	ICSA (%)	
Tangier	200	**155**	108	120	**163**	1
Casa	200	**118**	92	95	**126**	2

(d) Moroccan case study results

Classification datasets	XOR	Baloon	Iris	Cancer
Number of attributes	3	4	4	9
Number of training samples	8	16	150	255
Number of testing samples	8	16	150	100
Number of classes	2	2	3	2

(e) Classification datasets

Function-approximation datasets	Training samples	Test samples
Sigmoid: $y = 1/(1 + e^{-(x)})$	61: $x \in [-3 : 0.1 : 3]$	121: $x \in [-3 : 0.05 : 3]$
Cosine: $y = \left(\cos\left(\frac{x\pi}{2}\right)\right)^7$	31: $x \in [1.25 : 0.1 : 2.75]$	38: $x \in [1.25 : 0.04 : 2.75]$
Sine: $y = \sin(2x)$	126: $x \in [-2\pi : 0.1 : 2\pi]$	252: $x \in [-2\pi : 0.05 : 2\pi]$

(f) Function-approximation datasets

Datasets	Algorithm	AVG+STD	Classification
XOR	BBO	3.65e07 ± **0.00000**	**100**
	PSO	0.084050 ± 0.035945	37.50
	GA	0.000181 ± 0.000413	**100**
	ICSA	**5.03e–08 ± 7.94e–08**	**100**
Iris	BBO	0.019150 ± 3.66e18	**90.00**
	PSO	0.228680 ± 0.057235	37.33
	GA	0.089912 ± 0.123638	89.33
	ICSA	**0.01671 ± 0.00018**	**93.00**
Balloon	BBO	8.09e27 ± 1.51e42	**100.00**
	PSO	0.000585 ± 0.000749	**100.00**
	GA	5.08e24 ± 1.06e23	**100.00**
	ICSA	**1.04e–33 ± 2.56e–33**	**100.00**
Cancer	BBO	0.002807 ± 0.000000	95.00
	PSO	0.034881 ± 0.002472	11.00
	GA	0.003026 ± 0.001500	**98.00**
	ICSA	**0.001008 ± 9.38e–06**	**99.00**

(g) Data classification results

Datasets	Algorithm	AVG+STD	Test error
Sigmoid	BBO	1.33e05 ± **3.57e21**	**0.1438**
	PSO	0.022989 ± 0.009429	3.3563
	GA	0.001093 ± 0.000916	0.4496
	ICSA	**5.8796e–06** ± 3.3297e–06	**0.1400**
Cosine	BBO	0.013674 ± 1.83e18	1.4904
	PSO	0.058986 ± 0.021041	2.0090
	GA	0.010920 ± 0.006316	**0.7105**
	ICSA	**0.010150 ± 0.003834**	**0.6975**
Sine	BBO	0.102710 ± 0.000000	64.261
	PSO	0.526530 ± 0.072876	124.89
	GA	0.421070 ± 0.061206	111.25
	ICSA	**0.063484 ± 0.00069228**	**51.27**

(h) Functions datasets results

3 Hybrid Framework

3.1 Data Envelopment Analysis (DEA)

Measuring and improving the efficiency are two challenging task for all companies. Data envelopment analysis (DEA) is a nonparametric tool based on linear programming proposed by Charnes et al. [13] to measure the relative efficiency of a decision-making unit (DMU) and provide DMUs with relative performance assessment on multiple inputs and outputs. In this paper, an input-oriented model CCR is used to evaluate the efficiency of terminal port. Let us suppose that there are n DMUs, m inputs and s outputs. Suppose $x_{i,j}(i = 1, \ldots, m, j = 1, \ldots, n)$ is quantity of input i consumed by DMU_j and $y_{r,j}(r = 1, \ldots, s, j = 1, \ldots, n)$ is quantity of output r produced by DMU_j, u_r the weight of rth output element, v_i the weight ith input item. As suggested by the CCR model, the efficiency of DMU_k denoted as P_k can be measured by solving the linear equations below [13]

$$\begin{cases} \max P_k = \sum_{r \in s} u_r \times y_{r,k} \\ \text{Subjcct to:} \\ \sum_{i \in m} v_i \times x_{i,k} = 1 \\ \sum_{r \in s} u_r \times y_{r,j} - \sum_{i \in m} v_i \times x_{i,j} \le 0, \forall j \\ u_r, v_i \ge 0, \forall r, i \end{cases} \tag{1}$$

The DEA method determines the positive weight sets to maximize P_k, with the constraint of efficiency scores being between 0 and 1. The process continues to find efficiency scores of each DMU by solving n linear programs. A DMU is fully efficient if and only if it is impossible to improve any input or output without worsening some other inputs or outputs [13].

3.2 Feedforward Neural Network (FNN)

In the artificial neural network, the feedforward neural network (FNN) was the simplest type that consists of a set of processing elements called "neuron". In this network, the information moves in only one direction, forward, from the input layer, through the hidden layer and to the output layer. There are no cycles or loops in the network. An example of a simple FNN with a single hidden layer is shown in Fig. 1.

As shown, each neuron computes the sum of the inputs weight at the presence of a bias and passes this sum through an activation function (like sigmoid function) so that the output is obtained. This process can be expressed as (2) and (3).

$$H_j = \sum_{i \in R} \omega_{j,i} \times I_j + b_j \tag{2}$$

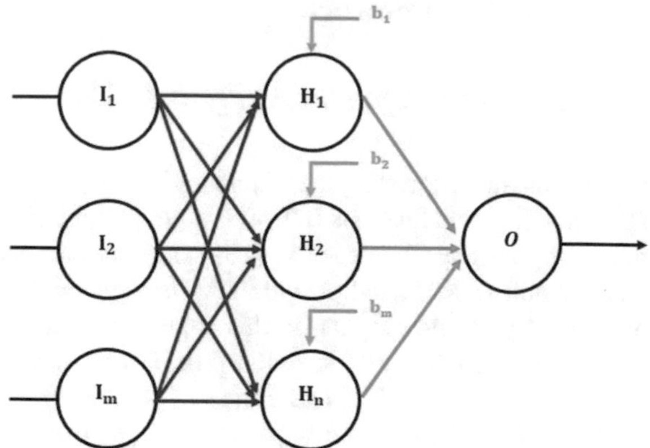

Fig. 1 Solution representation

where $w_{j,i}$ is the weight connected between neurons $i = (1, 2, \ldots, R)$ and $j = (1, 2, \ldots, N)$, b_j is a bias in hidden layer, R is the total number of neurons in input layer, N and I_j is the corresponding input data. Here, the S-shaped curved sigmoid function is used as the activation function, which is shown in

$$f_{(x)} = \frac{1}{(1 + e^{-(x)})} \tag{3}$$

In the output layer, the output of the neuron is shown in

$$y_k = f_k\left(\sum_{j \in N} \sigma_{k,j} \times H_j + b_k\right) \tag{4}$$

where is $\sigma_{k,j}$ the weight connected between neurons $j = (1, 2, \ldots, N)$ and $k = (1, 2, \ldots, S)$, b_k is a bias in output layer, N is the total number of neurons in hidden layer, and S is the total number of neurons in output layer.

3.3 *Improved Crow Search Algorithm*

Crow search algorithm is a new population-based stochastic search algorithm recently proposed by Askarzadeh [14], CSA is a newly developed optimization technique to solve complex engineering optimization problems [15]. It is inspired by the intelligent behaviors of the crows. The core mechanism of CSA algorithm consists of three basic phases, namely initialization, generate new position and updating the memory of crows. At first, the initial population of crows represented by n dimension is randomly generated. At iteration t, the position of crow is specified by $X^{i,t} = \left[X_1^{i,t}, X_2^{i,t}, \ldots, X_n^{i,t}\right]$ and it is assumed that this crow has memorized its best experience found so far in its memory $m^{i,t} = \left[m_1^{i,t}, m_2^{i,t}, \ldots, m_n^{i,t}\right]$. To generate a new position, the crow i select randomly a crow say j from the population and attempts to follow it to find the position of its hiding place (m^j). In this case, the position of the crows is updated as follows:

$$X^{i,t+1} = \begin{cases} X^{i,t} + r_1 \times fl^{j,t}(m^{i,t} - X^{i,t}, & r_j \geq AP_{t,j} \\ LB + rand * (UB - LB), & \text{Otherwise} \end{cases} \tag{5}$$

where r_j is uniformly distributed random number from [0, 1] and $AP^{j,t}$ denotes the awareness probability of crow j at iteration $iter$. Finally, the crows update their memory as follows $X^{i,t+1} = X^{i,t}$, if $f(X^{i,t})$ is better than $f(m^{i,t})$, $m^{i,t}$ otherwise, where $f(-)$ denotes the objective function value. It is seen that if the fitness function value of the new position of a crow is better than the fitness function value of the memorized position, the crow updates its memory by the new position. The above process is repeated until a given termination criterion ($t_{\max}$) is met. Finally, the best solution of the memories is returned as the optimal solution found by CSA.

It is well known that the balance between exploration and exploitation is the keys of success of any population-based optimization algorithms, such as GA, PSO, DE and so on. In conventional CSA, it may converge prematurely without enough exploration of search space. In order to increase the diversity of population against premature convergence and accelerate the convergence speed, this paper proposes an improved crow search algorithm based on Levy flights.

Levy flights represent a kind of non-Gaussian stochastic process whose step sizes are distributed based on a Levy stable distribution to generate new solutions. When a new solution is produced, the following Levy flight is applied:

$$X_i^{t+1} = X_i^t + \alpha \oplus Levy(\lambda) \tag{6}$$

Here, α is the step size that is relevant to the scales of the problem. The product $\oplus$ means entry-wise multiplications. Levy flights essentially provide a random walk while their random steps are drawn from a Levy distribution for large steps:

$$Levy(\lambda) = u = t^{-\lambda}, \quad 1 \leq \lambda \leq 3 \tag{7}$$

In this paper, we will use the algorithm proposed by Mantegna [16], which is one of the most efficient algorithms used to implement Levy flights.

Algorithm 1 Improved Crow Search Algorithm

Randomly initialize the position of a flock of (N_p) crows in the search space and parameters
Evaluate the position of the crows
Initialize the memory of each crow
while $(t \leq t_{max})$ **do**
  for $i = 1$ to $\frac{N_p}{5}$ **do**
    $X_i^{t+1} = X_{i,worst}^t + \alpha \oplus Levy(\lambda)$
    if $(F(X_i^{t+1}) < F(X_{i,worst}^t))$ **then**
      $X_{i,worst}^t = X_i^{t+1}$
      $F(X_{i,worst}^t) = F(X_i^{t+1})$
    end if
  end for
  for $i = 1$ to $\frac{4*N_p}{5}$ **do**
    Randomly choose one of the crows to follow (for example j)
    Define an awareness probability
    if $(r_j \geq AP^{j,t})$ **then**
      $X^{i,t+1} = X^{i,t} + r_i * fl^{i,t} * (m^{i,t} - X^{i,t})$
    else
      $X^{i,t+1}$ = a random position of search space
    end if
  end for
  Check the feasibility of new positions
  Evaluate the new position of the crows
  Update the memory of crows
end while

3.4 ICSA Trainer for FNN

In ICSA, every crow represents a candidate NN. Figure 2 demonstrates an example of encoding strategy of ICSA for the proposed framework. The evaluation of each crow is done by passing the vector of weights and biases to FNNs; then the mean squared error (MSE) criterion is calculated based on the prediction of the neural network using the training dataset. Through continuous iterations, the optimal solution is finally achieved, which is regarded as the weights and biases of a neural network. The MSE criterion is given in Eq. (8) where y and $\overline{y}$ are the actual and the estimated values based on proposed model and R is the number of samples in the training dataset.

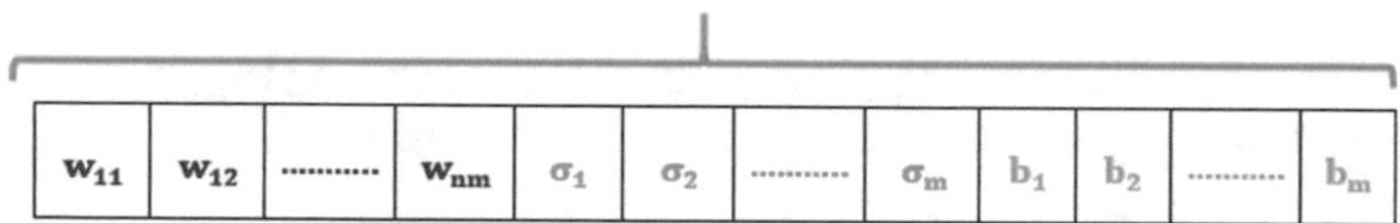

Fig. 2 Solution representation

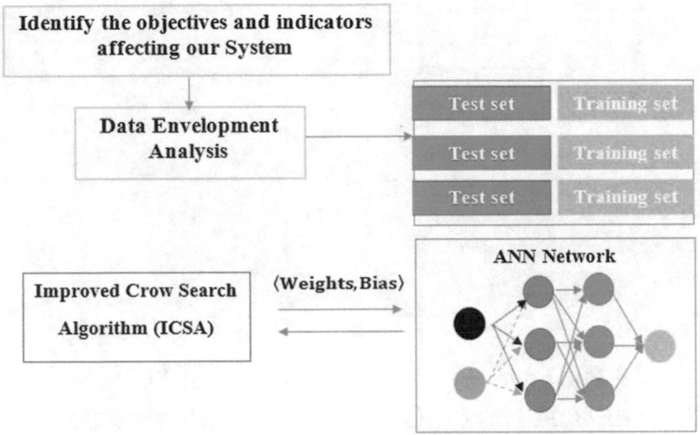

Fig. 3 Proposed hybrid framework

$$\text{MSE} = \frac{1}{R} \times \sum_{r \in R} (y - \overline{y})^2 \tag{8}$$

The training process is carried out to adjust the weights and bias until some error criterion is met (Fig. 3).

3.5 The Proposed Hybrid Algorithm

The proposed framework is based on three steps. In the first step, we identify the performance measures objectives and the indicators, the sub-indicators affecting our system. In the second step, we compute the efficiency scores by the CCR model (oriented inputs). In the last step, an improved crow search algorithm (ICSA) is employed as a new method for training FNNs to determine the efficiency scores. The steps of the proposed framework are given in Fig. 4.

4 Numerical Results

In this section, we investigate the efficiency of the proposed hybrid framework for evaluating the performance of operations in port container terminal. The experiments were done using a PC with a 3.30 GHz Intel(R) Core (TM) i5 processor, 4GB of memory. The entire algorithm was programmed in MATLAB R2014a.

The dataset of port container terminals is shown in Table 2a which is collected based on using the Delphi method developed by Linstone et al. [17]. This method consists of four properties: (i) the anonymity of the Delphi participants which gives

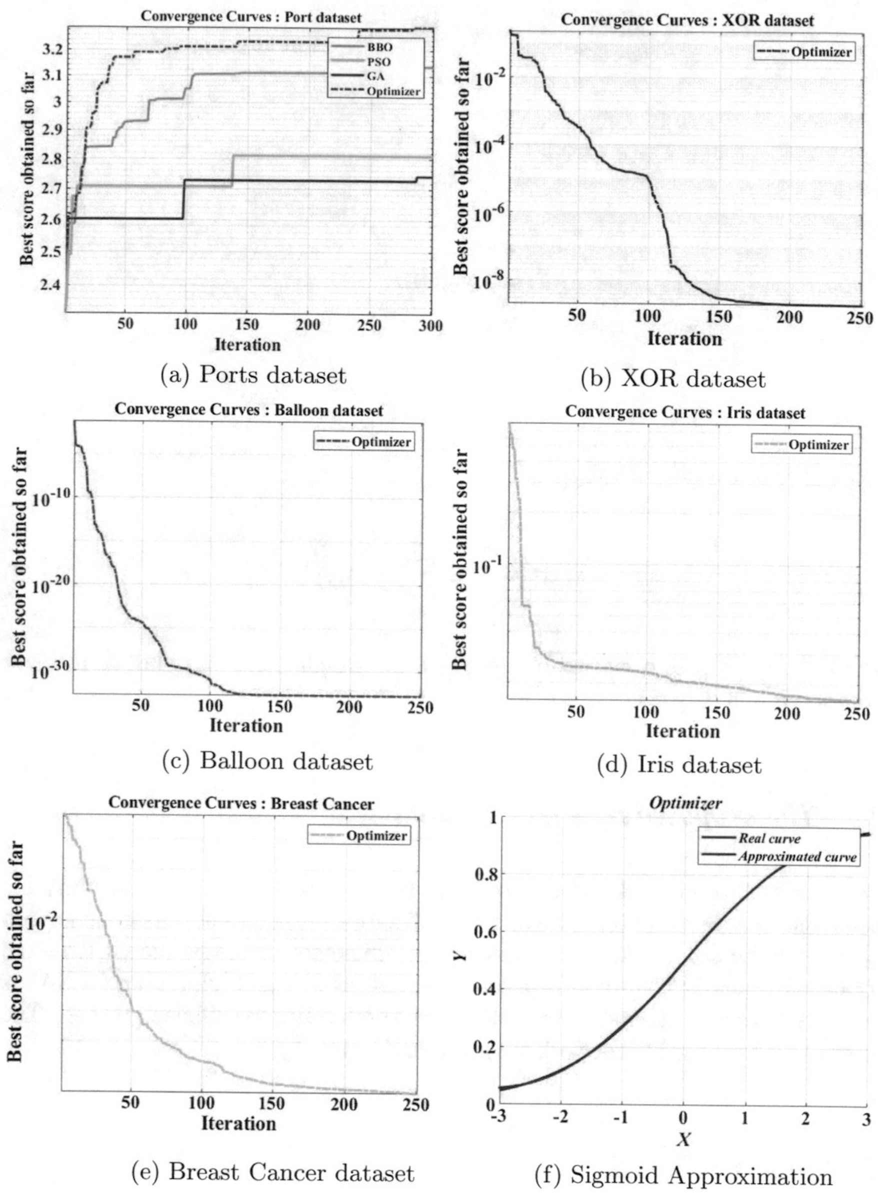

(a) Ports dataset (b) XOR dataset

(c) Balloon dataset (d) Iris dataset

(e) Breast Cancer dataset (f) Sigmoid Approximation

Fig. 4 Overall convergence curves

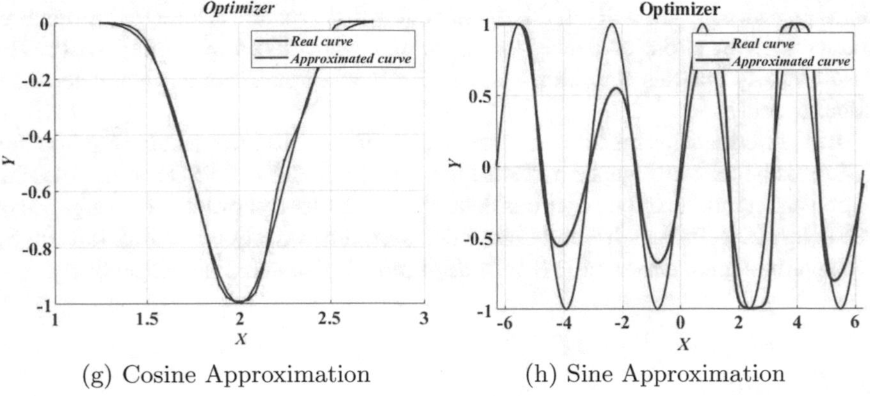

(g) Cosine Approximation (h) Sine Approximation

Fig. 4 (continued)

the participant more liberty to judge the company; (ii) steps iteration which allows participants to refine their work; (iii) feedback and control which allow participants to access the response of others; and (iv) statistical acceptance of answers which allows for quantitative analysis and interpretation of data. Due to confidential issue, the initial data was modified with uniform random variable between [0.15; 1] and partitioned based on DEA into 66% for training and 34% for testing. All experiments are executed for 20 different runs and for a given set of parameters presented on Table 2b.

The results of port datasets are reported on Table 2c based on the average (AVE) and standard deviation (STD). The main scope to employ these two measures is to indicate the ability of algorithms to avoid local minima. By analyzing Table 2c, the first thing that can be observed in the results is the highest performance obtained by the proposed method, this behavior is due to great ability to avoid local optima, significantly better than other algorithms.

In addition, a convergence comparative experiment was carried out to confirm that ICSA has better convergence performance than GA and PSO. Figure 4a shows the convergence of ICSA, BBO, PSO and GA. To validate our model, we conducted a case study based on two ports container terminals, namely: Tangier and Casablanca. Table 2d shows the obtained results.

Comparison with Others Algorithms

To show the efficacy of the proposed method, we are benchmarked it on four selected standard classification datasets from the University of California at Irvine (UCI) Machine Learning Repository [18]: XOR, balloon, Iris, breast cancer and on three function approximation datasets obtained from [18]: sigmoid, cosine and sine. The datasets are shown in Table 2e, f. Regarding the results shown in Table 2g, h. In those tables, the ICSA-FNN outperforms other algorithms in terms of not only the minimum MSE but also the maximum classification accuracy and test error. The results of ICSA-FNN follow by those of BBO-FNN, GA-FNN and PSO-FNN for all datasets. Since the difficulty of this dataset and MLP structure is high for this dataset, these results are strong evidences for the efficiencies of ICSA in training MLPs. The results testify that this algorithm has superior local optima avoidance and accuracy simultaneously.

In addition, a convergence comparative experiment was carried out to confirm that ICSA has better convergence performance than BBO, GA and PSO for data classification and approximation functions datasets. Figure 4b–e shows the convergence of ICSA for XOR, Baloon, Iris and Cancer datasets, respectively, while Fig. 4f–h shows the approximation curves of ICSA for Sigmoid, Cosine and Sine, respectively.

5 Conclusion

This paper employs the well-regarded FNN and DEA to overcome the drawbacks of DEA: statistical noise. After identifying the objectives and indicators affecting our system, the CCR model (oriented inputs) is used to evaluate the scores. As a result a crow search algorithm is employed as new trainer for ANN to improve the performance of the system. In addition, the algorithm was compared to the well-known benchmarks. The reported results demonstrated that the ICSA algorithm is able to outperform the BBO, GA and PSO on the collected port datasets and on the majority of classification and approximation datasets.

References

1. Taylor, I., Smith, K.: United Nations Conference on Trade and Development (UNCTAD). Routledge (2007)
2. Bentaleb, F., Mabrouki, C., Semma, A.: Key performance indicators evaluation and performance measurement in dry port-seaport system 58; a multi criteria approach. J. ETA Maritime Sci. **3**(2), 97–116 (2015)
3. Bentaleb, F., Fri, M., Mabrouki, C., Semma, E.: Dry port-seaport system development: application of the product life cycle theory. J. Transp. Logistics **1**, 116–128, 10 (2016)
4. Athanassopoulos, A.D., Curram, S.P.: A comparison of data envelopment analysis and artificial neural networks as tools for assessing the efficiency of decision making units. J. Oper. Res. Soc. **47**(8), 1000–1016 (1996)
5. Costa, Á., Markellos, R.N. : Evaluating public transport efficiency with neural network models. Transp. Res. Part C Emerg. Technol. **5**(5), 301–312 (1997)
6. Bauer, P.W.: Recent developments in the econometric estimation of frontiers. J. Econ. **46**(1–2), 39–56 (1990)
7. Wang, S.: Adaptive non-parametric efficiency frontier analysis: a neural-network-based model. Comput. Oper. Res. **30**(2), 279–295 (2003)
8. Rumelhart, D.E., Hinton, G.E., Williams, R.J.: Learning representations by back-propagating errors. Nature **323**(6088), 533 (1986)
9. Mirjalili, S., Mirjalili, S.M., Lewis, A.: Let a biogeography-based optimizer train your multi-layer perceptron. Inf. Sci. **269**, 188–209 (2014)
10. Zhu, J.: Quantitative models for performance evaluation and benchmarking: data envelopment analysis with spreadsheets, vol. 213. Springer (2014)

11. Mabrouki, C., Bentaleb, F., Mousrij, A.: A decision support methodology for risk management within a port terminal. Saf. Sci. **63**, 124–132 (2014)
12. de Lima, E.P., da Costa, S.E.G., de Faria, A.R.: Taking operations strategy into practice: developing a process for defining priorities and performance measures. Int. J. Prod. Econ. **122**(1), 403–418 (2009)
13. Charnes, A., Cooper, W.W., Rhodes, E.: Measuring the efficiency of decision making units. Eur. J. Oper. Res. **2**(6), 429–444 (1978)
14. Askarzadeh, A.: A novel metaheuristic method for solving constrained engineering optimization problems: crow search algorithm. Comput. Struct. **169**, 1–12 (2016)
15. Oliva, D., Hinojosa, S., Cuevas, E., Pajares, G., Avalos, O., Gálvez, J.: Cross entropy based thresholding for magnetic resonance brain images using crow search algorithm. Expert Syst. Appl. **79**, 164–180 (2017)
16. Mantegna, R.N.: Fast, accurate algorithm for numerical simulation of levy stable stochastic processes. Phys. Rev. E **49**(5), 4677 (1994)
17. Linstone, H.A., Turoff, M., et al.: The Delphi Method. Addison-Wesley Reading, MA (1975)
18. Hecht-Nielsen, R.: Kolmogorov's mapping neural network existence theorem. In: Proceedings of the IEEE International Conference on Neural Networks III, pp. 11–13. IEEE Press (1987)

A Comparative Study of HMMs and CNN Acoustic Model in Amazigh Recognition System

Meryam Telmem and Youssef Ghanou

Abstract In this paper, we apply two acoustic models to build the Amazigh speech recognition system; the first system based on hidden Markov models (HMMs) using the open-source CMU Sphinx-4, from the Carnegie Mellon University, the second system based on the convolution neural network CNN is a particular form of neural network implementing in TensorFlow and GPU computation. The two systems evaluated use mel frequency cepstral coefficients to extract the MFFc features. The corpus consists of 9900 audio files. The system obtained the best results when trained using CNN produced 92% of accuracy.

Keywords Speech recognition · Amazigh language · Deep learning · HMMs · CNN · CMU Sphinx-4 · TensorFlow · GPU · MFFc

1 Introduction

Automatic speech recognition is a branch of natural language processing. It identified the that words a person how has spoken using computer hardware and software techniques. This importance is explained by the privileged position of speech as a vector of human information. A remarkable change in the state of the art makes the systems more and more effective and used in many applications: assistance to independent living of people, voice control, language learning and translation, indexing large audiovisual databases etc.

However, the speech signal is characterized by many parameters, which make the task of a RAP system difficult [1]. This complexity of the speech signal originates from the combination of many factors, the redundancy of the acoustic signal, the inter and intra-speaker variability, the effects of the continuous speech co articulation, and the recording conditions. To overcome these difficulties, many mathematical methods

M. Telmem (✉) · Y. Ghanou
Team TIM, High School of Technology Moulay Ismail University, Meknes, Morocco
e-mail: meryamtelmem@gmail.com

Y. Ghanou
e-mail: youssefghanou@yahoo.fr

V. Bhateja et al. (eds.), *Embedded Systems and Artificial Intelligence*,
Advances in Intelligent Systems and Computing 1076,
https://doi.org/10.1007/978-981-15-0947-6_50

and models have been developed, including dynamic comparison, Vector Machine SVM, neural networks [2], the HMMs [3, 4]. In this work, we will be interested in the last two models to build Amazigh speech recognition system; the hidden Markov models (HMMs), which for decades the HMMs stay the ideal solution to the problems of automatic speech recognition; and the convolutional neural networks (CNN) a particular form of deep learning [2]. From 2012, CNN becomes the flagship architecture of deep learning proved in many domains like image recognition. In this paper, we compare the convolutional neural networks (CNN) with HMMs for Amazigh speech recognition system.

2 Related Works

This section presents some of the reported works available in the literature that are similar to the presented work.

In our previous work [1], we are looking for the optimal value of number of hidden Markov models (HMMs) states, and number of Gaussian mixture density functions for Amazigh speech recognition system. The system obtained best performance of 90% when trained using 128 Gaussian Mixture models and 5 number of HMMs states.

Satori and El Haoussi [4] have developed the Amazigh ASR based on the CMU Sphinx tools. The system obtained best performance of 92.89% when trained using 16 Gaussian Mixture models.

Zhang et al. [5] have studied a series of neural networks based on acoustic models; time delay neural network (TDNN), CNN, and the long short-term memory (LSTM), applied them in the Mongolian speech recognition systems, and compared their performance. The result shows that the LSTM is the best model among them with 8.12% WER.

Abushariah et al. [6] have developed the Arabic speech recognition system based on Carnegie Mellon University (CMU) Sphinx tools. The corpus contains a total of 415 sentences has been collected from 40 Arabic native speakers. The proposed Arabic speech recognition system is based on the Carnegie Mellon University (CMU) Sphinx and HTK tools.

3 Amazigh Language

The Amazigh languages are a group of very closely related and similar languages and dialects spoken in Morocco, Algeria, Tunisia, Libya, and the Egyptian area of Siwa, as well as by large Amazigh communities in parts of Niger and Mali. In c, for example, Amazigh is divided into three regional varieties, with tariffs in the North, Tamazigh in Central and Southeast Morocco, and Tachelhite in the South-West and the High Atlas [7].

Fig. 1 Official table of the Tifinaghe alphabet as recommended by l'RCAM

Various orthographies have been used to transcribe the Amazigh languages. In antiquity, the Tifinagh was utilized to write Berber. A modernized form of the Tifinagh alphabet was made official in Morocco in 2003, and only the IRCAM defined a precise order described by the expression below (*a* < *b*, means that *a* is sorted before *b*) [8] (Fig. 1).

3.1 Phonetics

In linguistics, when we talk about the phonology or phonetics of a language, we are usually referring to the sound system of a language. The graphic system of the standard amazighe proposed by the IRCAM comprises [8]:

- 27 consonants of: labels (ⴼ, ⴱ, ⵎ), dental (ⵜ, ⴷ, ⵟ, ⴹ, ⵏ, ⵔ, ⵕ, ⵍ).
- the alveolar (ⵙ, ⵣ, ⵚ,ⵥ) (ⵛ, ⴳ) ⴽ, ⴳ, ⴽⵯ, ⴽⵯ, ⵀ, ⵃ, ⵀ, ⵃ, ⵀ, ⵀ ⵀ.
- 2 semi-consonants: ⵢand ⵡ.
- vowels: the full ones (ⴰ, ⵉ, ⵓ), neutral (ⴻ).

4 Automatic Speech Recognition

Automatic speech recognition (ASR) systems convert speech to text; the problem of speech recognition can be stated as follows. From acoustic observations X, the system looks for the sequence of words W^* maximizing the following equation [1–3]:

$$w^* = \arg\max_{w} P(W|X) \tag{1}$$

After applying the Bayes theorem, this equation becomes:

$$w^* = \arg\max_{w} \frac{P(X|W)P(W)}{P(X)} \tag{2}$$

P(X) is considered constant and removed from Eq. 2.

$$w^* = \arg\max_{w} P(X|W)P(W) \tag{3}$$

Two types of probabilistic models are used to search for the most probable sequence of words: Acoustic models that provide the value of P(X|W), and the language model that provides the value of P(W).

4.1 Acoustic Pre-processing

The relevant acoustic information of the speech signal is essentially in the frequency band (50 Hz–8 kHz). A signal parametrization system, also known as acoustic pre-processing, is required for signal shaping and calculation of coefficients. This step must be done carefully, as it contributes directly to the performance of the system.

The acoustic analysis is divided into three stages; an analog filtering, an analog/digital conversion and a calculation of coefficients.

4.2 The hidden Markov model

An HMM is specified by the following components $\{Q, X, \Pi, A, B\}$ [1]:

- Q: $\{q_1 \ldots q_N\}$ a set of N states
- X: $\{x_1 \ldots x_M\}$ a sequence of M observations
- $P = \{p_i\}$ $\Pi = \{\pi_i = P(q_i)\}$ an initial probability distribution
- $A = \{a_{ij}\}$ a transition probability
- $B = \{b_{ik}\}$ emission probabilities, Each state of an HMM is represented by a set of Gaussian Mixture density functions.

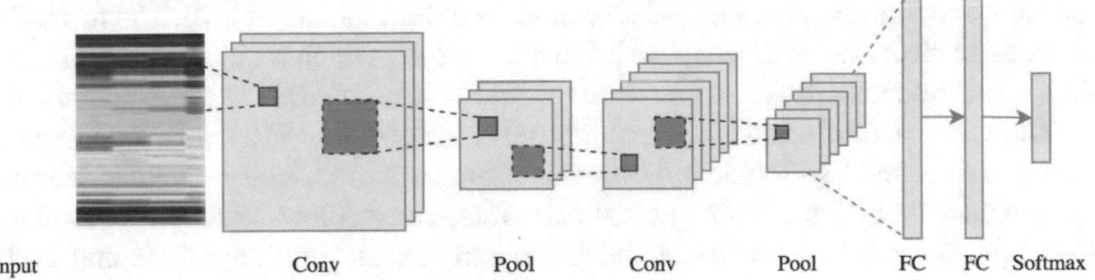

Fig. 2 Illustration of the architecture used for the CNN with many layers

The basic problems of HMM specified as following:

Evaluation: Our first problem is to Compute Probability of observation sequence given a model, calculated by forward algorithm and Viterbi algorithm.

Decoding: Find state sequence which maximizes probability of observation sequence, calculated by Viterbi algorithm.

Training: The third problem for HMMs is learning the parameters of an HMM: adjust model parameters to maximize probability of observed sequences, calculated by forward-backward algorithm.

4.3 Convolutional Neural Networks (CNN)

The CNN is a powerful neural network take input images, can also be applied to sound when it is represented visually as a spectrogram. CNN inspired by biological processes precisely the cat visual cortex [9]. The CNN architecture has two components:

- The convolutive part/feature extraction part: Consist of a series of convolutions and pooling to extract the MFFc features.
- The classification part: The fully connected layers leading into a softmax classifier (Fig. 2).

5 Experiments and Results

5.1 Environment

In this work, we build the HMMs acoustic model with CMU Sphinx-4; CMU Sphinx is an ASR tool from the Carnegie Mellon University can be download with the ability to modify the source code. The first versions of Sphinx (1, 2, and 3) are written in C language, but the recent version of Sphinx 4 is written in java. The Sphinx 4 decoder is very flexible in its configuration, with a modular architecture consists of the front end, the decoder, a knowledge base, and the application [10]. The formation of the acoustic model is done by the SphinxTrain tool, which requires the installation ActivePerl to edit scripts and Microsoft Visual Studio to produce the executables.

The CNN acoustic model is building with TensorFlow [11] is an open-source library developed by Google's AI organization, as a proprietary machine learning system based on deep learning neural networks, TensorFlow written in python and C++ with a model and robust architecture can run on multiple CPUs and GPUs [12]. It is necessary to be installed properly: pip and Virtualenv, CUDA Toolkit 9.0, GPU card with CUDA Compute Capability 3.0, GPU drivers, and cuDNN SDK v7 [11, 13].

5.2 *Training*

In order to compare the HMMs and CNN recognition system for Amazigh language, we used the same corpus [4] consists of 9900 audio files (33 letters × 10 repetitions × 30 speakers), we used mel frequency cepstral coefficients to extract the MFFc features.

Recording files are in MS WAV format with a specific sample rate—16 kHz, 16 bit, mono.

The language model predict P(W) captures the constraints of natural language in order to guide acoustic decoding. We use a statistical language models for two systems.

The acoustic model predicts P(X|W) the probability of observing X when W is pronounced:

- For HMMs model, it is a mixture of Gaussians, we use 8 Gaussian Mixture, 3 HMMs states. The training focus about estimate the parameters model to looking for the most probable sequence, calculated by forward-backward algorithm (Fig. 3).
- For CNN model, it is a convolution neural network, we build a CNN model use two convolutional layers, the number of parameters below 250 K, with 1280 units, the spectrogram creation parameters are: –window_stride_ms = 10.0, —dct_coefficient_count = 40, –clip_duration_ms = 10,000. Once we fixed the CNN architecture, we are going to train the model. During the training phase, in order to minimize the error and improve the prediction of the model, we optimize the weights or network parameters by gradient backpropagation (Table 1).

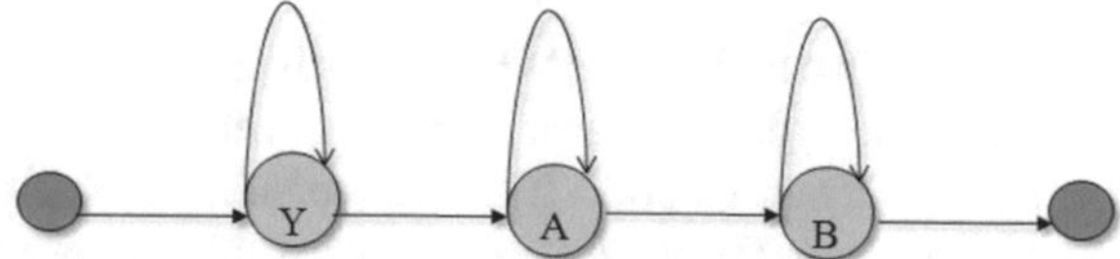

Fig. 3 A three-state left-right HMM for the word YAB

Table 1 Recognition accuracy

Model	Accuracy (%)
HMMs	90
CNN	92

These results show that the system obtained the best performance of 92% of accuracy when trained using CNN model.

The presented work has been compared with the existing similar works. In our previous work [1], the system obtained best performance of 90% when trained using 128 Gaussian Mixture models and 5 number of HMMs states. Zhang et al. [5] have studied a series of neural networks based acoustic models; time delay neural network (TDNN), CNN, and the long short-term memory (LSTM), applied them in the Mongolian speech recognition systems, and compared their performance. The result shows that the LSTM is the best model among them with 8.12% WER. Liu [14] has proposed analyzed and compared CNN and LSTM networks performance with that of DNNs. The results show that CNN and LSTM significantly improve ASR accuracy for various tasks.

6 Conclusion

This paper presents a comparative study of HMMs and CNN model of ASR tasks. For decades, the HMMs stay the solution by excellence to the problem of speech recognition system. From 2012, CNN becomes the flagship architecture of deep learning proved in many domains like image recognition. In this paper, the convolutional neural networks (CNN) have been used for improving automatic speech recognition (ASR) for the Amazigh language with 92% of accuracy.

In perspective, we can compare the regularization technique for CNN and HMMs in Amazigh speech recognition to reduce overfitting on recognition tasks.

References

1. Telmem, M., Ghanou, Y.: Estimation of the optimal HMM parameters for Amazigh speech recognition system using CMU-Sphinx. Proc. Comput. Sci. **127**, 92–101 (2018)
2. Bhandare, A., Bhide, M., Gokhale, P., Chandavarkar, R.: Applications of convolutional neural networks. Int. J. Comput. Sci. Inf. Technol. **7**(5), 2206–2215 (2016)
3. Telmem, M., Ghanou, Y.: Amazigh speech recognition system based on CMU-Sphinx. In: Proceedings of the Mediterranean Symposium on Smart City Applications, pp. 397–410. Springer, Cham (2017)
4. Satori, H., El Haoussi, F.: Investigation Amazigh speech recognition using CMU tools. Int. J. Speech Technol. **17**(3), 235–243 (2014)

5. Zhang, H., Bao, F., Gao, G., Zhang, H.: Comparison on neural network based acoustic model in Mongolian speech recognition. In: 2016 International Conference on Asian Language Processing (IALP), pp. 1–5. IEEE (2016)
6. Abushariah, M.A., Ainon, R.N., Zainuddin, R., Elshafei, M., Khalifa, O.O.: Natural speaker-independent Arabic speech recognition system based on Hidden Markov models using Sphinx tools. In: International Conference on Computer and Communication Engineering (ICCCE '10), pp. 1–6. IEEE (2010)
7. https://en.wikipedia.org/wiki/Berberlanguages
8. Amer, M., Bouhjar, A., Boukhris, F.:Initiation la langue amazigh. Institut Royal de la Culture Amazighe (2004)
9. Hubel, D.H., Wiesel, T.N.: Receptive fields, binocular interaction and functional architecture in the cat's visual cortex. J. Physiol. **160**(1), 106–154 (1962)
10. Oualid, M.D.: Reconnaissance Automatique De La Parole Arabe Par Cmu Sphinx 4. Doctoral Dissertation, Université Ferhat Abbas de Sétif 1 (2013)
11. https://www.tensorow.org/tutorials/sequences/audiorecognition
12. Wang, Q., Chu, X.: GPGPU performance estimation with core and memory frequency scaling. In: 2018 IEEE 24th International Conference on Parallel and Distributed Systems (ICPADS), pp. 417–424. IEEE (2018)
13. https://www.tensorow.org/install
14. Liu, X.: Deep Convolutional and LSTM Neural Networks for Acoustic Modelling in Automatic Speech Recognition (2018)

Region-Based Background Modeling and Subtraction Method Using Textural Features

Samah El kah, Siham Aqel, Abdelouahed Sabri and Abdellah Aarab

Abstract In this paper, we propose a region-based method using texture features for background modeling and subtraction in video sequences. To ensure the robustness of the proposed approach illumination changes and noise, we propose to combine the worthy properties of the texture analysis, particularly the local binary patterns and the region-based approaches. Experiments on outdoor videos attained satisfactory results in terms of both qualitative and quantitative results.

Keywords Motion detection · Background modeling · Background subtraction · Texture features · Local binary patterns

1 Introduction

Background subtraction is a fundamental component in many computer vision applications. In fact, the output data by this module can be considered as a valuable low-level visual cue to perform further processing such as tracking targets and event understanding. To meet this objective, one has to create a reliable representation of the static scene that should be robust against illumination changes, dynamic background, shadows to name a few, keep it updated over time, and compare it with each incoming frame to determine areas of divergence. The literature is replete of background modeling methods; however, there remain some challenging problems, due to the complex nature of real-world scenes. Changes in the scene can take place

S. El kah (✉) · A. Aarab
Laboratory LESSI, Department of Physics, Faculty of Sciences, BP 1796, 30000 Fes-Atlas, Morocco
e-mail: samah.elkah@gmail.com

S. Aqel
Higher Institute of Engineering and Business (ISGA_Fez), 38, Avenue des FAR - Ville Nouvelle, Fes, Morocco

A. Sabri
Laboratory LIIAN, Department of Computer Science, Faculty of Sciences, BP 1796, 30000 Fes-Atlas, Morocco

V. Bhateja et al. (eds.), *Embedded Systems and Artificial Intelligence*, Advances in Intelligent Systems and Computing 1076,
https://doi.org/10.1007/978-981-15-0947-6_51

owing to illumination changes during the sequence (abrupt or gradual), or changes in the geometry of the scene caused by the introduction/displacement of objects, or when an object remains still for a long period. The challenging nature of the issue has generated numerous papers tackling one or many of its aspects. Some interesting surveys can be found in [1–3].

Consequently, different types of features have been investigated to construct as much reliable background representation of the scene as possible. They can be of low level directly obtained from the image as color/ intensity features, or hand-crafted features like texture features, or calculated like edge features. Moreover, they can be computed for and from different image levels: pixel, block, or region. A detailed paper has addressed this topic [4].

Pixel-based background modeling methods generally proceed by analyzing the temporal histogram of intensities of each pixel over time in video sequence images. In this context, statistical methods provide a good framework for background modeling. The most famous pixel-based parametric method is the Gaussian model. Wren et al. introduced a statistical method, they proposed to model the history over time of intensity values of a pixel by a single Gaussian [5], and this model is powerless to handle multimodal backgrounds. Stauffer and Grimson proposed the Gaussian mixture model [6] method that is undoubtedly a pioneering work, where each pixel in a video frame is modeled by a mixture of Gaussians that ranges from three to five. However, this technique has a considerable computational cost besides it often fails to adapt fast temporal changes in the background. Since then, many variants have been suggested; some authors propose to automatically and dynamically determine the number of Gaussians to be more robust to dynamic environments [7, 8].

In contrast to pixel-based methods, region-based methods exploit the spatial information to segment the images. The image frames of the video sequence are usually divided into blocks, and according to specific characteristics, the selected blocks cooperate to form the background model. At authors, to model the background method named Rectgauss at different scales based on color histogram and texture measure criteria has been proposed [9]. In [10], Riahi et al. have slightly modified the original method by determining automatically the block size based on the image resolution. In [11, 12], Aqel et al. proposed a background modeling approach combining the entropy measure and the quadtree decomposition. The method splits each frame into regions and based on the entropy value to capture the variations in the scene that are occurred by moving objects. Heikkila et al. [13] proposed to model the background as set of adaptive local binary patterns histograms that are calculated on a circular area around the pixel. On the basis of these methods, it is reported that region-based approaches mitigate significantly the aforementioned phenomena.

In fact, the suitable choice of a feature or a combination of features in background modeling can improve the segmentation of moving objects. In this paper, we propose to join the merits of the region-based and texture-based methods. The texture of frames is estimated based on the local binary patterns (LBP) as they have shown great invariance to monotonic illumination changes, do not need many parameters to be set, and offer high discriminative power. The texture characteristic is used as a criterion to evaluate the presence of moving objects in regions obtained by iteratively partitioning the frames of the video sequence by quadtree decomposition.

The remainder of this paper is organized as follows: Sect. 2 presents our background modeling approach, experimental results are given in Sect. 3, and Sect. 4 concludes the paper.

2 Proposed Method

In this section, we present our proposed method that aims to model the background of outdoors scenes containing moving objects, captured by a fixed camera.

The proposed approach implements the main usual steps of background subtraction-based methods: background modeling and foreground detection. After an initialization phase, each frame of the video sequence is divided according to a quaternary tree production algorithm. First, each frame of the video sequence is divided into four regions of the same size (quadrant). For each quadrant resulting from the partitioning of the video, the texture descriptor based on the LBP is extracted. Based on the corresponding histograms of these descriptors, it is possible to retain the quadrant that presents the smallest distance in comparison with its adjacent quadrant by means of the Bhattacharyya measure. The selected quadrant is subjected to a moving object test based on temporal differentiation with the consecutive image. If the image is considered without movement, it is immediately retained to be part of the final background. Otherwise, the algorithm goes on toward a new partitioning where this region is again divided into four equal subquadrants. This procedure is reiterated until we reach the adopted minimum size of 16×16 pixels for the quadrant. Hence, the final background is obtained by the combination of quadrants of non-moving quadrants obtained at each iteration. Figure 1 explains in a general way the followed step of procedure. Then more details will be given in the next subsections.

2.1 Partitioning Mode

The adopted partitioning mode is of type quadtree decomposition that consists of splitting the image into equally sized regions and then browses through each region to check whether it meets a certain constraint. Otherwise, those regions are in turn divided into subblocks. The process of splitting is reiterated until a stopping criterion is met. In our proposed method, the constraint used is based on the texture features measurement. The quadtree technique allows direct access to the desired block regions, what would inevitably alleviate the computation time throughout the process. Figure 2 illustrates the partitioning mode principle.

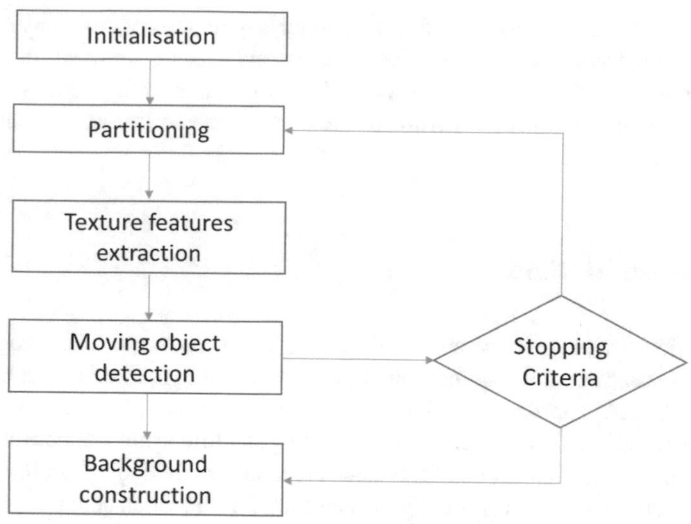

Fig. 1 Flowchart of the background modeling proposed method

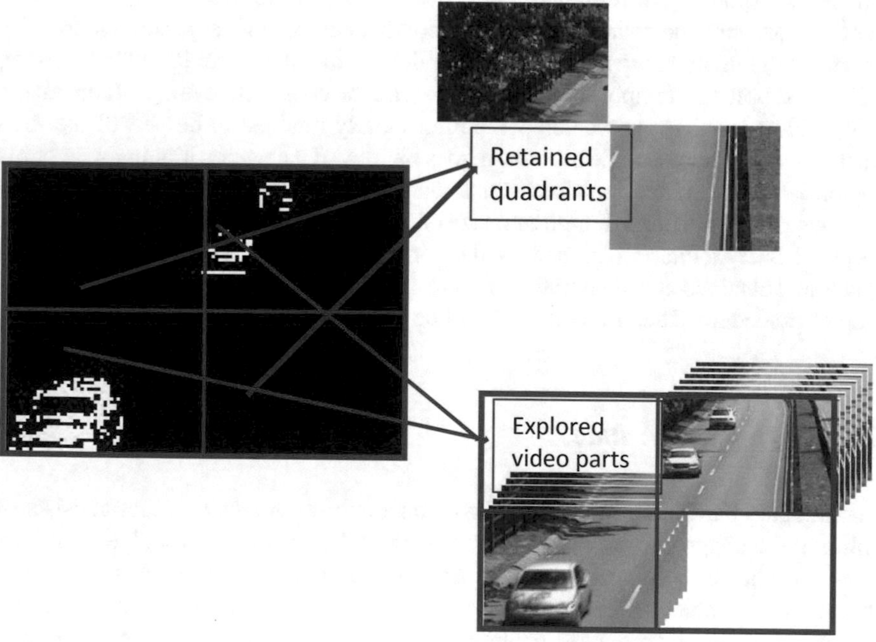

Fig. 2 Partitioning mode illustration

2.2 Texture Descriptor

In order to analyze the texture of the video frames, we used the local binary pattern that was first introduced by Ojala and Maenpaa [14] and has other variants as proposed in [15, 16]. It is a powerful graylevel invariant texture primitive. The operator describes each pixel by means of its relative neighboring pixels. If the graylevel of the neighboring pixels is higher or equal, the value is set to one, otherwise to zero. The descriptor returns the results over the neighborhood as a binary number. Figure 3 illustrates how a center pixel g_c is encoded by a series of bits relatively to its 3×3-neighborhood. The LBP operator applied to the $c(x_c, y_c)$ can be expressed as follows:

$$\mathrm{LBP}(x_c, y_c) = \sum_{p=0}^{P-1} S(g_i - g_c)2^p \tag{1}$$

g_c is the gray value of the center pixel c, whereas gp represents the gray values of the P neighborhood pixels of a center pixel. The thresholding function $S(x)$ is defined as follows:

$$S(x) = \begin{cases} 1 \text{ if} \geq 0; \\ 0 \text{ otherwise}; \end{cases} \tag{2}$$

The histogram is then calculated from the generated LBP codes as the LBP feature.

2.3 Distance Measure

Bhattacharyya distance measures the similarity of two probability distributions. It has a computational complexity $O(n)$.

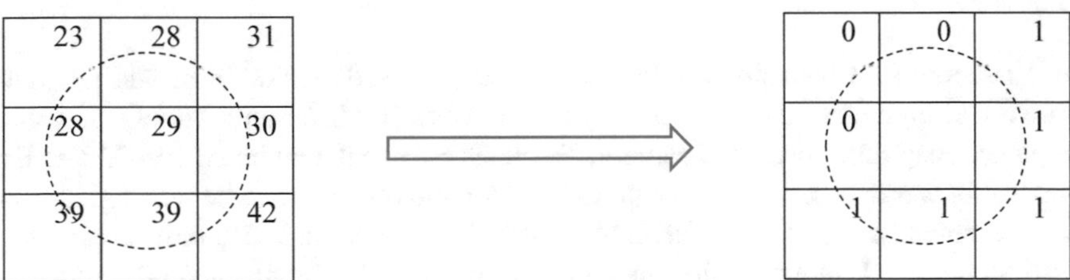

Fig. 3 LBP encoding illustration

$$D(x, y) = 1 - \sqrt{\sum_{i=1}^{n} \frac{\sqrt{x_i y_i}}{\sqrt{\sum_{i=1}^{n} x_i \sum_{i=1}^{n} y_i}}} \quad (3)$$

Once we characterize the spatial structure of the texture of the frames using the LBP operator, we calculate the LBP histogram with 32 bins, instead of 256. Thus, the size is reduced without losing too much useful information. After that, we calculate the distance between the bins of consecutive frames. Consecutive frames having the lowest distance value are the most probable to contain most of the background parts.

2.4 Foreground Detection

After generating the background model, the next step is to use background subtraction for foreground/background pixel classification using the current frame and the background model. The background difference frame is defined as follows:

$$\Delta_t(i, j) = |B_t(i, j) - I_t(i, j)| \quad (4)$$

A foreground mask representing regions of significant motion in the scene is generated by comparing the background difference to a threshold *Th* set experimentally. Hence, the foreground mask is obtained according to:

$$D_t(i, j) = \begin{cases} 255 \text{ if } \Delta_t(i, j) > Th \\ 0 \quad \text{else} \end{cases} \quad (5)$$

Figure 4 illustrates the foreground detection process.

3 Experimental Results

In this section, we evaluate the relevance of our proposed method by providing qualitative and quantitative results obtained by performing tests on a variety of video sequences captured by a fixed camera. The tests are carried on both 2014DATASET and wallflower data sets and the obtained foreground objects of some selected frames are presented in Fig. 5. The 2014DATASET is a benchmark database commonly used for testing different motion detection approaches. It contains several sequences arranged in categories, each representing a typical challenge scenario for video surveillance systems. We selected the following categories to conduct the tests on:

- Baseline Category: is considered as the reference data set to test motion detection schemes. It contains a variety of average challenges that range from background

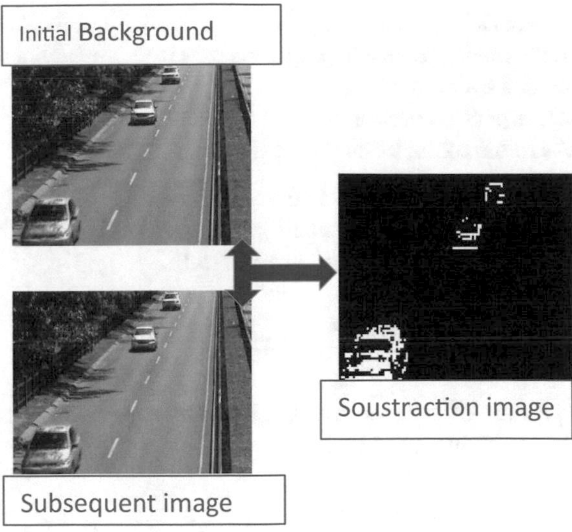

Fig. 4 Foreground detection illustration

Fig. 5 Foreground detection results for (from left to right): highway, pedestrians, bus station, traffic, lightswitch sequences

motions, shadows, to abandoned objects and pedestrians stopping for a while and then move away.

- Intermittent Object Motion: this category contains videos in which every once and again, an object appears or disappears from the scene, what causes the "ghost effect." This category aims at testing how the algorithm adapts to changing background.

- Shadow: contains sequences where shadowed moving objects are recurrent.
- Camera Jitter: contains sequences captured by a trembling camera due to natural or environmental event.
- Light Switch: A person enters a lighted room and turns off the lights. Afterward, person walks in the room, turns on the light, and moves the chair.

Concerning qualitative results of the detected moving objects, we used the recall, precision F-measure and similarity. Recall is the ratio of detected true positives to the total number of pixels present in the ground truth.

$$\text{Recall} = \frac{\text{TP}}{\text{TP} + \text{FN}} \tag{6}$$

Precision is the ratio of detected true positives as compared to the total number of pixels detected by the proposed method.

$$\text{Precision} = \frac{\text{TP}}{\text{TP} + \text{FP}} \tag{7}$$

F-measure is the weighted harmonic mean of precision and recall.

$$\text{F-measure} = \frac{2 * \text{recall} * \text{precision}}{\text{recall} + \text{precision}} \tag{8}$$

Finally, similarity measure is defined as:

$$\text{Similarity} = \frac{\text{TP}}{\text{TP} + \text{FN} + \text{FP}} \tag{9}$$

where true positive (TP) is the number of pixels correctly classified as belonging to the foreground, true negative (TN) represents the number of background pixels classified correctly. False positive (FP) is the number of background pixels that are incorrectly classified as foreground and false negative (FN) is the number of foreground pixels that are incorrectly classified as background. Table 1 reports the average evaluation metrics obtained by the suggested sequences.

In highway sequence, the GMM can detect well the moving objects but there are still some holes in the segmented image, also there exist some noisy pixels. In contrast to our proposed method, that detects the whole shape of cars without holes. In the Bus Station sequence, the GMM does not provide a well-detected object, with the presence of some noisy pixels. However, our method provides a good representation of the walking people without false negatives what is proved in the quantitative results that show that it handles well the shadow effect. In pedestrians and light switch sequences, the GMM detects only the outline of the dynamic objects, while the proposed method provides a full object representation.

Table 1 Average evaluation metrics of the proposed sequences

Sequence	Metrics	GMM	Proposed method
Highway	Similarity	0.5993	0.7850
	F-measure	0.4278	**0.8795**
Pedestrians	Similarity	0.5881	0.4293
	F-measure	0.4165	**0.6707**
Bus station	Similarity	0.5231	0.7560
	F-measure	0.6869	**0.8610**
Light switch	Similarity	0.2651	0.6146
	F-measure	0.4191	**0.7613**
Traffic	Similarity	0.4913	0.4250
	F-measure	0.3257	**0.5965**

4 Conclusion

This paper presents background modeling and moving object detection approach in a video sequence. The background model is generated using the texture features analysis; particularly, we used the local binary patterns due to not only their simplicity and their computational speed, but also because they are robust against illumination changes and have highly discriminative power. From the visualized results and the metrics values, we can deduce that our approach attained satisfactory results in terms of both qualitative and quantitative results.

References

1. Elhabian, S., El-Sayed, K., Ahmed, S.: Moving object detection in spatial domain using background removal techniques—state-of-art. Recent Patents Comput. Sci. Art 32–54 (2008)
2. Bouwmans, T.: Recent advanced statistical background modeling for foregroun detection—a systematic survey. Recent Patents Comput. Sci. Bentham Sci. Publ. **4**(3), 147–176 (2011)
3. El Kah, S., Aqel, S., Sabri, M.A., Aarab, A.: Background modeling method based on quad tree decomposition and contrast measure. In: The Second International Conference on Intelligent Computing in Data Sciences (ICDS '2018), 03–05 Oct 2018. Fez-Morocco. Proc. Comput. Sci. **148**, 610–617 (2019)
4. Bouwmans, T., Silva, C., Marghes, C., Zitouni, M., Bhaskar, H., Frelicot, C.: On the role and the importance of features for background modeling and foreground detection, to be submitted to Comput. Sci. Rev. (2016)
5. Wren, C., Azarbayejani, A., Darrell, T., Pentland, A.: Pfinder: real-time tracking of the human body. IEEE Trans. Pattern Anal. Mach. Intell. **19**(7), 780–785 (1997)
6. Stauffer, C., Grimson, W.: Adaptive background mixture models for real time tracking. Comput. Vis. Pattern Recognit. 246–252 (1999)
7. Zivkovic, Z.: Improved adaptive Gaussian mixture model for background subtraction. In: Pattern Recognition. ICPR 2004. Proceedings of the 17th International Conference on IEEE, pp. 28–31 (2004)

8. Carminati, L., Benois-Pinau, J.: Gaussian mixture classification for moving object detection in video surveillance environment. In: IEEE International Conference on Image Processing (ICIP 2005), Sept 2005, pp. 113–116
9. Bourezak, R., Bilodeau, G.A.: Iterative division and correlograms for detection and tracking of moving objects. In: The International Workshop on Intelligent Computing in Pattern Analysis/Synthesis, (IWICPAS 2006), Aug 2006, pp. 45–56. Xi'an, China
10. Riahi, D., Riahi, D., Bilodeau, G.: RECTAUSS-Tex: block based background subtraction. Technical Report EPM-RT-2012-03, Ecole Polytechnique de Montreal, pp. 1–9 (2012)
11. Aqel, S., Sabri, M.A., Aarab, A.: Background modeling algorithm based on transitions intensities Int. Rev. Comput. Softw. (IRECOS), 387–392 (2015)
12. Aqel, S., Hmimid, A., Sabri, M.A., Aarab, A.: Road traffic: vehicle detection and classification. In: 2017 Intelligent Systems and Computer Vision (ISCV), 17–19 Apr 2017, pp. 1–7. Fez, Morocco. Print on Demand (PoD). ISBN: 978-1-5090-4063-6. Electronic ISBN: 978-1-5090-4062-9. https://doi.org/10.1109/isacv.2017.8054916
13. Heikkilä, M., Pietikäinen, M., Schmid, C.: Description of interest regions with local binary patterns. Pattern Recognit. **42**, 425–436
14. Ojala, M.P., Maenpaa, T.: Multiresolution gray-scale and rotation invariant texture classification with local binary patterns. IEEE Trans. Pattern Anal. Mach. Intell. **24**(7), 971–987 (2002)
15. Yang, Y., Yang, J., Liu, L., Wu, N.: High-speed target tracking system based on a hierarchical parallel vision processor and gray-level LBP algorithm. IEEE Trans. Syst. Man Cybern. Syst. **47**(6) (2017)
16. Kim, H., Kim, H.-K., Lee, S.-J., Park, W.-J., Ko, S.-J.: Kernel-based structural binary pattern tracking. IEEE Trans. Circuits Syst. Video Technol. **24**(8) (2014)

Fuzzy Keyword Matching Using N-Gram and Cryptographic Approach Over Encrypted Data in Cloud

V. Lavanya, Somula Ramasubbareddy and K. Govinda

Abstract Due to widespread use of cloud computing, there is more and more data is being stored over cloud by the users every day in such case finding a solution for easy, secure and safe way to handle the data is necessary. So I have given the solution for such problem using some algorithms like AES encryption and similarity calculating algorithms. So, instead of searching directly using keyword here in this paper the approach is different. Here, I am using searching over the encrypted data which is much secure than the normal searching. In this project, I will be using AES encryption algorithm for encryption and Jaccard coefficient to find the similarity between the keywords. So after giving keyword, the encrypted keywords can be searched and user will be able to find the file over cloud. In this project, I have implemented and tested the solution to the problem of secure cloud computing. By using the fuzzy keyword searching over encrypted cloud computing and taking care of safety and secrecy of data files. This fuzzy keyword searching significantly increases the efficiency and safety over cloud. This is user friendly and easy to manage and require less resource. And the results are accurate enough to get exact files searched by the user. The efficiency is increased by approximated computing and it works great. In this solution, Jaccard coefficient is being used to calcite similarity and used two advanced techniques to make fuzzy keyword set, which achieves the great results. Where n-gram algorithm is used to generate the set of different length n-grams of any keyword which is going to be encrypted and is being stored over cloud. I have done security analysis and found out that this system is secure and reliable for modern cloud computing operations and file storage. This system achieves its proposed goal efficiently and results are shown for same.

Keywords Fuzzy · Encryption · Cloud · N-gram · Security

V. Lavanya · S. Ramasubbareddy (✉)
IT, VNRVJIET, Hyderabad, Telangana, India
e-mail: svramasubbareddy1219@gmail.com

K. Govinda
Scope, VIT University, Vellore, Tamilnadu, India

V. Bhateja et al. (eds.), *Embedded Systems and Artificial Intelligence*, Advances in Intelligent Systems and Computing 1076,
https://doi.org/10.1007/978-981-15-0947-6_52

1 Introduction

With the development in cloud computing, cloud servers are extensively being used for loading data centrally. This includes numerous communal financial records, game data, Web site login and more type of data. The cloud services deliver respite to user as it decreases to ring expenses and risk of behind the information due to hardware disappointments, i.e., it force occurs the hard disk of our system or due to nasty activity and we would finish up losing all the important data. The other problems may be deprived of maintenance and low formation facility as likened to cloud configuration services. On the other hand, cloud also has some disadvantages because cloud servers cannot be reliable by the data owners so it is the user's duty to encrypt the data before upload. By a plying data encryption, there is above the data utilization in more efficient manner as the data is secured and cannot be accessed by unauthenticated users. Also, in cloud computing, data owners share their subcontracted data with large amount of users due to which privacy of the data is not ensured. Thus, it is required that every discrete should save specific data files which they are looking for within a session. To apply this type of system, we need to deal with keyword search that retrieve the required files in its place of saving all the encrypted files. In plain text search situations such as Google search, the keyword search method is used which allows users to selectively recover the compulsory files. Unsuitably, encoded data limits user's ability to use the keyword search method and thus makes the plaintext search methods no use for cloud computing. Apart from this, encrypted data files which consist of file name needs to be protected as it may also describe the quality and sensitivity of information related to the data files. But by encrypting file name the outdated plain text practice gets totally useless as it is only able to search over plain text.

2 Literature Survey

Nowadays with the increasing use of cloud computation, it is necessary to make it safe and secure for the computation and easy to use [1]. They proposed this problem where the searching over the cloud is needed to be more secure and available. At starting, it seems easy that it can be achieved easily by constructing trap doors in the string. However, it suffers from several attacks like statistics attacks and it fails in privacy [1].

And in such systems, it is necessary to keep the time factors and cost should be cheap. Once these things are done then later the system's efficiency will come into play [2]. And we can use the Web sites and other sources for the coding and writing the algorithm. After all the considerations the system needs to be tested again later.

I have to investigate the cloud computing outline survey.

2.1 Cloud Computing

Cloud computing is the field in which the common platform is there for data storage while computation is done over different machine and it requires large security and safety. So it can be easily used by the users with no privacy issues [3–8].

2.2 Benefits of Cloud Computing

- Minimization of capital expenditure
- Independence location and device
- Improved utilization and efficiency
- Scalability very high.

2.3 Information Security

It is the field of Internet in which the common threats are analyzed and solution for them is researched. In this field, all things related to data transfer and its computation is studied.

3 Proposed Method

Algorithms used:

- AES encrypt rijndael 256
- N-grams generation
- Jaccard coefficient calculation.

3.1 Gram-Based Technique

It is the way for constructing the *n*-grams of the given keyword. The substrings of keyword are used as the gram and the permutations of such substrings are used to generate the grams.

While there are so many uses of *n*-gram algorithm; here, it is used to match the query with the fuzzy keyword so that the matching can be done easily and efficiently over cloud.

So, *n*-grams are formed and stored in the database after encryption and it is used for matching.

For example, the gram-based fuzzy set CASTLE, 1 for keyword CASTLE can be constructed as

{CASTLE, CSTLE, CATLE, CASLE, CASTE, CASTL, ASTLE}.

3.2 Jaccard Index

The Jaccard index is used to find the similarity between the set of keywords in the set. The intersection of the set is divided by the union of the sets. This is used in this project to calculate the similarity index between the search keyword and encrypted keyword in the stored dataset.

$$J(A, B) = |A \cap B|/|A \cup B| = |A \cap B|/(|A| + |B| - |A \cap B|)$$

(If A and B are both empty, we define $J(A, B) = 1$.)

$$0 \leq J(A, B) \leq 1.$$

3.3 AES Encrypt Rijndael 256

AES is the advanced method for encryption of the keys for safety purpose. Here, in this project, I am using it to encrypt the keywords and its *n*-grams and it will be stored in database. And matching will be done later with encrypted data. And files will be returned after matching the same keyword or its *n*-gram.

AES uses bytes for its operation rather than bits, so 128 bits word will be considered as the 16 byte block.

And the number of rounds is not constant and it varies as the number of blocks changes.

The schematic of AES structure is given in Fig. 1 (Fig. 2).

4 Experimental Results

This is the home page of cloud Web site in which we can see the various models, including user, registration and admin department. Through admin page, he/she can login by giving credentials which is fully secured mechanism. Here, is the demo of registration process in which user enter the details and register for the cloud and can login and upload the encrypted files and search and modify them (Fig. 3).

This is how user can upload the file using upload button (Fig. 4).

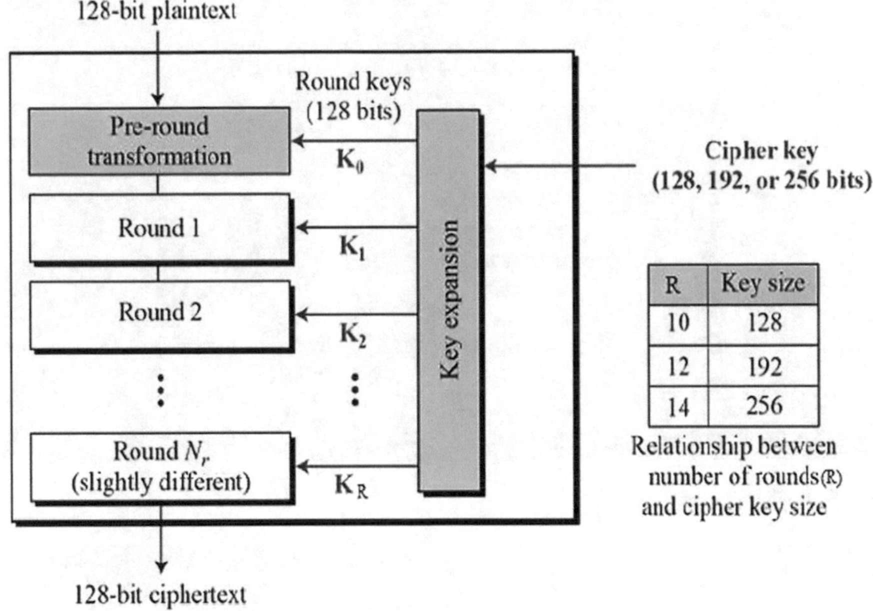

Fig. 1 AES structure

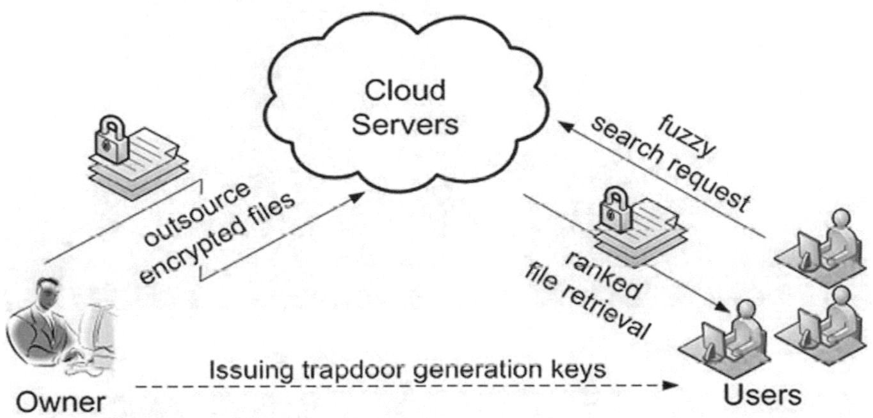

Fig. 2 Framework of project

After choosing file and giving proper keywords during uploading (all these keywords will be encrypted and saved in database over cloud).

And uploading the file he will be informed that file is being uploaded and it will be shown here (Fig. 5).

This is result of searching the files over cloud which is efficient and secure. In admin page, he/she can modify the status of files and search them (Fig. 6).

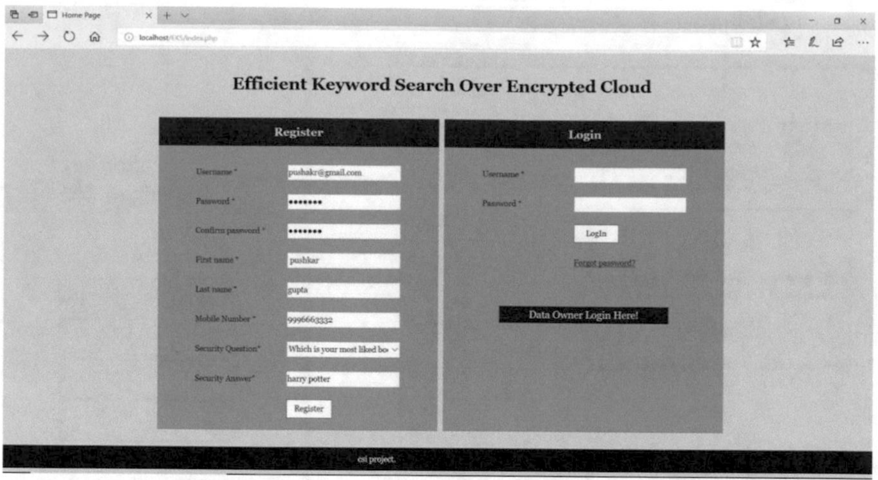

Fig. 3 Registration page

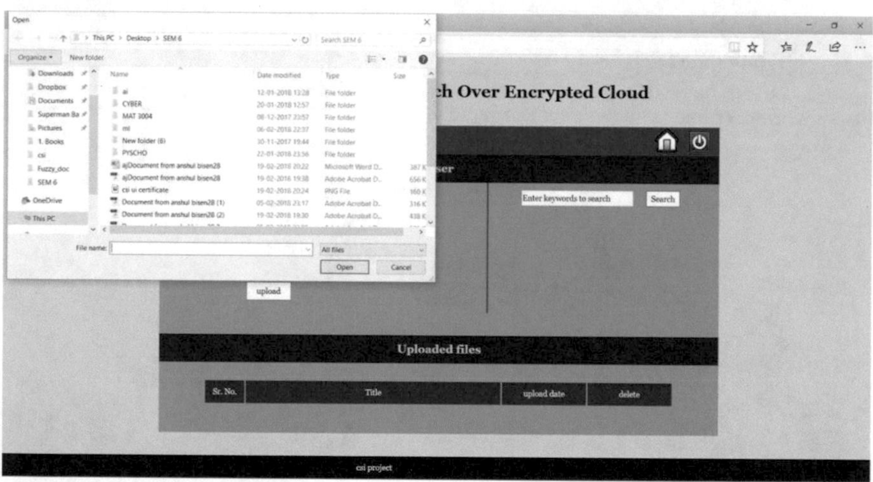

Fig. 4 File uploading page

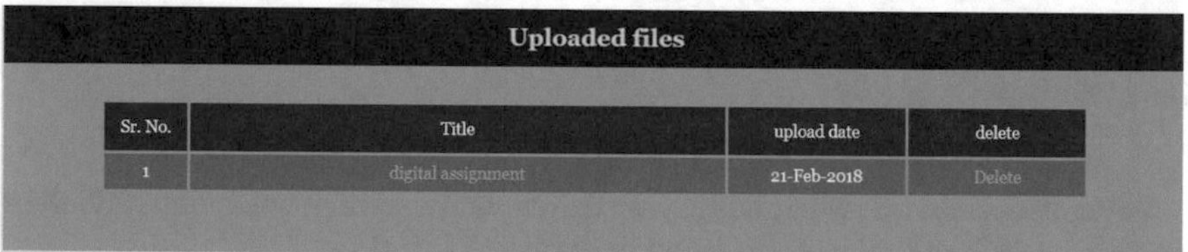

Fig. 5 File in cloud

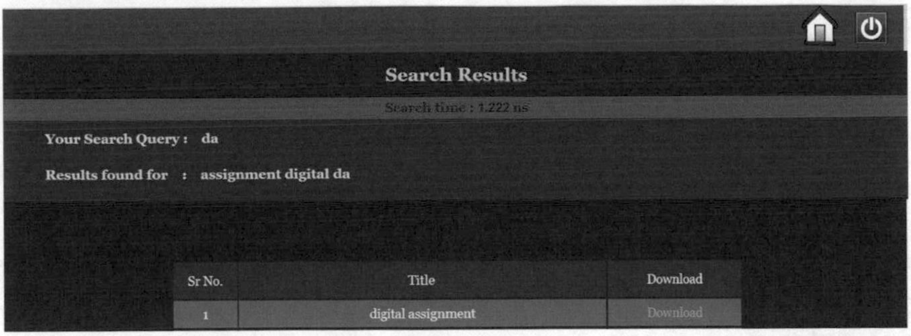

Fig. 6 Result of searching file in cloud

Figure 7 for time versus n-gram searching. In Figure 7, the time complexity of searching is shown where time is on y-axis and n-gram is on x-axis. So, we see that the time is decreasing as the size of n increases and the time fluctuates very less in the same set of n-grams.

In Fig. 8, the time complexity for 4 g for digital as a keyword is shown and as we can see the difference is very less between the same classes of grams. Like for 4 g the time taken is around the 0.3–0.4 ns time range so like this we can see the time need to search the keyword is very less secure as compare to other techniques.

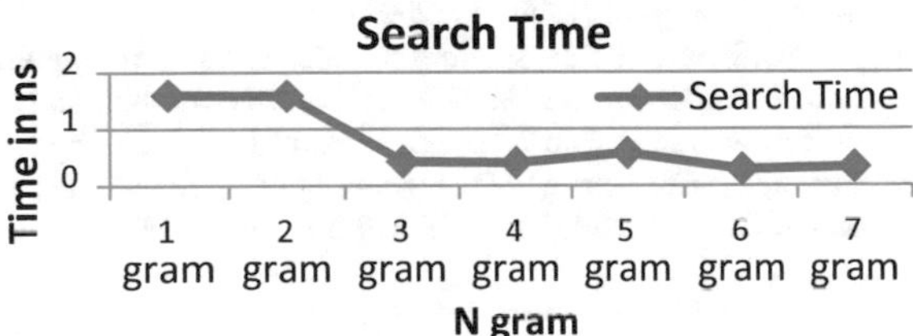

Fig. 7 Searching time complexity

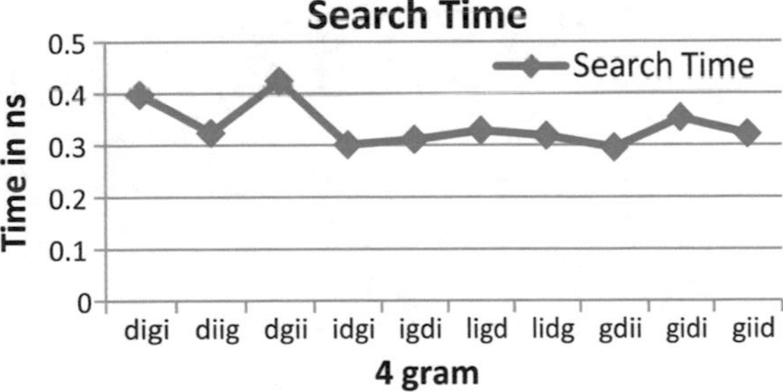

Fig. 8 Time complexity for 4 g

5 Conclusion

Here, in this article, I have given the solution for the problem of security issues over cloud using cryptographic methods to encrypt the data and search them over cloud so it will be easily used by the users. I have also given the facility for admin to manipulate the files over cloud for different users. My system also shows the time to search for different length of n-grams of the keywords given during file upload. There we can see the timing of n-gram searching algorithm. I have used AES algorithm to encrypt the data and Jaccard coefficient to find the similarity of the n-gram and encrypted keyword.

References

1. Li, J., Wang, Q., Wang, C., Cao, N., Ren, K. Lou, W.: Fuzzy keyword search over encrypted data in cloud computing. Department of ECE, Illinois Institute of Technology, Worcester Polytechnic Institute. Email: {jinli, qian, cong, kren}@ece.iit.edu, {ncao, wjlou}@ece.wpi.edu
2. Wasnakar, A.: Implementation of Fuzzy Keyword Search Over Encrypted Data in Cloud Computing
3. Song, D., Perrig, A.: Practical techniques for searches on encrypted data. In: IEEE (2000)
4. Curtmola, R., Garay, J.A., Kamara, S., Ostrovsky, R.: Searchabl symmetric encryption: improved definitions and efficient constructions. In: Proceedings of ACM CCS '06 (2006)
5. Li, C., Lu, J., Lu, Y.: Efficient merging and filtering algorithms for approximate string searches. In: Proceedings of ICDE '08 (2008)
6. Google: Britney spears spelling correction. http://www.google.com/jobs/britney.html (2009)
7. Bellare, M., Boldyreva, A., O'Neill, A.: Deterministic and efficiently searchable encryption. In: Proceedings of Crypto 2007, vol. 4622 of LNCS. Springer (2007)
8. Boneh, D., Crescenzo, G.D., Ostrovsky, R., Persiano, G.: Public key encryption with keyword search. In: Proceedings of EUROCRYP '04 (2004)

Stock Market Analysis and Prediction Using Artificial Neural Network Toolbox

Anand Mahendran, Kunal A. Vasavada, Rohan Tuteja, Yashika Sharma and V. Vijayarajan

Abstract Stock exchange is an open market for the exchanging of organization chart. Shares will be considered as minor entities of a company. Such small entities can be bought by different stake holders. In general, the company holds majority of the shares. Based on different factors the price of such shares is fixed. Stock brokers may be used by some companies to buy/sell the shares. The stock holders generally give advice to companies based on the public opinion. This entails the fact that a lot of factors cause variation in prices of the stock, making assumed prediction based on insignificant set of factors often inaccurate. In this paper, we explore an approach using machine learning techniques (in particular neural networks). Stock market closing prices of varied stocks are predicted using different algorithms on the ANN toolbox in MATLAB and correlation of predicted and acquired value analysis is performed pertaining to a NARX network which is nonlinear and autoregressive, amongst other accuracy metrics and performance criteria to draw out the precision and reliability of the respective training function.

Keywords Stock market · Prediction · Machine learning · Artificial neural network (ANN) · Analysis · MATLAB

1 Introduction

A good exemplar of evolving times is application of modern technologies to everyday tasks delivering better computation and boosting efficiency. In the past two or three decades, researchers made a lot of research of stock market prediction which may lead to consistent outputs. Accuracy is considered as one of the important factors to be considered in stock market prediction [1]. Finding the right timing to buy/sell stocks will be considered as one of the important tasks for the investors. Under these circumstances predicting stock market is the fundamental research problem.

A. Mahendran (✉) · K. A. Vasavada · R. Tuteja · Y. Sharma · V. Vijayarajan
School of Computer Science and Engineering, Vellore Institute of Technology (VIT), Vellore 632014, India
e-mail: manand@vit.ac.in

V. Bhateja et al. (eds.), *Embedded Systems and Artificial Intelligence*,
Advances in Intelligent Systems and Computing 1076,
https://doi.org/10.1007/978-981-15-0947-6_53

Such a stock market prediction can be achieved precisely by using some machine learning approaches. In this regard, we decided to use three specific algorithms to achieve the objective and publish the details about the same [2]. Knowing in what direction the market is going would be a great advantage. If we could predict the exact future outcome of a stock market, then it would give a great advantage and a chance to maximize return on investments and it would also aid in predicting economic instability and allow more time to prepare for such changes and reduce their impact on the economy.

In this paper, we have used MATLAB [3] for simulation purpose. In general, MATLAB will be used for technical computation(s). MATLAB can be used to solve the real time and mathematical problems by integrating and programming. In particular, we will be exploring all the applications on the ANN toolbox which allows us to run the simulation for a lot of different algorithms with different performance metrics.

In this paper, we use artificial neural network tool box for the simulation purpose. By using the artificial neural networks, the simulation problem can be solved by creating the neural network model. The next step in the simulation process is to train and visualize the created models. In particular, deep neural networks can be used to alter to whatever application necessary for training and testing various different data sets and the kind of operations intended to perform the same. In the simulation process, the following operations can be carried such as classification and clustering of patterns, forecasting, reduction of dimensionality, modelling of the systems in a dynamic manner. The nonlinear autoregressive network with exogenous inputs (NARX) is considered as a dynamic network which may function in a recurrent fashion. In addition to that NARX is a several layer network which has a feedback connection in nature. Time series modelling is an important variant in NARX. Such time series modelling will be functioned on the linear fashion. Such models are called as linear ARX model.

The equation which determines the characteristics of NARX model is given in [4]

$$y(t) = f(y(t-1), y(t-2), \ldots, y(t-ny), u(t-1), u(t-2), \ldots, u(t-nu))$$

The successor value of the output $y(t)$ mainly depends on the predecessor value of the output signal in addition to that with the value of the input signal also. A function f NARX model can be implemented by using NARX model in feedforward neural network mode. By using such model, the value of function will be an approximate value. Figure 1 shows such configuration. The configuration in NARX is a two-layer feedforward network. By using such a feedforward, it allows us to choose the input and output in a multidimensional manner. In addition to that, using NARX configuration model the input values can be predicted. The above-mentioned model suggests that NARX can be used in simulation mode which can act as a predictor so that the predecessor value of the input signal can also be predicted.

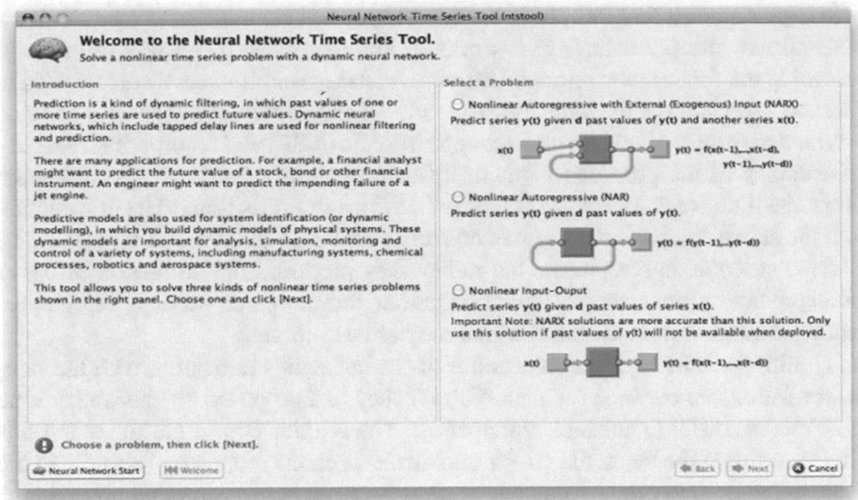

Fig. 1 NARX time series in ANN toolbox

1.1 NARX Setup

The NARX neural network model can be used as the initial predictor for testing any values. First, we need to train the network with a portion of historical data using one of the training algorithms. In the simulation setup, the initial for the feedback delay is fixed as 1:2. In addition to that the size of hidden layer is fixed to 10 [5]. Before any predictions the model was set as a closed loop as for the external feedback function to consist of passed predicted values instead of some external function. Predictions were then done in the closed loop NARX network as such. In accordance with the test scenarios, the neural networks are retrained and with new portions of training data before the next predictions are done.

2 Literature Survey

[5] performed a survey on writing about the utilization of forecasting the stock exchanges using artificial neural network. The authors concluded that the process of predicting the stock list is difficult with time related investigations. Whereas the authors concluded that artificial neural networks would be good choice to solve the problem. A neural network can remove helpful data from huge arrangement of information. Based on the literature survey the authors found that the neural networks seem to be the best machine learning technique in predicting the stock exchanges. The authors proposed there will be a significantly substantial degree for the utilization of artificial neural network to forecast of securities exchange file in a precise manner.

In [6], the stock returns are predicted using the neural network based methodology. A neural network predictor which works in autoregressive mode was utilized for foreseeing the future stock returns. Different mistake measurements were utilized to assess the execution of the indicator. The authors tried different things with genuine information from National Stock Exchange of India (NSE) was incorporated to study the accuracy of the proposed method. Information was taken from the following companies Tata Steel, Maruthi, TCS, BHEL, Wipro and Axis Bank. The information from the above companies was obtained from date 02-01-2007 till 22-03-2010. The obtained outcome is not precise but rather they proposed the utilization of better neural predictive frameworks. In addition to that, they prepared various strategies for limiting the expectation mistakes which may arise in future.

[7] utilized neural system on account of the capacities in dealing with the non-linear relationship between the data. Further they had proposed another fuzzy time arrangement model to enhance anticipating. The relationship in the fuzzy logic is utilized to figure the stock file (in particular the contents related to Taiwan). In the neural network, fuzzy time arrangement demonstrated in test perceptions is utilized for preparing and out-example perceptions are utilized for determining. The disadvantage of taking all the level of enrolment for preparing and anticipating may influence the execution of the neural systems.

[1] discuss the information mining strategies which have to be connected to outline the market capital framework for exchanging firms. The authors indicate how the neural systems can be used in blend with one of the important features in MATLAB (i.e. graphical user interface) which may guide to meet the necessary expectations. The prepared framework can be utilised to conjecture the capital (market) for a specific blend of information parameters. Exactness of the proposed strategy is high in light of the fact that the outcomes acquired were observed to be practically identical to the yield anticipated.

[8] deals about the domain where the stake holders can buy/sell stocks based on prices. Such a domain is called stock exchange. The prices to buy/sell such stocks cannot be done on random basis. Under these circumstances, prediction has to be done on fixing the stock prices. The prediction may be influenced by many external parameters such as sentiment analysis and opinion mining techniques. Such sentiment prediction can be recognized by means of opinion summary. The opinion summary can be collected by means of sentiment scores. In the past three decades, researchers used standard methods for sentiment analysis. But in recent years, the opinion mining has been improved by adding lot of new features such as visualization techniques (both spatial and temporal), different summary of textual data, opinion in a structural manner. In opinion mining, the level of understanding plays a crucial role. Such level of understanding can be obtained by complementing the summaries provided by the users. For example, for a product x, different people may provide different opinion based on their level of understanding about the product x. The opinion given by the user(s) may be of different types such as topic relevant or text based on the product chosen by the user. The main aim of the opinion mining is to facilitate the users about the negative and positive comments about the product which may help the customers. The approaches used in gathering the above task(s)

include lot of research problems such as natural language processing analysis, data clustering and sentiment analysis. The above-mentioned research problems can be solved by using heuristics or statistical approaches.

In [9], a framework was proposed that works by consolidating the two techniques. Initial one is the back propagation using ANN and second is regression. Relapse is one of the systems for anticipating estimations of any element. In any case, it is a measurable strategy and less exact than artificial neural network. Another technique is using back propagation algorithm utilizing artificial intelligence. This technique is significantly more precise than factual strategy. In this way, we are joining over two strategies utilizing verifiable information. The acquired by consolidating two strategies will be certainly precise and amend.

3 Methodology

In this paper, we explore the different alternate simulations that can be run on stock price data so as to actively and accurately predict future stock prices while also testing out various parameter values and functions which are evaluated over a mapping of the actual prices in order to validate the same. Hence, the first step in our progressive approach is collection of data in form of opening and closing stock prices. In this study, we have retrieved all our stock price data for the duration of one year and predicted the price data for the next five days, keeping the timeline, closest possible to the date of submission, specifics of which are as mentioned below:

- Test Data: Stock prices of five stocks
- Test data.

An Excel add-in allows access to data, including stock price information.

- Duration of a Test data: 27th April 2016–27th April 2017
- Duration of a Predicted data: 28th April 2017–2nd May 2017.

The first important step is to attain the data. The next step is to develop and execute a neural networking algorithm. The future stock price can be predicted by the constructed algorithm. The final step is to train the different combinations of data by developing suitable neural network map with the help of MATLAB (in particular using the neural network toolbox option).

To the same purpose, we use a lot of functions at different junctures throughout the course of action and methodology. It is crucial to note that, the model which we have designed mainly relies on the following components

- Gradient descent method
- Adaptive rate learning method
- Traingdx—training function.

The traingdx function is a combination method between momentum training and adaptive learning. In addition to the existing parameters the momentum co-efficient

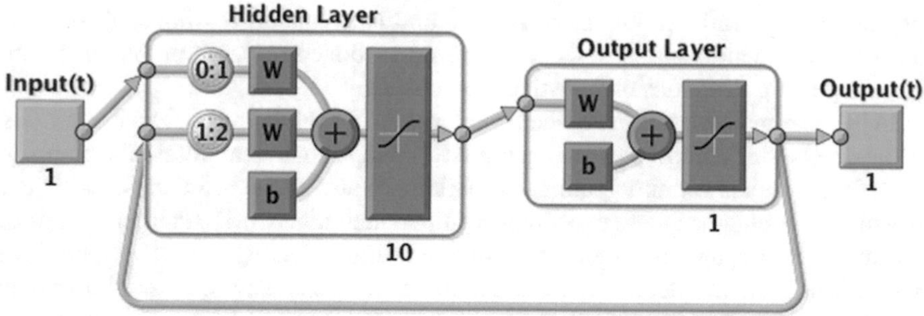

Fig. 2 Neural network map

mc is considered as an additional parameter. If the network has the derivative functions the traingdx can train any network if the derivative functions are associated with weight, net input and transfer functions.

- Upper limit on the maximum number of epochs.
- The maximum amount of time is reached.
- Goal centric performance.
- The lower limit of the performance gradient falls below min_grad.

The inputs for the NARX model are exogenous inputs. In particular, it is a several layers of network with multiple feedback connections. The constructed NARX model is a recurrent dynamic network. Another important parameter is the performance function i.e. mean square error (MSE) value which helps us evaluate how accurately the predicted and actual values are, and the degree of correlation between the expected and the predicted values. Figure 2 shows basic neural network map. Figure 3 depicts the loading of inputs and setting the parameter values.

Creating the above network for a sample stock of Bank of Baroda with the external input as the open prices of all the stock data for an entire year and the target as the closing prices of the same stock since the predicted data will be a function of the same target value and we will obtain the closing prices of the next five days as decided by setting the training function as *traingdx* function while the learning function is set as *learngdm.* The number of hidden layers is two and the performance function to evaluate will be the mean square error or the *mse.*

Figure 4 is an in-process training of the neural network that is built to test and achieve the predicted target data for five days and run, as pre decided, 1000 iterations for repetitive and recursive approach as is the NARX neural network with the training algorithm set as gradient descent with momentum and adaptive LR and the best training performance, in the validation phase once the data has been trained and measured by the performance function mapping of the MSE, is achieved at the value of 2.7836 at Epoch value of iteration 997 which has varied as opposed to since Epoch value 693 when the MSE function was valued at 2.98 and hence the variation is mapped in a graph as shown in Fig. 5.

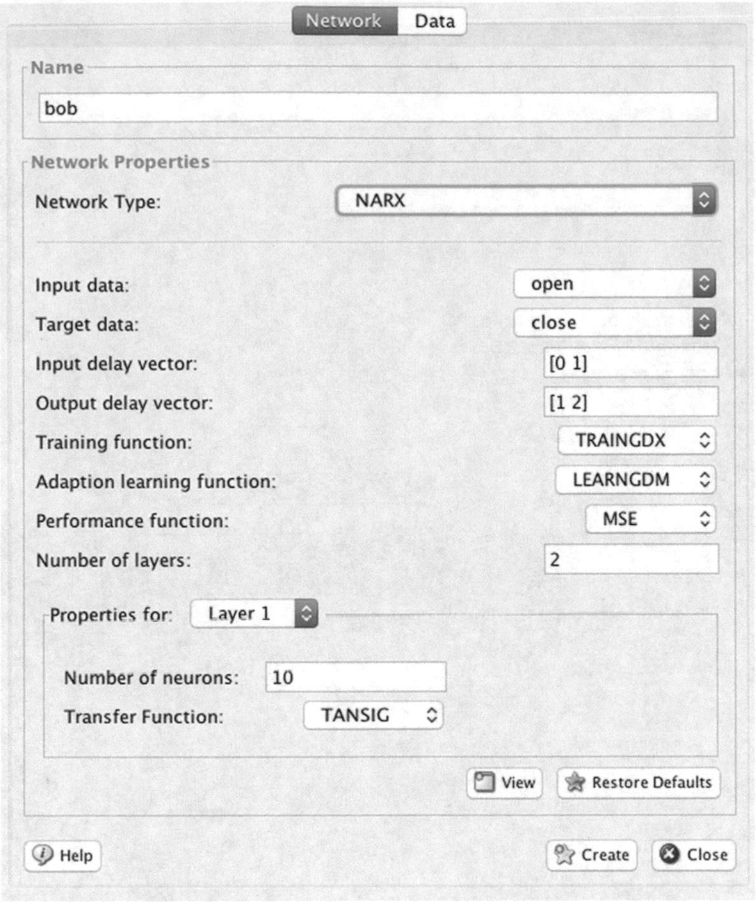

Fig. 3 Loading inputs and setting parameters

Once this is done and we have the predicted five closing stock prices of the particular stock, for validation purposes we are the liberty to plot a graph between the predicted generated five day values and the actual stock direction influenced closing prices at the end of the same five days in order to find out the amount of variation and difference in the two plotted lines and to critically examine the accuracy and efficiency of the predicting model and if it holds.

Figure 6 depicts the relationship between the actual and predicted values. In Fig. 6, blue line shows the predicted value, whereas the red line shows the actual values.

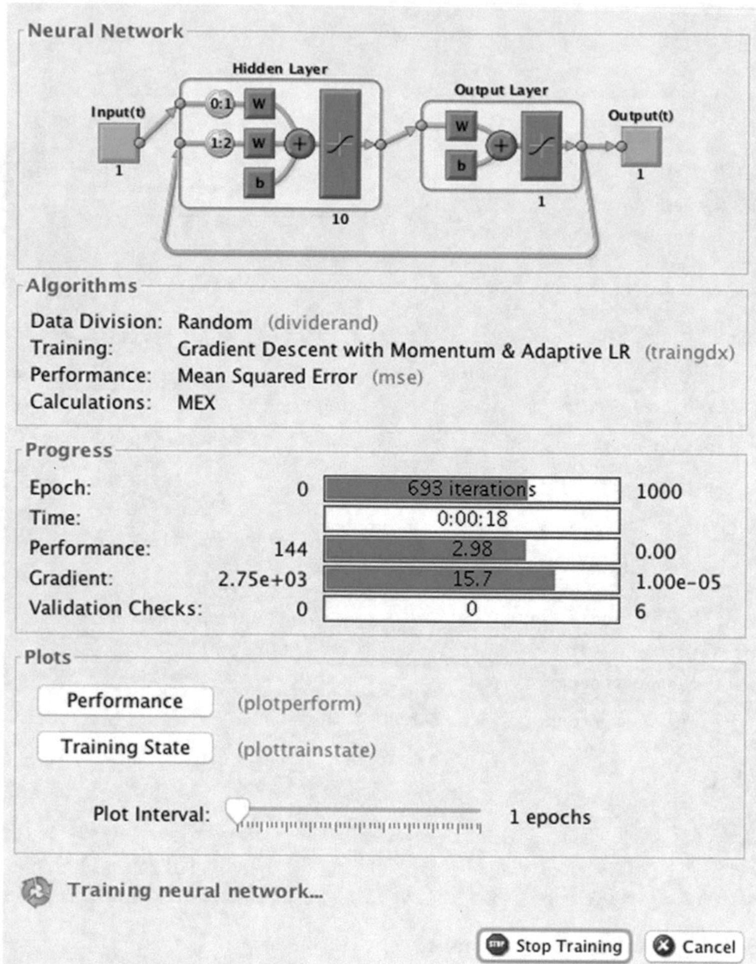

Fig. 4 In progress training neural network

4 Results

For different stock price data, the different results and outputs based on data obtained from aforementioned sources are shown in Table 1.

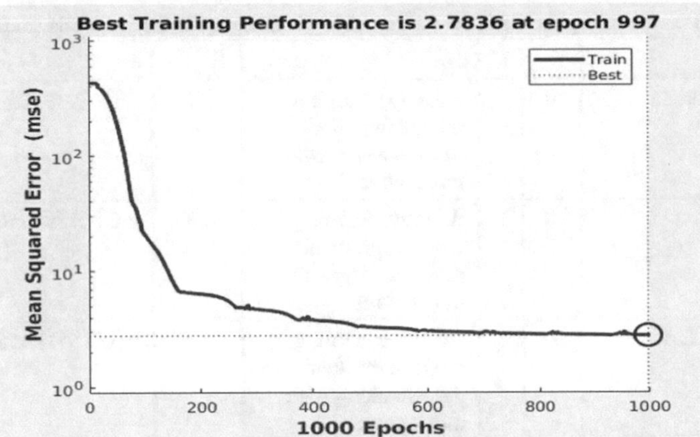

Fig. 5 Best training performance

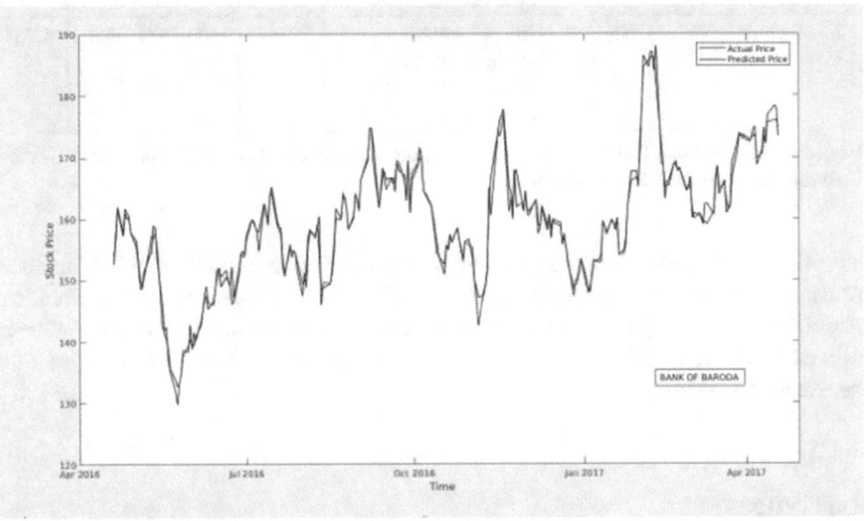

Fig. 6 Output curve for Bank of Baroda stock

5 Conclusion

The neural network models for anticipating securities exchange are at a developmental stage and there are future potential outcomes of change in the expectation precision and unwavering quality of the neural system based determining. In this paper, we have explored the fundamental part of the forecast issue of a securities exchange with simulated neural system. In this paper, we have just utilized the historic costs of the index values for expectation. Other macroeconomic components

Table 1 Different stock price data, the different results and outputs from various sources

A	B	C	D	E	F	G
Bank of Baroda	10	1	Adaptive learning mode in gradient descent along with momentum	2.78	0.029337598	1.706
Wipro Ltd	10	1	Adaptive learning mode in gradient descent along with momentum	10.3	−0.007899446	3.072
Coal India Ltd	10	1	Adaptive learning mode in gradient descent along with momentum	6.07	−0.015519456	2.468
Power Grid Corporation of India Ltd	10	1	Adaptive learning mode in gradient descent along with momentum	3.39	−0.086039305	1.845
HCL Technologies Ltd	10	1	Adaptive learning mode in gradient descent along with momentum	22.6	0.075893882	4.764

A—stock name; B—no of input neurons; C—no of output neurons; D—training function; E—MSE; F—mean error; G—standard deviation (error)

and other worldwide securities exchange information can likewise be utilized as information factors with a specific end goal which is to enhance the precision of the model. The different pattern markers of the specialized examination can likewise be utilized as a port of the information factors and can be checked for development in the execution system.

References

1. Nawani, A., Gupta, H., Thakur, N.: Prediction of market capital for trading firms through data mining techniques. Int. J. Comput. Appl. **70**(18) (2013)
2. Patel, M.B., Yalamalle, S.R.: Stock price prediction using artificial neural network. Int. J. Innov. Res. Sci. Eng. Technol. **3**(6), 13755–13762 (2014)
3. www. Mathworks.com
4. Xiaodong Li, H., Chen, L., Wanga, J., Deng, X.: News impact on stock price return via sentiment analysis. Knowl. Syst. Elsevier **69**, 14–23 (2014)
5. Pawar, D.D., Pawar, R.K.: Application of artificial neural network for stock market predictions: a review of literature. Int. J. Mach. Intell. **2**(2), 14–17 (2010). ISSN: 0975–2927
6. Rather, A.M.: A prediction based approach for stock returns using autoregressive neural networks. In: IEEE Conference of 2011 World Congress on Information and Communication Technologies, pp. 1271–1275 (2011)

7. Hui-Kuangyu, T., Huarng, K.: A neural network-based fuzzy time series model to improve forecasting. J. Exp. Syst. Appl. Elsevier **37**(4), 3366–3372 (2007)
8. Rajput, V., Bobde, S.: Stock market forecasting techniques—a literature survey. Int. J. Comput. Sci. Mob. Comput. **5**(6), 500–506 (2016)
9. Wadghule, Y., Sonawane: Stock market prediction and forecasting techniques—a survey. Int. J. Eng. Sci. Res. Technol. 39–42 (2007)

An Evaluation of Contrast Enhancement of Brain MR Images Using Morphological Filters

Mansi Nigam, Vikrant Bhateja, Anu Arya and Anuj Singh Bhadauria

Abstract Tumor is the uncontrollable growth of abnormal cells in the brain which can be screened using magnetic resonance imaging (MRI). But, MRI is prone to poor contrast and noise during acquisition. This might affect the visibility of the tumor in the image which makes contrast enhancement an essential part of MR image analysis for tumor detection. In this method, a disk-shaped flat structuring element is applied with morphological operators consisting of bottom-hat, dilation and erosion for the purpose of noise controlled enhancement of MRI tumors. The outcomes of the proposed method are validated by image fidelity assessment parameters like: contrast improvement index (CII) and peak signal-to-noise ratio (PSNR).

Keywords Contrast · Erosion · Dilation · Morphological filtering · MRI

1 Introduction

Over the past years, there has been an increase in number of diseases which have led to continuous improvement in biomedical field. Amongst these diseases, brain tumors due to their highly precarious nature have been a bane to the society [1]. Tumors in the brain lead to unnatural internal growth of peculiar cells; causing heavy pressure

M. Nigam · V. Bhateja (✉) · A. Arya · A. S. Bhadauria
Department of Electronics and Communication Engineering, Shri Ramswaroop Memorial Group of Professional Colleges (SRMGPC), Tiwari Ganj, Faizabad Road, Lucknow, UP 226028, India
e-mail: bhateja.vikrant@gmail.com

M. Nigam
e-mail: mansi2996@gmail.com

A. Arya
e-mail: 21anuarya@gmail.com

A. S. Bhadauria
e-mail: singh.anuj842@gmail.com

V. Bhateja
Dr. A.P.J. Abdul Kalam Technical University, Lucknow, Uttar Pradesh, India

V. Bhateja et al. (eds.), *Embedded Systems and Artificial Intelligence*,
Advances in Intelligent Systems and Computing 1076,
https://doi.org/10.1007/978-981-15-0947-6_54

and stress to an individual. The two main types of tumor based on their lifetime are Grade I and II which are initial stages of growth, whereas Grade III and IV are final stages of tumor where the cells become highly active and metastasize to other parts of the central nervous system [1–3]. Screening tumor is a challenging task which requires a mechanism that could take image by percolating through the skull amongst which MRI is the most docile and resourceful technique and can be expressed by various modalities each having different contrast characteristics [4, 5]. However, background noise and low visibility brightness are major challenges associated with the MR images [6]. Therefore, contrast enhancement is an expedient stage, necessary for pre-processing such images before undergoing further medical analysis. Techniques based on histogram equalization [7] were very common with their adaptive modifications [8]. These methods had a pervasive limitation of amplifying noise along with useful information due to which Lidong et al. [9] presented a combined technique of contrast limited adaptive histogram equalization (CLAHE) and discrete wavelet transform (DWT) which suppressed noise. But all histogram-based techniques altered the brightness value of the pixels, distorting the useful image features. Shortcomings of histogram-based techniques were overcome by using mathematical morphology-based contrast enhancement [10–13]. The proposed method of contrast enhancement consists of a combination of morphological top-hat and bottom-hat operators using a disk-shaped structuring element which is most suitable shape for medical image analysis [12]. The organization of this paper has been done as follows: Sect. 2 gives general view of morphology-based filters for MRI enhancement; Sect. 3 put forwards a detailed analysis of simulation results, computation of image quality assessment (IQA) parameters and outcomes of the methodology. Lastly, Sect. 4 summarizes the inferences derived at end of the work.

2 Methodology for Morphology-Based MRI Enhancement

2.1 An Outline of Mathematical Morphology

The structural form or features of anything is termed as morphology. MR image possesses a collection of such diagnostic objects (features) that have variable shapes and structures. To process these objects in the images, they are subjected to a set of morphological operations. These operators are non-linear filters that depend on the structure-based features of the object under test. They operate with two inputs: of which one is the input image and other one is the structuring element. A structuring element is a matrix that is used for the identification of the pixel that is being processed and also defines its neighborhood pixels. The selection of type of structuring element depends on the information being dealt [14]. Erosion, dilation, opening and closing [14–18] are the conventional and simple morphological operators. Noise suppression and edge enhancement are the primary requirements of MRI in medical diagnosis for any of its applications. Techniques based on conventional mathematical morphology are simpler to execute, but these operators have certain constraints which make their usage difficult. Firstly, they do not produce good results for highly complex images and secondly, there may be loss in prominent features of image due to improper choice of order of the structuring element [19, 20].

2.2 *Proposed MRI Enhancement Method Using Morphological Filters*

The method presents a computationally simple approach for brain MR image enhancement with the use of erosion and dilation along with morphological bottom-hat operator (which is obtained from mathematical combination of conventional opening and closing operators) [20]. Bottom-hat operator as represented by Eq. (1) is obtained by subtracting the original image form the closed image [21, 22]. This helps to focus upon the darker regions of the image which are around the brighter areas. The step-by-step procedure of the proposed methodology has been shown in Algorithm 1. In Algorithm 1, firstly the input MR image of T2 modality (I) is taken and then for software compatibility is converted from RGB to grayscale profile (f). Then, two disk-shaped [23] structuring elements (x_1 and x_2) of size 35 and 7 are defined, respectively, as shown in Fig. 1. The structuring element x_1 is used for bottom-hat transform of the grayscale image (f), whereas x_2 is used for the dilation of grayscale image (f). The dilated image is then subjected to erosion operator using the same structuring element (x_2). Dilation followed by erosion helps to connect the thinly separated objects of interest in the image that are needed to be converted after enhancement. Finally, the bottom-hat image is subtracted from the eroded image which will minimize the unnecessary dull background tissues within the brain. Hence, the resulting image is contrast enhanced with distinctly visible tumor region.

$$B_{\text{hat}}(f) = (f \cdot x_1) - f \tag{1}$$

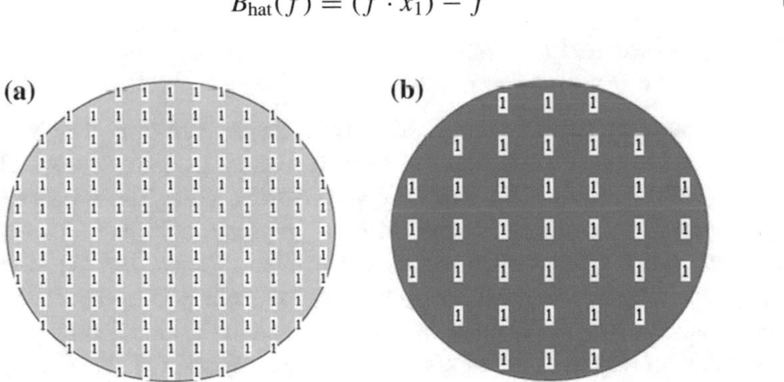

Fig. 1 Disk shape structuring elements of **a** size 35 (x_1) and **b** size 7 (x_2) (The dimension of this figure has been scaled down to one by six and one by two, respectively)

$$D(f) = f \oplus x_2 = \{z|(x_2)_Z \cap f \neq \varnothing\} \tag{2}$$

$$E(f) = f \Theta x_2 = \left\{z \middle| (x_2)_Z \cap f^C = \varnothing\right\} \tag{3}$$

The darker objects on a brighter background are highlighted via low pass filtering using bottom-hat transform. This also helps in discerning the visibility of tumor region which originally was obscured with the background tissues. Bottom-hat transform also helps in solving the non-uniformity in intensity that is prevalent constraint in MRI [24]. Difference of bottom-hat transformed image from the eroded one (as shown in Algorithm-1) is carried out to obtain only the relevant features by minimizing the undesirable background features.

Algorithm-1 *Proposed MRI Enhancement Method using Morphological Filters.*

BEGIN
Step 1: *Read* MRI as (I).
Step 2: *Convert* (I) from RGB to Grayscale (f).
Step 3: *Initialize* Structuring Sub-Matrix (x_1 and x_2) of size 35 and 7 respectively (as in Fig. 1).
Step 4: *Apply* Bottom-Hat transform of (f) with Sub-Matrix (x_1) using Eq. (1).
Step 5: *Compute* Dilation followed by Erosion of (f) with Sub-Matrix (x_2) using Eqs. (2) and (3).
Step 6: *Compute Difference* of outputs of Step 4 from Step 5.
Step 7: *Compute CII and PSNR* [25].
END

3 Results and Discussions

3.1 Experimental Results

Two publicly open research repositories available as Web-link: The Whole Brain Atlas [26] and Internet Brain Segmentation Repository [27] are deployed herein for providing MR test images for simulations in current work. The processing of MR image begins by converting it from RGB to grayscale colormap. The structuring element chosen is of disk shape as shown in Fig. 1. Then, Eq. (1) is applied on the grayscale image for bottom-hat transformation having 35 as the size of structuring element. The grayscale image is also subjected to dilation followed by erosion operations with disk element of size 7 using Eqs. (2) and (3). This is followed by subtracting the bottom-hat image from the eroded image to yield the output image. The enhanced image has distinctly visible tumor regions with nullified background objects. Simulations responses of Algorithm-1 with original and enhanced images are shown in Fig. 2 along with the computed values of quality parameters in Table 1.

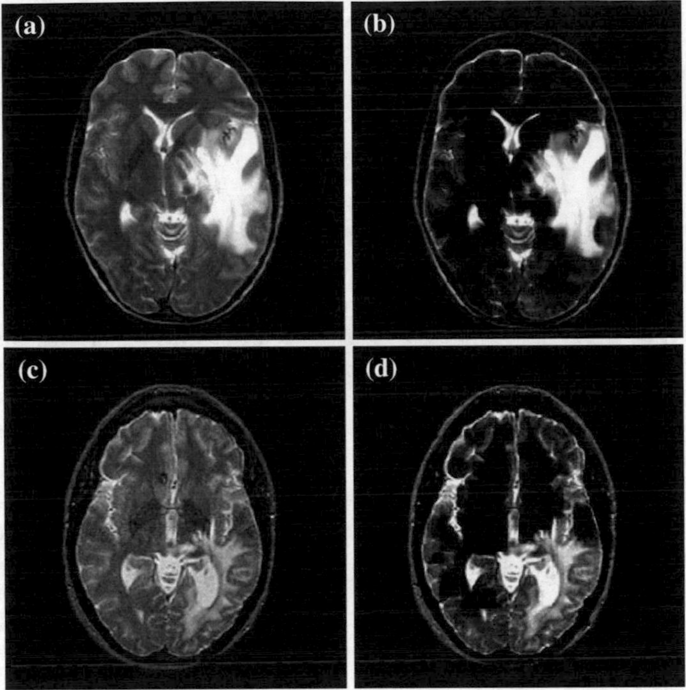

Fig. 2 **a** Original Test_Case#1. **b** Enhanced Test_Case#1. **c** Original Test_Case#2. **d** Enhanced Test_Case#2

Table 1 Quality assessment metrics for proposed morphology-based enhancement method

Images	CII	PSNR	
		Original	Enhanced
Test_Case#1	1.7694	2.3679	3.5280
Test_Case#2	1.6973	2.9024	3.5534
Test_Case#3	1.5811	2.1862	3.6377
Test_Case#4	1.4934	2.1528	3.3879

3.2 *Discussions*

Figure 2 highlights two different MR images that have been enhanced using the proposed morphology-based method as outlined in Algorithm 1. The original image Test_Case#1 consists of the main tumorous region of irregular shape on the left hemisphere of bright intensity but has its edema (region surrounding the tumor) tissues highly merged with the background hiding the useful tumor edges. Upon enhancement, the unnecessary background tissues have been removed which makes the tumor distinctly visible with high sharpness. Similarly, Test_Case#2 has tumorous tissue on the left occipital lobe which upon enhancement has been retained successfully without loss of useful information and is easily distinguishable. Satisfactory values of CII and PSNR in Table 1 indicate successful contrast improvement without increment in the noise levels.

4 Conclusion

A timely diagnosis and medication of brain tumor require a detailed analysis of its MRI which is felicitated by contrast enhancement. This paper presents contrast improvement technique based on morphological filtering which is computationally simple for implementation. The proposed method uses bottom-hat, dilation and erosion morphological operators to enhance the ROI thereby suppressing the background. A flat structuring element has been applied of size similar to that of tumor region for various morphological operations involved. The results obtained ensure improvement in contrast along with simultaneous noise suppression as observed from image fidelity assessment parameters.

References

1. Types of Brain Tumors. https://www.abta.org
2. Alankrita, A.R., Shrivastava, A., Bhateja, V.: Contrast improvement of cerebral MRI features using combination of non-linear enhancement operator and morphological filter. In: Proceedings of IEEE International Conference on Network and Computational Intelligence (ICNCI), vol. 4, pp. 182–187, Zhengzhou, China (2011)
3. Bahadure, N.B., Ray, A.K., Thethi, H.P.: Comparative approach of MRI-based brain tumor segmentation and classification using genetic algorithm. Int. J. Digit. Imaging 1–13 (2018)
4. Somasundaram, K., Kalaiselvi, T.: Automatic brain extraction methods for T1 magnetic resonance images using region labeling and morphological operations. Comput. Biol. Med. **41**(8), 716–725 (2011)
5. MRI Sequences (Overview). https://radiopaedia.org/articles/mri-sequences-overview
6. Akram, M.U., Usman, A.: Computer aided system for brain tumor detection and segmentation. In: IEEE International Conference on Computer Networks and Information Technology (ICCNIT), pp. 299–302 (2011)
7. Yeganeh, H., Ziaei, A., Rezaie, A.: A novel approach for contrast enhancement based on histogram equalization. In: IEEE International Conference on Computer and Communication Engineering (ICCCE), pp. 256–260 (2008)
8. Stark, J.A.: Adaptive image contrast enhancement using generalizations of histogram equalization. IEEE Trans. Image Process. **9**(5), 889–896 (2000)
9. Lidong, H., Wei, Z., Jun, W., Zebin, S.: Combination of contrast limited adaptive histogram equalisation and discrete wavelet transform for image enhancement. IET Image Process. **9**(10), 908–915 (2015)
10. Moses, C., Prasad, P.M.K.: Image enhancement using stationary wavelet transform. Int. J. Comput. Math. Sci. (IJCMS) **6**(9), 84–88 (2017)
11. Kharrat, A., Benamrane, N., Messaoud, M.B., Abid, M.: Detection of brain tumor in medical images. In: 3rd IEEE International Conference on Signals, Circuits and Systems (SCS), pp 1–6 (2009)

12. Hassanpour, H., Samadiani, N., Salehi, S.M.: Using morphological transforms to enhance the contrast of medical images. Egypt. J. Radiol. Nucl. Med. **46**(2), 481–489 (2015)
13. Bhadauria AS et al.: Skull stripping of brain MRI using mathematical morphology. Smart Intelligent Computing and Applications, pp. 775–780. Springer, Singapore (2020)
14. Raj, A., Srivastava, A., Bhateja, V.: Computer aided detection of brain tumor in magnetic resonance images. Int. J. Eng. Technol. **3**(5), 523–532 (2011)
15. Verma, R., Mehrotra, R., Bhateja, V.: A new morphological filtering algorithm for pre-processing of electrocardiographic signals. In: Proceedings of the Fourth International Conference on Signal and Image Processing (ICSIP), pp. 193–201. Springer, India (2013)
16. Tiwari, D.K., Bhateja, V., Anand, D., Srivastava, A., Omar, Z.: Combination of EEMD and morphological filtering for baseline wander correction in EMG signals. In: Proceedings of 2nd International Conference on Micro-Electronics, Electromagnetics and Telecommunications, pp. 365–373. Springer, Singapore (2018)
17. Bhateja, V., Urooj, S., Mehrotra, R., Verma, R., Lay-Ekuakille, A., Verma, V.D.: A composite wavelets and morphology approach for ECG noise filtering. In: International Conference on Pattern Recognition and Machine Intelligence, pp. 361–366. Springer, Heidelberg, Berlin (2013)
18. Bhateja, V., Devi, S.: A novel framework for edge detection of microcalcifications using a non-linear enhancement operator and morphological filter. In: IEEE 3rd International Conference on Electronics Computer Technology (ICECT), vol. 5, pp. 419–424 (2011)
19. Bhateja, V., Urooj, S., Verma, R., Mehrotra, R.: A novel approach for suppression of powerline interference and impulse noise in ECG signals. In: IEEE International Conference on Multimedia, Signal Processing and Communication Technologies, pp. 103–107 (2013)
20. Arya, A., Bhateja, V., Nigam, M., Bhadauria, A.S.: Enhancement of brain MR-T1/T2 images using mathematical morphology. In: 3rd International Conference on ICT for Sustainable Development, pp. 1–8, Panaji, Goa (2018)
21. Chaddad, A., Tanougast, C.: Quantitative evaluation of robust skull stripping and tumor detection applied to axial MR images. Brain Inf. **3**(1), 53–61 (2016)
22. Verma, R., Mehrotra, R., Bhateja, V.: An integration of improved median and morphological filtering techniques for electrocardiogram signal processing. In: IEEE 3rd International Conference Advance Computing, pp. 1223–1228 (2013)
23. Gonzalez, R.C., Woods, R.E.: Digital Image Processing, Chapter 10, pp. 689–794. Pearson Education (2009)
24. Das, S.S., Mohan, A.: Medical image enhancement techniques by bottom-hat and median filtering. Int. J. Electron. Commun. Comput. Eng. **5**(4), 347–351 (2014)
25. Bhateja, V., Nigam, M., Bhadauria, A.S., Arya, A., Zhang, E.Y.D.: Human visual system based optimized mathematical morphology approach for enhancement of brain MR images. J. Ambient Intell. Hum. Comput. 1–9 (2019)
26. The Whole Brain Atlas. http://www.med.harvard.edu/aanlib/home.html
27. The Internet Brain Segmentation Repository. https://www.nitrc.org/projects/ibsr/

Information and Decision Sciences

A New Color Image Encryption Algorithm Using Random Number Generation and Linear Functions

Mohammed Es-sabry, Nabil El Akkad, Mostafa Merras, Abderrahim Saaidi and Khalid Satori

Abstract In this paper, we introduce a new method of color image encryption. The principle of this approach is to encrypt any color image (called the original image), into a sequence of N images whose $(N-1)$ of them are generated randomly, and the Nth image is calculated from the original image and the $(N-1)$ generated images using a linear equation. This process will be repeated for each color image channel (red, green, and blue). Those $(N-1)$ generated images are used as a symmetric key in the decryption phase; therefore, they should be shared with the receiver privately or via a secure channel. Only the Nth image will be sent. To obtain the original image, we need all N images in the decryption phase. Our encryption system can resist brute-force attacks as well as statistical attacks. The results are justified by applying several safety criteria, such as correlation coefficient and histogram.

Keywords Color image encryption · Color image decryption · Security · Coding

1 Introduction

Nowadays, information security is becoming more important in data storage and transmission, and the images have become widely used in the different field, since

M. Es-sabry (✉) · N. El Akkad · M. Merras · A. Saaidi · K. Satori
LIIAN, Department of Mathematics and Computer Science, Faculty of Sciences Dhar-Mahraz, Sidi Mohamed Ben Abdellah University, Atlas, B.P 1796 Fez, Morocco
e-mail: mohammed.es.sabry@usmba.ac.ma

N. El Akkad
Department of Electrical and Computer Engineering, National School of Applied Sciences (ENSA), Sidi Mohamed Ben Abdellah University, Fez, Morocco

M. Merras
Department of Computer Science, High School of Technology, Moulay Ismail University, Meknes, Morocco

A. Saaidi
LSI, Department of Mathematics, Physics and Informatics, Polydisciplinary Faculty of Taza, Sidi Mohamed Ben Abdellah University, Taza, Morocco

V. Bhateja et al. (eds.), *Embedded Systems and Artificial Intelligence*, Advances in Intelligent Systems and Computing 1076,
https://doi.org/10.1007/978-981-15-0947-6_55

they have their own characteristics like the large amount of information (Pixels). As a result, image encryption is given much attention in research of information security and a lot of image encryption algorithms [1–7] have been introduced in order to make the decryption phase for a malicious user very difficult. The image encryption (Fig. 1) is different from that of text. Therefore, it is difficult to handle them by traditional encryption methods [8] such as AES or DES, because they take a lot of time in computing phase.

In this field, we can mention several encryption systems based on the random generation of numbers such as true random number generators (TRNGs), pseudo-random number generators (PRNGs), and chaotic random number generators (CRNGs).

The first category (TRNGs) [2, 3] makes the prediction of the next number impossible because it is based on the use of unpredictable and non-deterministic sources such as natural processes or physical phenomena; it does not matter if you know the previous numbers. Therefore, TRNGs are particularly very useful in cryptography and much more in the production of keys. The second category (PRNGs) is algorithmic generators of numbers that have the appearance of random but with periodic digit sequences; nevertheless, their results are predictable. Good random number generators produce very long sequences that seem random, in the sense that no efficient algorithm can guess the next number according to the prefix of the sequence. The third category (CRNGs) is based on the chaotic phenomenon; it is a multiparametric system; these parameters must be refined to improve its performances; although they are periodic, by applying spatiotemporal techniques, their period can last several years, so practically inaccessible.

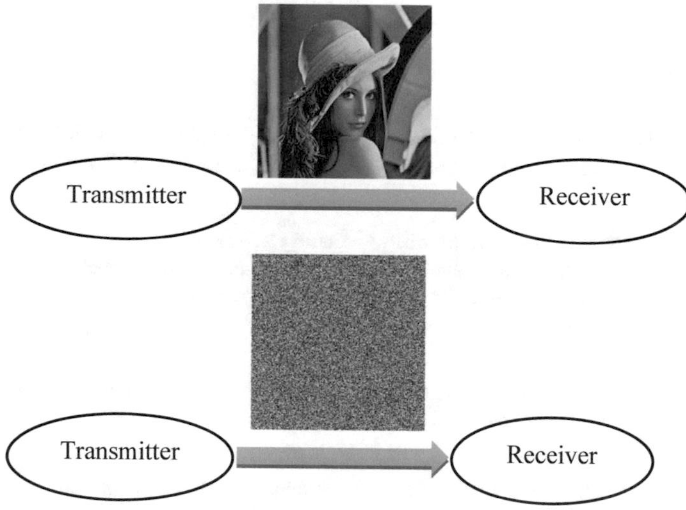

Fig. 1 Transmission of an encrypted color image

The rest of this work is organized as follows: The second part presents the proposed method. Experimentation and performance analysis of the method is covered in the third part. A conclusion of this work is presented in the fourth part.

2 Proposed Method

2.1 Image Encryption

Our system is initialized by a color image; after that, we extract the decimal matrix corresponding to each channel of image (red, green, and blue) whose values are between 0 and 255; then, we generate $N - 1$ matrices randomly for each channel and we use a linear function to calculate the output matrix from the original image and the $N - 1$ generated matrices; this function will be described below, and it must be bijective in order to do the decryption phase; finally, we merge the three output matrices for each channel into one single encrypted color image. The flowchart (Fig. 2) illustrates the various steps used to encrypt the original image.

The first phase of our algorithm consists of decomposing the color image that we want to encrypt in three channels; then, we extract the corresponding decimal matrix for each of them. After that, we generate randomly $(N - 1)$ matrices of the same size as the original image for each channel (red, green, and blue). The values of these matrices will be between 0 and 255. Those matrices are used as the key in the encryption and decryption phase; they should be shared privately with the receiver. Then, we will use our linear function to determine the Nth matrix (Nth image); this function is defined as follows:

$$F\left(x_{i,j}, y_{i,j}\right) = \begin{cases} x_{i,j} - y_{i,j} \text{ if } x_i > y_i \\ y_{i,j} - x_{i,j} \text{ if } y_i \geq x_i \end{cases} \quad (1)$$

$x_{i,j}$ and $y_{i,j}$ represent the values of the pixels located at row i and column j for the first and second matrices of the $(N - 1)$ matrices that we generate randomly. This procedure will be repeated two by two until the end of the matrix sequence.

At the end, we will obtain the Nth matrix of each channel (red, green, and blue). Then, we merge them into one single color image, and we can now send it to the receiver.

2.2 Image Decryption

In the decryption phase, we just follow these steps:

Step 1: Decompose the encrypted color image into three channels (red, green, and blue).

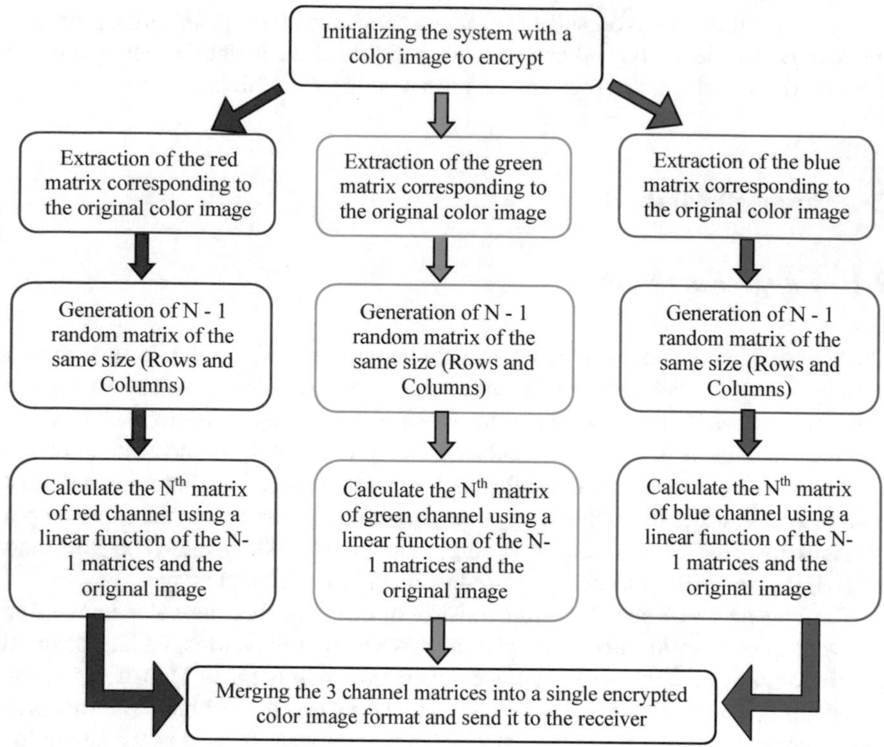

Fig. 2 Flowchart of the steps used to encrypt the original color image

Step 2: Extract the corresponding matrix of each channel (red, green, and blue) of the encrypted color image.
Step 3: Calculate the F^{-1} of extracted matrix and all the $(N - 1)$ matrices used as the key two by two until the end of all matrices.

$$F^{-1}\left(x_{i,j}, y_{i,j}\right) = \begin{cases} x_{i,j} + y_{i,j} \text{ if } x_i > y_i \\ y_{i,j} - x_{i,j} \text{ if } y_i \geq x_i \end{cases} \tag{2}$$

Step 4: Repeat step 3 for each channel (red, green, and blue).
Step 5: Merge the output matrix of each channel into one single image which is the original color image.

3 Experimentation

3.1 Histograms

The first criterion that will be tackled is the histogram analysis. We will make a comparison between the original image and encrypted image histograms; a good encryption algorithm must completely change the pixel distribution of the original image. And that is what our algorithm did.

We took 10 color images with different size (256 × 256, 512 × 512, and 1024 × 1024); we applied our own encryption system on those images using java programming language. Figure 3 shows the histogram of Lena image as an example before and after using our approach.

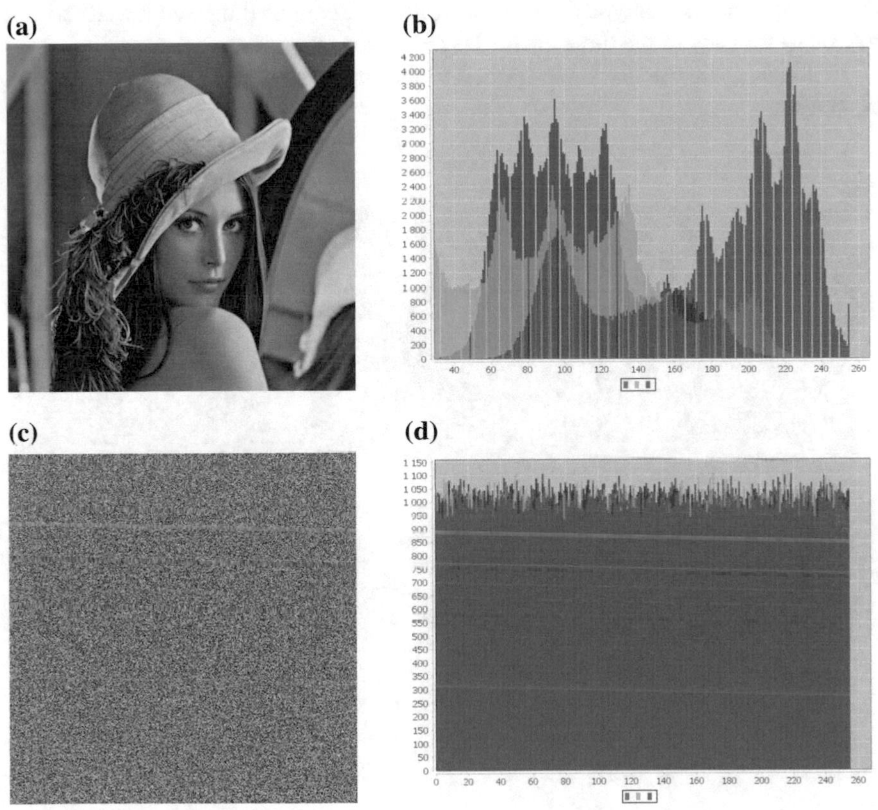

Fig. 3 **a** Lena image, **b** histogram of Lena image, **c** encrypted Lena image, and **d** histogram of encrypted Lena image

From Fig. 3, we can note that not only the encrypted Lena image histogram is completely different from that of the original Lena image, but also the pixel distribution is almost uniform; therefore, our approach can resist any statistical attacks and increases the protection of the information included in the images.

3.2 Correlation Analysis of Two Adjacent Pixels

The correlation coefficient of adjacent pixels is crucial internal information of the image; in order to decrypt images, a lot of statistical attacks use this kind of information; consequently, a good encryption system must clear this information by minimizing the correlation coefficient as much as possible.

To do so, we took 4000 pairs of randomly selected adjacent pixels from the original color image and the encrypted image along three directions (horizontal, vertical, and diagonal) for each of the three channels (red, green, and blue).

In this test, the calculation is applied to the image of Fig. 4 using the following equations:

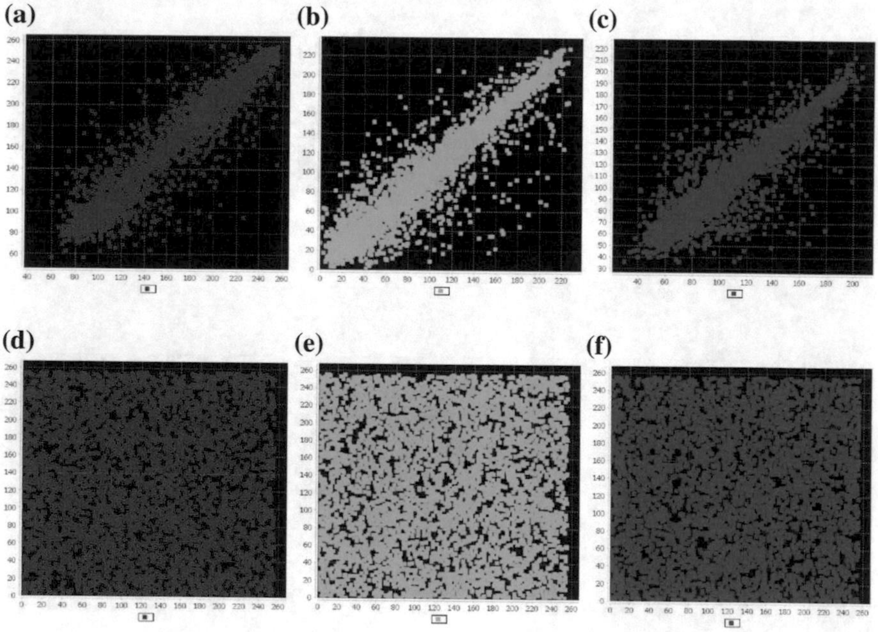

Fig. 4 Pixel distribution in the horizontal direction of Lena image: **a** original red channel, **b** original green channel, **c** original blue channel, **d** encrypted red channel, **e** encrypted green channel, and **f** encrypted blue channel

Table 1 Correlation coefficient of original Lena image and encrypted Lena image

Image	Channel	Original image			Encrypted image		
		Horizontal	Vertical	Diagonal	Horizontal	Vertical	Diagonal
Lena	Red	0.974259	0.951737	0.920299	0.044035	0.02504	0.011318
	Green	0.971863	0.942889	0.921788	0.016395	0.013671	−0.020847
	Blue	0.955128	0.906188	0.866483	−0.006216	−0.004295	−0.015779
	Average	0.967083	0.933605	0.902857	0.018071	0.011472	−0.008436

$$r_{xy} = \frac{\mathrm{Cov}(x, y)}{\sqrt{D_x}\sqrt{D_y}} \tag{3}$$

$$\mathrm{Cov}(x, y) = \frac{1}{N}\sum_{i=1}^{N}(x_i - E_x)\big(y_i - E_y\big) \tag{4}$$

$$D_x = \frac{1}{N}\sum_{i=1}^{N}(x_i - E_x)^2 \tag{5}$$

$$E_x = \frac{1}{N}\sum_{i=1}^{N}x_i \tag{6}$$

where N represents the number of pixels of the original image.

Table 1 shows the results obtained after calculating the correlation coefficient of adjacent pixels in the three channels (red, green, and blue) and following the three directions (horizontal, vertical, and diagonal).

In the case of original images, we notice that the correlation coefficient (CC) in the three channels (red, green, and blue) including the three directions (horizontal, vertical, and diagonal) takes values very close to 1. On the other hand, it takes values close to 0 in the case of encrypted images, which means that the degree of dependence between the two adjacent pixels is very strong before the application of our method, and this dependence is destroyed after having applied our encryption algorithm.

Figure 4 shows the pixel distribution in three channels (red, green, and blue) and following horizontal direction of the Lena image. We can see that the distribution before the application of our method follows one line, which means that the correlation is strong between the pixels. On the other hand, this correlation becomes weak after applying our algorithm and the pixels are scattered everywhere; therefore, our proposed method gives good results.

Table 2 represents the correlation coefficients (CC) obtained by our approach compared with method of Tong [9], method of Borujeni [10], and Hua [11].

The results show that the correlation existing between the adjacent pixels has been destroyed by our algorithm, which shows the strong performance of our approach compared to other methods.

Table 2 CC comparison of two adjacent pixels of our method with other methods

Encrypted image	Directions			Average
	Horizontal	Vertical	Diagonal	
Tong [9]	0.017188	0.009852	0.033045	0.020028
Borujeni [10]	0.004100	0.030800	0.005300	0.013400
Hua [11]	0.002383	0.008576	0.040242	0.017067
Proposed method	0.018071	0.011472	−0.008436	0.007035

4 Conclusion

In this work, we proposed a new method that encrypts color images in a very secure way. The method is based on the random number generation of $(N - 1)$ matrices; those matrices are used as a key and shared privately with the receiver. Then, we calculate the Nth matrix of each channel (red, green, and blue) using a linear function; at the end, we merge the three matrices into one single image format. The resistance of our method against statistical attacks has been approved using several criteria such as histogram and correlation coefficient.

References

1. Es-sabry, M., El Akkad, N., Merras, M., Saaidi, A., Satori, K.: Grayscale image encryption using shift bits operations. In: International Conference on Intelligent Systems and Computer Vision, ISCV (2018)
2. Wu, X., Hu, H., Zhang, B.: Parameter estimation only from the symbolic sequences generated by chaos system. Chaos Solitons Fractals **22**, 359–366 (2004)
3. Arroyo, D., Rhouma, R., Alvarez, G., Li, S., Fernandez, V.: On the security of a new image encryption scheme based on chaotic map lattices. Chaos Interdiscip. J. Nonlinear Sci. **18** (2008)
4. Norouzi, B., Seyedzadeh, S.M., Mirzakuchaki, S., Mosavi, M.R.: A novel image encryption based on hash function with only two-round diffusion process. Multimedia Syst. **20**, 45–64 (2013)
5. Zhu, C.: A novel image encryption scheme based on improved hyperchaotic sequences. J. Opt. Commun. **285**, 29–37 (2012)
6. Tong, X.G., Wang, Z., Zhang, M., Liu, Y., Xu, H., Ma, J.: An image encryption algorithm based on the perturbed high-dimensional chaotic map. Nonlinear Dyn. **80**, 1493–1508 (2015)
7. Wang, X.Y., Guo, K.: A new image alternate encryption algorithm based on chaotic map. Nonlinear Dyn. **76**, 1943–1950 (2014)
8. Es-sabry, M., El Akkad, N., Merras, M., Saaidi, A., Satori, K.: A novel text encryption algorithm based on the two-square Cipher and Caesar Cipher. In: International Conference on Big Data, Cloud and Applications, BDCA, vol. 872, pp. 78–88 (2018)
9. Tong, X., Cui, M., Wang, Z.: A new feedback image encryption scheme based on perturbation with dynamical compound chaotic sequence cipher generator. J. Opt. Commun. **282**, 2722–2728 (2009)
10. Borujeni, S.E., Eshghi, M.: Chaotic image encryption system using phase-magnitude transformation and pixel substitution. J. Telecommun. Syst. **52**, 525–537 (2011)
11. Hua, Z.Y., Zhou, Y.C., Pun, C.M., Philip Chen, C.L.: 2D Sine logistic modulation map for image encryption. Inf. Sci. **297**, 80–94 (2015)

A New Encryption Approach Based on Four-Square and Zigzag Encryption (C4CZ)

Fouzia Elazzaby, Nabil El Akkad and Samir Kabbaj

Abstract With the evolution and the development of data-processing tools and their pervasive scopes of applications, the need for data protection today has become a crucial necessity in order to reduce the number of incidents caused by non-authorized data accesses and to prevent malicious usage. With the purpose of maintaining a significantly high level of security, cryptography englobes the entirety of methods as well as a couple of techniques and tools aimed to assure data confidentiality. Indeed, this work's main focus is a new approach based on a four-square encryption accompanied by a zigzag transformation. This approach allows us to provide the appropriate settings for the four-square technique and the new conception of a transformation that manages future plans. These last are immune to brute attacks and statistic attacks. Considering all of that, this exhaustive study will emphasize the reliability of this approach.

Keywords Text encryption · Four-square encryption · Zigzag transformation · Static attack · Brute attack

1 Introduction

Thanks to the internet, virtual communication [1] is constantly spreading day after day. It's mainly due to the important role it plays when it comes to the exchange of information, knowledge sharing and entertainment with our loved ones. Many threats affect this exchange. This lack of security is the main source of various critical

F. Elazzaby (✉) · S. Kabbaj
Department of Mathematics, Faculty of Sciences, IBN TOFAIL University, 14000 Kenitra, Morocco
e-mail: fouzia_099@hotmail.com

S. Kabbaj
e-mail: samkabbaj@yahoo.fr

N. El Akkad
Electrical and Computer Engineering Department, National School of Applied Sciences of Fez, Sidi Mohamed Ben Abdellah University, Fez, Morocco

V. Bhateja et al. (eds.), *Embedded Systems and Artificial Intelligence*,
Advances in Intelligent Systems and Computing 1076,
https://doi.org/10.1007/978-981-15-0947-6_56

Fig. 1 Encryption and decryption process

problems during the communication process. That drives a lot of researchers to conceive some of the simplest security mechanisms while going through the process of encryption and decryption as shown in Fig. 1.

In order to fulfill this need, cryptography [2–6] aims to search for the best way to send data in a confidential manner by plainly encrypting information into coded information (incomprehensible). In addition to that, this process has the privilege to restore coded data into the original texts through a confidential key. Despite the variety of these techniques, cryptography still is a hot topic that draws more and more researchers to create even more effective and new approaches [13–24] aimed to deal with unauthorized attacks. Indeed, this work focuses on a new encryption method based on the four-square encryption and a zigzag transformation.

2 Suggested Method

This new approach is based on the exceptional four-square encryption [7] fundamentals convoluted with a zigzag transformation that will highlight the reliability of this approach when it comes to attacks.

2.1 Principle of Encryption

Regarding our approach, the encryption of a message is based on the existence of four grids of the four-square encryption method (Fig. 2). Each of these grids consists of a matrix of 25 cells occupied by letters of the alphabet (as illustrated in Fig. 2), except the letter W which has a low frequency in the French language (we eliminate the letter V in English). The two grids numbered 1 and 4 represent the plain text; they are filled in alphabetical order. While to fill the other two grids that represent the digit text, this technique uses two different keys—one is used to initialize the grid 2 and the other for the grid 3. These last two grids contain the distinct alphabets of these keys and the empty cells are completed with the letters that do not appear in

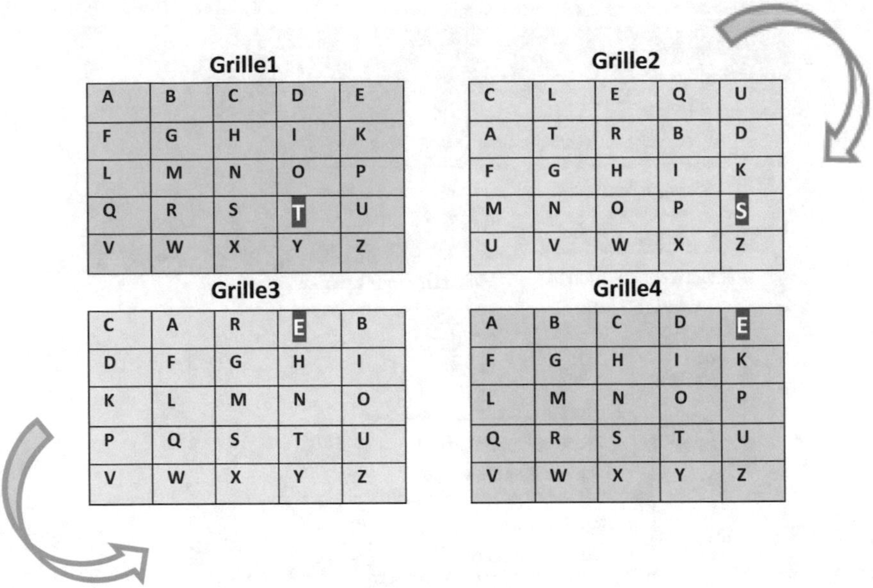

Grille1

A	B	C	D	E
F	G	H	I	K
L	M	N	O	P
Q	R	S	T	U
V	W	X	Y	Z

Grille2

C	L	E	Q	U
A	T	R	B	D
F	G	H	I	K
M	N	O	P	S
U	V	W	X	Z

Grille3

C	A	R	E	B
D	F	G	H	I
K	L	M	N	O
P	Q	S	T	U
V	W	X	Y	Z

Grille4

A	B	C	D	E
F	G	H	I	K
L	M	N	O	P
Q	R	S	T	U
V	W	X	Y	Z

Fig. 2 Four-square and zigzag encryption

these keys. Once the 4 grids are generated, the method removes all non-alphabetic characters (punctuation, number, etc.). Then, it transforms the message into capital letters; then she slices the message in pairs of two letters. In case the number of letters of the text is odd, it will be completed with a neutral letter x.

After these steps, for each digraph, we first choose the index of the line and the column of the first letter of the message in matrix 1 then we do the same thing for the second letter in matrix 4 then, we look in matrices 3 and 4 for the letters that complete the rectangle and at the end, this process is repeated at each pair of the message to be encrypted with a zigzag rotation to the right for the grid 2 and a rotation to the left of the grid 3 in order to overcome the fact that a letter can be coded in one way.

2.2 *The Flowchart of the New Approach*

See Fig. 3.

3 Results Experimental

In this section, we are going to see a concrete example of an easy to encrypt a message with (C4CZ).

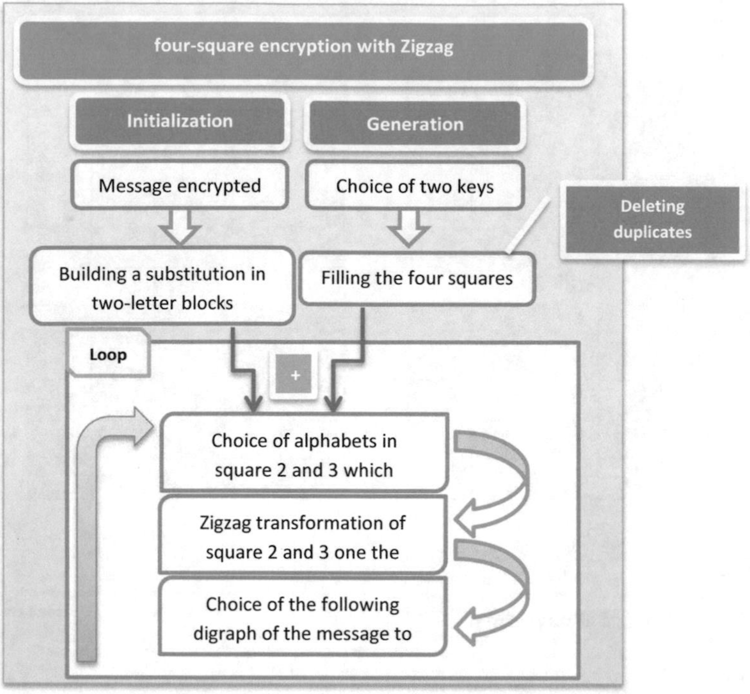

Fig. 3 The process encryption at four-square and zigzag

The chosen message is: «**EXEMPLE DE CRYPTAGE A QUATRE CARREES ET ZIGZAG**». It's composed of 38 letters. The two keys used to fill the second and third cells (Fig. 4) are Key 1 **CLEQUATRE** and key 2 **CARREE**.
The second message is: «**LA METHODE DECOUPE LE MESSAGE EN CLAIR EN DEUX LETTRES**» composed of 45 Letters. To know that this time the proposed keys are key 1 «**ELAZZABY**» and key 2 «**FOUZIA**».

3.1 The Encryption

a. **Example 1**

This schema (Fig. 4) clarifies the principle of encryption (as you can see down below) and puts in place the four cells with the two keys without forgetting to delete duplicates and punctuation while splitting the text into digraphs:

«EX EM PL ED EC RY PT AG EA QU AT RE CA RR ES ET ZE GZ AG»

First of all, let's encrypt the digraph EX:

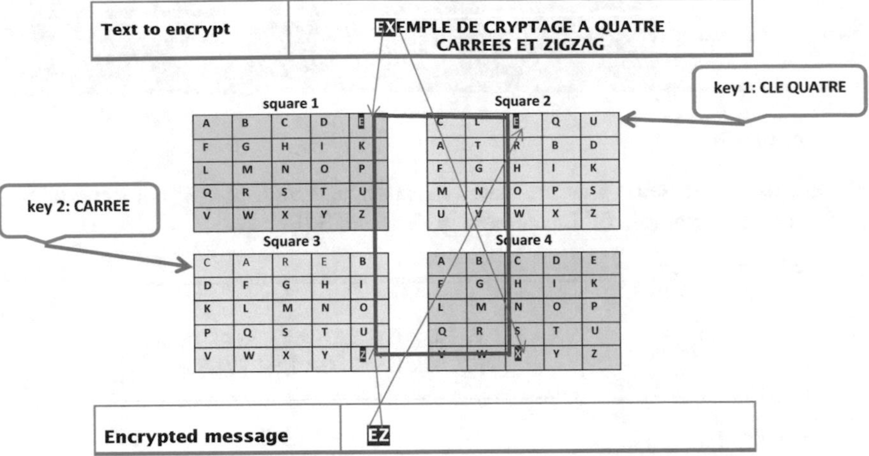

Fig. 4 Step 1 of encryption a clear text

- We find the E in the square on the left, the X in the square on the right, then we search in these squares for letters that complete the rectangle:
  In our example, the E in the left cell and the Z in the right cell. Then EX becomes EZ. It's worth mentioning that the first letter out of the first encrypted two is aligned with the first clear letter.
- Then, we are going to enforce the four-square methods with some zigzag transformations on grid number 2 and 3. Indeed, in every couple of the message, we rotate matrix 1 to the right and matrix 2 to the left which prevents an attack with probable words, which means blocking the fact that two couples of letters can be coded by the same digraph, which makes the dismantling of the number by frequency analysis [8–12] more difficult. Figure 5 illustrates these steps.

And so on, in every block of two letters from the clear text, we redo the zigzag transformation the right and on the left, respectively, to grid number 2 and 3, here's the result of the text's encryption with this approach.

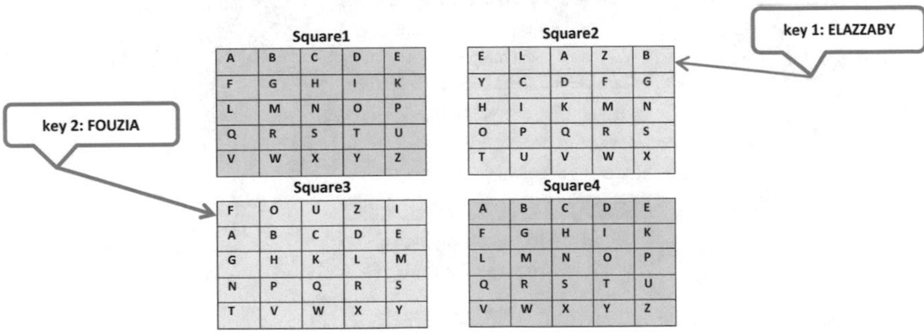

Fig. 5 Four squares of example 2

Encrypted message	EZCOBPCFYGIARCSMOLFZPCBLINQS MTUQQNKCEXI

b. **Example 2**

We proceed in the same way presented in example 1 and except that the letters of the keys in this example fill the cells of grids 2 and 3 (Fig. 6).

Here is the result:

PLAIN TEXT	LA METHODE DECOUPE LE MESSAGE ENCLAIR EN DEUX LETTRES
keys :	key 1 « ELAZZABY » key 2 « FOUZIA ».
CIPHER TEXT	HF MUOAHBZ DWEFYDH CD YGDUOPQ PZ NUMPD IA KTZQ TSXLXWTS

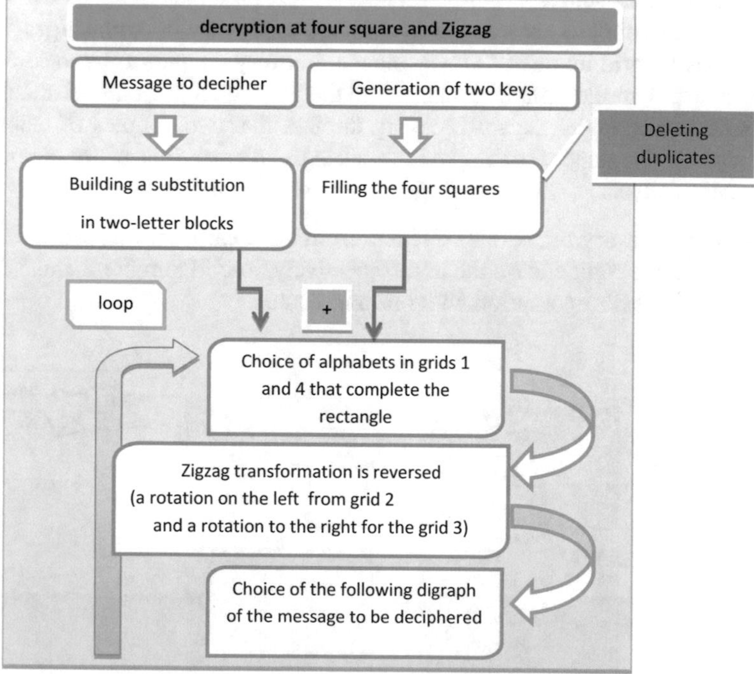

Fig. 6 The decryption process at four-square and zigzag

3.2 The Decryption

We proceed in the same way presented in example 1 and except that the letters of the keys in this example fill the cells of grids 2 and 3 (Fig. 6).

Decryption Example

EXAMPLE1	**CIPHER TEXT**	**EZCOBPC FY GIARCSMO L FZPCBL INQSMTU QQ NKCEXI**
	Keys :	**key 1 :CLEQUATRE key 2 : CARREE.**
	PLAIN TEXT	**EXEMPLE DE CRYPTAGE A QUATRE CARRES ET ZIGZAG**

EXAMPLE2	**CIPHER TEXT**	**HF MUOAHBZ DWEFYDH CD YGDUOPQ PZ NUMPD IA KTZQ TSXLXWTS**
	Keys :	**key 1 « ELAZZABY » key 2 « FOUZIA ».**
	PLAIN TEXT	**LA METHODE DECOUPE LE MESSAGE EN CLAIR EN DEUX LETTRES**

The experimental part clearly shows that our approach gives good results. Encrypted texts are very different from the original texts. Indeed, the convolution between the 4-square encryption method and the translation of the cells made right in Grid 2 and translating cells made on the left (the zigzag transformation) prevents the encryption of each digraph in the same way. This results in high-security results slowing down the weather decryption of the message and decreasing the risk of attack considerably.

4 Conclusion

To conclude, this work focuses on a new encryption approach that is based on four-square encryption accompanied by a zigzag transformation. The latter has the advantage of encrypting the same digraph of 26 pairs of different letters which further slows down the time of each unauthorized access. The added value of our approach is that it resists cryptanalysis attacks and that it delays the temptations of decryption by diversifying the encryption of each digraph differently for each pair of the message clear efficiency of the method is relatively safe when the message has a number is short enough.

Despite the good properties of our approaches including its ability to defend against attacks, it still has limits to know: once we repeat in a message the same digraph more than 26 times the encryption process is the original couple. So this flaw is clearly visible especially when the size of the message is considerable. Our goal is to recognize this constraint and to analyze it.

References

1. Tong, X., Cui, M., Wang, Z.: A new feedback image encryption scheme based on perturbation with dynamical compound chaotic sequence cipher generator. J. Opt. Commun. **282**, 2722–2728 (2009)
2. Borujeni, S.E., Eshghi, M.: Chaotic image encryption system using phase-magnitude transformation and pixel substitution. J. Telecommun. Syst. (2011)
3. Wang, X.Y., Guo, K.: A new image alternate encryption algorithm based on chaotic map. Nonlinear Dyn. **76**, 1943–1950 (2014)
4. Hua, Z.Y., Zhou, Y.C., Pun, C.M., Philip Chen, C.L.: 2D Sine logistic modulation map for image encryption. Inf. Sci. **297**, 80–94 (2015)
5. Tong, X.G., Wang, Z., Zhang, M., Liu, Y., Xu, H., Ma, J.: An image encryption algorithm based on the perturbed high-dimensional chaotic map. Nonlinear Dyn. **80**, 1493–1508 (2015)
6. Zhu, C.: A novel image encryption scheme based on improved hyperchaotic sequences. J. Opt. Commun. **285**, 29–37 (2012)
7. Es-Sabry, M., El Akkad, N., Merras, M., Saaidi, A., Satori, K.H.: A novel text encryption algorithm based on the two-square cipher and Caesar cipher, pp. 78–88 (2018)
8. Ling, C., Wu, X., Sun, S.: A general efficient method for chaotic signal estimation. IEEE Trans. Signal Process. **47**, 1424–1428 (1999)
9. Wu, X., Hu, H., Zhang, B.: Parameter estimation only from the symbolic sequences generated by chaos system. Chaos Solitons Fractals **22**, 359–366 (2004)
10. Skrobek, A.: Cryptanalysis of chaotic stream cipher. Phys. Lett. A **363**, 84–90 (2007)
11. Patidar, V., Pareek, N., Sud, K.: Communications in nonlinear science and numerical simulation **14**, 3056 (2009)
12. Parvaz, R., Zarebnia, M.: A combination chaotic system and application in color image encryption. Optics Laser Technol. **101**, 30–41 (2018)
13. Norouzi, B., et al.: A novel image encryption based on hash function with only two-round diffusion process. Multimedia Syst. (2013)
14. Weidenmuller, H.A., Mitchell, G.E.: Random matrices and chaos in nuclear physics: nuclear structure. Rev. Mod. Phys. **81**(2), 539–589 (2009)
15. Lu, L., Luan, L., Meng, L., Li, C.R.: Study on spatiotemporal chaos tracking synchronization of a class of complex network. Nonlinear Dyn. **70**(1), 89–95 (2012)
16. Lesne, A.: Chaos in biology. Riv. Biol. **99**(3), 467–481 (2006)
17. Liu, C.X., Lu, J.J.: A novel fractional-order hyperchaotic system and its circuit realization. Int. J. Mod. Phys. B **24**(10), 1299–1307 (2010)
18. Hilborn, R.C.: Chaos and nonlinear dynamics: an introduction for scientists and engineers, 2nd edn. Oxford University Press, USA (2001)

19. Arroyo, D., Rhouma, R., Alvarez, G., Li, S., Fernandez, V.: On the security of a new image encryption scheme based on chaotic map lattices. Chaos Interdiscip. J. Nonlinear Sci. **18** (2008)
20. Hayat, U., Azam, N.A.: A novel image encryption scheme based on an elliptic curve. Signal Process. **155**, 391–402 (2019)
21. Koppula, V., Waters, B.: Realizing chosen ciphertext security generically in attribute-based encryption and predicate encryption. In: IACR Cryptology ePrint Archive, p. 847 (2018)
22. Håstad, J.: Solving simultaneous modular equations of low degree. SIAM J. Comput. **17**(2), 336–341 (1988)
23. Goldwasser, S., Micali, S.: Probabilistic encryption. J. Comput. Syst. Sci. **28**, 270–299 (1984)
24. Es-sabry, M., El Akkad, N., Merras, M., Saaidi, A., Satori, K.: Grayscale image encryption using shift bits operations. In: International Conference on Intelligent Systems and Computer Vision, ISCV (2018)

From Marker to Markerless in Augmented Reality

Zainab Oufqir, Abdellatif El Abderrahmani and Khalid Satori

Abstract Our article discusses existing methods in marker-based and markerless augmented reality and their evolution over time. Markers are the most optimal solution to solve the problem of calculating the pose for augmented reality and do not require powerful devices. Its simplicity, robustness and efficiency have a great advantage; the objects to detect are provided to the application and require to be always visible by the sensor. On the other hand, markerless augmented reality detects objects or characteristic points of a scene without any prior knowledge of the environment. This mechanism is more difficult to implement because it implements algorithms that are expensive in terms of computation time. The appearance and improvement of new devices have made it possible to exploit this technology and to detach from the presence of a marker in a scene and emerge to markerless methods.

Keywords Augmented reality · Marker · Markerless · Camera pose

1 Introduction

Augmented reality is one of the recent technologies that enrich the real world with additional information in the form of digital virtual objects [1]. This combination is possible by using electronic devices containing a camera sensor [2] (glasses, smartphone, tablet, etc.). This technology is used in several fields; for example, for medicine, a doctor can display organs on the human body and perform interactive simulations of surgical procedures [3]. In education, it has revolutionized the

Z. Oufqir (✉) · K. Satori
LIIAN, Department of Mathematics and Informatics, Faculty of Sciences Dhar-Mahraz, Sidi Mohamed Ben Abdellah University, Atlas, P.O.Box 1796, 30000 Fes, Morocco
e-mail: zayna.oufkir@gmail.com

K. Satori
e-mail: khalidsatori@gmail.com

A. El Abderrahmani
LSTA, Department of Computer Science, Larache Polydisciplinary School, Larache, Morocco
e-mail: elabderrahmani@yahoo.fr

V. Bhateja et al. (eds.), *Embedded Systems and Artificial Intelligence*,
Advances in Intelligent Systems and Computing 1076,
https://doi.org/10.1007/978-981-15-0947-6_57

world of learning [4]. The calculation of the camera position in the real world is the problem studied in augmented reality, and it means finding the translation and orientation of the camera in a reference in the real world. The solution to this problem is to provide the position of the 3D points of the scene in the real-world reference and the 2D position of their projection in the image sequence acquired in real time [5]. In computer vision, the process of calculating the pose is summarized in two steps: detection and tracking; the detection phase consists in detecting the points that most characterize a scene, and there are several algorithms to detect them as SIFT [6], SURF [7], BRIEF [8], etc. The tracking phase matches the detected key points and finds their position in a sequence of images taken from different points of view. By obtaining the 3D position of the key points of the scene through a 3D reconstruction [9] or provide it based on real-world references such as markers [10], it is possible to calculate a user's point of view in a scene. Using markers that optimize the process of calculating the pose because they are detected easily and quickly, unlike 3D reconstruction, which is costly in terms of calculation time. The detection phase and matching of 2D points in images acquired in real time are the most important phase in the pose calculation process because it correctly aligns a virtual object in a real scene so that they coexist together [11]. In the following sections, we will study in detail marker-based and markerless methods and we will present some existing simulations built in different fields to exploit the richness of these methods.

2 Related Work

Markers are planar elements placed in a scene and are designed specially to facilitate the calculation of the camera position; the process consists in detecting the marker and inserting a virtual object on it. The most popular type of marker is the black and white square marker, and corners are easily detected and tracked in real time. Hirzer [12] provides a rapid detection algorithm using an edge-based approach that increases robustness in front of lighting conditions and occlusions. Several libraries that implement this type of marker as ARToolKit [13], ARTag [14], Aruco [15, 16] models are saved in a database and compared by the markers placed in the scene.

For markerless augmented reality, they are based on the natural characteristics of the scene that are captured by the camera to calculate its position. Markerless methods use natural characteristics such as planes, edges or corner points that are present in the scene, and these characteristics are extracted from the environment and have been demonstrated that are suitable for inserting virtual objects into the real world [17, 18, 19]. When it is a planar scene, the calculation of a homography between frames provides an effective detection and tracking mechanism to make augmentations [20]. The challenge of markerless augmented reality is to overcome the requirement of powerful materials to implement its robust algorithms such as the SURF descriptor, make a 3D reconstruction and optimize the result with the RANSAC algorithm [21].

Marker-based and markerless methods use the same mathematical formula to calculate the camera position. Estimating the pose means calculating the translation and orientation of the camera in the real world, using its internal parameters, as well as the knowledge of the position of 3D reference points and their projections in the image (Fig. 1):

$$s\begin{bmatrix} x_c \\ y_c \\ 1 \end{bmatrix} = \begin{bmatrix} P_{11} & P_{12} & P_{13} \\ P_{21} & P_{22} & P_{23} \\ P_{31} & P_{32} & P_{33} \end{bmatrix} \begin{bmatrix} R_{11} & R_{12} & R_{13} & t_1 \\ R_{21} & R_{22} & R_{32} & t_2 \\ R_{31} & R_{32} & R_{33} & t_3 \\ 0 & 0 & 0 & 1 \end{bmatrix} \begin{bmatrix} X_c \\ Y_c \\ Z_c \end{bmatrix} = PA \begin{bmatrix} X \\ Y \\ Z \end{bmatrix}$$

(X_c, Y_c, Z_c) Camera coordinate
(X, Y, Z) 3D point coordinate
R Rotation
t Translation
A Pose matrix
P The intrinsic parameters of thc camera
(x_c, y_c) The coordinates of the projection on the camera screen.

For marker-based methods, the marker is easily recognized, and the 3D position of its points is already known, which facilitates the process of calculating the pose. On the other hand, for methods based without markers, it is necessary to extract the characteristic points, finds their correspondent in the different images acquired in real time with a robust descriptor and estimates their 3D position in the real world.

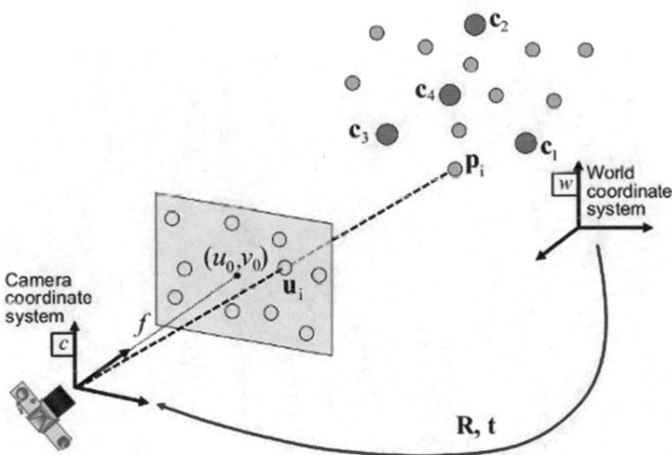

Fig. 1 Rotation and translation between the camera and the world coordinate

3 Marker Augmented Reality

The history of markers begins in 1998, and Jun Rekimoto introduces a new method to build an augmented reality system using printed 2D matrix codes that are applied on a video image [21]. In 2000, Kato and Billinghurst developed the ARToolKit open-source library that creates real-time augmented reality applications using a black and white square marker [22]. This library calculates the user's point of view, which allows a virtual object to be correctly aligned on the marker. This technology has been successfully integrated into mobile phones [23], and this experimentation has allowed this technology to be used by many users and to be easily accessible via their device. Markers are also used in the world of tourism; for example, ARCO uses square-shaped markers containing specific letters or characters and is placed in museums, and the visitor displays virtual objects on the marker and can interact with [24]. In the world of games, markers have shown their feasibility and produced graphic overlays on the physical components of the game [25].

To correctly align a virtual object on the marker, it is necessary to calculate the position of the camera in a real-world reference. For this, we must have at least the 3D position of four points in the real world and their 2D position in their projection in the image sequence acquired in real time [26, 27]. A marker represents a reference and landmark in the real world and easily detects its points of interest that represents its corners [28] as illustrated in Figs. 2 and 3.

Markers differ from one application to another; planar markers are popular for their simplicity of detection; they must be visible by the sensor all along the processing to ensure a good insertion quality. They allow finding the translation and orientation of the camera relative to the marker landmark in real time when the images are acquired

Fig. 2 Augmented reality with markers

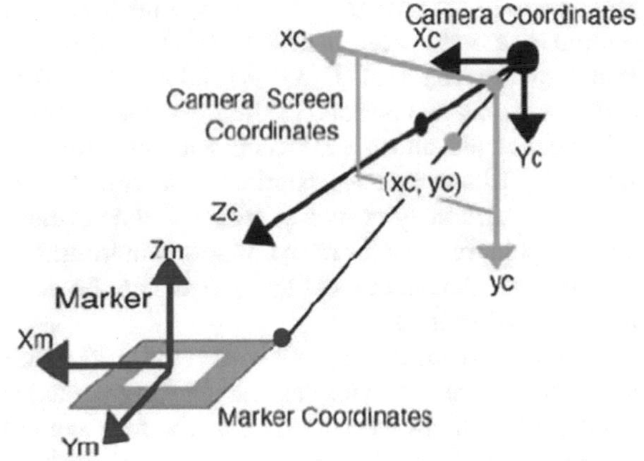

Fig. 3 Representation of the camera landmark and marker landmark

by the sensor. A large part of these markers are represented in the form of a white and black square or a QR code, and the marker must first be detected and then identified between the different markers loaded in the application's database.

4 Markerless Augmented Reality

On the other hand, markerless augmented reality applications are based on the points that characterize a scene and use them as markers, and no objects are provided or added to the real or loaded in the application. Today's applications are mostly oriented towards this mode of operation, and the potential of augmented reality increases with the emergence of new technologies and their high performance; so the recent emergence of better cameras and more accurate sensors has opened new opportunities to develop markerless augmented reality experiences without prior knowledge of the user's environment to place virtual content in a 3D scene.

The first works aimed to detect objects in square rectangle or other form that already exists in the scene and uses them as markers to calculate the pose [17, 29, 30], the disadvantage of this method is the computation time to find these objects with the possibility of finding several objects with the same characteristics which can bring ambiguities in the calculation of the pose.

The technology that has revolutionized markerless augmented reality is the method of simultaneous location and mapping (SLAM), which is used in planar surfaces. It reconstructs the real environment by aggregating key points between two consecutive images. These approaches do not therefore locate a specific object in a real scene, and it calculates the position and orientation of the point of view basing on key points of the scene; it estimates the position of a camera in a scene at the same time as this scene is reconstructed [31]. This technology has been integrated into mobiles [32], and the information captured by the camera is analysed and filtered as key images to optimize the calculation of the pose. Other technologies are combined with SLAM such as the Kinect sensor which allows us to calculate the depth represented by Newcombe et al. [33, 34] and test its performance in augmented reality to estimate the camera position [35]. SLAM has also contributed to the construction of a framework that allows a 3D reconstruction in real time of the scene in combination with a depth camera and an inertial measurement unit [36]. SLAM is also combined with other technologies such as compass and accelerometer; the aim is to calculate a user's point of view for in an unknown environment [37]; the role of the compass is to indicate the direction of the device; and the accelerometer is used to calculate the user's orientation.

The markerless AR system also uses the GPS function integrated into the devices to locate and interact with the available augmented reality resources. Feiner et al. made the first application based on this technology in 1997 [38], and they used GPS tracking to build augmented reality interfaces and demonstrated its feasibility to display additional information in the real world. To get a good estimate of the pose, GPS is often combined with inertial sensors and uses vision-based tracking [39]. Smartphones and tablets have opened new opportunities and integrate the technologies needed to implement augmented reality applications [40]. The popular Pokémon Go application implements GPS to estimate the user's position and displays 3D cartoons in the real world through the device's camera [41]; it popularized the concept of augmented reality and showed the general public this technology which consists in displaying a virtual object in the real environment through a screen. These methods and devices offer many advantages for the insertion of 3D virtual objects into 2D images.

Another technology that has revolutionized the world of augmented reality is holography, which consists in projecting a 3D hologram into space that can be viewed from different angles [42], bypassing it without the need for a headset or 3D glasses as shown in Fig. 4a. Using computer vision techniques, it is used in HoloLens devices (Fig. 4b), and they are glasses based on a visor composed of layers that provide real-time 3D stereoscopic vision, using techniques based on holography [43].

Fig. 4 Markerless augmented reality

5 Simulation

We will present some simulations of some existing work in different fields exploiting the richness of augmented reality and implementing marker-based and markerless methods to display virtual objects in the real world.

5.1 Marker Augmented Reality

Markers are implemented in different fields; for example, in education, a study was carried out on students to exploit the influence of augmented reality on their learning, and a marker is used to visualize 3D microparticles on markers added to the scene as shown in Fig. 5. This experiment had a relatively greater influence on students with low scores [44]. The software is programmed in Java, and the extra packages used include NyARToolkit, Java3D and JMF.

Marker technology has been integrated into veterinary medicine, and it has been used as a useful training tool for students before they attempt to intravenous injection into dogs [45]. A marker is added to the scene to display the correct injection area as shown in Fig. 6. The simulator was developed using ARToolKit software, an open-source library for augmented reality applications.

The markers have also proven their effectiveness in the gaming world. As shown in Fig. 7, the user manipulates the marker on a surface to move the ball, and the virtual representation of the game is displayed on a screen of a device [46]. The system was developed using Microsoft's XNA platform (C#) and works with any DirectShow-compatible webcam using ARToolKitPlus library.

(a) Hologramme

(b) HoloLens smart glass display

Fig. 5 Augmented reality with holography

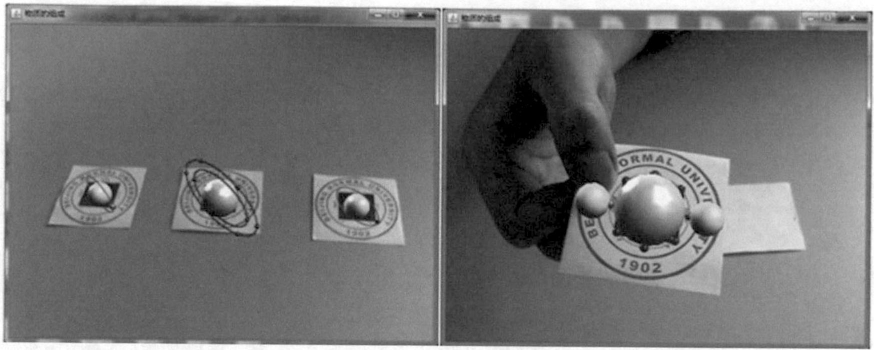

Fig. 6 Micro-particles augmented on markers

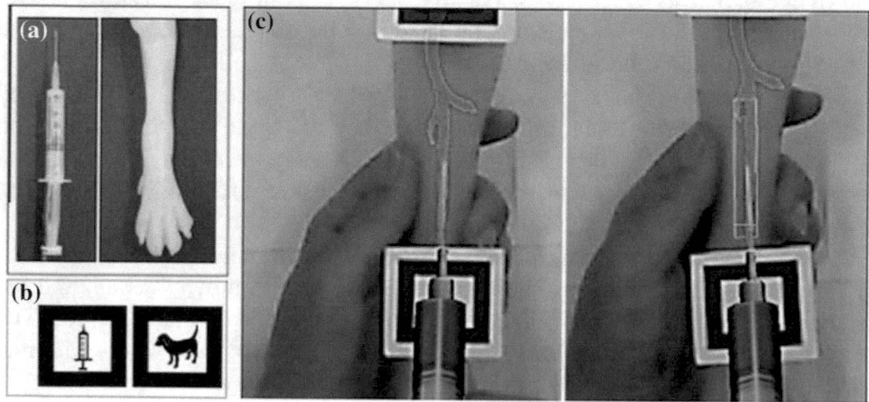

Fig. 7 Injection experience using markers

5.2 Markerless Augmented Reality

Markerless augmented reality has also been implemented in several areas. For example, in tourism, a mobile phone is used as a guide to increase the painting table with additional information, and this method has been very efficient because the addition of markers in the museum has been disruptive [47]. The augmented table is considered a giant marker that can be reliably tracked on a wide range of paints as shown in Fig. 8. They used a handheld augmented reality system that is built around a Samsung Q1 Ultra Premium UMPC.

Pokémon Go is a markerless augmented reality game developed by the Japanese company Nintendo, and it swept the world in the first week of its release and became the most downloaded application in its time [48]. It uses GPS technology to align cartoons in the real world through the screen of the device as shown in Fig. 9.

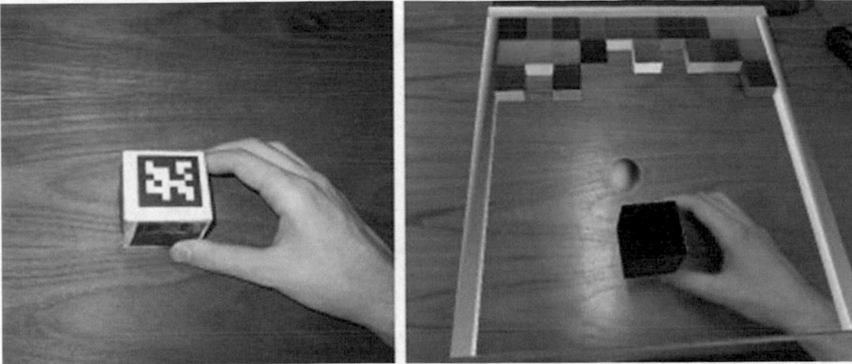

Fig. 8 Game with marker augmented reality

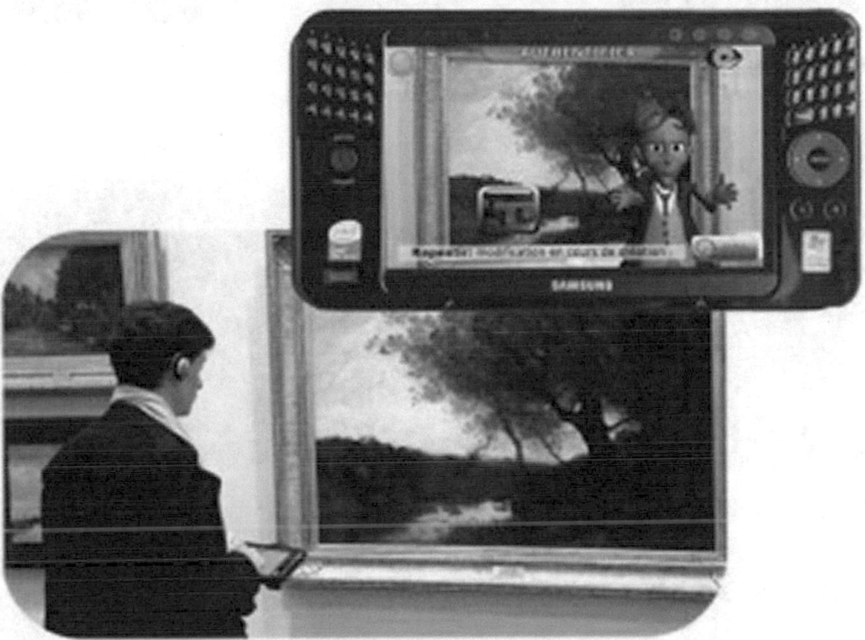

Fig. 9 Markerless augmented reality in the museum

In medicine, it is becoming more and more useful and helps doctors in diagnosis and surgical simulation. For example, iRay is a markerless mobile application that uses the SLAM method to calculate the pose and correctly display the virtual part as shown in Fig. 10 and can be used in learning and medical interventions [49, 50]. The hardware devices used for this work are a structure sensor [51] that provides 3D information about the scene and an Apple iPad (Fig. 11).

Fig. 10 Pokémon Go application

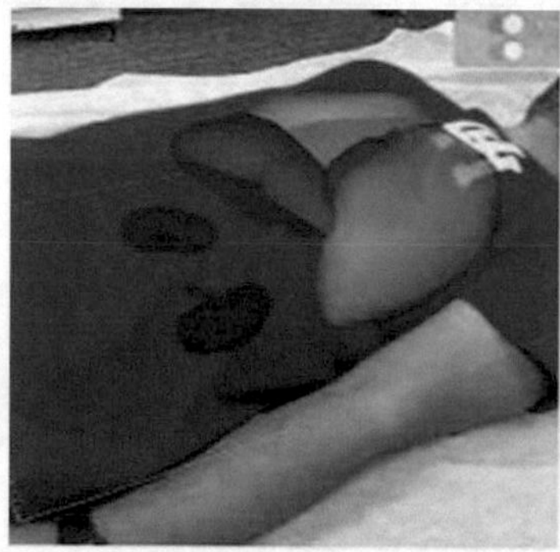
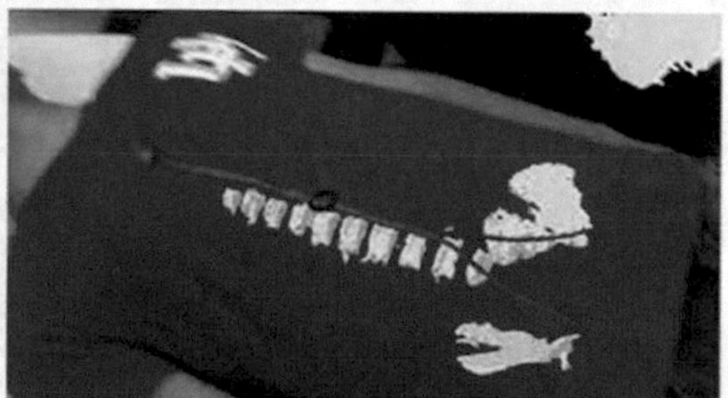

Fig. 11 Overview of iRay application

6 Interpretation

For marker-based methods, it is necessary to place an image reference in the real world and must be visible by the sensor during the entire process in order to estimate the camera pose and to correctly align the virtual object on the marker. This method is widely used in the augmented reality world because the markers are easy and fast to recognize and do not require high power equipment. Markerless methods combine several technologies such as SLAM, GPS, accelerometers or compass to build an augmented reality system and calculate the user's point of view based on the information taken by the camera sensor without having prior knowledge of the environment, and it offers flexibility and detaches the user from adding a component to the scene. Today's applications are mostly oriented towards markerless methods due to rapid technological advances.

7 Conclusion

This article discusses marker-based and markerless methods and their use in different fields, markerless augmented reality depends on the natural characteristics of an environment while markers are added in a scene and are easy to recognize and track and do not require powerful devices. The markerless augmented reality system is now the preferred image recognition method over its marker-based counterpart due to the power of new technologies. The mobile device industry is improving the quality of its cameras every year and according to it will have mobile devices that integrate a 3D camera, which will revolutionize and enrich the world of augmented reality.

References

1. Azuma, R.T.: A survey of augmented reality **48** (1997)
2. Zhou, F., Duh, H.B., Billinghurst, M.: Trends in augmented reality tracking, interaction and display: a review of ten years of ISMAR. In: 2008 7th IEEE/ACM International Symposium on Mixed and Augmented Reality, pp. 193–202. Presented at the 2008 7th IEEE/ACM International Symposium on Mixed and Augmented Reality (2008). https://doi.org/10.1109/ismar.2008.4637362
3. Kutter, O., Aichert, A., Bichlmeier, C., Traub, J., Michael, S., Ockert, B., Euler, E., Navab, N.: Real-time volume rendering for high quality visualization in augmented reality **10** (2008)
4. Cabero Almenara, J., Barroso Osuna, J.: The educational possibilities of augmented reality. J. New Approaches Educ. Res. **6**(1), 44–50 (2016). https://doi.org/10.7821/naer.2016.1.140
5. Marchand, E., Uchiyama, H., Spindler, F.: Pose estimation for augmented reality: a hands-on survey. IEEE Trans. Visual Comput. Graph. **22**(12), 2633–2651 (2016). https://doi.org/10.1109/TVCG.2015.2513408
6. Lowe, D.G.: Object recognition from local scale-invariant features. In: Proceedings of the Seventh IEEE International Conference on Computer Vision, vol. 2, pp. 1150–1157. Presented at the Proceedings of the Seventh IEEE International Conference on Computer Vision (1999). https://doi.org/10.1109/iccv.1999.790410
7. Bay, H., Tuytelaars, T., Van Gool, L.: SURF: speeded up robust features. In: Leonardis, A., Bischof, H., Pinz, A. (eds.) Computer Vision – ECCV 2006, vol. 3951, pp. 404–417. Springer, Berlin, Heidelberg (2006). https://doi.org/10.1007/11744023_32
8. Calonder, M., Lepetit, V., Strecha, C., Fua, P.: BRIEF: binary robust independent elementary features. In: Daniilidis, K., Maragos, P., Paragios, N. (eds.) Computer Vision – ECCV 2010, vol. 6314, pp. 778–792. Springer, Berlin, Heidelberg (2010). https://doi.org/10.1007/978-3-642-15561-1_56
9. Annich, A., El Abderrahmani, A., Satori, K.: Fast and easy 3D reconstruction with the help of geometric constraints and genetic algorithms. 3D Res. **8**(3) (2017). https://doi.org/10.1007/s13319-017-0139-6
10. Li, Y., Wang, Y.-T., Liu, Y.: Fiducial marker based on projective invariant for augmented reality. J. Comput. Sci. Technol. **22**(6), 890–897 (2007). https://doi.org/10.1007/s11390-007-9100-0

11. Bergamasco, F., Albarelli, A., Torsello, A.: Pi-Tag: a fast image-space marker design based on projective invariants. Mach. Vis. Appl. **24**(6), 1295–1310 (2013). https://doi.org/10.1007/s00138-012-0469-6
12. Hirzer, M.: Marker detection for augmented reality applications **28** (n.d.)
13. Matsuoka, H., Onozawa, A., Hosoya, E.: Environment mapping for objects in the real world: a trial using ARToolKit. In: The First IEEE International Workshop Augmented Reality Toolkit, p. 2. Presented at the First IEEE International Workshop Augmented Reality Toolkit (2002). https://doi.org/10.1109/art.2002.1107006
14. Fiala, M.: ARTag, a fiducial marker system using digital techniques. In: 2005 IEEE Computer Society Conference on Computer Vision and Pattern Recognition (CVPR'05), vol. 2, pp. 590–596. Presented at the 2005 IEEE Computer Society Conference on Computer Vision and Pattern Recognition (CVPR'05). IEEE, San Diego, CA, USA (2005). https://doi.org/10.1109/cvpr.2005.74
15. Romero-Ramirez, F.J., Muñoz-Salinas, R., Medina-Carnicer, R.: Speeded up detection of squared fiducial markers. Image Vis. Comput. **76**, 38–47 (2018). https://doi.org/10.1016/j.imavis.2018.05.004
16. Garrido-Jurado, S., Muñoz-Salinas, R., Madrid-Cuevas, F.J., Medina-Carnicer, R.: Generation of fiducial marker dictionaries using mixed integer linear programming. Pattern Recogn. **51**, 481–491 (2016). https://doi.org/10.1016/j.patcog.2015.09.023
17. Ferrari, V., Tuytelaars, T., Van Gool, L.: Markerless augmented reality with a real-time affine region tracker. In: Proceedings IEEE and ACM International Symposium on Augmented Reality, pp. 87–96. Presented at the IEEE and ACM International Symposium on Augmented Reality. IEEE Computer Society, New York, NY, USA (2001). https://doi.org/10.1109/isar.2001.970518
18. Simon, G., Fitzgibbon, A.W., Zisserman, A.: Markerless tracking using planar structures in the scene. In: Proceedings IEEE and ACM International Symposium on Augmented Reality (ISAR 2000), pp. 120–128. Presented at the IEEE and ACM International Symposium on Augmented Reality (ISAR 2000). IEEE, Munich, Germany (2000). https://doi.org/10.1109/isar.2000.880935
19. Skrypnyk, I., Lowe, D.G.: Scene modelling, recognition and tracking with invariant image features. In: Third IEEE and ACM International Symposium on Mixed and Augmented Reality, pp. 110–119. Presented at the Third IEEE and ACM International Symposium on Mixed and Augmented Reality. IEEE, Arlington, VA, USA (2004). https://doi.org/10.1109/ismar.2004.53
20. Lee, T., Hollerer, T.: Viewpoint stabilization for live collaborative video augmentations. In: Proceedings of the 5th IEEE and ACM International Symposium on Mixed and Augmented Reality, pp. 241–242. IEEE Computer Society, Washington, DC, USA (2006). https://doi.org/10.1109/ismar.2006.297824
21. Rekimoto, J.: Matrix: a realtime object identification and registration method for augmented reality (1998)
22. Kato, H., Billinghurst, M.: Marker tracking and HMD calibration for a video-based augmented reality conferencing system. In: Proceedings 2nd IEEE and ACM International Workshop on Augmented Reality (IWAR'99), pp. 85–94. Presented at the Proceedings 2nd IEEE and ACM International Workshop on Augmented Reality (IWAR'99) (1999). https://doi.org/10.1109/iwar.1999.803809
23. Henrysson, A., Ollila, M.: UMAR: ubiquitous mobile augmented reality. In: Proceedings of the 3rd International Conference on Mobile and Ubiquitous Multimedia - MUM'04, pp. 41–45. Presented at the 3rd International Conference. ACM Press, College Park, Maryland (2004). https://doi.org/10.1145/1052380.1052387
24. Wojciechowski, R., Walczak, K., White, M., Cellary, W.: Building virtual and augmented reality museum exhibitions. In: Proceedings of the Ninth International Conference on 3D Web Technology, pp. 135–144. ACM, New York, NY, USA (2004). https://doi.org/10.1145/985040.985060
25. Rohs, M., Laboratories, D.T., Berlin, T.: Marker-based embodied interaction for handheld augmented reality games. J. Virtual Reality Broadcast. (2007)

26. Neumann, U., Park, J.: Extendible object-centric tracking for augmented reality. In: Proceedings. IEEE 1998 Virtual Reality Annual International Symposium (Cat. No.98CB36180), pp. 148–155. Presented at the Proceedings. IEEE 1998 Virtual Reality Annual International Symposium (Cat. No.98CB36180), IEEE Computer Society, Atlanta, GA, USA (1998). https://doi.org/10.1109/vrais.1998.658482
27. Fischler, M.A., Bolles, R.C.: Random sample consensus: a paradigm for model fitting with applications to image analysis and automated cartography. Commun. ACM **24**(6), 381–395 (1981). https://doi.org/10.1145/358669.358692
28. Harris, C., Stephens, M.: A combined corner and edge detector. In: Proceedings of the Alvey Vision Conference 1988, pp. 23.1–23.6. Presented at the Alvey Vision Conference 1988. Alvey Vision Club, Manchester (1988). https://doi.org/10.5244/c.2.23
29. Comport, A.I., Marchand, E., Chaumette, F.: A real-time tracker for markerless augmented reality. In: The Second IEEE and ACM International Symposium on Mixed and Augmented Reality, 2003. Proceedings, pp. 36–45. Presented at the Second IEEE and ACM International Symposium on Mixed and Augmented Reality. IEEE Computer Society, Tokyo, Japan (2003). https://doi.org/10.1109/ismar.2003.1240686
30. Simon, G., Berger, M.-O.: Reconstructing while registering: a novel approach for markerless augmented reality. In: Proceedings. International Symposium on Mixed and Augmented Reality, pp. 285–293. Presented at the IEEE and ACM International Symposium on Mixed and Augmented Reality. IEEE Computer Society, Darmstadt, Germany (2002). https://doi.org/10.1109/ismar.2002.1115118
31. Davison, A.J., Reid, I.D., Molton, N.D., Stasse, O.: MonoSLAM: real-time single camera SLAM. IEEE Trans. Pattern Anal. Mach. Intell. **29**(6), 1052–1067 (2007). https://doi.org/10.1109/TPAMI.2007.1049
32. Ondruska, P., Kohli, P., Izadi, S.: MobileFusion: real-time volumetric surface reconstruction and dense tracking on mobile phones. IEEE Trans. Visual Comput. Graph. **21**(11), 1251–1258 (2015). https://doi.org/10.1109/TVCG.2015.2459902
33. Newcombe, R.A., Izadi, S., Hilliges, O., Molyneaux, D., Kim, D., Davison, A.J., Kohli, P., Shotton, J., Hodges, S., Fitzgibbon, A.: Kinectfusion: real-time dense surface mapping and tracking **66** (2011)
34. Izadi, S., Davison, A., Fitzgibbon, A., Kim, D., Hilliges, O., Molyneaux, D., Newcombe, R., Kohli, P., Shotton, J., Hodges, S., Freeman, D.: KinectFusion: real-time 3D reconstruction and interaction using a moving depth camera. In: Proceedings of the 24th Annual ACM Symposium on User Interface Software and Technology - UIST'11, p. 559. Presented at the 24th Annual ACM Symposium. ACM Press, Santa Barbara, California, USA (2011). https://doi.org/10.1145/2047196.2047270
35. Bostanci, E., Kanwal, N., Clark, A.F.: Augmented reality applications for cultural heritage using Kinect. Human-Centric Comput. Inf. Sci. **5**(1) (2015). https://doi.org/10.1186/s13673-015-0040-3
36. Chen, C.-W., Chen, W.-Z., Peng, J.-W., Cheng, B.-X., Pan, T.-Y., Kuo, H.-C.: A real-time markerless augmented reality framework based on SLAM technique, pp. 127–132. IEEE, Exeter (2017). https://doi.org/10.1109/ispan-fcst-iscc.2017.87
37. Basori, A.H., Afif, F.N., Almazyad, A.S., AbuJabal, H.A.S., Rehman, A., Alkawaz, M.H.: Fast markerless tracking for augmented reality in planar environment. 3D Res. **6**(4) (2015). https://doi.org/10.1007/s13319-015-0072-5
38. Feiner, S., MacIntyre, B., Höllerer, T., Webster, A.: A touring machine: prototyping 3D mobile augmented reality systems for exploring the urban environment. Pers. Technol. **1**(4), 208–217 (1997). https://doi.org/10.1007/BF01682023
39. Azuma, R.T., Hoff, B.R., Iii, H.E.N., Sarfaty, R., Daily, M.J., Bishop, G., Chi, V., Welch, G., Neumann, U., You, S., Cannon, J.: Making augmented reality work outdoors requires hybrid tracking **6** (1998)
40. Ventura, J., Hollerer, T.: Wide-area scene mapping for mobile visual tracking. In: 2012 IEEE International Symposium on Mixed and Augmented Reality (ISMAR), pp. 3–12. Presented at the 2012 IEEE International Symposium on Mixed and Augmented Reality (ISMAR), IEEE, Atlanta, GA, USA (2012). 10.1109/ISMAR.2012.6402531

41. McCartney, M.: Margaret McCartney: game on for Pokémon Go. BMJ i4306 (2016). https://doi.org/10.1136/bmj.i4306
42. Ando, T., Matsumoto, T., Takahashi, H., Shimizu, E.: Head mounted display for mixed reality using holographic optical elements **6** (1999)
43. Evans, G., Miller, J., Iglesias Pena, M., MacAllister, A., Winer, E.: Evaluating the Microsoft HoloLens through an augmented reality assembly application. In: Sanders-Reed, J. (Jack) N., Arthur, J. (Trey) J. (eds.) Presented at the SPIE Defense + Security, p. 101970V, Anaheim, California, United States (2017). https://doi.org/10.1117/12.2262626
44. Cai, S., Wang, X., Chiang, F.-K.: A case study of augmented reality simulation system application in a chemistry course. Comput. Hum. Behav. **37**, 31–40 (2014). https://doi.org/10.1016/j.chb.2014.04.018
45. Lee, S., Lee, J., Lee, A., Park, N., Song, S., Seo, A., Lee, H., Kim, J.I., Eom, K.: Augmented reality intravenous injection simulator based 3D medical imaging for veterinary medicine. Vet. J. **196**(2), 197–202 (2013). https://doi.org/10.1016/j.tvjl.2012.09.015
46. Bur, J.W., McNeill, M.D.J., Charles, D.K., Morrow, P.J., Crosbie, J.H., McDonough, S.M.: Augmented reality games for upper-limb stroke rehabilitation. In: 2010 Second International Conference on Games and Virtual Worlds for Serious Applications, pp. 75–78. Presented at the 2010 2nd International Conference on Games and Virtual Worlds for Serious Applications (VS-GAMES 2010). IEEE, Braga, Portugal (2010). https://doi.org/10.1109/vs-games.2010.21
47. Tillon, A.B., Marchal, I., Houlier, P.: Mobile augmented reality in the museum: can a lace-like technology take you closer to works of art? In: 2011 IEEE International Symposium on Mixed and Augmented Reality - Arts, Media, and Humanities, pp. 41–47. Presented at the 2011 IEEE International Symposium on Mixed and Augmented Reality - Arts, Media, and Humanities (ISMAR-AMH). IEEE, Basel, Switzerland (2011). 10.1109/ISMAR-AMH.2011.6093655
48. Dorward, L.J., Mittermeier, J.C., Sandbrook, C., Spooner, F.: Pokémon Go: benefits, costs, and lessons for the conservation movement: conservation implications of Pokémon Go. Conserv. Lett. **10**(1), 160–165 (2017). https://doi.org/10.1111/conl.12326
49. Kakadiaris, I.A., Islam, M.M., Xie, T., Nikou, C., Lumsden, A.B.: iRay: mobile AR using structure sensor. In: 2016 IEEE International Symposium on Mixed and Augmented Reality (ISMAR-Adjunct), pp. 127–128. Presented at the 2016 IEEE International Symposium on Mixed and Augmented Reality (ISMAR-Adjunct). IEEE, Merida, Yucatan, Mexico (2016). https://doi.org/10.1109/ismar-adjunct.2016.0058
50. Xie, T., Islam, M.M., Lumsden, A.B., Kakadiaris, I.A.: Semi-automatic initial registration for the iRay system: a user study. In: De Paolis, Ld.T., Bourdot, P., Mongelli, A. (eds.) Augmented reality, virtual reality, and computer graphics, vol. 10325, pp. 33–42. Springer International Publishing, Cham (2017). https://doi.org/10.1007/978-3-319-60928-7_3
51. Official Structure Sensor Store - Give Your iPad 3D Vision.: Retrieved 22 Apr 2019, from https://structure.io/

Towards Eclipse Plug-ins for Automated Data Warehouse Design from an Ontology

Morad Hajji, Mohammed Qbadou and Khalifa Mansouri

Abstract The Semantic Web is experiencing a rise in recent years. As a result, ontologies have become ubiquitous in many areas while attracting growing interest from researchers. However, this knowledge representation model escapes to multidimensional analysis due to the lack of tools allowing the transition from the ontological model to the multidimensional model. In this contribution, we present our approach for the implementation of our multi-approaches model for the automatic generation of a Data Warehouse model from an ontology. In this context, we have first developed an Eclipse Plug-in for generating a conceptual model of a relational database from an ontology. Secondly, we have developed another Eclipse Plug-in for generating a Data Warehouse model from a database model. And finally, we have integrated these two plug-ins, based on model-to-model transformation and to build an automated design solution for a Data Warehouse model from an ontology.

Keywords Data Warehouse · Semantic Web · Ontology · Relational database · Plug-in Eclipse · Query/View/Transformation

1 Introduction

Beyond the prediction of consumer trends, decision support systems are proliferating to reach any kind of data collection system. As a result, Business Intelligence is operated in small businesses as in large firms mainly with the advent of open-source solutions. It is by studying the past, determining the present, and predicting the future

---

M. Hajji (✉) · M. Qbadou · K. Mansouri
Laboratory SSDIA, ENSET
Mohammedia, Hassan II University of Casablanca, Mohammedia, Morocco
e-mail: morad.hajji@gmail.com

M. Qbadou
e-mail: qbmedn7@gmail.com

K. Mansouri
e-mail: khmansouri@hotmail.com

V. Bhateja et al. (eds.), *Embedded Systems and Artificial Intelligence*,
Advances in Intelligent Systems and Computing 1076,
https://doi.org/10.1007/978-981-15-0947-6_58

that an organization can make good decisions aligned to its strategy to achieve its intended goals.

This has a huge impact on the evolution of an organization over time and can outpace these competitors. In the information age, having the right data is a vital necessity for the survival of today's business. According to Inmon [1], the distribution of operational and informational databases occurs for several reasons:

- Data used to meet operational requirements are physically different from those used for information or analysis purposes.
- Support technology for operational processing is fundamentally different from the technology used to meet information or analysis needs.
- The user community for operational data is different from that served by informative or analytic data.
- The processing characteristics of the operational environment and the information environment are fundamentally different.

The Data Warehouse is the main element of the difference between a transactional system and a decision-making system. While the purpose of an operational system is to safeguard transactional operations, the decision-making system aims at providing an appropriate and adequate solution to decision-makers to meet the specific needs of such users, and in this case [2]:

- "We have mountains of data in this company, but we cannot access it."
- "We need to slice and dice the data every which way."
- "You've got to make it easy for business people to get at the data directly."
- "Just show me what is important."
- "It drives me crazy to have two people present the same business metrics at a meeting, but with different numbers."
- "We want people to use the information to support more fact-based decision-making."

In other words, a Data Warehouse must allow easy access, greater consistency, adaptation, and flexibility of an organization's information, security, and improvement of an organization's decision-making. A Data Warehouse is considered by Inmon as being a collection of thematic, integrated, nonvolatile, and historiated data organized for decision-making [1]. It is one of the fundamental concepts on which the development of decision support systems is based, namely data mining, ETL, OLAP, etc.

Literature provides several multidimensional approaches dealing with ontology analysis. Many approaches have been proposed by researchers for the design of a Data Warehouse from an ontology, such as [3–5]. However, most of them ignore both the conceptual level and the end-user's needs and do not incorporate the customization and verification of results by an expert. Consequently, these approaches may not meet the specific needs of the end user. On the other hand, they promote the propagation of anomalies from the data source (ontology) to the Data Warehouse. Therefore, they can lead to inaccurate or inadequate analysis. In addition, each approach proposed in these works is intended for saving the treatment of ontologies expressed in a specific language. In this case, global transformation is favored over customizable transformation. In fact, this way of proceeding can duplicate the inherent flaws of the data sources in the resulting Data Warehouse. There is partial satisfaction for the ontology analysis. These approaches suffer from certain limitations and may lead to inadequate analyzes because the loading of data from an ontology into a multidimensional schema can lead to facts related to unspecified or multiple dimensions or measurements. Not to mention that most of them do not offer tools ready for employment or with a low or even without a level of abstraction.

To be able to implement our global model for the integration of textual resources into the decision-making process [6] and to overcome the aforementioned limitations, we present in this contribution our approach for the development of a solution composed of two Eclipse Plug-ins allowing together the generation of a Data Warehouse model from an ontology. Indeed, this contribution is a part of the general context of our research work in the fields of Semantic Web and Business Intelligence. In particular, it is part of the "ETL and Data Warehousing" phase of the model proposed in [6].

2 Designing a Database from an Ontology

In our case, the conceptual model of a relational database (RDB) refers to the Entity-Relationship. In fact, a conceptual model of a RDB includes entities with attributes, relationships with or without attributes in addition to links between entities and relationships, and links between attributes and their entity or relationship. An entity is represented by a rectangle; a relationship is represented by lozenge; an attribute is represented by a rounded rectangle; an attribute forming part of the identifier of an entity is represented by a rounded rectangle with the underlining of the text; and finally, a link between an entity and a relation by a line labeled with cardinality. These symbols and their meanings are shown in Fig. 1.

As illustrated in Fig. 2, we use the Ecore graphical editor in order to specify the metamodel of a RDB conceptual model. We consider that a conceptual model of a RDB is composed of elements and links. Entities, relationships, attributes, and attributes that are part of an identifier are elements. There are two types of links: the links that link an attribute to its entity or relationship, and the links that link entities and relationships.

As illustrated in Fig. 3, the user interface comprises principally a graphic editor. The graphic editor displays the schema resulting from the automatic generation of a RDB conceptual model from an ontology and also allows users to edit this model by making modifications according to needs. The properties' view is used to edit the properties of a model and its elements (entities, relationships, attributes, identifiers, and links). Thus, the elements of a model are represented graphically at the editor level and their properties are manageable in the properties view. The editor has a palette that groups all the types of graphic elements for the design of RDB models,

Concept	Representation & Example	
Entity	Student	
Relationship	Lives in	
Attribute	Identifier (key)	Student-id
	Descriptor (non-key)	Student-name
Cardinality	0 to 1	0,1
	1 to 1	1,1
	0 to n	0,n
	1 to n	1,n
	n to m	n,m

Fig. 1 Basic ER model symbol

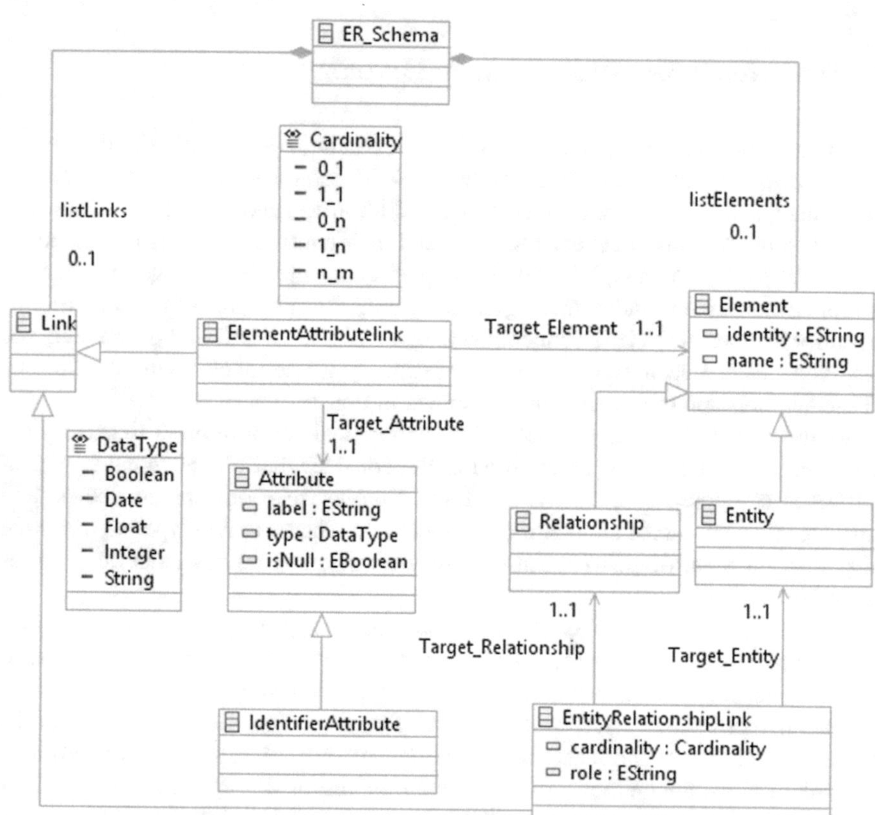

Fig. 2 Ontology to RDB plug-in metamodel

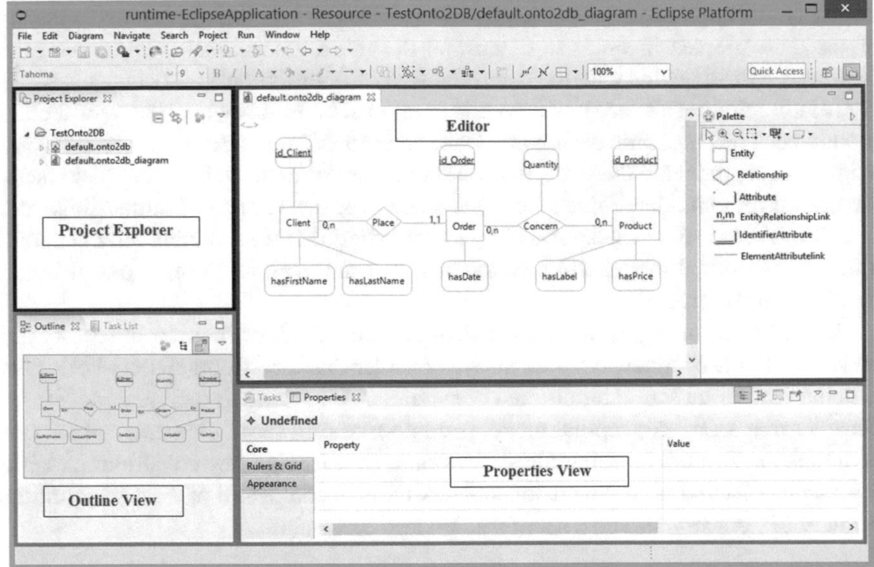

Fig. 3 Ontology to RDB plug-in preview

such as entity, relation, attribute, and link between attribute and its entity or relation. A user can therefore modify an automatically generated model from an ontology.

This Plug-in is a visual modeling environment for generating a conceptual model of a RDB from an ontology. It gives designers the ability to customize the generated conceptual model by interacting with graphical user interface widgets. The development of this Plug-in is based on the Eclipse platform because of the advantages such as its extensibility and the wide range of tools available as a framework. In fact, this Plug-in is implemented as an Eclipse Plug-in, adding capabilities to generate and visualize conceptual models of RDBs from ontologies, and manipulate these models using graphical user interface, such as entities, relationships, attributes. The stacking relationship between the Plug-in and Eclipse in a layered architecture. Indeed, it is based on the exploitation of the services offered by GMF.

3 Designing a Data Warehouse from a Database

Approaches to multidimensional modeling, according to Winter and Strauch [7], El Moukhi et al. [8], Romero and Abello [9] and Feki et al. [10], can be classified into three classes: demand-driven (also called requirement-driven or goal-driven), supply-driven (also called data-driven), and hybrid approaches of the two previous classes. The approaches that form part of the "demand-driven" class begin with the determination of the specific needs of the multidimensional model end users.

Inversely, the approaches of the supply-driven class begin with the detailed analysis of the data sources in order to determine the multidimensional model concepts. However, hybrid begins with the determination of multidimensional concepts while confronting them with the specific needs of end users. In this work, the hybrid class is adopted. The automatic design of a Data Warehouse from a RDB involves several steps. Among them are the search for candidate measures as well as candidate facts, the search for candidate dimensions, the search for hierarchies of dimensions, etc. In this work, we adopt the hybrid class to the extent that we automatically generate a multidimensional model and give the designer the hand to closely control it and intervene in the process.

As illustrated in Fig. 4, we use the Ecore graphical editor in order to specify the metamodel of DW conceptual model. We consider that a conceptual model of a DW is comprised of tables, columns, and constraints. According to this metamodel, we consider that a DW conceptual model is composed of a set of elements, which can be a table, a column, a primary key constraint, or a foreign key constraint. A table represents a fact table or dimension table. A table is comprised of a set of columns, primary key constraint, and a set of foreign key constraints.

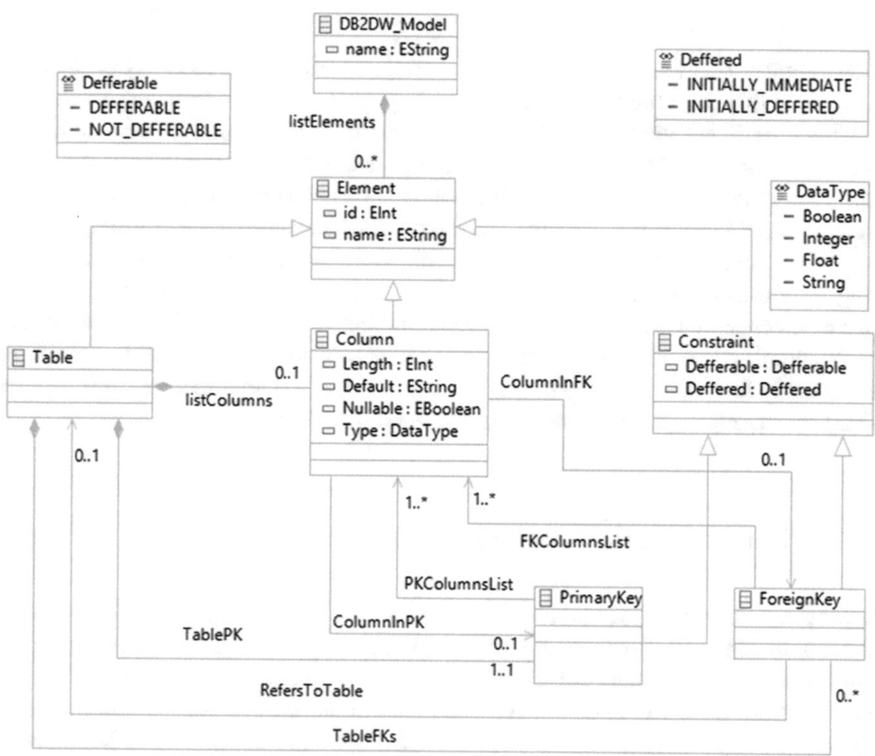

Fig. 4 DW plug-in metamodel

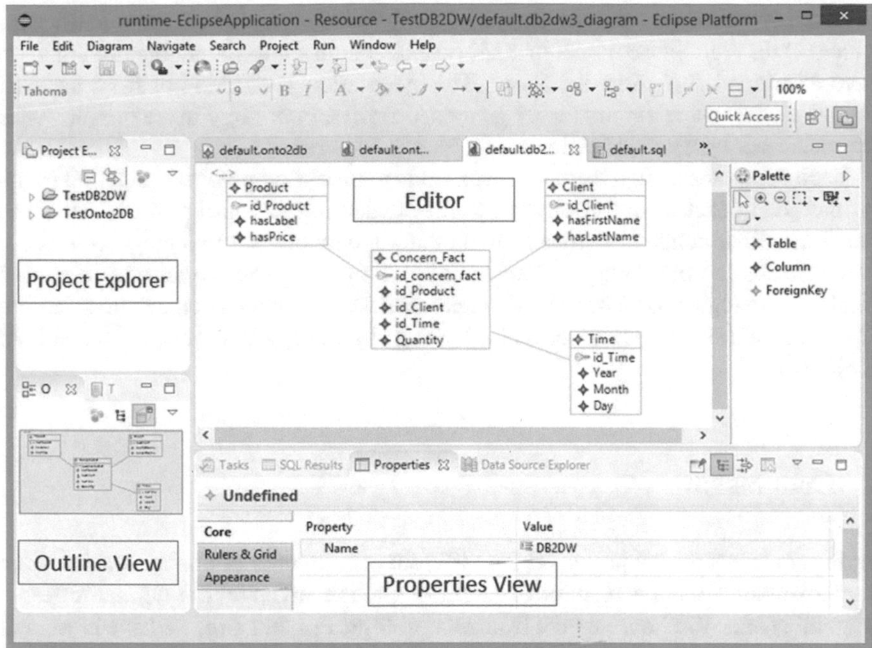

Fig. 5 RDB to Data Warehouse plug-in preview

As illustrated in Fig. 5, this Plug-in provides a visual modeling environment for generating a model of a Data Warehouse from a RDB model. It gives designers the ability to customize this generated model via the manipulation of graphical user interface widgets.

4 Model to Model

A model is a kind of instance of a metamodel like objects with respect to classes in the object-oriented programming language. In our case, a model of a RDB is an instance of the metamodel presented in Fig. 2 and a model of a Data Warehouse is an instance of the metamodel presented in Fig. 4. Automating the generation of a Data Warehouse model from a RDB model consists of automating the transformation of this RDB model to a Data Warehouse model. This transformation refers to the operation of the semantic translation of the model components into components of another model. To our knowledge, there are at least two types of languages in the Eclipse Modeling Project, namely Atlas Transformation Language (ATL) and Query/View/Transformation (QVT). The QVT language includes three sub-languages, namely QVT Operational Mapping Language (QVT OML or QVTO), QVT Relations Language, and QVT Core Language.

In our contribution, we have exploited the QVTO language to automatically generate a Data Warehouse model in XMI format according to the metamodel that we have developed and presented above. This generation is carried out from an XMI format model that is automatically generated from an ontology in accordance with the metamodel presented in Fig. 2.

Each automatic design method corresponds to the development of a QVTO transformation. In fact, a transformation is the realization of an algorithm. All these transformations constitute a means of integration between our two plug-ins. Indeed, they offer us the possibility of transition from a conceptual model of a relational database to a logic model of a Data Warehouse. These methods come from scientific literature such as Phipps and Davis [11], Song et al. [12], Nazri et al. [13], and Feki et al. [10].

5 Conclusion

In this contribution, we presented the different constituents of our approach for the implementation of our perception of a multi-approach model for the automatic generation of a Data Warehouse model from an ontology. For that aim, we have presented the developed Eclipse Plug-ins and their integration through QVTO transformations. This integration is a suite of tools for transposing the ontological model to the multidimensional model at the conceptual level, which offers a high level of abstraction. Our solution has the advantage of offering to the end users, in addition to automatic generation, the ability to customize generated models. In future work, we will improve the quality of development of these plug-ins and incorporate constraints using the Object Constraint Language (OCL), evaluate them on a large scale, and exploit them in real cases.

References

1. Inmon, W.H.: Building the Data Warehouse, 2nd edn, p. 401. Wiley, New York (1996)
2. Kimball, R., Ross, M.: The Data Warehouse Toolkit: The Definitive Guide to Dimensional Modeling. Wiley, London (2013)
3. Nebot, V., Berlanga, R.: Building data warehouses with semantic web data. Decis. Support Syst. **52**(4), 853–868 (2012)
4. Colazzo, D., Manolescu, I., Roatis, A., Roati, A.: Warehousing RDF Graphs To cite this version: Warehousing RDF Graphs*. Hal (2013)
5. Berkani, N., Bellatreche, L., Benatallah, B.: A Value Added Approach to Design BI Applications, vol. 9263, pp. 257–269 (2015)
6. Hajji, M., Qbadou, M., Mansouri, K.: Proposal for a new systemic approach of analitical processing of specific ontology to documentary resources: case of educational documents. J. Theor. Appl. Inf. Technol. **89**(2), 481–488 (2016)
7. Winter, R., Strauch, B.: A method for demand-driven information requirements analysis in data warehousing projects. In: Proceedings of the 36th Hawaii International Conference on System Sciences (HICSS'03), p. 231.1 (2003)

8. El Moukhi, N., El Azami, I., Mouloudi, A., El Mounadi, A.: Requirements-driven modeling for decision-making systems. In: 2018 International Conference on Electronics, Control, Optimization and Computer Science (ICECOCS), pp. 1–7 (2018)
9. Romero, O., Abello, A.: A survey of multidimensional modeling methodologies. Int. J. Data Warehouse. Min. **5**(2), 1–23 (2009)
10. Feki, J., Nabli, A., Ben-Abdallah, H., Gargouri, F.: An automatic data warehouse conceptual design approach. In: Encyclopedia of Data Warehousing and Mining, John Wang edn (2008)
11. Phipps, C., Davis, K.C.: Automating data warehouse conceptual schema design and evaluation. In: Proceedings of the International Workshop on Design and Management of Data Warehouses (DMDW'2002), vol. 58, pp. 23–32
12. Song, Y., Khare, R., Dai, B.: SAMSTAR: a semi-automated lexical method for generating star schemas from an entity-relationship diagram. In: Proceedings of the ACM Tenth International Workshop on Data Warehousing and OLAP, pp. 9–16 (2007)
13. Nazri, M.N.M., Noah, S.A.M., Hamid, Z.: Automatic data warehouse conceptual design. In: 2008 International Symposium on Information Technology, Kuala Lumpur, Malaysia, pp. 1–7 (2008)

Towards a Semantic Annotation of Pedagogical Unstructured Documents

Intissar Salhi, Hanaa El Fazazi, Mohammed Qbadou and Khalifa Mansouri

Abstract The proliferation of information in the form of unstructured documents has steadily increased in different areas. However, in the education sector, most e-learning systems do not include any mechanism that can support this diversity. The interest of ontologies in the field of education has been emphasized because it allows more efficient description and retrieval of content, thus facilitating exchange and sharing between teachers and learners. To achieve this goal, a scalable architecture for the educational field becomes paramount, based on the standardization of data to transform them into FAIR data using semantic annotations. In this paper, we examine key contributions related to the development and use of ontologies in e-learning systems. We provide a critical framework for these studies, explaining how ontologies can be used to describe learning objects, and we propose a new ontology that will annotate unstructured documents, and we will talk towards the end of our future work.

Keywords Unstructured documents · Standardization · E-learning · Ontology · Learning objects · Semantic annotations

I. Salhi (✉) · H. El Fazazi · M. Qbadou · K. Mansouri
Laboratory Signals, Distributed Systems and Artificial Intelligence ENSET, University Hassan II, Mohammedia, Morocco
e-mail: salhi9477@gmail.com

H. El Fazazi
e-mail: elfazazi.hanaa@gmail.com

M. Qbadou
e-mail: qbmedn7@gmail.com

K. Mansouri
e-mail: khmansouri@hotmail.com

V. Bhateja et al. (eds.), *Embedded Systems and Artificial Intelligence*,
Advances in Intelligent Systems and Computing 1076,
https://doi.org/10.1007/978-981-15-0947-6_59

1 Introduction

With the WWW, learners can simply open their browser and immerse themselves in a learning environment that allows them to gather the information they are looking for. In addition, this generic interface provides teachers with an effective and convenient way to host their teaching materials. Several formats have been proposed by various international organizations, including the Sharable Content Object Reference Model (SCORM). Given these standards, learning objects from different learning management systems (LMS) can not only be shared but also reused or even recombined. Among these international standards, we found The Advanced Distributed Learning SCORM 2004 that was designed to facilitate adaptive learning through declarative, rules-based descriptive means. Although training content is typically composed as directed graphs, it is essential to annotate them with metadata according to SCORM packages so that they can work well with other standard-compliant learning systems. In this regard, the W3C proposed standardizations of its annotations. The most interesting are the Resource Description Framework (RDF) and the RDF Schema, which are currently used together to exchange knowledge and represent information on the Web. However, these annotations are limited unless their specific meaning is normally understood.

Metadata is constructed from a set of ordered terms to describe learning objects (LO). They include a number of “descriptors” that make them more easily identifiable (accessible) and easier to handle (interoperable, reusable, durable, and adaptable) from a content repository. Ontologies can indeed help meet this requirement by providing “an explicit specification” of a particular area of interest conceptualization that can be easily manipulated by teachers because they represent a structured set of concepts organized in a graph whose relationships can be semantics and/or relationship of composition and inheritance (in the object sense). In addition, in terms of creation, learning objects are more likely to rely on conceptual models in the form of ontologies that offer new possibilities for individuals to move from object-oriented to content-oriented learning, but it is clear that learning objects are structured, interconnected, combined, and used, thus facilitating interaction between teachers. In this way, the objectives of the content framework of the teaching material can be achieved.

Semantic Web, which contains ontologies and relational metadata that can be processed by a machine, can greatly facilitate the construction of a course unit. The key to achieving the Semantic Web vision is the provision of metadata and the association of metadata with web resources. This process is called a semantic annotation.

In this study, we apply OWL language techniques to explore the best way to create Web ontology metadata aggregation for educational unstructured documents based on a specific content repository. We start with a review of the literature and follow with a discussion of the methodology that we used to create an ontology that we will integrate later in a process of annotations of unstructured documents and we concluded with some recommendations for future studies.

2 Literature Review

One of the disadvantages of traditional information retrieval methods is their limited ability to respond to the complex needs of users, especially on the Web, with this proliferation of raw data under different formats. To cope with this, several research works bring different contributions especially with the appearance of the Semantic Web. Several general-purpose tools have been developed to support the annotation process starting with preprocessing, query translator, annotation, indexing, content retrieval, and classification [1]. Research groups have also proposed ontologies and specific knowledge bases. Several studies explain how ontologies can be put into practice to describe learning objects and how these semantic techniques can facilitate the sharing and retrieval of learning resources in learning repositories, that we studied in order to be abled to compare them based on several aspects, namely the objectives that they wish to reach, the standards used, the disciplines taken into account, etc., in order to draw advantages, disadvantages, and limitations of each of them taking into account the annotation of documents. Table 1 presents the result of this study.

2.1 Critical Ontologies

In this part, we have chosen five ontologies that we have judged better since they take into account the annotation of the pedagogical objects. The five ontologies are described below:

PeOnto & PedaOnto: the authors proposed a conceptual framework for personalized education for English language learning as a second language (ESL) and they demonstrated the necessary attributes required in delivering personalized education services by integrating five interdependent ontologies: Target Audience Ontology, Language Ontology, Pathology Ontology, Ontology of Pedagogy, and Ontology of Objectives.

LOSON: the author focused mainly on the semantic aspects of the learning object repository for sharing pedagogical knowledge and showed how to support learning objects in the Semantic Web era. Therefore, he proposed an ontology that can practically be used for educational development software. Except that users must clearly understand the ontological structure of the repository of organizational learning objects to be able to transform unstructured actions into structured tasks by applying the repository of learning objects on the basis of their own knowledge a priori in order to develop their new knowledge.

Universia: the authors proposed an approach to classify learning material based on a set of categories automatically extracted from DBpedia. In order to connect the LOM textual fields describing the subjects of the LO contained in Universia digital library [14]. Except that in LOM the pedagogical objects are not central in the learning process but rather the activities associated with them. Consequently,

Table 1 Different ontologies proposed to annotate pedagogical objects and documents

Ontology	Objectives	Standard	Discipline	Annotating documents
PeOnto & PedaOnto [2]	Describes pedagogical approaches, pedagogical design procedures and relationships between pedagogical resources and activities and helps to identify the usability of various resources to discover teaching, learning preferences and styles	IMS SCORM LOM	English	Yes
SOLR [2]	An approach towards semantic LO repositories by designing a prototype using an ontology scheme for the description of its entities	LOM	ALL	No
LOSON [3]	Transform unstructured to structured tasks action by applying the learning object repository based on their own knowledge a priori	DCMI IMS	ALL	Yes
Universia [4]	Classification of educational objects according to a set of categories provided by Wikipedia via the DBpedia ontology	DBpedia LOM SKOS	ALL	Yes
MSLF [5]	Intelligent discovery of learning objects using Semantic Web technologies	LMS SCORM	ALL	No

(continued)

Table 1 (continued)

Ontology	Objectives	Standard	Discipline	Annotating documents
LOM Ontology [6]	Proposes a LOM ontology that models the LOM standard. They then create a "wizard" prototype that helps users to load metadata through automation	LOM	ALL	Yes
EduProgression [7]	Proposes ontologies built from official texts in French describing the curriculum and completed this ontology	LOM DBpedia	ALL (primary school)	Course
LOOR [8]	Provides a learning object ontology and explains how to integrate pedagogical objects with other semantic standards, namely SKOS, to facilitate their description and discovery	LOM SKOS DCMI	Medicine Math's	Yes
Wong et al. [9]	Development of a new ontology for knowledge management (KM) technologies and determination of the relationship between these technologies and their classification	None	ALL	Yes

(continued)

Table 1 (continued)

Ontology	Objectives	Standard	Discipline	Annotating documents
Pech et al. [10]	Proposes semantic annotation strategy for unstructured documents as part of a semantic search engine	None	ALL	Yes
Obeid et al. [11]	Proposes a semantic recommendation system for universities	None	ALL	No
Educlever [12]	Proposes and evaluates a semantic Web approach to support the functionalities and the interoperability of a real e-education system	LMS SCORM	ALL	Yes
ROO [13]	Publication of related data in radiation oncology using Semantic Web techniques and ontologies	None	Medicine	Yes

this calls into question certain aspects of the LOM, which is not very suitable for annotation, since objects cannot be defined a priori apart from their use in "learning units".

LOOR: the authors want to lead to semantification of teaching resources by exposing their metadata through ontologies and associating their educational resources with SKOS thematic terminologies to support integration with other discovery mechanisms, Digital repositories and the Web of linked open data (LOD). LOOR is based on standard thematic taxonomies and well-known semantic standards that combines terminology with the specification of Dublin Core metadata terms (DCMI). This allowed them to retrieve objects from additional support, perform federated searches, and collect material from different online repositories and benefit from the use of knowledge in the form of SKOS vocabularies maintained by curators and librarians to improve content annotations and queries.

ROO: the authors successfully demonstrated that it was possible to convert clinical data according to the principles of FAIR using the combination of ontologies and Semantic Web technologies and showed how Semantic Web technologies based on ontologies developed allow to easily and efficiently query data from relational data sources without knowing a priori their structures except that this ontology does not address the pedagogical field but this technique could be beneficial for us if we adapt it to the educational field.

3 Proposed Architecture

In this section, we present a new semantic annotation approach based on ontologies for to improve information retrieval in unstructured pedagogical documents.

3.1 XLOMAPI Ontology

Several models of metadata have been proposed to describe digital documents. Among these models, we found the DCMI [15] which responds, through its fifteen elements, to a description of a digital resources in a simple and standardized way. However, there are much more precise needs that require a more detailed description. The LOM [16] finds its place as a metadata model for the description of resources for educational purposes. It defines a schema consisting of nine categories to represent metadata. SCORM is a second language developed by the AICC and then taken over by the ADLNeT consortium [17]. These proposals mainly concern online content. It was to make training totally free through the Web. SCORM enriches the LOM standard with a somewhat different aggregation model and a runtime environment that allows to monitor the activity of a learner in an LMS.

Critics of the SCORM standard recognize that this is by far the most widely used standard for specifying pedagogical metadata, but it also lists a number of shortcomings, including the fact that it is not possible to collect certain data on learners' learning experiences, that SCORM content has to work in a browser and therefore, it is impossible to follow the activities of learners offline. Following this, ADL realized that the specification did not take into account technological advances, and that it needed to be updated. So, he proposed a new comprehensive xAPI model that stores and retrieves data (mobile apps, games, ITS experiments, and virtual worlds, team/group activities) not just the courses.

In this study, we have proposed a new XLOMAPI ontology which will allow us to annotate pedagogical documents. To this end, we have been required to translate the LOM specification into an ontology based on the ontology built by Rajabi et al. [18]. As well as XAPI with some modifications to deal with limitations of LOM and to better annotate documents. To do this, we followed the methodology below:

1. Add two fields (SingleRequirement, MultiRequirement) describing the conditions that must be used for pedagogical object. Since LOM stores all knowledge but does not really specify the prerequisites that must be used to take into account this OL.
2. Place the langString field among the categories to process the multilingual.
3. Similarly for Taxon and TaxonPath, to highlight the classification of pedagogical objects from the general to the specific.
4. Match the modified LOM ontology with that of XAPI ontology to have visibility on the course and activity of the student either online or offline.

Figures 1 and 2 show the structure of the resulting Ontology XLOMAPI under Protege and WebVOWL [19].

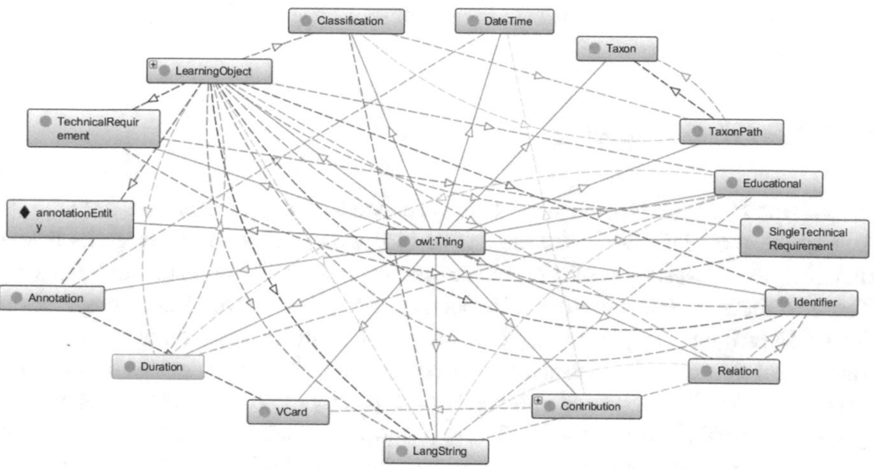

Fig. 1 XLOMAPI ontology under protege

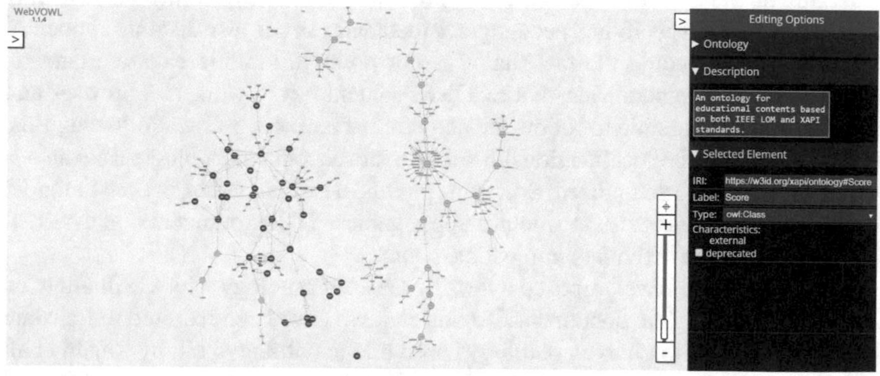

Fig. 2 XLOMAPI ontology under WebVOWL

3.2 The Proposed Model for the Annotation of Unstructured Teaching Documents

Figure 3 shows an overview of our proposed solution for semantic annotation.

Pretreatment of documents: divides texts into smaller units (tokens), filters them, performs normalization (Tokinization, lemmatization), creates n-grams tokens, and associates them with corresponding grammatical information as part of the speech (POS). The steps of the analysis are applied sequentially.
Indexing of documents: a generation of an inverted index in the form of a knowledge base containing all the terms to be compared to ontology entities.
Semantic extraction of instances: extraction of textual descriptions from the document and the knowledge base. When an entity is identified (has the highest frequency), a check is made to see if it exists as an instance in the knowledge base and the domain ontology, as well as its variants, its common abbreviations, and its synonyms.
Disambiguation of entities: it is possible that there are several references in the knowledge base for an identified entity, so here, it is necessary to identify different references corresponding to the same entity of the real world. Disambiguation of entities is crucial for basic functionality such as database/ontology integration. There is a multitude of approaches to disambiguation of entities depending on the nature of the data source and the level of precision required. References [20, 21, 22] represent a small sample of the literature.
Semantic annotation: after the disambiguation of the entities (in the presence of ambiguities), comes the step of annotation which consists of associating semantic metadata with the entities of the document via the annotation process. Usually intended for use by users and agents.

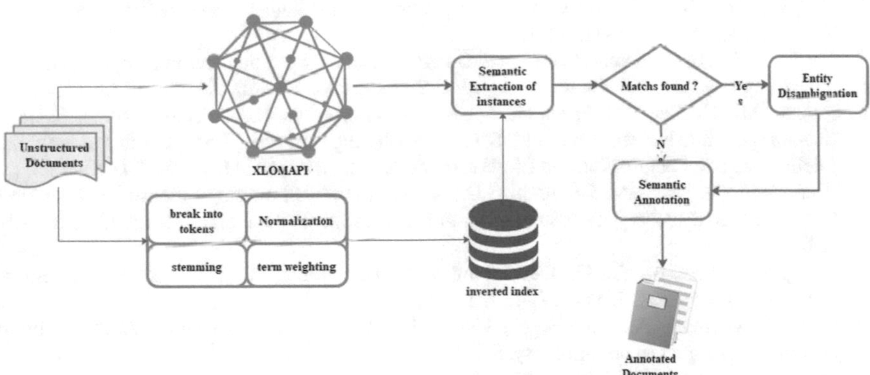

Fig. 3 Pipeline of semantic annotation process using XLOMAPI ontology

4 Conclusion

During this study, we found that it is not the pedagogical objects that are central in the learning process but the activities associated with them. Consequently, this calls into question certain aspects in the standards, namely LOM which is not very adapted to this representation, since objects cannot be defined a priori apart from their use in "learning units". The aim was to provide models adapted to the pedagogical design of diversified learning situations and to annotate unstructured educational documents. To deal with this, we presented a number of ontologies that we considered critical for the semantic annotation and after evaluation and identification of several limitations, we released a new ontology XLOMAPI that we integrated in a process of semantic annotations. In our future work, we want to test our ontology and then use the semantic annotation of unstructured educational documents to feed a pedagogical recommendation system to personalize e-learning.

References

1. Wei, W., Barnaghi, P.: The anatomy and design of a semantic search engine. Int. J. Commun. SIWN **3**(March), 76–82 (2007)
2. Fok, A.W.P., Ip, H.H.S.: Educational ontologies construction for personalized learning on the web. Stud. Comput. Intell. **62**, 47–82 (2007)
3. Wang, S.: Ontology of learning objects repository for pedagogical knowledge sharing. Interdiscip. J. e-Skills Lifelong Learn. **4**, 001–012 (2017)
4. Lama, M., et al.: International Forum of Educational Technology & Society Semantic Linking of Learning Object Repositories to DBpedia Published by: International Forum of Educational Technology & Society Linked references are available on JSTOR for this article: Semantic, vol. 15, no. 4 (2012)
5. Hsu, I.C.: Intelligent discovery for learning objects using semantic web technologies. Educ. Technol. Soc. **15**(1), 298–312 (2012)
6. Casali, A., Deco, C., Romano, A., Tomé, G.: An assistant for loading learning object metadata: an ontology based approach. Interdiscip. J. e-Skills Lifelong Learn. **9**, 077–087 (2013)
7. Rocha, O.R., Zucker, C.F., Pelap, G.F.: A formalization of the French elementary school curricula. In: Lecture Notes in Computer Science (including Subseries. Lecture Notes in Artificial Intelligence and Lecture Notes in Bioinformatics), vol. 10180 LNAI, pp. 82–94 (2016)
8. Koutsomitropoulos, D.A., Solomou, G.D.: A learning object ontology repository to support annotation and discovery of educational resources using semantic thesauri. IFLA J. **44**(1), 4–22 (2017)
9. Wong, M., Wong, S., Ke, G.: Developing a domain ontology for knowledge management technologies. Asia Pac. J. Mark. Logist. 1–55 (2016)
10. Pech, F., Martinez, A., Estrada, H., Hernandez, Y.: Semantic annotation of unstructured documents using concepts similarity. Sci. Program. **2017**, 1–10 (2017)
11. Obeid, C., Lahoud, I., El Khoury, H., Champin, P.-A.: Ontology-based recommender system in higher education, vol. 2, pp. 1031–1034 (2018)
12. Pelap, G.F., Zucker, C.F., Gandon, F.: Semantic Models in Web Based Educational System Integration, pp. 78–89 (2018)
13. Traverso, A., van Soest, J., Wee, L., Dekker, A.: The radiation oncology ontology (ROO): publishing linked data in radiation oncology using semantic web and ontology techniques. Med. Phys. **45**(10), e854–e862 (2018)

14. "Universia.net." [Online]. Available: https://www.universia.net/. Accessed: 14 Mar 2019
15. DublinCore: Dublin core metadata initiative encoding guidelines, DCMI (2010) [Online]. Available: http://dublincore.org/
16. Duval, E.: Learning Object Metadata, DCC Digital Curation Manual, no. 05-06-2007 (2007)
17. ADL: The Advanced Distributed Learning Initiative [Online]. Available: https://adlnet.gov/scorm. Accessed: 19 Mar 2019
18. Rajabi, E., Sicilia, M.-A., Ebner, H., Palmer, M., Sanchez, S.: Recommendation on Exposing IEEE LOM as Linked Data 1.0 (second version). ODS Recommendation Draft (2014) [Online]. Available: http://data.opendiscoveryspace.eu/ODS_LOM2LD/ODS_SecondDraft.html. Accessed: 21 Mar 2019
19. Negru, S., Lohmann, S., Haag, F.: WebVOWL, no. January 2013, pp. 1–9 (2013)
20. Kalashnikov, D.V., Mehrotra, S.: A probabilistic model for entity disambiguation using relationships* (2004)
21. Chai, M., Li, D., Zhuang, T., Yang, S.: Named entity disambiguation based on classified and structural semantic relatedness. Chin. J. Electron. **27**(6), 1176–1182 (2018)
22. Karimzadeh, M., Pezanowski, S., MacEachren, A.M., Wallgrün, J.O.: GeoTxt: a scalable geoparsing system for unstructured text geolocation. Trans. GIS **23**(1), 118–136 (2019)

A Follow-Up Survey of Audiovisual Speech Integration Strategies

Ilham Addarrazi, Hassan Satori and Khalid Satori

Abstract The automatic speech recognition (ASR) systems benefit from visual modality to improve its performance especially in noisy environments. By combining acoustic features with the visual features, audiovisual speech recognition (AVSR) system could be implemented. This paper presents a review on various existing and recent techniques for AVSR. A special emphasis was placed on recent AVSR system fusion technique, where the AVSR systems fusion stages (early, intermediate and late integration) are discussed with their corresponding models. The aim of this study is to discuss different AVSR approach and compare the existing AVSR techniques.

Keywords Audiovisual speech recognition · Automatic speech recognition · Lip reading · Late integration · Early integration · Features extraction

1 Introduction

Human speech perception is naturally a bimodal process that combines information coming from different sensory modalities to decide what has been spoken. According to the McGurk effect [1], the recognition of speech can be determined by both auditory and visual signals. Audiovisual speech recognition (AVSR) systems aim to assume the bimodal nature of human speech perception by integrating audio and visual information. The purpose of this integration is to improve robustness of automatic speech recognition systems especially in difficult condition. When an audio cue is degraded by acoustic noise, visual information such as speaker's lip movement enhances the accuracy of speech recognition.

Traditional AVSR systems consist of two stages, feature extraction from each modality (image and audio signals) and the combination of the information delivered by modalities. A key open challenge is to find effective approaches to combine audio and visual information because the performance of this system depends on fusion strategies. Our goal in this work is to describe the integration problems by considering

I. Addarrazi (✉) · H. Satori · K. Satori
Department of Mathematics and Computer Science FSDM, University of Sidi Mohamed Ben Abdllah, Fez, Morocco

V. Bhateja et al. (eds.), *Embedded Systems and Artificial Intelligence*, Advances in Intelligent Systems and Computing 1076,
https://doi.org/10.1007/978-981-15-0947-6_60

the exiting audiovisual systems. In this survey, we first give detailed description of AVSR steps. We also introduce the main concepts and review recent work on the challenging AVSR information fusion strategies. There are a number of review works on the multimodal fusion technique (i.e., [2–4]). In particular, we concentrate on fusion for the AVSR system.

The paper is organized as follows. Section 2 presents the architecture of AVSR system. Section 3 gives an overview of fusion methods. A summary on audiovisual system integration is presented in Sect. 4.

2 Audiovisual Speech Recognition System Processing

The concept of AVSR system is based on use of both acoustic and visual information during speech to perform the recognition task [5]. As it was briefly mentioned in the introduction, AVSR system architecture consists of two components: the multimodal features extraction and the multimodal integration. Firstly, the signals coming from different multimodal input (audio and video channels) are analyzed to extract the significant multimodal features. Therefore, the audio features are extracted from audio signal, and the visual features are extracted from the region of the speaker's mouth in video cues. Finally, the integration of the two streams signify that the audio and visual information are joint conjointly in order to attain the recognized word.

2.1 Multimodal Feature Extraction

Toward catching a signal for each spoken word, it is necessary to determine an appropriate method to extract the most relevant features. There are many features that can be used to represent audio and visual information. Audio feature extraction is an operation that aims to convert audio signal to a set of features. Those features contain the meaningful information from audio data in order to get a higher-level understanding of the audio. Many feature extraction techniques are available such as Mel-frequency cepstral coefficients (MFCC) [6–8], linear predictive coding (LPC) [9], perceptual linear prediction (PLP) [10] and relative spectra filtering of log domain coefficients (RASTA) [9]. These techniques have their own strengths and drawbacks in extracting the audio signals.

However, the extraction of visual feature from region mouth for speech recognition is a critical step in AVSR. Three main types of features are used for visual speech recognition:

- **The appearance-based visual features** that include image transformation-based feature extraction methods such as discrete cosine transform (DCT), discrete wavelet transform (DWT) and principal component analysis (PCA) [11]. These algorithms are usually applied to dimensionality reduction and to eliminate redundant data.

- **The geometric-based approaches** that aim to geometric parameters from the mouth region such as the mouth shape, height, width and area. There are many methods that are based on this concept like active shape models (ASM) [12], snakes [13] or AAM [14].
- **Hybrid approaches** are the combination of the appearance-based and the geometric-based approaches.

2.2 *Multimodal Integration Categories*

The actual audiovisual recognition systems combine the auditory and visual information from the acoustic in various forms. However, the more important issue is how to fuse these two modalities' information. Three main integration models have been described in the literature: early integration, intermediate integration and late integration.

Early integration: In the early integration EI (also called features fusion), thc fusion is realized after the unimodal features extraction process. Figure 1 displays this integration strategy.

Once the features of each modality (audio and visual) are extracted, the vectors of audio features $O_{a;t}$ and video features $O_{v;t}$, with dimension d_a and d_v, respectively, are combined at each instant t to obtain a single combined audiovisual vector.

$$O_{av;t} = \left[O_{a;t};\, O_{v;t}\right] \in R^D \tag{1}$$

where

$$D = d_a + d_v \tag{2}$$

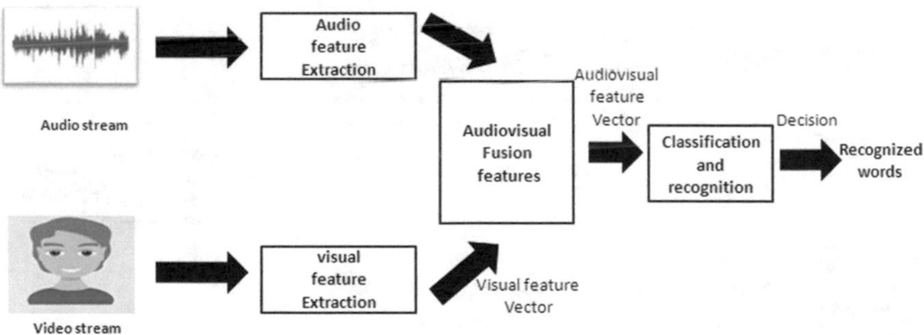

Fig. 1 Early integration

This combined vector can be used in the classification and recognition stage to obtain the final result. Moreover, the dimension of the audiovisual vector (O_{av}) is large. Thus, the recognizer needs a large number of parameters which make the modeling process difficult. To overcome this challenge, some feature reduction techniques such as principal component analysis (PCA) and linear discriminant analysis (LDA) are used. The implementation of early integration is simple, and the synchronization of the acoustic and visual features is not required. For that reasons, this type of fusion is widely used in the literature. Nevertheless, the main disadvantage of early integration is that the modalities features can be incompatible, and the corrupt of acoustic or visual modalities can influence on the entire speech modalities.

Intermediate integration: In intermediate integration (called also classifier-level fusion), the information from each modality is combined inside the classifier using separated audio and visual cues. In comparison with the feature fusion level, which does not distinguish between the modalities features, the output of the intermediate integration includes sufficient amount information to differentiate between the features from audio and visual modalities. Consequently, to treat the audio and visual streams, a composite classifier is used. Considering the correlation between the two modalities, a modeling process is done (Fig. 2).

The ability of a classifier-level fusion to treat the asynchrony between the two modalities (audio and visual) features is an advantage. In contrast, the main disadvantage of this type of fusion is that the modeling techniques had to be built particularly. Thus, there is limitation in modeling process.

Late integration: In late Integration (LI) (or what is called decision fusion), each modality can be trained independently. Usually, an audiovisual speech recognition system based on the decision fusion performs as follows: each subsystem (audio only and visual only) accepts, as input, the extracted features to produce a decision score. Subsequently, these decisions are integrated to get the final decision for recognition (D) (Fig. 3).

The straightforward techniques S such as weighting [15], summation [16] and voting [17] are used to fuse the decision values. Suppose that the models M_a and M_v are used on audio and video modalities, respectively:

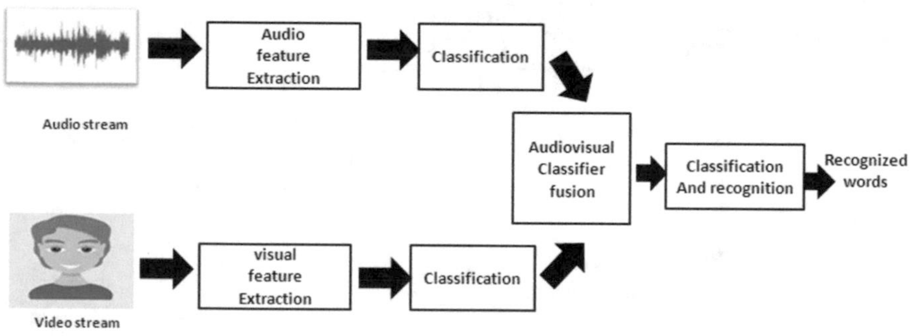

Fig. 2 Intermediate integration

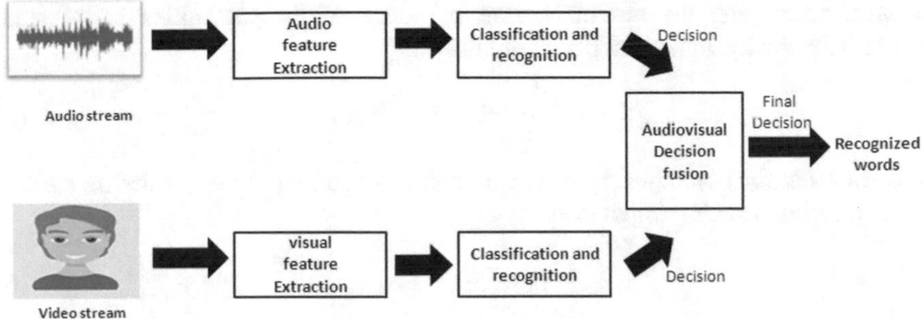

Fig. 3 Late integration

$$D = S(M_a(O_a), M_v(O_v)) \tag{3}$$

In this kind of integration, any dependencies revolving between the two modalities are absent. Moreover, the decision outputs of each modality have the same representation which makes the combination easier than in features fusion. Therefore, the systems based on late integration use recognizers that offer scalability in the number of inputs modes and vocabulary. Another advantage of this strategy is that it is possible to use different methods for modeling each modality (i.e., hidden Markov model (HMM) for audio modality and neural network models (NNM) for visual modality). Contrarily, the main disadvantage of the LI scheme is that the modalities features are not correlated. Additionally, the final decision is based on multi classifiers, which the training of each of them needs time.

3 Overview of Fusion Methods

3.1 Rule-Based Methods

The rule-based fusion method includes a variety of basic rules of combining multimodal information. Suppose that S_f is the fused score, and S_{MA} and S_{MV} are the scores of audio and visual modalities, respectively:

- **AND, OR fusion**: in AND fusion, an accepted decision is done when all classifiers agree, whereas in OR fusion, a positive decision is made as soon as one of the classifiers makes an acceptance decision.
- **Maximum rule**: the maximal value of score of the modalities is selected. Mathematically, maximum rule is defined as

$$S_f = \max(S_{MA}, S_{MV}) \tag{4}$$

- **Minimum rule**: the minimal value of score of the modalities is selected. Mathematically, minimum rule is defined as

$$S_f = \min(S_{MA}, S_{MV}) \tag{5}$$

- **Sum rule**: the combined score is calculated by adding values of the modalities scores. Sum rule is computed as

$$S_f = S_{MA} + S_{MV} \tag{6}$$

- **Product rule**: the combined score is calculated by multiplying values of the modalities scores. Product rule is computed as

$$S_f = S_{MA} * S_{MV} \tag{7}$$

- **Majority voting rule**: the final decision is obtained when the most of the classifiers achieve the similar decision. In case of two-class classification task (audio and visual), the number of classifiers should be odd and more than two.
- **Adaptive weighting**: firstly, each of the audio or the visual modality score is multiplied by its weight. Secondly, the multiplications, obtained in first, for each modality are added jointly. Mathematically, adaptive weighting is described as

$$S_f = (W_{MA} * S_{MA}) + (W_{MV} * S_{MV}) \tag{8}$$

where W_{MA} and W_{MV} are the weights of audio and visual modalities, respectively:

$$W_{MA} + W_{MV} = 1 \tag{9}$$

3.2 *Classification Fusion Methods*

Hidden Markov Models

Hidden Markov models (HMMs) are statistical tools used to model tasks such as speech recognition and video processing. As shown in Fig. 4, early, intermediate or late integration schemes can be used to combine the audio and visual cues using HMMs. In Fig. 4, S_1, S_2, S_3, S_4 define the hidden state variables, and O_a, O_v and $O_{a,v}$ represent, respectively, the sequence of audio, visual and audiovisual feature vectors. For the early integration, the modeling process is established using a single HMM with combined audiovisual vectors (Fig. 4a). Against this background, in late integration, two single HMM were used for modeling process with each HMM model corresponds to one of the modalities. The outputs of each HMM are fused (Fig. 4b). Most of intermediate integration is realized using coupled hidden Markov models

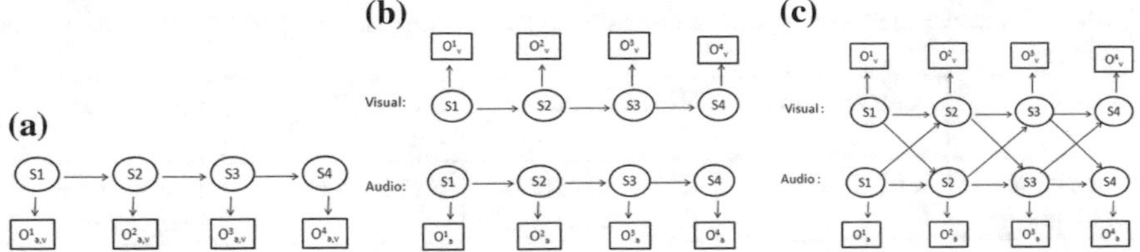

Fig. 4 HMM topologies in: **a** early integration; **b** late integration; **c** intermediate integration using CHMM

(CHMM). As shown in Fig. 4c, a CHMM couples two HMMs, one for each modality streams, with temporal, asymmetric conditional probabilities [18].

Support Vector Machines
Support vector machines (SVMs) are the most popular classification method. To be more precise, SVMs can be used in all steps of the audiovisual speech recognition, including features classification, face detection, audiovisual integration, etc. Basically, the features vector (the combination of audio and visual features) obtained from early integration will be used for SVM classification to achieve the recognized words. However, SVMs have also been used for late integration by modeling each modality separately, and the outputs are integrated by another SVM.

Deep Neural Networks (DNN)
Actually, various audiovisual speech recognition systems are created through diverse neural network-based architectures. Deep neural networks (DNN) are a type of neural networks that characterized by an input layer, hidden layers (at least one) and an output layer. DNN can be classified into convolution neural network (CNN), multilayer perceptron (MLP).

In the early fusion approach, the extracted audio and visual features are joined at the input level. The concatenated features become the input vector to a DNN. In late integration, the DNNs unimodal decisions are combined for modeling the audio and visual modalities.

Summary on Audiovisual Systems Integration
Table 1 shows a list of audiovisual speech recognition applications.

4 Conclusion

We have surveyed the recent studies and works in AVSR fusion techniques published in the literature. In addition, we have compared the characterization and benefits of the fusion schemes: early, intermediate and late integration. Regarding early integration, different techniques have been presented to fuse the audio and visual features to obtain

Table 1 Various audiovisual speech recognition systems using different fusion level and fusion technique

System	Audio features	Visual features	Fusion techniques	Fusion level
Subashini et al. [19]	MFCC	Color histogram coefficients	SVM-AANN	Intermediate integration
Paleček et al. [14]	MFCC	AAM	HMM	Intermediate integration
Ibrahim et al. [20]	MFCC	Lip geometrical-based	HMM	Early integration
Makhlouf et al. [15]	RASTA-PLP	DCT	GA/HMM	Late integration
Chelali et al. [21]	MFCC + PLP	DCT + DWT	ANN (MLP + RBF)	Early integration
Rahmani et al. [22]	MFCC	DBNF	DNN-HMM	Early integration

the best result. When it came to late integration, we have discussed the methods used to combine the decisions provided by the audio and visual modalities.

Practically, we will compare the early, intermediate and late integration by using different visual and acoustic modalities.

References

1. McGurk, H., MacDonald, J.: Hearing lips and seeing voices. Nature **264**, 746 (1976)
2. Aleksic, P.S., Katsaggelos, A.K.: Audio-visual biometrics. Proc. IEEE **94**, 2025–2044 (2006)
3. Atrey, P.K., Hossain, M.A., El Saddik, A., Kankanhalli, M.S.: Multimodal fusion for multimedia analysis: a survey. Multimedia Syst. **16**, 345–379 (2010)
4. Katsaggelos, A.K., Bahaadini, S., Molina, R.: Audiovisual fusion: challenges and new approaches. Proc. IEEE **103**, 1635–1653 (2015)
5. Addarrazi, I., Satori, H., Satori, K.: Amazigh audiovisual speech recognition system design. In: 2017 Intelligent Systems and Computer Vision (ISCV), pp. 1–5. IEEE (2017)
6. Satori, H., El Haoussi, F.: Investigation Amazigh speech recognition using CMU tools. Int. J. Speech Technol. **17**, 235–243 (2014)
7. Satori, H., Zealouk, O., Satori, K., ElHaoussi, F.: Voice comparison between smokers and non-smokers using HMM speech recognition system. Int. J. Speech Technol. **20**, 771–777 (2017)
8. Zealouk, O., Satori, H., Hamidi, M., Laaidi, N., Satori, K.: Vocal parameters analysis of smoker using Amazigh language. Int. J. Speech Technol. **21**, 85–91 (2018)
9. Gupta, K., Gupta, D.: An analysis on LPC, RASTA and MFCC techniques in automatic speech recognition system. In: 2016 6th International Conference-Cloud System and Big Data Engineering (Confluence), pp. 493–497. IEEE (2016)
10. Dave, N.: Feature extraction methods LPC, PLP and MFCC in speech recognition. Int. J. Adv. Res. Eng. Technol. **1**, 1–4 (2013)
11. Upadhyaya, P., Farooq, O., Abidi, M.R., Varshney, P.: Comparative study of visual feature for bimodal Hindi speech recognition. Arch. Acoust. **40**, 609–619 (2015)

12. Morade, S.S., Patnaik, S.: A novel lip reading algorithm by using localized ACM and HMM: tested for digit recognition. Optik **125**, 5181–5186 (2014)
13. Aleksic, P.S., Williams, J.J., Wu, Z., Katsaggelos, A.K.: Audio-visual continuous speech recognition using MPEG-4 compliant visual features. In: Proceedings. International Conference on Image Processing, vol. 1, pp. I–I. IEEE (2002)
14. Paleček, K., Chaloupka, J.: Audio-visual speech recognition in noisy audio environments. In: 2013 36th International Conference on Telecommunications and Signal Processing (TSP), pp. 484–487. IEEE (2013)
15. Makhlouf, A., Lazli, L., Bensaker, B.: Evolutionary structure of hidden Markov models for audio-visual Arabic speech recognition. Int. J. Signal Imaging Syst. Eng. **9**, 55–66 (2016)
16. Lucey, S., Chen, T., Sridharan, S., Chandran, V.: Integration strategies for audio-visual speech processing: applied to text-dependent speaker recognition. IEEE Trans. Multimedia **7**, 495–506 (2005)
17. Sanderson, C., Paliwal, K.K.: Information fusion and person verification using speech and face information. Research Paper IDIAP-RR, pp. 02–33 (2002)
18. Amarnag, S., Gurbuz, S., Patterson, E., Gowdy, J.N.: Audio-visual speech integration using coupled hidden markov models for continuous speech recognition. In: Student Forum Paper at ICASSP (2003)
19. Subashini, K., Palanivel, S., Ramalingam, V.: Audio-video based classification using SVM and AANN. Int. J. Comput. Appl. **44**(6), 33–39 (2012)
20. Ibrahim, M.Z., Mulvaney, D.J., Abas, M.F.: Feature-fusion based audio-visual speech recognition using lip geometry features in noisy enviroment. ARPN J. Eng. Appl. Sci. **10**, 17521–17527 (2015)
21. Chelali, F., Djeradi, A.: Audiovisual speaker identification based on lip and speech modalities. Int. Arab J. Inf. Technol. (IAJIT) **14** (2017)
22. Rahmani, M.H., Almasganj, F., Seyyedsalehi, S.A.: Audio-visual feature fusion via deep neural networks for automatic speech recognition. Digit. Signal Proc. **82**, 54–63 (2018)

Handwritten Text Segmentation Approach in Historical Arabic Documents

Noureddine El Makhfi

Abstract The sharp increase in the number of historical documents available as images in national libraries only increases their storage capacity. The need for access to content of this cultural heritage is also increasing. The methods of indexing and searching in image content are still very limited. Text segmentation of digital historical documents is an important step in recognizing content as images. In this paper, we present an original method of segmentation of lines and pseudo-words in historical documents based on Gaussian filters. Our method consists in detecting elliptical blobs in scales formed by Gaussian filters. Experimental tests are performed on hundreds of pages of historical documents. The experimental results showed with this method are excellent in front of other methods of text segmentation in manuscript images.

Keywords Arabic manuscripts · Cultural heritage · Digital libraries · Gaussian filters · Historical documents · Image/Text · Segmentation · Recognition

1 Introduction

The automatic extraction process of regions of interest is called segmentation [1, 2]. It is the most important process in computer vision [3]. Segmenting an image containing Arabic handwritten text is not a simple process, unlike the segmentation of printed text that has intrinsic properties of regions easily identified by segmentation algorithms. Most of the existing research deals with the word segmentation of Arabic printed text or the handwritten text containing regular spacing between these words. In general, in the case of handwriting Arabic, two consecutive lines are close and touch. Fluctuations in the baseline result in overlapping. The massive

N. El Makhfi (✉)
Laboratory of Physics, Mathematics, Computer Science and Communication, Department of Mathematics and Computer Science, Faculty of Science and Technology Al-Hoceima, Abdelmalek Essaadi University, Tetouan, Morocco
e-mail: n.elmakhfi@gmail.com

V. Bhateja et al. (eds.), *Embedded Systems and Artificial Intelligence*, Advances in Intelligent Systems and Computing 1076,
https://doi.org/10.1007/978-981-15-0947-6_61

presence of diacritical symbols, exaggerated closeness and the possibility of contacts between lines make it impossible to segment Arabic handwritten text. In addition to the problem of words segmentation, the cursive nature of Arabic writing presents an obstacle for segmentation.

2 Related Works

Most studies on the segmentation of the cursive writing of a page in lines are based on the segmentation of the binary images. Only the scale-space segmentation method [4] and the projection method use grayscale images. The most common segmentation process is used in the literature:

Segmentation of lines is based on the related components. The related components are grouped in alignment. Conflicts due to overlapping lines are solved by local analysis.

In the case of line and word segmentation in Latin manuscript images, the best-known algorithms are the Run Length Smoothing [5, 6] and Hough transform [8, 9] algorithms.

Segmentation of lines and words is based on the projection [7, 9, 10]. This method is also used by OCR for the segmentation of printed text. Note that this method is applicable for binary images using Otsu thresholding in gray levels. Despite the existence of a method [11] which is based on the analysis of local minima and models derived from maxima of the vertical projection profile of the image in gray levels, it cannot be applicable in the case of Arabic manuscripts. The performance of this approach can be seriously affected by the contacts of text lines as well as by the existence of distorted or asymmetric text lines.

The authors of [11] propose an image analysis based on the scale space technique in order to segment words in images of Latin manuscripts. In this context, a recent study [12] highlighted the segmentation of multi-font text in Arabic printed images. However, in the case of Arabic manuscripts, these methods are not applicable because of overlapping characters.

3 Proposed Methodology

In this paper, we present a method of text segmentation for manuscripts from image. Our method is to select lines using the elliptical blobs detected in the image. These elliptical blobs are generated horizontally at a given scale in order to aggregate the appearance of text lines. After that, we use the horizontal projection histogram to favor the appearance of lines. With thresholding, we detect the start and end of lines. The generation of blobs is therefore done by smoothing the image with the convolution of a Gaussian. The Laplacian of the Gaussian of the image makes it possible to bring out these blobs. In a final step, we exploit these blobs which serve

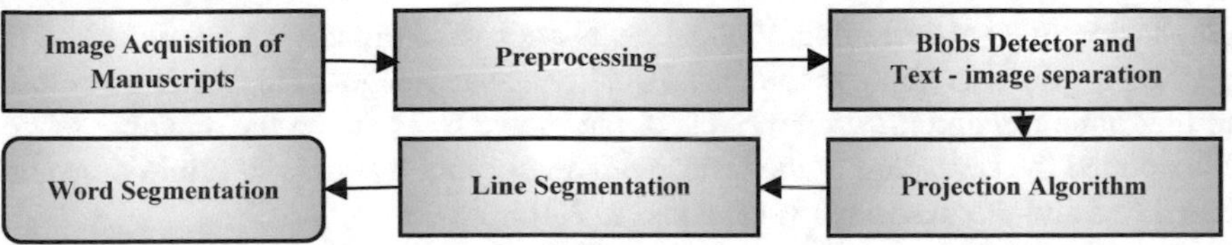

Fig. 1 Block diagram of the proposed method

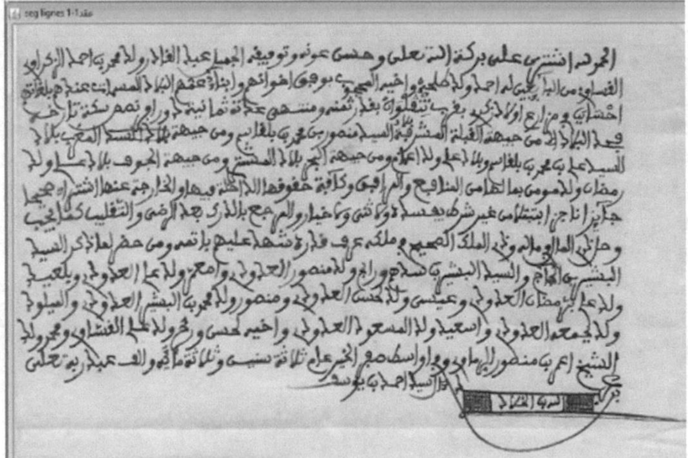

Fig. 2 Original image of an Arabic manuscript dated (1303 Hijri-1881)

as related components. We frame the blobs in rectangles of interest to deduce the locations of the words in the image. Figure 1 illustrates the synoptic diagram of the proposed method for text segmentation.

3.1 Acquisition

Digitization is the first step when that allows the conversion of a manuscript into digital images. We used a professional scanner [13] to scan our manuscripts in different formats and sizes. Scanned images are stored in brut format in folders. This gives very voluminous images. The scanner integrates algorithms such as compression to reduce the capacity of the images. It also incorporates calibration and resolution settings (Fig. 2).

3.2 Preprocessing

Manuscripts scanned and stored in brut format images require a series of preprocessing such as contrast enhancement, straightening, curvature correction, detail emphasis and level spreading. Pretreatments are image processing operations to be applied to pages to improve the visual quality of the images. This function adjusts to the minimum and maximum levels of the image to optimize the dynamic range. We applied the spreading of the levels with an optimal dynamic of minimum value equal to 5 and the maximum value equal to 250.

3.3 Blobs Detection

We adopted the multi-scale-space theory for the selection of lines in grayscale images containing Arabic handwritten text. Our main interest is the line segmentation in manuscripts at different scales. The theory of multi-scale space has been formalized by Linderberg [14]. It consists in a simulation of the functioning of the human eye. When we look at an image by placing it far from the eye, there is the appearance of the blur. This appearance is represented by smoothing of image via a Gaussian filtering in scales σ.

The image is represented by the intensity of gray-level $I(x, y)$ at the coordinates (x, y) of pixels. The Gaussian function $G(x, y, \sigma)$ represents the filter of kernel σ. The convolution is shown by the following formula:

$$I(x, y, \sigma) = I(x, y) * G(x, y, \sigma) \tag{1}$$

where

$$G(x, y, \sigma) = \frac{1}{2\pi\sigma^2} \exp\left(-\frac{x^2 + y^2}{2\sigma^2}\right) \tag{2}$$

- Detection of circular blobs

Laplacian of Gaussian LOG is a way to aggregate circular blobs in filtered image [11]. These blobs are the result of Gaussian blurs generated by the image filtering with the Gaussian function of kernel σ.

$$\text{LOG}(x, y, \sigma) = I_{xx}(x, y, \sigma) + I_{yy}(x, y, \sigma) \tag{3}$$

In general, the Arabic script is written from right to left along the horizontal line direction. Indeed, circular blobs in an image observed at different scales do not represent lines or words or other characters of the text.

- Detection of elliptical blobs

The LOG at the scale σ applied to an image $I(x, y)$ gives the circular blobs. Circular blobs do not work for this case. As a solution, we propose other forms of blobs that really represent lines and words. This solution is the elliptical blobs that can accomplish this task. An elliptical blob can be defined by LOG at the scale (σ_x, σ_y)

applied to the image $I(x, y)$. With σ_x follows the x-axis and σ_y in turn follows the y-axis, the elliptical filter for Gaussian kernel is as follows:

$$G(x, y, \sigma_x, \sigma_y) = \frac{1}{2\pi\sigma_x\sigma_y}\exp\left(-\frac{1}{2}\left[\frac{x^2}{\sigma_x^2} + \frac{y^2}{\sigma_y^2}\right]\right) \tag{4}$$

The convolution of the image $I(x, y)$ with a Gaussian kernel filter (σ_x, σ_y) is as follows:

$$I(x, y, \sigma_x, \sigma_y) = I(x, y) * G(x, y, \sigma_x, \sigma_y) \tag{5}$$

Hence, the Laplacian of Gaussian formula is the following:

$$\mathrm{LOG}(x, y, \sigma_x, \sigma_y) = I_{xx}(x, y, \sigma_x) + I_{yy}(x, y, \sigma_y) \tag{6}$$

3.4 Text–Image Separation

The choice of Gaussian kernels (σ_x, σ_y) is a way to define the scales for the selection of elliptical blobs determining the text in images from Arabic manuscripts. A convolution of the image of intensity $I(x, y)$ with a Gaussian filter at scale (σ_x, σ_y) is necessary before producing blobs with LOG. The first step in segmentation is detection of lines in the manuscript image. The Gaussian kernel σ_x must be very large in front of σ_y to aggregate blobs in the direction of the ordinate axis. Nevertheless, the σ_y kernel must be infinitely small in order to reduce the contact effects of text line blobs. An article [15] that we have been published deals with the character segmentation of Arabic printed documents using also the space-scale technique. After several experimental tests of σ_x and in order to aggregate the elliptical blobs of lines, we noticed that the σ_x scale is equal to the square of σ_y as shown in the following formula:

$$\sigma_x = (\sigma_y)^2 \tag{7}$$

For this purpose, we have chosen eight Gaussian kernel filters (σ_x, σ_y) with a step of 0.8 for switching from a lower filter to a higher filter. The choice of these filters is based on several methods in the literature that use scale space such as Harris-Laplacian, Hessian-Laplacian [16] and SURF [17]. The smallest scale we have chosen is $(\sigma_x, \sigma_y) = (1.44, 1.20)$. However, the large scale is $(\sigma_x, \sigma_y) = (46.24, 6.8)$ which represents the largest filter to use. Table 1 shows the different scales used.

Figure 3 displays the scales (σ_x, σ_y) in order to give the correct line segmentation. Note that when the ladder increases the blobs begin to disappear.

Table 1 Gaussian filters for scale-space selection

σ_x	1.44	4	7.84	12.96	19.36	27.04	36	46.24
σ_y	1.2	2	2.8	3.6	4.4	5.2	6	6.8

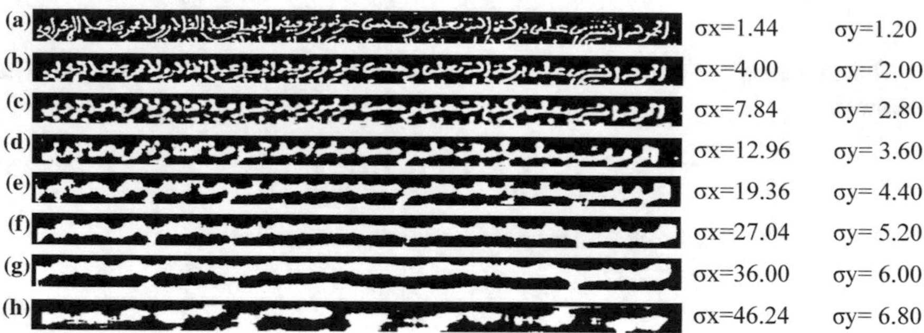

Fig. 3 Elliptical lines blobs result for each scale

The method of detecting text in the grayscale image of the manuscript is based on the following observations: The texture represents the physical medium whose luminous variations are extremely small. On the other hand, the text presents wide variations. The LOG sign is useful for separating the text from the image as shown in Fig. 4 (*left*). The positive sign corresponds to the handwriting. On the other hand, the negative sign corresponds to the background of the image. The writing presents strong variations of the luminous intensity, which generates the positive blobs. The filtering can be done with mathematical morphological operations, such as dilatation and morphological closing (Fig. 4 *right*).

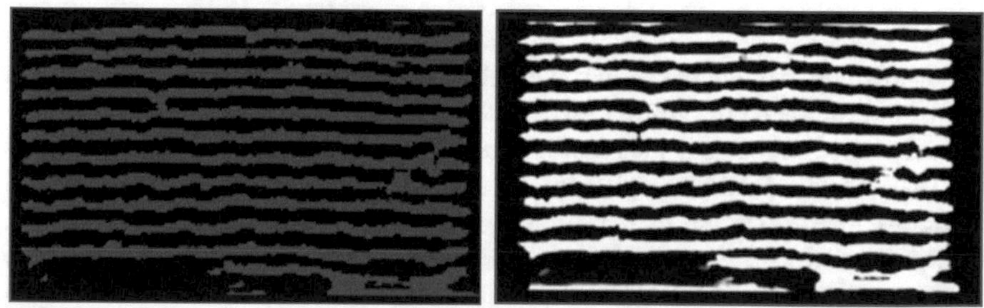

Fig. 4 Elliptical lines blobs selection result (*left*) and morphological dilation and closing by a structuring element 3 * 3 (*right*)

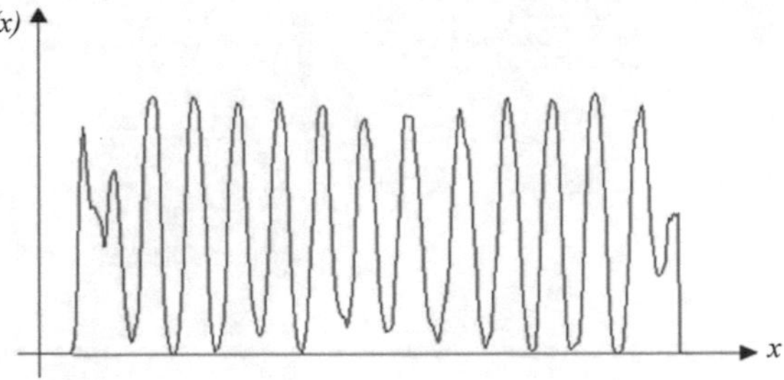

Fig. 5 Vertical projection profile

3.5 *Image Projection*

The projection function of an image is defined by:

$$f(y) = \sum_{x=0}^{w} I(x, y) \tag{8}$$

where $f(y)$ is the horizontal projection of the image $I(x, y)$ represented at Fig. 4. We thus obtain a 1D function facilitating the extraction of lines.

$$\boldsymbol{Minimas}: \begin{cases} \text{Minimas} = 0 \rightarrow \text{Background of the image} \\ \text{Minima} \neq 0 \rightarrow \text{Background image plus overlapping lines} \end{cases}$$

$\boldsymbol{Maxmas}$: baselines

Figure 5 shows the transformation of image in Fig. 4 into a 1D signal. In general, the words on the line are characterized by a remarkable space. For computing the threshold, we used the following function:

$$f(x) = (\text{Maximas} + \text{Minimas})/2 \tag{9}$$

4 Line Segmentation

The first processing consists in filtering the segmented lines of the remaining components of the adjacent lines, these components being detectable on the clipping line as shown in Fig. 6. A filtering based on mathematical morphologies is a solution allowing the suppression of these components which are considered as additional noise. The second treatment is devoted to the images of the scanned manuscripts. Acquisition sensors can generate noise. However, these images are generally noisy by the presence of the artifacts of the compression process. Gaussian filtering is based on the principle that the information contained in an image is redundant. This redundancy is used to suppress noise by smoothing the image. For that, we define a

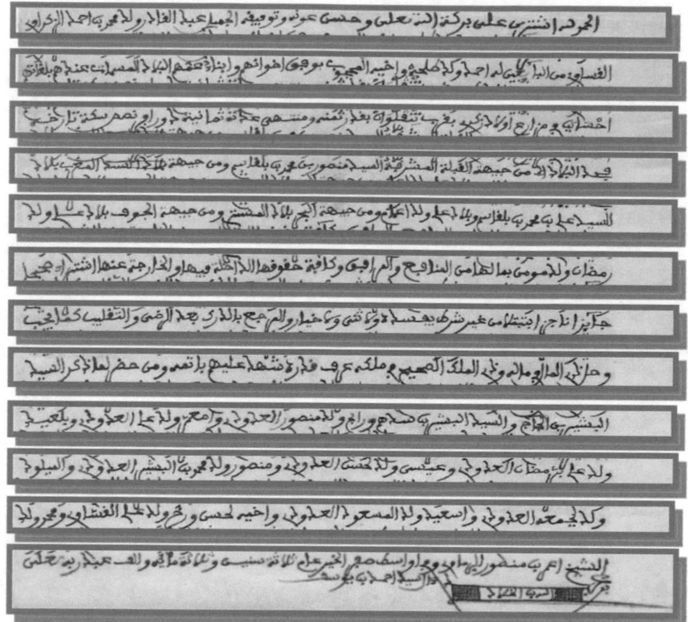

Fig. 6 Lines segmentation results with scale-space method

convolution core whose weights follow a 2D Gaussian curve centered in the middle of the matrix. This core is associated with a scale factor corresponding to the sum of the coefficients in order to keep the luminous intensities globally.

5 Pseudo-Word Segmentation

The principle of the method is like the line segmentation by the blob detection at each line. The connected components represent an element of the detected pseudo-word. Our aim is to extract all the useful information in each pseudo-word to facilitate the task with the recognition algorithms. For this purpose, we apply the image labeling to the extraction of the connected components of all the blobs at each line. Figure 7 shows an example of pseudo-words segmentation.

Fig. 7 Pseudo-word segmentation result

6 Discussion of Results

We applied our method to images containing Arabic handwriting. The results are encouraging. We found some segmentation problems for manuscripts that have unregulated lines. Our multi-scale segmentation algorithm detects lines even in the case of low resolutions. We have achieved a good rate of line segmentation of around 95%. We have surpassed rate of 98% for line segmentation in the absence of overlapping figures with text. Regarding the detection of words for complex manuscripts, we can segment the related components for each line. These components are elements of the search words and pseudo-words.

7 Conclusion

In this paper, we have developed a method of segmenting the lines and pseudo-words in Arabic manuscripts. We used scale space as a selector for horizontal elliptic blobs. Our method deals the Handwritten Text Segmentation in two essential steps: The first is to apply scale-space segmentation for Arabic manuscripts, to locate significant variations in intensity in different scales and to select horizontal elliptic blobs for lines, and the second is the selection of the elliptical blobs in the horizontal direction to favor the extraction of the lines and to reduce the effects of overlapping. About pseudo-words are detected on each segmented line.

The results obtained are encouraging compared to other segmentation methods. The error rate is mainly due to the massive presence of diacritical symbols in some Arabic manuscripts and to the problems of overlapping lines.

References

1. Pal, N., Pal, S.: A review on image segmentation techniques. Pattern Recogn. **26**, 1277–1294 (1993)
2. Sahoo, P.K., Soltani, S.A., Wong, K.C., Chen, Y.C.: A survey of thresholding techniques. Comput. Vision Graph. Image Process. **41**(2), 233–260 (1988)
3. Haralick, R., Shapiro, L.: Computer and Robot Vision, vol. 1, 2. Addison-Wesley Inc., Reading, Massachusetts (1992, 1993)
4. Ter Haar Romeny, B.M.: Introduction to scale-space theory: multiscale geometric image analysis. Technical Report ICU-96-21, Utrecht University, Netherlands, p. 26, Sep 1996
5. Konidaris, T., Gatos, B., Ntzios, K., Pratikakis, I., Theodoridis, S., Perantonis, S.J.: Keyword guided word spotting in historical printed documents using synthetic data and user feedback. Int. J. Doc. Anal. Recogn. **9**(2–4), 167–177 (2007)
6. Shi, Z., Govindaraju, V.: Line separation for complex document images using fuzzy runlength. In: Proceedings—First International Workshop DIAL 2004, pp. 306–312 (2004)
7. He, J., Downton, A.C.: User-assisted archive document image analysis for digital library construction. In: 7th International Conference (ICDAR), pp. 498–502 (2003)
8. Likforman-Sulem, L., Hanimyan, A., Faure, C.: A Hough based algorithm for extracting text lines in handwritten document. In: Proceedings of ICDAR'95, pp. 774–777 (1995)
9. Antonacopoulos, A., Karatzas, D.: Semantics-based content extraction in typewritten historical documents. In: 8th ICDAR, pp. 48–53 (2005)
10. Pechwitz, M., Maergner, V.: Baseline estimation for arabic handwritten words. In: Frontiers in Handwriting Recognition, V, pp. 479–484 (2002)
11. Manmatha, R., Rothfeder, J.L.: A scale space approach for automatically segmenting words from historical handwritten documents. IEEE Trans. Pattern Anal. Mach. Intell. **27**(8), 1212–1225 (2005)
12. Zoizou, A., Zarghili, A., Chaker, I.: A new hybrid method for Arabic multi-font text segmentation. JKSU CIS J. (2018) ISSN 1319-1578
13. CopiBook, i2s-DigiBook. https://www.i2s.fr/en/product/copibook-cobalt-hd. 25 Mar 2019
14. Lindeberg, T.: Feature detection with automatic scale selection. Technical Report ISRN KTH NA/P-96/18-SE, Royal Institute of Technology, Stockholm, Sweden, S-100 44, May 1996
15. El Makhfi, N., El Bannay, O.: Scale-space approach for character segmentation in scanned images of Arabic documents. JATIT J. **94**(1) (2016) ISSN 1992-8645
16. Mikolajczyk, K., Schmid, C.: Indexing based on scale invariant interest points. In: ICCV, vol. 1, pp. 525–531 (2001)
17. Bay, H., Tuytelaars, T., Gool, L.V.: SURF: speeded up robust features. In: 9th European Conference on Computer Vision, Graz Austria, pp. 404–417, May 2006

Mining Learners' Behaviors: An Approach Based on Educational Data Mining Techniques

Ouafae El Aissaoui, Yasser El Alami El Madani, Lahcen Oughdir, Ahmed Dakkak and Youssouf El Allioui

Abstract Educational data mining is a research field that aims to apply data mining techniques in educational environments. Many data mining techniques such as clustering, classification, and prediction can be performed on educational data in order to analyze the learner behaviors. In this work, we have used the clustering and classification techniques to predict the learners' learning styles. The students' behaviors while using the e-learning system have been captured from the log file and given as an input of a clustering algorithm to group them into 16 clusters. The resulted clusters were labeled with learning styles combinations based on the Felder and Silverman learning style model. Then the labeled behaviors were given as input to four classifiers: naive Bayes, Cart, Id3, and C4.5 to compare their performance in predicting students' learning styles. The four classifiers were performed using Weka data mining tool, and the obtained results showed that Id3 yielded better results than the other classifiers.

Keywords Educational data mining · Data mining techniques · Clustering · Classification · E-learning system · Felder and Silverman learning style model

1 Introduction

In recent years many technologies have been developed to improve the efficiency of E-learning environments. Using those technologies, the learners' behaviors and preferences can be detected and used to enhance the learning process. In this context, the Educational data mining field (EDM) has emerged.

O. El Aissaoui (✉) · L. Oughdir · A. Dakkak
LSI, FPT, University of Sidi Mohammed Ben Abdellah, Taza, Morocco
e-mail: ouafae.elaissaoui@usmba.ac.ma

Y. El Alami El Madani
ENSIAS, Mohammed V University, B.P.: 713, Agdal, Rabat, Morocco

Y. El Allioui
LS3M, FPK, USMS University, B.P.: 145, 25000 Khouribga, Morocco

V. Bhateja et al. (eds.), *Embedded Systems and Artificial Intelligence*,
Advances in Intelligent Systems and Computing 1076,
https://doi.org/10.1007/978-981-15-0947-6_62

EDM is a research field that aims to apply data mining techniques on educational data to extract useful information about the students' behaviors and then make enhanced learning strategies based on the extracted information. Various data mining techniques can be used in the EDM field to achieve different purposes. One of the important purposes of the EDM is the enhancement of the student model by predicting the learners' characteristics.

The learning style is one of the vital characteristics that construct a student model. Knowing learners' learning styles helps adaptive E-learning systems to provide customized contents that fit the learners' preferences. Many solutions have been proposed to identify students' learning styles, the traditional one consists in asking the students to fill in a questionnaire, this solution has many disadvantages. Firstly, filling in a questionnaire is a boring task that consumes a lot of time. Secondly, students aren't always aware of their learning styles and the importance of the further use of questionnaires, which can lead them to give arbitrary answers, therefore, the results obtained from the questionnaires can be inaccurate and might not reflect the real learning styles of the students. Thirdly, the results obtained from the questionnaires are static, while the learning styles can be changed during the learning process.

This paper presents an approach to detect the learning style automatically using the existing learners' behaviors and based on the This paper presents an approach to detect the learning style automatically using the existing learners' behaviors and based on the Felder and Silverman Learning Style Model (FSLSM). Four classifiers have been performed to detect the learners' learning styles which are: Naive Bayed, Id3, CART, and C4.5. We have compared the performance of the four classifiers to get the most efficient in detecting students' learning styles.

The rest of this paper is organized as follows. Section 2 gives a brief definition of EDM and describes the FSLSM and the used algorithms. Section 3 introduces a literature review of related work. Section 4 describes the methodology of our approach. The results are presented in Sect. 5; Finally, Sect. 6 presents our conclusions and future works.

2 Preliminary

2.1 Application of Data Mining in Educational Data

Data mining, also known as Knowledge Discovery in Data (KDD), is the process of finding hidden and useful information from large volumes of data. Many fields have exploited the efficiency of data mining techniques to make decisions such as e-commerce, bioinformatics, and e-learning. Applying data mining techniques to educational data is commonly known as educational data mining (EDM). One of the major utilities of EDM is the improvement of the student model. According to [1] a student model contains many characteristics which are: Knowledge and skills, misconception and errors, affective factors, Cognitive features, Meta-cognitive skills, and learning style. Many researches have been focusing on detecting the learner' learning style, various techniques have been used and many approaches have been proposed. Among the techniques that were widely used there are: Neural Network, Decision Tree, Bayesian Networks, and Association Rule.

2.2 *The Felder and Silverman Learning Style Model*

Felder and Silverman learning style model was developed by Felder and Silverman in 1998 [2]. This model uses the notion of dimensions where each dimension contains two opposite categories, and each student has a dominant preference in each dimension's category. The four dimensions of the FSLSM are: processing (active/reflective), perception (sensing/intuitive), input (visual/verbal), understanding (sequential/global). A learning style is defined by combining one category from each dimension.

Active learners tend to retain and understand information best by doing something with the learning material. Reflective learners prefer to think about the learning material quietly first. Active learners also prefer to study in group, while the reflective learners prefer to work individually.

Sensing learners like courses that deal with the real-world facts. They tend to be more practical by doing hands-on (laboratory) works and they are more competent than intuitive at memorizing facts. Intuitive learners don't like materials that contain a lot of memorization and routine calculations, and they prefer the ones involving abstractions and mathematical formulations. They may be better than sensors at innovating and grasping new concepts.

The visual learners prefer to see what they learn by using visual representations such as pictures, diagrams, and charts. While the verbal learners like information that are explained with words; both written and spoken.

Sequential learners tend to go through the course step by step in a linear way, with each step followed logically by the next one. Global learners tend to learn in large jumps, by accessing courses randomly without seeing connections.

In this work, we have based on the FSLSM for many reasons. Firstly; this model uses the concept of dimensions and describes thoroughly the learning styles preferences. This description mentions the types of learning objects that can be included in each learning style preference, and this is an interesting characteristic to our work because knowing the learning styles preferences of learning objects helps us to determine the learning style of learners 'sequences. Secondly; according to our approach a LS can be changed over time, and then updated using a classifier technique. The FSLSM provides this possibility by considering LS as tendencies and students can act in a non-deterministic way as pointed by [3]. Thirdly, the FSLSM is the most used in adaptive e-learning systems and the most appropriate to implement them as mentioned by [4, 5].

According to FSLSM, each learner prefers a specific category in each dimension. Thus a learning style is defined by combining one preference from each dimension. As a result, we will obtain sixteen combinations:

```
Learning Styles Combinations (LSCs) = {(A,Sen,Vi,G),
(A,Sen,Vi,Seq), (R,Sen,Vi,G), (A,Sen,Ve,Seq),
(A,Sen,Ve,G), (R,Sen,Ve,Seq), (R,Sen,Ve,G), (A,I,Ve,G),
(A,I,Vi,Seq), (A,I,Vi,G), (R,I,Vi,Seq), (R,I,Vi,G),
(R,S,Vi,Seq), (A,I,Ve,Seq), (R,I,Ve,Seq), (R,I,Ve,G)}.
```

2.3 Used Algorithms

K-modes clustering algorithm

The k-modes is a clustering algorithm that aims to group similar categorical objects into k clusters. This algorithm is an extension of k-means clustering algorithm with three modifications: replacing the Euclidean distance function with the simple matching dissimilarity measure. Instead of means, it uses modes, and instead of updating the centroids it updates modes using a frequency based method.

The following steps illustrate how to cluster a categorical data set X into k clusters:

Step 1: Randomly Select k initial modes, one for each cluster.

Step 2: compute the distance from each object to each mode using the dissimilarity Measures, and then associate each object to the cluster whose mode is the nearest to it. This association defines the first k clusters.

Step 3: Update the modes of the newly defined clusters from Step 2 using Theorem (1) described below and then retest the dissimilarity between the objects and the updated modes. If an object is found such that its nearest mode belongs to another cluster rather than its current one, reallocate the object to that cluster and update the modes of both clusters.

Step 4: repeat step 3 until no object has moved to another cluster after a full cycle test of the whole data set.

The dissimilarity Measures and the theorem (1) that are used in the previous steps; are described below:

- Dissimilarity Measures

To compute the dissimilarity between two objects X and Y described respectively by m categorical attributes values $(x_1, x_2, \ldots, x_m)$ and $(y_1, y_2, \ldots, y_m)$, the two following functions can be used:

$$d(X, Y) = \sum_{j=1}^{m} \delta(x_j, y_j) \text{ where } \delta(x_j, y_j) = \begin{cases} 0, x_j = y_j \\ 1, x_j \neq y_j \end{cases} \quad (1)$$

$$d_{x^2}(X, Y) = \sum_{j=1}^{m} \frac{(n_{x_j} + n_{y_j})}{n_{x_j} n_{y_j}} \delta(x_j, y_j) \tag{2}$$

where n_{x_i} and n_{y_j} are respectively the numbers of objects that have the categories x_j and y_j for attribute *j*.

- How to select a mode for a set

Let $X = \{X_1, X_2, \ldots, X_n\}$ a mode of X is a vector $Q = [q_1, q_2, \ldots, q_m]$ that minimizes the following equation:

$$D(Q, X) = \sum_{i=1}^{n} d(X_i, Q) \tag{3}$$

where *D* can be either Eqs. (1) or (2).

Theorem 1 *The function* $D(Q, X)$ *is minimized if* $f_r(A_j = q_j \backslash \boldsymbol{X}) \geq f_r(A_j = C_{k,j} \backslash \boldsymbol{X})$ *for* $q_j \neq C_{k,j}$ *for all* $j = 1, \ldots, m$ *where* $f_r(A_j = C_{k,j} \backslash \boldsymbol{X}) = \frac{n_{C_{k,j}}}{n}$ *is the relative frequency of category* $C_{k,j}$ *in X.*

Naive Bayes

The Naive Bayes is a supervised classifier and it is an extension of the Bayes theorem with two simplifications [6, 7].

The first simplification is to use the conditional independence assumption. That is, each attribute is conditionally independent of every other attribute given a class label C_i. See the following equation:

$$\begin{aligned} P(C_i \backslash A) &= P(C_i) * \frac{P(a_1, a_2, \ldots, a_m \backslash C_i)}{P(a_1, a_2, a_3, \ldots, a_m)} \\ &= P(C_i) * \frac{P(a_1 \backslash C_i) P(a_2 \backslash C_i), \ldots, P(a_m \backslash C_i)}{P(a_1, a_2, a_3, \ldots, a_m)} \\ &= P(C_i) * \frac{\prod_{j=1}^{m} P(a_j \backslash C_i)}{P(a_1, a_2, a_3, \ldots, a_m)} \end{aligned} \tag{4}$$

The second simplification is to ignore the denominator $P(a_1, a_2, a_3, \ldots, a_m)$. Because $P(a_1, a_2, a_3, \ldots, a_m)$ appears in the denominator of $\boldsymbol{P(C_i \backslash A)}$ for all values of *i*, removing the denominator will have no impact on the relative probability scores and will simplify calculations. After applying the two simplifications mentioned earlier, we obtain the Naive Bayes classifier which consists in labeling the object A with the class label C_i that maximizes the following equation:

$$P(C_i \backslash A) \sim P(C_i) \cdot \prod_{j=1}^{m} P(a_j \backslash C_i),\ i = 1, 2, \ldots, n \tag{5}$$

Decision Tree Algorithms

Many algorithms exist to implement a decision tree, where each algorithm use a specific method to construct a tree. Some popular algorithms include ID3 [8], C4.5 [9], and CART [10].

- ID3 Algorithm:

ID3 (or Iterative Dichotomiser 3) is one of the first decision tree algorithms that Handles only Categorical value, and it was developed by John Ross Quinlan. Let S be a set of categorical input variables where each input variable y belongs to an attribute Y, x be the output variable (or the predicted class), The ID3 requires the following steps:

Step 1: compute the entropy $H(S)$ for data-set
Step 2: for every attribute:

Step 2.1: Compute the conditional entropy $H_{x/y}$. Given an attribute Y, its value y, its outcome X, and its value x.
Step 2.2: compute the information gain for the current attribute.
Step 2.3: choose the attribute with the largest information gain for the first split of the decision tree.
Step 2.4: Repeat until we get the desired tree.

The entropy $H(S)$, the conditional entropy $H_{x/y}$, and the information gain are computed respectively as follows:

$$H(S) = -\sum_{\forall x \in X} p(x) \cdot \log_2 p(x) \tag{6}$$

$$H_{x/y} = -\sum_{\forall y \in Y} p(y) \cdot \sum_{\forall x \in X} p(x/y) \cdot \log_2 p(x/y) \tag{7}$$

$$\text{InfoGain}_A = H_S - H_{S/A} \tag{8}$$

- C4.5 Algorithm

The C4.5 algorithm is an improved version of the original ID3 algorithm. The C4.5 algorithm can handle missing data. If the training set contains missing attribute values, the C4.5 evaluates the gain for an attribute by considering only the records where the attribute is defined.

The C4.5 algorithm Handles both Categorical and continuous attributes. Values of a continuous variable are sorted and partitioned. For the corresponding records of each partition, the gain is calculated, and the partition that maximizes the gain is chosen for the next split.

The C4.5 algorithm addresses the overfitting problem in ID3 by using a bottom-up technique called pruning to simplify the tree by removing the least visited nodes and branches.

- CART Algorithm

CART (or Classification And Regression Trees) can handle both Categorical and continuous attributes like the C4.5 algorithm. Whereas C4.5 uses entropy based criteria to rank tests, CART uses the Gini diversity index defined in the following Equation

$$\mathrm{Gini}_X = 1 - \sum_{\forall x \in X} P(x)^2 \tag{9}$$

Similar to C4.5, CART can handle missing values and overcome the overfitting problem by using the Cost-Complexity pruning strategy.

3 Related Works

Many researchers have focused on detecting students' learning styles using educational data mining techniques. The EDM uses algorithms from the field of data mining and machine learning to build a model from the existing students' behaviors and then the constructed model is used to determine the learning style of a new student. Among the popular algorithms being used, there are the neural network, Decision tree, and Bayesian network. Reference [11] introduces an approach which uses artificial neural networks to identify students' learning styles. The approach has been evaluated with data from 75 students and found to outperform the current state of the art approaches. Reference [12] presents an approach to recognize automatically the learning styles of individual students according to the actions that they have performed in an e-learning environment. This recognition technique is based upon feed-forward neural networks.

Beside the neural network, the decision tree is also a widely used technique to determine LSs. In [13] the authors compare the performance of J48, NBTree, RandomTree and CART in inferring LSs. Authors in [14] used NBTree classification algorithm in conjunction with Binary Relevance classifier, the learners are classified based on their interests. Then, learners' learning styles are detected using these classification results.

According to [15] the Bayesian networks are ones of the most widely adopted classifiers to infer the learning style [16]. Used a Bayesian network (BN) in order to identify students' learning styles. They identified various behaviors that may be relevant to detect learning styles in a given E-learning system. Then, a BN was trained with data from 50 students, using initial probabilities based on expert knowledge. The trained BN was then evaluated using 27 students. For each student, the BN provides a probability that the student has a particular learning style preference. As a result, the approach obtained an overall precision of 58% for A/R, 77% for S/I a 63% for S/G (the V/V dimension was not considered) [17]. Presents a work that aims to enhance the capability of such systems by introducing adaptivity in the way the information is presented to the online learner. The adaptivity takes into account the learners' learning styles by modeling them using a Bayesian Network.

Most of the proposed approaches aim to classify learners according to the eight learning styles preferences of FSLSM. This study differs from others by considering 16 learning styles combinations instead of eight.

4 Methods

The EDM techniques consist in applying data mining techniques on educational data. One of the sources of educational data is the log file since most of the E-learning environments have their own log files that record the interaction history of students. The data recorded in log files can be analyzed by using web usage mining techniques to investigate the learners' behaviors. The basic steps of the web usage mining process are; data collection, preprocessing and pattern discovery.

4.1 Data Collection

In our work, the learners' behaviors were collected from the E-learning platform of Sup 'Management Group. The online educational system was developed based on Moodle platform. Moodle is a free Open Source software used to help educators and students to enhance the learning process. We collected 1235 sequences from the E-learning platform, which represent the learners' behavior.

4.2 Data Preprocessing

In This step, we cleaned the collected data from all unnecessary information and we kept just the ones that seemed relevant for our study. Thus, for each learner's sequence, we kept the following information: the sequence id, session id, learner id, and the set of learning objects accessed by the learner in a session.

The accessed learning object can be used to determine the learning style of a learner's sequence, because each student prefers to use some specific types of learning objects while learning, so there is a relationship between the learner's learning styles and the accessed learning objects, this relationship has to be analyzed by matching each LO with its relevant learning style preferences.

Based on the matching table presented in our previous work [18]; we have obtained the following table, where each LO is matched with its appropriate LSC.

4.3 Pattern Discovery

This step aims to apply data mining techniques to the data preprocessed in the previous step in order to predict the students' learning styles. In this study we have used two data mining techniques:

- The clustering to group the students into 16 clusters where each cluster corresponds to LSC.
- The classification to predict the learning style combination for a new sequence.

The following schema resumes our proposed approach (Fig. 1).

Clustering

This step aims to group the learners' sequences into 16 LSCs. We have used the *k*-modes clustering algorithm because of its ability to deal with categorical objects, since the learners' sequences which were considered as the input to that algorithm have categorical attribute values.

After extracting and preprocessing the learners' behaviors from the log file using web usage mining techniques, we can use them as an input to the *K*-modes by turning

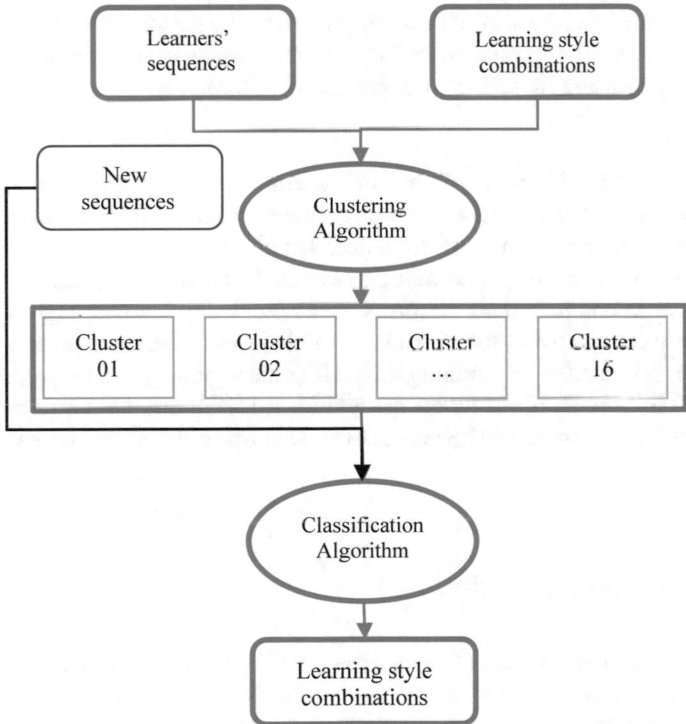

Fig. 1 Our approach

them into a matrix with M rows corresponding to M sequences and sixteen columns to store the attribute values where the attributes of each sequence correspond to the sixteen Los presented in the previous mapping table (Table 1).

Let $S = \{S_1, \ldots, S_i, \ldots, S_n\}$ be a set of n sequences (categorical objects), each sequence S_i is defined by 16 attribute values $(s_{i1}, \ldots, s_{ij}, \ldots, s_{i16})$.

And let $A_1, \ldots, A_j, \ldots, A_{16}$ be the 16 attributes describing the n sequences where: $\boldsymbol{A}_1$ = video, $\boldsymbol{A}_2$ = PPTs, $\boldsymbol{A}_3$ = demo, $\boldsymbol{A}_4$ = Exercise, $\boldsymbol{A}_5$ = Assignments, $\boldsymbol{A}_6$ = PDFs , $\boldsymbol{A}_7$ = Announcements, $\boldsymbol{A}_8$ = References, $\boldsymbol{A}_9$ = Examples, $\boldsymbol{A}_{10}$ = Practical Material, $\boldsymbol{A}_{11}$ = Forum, $\boldsymbol{A}_{12}$ = Topic list, $\boldsymbol{A}_{13}$ = Images, $\boldsymbol{A}_{14}$ = Charts, $\boldsymbol{A}_{15}$ = Email, $\boldsymbol{A}_{16}$ = Sequential.

For each sequence, Each attribute has two possible categorical attribute values: {yes or no, yes if the jth learning object exists in the sequence, No if it doesn't exist.

Therefore, each sequence is presented as $(s_{i1}, \ldots, s_{ij}, \ldots, s_{i16})$ where s_{ij} = {yes or no}.

To perform the *k*-modes algorithm we used the *R* software framework for statistical analysis and graphics. The dataset employed in our approach was extracted from the E-learning platform's log file of Sup' Management Group. This dataset records 1235 learners' sequences. The following figure displays the results obtained after running the *k*-modes with the *R* (Fig. 2).

The above figure shows the size of each cluster. The obtained clusters were labeled with the LSCs based on the minimum distances between the clusters' Modes and the LSCs' vectors presented in Table 1. The distances are computed using the dissimilarity measure (described in subsection II. C. 1).

Classification

After applying the *K*-Modes algorithm and labeling the sequences with the LSCs, the labeled sequences can be used as a training dataset to train a classification algorithm, and then use it to predict the LSC for a new sequence.

In this work, we have applied four classifiers, the naïve Bayes, CART, C4.5, and ID3 in order to compare their performance in predicting learning styles. The four algorithms were run using the Weka data mining tool. Weka implements machine learning algorithms for data mining tasks. It contains tools for data preprocessing, classification, regression, clustering, association rules mining, and visualization. The performance of the four classifiers was evaluated using the *K*-fold cross validation technique.

5 Results and Discussion

To compare the performance of the four applied algorithms in predicting LSC, we used the 10-fold cross validation for every classifier as suggested by Weka. The results of the experiment are shown in Fig. 3, Tables 2 and 3. Table 2 presents for each classifier the results of the Correctly Classified Instances, the Incorrectly Classified Instances, the Kappa statistic and the Time taken to build the model.

Table 1 Matching learning objects to LSCs

Cluster ID	Cluster meaning	Videos	PPTs	Demo	Exercise	Assignments	PDFs	Announcements	References
C01	Reflective-intuitive-verbal-global	Yes	Yes	No	Yes	Yes	Yes	Yes	Yes
C02	Active-intuitive-verbal-global	Yes	Yes	Yes	Yes	Yes	Yes	Yes	Yes
C03	Reflective-sensing-verbal-global	Yes	Yes	No	Yes	Yes	Yes	Yes	Yes
C04	Active-sensing-verbal-global	Yes	Yes	Yes	Yes	Yes	Yes	Yes	Yes
C05	Reflective-intuitive-visual-global	Yes	Yes	No	Yes	Yes	Yes	Yes	Yes
C06	Active-intuitive-visual-global	Yes	Yes	Yes	Yes	Yes	Yes	No	Yes
C07	Reflective-sensing-visual-global	Yes	Yes	No	Yes	Yes	Yes	Yes	Yes
C08	Active-sensing-visual-global	Yes	Yes	Yes	Yes	Yes	Yes	No	Yes
C09	Reflective-intuitive-verbal-sequential	Yes	Yes	No	Yes	Yes	Yes	Yes	Yes
C10	Active-intuitive-verbal-sequential	Yes	Yes	Yes	Yes	Yes	Yes	Yes	Yes
C11	Reflective-sensing-verbal-sequential	Yes	Yes	No	Yes	Yes	Yes	Yes	Yes
C12	Active-sensing-verbal-sequential	Yes	Yes	Yes	Yes	Yes	Yes	Yes	Yes
C13	Reflective-intuitive-visual-sequential	Yes	Yes	No	Yes	Yes	Yes	Yes	Yes
C14	Active-intuitive-visual-sequential	Yes	Yes	Yes	Yes	Yes	Yes	No	Yes
C15	Reflective-sensing-visual-sequential	Yes	Yes	No	Yes	Yes	Yes	Yes	Yes
C16	Active-sensing-visual-sequential	Yes	Yes	Yes	Yes	Yes	Yes	No	Yes

(continued)

Table 1 (continued)

Cluster ID	Cluster meaning	Examples	Practical Material	Forum	Topic list	Images	Charts	Email	Sequential
C01	Reflective-intuitive-verbal-global	No	No	Yes	Yes	No	No	Yes	No
C02	Active-intuitive-verbal-global	No	No	Yes	Yes	No	No	Yes	No
C03	Reflective-sensing-verbal-global	Yes	Yes	No	Yes	No	No	Yes	No
C04	Active-sensing-verbal-global	Yes	Yes	No	Yes	No	No	Yes	No
C05	Reflective-intuitive-visual-global	No	No	Yes	Yes	Yes	Yes	No	No
C06	Active-intuitive-visual-global	No	No	Yes	Yes	Yes	Yes	No	No
C07	Reflective-sensing-visual-global	Yes	Yes	No	Yes	Yes	Yes	No	No
C08	Active-sensing-visual-global	Yes	Yes	No	Yes	Yes	Yes	No	No
C09	Reflective-intuitive-verbal-sequential	No	No	Yes	Yes	No	No	Yes	Yes
C10	Active-intuitive-verbal-sequential	No	No	Yes	Yes	No	No	Yes	Yes
C11	Reflective-sensing-verbal-sequential	Yes	Yes	No	No	No	No	Yes	Yes
C12	Active-sensing-verbal-sequential	Yes	Yes	No	No	No	No	Yes	Yes
C13	Reflective-intuitive-visual-sequential	No	No	Yes	Yes	Yes	Yes	No	Yes
C14	Active-intuitive-visual-sequential	No	No	Yes	Yes	Yes	Yes	No	Yes
C15	Reflective-sensing-visual-sequential	Yes	Yes	No	No	Yes	Yes	No	Yes
C16	Active-sensing-visual-sequential	Yes	Yes	No	No	Yes	Yes	No	Yes

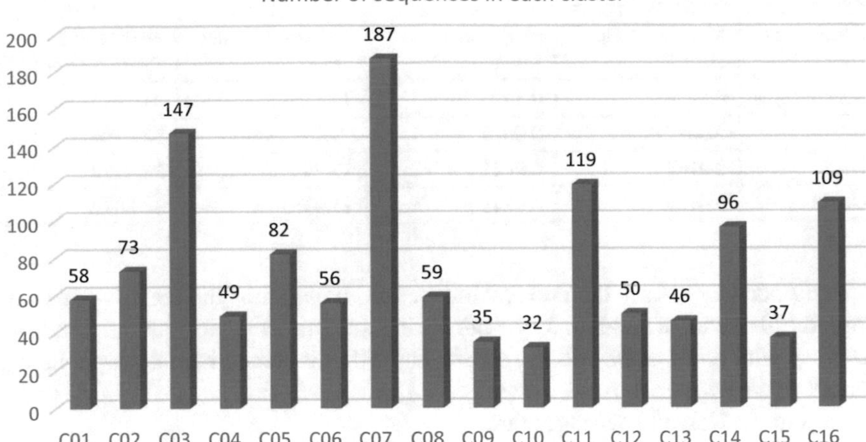

Fig. 2 Result of the *K*-modes algorithm

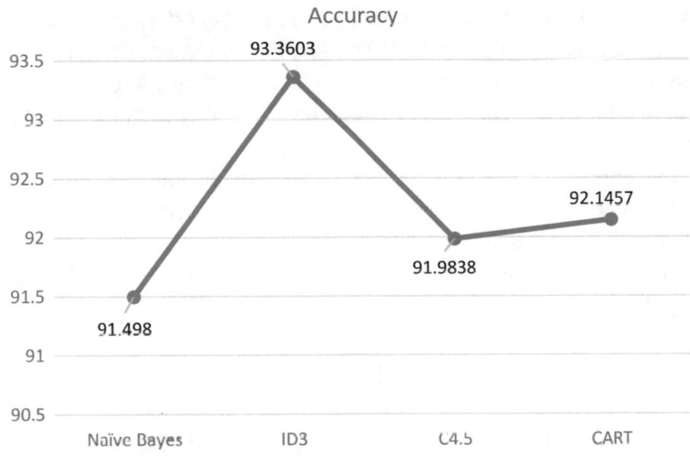

Fig. 3 Accuracy result by classifier

Table 2 Validation metrics by classifier

Algorithms	Correctly classified instances (%)	Incorrectly classified instances (%)	Kappa statistic	Time taken (s)
Naïve Bayes	91.498	8.502	0.9093	0.01
ID3	93.3603	6.6397	0.9292	0.03
C4.5	91.9838	8.0162	0.9145	0.05
CART	92.1457	7.8543	0.9162	0.52

Table 3 Error metrics by classifier

Algorithms	Mean absolute error	Root mean squared error	Relative absolute error (%)	Root relative squared error (%)
Naïve Bayes	0.0186	0.1005	15.8854	41.5141
ID3	0.0087	0.0679	7.4314	28.0504
C4.5	0.0163	0.0921	13.895	38.0679
CART	0.0129	0.0843	11.0151	34.8072

The Accuracy defines the rate at which a model classifies the records correctly. From the above table and Fig. 3, we can notice that the ID3 algorithm has the heist accuracy with a rate of 93.3603%, so this algorithm is more efficient in predicting correct instances than the others classifiers. The lowest Accuracy is 91.498% and it belongs to the Naïve Bayes algorithm.

The Kappa statistic is a measure of agreement between the predictions and the actual labels. It can also be defined as a comparison of the overall accuracy to the expected random chance accuracy. The higher the Kappa metric is, the more efficient a classifier to be as a random chance classifier. The kappa statistic results of the used algorithms are between 0.9093 and 0.9292, so all of them have a good accuracy.

The time taken to build the model is also an important factor to be considered. From the above table, we can observe that the naive Bayes is the fastest one with 0.01 s, while the slowest on is the CART with 0.52 s

Table 3 presents the results of some statistical metrics that are computed based on the differences between forecast and the corresponding observation. Those metrics can be computed using the following formula.

Let's denote the true value of interest as Y and the value estimated using some algorithm as X.

$$\mathrm{MAE} = \frac{1}{N}\sum_{i=1}^{N} |X_i - Y_i| \tag{10}$$

$$\mathrm{RMSE} = \sqrt{\frac{1}{N}\sum_{i=1}^{N} (X_i - Y_i)^2} \tag{11}$$

$$\mathrm{RAE} = \frac{\sum_{i=1}^{N} |X_i - Y_i|}{\sum_{i=1}^{N} \left|\overline{Y} - Y_i\right|} \tag{12}$$

$$\mathrm{RRSE} = \sqrt{\frac{\sum_{i=1}^{N} (X_i - Y_i)^2}{\sum_{i=1}^{N} \left(\overline{Y} - Y_i\right)^2}} \tag{13}$$

As we can notice, all the above metrics compare true values to their estimates, but each one do it in a slightly different way. They all tell us "how far away" are our estimated values from the true value of Y. The model with the smaller metrics values is the more performant, in our case, the ID3 has the lowest values so it performs better than the other algorithms in predicting correctly.

6 Conclusion

In this paper, we have used two educational data mining techniques to predict the learners' learning styles. The first technique is the clustering and it has been applied in order to group the learners into 16 LSCs using the K-modes algorithm and based on the FSLSM. The second one is the classification that aimed to use the result of the clustering algorithm as a training set to fit a classifier on it and then predict LSCs of new learners' sequences.

In this work, we have compared the performance of four classifiers in predicting LSCs using Weka data mining tool. The comparison was done based on the correctly and incorrectly classified sequences, the time taken by each classifier, the Kappa statistic and the mean error.

The obtained results showed that the ID3 algorithm yielded the highest accuracy and Kappa statistic and the lowest mean error, so it is the most efficient in classifying correct sequences. The naive Bayes is less performant in predicting correct instances, but it is the fastest one.

In the future work, we'd like to have determined automatically others learner's characteristics such as the affective factors in order to create a more efficient learning style model.

References

1. Chrysafiadi, K., Virvou, M.: Student modeling approaches: a literature review for the last decade. Expert Syst. Appl. **40**(11), 4715–4729 (2013)
2. Felder, R.M., Silverman, L.K.: Learning and teaching styles. J. Eng. Educ. **78**, 674–681 (1988)
3. Kinshuk, Liu, T.C., Graf, S.: Coping with mismatched courses: students' behaviour and performance in courses mismatched to their learning styles. Educ. Technol. Res. Dev. **57**(6), 739–752 (2009)
4. Graf S., Kinshuk, Advanced adaptivity in learning management systems by considering learning styles. In: Proceedings—2009 IEEE/WIC/ACM International Conference on Web Intelligence and Intelligent Agent Technology—Workshops, WI-IAT Workshops 2009, vol. 3, pp. 235–238 (2009)
5. Kuljis J, Liu, F.: A comparison of learning style theories on the suitability for elearning. In: Conference on Web Technologies, Applications, and Services, pp. 191–197 (2005)
6. Dietrich, D., Heller, R., Yang, B., EMC Education Services: Data Science and Big Data Analytics: Discovering, Analyzing, Visualizing and Presenting Data
7. Richert W., Coelho, L.P.: Building Machine Learning Systems with Python (2013)

8. Quinlan, J.R.: Induction of decision trees. Mach. Learn. **1**(1), 81–106 (1986)
9. Salzberg, S.L.: Book review: C4.5: programs for machine learning by J. Ross Quinlan. Morgan Kaufmann Publishers Inc., 1993. Mach. Learn. **16**(3), 235–240 (1994)
10. Breiman, L., Friedman, J., Stone, C.J., Olshen, R.A.: Classification and Regression Trees, 1st edn, Chapman & Hall (1993)
11. Bernard, J., Chang, T.-W., Popescu, E., Graf, S.: Using Artificial Neural Networks to Identify Learning Styles, pp. 541–544. Springer, Cham (2015)
12. Villaverde, J.E., Godoy, D., Amandi, A.: Learning styles' recognition in e-learning environments with feed-forward neural networks. J. Comput. Assist. Learn. **22**(3), 197–206 (2006)
13. Ahmad N.B.H., Shamsuddin, S.M.: A comparative analysis of mining techniques for automatic detection of student's learning style. In: 2010 10th International Conference on Intelligent Systems Design and Applications, 2010, pp. 877–882
14. Özpolat E., Akar, G.B.: Automatic Detection of Learning Styles for an E-Learning System, vol. 53(2). Pergamon (2009)
15. Feldman, J., Monteserin, A., Amandi, A.: Automatic detection of learning styles: state of the art. Artif. Intell. Rev. **44**(2), 157–186 (2015)
16. García, P., Amandi, A., Schiaffino, S., Campo, M.: Evaluating Bayesian networks' precision for detecting students' learning styles. Comput. Educ. **49**(3), 794–808 (2007)
17. Alkhuraiji, S., Cheetham, B., Bamasak, O.: Dynamic adaptive mechanism in learning management system based on learning styles. In: 2011 IEEE 11th International Conference on Advanced Learning Technologies, 2011, pp. 215–217
18. El Aissaoui, O., El Madani El Alami, Y., Oughdir, L., El Allioui, Y.: Integrating web usage mining for an automatic learner profile detection: a learning styles-based approach. In: 2018 International Conference on Intelligent Systems and Computer Vision (ISCV), 2018, pp. 1–6

Adaptive Local Gray-Level Transformation Based on Variable S-Curve for Contrast Enhancement of Mammogram Images

Hamid El Malali, Abdelhadi Assir, Mohammed Harmouchi, Mourad Rattal, Aissam Lyazidi and Azeddine Mouhsen

Abstract Mammogram image enhancement plays an important role in the accuracy of the diagnosis. Resulting images must allow a good contrast and better appearance of the limiting edges between different areas in the image, while preserving both information and brightness. In this paper, an adaptive local gray-level s-curve transformation is proposed. The main aim is to improve as much contrast of mammogram images as possible. The principle is to find all the parameters of the local gray-level transformation for each image which will allow for a better improvement using the genetic algorithm that is among global optimization methods. The evaluation of the results found is done based on image quality assessment (IQA) metrics and visual inspection in comparison with three existing techniques based on histogram equalization.

Keywords Mammogram image enhancement · Global optimization · Sigmoid function

1 Introduction

Mammography is among the reliable screening procedures that can detect breast cancer early [1–4]. However, many factors led to a decrease in the mammogram image quality, such as the high-density breast tissue and the limitation of X-ray hardware [4]. Moreover, contrast between abnormal tissues and normal glandular is not sufficient because of low attenuation between tissues in the image [1, 5]. Thus, the decision taken by the radiologist or computer-aided diagnosis (CAD) system would not be precise [1, 3–5]. The image enhancement is used to overcome this

H. El Malali (✉) · A. Assir · M. Harmouchi · Az. Mouhsen
Faculty of Science and Technology, Laboratory of Radiation-Matter and Instrumentation, University Hassan 1st, Settat, Morocco
e-mail: elmalalihamid@gmail.com

M. Rattal · A. Lyazidi
Higher Institute of Health Sciences, University Hassan 1st, Settat, Morocco

V. Bhateja et al. (eds.), *Embedded Systems and Artificial Intelligence*, Advances in Intelligent Systems and Computing 1076,
https://doi.org/10.1007/978-981-15-0947-6_63

problem. Its goal is to improve the contrast and the appearance of the limiting edges between areas while maintaining as much brightness and information contained in the image as possible [2, 4–6]. The objective assessment is preferred in this case of the improvement of medical images using the image quality assessment (IQA) measures [2, 5, 7, 8]. Indeed, the increase of the values of the Effective Measure of Enhancement (EME) parameter, the Edge Content (EC) parameter, and the Feature Similarity Index Measure (FSIM) parameter, and the decrease of the value of the Absolute Mean Brightness Error (AMBE) parameter indicate that the improvement is better [5, 9, 10]. Gandhamal et al. proposed a parametric local gray-level s-curve transformation for medical image enhancement. The subdivision of all images into blocks is the same according to their size (10%), and the intensities of the pixels of images are transformed using the same s-curve [10]. Our proposed approach aims to find a subdivision in non-overlapping blocks and an s-curve for each mammogram image that will allow a better improvement. The search process for the parameters of the local transformation is managed by the genetic algorithm to optimize all the parameters mentioned above. The tested images are mdb209, mdb210, mdb211, and mdb212 from Mini-MIAS database [11]. The performance of the proposed method is evaluated by comparing its results with those results using three techniques that are widely used namely histogram equalization (HE), contrast limited adaptive histogram equalization (CLAHE), and brightness preserving bi-histogram equalization (BBHE).

2 Problem Formulation

In the literature, several techniques for improving medical image quality have been used. Point-to-point transformations are among the most used techniques [12]. These transformations correspond to each pixel of the input image a new value of the output one that can be expressed by:

$$s = T(r) \tag{1}$$

where r is the gray level of the current pixel, s is its new value in the output image, and T is the gray-level transformation function. Figure 1 summarizes some types of these transformations from which we quote linear, power law, logarithmic, exponential, and s-curve [10, 12].

The current study exploits the s-curve transformation to enhance the contrast of mammogram images. It makes it possible to exploit both the properties of the exponential and logarithmic transformations. This combination makes it possible to improve the contrast and the appearance of edges [10]. In fact, this transformation reduces both the range of lower and higher gray levels and increases the difference between them in each block (Fig. 1) [10]. These differences in optical densities are necessary to improve the visual contrast and the sharpness of edges which allow the enhancement of human visual perception for an accurate interpretation [10, 13, 14]. Moreover, the s-curve transformation shows its performance in brightness

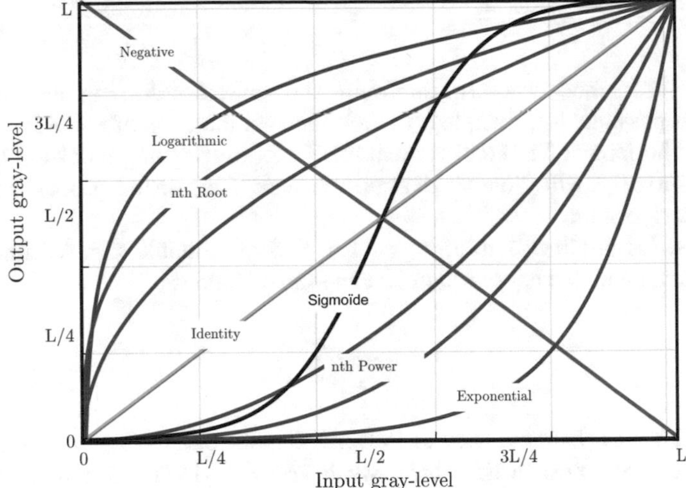

Fig. 1 Gray-level transformations

preservation as well as the information contained in the image [10]. The equation of this transformation is given by the following expression [10, 15, 16]:

$$s = \frac{1}{1 + e^{(\frac{-(r-\alpha)}{\beta})}} \tag{2}$$

where r is the normalized gray level of the current pixel in the range of [0, 1] given by:

$$r = \frac{I - I_{\min}}{I_{\max} - I_{\min}} \tag{3}$$

I is the gray level of the current pixel, $I_{\min}$ and $I_{\max}$ are the minimum and maximum gray level, respectively, in the input image. α and β are, respectively, the center and width of sigmoid function in the range of]0, 1]. s is the new normalized gray-level value, which must be de-normalized to find the new gray-level value in the range $[I_{\min}, I_{\max}]$ using the following expression:

$$I_{\text{out}} = s.(I_{\max} - I_{\min}) + I_{\min} \tag{4}$$

Our aim is to find the parameters of the local transformation for each mammogram image by a variable sigmoid function, which will give a better improvement. The quality of this improvement is objectively evaluated using image quality assessment parameters that are recommended in the case of medical image enhancement [4, 7].

3 Proposed Approach

The proposed method consists of improving the contrast of the mammogram images as much as possible. Its principle is to look for a combination of a subdivision into a number of horizontal NBH and vertical NBV non-overlapping blocks of the image and the s-curve that will allow a better improvement. The research process is managed by genetic algorithm.

Proposed sigmoid function: in order to find the suitable sigmoid function for each mammogram, we have written its equation as follows:

$$s = \frac{1}{1 + B^{\left(\frac{-(r-\alpha)}{\beta}\right)}} \tag{5}$$

where s, r, α, and β are the same as in Eq. 2; B is the basis of sigmoid function.

Genetic algorithm: In this study, we have used MATLAB genetic algorithm toolbox to manage the process of finding the parameters (NBH, NBV, B, α, and β) of the local transformation for each mammogram image. The main options of the genetic algorithms chosen are: the population number is 50, the number of generation is 100, the crossover function used is single point, and the mutation function is adaptive feasible. The rest of the options are the default toolbox options.
For the fitness function, we have employed the famous function that is much used to optimize image contrast enhancement. Its equation is as follows:

$$F(X) = \log(\log(E(\mathrm{Im}(X)))) \times \frac{n_{\mathrm{edgels}}(\mathrm{Im}(X))}{M \times N} \times H(\mathrm{Im}(X)) \tag{6}$$

where $F(X)$ is the fitness function, $X =$ (NBH, NBV, B, α, and β) is the decision vector in the objective space, $\mathrm{Im}(X)$ is the enhanced image according to the local transformation using NBH $\times$ NBV non-overlapping blocks, and the s-curve whose equation is represented in Eq. 5. $E(\mathrm{Im}(X))$ and $n_{\mathrm{edgels}}(\mathrm{Im}(X))$ are the intensity and the number of edgel pixels, respectively, detected with a Sobel edge detector for the enhanced image $\mathrm{Im}(X)$. M and N are the number of rows and columns of the image. Lastly, $H(\mathrm{Im}(X))$ measures the entropy of the image $\mathrm{Im}(X)$.

Assessment Methodology: To objectively evaluate the performance of the proposed method, we have assessed, on the one hand, the enhancement of contrast and edges using, respectively, the parameters EME and EC. On the other hand, the conservation of brightness and information contained in the image using, respectively, the parameters AMBE and FSIM. The expressions of the parameters mentioned above are as follows [8, 10, 17]:

$$\mathrm{EC} = \frac{1}{M.N} \sum_{i=1}^{M} \sum_{j=1}^{N} \sqrt{|G_x^2 + G_y^2|} \tag{7}$$

where G_x and G_y are the magnitude of gradient in the horizontal and vertical direction. M and N represent the size of the image.

$$\text{EME} = \frac{1}{K_1.K_2}.\sum_{i=1}^{k_1}\sum_{j=1}^{k_2} 20.\log\frac{Imax(i,j)}{Imin(i,j)} \tag{8}$$

K_1 and K_2 are non-overlapping blocks of size 5 × 5 pixels in the horizontal and vertical direction; $I_{\max}(i, j)$ and $I_{\min}(i, j)$ are, respectively, the maximum and minimum intensity values in each block.

$$\text{AMBE} = |E(S) - E(R)| \tag{9}$$

where $E(R)$ and $E(S)$ are the mean of the intensity for original and enhanced image, respectively.

$$\text{FSIM} = \frac{\sum_{X\in\Omega} S_L(x).\text{PC}_m(x)}{\sum_{X\in\Omega} \text{PC}_m(x)} \tag{10}$$

where $\text{PC}_m(x)$ is the feature perceived at a point where the Fourier components reach maximum in phase, $S_L(x)$ is the global similarity between original image and the enhanced one, and Ω is the entire image spatial domain.

Finally, a comparison between the results found and those given using histogram equalization HE, brightness preserving bi-histogram equalization BBHE, and contrast limited adaptive histogram equalization CLAHE is done.

4 Results and Discussion

In this section, quantitative and qualitative results of the applied method and the comparison techniques are presented. Table 1 shows the resulting decision vectors for all tested images. The results prove that each image has its own local transformation parameters (NBH, NBV, B, α, and β) which allows better enhancement. The quantitative results of parameters EC, EME, AMBE, and FSIM are recapitulated in Table 2 where it is clearly seen that the proposed method improves all considered

Table 1 Results of decision vectors

Image	NBH	NBV	B	α	β
mdb209	52	51	8.746	0.496	0.00013
mdb210	53	58	6.430	0.472	0.00330
mdb211	61	52	9.396	0.425	0.0015
mdb212	52	65	4.941	0.497	0.00081

Table 2 Quantitative results of all tested mammogram images

Images		EC	EME	AMBE	FSIM
MDB209	Original image	1.345	6.781	0	1
	HE	1.604	0.929	102.834	0.857
	BBHE	1.939	2.784	25.539	0.907
	CLAHE	4.297	5.679	20.672	0.796
	Proposed method	2.286	9.236	0.271	0.9167
MDB210	Original	1.326	6.3546	0	1
	HE	1.637	0.926	97.110	0.857
	BBHE	2.151	2.624	26.265	0.883
	CLAHE	4.564	5.822	18.029	0.765
	Proposed method	2.323	9.539	0.546	0.924
MDB211	Original	1.119	4.786	0	1
	HE	0.792	0.367	112.561	0.868
	BBHE	1.459	1.682	27.036	0.912
	CLAHE	3.543	4.493	6.858	0.809
	Proposed method	1.901	6.528	1.193	0.946
MDB212	Original	0.954	5.267	0	1
	HE	0.933	0.4494	114.291	0.841
	BBHE	1.371	1.725	24.636	0.900
	CLAHE	3.464	4.625	8.4205	0.803
	Proposed method	1.716	7.556	0.198	0.951

IQA metrics, high values for EC, EME, and FSIM and low values for AMBE. The EC parameter is increased from about 1.3 to about 2.29 for both mdb209 and mdb210 and from about 1 to about 1.8 for both mdb211 and mdb212. The EME parameter is increased for all tested images from an average of 5.795 to an average of 8.214 for all tested images. As for the two parameters AMBE and FSIM, proposed method is characterized by a low AMBE value with an average of 0.55 and a high FSIM value with an average of 93.44%. However, the techniques HE, BBHE, and CLAHE have varied results from an image to another and from a parameter to another. HE, BBHE, and CLAHE have improved the parameter EC for all images except for the images mdb211 and mdb212 whose EC has been decreased by HE. Yet they have decreased EME for all images. AMBE is too high for HE (from 97.11 for mdb210 to 114.291 for mdb212) and high for both BBHE (from 24.636 for mdb212 to 27.036 for mdb211) and CLAHE (from 6.858 for mdb211 to 20.672 for mdb209). For FSIM, its average is 79.3%, 85.5%, and 90% for CLAHE, HE, and BBHE, respectively. Visual inspection from Fig. 2 shows that the resulting images using contrast enhancement techniques HE, BBHE, and CLAHE (Fig. 2b–d) are too bright due to the high value of the parameter AMBE which is too high for HE and high for BBHE and CLAHE. In addition, some details are lost in the resulting image with the techniques of comparison; this can be justified by the low values of FSIM. However, improved images with

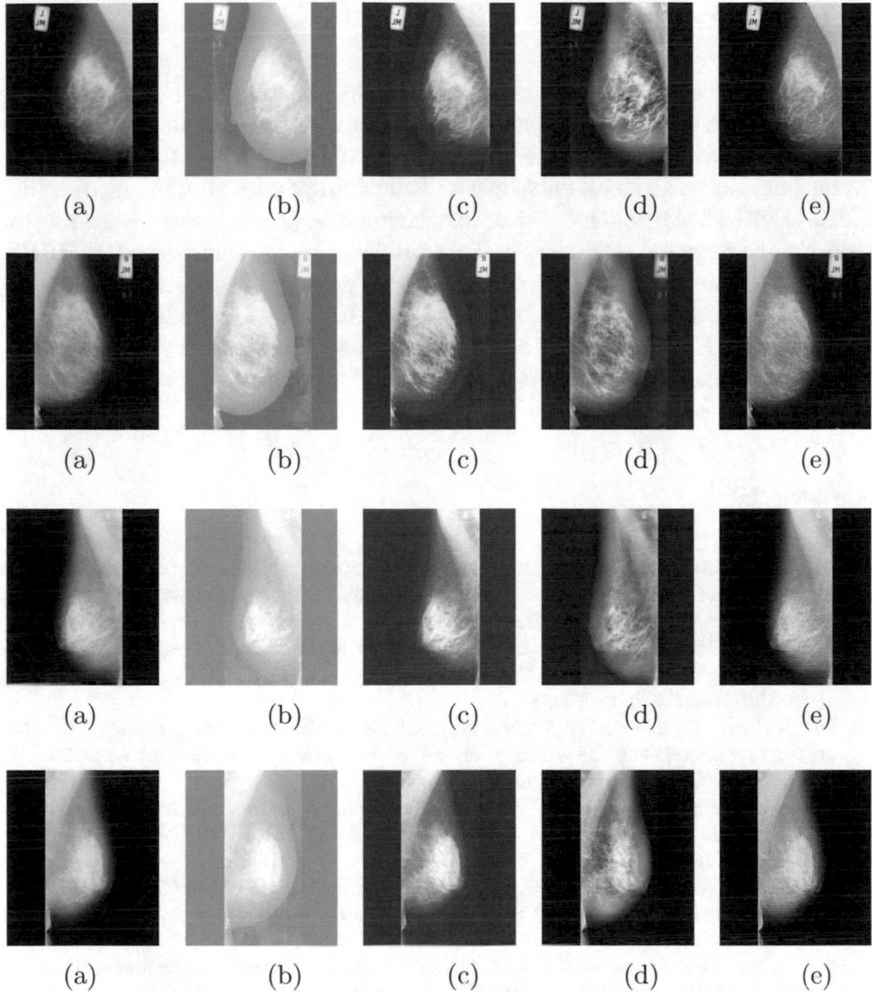

Fig. 2 Qualitative results for all tested images: **a** Original image, **b–e** enhanced image using: HE, BBHE, CLAHE, and proposed method

proposed method (Fig. 2e) are more prominent and faithful to the original images as far as brightness is concerned. The parameter AMBE is very low as shown in Table 2. Moreover, the resulting images show that they have more visual clarity and more clear details. The suggested method allows a better enhancement of the edges and a better contrast among the regions of the image. Furthermore, it enhances the image features compared to the other techniques.

5 Conclusion

In this paper, an adaptive local gray-level transformation based on variable s-curve for contrast enhancement of mammogram images was presented. The method is based on genetic algorithm to look for the best combination of subdivision into non-overlapping blocks and the s-curve that will allow to optimize all IQAs metrics (EC, EME, AMBE, FSIM) which indicates that the improvement is better. Comparing the results of the proposed method with those results using the techniques HE, BBHE, and CLAHE demonstrated the effectiveness of proposed approach in improving the contrast and the edges while maintaining the brightness and information of the original image. Future work will be devoted to study the effect of this improvement on the segmentation and precise extraction of regions of interest.

References

1. Akila, K., Jayashree, L.S., Vasuki, A.: Mammographic image enhancement using indirect contrast enhancement techniques—a comparative study. Proc. Comput. Sci. **47**, 255–261 (2015). https://doi.org/10.1016/j.procs.2015.03.205
2. Srivatava, H., Mishra, A.K., Bhateja, V.: Non-linear quality evaluation index for mammograms. In: 2013 Students Conference on Engineering and Systems (SCES) 2013. https://doi.org/10.1109/SCES.2013.6547534 (2013)
3. Jain, A., Singh, S., Bhateja, V.: A robust approach for denoising and enhancement of mammographic images contaminated with high density impulse noise. Int. J. Converg. Comput. **1**, 38 (2013). https://doi.org/10.1504/ijconvc.2013.054659
4. Panetta, K., Samani, A., Agaian, S.: Choosing the optimal spatial domain measure of enhancement for mammogram images. Int. J. Biomed. Imaging **2014**, 1–8 (2014). https://doi.org/10.1155/2014/937849
5. Patel, B.C., Sinha, G.R.: Gray level clustering and contrast enhancement (GLC-CE) of mammographic breast cancer images. CSI Trans. ICT **2**, 279–286 (2015). https://doi.org/10.1007/s40012-015-0062-z
6. Yao, Z.: Image enhancement based on bi-histogram equalization with non-parametric modified technology (2016). https://doi.org/10.1109/ICPADS.2016.160
7. Jaya, V.L., Gopikakumari, R.: IEM: A new image enhancement metric for contrast and sharpness measurements. **79**, 1–9. https://doi.org/10.5120/13766-1620 (2013)
8. Zhang, L., Zhang, L., Mou, X., Zhang, D.: Correspondence FSIM: A feature similarity index for image. IEEE Trans. IMAGE Process **20**, 2378–2386 (2011). https://doi.org/10.1109/TIP.2011.2109730
9. Liu, S., Zeng, J., Gong, H., et al.: Quantitative analysis of breast cancer diagnosis using a probabilistic modelling approach. Comput. Biol. Med. **92**, 168–175 (2017). https://doi.org/10.1016/j.compbiomed.2017.11.014
10. Gandhamal, A., Talbar, S., Gajre, S., et al.: Local gray level S-curve transformation—A generalized contrast enhancement technique for medical images. Comput. Biol. Med. **83**, 120–133 (2017). https://doi.org/10.1016/j.compbiomed.2017.03.001
11. Astley, S., Betal, D., Cerneaz, N., Dance, D.R., Kok, S.-L., Parker, J., Ricketts, I., Savage, J., Stamatakis E., Taylor, P.: the mini-MIAS database of mammograms (1994). http://peipa.essex.ac.uk/info/mias.html. Accessed 1 Mar 2019
12. Gonzalez, R.C., Woods, R.E.: Digital Image Processing, 2nd edn. Prentice-Hall Press, ISBN 0-201-18075-8 (2002)

13. Bhateja, V., Misra, M., Urooj, S.: Human visual system based Unsharp masking for enhancement of mammographic images. J. Comput. Sci. (2016). https://doi.org/10.1016/j.jocs.2016.07.015
14. Bhateja, V., Misra, M., Urooj, S.: Unsharp masking approaches for HVS based enhancement of mammographic masses: A comparative evaluation. Futur. Gener. Comput. Syst. (2017). https://doi.org/10.1016/j.future.2017.12.006
15. Lal, S., Chandra, M.: Efficient Algorithm for Contrast Enhancement of Natural Images. **11**, 95–102 (2014)
16. Kciuk, M., Chwastek, K., Kluszczynski, K., Szczyglowski, J.: A study on hysteresis behaviour of SMA linear actuators based on unipolar sigmoid and hyperbolic tangent functions. Sens. Actuators, A Phys. **243**, 52–58 (2016). https://doi.org/10.1016/j.sna.2016.02.012
17. Gupta, S., Porwal, R.: Appropriate contrast enhancement measures for brain and breast cancer images (2016). https://doi.org/10.1155/2016/4710842 (2016)

New Objective Function for RPL Protocol

Soukayna Riffi Boualam and Abdellatif Ezzouhairi

Abstract Nowadays, connected objects are considered as the most used network for short-range services. As IPv6 is no longer appropriate to support this kind of communication, the IPv6 routing protocol for LLNs (RPL) is proposed by IETF as an alternative to overcome the aforementioned drawback. Accordingly, researchers are focused on RPL protocol to adapt LLN requirements to the IoT context. The selection of optimal routes is based on predefined objective functions. Nevertheless, these objective functions lack to use the right combination of parameters which limits drastically the energy consumption and the quality of data transmission. In this paper, we propose a new FO that aims to improve the quality of selected routes by using combined metrics as well as fuzzy logic. The obtained results show that our solution performs better than standard RPL in terms of stability, energy consumption, and packet delivery.

Keywords IOT · RPL · Objective function · Energy consumption · ETX · HC · Packet delivery · Fuzzy logic

1 Introduction

The low power and lossy networks (LLN) are composed of several interconnected nodes that have low power, and their batteries are limited with a low lifetime and a short transmission range. The big challenge of the IoT is to provide reliable communications with minimum packet loss and low power consumption. In addition, the instability of IoT networks leads to frequent loss of wireless link as well as depletion of battery. Hence, existing routing protocols such as ad hoc and dynamic source routing (DSR) are not suitable for this new connected objects context. Recently, this issue has attracted the attention of several researchers to find a protocol that meets

S. R. Boualam (✉)
Faculty of Sciences and Technique, University Sidi Mohammed Ben Adbellah, Fez, Morocco

A. Ezzouhairi
Renewable Energies Laboratory and Intelligent Systems, Fez, Morocco
e-mail: abdellatif.ezzouhairi@polymtl.ca

V. Bhateja et al. (eds.), *Embedded Systems and Artificial Intelligence*,
Advances in Intelligent Systems and Computing 1076,
https://doi.org/10.1007/978-981-15-0947-6_64

the requirements of this new tendency. To meet these features, the routing over low power and lossy networks(ROLL) group relevant to the Internet Engineering Task Force (IETF) develop a routing architecture for 6LoWPAN networks called RPL [1] that allows to build a routing topology on constrained networks and leaves a large number of possible settings in order to adapt it to a specific environment.

RPL is a proactive protocol based on distance vectors for the low power and lossy networks specified by the IETF ROLL working group as a solution to allow small connected objects to be integrated into a large-scale network by using easy mechanisms and metrics. RPL supports three types of communication: point-to-point, point-to-multipoint, and multipoint-to-point.

RPL builds a destination-oriented direct acyclic graph (DODAG) [2] to route data to the base station. The built DODAG allows each DODAG node to transmit the data that has collected up to the root DODAG. Each node in the DODAG selects a parent according to a given routing metric and a well-defined OF. The collected data is then forwarded from parent to root.

RPL responds to a number of limitations by varying the link and node metrics as well as the objective function. The OF allows to discover and determine the best solution to find an optimal route during data transmission. This key factor is based on the metrics to choose the adequate parent up to the root. The well-known standard OFs are based on a single metric. More specifically, the OF0 [3] and MRHOF [4] are based on the number of hops (HC) and the expected transmission number (ETX), respectively. The use of multiple metrics allows performance improvement over the use of a single metric by choosing the best parent without taking into account other criteria that may impact the nodes chosen during data transmission such as the choice of a link with a poor quality, and also the choice of the same route during each transmission will exhaust the energy of the node [5] which impacts the lifetime of the network.

2 Related Work

In the past few years, several researches are interested to low power and lossy networks (LLN) issue especially when applied in a difficult environments such as the one relevant to IoT. The main concern of these studies consists in improving the quality of the selected routes based, eventually, on a combination of metrics used by a predefined OF. Among these studies, we find the one based on a single metric like the OF0 [3] which uses the hop count as the only metric to select the best parent to transmit data. Each node chooses the neighbor who has the minimum rank during transmission. However, this proposal suffers from drawbacks such as choosing the same node whenever route selection is initiated which affects drastically energy consumption and link quality.

Another objective function commonly designated by MRHOF is proposed [4]. It is based on the expected transmission number ETX which is considered as the only criterion to select the appropriate parent without taking into account the residual energy. The use of this node metric allows a long network lifetime but risks a bad choice of the link that can impact the quality of the information transmitted.

The existing objective functions OF0 and RHMOF are based on a single metric that are ineffective [5] because the latter may not meet the requirements of the application at this time there will be a degradation of the performance of the DAG.

In [6, 7], authors have proposed a new objective OF-FL function for LLN networks which is based on the combination of four nodes and link metrics. They used fuzzy logic to choose the optimal route based on end-to-end delay, ETX, HC, and battery level. Nevertheless, this solution is only compared to the standard OF: OF0 and MRHOF, which limits its reliability compared to other OF.

In [8, 9], authors have proposed an improvement of the RPL routing protocol by improving the objective function. As this work proposes a combination of two metrics, ETX and residual energy using fuzzy logic to choose the preferred parent to transmit information to the root, also taking into account (HC = 1) and this OF is called OF-EC, it allows to equalize the distribution of energy consumption between all the nodes to avoid the death of the nodes in the case of use successively during the transmission. But, this OF that combine only two metrics is insufficient to satisfy all the requirements of an application, the purpose of each LLN application may differ from one application to another.

In [10, 11], authors have designed and implemented a proposal that uses of the OF based on the residual energy [10] to choose the best route. But this solution cannot guarantee the good link of transmitted data as well as transmission delay. After it to improve its objective function by creating a new objective function RPL combining several metrics "ETX–Délai–Energy" [11], this study has made it possible to take into account QoS considering more than one parameter of network performance such as packet loss rate, routing stability, and energy efficiency, and even end-to-end delay. But this solution in only compared to the standard OF: MRHOF, which limits its reliability compared to other OF.

In the rest of our paper, we will present a new objective function based on a combination of the two previous methods OF0 and MRHOF taking into account the residual energy of nodes during each operation, so in our case, we will use three metrics are ETX, HC, and the residual energy.

3 Proposed Objective Solution

1. Fuzzy logic method:

Fuzzy logic is a method that allows to formalize the inaccuracies due to a global knowledge of a complex system and the expression of the behavior of a system by words, as it allows to transform several input variables into only output variable, the fuzzy inference system takes into consideration four steps:

- Fuzzification: This step consists of characterizing the linguistic variables used in a system, it transforms input values into a fuzzy part defined on a representation space linked to the input.
- Fuzzy Inference: This step allows you to apply combination rules of input values to calculate a fuzzy output.
- Aggregation: This step consolidates all values into one.
- Defuzzification: Converts a fuzzy out to a net value.

In this article, we will use the fuzzy inference method called Mamdami because of its simplicity to be applied in a study.

2. Linguistic variables:

To deal with the aforementioned drawbacks, we identify three important metrics while choosing the preferred parent. The combination of these metrics is expected to create an adequate and reliable path to the root by using fuzzy logic.

More specifically, we consider ETX which refers the expected number of transmissions for a packet to be transmitted correctly to its destination. It is an indicator of the link quality between the nodes in a DAG. For this reason, we can say that ETX rate gives us an indication about link quality and reliability. More than the ETX value is high plus the quality of the link is not reliable for data transmission, for that to guarantee an adequate connection, it is necessary that the value ETX tends toward zero.

The second criterion considered in our proposal is the number of hops which refers to the number of hops between each neighboring node and the root. This metric is necessary to reach the destination where the number of hops is minimal. However, some shorter paths may have bad link quality than other paths and higher energy dissipation due to congestion/overload. Thus, the combination of this metric with ETX will be more appropriate for real-time applications by optimizing the preferred parent selection process.

The last considered is the remaining energy of the node. During each transmission, the energy of used nodes decreases. Hence, this metric is used to avoid choosing nodes with low energy.

In order to take benefit of the above-described parameters (ETX, HC and energy), we propose to use the fuzzy logic to improve the route decision-making process. To have an idea about the performance of nodes, in Fig. 1, we present the considered linguistic variables through a combination performed by a fuzzy inference engine.

3. Fuzzification process:

To illustrate the fuzzy process, we chose to do a two-step combination: in the first place, we will combine two metrics are ETX and HC which will be considered input variables, and after the calculation of the output value which is the QoS that will combine with the residual energy in the second stage.

- The linguistic variables used to represent ETX are: small, average, and high.
- The linguistic variables used to represent HC are: near, vicinity, and far.

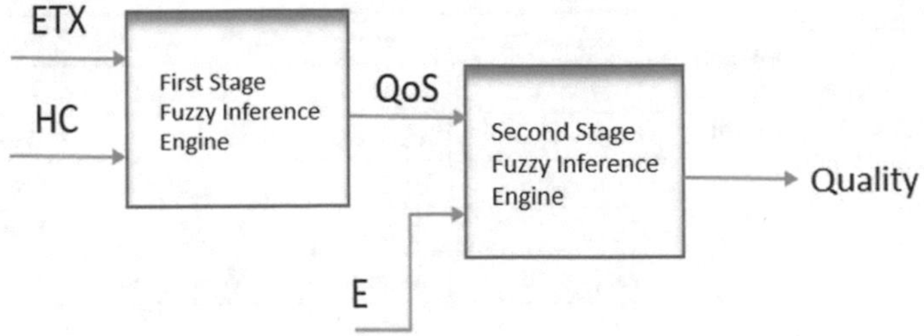

Fig. 1 Fuzzy inference engine

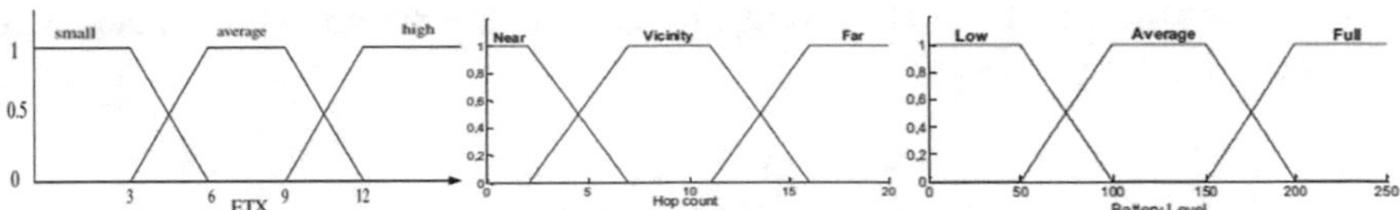

Fig. 2 Membership functions of ETX, battery level, and HC

Table 1 Illustrates the relationship between ETX and HC (we consider that the delay is fixed) to calculate the output variable

ETX/HC	Near	Vicinity	Far
Small	Very-good	Good	Average
Average	Good	Average	Bad
High	Average	Bad	Very-bad

- The linguistic variables used to represent the residual energy are: low, average, and high.

In Fig. 2, we illustrate the membership function relevant to each one of the considered parameters that will be used as our objective function (Table 1).

Three QoS functions are provided from these membership functions: Very-good, Good, and Average, the Formula 1 indicates how to compute average (OoS) fuzzy set from inputs.

$$\text{avg(QoS)}=\max\begin{pmatrix}\text{Min(High(ETX); Near(HC))}\\ \text{min(Avg(ETX); Vicinty(HC)}\end{pmatrix} \tag{1}$$

$$\text{min(Small(ETX); Far(HC))}$$

If we choose randomly ETX = 4 and HC = 60, the membership level will be calculated as follows.

Small (ETX) = 0.66; Average (ETX) = 0.33; High (ETX) = 0 and Near (HC) = 0.83; Vicinty (HC) = 0.16 and Far (HC) = 0. Knowing that the quality of service depends solely on ETX and HC, the Mamdani model [12] will allow us to calculate

Table 2 Illustrates the relationship between QoS and energy consumed by each node in order to calculate the output variable "quality"

QoS/energy	Low	Average	Full
Very-bad	Awful	Very-bad	Average
Bad	Very-bad	Bad	Average
Average	Bad	Average	Good
Good	Average	Good	Very-good
Very-good	Average	Very-good	Excellent

the fuzzy set (QoS) from these entries as follows: Very-good (QoS) = 0.83, Good (QoS) = 0.33, and Avg(QoS) = 0.66.

The defuzzification method converts all values into a single output value using the center of gravity.

$$CG = \frac{\sum_{i=1}^{k} \mu(x) * x_i}{\sum_{i=1}^{k} \mu(x)}$$

CG is the abscissa of the center of gravity of the polygon defined by the values of the variable linguistic output. *k* is the number of rules enabled in the engine of inference, $\mathbf{x_i}$ presents the relative value domain to the *i*th rule and $\mu(x)$ is the level of trustiness value according to corresponding domain.

The first stage of fuzzification: In the same way, the second combination is carried out, Table 2 shows the combination between the QoS and the residual energy.

$$CG = \frac{\sum_{i=1}^{k} \mu(x) * xi}{\sum_{i=1}^{k} \mu(x)} - \frac{0,2 * 60 + 0,66 * 72 + 0,8 * 83}{0,8 + 0,66 + 0,2} = 75,85$$

After the calculations, we get the level of quality as equal: **75.85%**.

4 Performance Analysis

To implement our new OF, we use the COOJA simulator [13] running on the Contiki operating system (version 2.7). Contiki OS is a lightweight open-source operating system designed for the Internet of things. Contiki has been focused on several platforms running on different types of processors [14]. The main advantage of the Contiki is that it works on a concept that is between multi-threading and event programming, this allows the processes to share the same execution context and thus improve the memory usage and energy. Table 3 shows the parameters and values that will be used during our simulation.

In this sub-section, we will evaluate the performance of our new OF by making a comparison with the standard OF and also other existing OF.

Table 3 COOJA parameters

Network simulator	COOJA under Contiki OS (2.7)
Number of node	20, 40, 60
Emulated nodes	Tmote Sky
Deployment type	Random position
Interference range	100 m
Total simulation time	2 J

Number of parents changed: This is the number of times that a mobile node changes its parent to transmit packets based on the OF, this parameter is an indicator of the stability of the environment that allows to choose the most appropriate and the best parent.

Figure 3 shows the number of parents changed on average for four OFs (OF0, MRHOF, OF-FL, and OF-EHE). This simulation measures the stability of our network, i.e. the number of parents exchanged based on the new OF. We can notice on the one hand that the number of parents changed for the two standard OFs is less than the two others OF, this result is due to the fact that they do not take into account the optimization of the road but aim only to minimize the rank between a node and the root is the choice of a route where the number of retransmission of the data is minimal which implies the rare change of the parent. On the other hand, the network that is based on our new OF illustrates a smaller number of parents traded than the OF-LF-based network which helps to reduce that our network is more stable by improving the quality of transmission.

The lifetime of the network: The OF has a direct impact on the lifetime of the network precisely the energy consumed by each node, the objective of our study is to track the energy consumption of the nodes during a communication. For this reason, we track node energy for 20 nodes to make a comparison of the network lifetime for each objective function used.

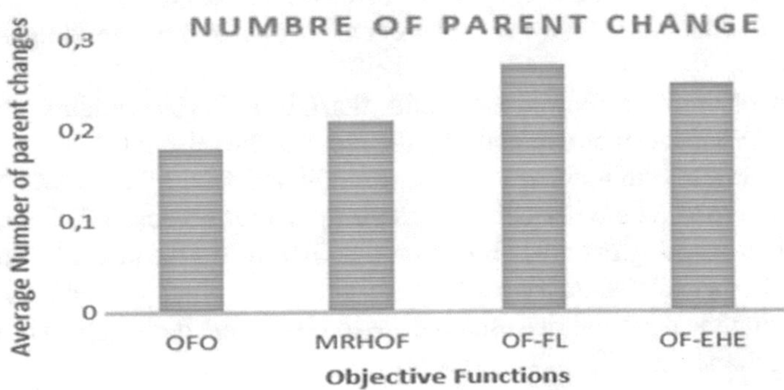

Fig. 3 Comparison between OF0, MRHOF, OF-FL, and OF-EHE in term of the average number of parent changes

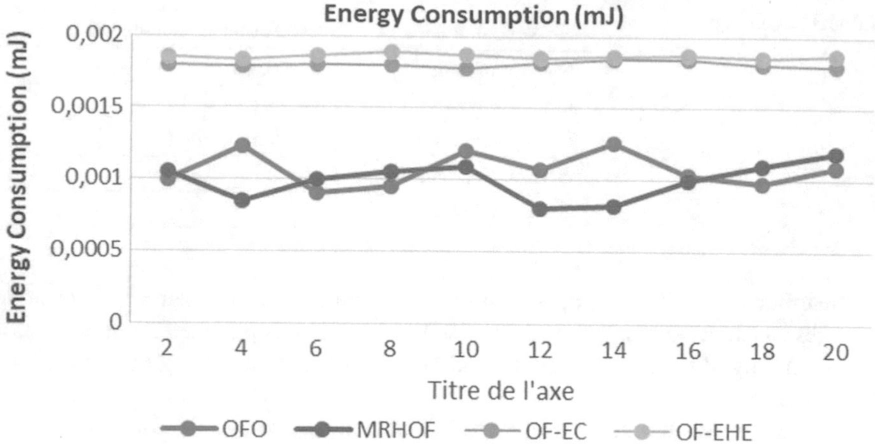

Fig. 4 Comparison between OFs in term of consumed energy

Figure 4 illustrates the distribution of the energy consumption of each nodes along the network, it is visible for the two OFs: OF0 and MRHOF that the energy consumption is low because they are based on a single metric and do not take into account the quality of the transmission link which will impact the performance of the network, also the distribution of energy is not balanced between the nodes which causes the rapid death of some of them and creates a disturbance and a loss of information during the transmission of the data, on the contrary, for the other two OF energy consumption is well distributed between the nodes which will delay the depletion of the battery of the first nodes. It is remarkable that OF-EC and OF-EHE consume more energy because of the crazy calculation that takes time to find the optimal route. The consumption of more energy will impact the survival of the network but not the failure of the nodes.

Packet delivered ratio: It is clear that to measure the reliability of our network to know the number of packets delivered successfully compared to the number of packets sent, and for this reason, we have varied the number of nodes on the network from 20 to 60 nodes in order to see the effect of successfully transmitting packets on the network.

Our simulations have allowed us to notice that OF0 offers less reliability compared to the other OFs that can be explained by the fact that the selected node and its parent taking into account the number of hops only without taking into account the link quality of the road. The other OF offer more reliability because ETX is taken into account when choosing the route; however, our OF remains the best OF which offers a large PDR especially when the network is more dense which justified the use of an OF combining the three metrics used before to choose the best links to transmit the data more reliably (Fig. 5).

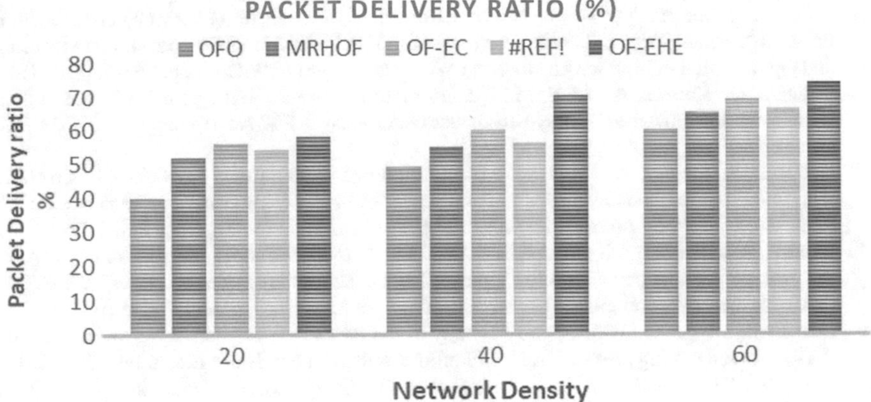

Fig. 5 Comparison between OFs in terms of PDR

5 Conclusion

In this paper, we presented a new FO for LLNs that improves the RPL protocol, comparing it with other existing OF, OF-EHE combines three important metrics of nodes and links are (ETX: number expected retransmission, HC: number of jumps between a node and root and consumed energy) using fuzzy logic. This comparison is made with some OFs like OF0, MRHOF, and OF-FL, the results of the simulation have approved that OF-EHE offers the best performance compared to other OFs in terms of energy consumption in a balanced way, the rate of delivered packages, and the stability of the environment. In this paper, we gave the same importance to each metric used.

In the future work, we will further improve protocol RPL by evaluating the performance of its metrics and ameliorate the quality of the transmission line by choosing other methods to do the combination to select the proper route for each application in short time.

References

1. Brachman, A.: RPL objective function impact on LLNs topology and performance. In: Internet of Things, Smart Spaces, and Next Generation Networking, vol. 8121, pp. 340–351 (2013)
2. Iova, O., Theoleyre, F., Noel, T.: Using multiparent routing in RPL to increase the stability and the lifetime of the network. Ad Hoc Netw. **29**, 45–62 (2015). Available at http://dx.doi.org/10.1016/j.adhoc.2015.01.020
3. Thubert, P.: Objective function zero for the routing protocol for low-power and lossy networks (RPL). RFC **6552**, 1–14 (2012)
4. Gnawali, O., Levis, P.: The ETX Objective Function for RPL. IETF Internet Draft: draft-gnawali-roll-etxof-00 (2010)

5. Gaddour, O., Koubaa, A., Baccour, N., Abid, M.: Of-fl: Qos aware fuzzy logic objective function for the RPL routing protocol. In: Proceeding of the 12th International Symposium on Modeling and Optimization in Mobile, Ad Hoc, and Wireless Networks (WiOpt), pp. 365–72 (2014)
6. Gaddour, O., Koubâa, A., Abid, M.: Quality-of-service aware routing for static and mobile IPv6-based low-power and lossy sensor networks using RPL. Ad Hoc Netw. **33**, 233–256 (2015)
7. Gaddour, O., Koubäa, A., Rangarajan, R., Cheikhrouhou, O., Tovar, E., Abid, M.: Co-RPL: RPL routing for mobile low power wireless sensor networks using corona mechanism. In: 9th IEEE International Symposium on Industrial Embedded Systems, 2014, pp. 200–209
8. Lamaazi, H., Benamar, N., Imaduddin, M.I., Jara, A.J.: Performance assessment of the routing protocol for low power and lossy networks. In: The International Workshop WINCOM (Wireless Networks and mobile COMmunications) in Marrakech, Morocco (2015)
9. Lamaazi, H., Benamar, N., Antonio J.: Study of the impact of designed objective function on the RPL-based routing protocol. Adv. Ubiquitous Netw. 2, Lect. Notes Electr. Eng. 397 (2016)
10. Kamgueu, P.O., Nataf, E., Ndié, T.D., Festor, O.: Energy-based routing metric for RPL. [Research Report], INRIA. 2013, pp. 14. http://hal.inria.fr/hal-00779519 (2013)
11. Kamgueu, P.O., Nataf, E., Djotio, T.N.: On design and deployment of fuzzy-based metric for routing in low-power and lossy networks. In: Proceedings—Conference Local Computer Networks, LCN, vol. 2015–December, pp. 789–795 (2015)
12. Mamdani, E.H.: Application of fuzzy logic to approximate reasoning using linguistic synthesis. IEEE Trans. Comput. **12**, 1182–1191 (1977)
13. Osterlind, F., Dunkels, A.: Contiki COOJA Handson crash course: session notes. Sics.Se (July) (2009)
14. Saad, L.B., Chauvenet, C., Tourancheau, B.: Simulation of the RPL routing protocol for IPv6 sensor networks. In: 5th International Conference on Sensor Technologies and Applications (2011)

A Smart Data Approach for Automatic Data Analysis

Fouad Sassite, Malika Addou and Fatimazahra Barramou

Abstract The purpose of this paper was to propose an automatic data analysis approach to achieve real-time processing for heterogeneous collected big data volumes from multiple sources. With this paradigm, which has limited traditional analytical capabilities to deal with the rapidly changing data storage capacities, the importance was focused on the value aspect of the smart data approach in order to accelerate data processing and decision-making. It was therefore important to find a method of extracting the most relevant data (smart data), the most useful in real time, from the massive quantities of data from the big data. A multilayer architecture and a projection of agents on this architecture were proposed for real-time big data processing using the smart data approach, with three layers: data acquisition, data management and processing and data services.

Keywords Automatic data analysis · Big data · Smart data

1 Introduction

Today, with the big data approach, we are witnessing an increase in data storage capacity and an explosion of collected data volumes. This becomes a problem in itself, given the traditional analysis capabilities of these masses of data.

Smart data is a new approach that responds to this problem by providing real-time data analysis by extracting the most relevant data from the large amount of data collected by big data. This paradigm will be very useful in the automatic analysis of these collected big data in order to optimize the processing time, especially with

F. Sassite (✉) · M. Addou · F. Barramou
Architecture, System and Networks Team (ASYR)—Laboratory of Systems Engineering (LaGeS), Hassania School of Public Works EHTP, BP 8108 Oasis Casablanca, Morocco
e-mail: fouad.sassite@gmail.com

M. Addou
e-mail: malika.addou@gmail.com

F. Barramou
e-mail: f.barramou@gmail.com

V. Bhateja et al. (eds.), *Embedded Systems and Artificial Intelligence*,
Advances in Intelligent Systems and Computing 1076,
https://doi.org/10.1007/978-981-15-0947-6_66

real-time data processing, this paper will be organized as follows: the section "big data vs smart data" presents the two paradigms and discuss the transition from big to smart data. The next section "automatic data analysis" will present some approaches to achieve automatic data analysis in real time. The last section "proposed approach" will present a proposition of a multilayer architecture for automatic analysis of big data using multi-agent approach.

2 Big Data Versus Smart Data

Big data is a data characterized by high volume, velocity, variety, does not match the constraints of database architecture and cannot be processed with conventional database systems and need an alternative way to process it to extract value from this data [1].

Smart data or intelligent data offers the value aspect by extracting relevant data from the massive big data quantities. By using semantics and intelligent processing, to exploit the **value** and overcome the challenges of the 4 V related to the big data: variety, velocity, veracity and volume [2], and providing such relevant and useful data can help to improve the real-time processing and the decision-making.

2.1 From Big Data to Smart Data

It is important to expose the main features in the process of transformation the big data to smart data [3] (Fig. 1):

Automate analysis. The process of decision-making needs to overcome the challenge of turning real-time data into valuable decisions this process should automate the operational decision-making in order to turn data dynamically and in the real-time to valuables decisions.

Forward looking. To have a predictive view to understand the impact of events on the organization.

Process oriented. The smart data should be process oriented in order to make the process itself smarter.

Scalable solutions. The aspect of scalability is critical and important in the process of transformation of the big data to smart data and should take into consideration the random evolution of events stream that can generate high data volumes.

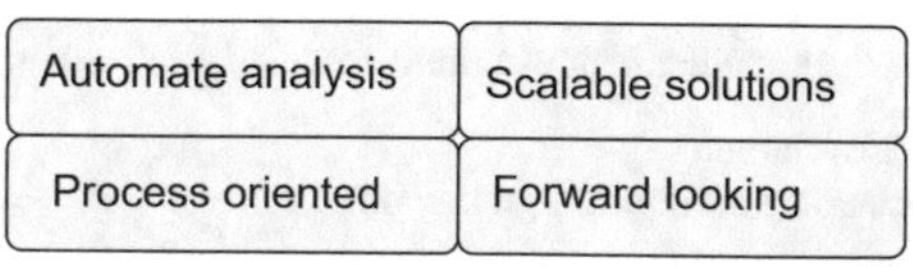

Automate analysis	Scalable solutions
Process oriented	Forward looking

Fig. 1 Important components to transform big data to smart data

2.2 *Models for Smart Data Management*

In order to build smart data architectures to extract valuable data from different sources, there are several approaches focused on:

Ontology. By adopting a progressive and continuous data integration approach for both static and streaming data, using a linked data-based approach to index and query data from multiples and heterogeneous sources based on semantic Web technologies [4].
Data quality and service optimization. This approach focuses on data services and introduces the concept of quality-aware service oriented in the process of data integration which takes into consideration multiples aspects such as licensing, service and data quality [5].
Data source characteristics. In order to exploit the unused generated data, due to difficulty to manage the access from multiple separated data sources, the aim of this approach is to give the possibility to combine multi-origin data in a smart and coherent dataset, through the definition of data source meta-models and data processing workflows. The data source meta-models give the possibility to describe the characteristics of each data source to propose an adaptive architecture that generate runtime integration workflows [6].

3 Automatic Data Analysis

This section presents some approaches in order to achieve an automatic data analysis.

3.1 *Lambda Architecture*

The main concept of lambda architecture is to deal with the challenge of processing big data streams in real time, especially when the system should respond to queries implies online and offline data based on history, the architecture is composed of three main components: speed, batch and serving layers.

An architecture designed for different implementations requiring real time and offline big data processing [7].

Figure 2 shows the main components of lambda architecture [7]: The batch layer is responsible for storing the master dataset. Thanks to MapReduce algorithms, it computes contiguously the views of this data available for the different applications. The serving layer is responsible for serving views calculated by the batch-processing layer. This process can be facilitated by additional data indexing to speed up readings. Moreover, the role of the speed layer is to calculate in real time the data that has just arrived and that not yet been processed by the batch layer. It serves this data in the form of real-time views, which are incremented as new data arrives and can be used with batch views for a complete view of the data.

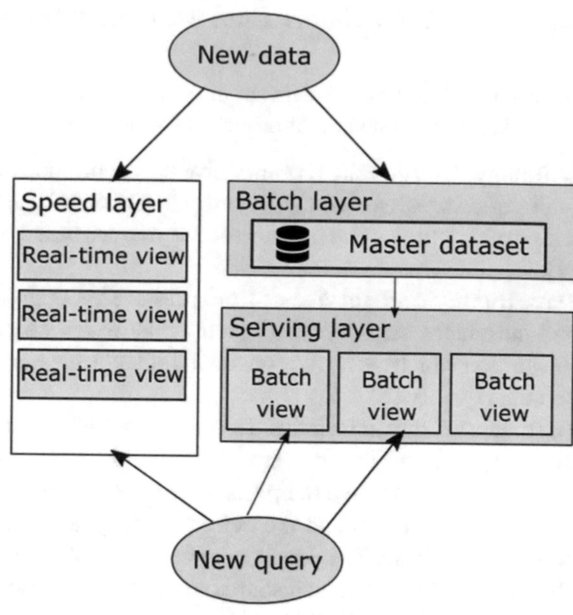

Fig. 2 Lambda architecture. Redrawn based on [7]

3.2 *Real-Time Big Data Processing Based on Multi-agent Approach*

Multi-agent systems are systems composed of a set of interacting computing elements, called agents. Agents are computer systems with an autonomous behavior and they are capable of autonomous actions and interacting with other agents like exchanging data, engaging in cooperation, coordination and negotiation [8].

The combination of the lambda architecture and the multi-agent system paradigms can be very useful for real-time big data processing by using the lambda architecture and autonomous agents to enhance the capabilities for robust data processing in real time [9]. This approach can be applied whenever the big data analysis needs quick responses with short latency.

3.3 *Reference Architecture for Big Data Solutions*

The big data solution reference architecture created to perform quantitative data analysis, and give the representations of the various elements, the reference architecture gives a guideline to build big data solutions and perform predictive analytics; the elements of the model are optional in the architecture [10]. It represents an abstraction of real architectures, a skeleton to adapt for multiples areas of expertise capable to work with big data. It is composed of two patterns the first is pipes and filters and the second is layers. Pipes and filters pattern is more suitable for processing and transforming stream-oriented data flow [11].

3.4 *Automated Real-Time Analytical Process*

Big data analytics process is composed of three process of collecting, organizing and analyzing collected data in the warehouse through data science technologies and data mining techniques, to deal with the challenge of processing the exponential increasing of data in terms of volume, velocity, variety and veracity. The most of real-time big data analytics done by manually accumulated tasks. The nested automatic service composition (NASC) [12] is a technology derived from service-oriented architectural used in this approach to automate big data analytics process in intelligent way. The development of this concept includes the following steps: defining types and instances development for the big data analytics, defining the workflow, developing service discovery algorithm; developing service selection algorithm and developing algorithm result for big data analytics [13].

This approach represents three main layers:

Infrastructure Layer: It essentially considers the data warehouse and the data mart layer.

Technology Layer: It implements the nested automatic service composition and defines multiple agents to assure an intelligent workflow automation facility.

Analytical Layer: This layer provides the data mining process to complete the automation of mining requirement. This approach can be useful to automate big data analytics process, by the separation of the workflow management and the functional modules.

Note: On the four approaches mentioned above, the first three deals with the analysis aspect of big data in real time but not with automation. The fourth deals with the automation aspect of the analysis process in an intelligent way. Our objective is to make a compromise between these approaches, namely the lambda architecture, multi-agent systems, big data flow processing and intelligent automation of this processing by adopting the smart data concept.

4 Proposed Approach

This section introduces the proposed approach of multilayer architecture for automatic big data analysis and exposes the organization of used agents.

4.1 Proposed Architecture

A multilayer architecture approach for the automatic analysis of big data using a multi-agent approach.

This architecture will be composed of three layers: The first is responsible for data collection and generation, the second of the management and processing which will be in charge of the automatic analysis of data by adopting a distributed multi-agent approach. In addition, the third layer will be data services layer capable of transmitting the calculation results to users, applications and services (Fig. 3).

Data with different formats are collected via data collectors from multiple data sources and then transmitted to the data management and processing layer to process and analysis data.

The collected data is then preprocessed and aggregated by a set of defined processing and calculation tasks. Complex processing and automatic analysis tasks are performed in a distributed and parallel manner by using the MapReduce [14] paradigm and an agent-based architecture.

Big data analysis can extract more interesting information than traditional data analysis, and to perform effective big data analysis, it is necessary to properly aggregate the data and perform sufficient preprocessing before data processing.

Data preprocessing performs some important tasks to prepare data to analysis and storage phases and it is responsible of data cleaning, reduction, transformation and normalization or standardization, which plays a very important role when it comes to the parameters of the different units and scales [15].

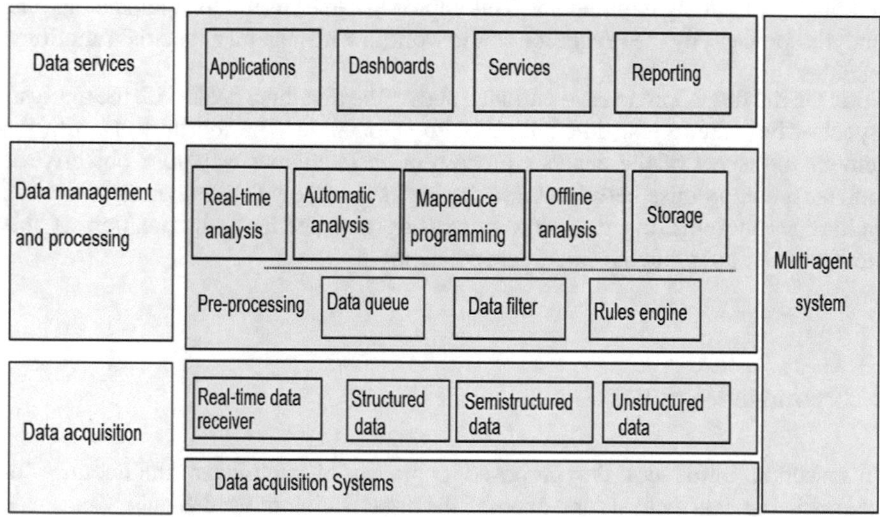

Fig. 3 Agent-based architecture for automatic big data analysis

Similarly, the Kalman filter can be used to speed up data processing by separating valuable and noisy data. Noisy data could be that kind of the data that is not important to process and does not affect real-time processing [16].

The proposed architecture is capable of autonomous data aggregation and preprocessing, intelligent real-time analysis and decision-making, and management of events and notifications. The data processing can be done using various methods and algorithms particularly data mining methods to extract knowledge, find and evaluate patterns [17]. The processing is followed by an associated generation of intelligent decisions.

4.2 Agents' Projection on the Proposed Architecture

The proposed approach is composed of several agents, in order to collaborate to perform automatic real-time big data analysis. The agents orchestrate the processing and dialog between the different layers through collaboration between them (Fig. 4).

The proposed architecture contains many agents, the collaboration between them can solve many complex problems, by providing a greater amount of flexibility and the ability of configuration the system [18]. The role of each agent is explained.

Starting with the receiver agent, which is a reactive agent, who will receive the data from the acquisition layer, in order to transfer them to the data management and processing layer. Mainly, in the processing layer, a cognitive agent and manager agent are responsible of coordination between preprocessing, storage and analysis agents, and the decision of delegate the task based on the query or the data type. The stream analysis agent is responsible for real-time data analysis and the offline analysis agent for massive data analysis in history and preparation of pre-computed data for stream analysis and then storage agent, which is a reactive agent capable of managing the

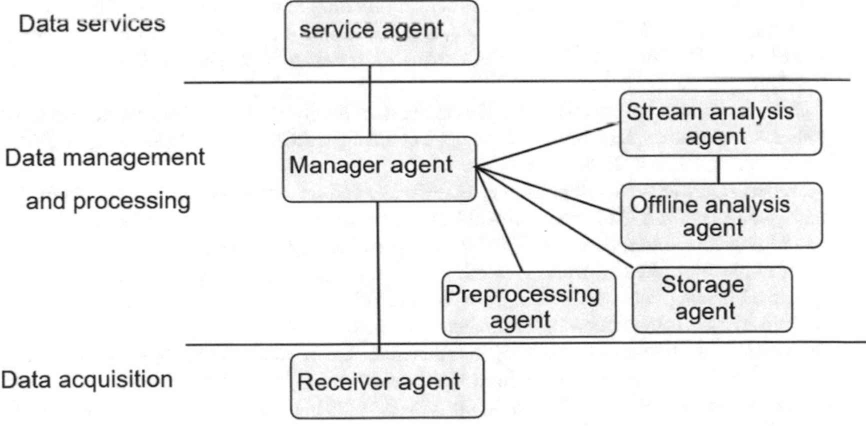

Fig. 4 Agents' projection on the proposed architecture

tasks of verification and storage of big data. The service agent can transfer queries from application to the processing layer to perform the requested analysis, it can also notify applications and services for results of an automated analysis tasks.

5 Conclusion

The paper presents a smart data approach for automatic big data analysis. A state of art was presented to motivate the transition from the big data and smart data and expose some approaches for automatic and real-time analysis. An agent-based multilayer architecture is proposed and the organization of the agents is explained.

Some related works [19] aim to present schemes and architectures for big data processing, this paper focuses on making an automatic data analysis by adopting a smart data approach, the approach of using smart data and a multi-agent system can enhance capabilities of analysis big data in real time. The next work is to test, implement and simulate this approach through a specific area.

References

1. Dumbill, E.: Making sense of big data. Big Data **1**, 1–2 (2013). https://doi.org/10.1089/big.2012.1503
2. Sheth, A.: Transforming big data into smart data: deriving value via harnessing volume, variety, and velocity using semantic techniques and technologies. In: 2014 IEEE 30th International Conference on Data Engineering. pp. 2–2 (2014). https://doi.org/10.1109/ICDE.2014.6816634
3. Iafrate, F.: A Journey from big data to smart data. In: Benghozi, P., Krob, D., Lonjon, A., Panetto, H. (eds.) Digital Enterprise Design & Management. pp. 25–33. Springer International Publishing (2014)
4. Lopez, V., Kotoulas, S., Sbodio, M.L., Stephenson, M., Gkoulalas-Divanis, A., Aonghusa, P.M.: QuerioCity: a linked data platform for urban information management. In: Cudré-Mauroux, P., Heflin, J., Sirin, E., Tudorache, T., Euzenat, J., Hauswirth, M., Parreira, J.X., Hendler, J., Schreiber, G., Bernstein, A., Blomqvist, E. (eds.) The Semantic Web—ISWC 2012, pp. 148–163. Springer, Berlin (2012)
5. Dustdar, S., Pichler, R., Savenkov, V., Truong, H.-L.: Quality-aware service-oriented data integration: requirements, state of the art and open challenges. ACM SIGMOD Rec. **41**, 11 (2012). https://doi.org/10.1145/2206869.2206873
6. De vettor, P., Mrissa, M., Benslimane, D.: Models and adaptive architecture for smart data management. In: 24th IEEE International Conference on Enabling Technologies Infrastructure for Collaborative Enterprises (WETICE 2015), Larnaca, Cyprus (2015)
7. Marz, N., Warren, J.: Big Data: Principles and Best Practices of Scalable Real-Time Data Systems. Manning Publications Co., New York (2015)
8. Michael, W.: An Introduction to MultiAgent Systems, 2nd Edn
9. Twardowski, B., Ryzko, D.: Multi-agent Architecture For Real-Time Big Data Processing. In: IEEE/WIC/ACM International Joint Conferences on Web Intelligence (WI) and Intelligent Agent Technologies (IAT), vol. 3, pp. 333–337 (2014). https://doi.org/10.1109/WI-IAT.2014.185

10. Geerdink, B.: A reference architecture for big data solutions introducing a model to perform predictive analytics using big data technology. In: 8th International Conference for Internet Technology and Secured Transactions (ICITST-2013). pp. 71–76. IEEE, London, United Kingdom (2013). https://doi.org/10.1109/ICITST.2013.6750165
11. Avgeriou, P., Zdun, U.: Architectural patterns revisited—a pattern language. In: In 10th European Conference on Pattern Languages of Programs (EuroPlop 2005), Irsee, pp. 1–39 (2005)
12. Paik, I., Chen, W., Huhns, M.N.: A scalable architecture for automatic service composition. IEEE Trans. Serv. Comput. **7**, 82–95 (2014). https://doi.org/10.1109/TSC.2012.33
13. Siriweera, T.H.A.S., Paik, I., Kumara, B.T.G.S., Koswatta, K.R.C.: Intelligent big data analysis architecture based on automatic service composition. In: 2015 IEEE International Congress on Big Data. pp. 276–280. IEEE, New York City, NY, USA (2015). https://doi.org/10.1109/BigDataCongress.2015.46
14. Dean, J., Ghemawat, S.: Commun. ACM **51**, 107 (2008). https://doi.org/10.1145/1327452.1327492
15. Jain, Y.K., Bhandare, S. K.: Min max normalization based data perturbation method for privacy protection. Int. J. Comput. Communication Tech. **3**(4), 45–50 (2014)
16. Li, D., Kar, S., Moura, J.M.F., Poor, H.V., Cui, S.: Distributed Kalman filtering over massive data sets: analysis through large deviations of random Riccati equations. IEEE Trans. Inf. Theory **61**, 1351–1372 (2015). https://doi.org/10.1109/TIT.2015.2389221
17. Chen, F., Deng, P., Wan, J., Zhang, D., Vasilakos, A.V., Rong, X.: Data mining for the internet of things: literature review and challenges. Int. J. Distrib. Sens. Netw. **11**, 431047 (2015). https://doi.org/10.1155/2015/431047
18. Wood, M.F., DeLoach, S.A.: An overview of the multiagent systems engineering methodology. In: Ciancarini, P., Wooldridge, M.J. (eds.) Agent-Oriented Software Engineering, pp. 207–221. Springer, Berlin (2001)
19. Babar, M., Arif, F.: Real-time data processing scheme using big data analytics in internet of things based smart transportation environment. J Ambient Intell. Hum. Comput. (2018). https://doi.org/10.1007/s12652-018-0820-5

For a Wide Propagation of Information on Social Networks

Sara Abas, Malika Addou and Zineb Rachik

Abstract Propagation of information happens over a defined period of time and through a quantified interactive audience. In order to contribute to a wider propagation, the following analysis was conducted. In this article, a system that generates primary network seeds that optimize diffusion was elaborated on a macroscopic level, as a result of certain input variables: the type of the network, the type of the information, the preferred duration of propagation time, and the number of target nodes. The key component of our algorithm was to extract nodes accordingly to the type of the information, within our fixed social network. As the other factors also majorly interfere with the selection of the primary seeds observed in the output of our main system, the other parameters were fixed with the exception of the type of information. An analysis took place, which described how the suggested approach generates efficient seeds.

Keywords Propagation of information · Social network · Wider propagation

1 Introduction

Social networks have led to an exponentially increasing propagation of information year after another over the last decade. Facebook, Instagram, Twitter, and LinkedIn are globally influential as they support wide information diffusion across the world, within seconds if not less. A social network is a network that contains nodes known as its members that interact with each other, therefore creating relations with one another called edges. Undeniably, the concept of a virtual life-like world combined

S. Abas (✉) · M. Addou · Z. Rachik
Architecture, System and Networks Team (ASYR)—Laboratory of Systems Engineering (LaGeS), Hassania School of Public Works EHTP, BP 8108 Oasis Casablanca, Morocco
e-mail: sara9abas@gmail.com

M. Addou
e-mail: malika.addou@gmail.com

Z. Rachik
e-mail: zineb.rachik@gmail.com

V. Bhateja et al. (eds.), *Embedded Systems and Artificial Intelligence*, Advances in Intelligent Systems and Computing 1076,
https://doi.org/10.1007/978-981-15-0947-6_67

with technology is a better transmission tool for information, in terms of information traveling speed and the size of its propagation. However, most phenomena that make it worldwide on these social networks cease to propagate at a certain point in time. It is very common to stop hearing about an issue or a topic after it is been widely diffused.

This disappearance of propagation, whether it is abrupt or gradual, represents an obstacle for the information diffusion, as it stops it from reaching more interactive audience, and from having a longer time span on the platform.

In this paper, the global system that aims to maximize influence as an end goal considers many factors: the type of the information, the chosen time span of its propagation, the size of the network to be reached, and the type of the network.

Type of information: The relevance of an information on social networks lies in its semantic aspect. The suggested system in the present work takes in entry the type of the information, in terms of its semantic meaning rather than just lexical.

Time span: When observing an information spread through a social network, at a certain point in time, nodes are no longer participating in its propagation. The time span considered in the input parameters is the time duration between the step in time when it starts spreading in terms of its increasing degree of propagation and the moment observed in time when the social network's activity is no longer related to it on a global scale.

Type of the network: Different types of networks relate to each other in terms of their randomness, heterogeneity, and modularity. Examples for type of networks are small-world networks, scale-free networks, and random graphs.

This study focuses on a method that represents a strong component of our final system, by extracting the best primary nodes, based on the type of information. The target primary nodes are the best influential users, which can guarantee a wide propagation of information through the network as soon as it lands on their feed. These are the basic nodes for a good propagation of information. In order to achieve that, all the input parameters are fixed with the exception of the type of the information. The compatibility of the type of the information with a node's interests makes the node's decision to spread the information more likely to happen and is our main focus.

The type of the information as previously defined is in terms of semantic meaning. Our approach relies on semantic similarity concepts.

After having introduced on a global scale the work of authors that targets a wide propagation of information, in the next chapter, we discuss the past studies of influence maximization. In the third chapter, most commonly used semantic similarity methods that achieved great results are discussed. In the fourth chapter, a semantic similarity-based method is developed in order to extract the most likely compatible nodes with the type of the information, using our proposed method. At last, we summarize our study on a macroscopic level, to show how it contributes to our final system.

2 Past Influence Maximization Studies

Today, many studies have been conducted in order to contribute to a more efficient propagation of information. Some of the aims were to maximize the influence of users on each other. Each of these researches considers different criteria to select the most influential users.

In 2014, [1] Neda Salehi Najaf Abadi and Mohammad Reza Khayyambashi suggest a model that maximizes influence in viral marketing on social media. The algorithm selects expert or influential users of the social network, which eventually efficiently spread the word on the platform. The approach consisted of four steps. The first one was to discover communities in relation with the commercialized product using *Tag Mining*, a text mining method with tags or keywords. A selection of the number of posts related to the product is calculated, using text mining. The third step is to identify users that are more popular than others. The intersection of the three steps is portrayed in the fourth one, comparing propagation with the nodes selected using The independent cascade model, with other models like the greedy algorithm and the high degree algorithm, is using random nodes. The independent cascade model is an information diffusion model commonly used in studies that starts with a series of active nodes, then activates the others in cascade mode step by step. This process continues until no further activation is possible. The greedy algorithm adopts the principle of making, step by step, an optimum local choice. The high degree algorithm considers most influential the nodes that have the most relationships with others. In this approach, the independent cascade model with the leader nodes provides a greater influence than the other methods.

The author focuses on extracting the best influencers through the extraction of keywords that would contribute to easily identifying nodes that spread certain words more often than others. Adding to that, a method that selects the popular nodes amongst that first set would make an efficient primary sample that spreads the information faster, as noticed in the independent cascade model compared to the other models. This method indeed helps spread the information faster. However, what our method suggests is to not only select based on keywords but also on concepts. For example, a user A can very well discuss alcoholic "drinks" very often and be targeted by a tea drink brand. The two concepts are not related; however, they share some identical words in their usual posts, and this makes the targeting inefficient. In our method, user A belongs to a set interested in the alcohol section of the social network, while the brand is herbal-related. Our method suggests targeting nodes that are potentially interested in herbal consuming products, like tea. In this paper, the popular nodes will not be part of the analysis, as the aim is to extract nodes accordingly to the type of the information, as similarity in interests can also be positively influential to the propagation of information.

In 2018, [2] Alizera Louni studied the detection of the source of rumor. The aim of the study is the key in identifying major propagators on social networks. The author tries in his work to determine the source in two phases. The first phase consists in selecting a candidate group through the method of Louvain. The Louvain method is a hierarchical algorithm for community extraction applicable to large networks. The second phase consists of selecting the candidate source using the concept of betweenness centrality. The betweenness centrality is a means of detecting the amount of influence a node has on the flow of information in a graph. It is often used to search for nodes serving as a bridge between one part of a graph and another. This work makes it possible to define the source to a distance of three nodes on a graph whose diameter does not exceed 13, supposing that the source has a high betweenness centrality. The author elaborates a model to detect the source of the rumor, by first selecting communities. The experience shows that the detection of the rumor is correct to an extent. However, many nodes can be sources of the rumor, and therefore, many influencers can emerge in a social network. In our suggested method, we address every possibly and strongly related node, with no elimination.

In 2011, [3] Yulia Tyshchuk studies propagation in the case of an alert. The first phase consists in identifying the key actors by calculating the centrality and the proximity factor. After identifying key actors, the author perceives reactive elements as cohesive groups, where a cohesive group is a set of individuals with similar contents. The author distributes the alert in six phases. The results conclude that the participation of key actors is done in all six phases, cohesive group formation is toward the final stages, and the main nodes are part of at least two cohesive groups. The author studies the formation of cohesive groups as an emergency information spreads through the social network and identifies the stages where it forms. Our approach focuses on identifying nodes that are interested in the information whether they eventually form a cohesive group or not.

Our end goal was to elaborate a system that aims—amongst other goals—to stretch the size of the reached audience, thus maximize influence as final result. In order to do that, the conceived system matches the type of the information with its efficient propagator nodes, which is our main focus in the rest of the paper.

3 Semantic Similarity Approaches

The key component of our method is to extract nodes that are compatible with the type of the information to be spread. If the node is compatible with the type of the information, it is more likely for that node to spread that information. This strategy makes information spread easier, as the specialty of its propagators is about the information, and thus, nodes are easily prompted to take action in spreading the word.

Using successful previous research in the field, semantic similarity approaches are efficient in our aim to propagate information, through specialized network nodes. Research conducted in the field of artificial intelligence to solve semantic problems relies on a few common concepts.

3.1 Basic Concepts

Semantic Similarity

Semantic similarity analysis is an important aspect in analyzing social networks activities. The resemblance of word structure is not enough to identify similarity. Several measuring approaches exist, from natural processing language to recent researches semantic approaches. In the following, basic concepts of semantic similarity are discussed, as well as few studies conducted in this sense.

NPL

Natural processing language (NPL) is a multidisciplinary field involving linguistics, computer science, and artificial intelligence. It aims to create tools for natural language processing for various applications. It supports artificial intelligence. NPL was initially used for machine translation and speech recognition. The basic idea of using NLP is analyzing grammatical structure at sentence level and then constructing grammatical rules for some useful information within sentence.

WordNET

WordNET is a frequent basic reference in semantic similarity researches. Its purpose is to catalog, classify, and link in various ways the semantic and lexical content of the English language. It groups English words into sets of synonyms called synsets. Each synset has a common semantic field it represents. Synsets are connected to each other with defined relations. These relations are hierarchically defined. For example, gymnastics is a sport, and thus, "gymnastics" is a hypernym of "sport." Words can also be connected to each other with antonomy, meronymy, hyponymy.

Many researchers rely on WordNet to come up with semantic similarity measuring techniques.

In 2009, [4] Zhongcheng Zhao developed a method that consists of calculating similarity between concepts, based on WordNET 3.0 using Java WordNet Library in the operating phase. In his work, the author introduced the "PART-OF" concept to minimize calculated similarity errors. The PART-OF relation is different than the "IS-A" relation. A trunk is part of a tree when the "IS-A" concept is expressed in hierarchy as the following: a tree is a plant. This work is innovative by adding the PART-OF link to his equation.

Information Retrieval

Information retrieval is the domain that studies how to extract information from a corpus. This is composed of documents from one to more databases, which are described by content or associated metadata. Information retrieval methods are used in the field of data analysis. It is the first step to achieve before implementing an analysis approach on significant data.

3.2 Semantic Similarity Measures

Although they adopt the same hierarchy principles, semantic similarity measures are used in different ways. Some of the following methods are more commonly used than others.

Cosine Similarity METHOD

The cosine similarity (or cosine) calculates the similarity between two vectors and its dimensions by determining the angle measure between them. This metric is frequently used in text mining. Given two vectors A and B, the θ is obtained by the scalar product and the norm of the vectors:

$$\cos\theta = \frac{A \cdot B}{\|A\| \cdot \|B\|} \quad (1)$$

In 2014, [5] Rawashdeh Ahmed created a unified similarity measure to study node profile similarity in social networks, using Facebook data. The author uses the cosine similarity approach, where each feature field is encoded as a vector.

Set-Based Similarity

Jaccard's index and distance are two metrics used in statistics to compare similarity and diversity (in) between samples. Jaccard's index is the relation between the cardinal (the size) of the intersection of the considered sets and the cardinal of the union of the sets. It enables an evaluation of the similarity between the sets.

$$J(A, B) = \frac{|A \cap B|}{|A \cup B|} \quad (2)$$

4 Proposed Semantic Similarity-Based Extraction Method

Many information retrieval methods in machine learning majorly contributed to semantic analysis through some breakthrough semantic similarity methods.

In our work, we develop a method that uses previously developed semantic similarity extraction approaches and applies it to a social network, combined with reality-inspired concepts and sociological ideologies of how information travels through the network.

4.1 Computerizing Sociological Aspects

Indirect Similarity

If information A belongs to the synset Sa, while information A is related to information B which belongs to synset Sb, Sa and Sb are not necessarily semantically related, but there is an implication relationship between the two concepts. Information B can therefore be efficient in the extraction of similar nodes that are actively spreading information A. For example, users who are interested in weightlifting, which belongs to the sports category/synset, are usually interested in diets which are hyponyms of nutrition. The nutrition synset is different than the sports synset. However, the two concepts are related. Therefore, a user that is interested in weightlifting can also be interested in diets. In this study, synsets that are different from each other, but strongly correlated, are matched.

Based on the type of the information provided in the entry of our globally conceived final model, a list of nodes whose interests are similar to the type of the information is extracted.

Fellow Membership Interests

In a social network, each node is linked to a number of fellow members of the platform. For example, they are called "friends" on Facebook or "followers/following" on Twitter and Instagram. In this study, we suppose that a node is more likely to share an information if the nodes related to it that we call friends in the rest of the paper are interested in that information.

4.2 Proposed Method

The first step of the method is to extract nodes that are semantically similar to the type of the information. In the second phase, we extract the common interest of the selection resulted in the first phase. The concept that most nodes obtained in phase one are commonly interested in is selected. In phase three, a selection based on friends' interests is put into place. Finally, the set resulting from the three phases is considered to be the proper one for the selection of primary nodes.

Phase one:

The first step of the algorithm is a procedure of an extraction of the concepts that node x is interested in, using tag mining [6]. Each post shared by user *x* is subject to tag mining, in order to obtain item features from tags. A set of keywords is extracted from the shared post. A concept C2 is obtained as a result of the set of keywords, using the seed concept extraction subtask of the ONTOEXTRACTOR'S method in Xuejun Nie's work [7].

Fig. 1 Frequency indicator calculation of recurring concepts

Extracted Concepts	Similarity	Indicator	I (C1)
C2	0.85	+1	1
C3	0.40	+0	1
C4	0.91	+1	2
C5	0.00	+0	2

Next, adapting Resnik's [8] information content approach, we calculate the similarity between the information's concept and the already existing content, as follows (3).

Where C1 is the concept of the type of the information parameter **in the input of our model**, and C2 is a concept expressed in one of node x's posts.

$$\mathrm{Sim_R}(C_1, C_2) = \mathrm{IC}(\mathrm{LCS}(C_1, C_2)) \tag{3}$$

where IC(C) is defined by

$$\mathrm{IC}(C) = -\log p(C) \tag{4}$$

And

$$p(C) = \frac{\mathrm{freq}(C)}{M} \tag{5}$$

where M is the total number of observed instances of nouns in the corpus, and LCS as expressed in Kishor Wagh's work [9] is the least common subsumer.

It is the most specific common ancestor of two concepts found in a given ontology. Semantically, it represents the commonality of the pair of concepts. For example, the LCS of moose and kangaroo in WordNet is mammal.

We apply Eq. (3) on all the content shared by node x. We consider that concept C1 is similar to concept C2 if the following statement is true (Fig. 1):

$$\mathrm{Sim_R}(\mathrm{C_1}, \mathrm{C_2}) > 0.8 \tag{6}$$

If concept C_1 is similar to concept C_2 in a shared post, an indicator $I(C_1)$ is incremented by 1.

Let $f(Ci)$ be the frequency of the concept Ci in node x's content. We define $f(Ci)$ as follows:

$$f(Ci) = \frac{I(Ci)}{S} \tag{7}$$

where S is a constant number that equals the number of concepts expressed in x's posts.

We conclude that node x is compatible with the concept C_1 if it supports a concept Cj, where Cj is most similar to concept C_1:

$$f(Cj) = \max(f(Ci))_{0<i<S+1} \\ \text{while } \mathrm{Sim_R}(C_1, C_j) > 0.8. \tag{8}$$

A final set of nodes $S_1 = \{n1, n2, n3\ldots\}$ is extracted from the fixed network, where node n_k supports a concept C_k where

$$f(Ck) = \max(f(Ci))_{0<i<S+1} \\ \text{and } \mathrm{Sim_R}(\mathrm{C_1}, \mathrm{C_k}) > 0.8. \tag{9}$$

Let S_1 be the result set of nodes resulted from phase one.

Phase two:

In phase two, the same approach is applied. In this step, we look for the second-best concept. Based on the first selection, we extract the concept C_x that they are mostly interested in. The idea is if S_1 is interested in C_1, then S_1 is interested in C_x. In this case, we can target nodes that are interested in C_x, since they can be potentially interested in C_1. We re-apply the method of phase one to the rest of the concepts while eliminating the first chosen one C_1. This will help target concepts that are not evidently similar but have an implication from one to another. Once the concept of phase two is selected, we proceed to an extraction of all nodes that are related to the concept of phase two.

The intersection of the second-best concepts, mentioned in all nodes that are compatible with the first concept, is the set obtained that represents the second big common interest of all nodes that are compatible with C_1.

At the end of phase two, a set S_2 is resulted.

Phase three:

In phase three, through an analysis that identifies node y's friends' interests, we extract nodes that are most likely to have friends that are interested in the concept. We apply the same approach of phase one to each of y's friends, to extract the friends that are interested in the concept. Let N be the number of friends obtained at the result set Fy.

Let

$$f(Fy) = \frac{N}{N'} \tag{10}$$

where N' is the total number of node y's friends.

If $f(Fy) > 1/2$, node y is a potential target node.

As we apply this analysis to the entire network, let S_3 be the final set of phase three.

Phase four:

In this phase, we gather the results as follows:

$$S = \{S_1 \cup S_2 \cup S_3\} \tag{11}$$

as the final set that we consider to be related to the type of information in the input of our algorithm.

The final set S gathers:

- Nodes that are interested in the information, based on its type.
- Nodes that can potentially be interested in the information based on its type, where the information represents a positive relation with their usual interests.
- Nodes that have friends who are interested in the information, based on its type.

5 Global System Perspectives

The suggested and developed method generates nodes that are compatible with the type of information as a result of the input chosen in the type of the information parameter. However, the global suggested method can be used for different purposes in information distribution and analysis on social networks. One of the aims is the following one: Having previously fixed the type of the network, the duration of time of propagation of information, the size of network, the algorithm mainly interferes so far with the selection of nodes that are most likely to have interest in the information based on its type, since it is compatible with them based on behavioral and sociological criteria we considered to be true in our work. As illustrated in the following figure, the first step of our global approach has been achieved (Fig. 2).

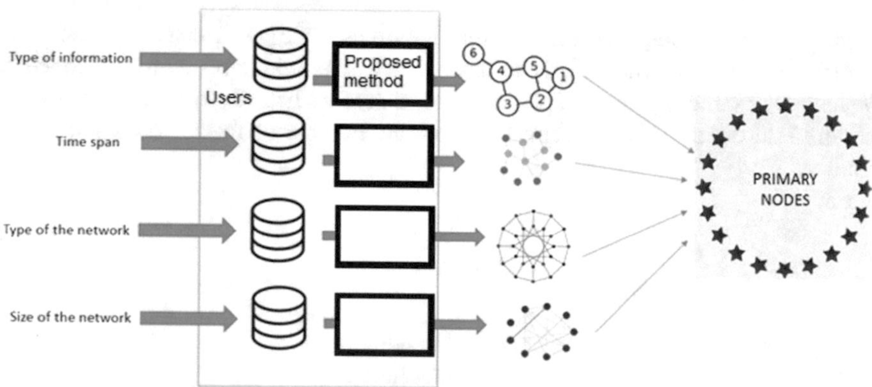

Fig. 2 Overall mechanism of primary nodes generation

6 Conclusion

In this paper, to contribute to the demand raised in data analysis in social networks due to the increasing diffusion of information, we formulated a new method, for the selection of nodes that are most likely to take action when they receive the information. We used the information content approach combined with principles of semantic similarity based on previous work in the field, to generate equations that optimize the efficiency of our method. We designed a general model for adequate propagation that takes into consideration several input factors. Our main focus in this study was the type of the information in entry; therefore, we fixed all the other parameters. As future work, we plan to vary another parameter in order to manage its output selection that is most likely to pair with it.

References

1. Abadi, N.S.N., Khayyambash, M.R.: Influence maximization in viral marketing with expert and influential leader discovery approach. In: 8th International Conference 24–25 April 2014, Mashhad, Iran on e-Commerce with focus on e-Trust (2014)
2. Alizera Louni: Who Spread That Rumor: Finding the Source of Information in Large Online Social Networks With Probabilistically Varying Internode Relationship Strengths in IEEE Transactions on Computational Social Systems (2018)
3. Tyshchuk, Y.: Actionable information during extreme events—case study: warnings and 2011 Tohoku Earthquake. In: 2012 International Conference on Privacy, Security, Risk and Trust and 2012 International Conference on Social Computing (2012)
4. Zhao, Z.: Measuring semantic similarity based on WordNet. In: Sixth Web Information Systems and Applications Conference (2019)
5. Rawashdeh, A., Rawashdeh, M.: Measures of semantic similarity of nodes in a social network. In: Information Processing and Management of Uncertainty in Knowledge-Based Systems (2014)
6. Rajaraman, A., Ullman, J.D.: Mining of Massive Datasets, Stanford University (2011)
7. Nie, X., Zhou, J.: A Domain adaptive ontology learning framework. In: 2008 IEEE International Conference on Networking, Sensing and Control (2008)
8. Resnik, P.: Using information content to evaluate semantic similarity in a taxonomy. In: Proc. 14th International Joint Conference on Artificial Intelligence (1995)
9. Wagh, K., Kolhe, S.: A new approach for measuring semantic similarity in ontology and its application in information retrieval. In: Multi-disciplinary Trends in Artificial Intelligence (2012)

New Algorithm for Aligning Biological Data

Wajih Rhalem, Mourad Raji, Ahmed Hammouch, Hassan Ghazal and Jamel El Mhamdi

Abstract The change in data processing conditions obtained from biological experiments, in particular, SHAPE (Selective 2′-Hydroxyl Acylation analyzed by the Extension Primer) technique, results in the time shift of the data which are in the form of signals. In this study, a SHAPE data alignment algorithm is proposed using a new pattern recognition approach based on the discrete-to-continuous transition of entities. The advantage of our algorithm lies in the ability to process the information concerned with a logarithmic complexity, therefore, powerful results have been obtained.

Keywords Time series · Signal alignment · SHAPE data · Logarithmic complexity · SHAPEfinder

1 Introduction

In new ribonucleic acid (RNA) structure probing technologies, accurate alignment of signals obtained from biological experiments is essential for assessing the quality of the results.

SHAPE (Selective 2′-Hydroxyl Acylation Analyzed by Primer Extension) [1] is considered one of the most used techniques for efficiently analyzing and detecting RNA structure modification zones. The SHAPE data in question are electropherograms in the form of time-shifted signals (see Fig. 1), which is a challenge to process them in a careful manner.

To solve the problem of alignment of time-shifted signals resulting from this technique, SHAPEFinder software [2] has been developed and became a tool widely

W. Rhalem (✉) · M. Raji · J. El Mhamdi
LRGE Laboratory, ENSET Mohammed V University, Rabat, Morocco
e-mail: Wajih.dgapr@gmail.com

A. Hammouch
Direction of Scientific Research and Innovation, Ministry of Higher Education, Rabat, Morocco

H. Ghazal
National Center for Scientific and Technical Research (CNRST), Rabat, Morocco

V. Bhateja et al. (eds.), *Embedded Systems and Artificial Intelligence*,
Advances in Intelligent Systems and Computing 1076,
https://doi.org/10.1007/978-981-15-0947-6_68

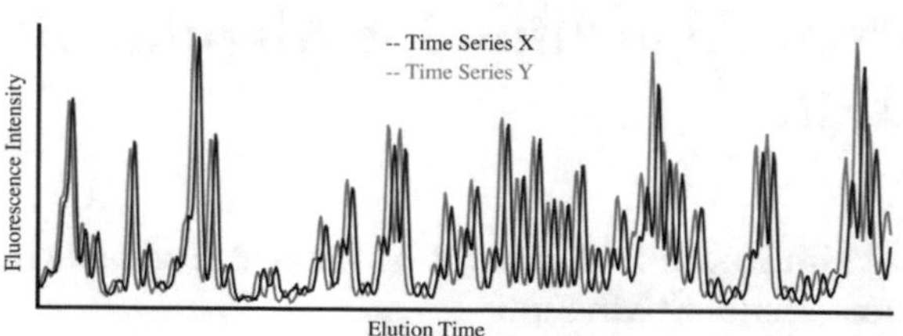

Fig. 1 Raw data obtained from two different capillaries. Time series X (black) and Time series Y (grey) have the same pattern but are out of phase

used by the scientific community. It is a modular and extensible software specialized in the processing and analysis of SHAPE data.

It was designed and launched primarily by Morgan Giddings and Jessica Severin at the University of Wisconsin-Madison, in Lloyd Smith's lab and was known as BaseFinder, then the refinement was carried out in 1998. Subsequently, the BaseFinder platform was extended to analyze the nucleic reactivity information. However, this software uses time offset correction tools by initiating the parameters manually. This process takes a long time (10–15 min) depending on the length of the signals.

The main objective of this work is to develop, adapt and apply a new algorithm in the field of fast and efficient processing of the shifted signals that suffers the data SHAPE, by implementing it in an automated system, in order to answer the problem of slowness in software using alignment methods based on manual processing (Shapefinder). Our alignment algorithm is based on the principle of switching from discrete to continuous, to optimally align the time series.

This approach was created in 1999 by Professor Raji Mourad [3], a member of our research team. Subsequently, it has been developed in collaboration with Professor Hammouch Hmed to apply it in other fields of application such as algorithmic geometry [4], fingerprint matching [5] online signature matching [6] and also in voice recognition [7].

To test the effectiveness, it should be noted that our algorithm signal processing, in this work, be validated by the use of Two offset signals for future use with SHAPE data.

This sense, we will proceed by generating two pseudo-sinusoidal offset signals A and B (see Fig. 2) shifted in x-axis with the following functions.

$A = \mathrm{Sin}(x)\sqrt{x}$ and $B = \mathrm{Sin}(x)\sqrt{x} + \mathrm{Tr}$ where Tr is a random shift (translation) of the signal A.

For our tests, a first series will concern an exact alignment, while the second will take into account, in addition to the offset, a slight perturbation of the B signal.

Generally, in order to solve the time shift problem, the "Dynamic Time Warping" algorithm (DTW) [8–10], is used as a reference method that has been proven in this field.

The temporal complexity of DTW is of the order $O(m.n)$. For the case of our discrete To continuous algorithm, the complexity is of the logarithmic order, which explains the performance of our approach.

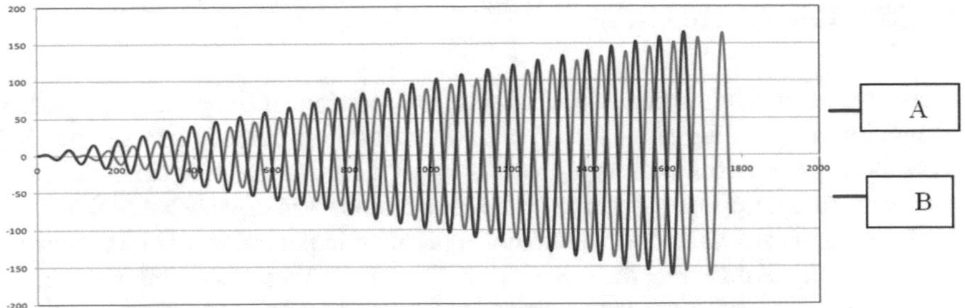

Fig. 2 Two offset signals *A* and *B*

This work is organized as follows. Section 2 is dedicated to related works. Section 3 presents the scientific contribution of the work. Section 4 details the alignment algorithm. The results obtained are presented and discussed in Sect. 5.

2 Related Works

Given the need to address the lag problems in the field of analysis of data from biological experiments, including the SHAPE Technical studies and several methods have been developed. Traditionally, for the resolution of this kind of problem we use the algorithm DTW, whose results are honorable. In addition to the classic DTW, a version with improvements [11] has been made to increase the efficiency of the latter, by implementing preprocessing tools. Still based on dynamic programming for the alignment of time series data, Folgado et al. [10] carried out in 2018 a work on the improvement of the performance of the time series alignment in the characterization of synthetic and real data during the human movement.

Other studies have been done to correct the drift of retention time. As an example, Gong et al. [12] used the combination of chemo-metric resolution and interpolation of cubic spline data to correct time offsets of chromatographic fingerprints of herbal drugs. Correlation optimized warping was used to align the chromatographic data, which is also time-shifted [13].

On the other hand, an approach to align gas chromatography-mass spectrometry has been proposed based on dynamic programming and peak similarity [14].

Generally, most of the existing methods are based on the DTW algorithm with a complexity of the order $O(m.n)$. In our alignment approach, DTC tries to find the best superposition between the input signal and a reference signal based on a logarithmic complexity seeking a transformation based on a Euclidean metric. This explains the performance of the results obtained.

It should be noted that our signal processing algorithm over time will, in this work, be validated by the use of two pseudo-sinusoidal shifted signals A and B with the following functions: $A = \mathrm{Sin}(x)\sqrt{x}$ et $B = \mathrm{Sin}(x)\sqrt{x} + \mathrm{Tr}$ for future use with SHAPE data.

3 Scientific Contribution

Generally, for making a final signal alignment decision in the context of pattern recognition, the raw signal at the input must undergo different pretreatment **and extraction of characteristics** steps, in order to mitigate the number of characteristics, and transforming the input signal into a more suitable form of decision making.

The alignment of SHAPE data remains a problem in pattern recognition because of the difficulty in matching the resulting traces that are very often time-shifted.

The SHAPE data, which is in the form of signals, is modeled as a set of minutiae. After successful extraction, the minutiae stored in a model gets the position of the minutia (x, y).

After the step of extraction of characteristics, comes the stage of recognition which is a principal task in the alignment systems. We chose to implement our DTC algorithm as an algorithm to find the correspondence between the entities to be treated.

While other algorithms try to find a good fit of certain pairs of features by comparing the composition of the two features (point by point), the DTC algorithm treats this problem as a point pattern matching problem. It brings the discrete representation of the signal in question on the continuous representation of the reference signal by using the cubic spline interpolation of the set of training data by considering the problem in its entity and not point-by-point. This makes it possible to compare the signals (Question and Reference) by seeking the inclusion of one form in another by searching for the best superposition.

4 Description of the Algorithm DTC

4.1 Formulation of the Problem

Let A and B be the stored signal and the input signal, respectively

$$A = \{m_1, m_2, m_3, \ldots, m_n\}. \tag{1}$$

where $m_i = \{x_i, y_i, \ldots\}$ and $i = 1, \ldots, m$

$$B = \left\{m'_1, m'_2, m'_3, \ldots, m'_n\right\}. \tag{2}$$

where $m'_j = \left\{x'_j, y'_j, \ldots\right\}$ and $j = 1, \ldots, n$.

n and m denote the number of minutiae in A and B, respectively. We consider that two minutiae points m and m' correspond if their positions are close, and this based on the approach of the spatial distance S_d.

$$S_d(m_i, m'_j) = \sqrt{\left(x_i - x'_j\right)^2 + \left(y_i - y'_j\right)^2} \leq r_0 \tag{3}$$

where r_0 is the tolerance that we can adjust.

4.2 Similarity Score

The similarity score is calculated by normalizing the corresponding number of minutiae (k):

$$\text{Score} = \frac{2k}{m+n} \tag{4}$$

A more efficient normalization considers the number of minutiae belonging to the intersection of the two signals after determining the optimal alignment.

A quantitative factor can be used, which specifies that at least a certain number of corresponding details must be found. In this article, the quantitative factor is used in the calculation of the similarity score.

Our method is a dot pattern matching technique. In this work, it applies to the alignment of offset signals, considering this problem as the biggest problem of the set of approximately common points (LACP: Largest Approximately Common Point Set). The problem is to decide whether or not there is an authorized transformation T, such that $T(B) \subset A$.

The origin of the difficulties is clearly due to the discrete nature of the sets of points to be matched. To overcome this difficulty, a transition from this discrete representation to a continuous representation has been made in this approach.

At first, instead of matching the two sets of points; we will try to bring one of them on the continuous representation of the other. After this pretreatment, an isomorphism is proposed.

A transformation T encounters difficulties if it is performed on a discrete representation without prior knowledge of the correspondence. We propose, for our part, an intermediate step, where we compute a transformation T' which brings the set of discrete data (B) on the continuous representation of the set of models (A).

The proposed algorithm aligns the signals in question and finds the correspondence between them. At first, the points of (A) are interpolated using a polynomial

$$\forall(x, y)\ P(x_i) = y_i. \tag{5}$$

If (x_i, y_i) is a point of B and if $T'(x_j, y_j) = (x'_j, y'_j)$ then (x'_i, y'_i) must be included in the polynomial representation of A.

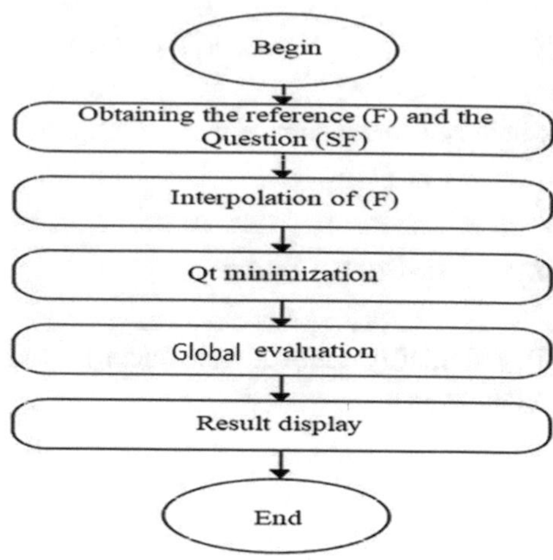

Fig. 3 DTC process

This means that:

$$P(x'_j) = y'_j. \tag{6}$$

The parameters T' are achieved by minimizing the QT cost function applied to points B.

$$QT(ty) = \sum_{i=1}^{n} (P(x''_j) - y''_j)^2. \tag{7}$$

After the step of adjusting the points B on the continuous representation of A by means of T', we associate each minutia of B with the nearest point form in B according to the nearest neighbor rule. The last step is to calculate the distance measurement, as indicated in (3), and the score is calculated as in (4).

4.3 Algorithm Process

Figure 3 gives an overview of the DTC process.

5 Results and Discussion

In order to solve the problem of time shift, often encountered in SHAPE data processing, a new algorithm, based on the principle of switching from discrete to continuous (DTC) has been successfully tested.

While waiting for the preparation of the SHAPE data, which requires a very specific treatment in specialist providers in this field, we proceeded, in the first place, to the generation of two offset signals *A* and *B* to test the effectiveness of our approach in the treatment offset signals.

The test procedure is as follows:

Generation of the signal *A* (reference) with the function $A = \mathrm{Sin}(x)\sqrt{x}$ of 1000 points with a step of 0.1 for the variable *x*.

Signal generation (Question) *B* with the function $B = \mathrm{Sin}(x)\sqrt{x} + \mathrm{Tr}$

Tr is the offset of the signal which would be generated randomly with the same interval and the same pitch.

These signals have the same pattern but out of phase as shown in Fig. 2.

For our tests, a first series will concern an exact alignment, while the second will take into account, in addition to the offset, a slight random disturbance of the signal *B* simulating the possible background noise encountered in the experiments.

The DTC algorithm has been implemented in the Java programming language. The machine used was a 2.40 GHz Intel Core i7 processor with 8 GB of RAM.

The results obtained, as shown in Fig. 4, have shown that the proposed procedure for signal alignment can correct the time shift problem, often encountered in SHAPE data processing, in an efficient and accurate manner.

The execution time of the developed signal alignment algorithm is less than 2 s with accuracy greater than 98%. Knowing that SHAPEFINDER requires treatment of about 10 min.

Figure 5 corresponds to the temporal complexity in the case of an exact alignment.

Figure 6 corresponds to the time complexity of the case of alignment with disturbance along the *x*-axis.

The results obtained demonstrate that the temporal complexity of our algorithm is of the logarithmic order $O(m.\log(n))$ unlike that of DTW which is of the order $O(m.n)$.

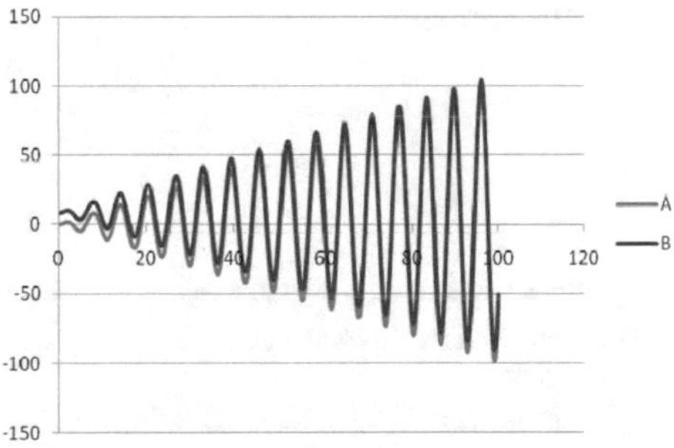

Fig. 4 Superposition of the two signals using the proposed algorithm DTC

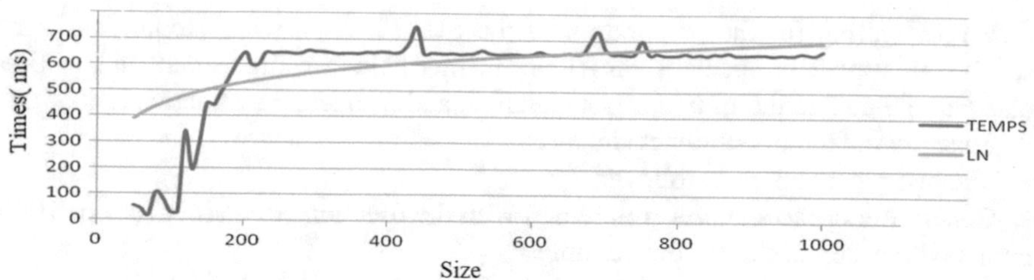

Fig. 5 Time complexity in the case of an exact alignment

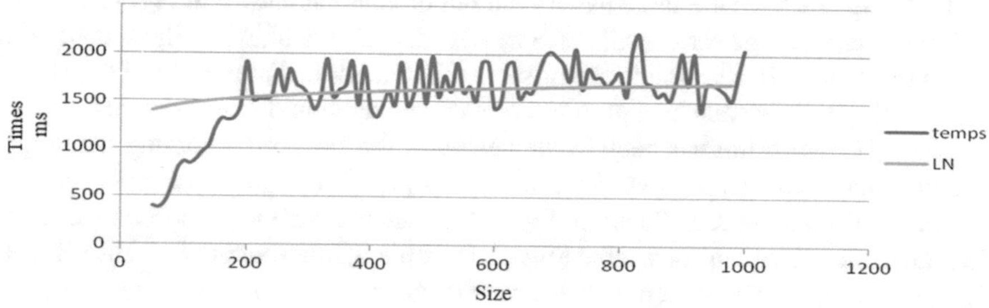

Fig. 6 Temporal complexity in the case of an alignment with disturbance

6 Conclusion

Time shift issues, which are one of the most important components in SHAPE data analysis, are solved correctly using the developed algorithm.

The obtained results encourage the application of our algorithm on SHAPE data by carrying out a comparative study with the DTW algorithm, which is often used to solve this kind of problem. As well as studying the opportunity to implement our algorithm in SHAPEFINDER software.

References

1. Wilkinson, K.A., Merino, E.J., Weeks, K.M.: Selective 2′-hydroxyl acylation analyzed by primer extension (SHAPE): quantitative RNA structure analysis at single nucleotide resolution. Nat. Protoc. **1**, 1610–1616 (2006)
2. Vasa, S.M., Guex, N., Wilkinson, K.A., Weeks, K.M., Giddings, M.C.: ShapeFinder: a software system for highthroughput quantitative analysis of nucleic acid reactivity information resolved by capillary electrophoresis. RNA **14**, 1979–1990 (2008)
3. Raji, M., Cossé-Barbi, A.: Shape recognition and chirality measure: reestablishing the link between similarity and dissimilarity in discrete space. Chemom. Intell. Lab. Syst. **47**, 219–225 (1999)
4. Aqili, N., Raji, M., Hammouch, A.: PPM translation, rotation and scale in D-dimensional space by the discrete to continuous approach. Int. Rev. Comput. Softw. IRECOS **11**, 270–276 (2016)

5. Aqili, N., Maazouzi, A., Raji, M., Jilbab, A., Chaouki, S., Hammouch, A.: Fingerprint matching algorithm based on discrete to continuous approach. In: International Conference on Electrical and Information Technologies (ICEIT), pp. 414–417 (2016)
6. Aqili, N., Maazouzi, A., Raji, M., Jilbab, A., Chaouki, S., Hammouch, A.: On-line signature verification using point pattern matching algorithm. In: 2016 International Conference on Electrical and Information Technologies (ICEIT), pp. 410–413 (2016)
7. Maazouzi, A.-E., Aqili, N., Raji, M., Hammouch, A.: A speaker recognition system using power spectrum density and similarity measurements. In: Third World Conference Complex Systems (WCCS), pp. 1–5 (2015)
8. Itakura, F.: Minimum prediction residual principle applied to speech recognition. IEEE Trans. Acoust. Speech Signal Process ASSP **23**, 52–72 (1975)
9. Kruskall, J.B., Liberman, M.: The symmetric time warping algorithm: from continuous to discrete. In: Time Warps, String Edits and Macromolecules: The Theory and Practice of String Comparison. Addison-Wesley (1983)
10. Folgado, D., Barandas, M., Matias, R., Martins, R., Carvalho, M.: Time alignment measurement for time series. Patt. Recognit. **81**, 268–279 (2018)
11. Karabiber, F.: An automated signal alignment algorithm based on dynamic time warping for capillary electrophoresis data. Turk. J. Elec. Eng. Comp. Sci. **21**, 851–863 (2013)
12. Gong, F., Liang, Y.Z., Fung, Y.S., Chau, F.T.: Correction of retention time shifts for chromatographic fingerprints of herbal medicines. J. Chromatogr. A, **1029**, 173–83 (2004)
13. Tomasi, G., Van Den Berg, F., Andersson, C.: Correlation optimized warping and dynamic time warping as preprocessing methods for chromatographic data. J. Chemom. **18**, 231–241 (2004)
14. Robinson, M.D., De Souza, D.P., Keen, W.W., Saunders, E.C., McConville, M.J., Speed, T.P., Likić, V.A.: A dynamic programming approach for the alignment of signal peaks in multiple gas chromatography-mass spectrometry experiments. BMC Bioinf. **8**, 419 (2007)

Data Mining Approach for Employability Prediction in Morocco

Saouabi Mohamed and Ezzati Abdellah

Abstract Finding a career opportunity for a young job seeker is a challenge to every graduate; employability is a factor to maintain social cohesion, and political stability, to reduce the crisis of confidence in the education system. Answering the employability problematic is a priority in the strategy of most countries, and which quite often drives the public and scientific debates. The era of Big Data is now in full force because the world is changing, thanks to advances in communication technologies, people and objects are increasingly interconnected, not just a part of the time, but almost all the time. People are starting to use many connected objects, such as cars, phones, they are becoming more involved in social networks. Data is generated almost all the time, giving this huge amount of data ready to be used and stored. Storing this data is not the major problem; extraction of the hiding patterns between this data is actually the objective. The use of big data and data mining will give the opportunity to decision-makers to make a lot of improvements in the employability domain. In this paper, we present a data mining approach to predict employability in Morocco; we present it in a form of a graph, with phases, explaining every phase and what it includes in details, from data collection till the visualization of the results.

Keywords Big data · Data mining · Employability · Prediction

1 Introduction

The explosion [1] of the data has forced researchers to find new ways of seeing and analyzing the world. It's about discovering new orders of magnitude for capturing, searching, sharing, storing, analyzing, and presenting data. So was born the Big Data [2]. This is a concept for storing and processing an unspeakable amount of

---

S. Mohamed (✉) · E. Abdellah
LAVETE Laboratory, University Hassan the 1st, Fst, Settat, Morocco
e-mail: mohamed.saouabi@gmail.com

E. Abdellah
e-mail: abdezzati@gmail.com

V. Bhateja et al. (eds.), *Embedded Systems and Artificial Intelligence*,
Advances in Intelligent Systems and Computing 1076,
https://doi.org/10.1007/978-981-15-0947-6_69

information on a digital basis. This amount of data cannot but stored or managed by traditional database systems [3]. But storing this data is not the only process, we need to use this data, to discover the hidden patterns from it and use it for decision making; and it is the power of data mining. Data mining [4] nowadays became more and more important; it's used to extract knowledge from a large amount of data to help decision-makers make decisions. Data mining is about using data analysis tools to find out new unknown knowledge, hidden relationships between the large data set we have on hands.

These tools can handle mathematical algorithms, statistical models and machine learning methods. Data mining is not just about the collection and managing the data, it includes data analysis and prediction. These technologies were used in several domains, a lot of domains are taking advantage of their power, and employability is one of these domains. The concept of employability appeared in the 1980s [5]. It was introduced by corporations, marketed as a response to the need to be flexible in the face of global competition, adapting to the unstable economic environment. It represents a serious problem to graduates, finding a career opportunity for a young job seeker is a challenge to every one of them. Answering the employability problematic is a priority in the strategy of most countries, and which quite often drives the public and scientific debates. We present a data mining approach to predict employability in Morocco; we present it in a form of a graph, with phases, explaining every phase and what it includes in details, from data collection till the visualization of the results.

2 Related Works

Many domains are now taking advantage of the powerful use of data mining and the opportunities to improve a particular domain. We present below some works using data mining techniques for the prediction of employability.

Kaur et al. [6] proposed an experiment to identify students with slow learning in the field of education using data mining classification techniques using world data collected in secondary schools in order to provide solutions and ensure good employability for these students. Using Weka, they applied several algorithms to find the best classifier model for employability prediction.

Mishra et al. [7] conducted a survey to explain that higher education has become an exciting field of research and that employability prediction using data mining techniques is beneficial for institutions. Then they discussed the work done in both areas of prediction. They concluded that the employability forecast has progressed significantly.

Mohd tajul rizal and yuhanis Yusof proposed a graduate employment forecast experiment, their experience included the use of five data extraction techniques, namely Naïve Bayes, logistic regression, multilayer perceptron, k-Nearest-Neighbor, and the decision tree. The results showed that logistic regression is the best classifier for the dataset.

Thakar et al. [8] proposed an empirical study that compares varied classification algorithms on two datasets of MCA (Masters in Computer Applications) students collected from various affiliated colleges of a reputed state university in India.

3 The General Architecture of the Data Mining Approach for Employability Prediction in Morocco

Employment is the main form of social integration, a factor for improving living conditions and preventing risks of poverty and vulnerability, and the most appropriate indicator for assessing the level of social cohesion in a country. We present an approach that describes how to use data mining and big data technologies in order to answer the employability problem and propose solutions (Fig. 1).

3.1 Data Collection

In this phase, we collect the data, this data probably collected through a survey, from an internet website or social media, the important is that this data is answering the problem in hand and could be used for the experiment. In our experiment, we collected the data from a survey of employability conducted by Hassan the 1st University in partnership with the National Evaluation Office (NEO) under the Higher Council for Education.

3.2 Entering the Data

After data collection, we need to input this data into a database; we can use a PHP website with MYSQL database so the data can be ready for the next phase to be ingested into the Hadoop ecosystem. Data Ingestion means taking data from various databases and files and putting it into Hadoop. Hadoop offers so many technologies and tools for ingesting the data, it depends on the type of the data we have, so we can choose the right tool, for example, if we have a structured data from a relational database, in this case, we'll use SQOOP [9] as a tool for data ingestion into the distributed file system of Hadoop HDFS.

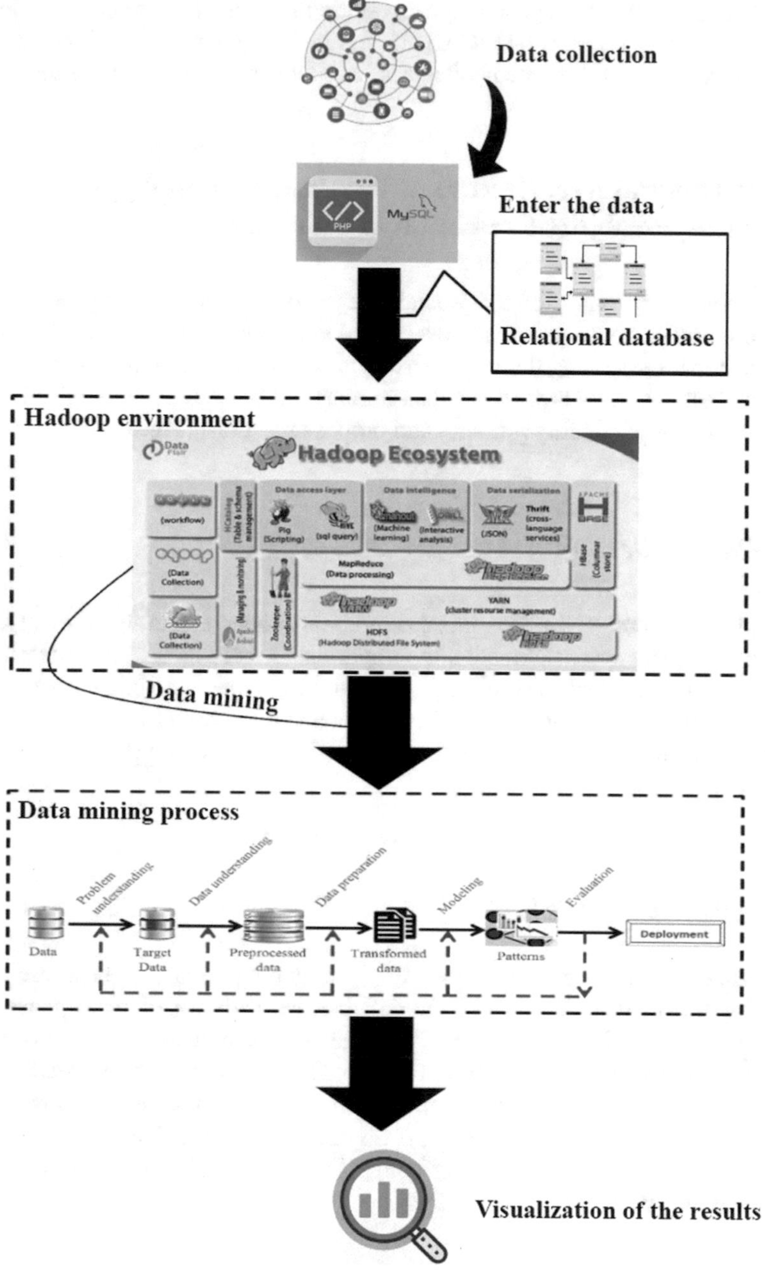

Fig. 1 The global architecture describing the phases of the data mining approach

3.3 Big Data Environment: Apache Hadoop

Hadoop [10] is one of the driving forces behind the Big Data revolution. It allows the distributed processing of large datasets on standard server clusters. It is designed to go from a single server to thousands of machines with a very high degree of fault tolerance. Rather than relying on high-end hardware, the resilience of these clusters stems from the software's ability to detect and manage failures. Hadoop has changed the data and the dynamics of computing on a large scale by building high computing capacity at a very low cost. This is why we chose to experiment in a Hadoop environment. In order to take advantage of all the solutions and technologies that Hadoop offers.

3.4 Data Mining

The goal of data mining is to extract new knowledge and give an understanding of a given set of data, which can be used for decision making. Data mining overlaps with other disciplines, including statistics. Several techniques offered by data mining and prediction are one of them. As we explained before, data mining will be used for employability prediction in order to predict either a graduate is classified as working or not working. Data mining involves several phases [11] starting from problem understanding until the deployment of the found results, as described below.

The problem understanding: Get a clear understanding of the problem we want to solve. Tasks in this phase include:

- Identifying the objectives for answering the problem;
- Defining the data mining goals;
- Producing the objectives plan.

This is a major important step because it guides us towards the objectives we want to achieve, and it gives us a clear view of the problem we want to answer.

Data understanding: In this phase, we review the data that we have, documented it; identify data management and data quality issues. Tasks for this phase include:

- Gathering data;
- Describing;
- Exploring;
- Verifying quality.

After this step, we obviously understood the data more, and we have a clear view of how data need to be and how the structure will look like, all we have to do now is to prepare the data in the next step.

Data preparation: We make the data ready to use for modeling. Tasks for this phase include:

- Selecting data;
- Cleaning data;
- Constructing;
- Integrating;
- Formatting.

Now that the data is prepared, we will apply the models and the data mining algorithms, depends on the data, several algorithms may be applied.

Modeling: We use mathematical techniques to identify patterns within the data. Tasks for this phase include:

- Selecting techniques;
- Designing tests;
- Building models;
- Assessing models.

In this phase, we apply the data mining algorithms, depends on the problem of the data, we choose which type of algorithms we'll apply, and in the next step to be assessed.

Evaluation: We review the patterns we have discovered and assess their potential for use. Tasks for this phase include:

- Evaluating results;
- Reviewing the process;
- Determining the next steps.

In this step, we assess each algorithm and we choose which one is best suited for our data and which provide the best model for our data.

Deployment: We put the discoveries founded into use. Tasks for this phase include:

- Planning deployment (the methods for integrating data mining discoveries into use);
- Reporting the final results;
- Reviewing the final results.

This is obviously the last step of the process, in which we take advantage of the results and we deploy it practically in order to solve the problem in hand.

3.5 Visualization of the Results

Data visualization [12] is one of the most powerful and important phases. Visualization presents an interesting challenge for computer systems; it is the conversion of data into a visual or tabular format so that data characteristics and relationships between data elements or attributes can be analyzed or reported. As we finish the previous phases, results such as models, variables, and graphs will be presented in order to give insight and solutions for decision-makers.

4 Conclusion

The exponential growth in the amount of data collected by IT systems poses new problems in the analysis and exploitation of IT systems. Big data brings together a set of techniques that make it possible to cope with the storage challenge and the difficulties of heterogeneous data processing. Database management systems are no longer based just on relational database architecture, but big data technologies now can handle both structured and unstructured data. Mining this data and extract meaningful information and hidden patterns in order to predict the future and propose solutions to decision-makers is the goal. It's used in several domains, and employability is one of them. In this paper, we have proposed an approach using data mining and big data in order to predict employability and offer answers and solutions for future adjustments, we proposed the in a form of architecture the general approach and the phases describing every phase aside in details.

References

1. Zhu, Y., Zhong, N., Xiong, Y.: Data explosion, data nature and dataology. In: Brain Informatics 2 (2009)
2. Elgendy, N., Elragal, A.: Big Data Analytics: A Literature Review Paper, Springer Lecture Notes in Computer Science, pp. 214–227 (2014)
3. Lungu, I., Velicanu, M., Botha, I.: Database Systems—Present and Future, Informatica Economică, **13**(1), 84–99 (2009)
4. Haripriya, P., Porkodi, R.: A survey paper on data mining techniques and challenges in distributed DICOM. Int. J. Adv. Res. Comput. Commun. Eng. **5**(3), 741–747 (2016)
5. McQuaid, R.W., Lindsay, C.: The concept of employability. Urban Stud. **42**(2), 197–219 (2005)
6. Kaur, P., Singh, M., Josan, G.S.: Classification and prediction based data mining algorithms to predict slow learners in education sector. Procedia Comput. Sci. **57**, 500–508 (2015)
7. Mishra, T., Kumar, D., Gupta, S.: Students' performance and employability Prediction through data mining: a survey. Indian J. Sci. Technol. **10**(24), 1–6 (2017)
8. Thakar, P., Mehta, A., Manisha: Role of secondary attributes to boost the prediction accuracy of student's employability via data mining, IJACSA (2015). https://doi.org/10.14569/IJACSA.2015.061112
9. Aravinth, S., Haseenah Begam, A., Shanmugapriyaa, S., Sowmya, S.: An efficient HADOOP frameworks SQOOP and Ambari for big data processing. Int. J. Innov. Res. Sci. Technol. **1**(10), 252–255 (2015)
10. Dhyani, B., Barthwal, A.: Big data analytics using Hadoop. Int. J. Comput. Appl. **108**(12), 1–5 (2014)
11. Danubianu, M.: Step by step data preprocessing for data mining. A case study. In: International Conference on Information Technologies, pp. 1–8 (2015)
12. Wang, L., Wang, G., Alexander, C.A.: Big data and visualization: methods, challenges and technology progress. Digital Technol. **1**(1), 33–38 (2015)
13. Bharathi, A., Deepankumar, E.: Survey on classification techniques in data mining. Int. J. Recent Innov. Trends Comput. Commun. **2**(7), 1–4

System Service Provider–Customer for IoT (SSPC-IoT)

Abderrahim Zannou, Abdelhak Boulaalam and El Habib Nfaoui

Abstract Internet of things (IoT) is a new promising paradigm based on the interconnection of heterogeneous objects, which aims to provide specific services anywhere, anytime by anything and anyone. IoT systems include different devices and use different protocols and technologies. Whereas IoT is now a reality, many IoT challenges are not resolved such as horizontal silos, defined level communication services in the case of constraint things, trust object services, etc. Some IoT architectures for global communication are proposed for resolving those challenges, but they are deployed on specific applications for resolving some constraints such as power energy, capacity of calculation, etc. This paper proposes a System Service Provider–Customer (SSPC) to manage and control communication of things and related services in IoT systems. SSPC is a system that consists of discovering services by considering constraints of objects. We provide all details about the SSPC algorithms, their features, and the SSPC workflow. We discuss the proposed system in a use case and we give a functional comparison with some recent related work.

Keywords Internet of things · Dynamic service creation · Cloud computing · Semantic service · Horizontal and vertical communication · Ontologies

1 Introduction

Internet of Things (IoT) has been recognized as one of the major technological revolutions of this century [1, 2]. The IoT concept aims to make the internet more immersive by allowing, anything to be connected in anytime and anyplace. In the

A. Zannou (✉) · E. H. Nfaoui
LIIAN Laboratory, University Sidi Mohamed Ben Abdellah, Fez, Morocco
e-mail: abderrahim.zannou@usmba.ac.ma

E. H. Nfaoui
e-mail: elhabib.nfaoui@usmba.ac.ma

A. Boulaalam
LSI Laboratory, University Sidi Mohamed Ben Abdellah, Fez, Morocco
e-mail: abdelhak.boulaalam@usmba.ac.ma

V. Bhateja et al. (eds.), *Embedded Systems and Artificial Intelligence*, Advances in Intelligent Systems and Computing 1076,
https://doi.org/10.1007/978-981-15-0947-6_70

IoT vision, there is a large number of devices and things gather information from diverse physical or virtual environments and deliver the information to a variety of intelligent applications and services, where defined as the internetworking of physical devices, vehicles, machines and other objects, embedded with sensors, electronics and network connectivity that enable them to collect and transfer data.

There are variety of applications and platforms [3–6] of IoT oriented to realize communication between different devices, and execute these services in real-time, the benefits can be derived from analyzing the applications of IoT in each field aren't limited, we can find a set of applications like smart home (smart thermostat, connected lights, smart fridge, smart door lock), wearable (smartwatch, activity tracker, smart glass), smart city (smart light, smart parking), smart grid, industrial internet, connected car, etc. From these applications, to make the Internet of Things useful, we need analytics of things. This will mean new data management and integration approaches, and new ways to analyze streaming data continuously.

There are many objects, e.g., sensor (detectors), actuators, router (s), etc. Also, every object connects to internet or constructs a network between them and has many constraints. On other side, objects (sensors, actuators, RFID, etc.) generally constructed by three elements: input, output, and processing. Behind each element there are many mechanical operations, chemical, electronic, etc. As result, object from input to output passes through with many operations and instructions and also every object has its own function or more functions. Each object has many constraints programming and hardware. Thus, for resolving the heterogeneity of devices, there has been much effort to create open platforms tried to resolve these problems but rest incomplete. Furthermore, there isn't a standard communication for connecting constraints objects and global communication services.

To overcome these limitations, we propose SSPC-IoT, it is a system that consists of two conceptions, the management of services and constraints related to devices, it may exist in the form of plugin/software, each object may integer SSPC inside (virtual object) or outside (exists in another real object). This system is composed of two parts System Service Provider (SSP) is responsible for response service for service client/customer and System Service Customer (SSC) is responsible to demand service.

The remainder of the paper is organized as follows: In Sect. 2, we present some related works. In Sect. 3, we present systems vision SSPC for IoT. In Sect. 4 we give a use case and a comparison functional of SSPC, respectively. Finally, we conclude the paper.

2 Related Works

Several IoT proposals have been investigated to define an architectural model for IoT that are usually applicable to a specific application domain [7–14]. As far as we know, there is no suitable unified architecture till date that is appropriate for a global IoT infrastructure.

In [15], a new mechanism has been proposed in order to resolve some limitations (e.g., data distribution, tractability of devices and resources, and security.) of architectures designed for IoT. The mechanism is consist of data management via information-centric networking, configured objects accordingly with services contracts, configured objects accordingly with services requirements, resolution name of the physical and virtual objects, and routing configuration (name-based and networking caching).

To facilitate the service discovery of IoT services, a scalable and self-configuring architecture has been proposed [16]. It is based on peer to peer technologies to provide an automated service and resource discovery without any human intervention. While increasing the device's capacity, the participants in network increasing robust systems to create high availability, and a mechanism manage to join and leave of objects automatically, e.g., the device's system reorganizes themselves without any intervention.

To minimize transmission size between devices in IoT, e.g., object size and power energy, a global generic architecture for the future IoT (GGIoT) [17] framework is proposed, the principals concepts of GGIoT framework are object description, object virtualization, building ontologies, virtualization system, and service coordination.

In [18], proposed architecture (DIAT), the functionalities of IoT infrastructure are grouped into three layers Virtual Object Layer (VOL), Composite Virtual Object Layer (CVOL), and Service Layer (SL). The three layers are responsible for object virtualization, service composition and execution, and service creation and management, respectively.

In [19], proposed a unified semantic base for IoT that uses ontologies, resource, location, context and domain, policy and service ontologies. All these ontologies applied to architecture DIAT. In [20], derived from [18], this paper proposes an architecture that supports Web Objects based energy efficiency for smart home IoT services.

The works presented in [15–17], resolve self-management, self-configuring, stability and reliability and power consumption in case non-global and can't apply in lossy network. Also, in [18–20], there are three biggest limitations:

- It is very difficult to apply the architecture DIAT in constraints objects and the workflow of management of service in lossy network objects not defined.
- To get services from the lossy network, we must have devices of high capacity, we present our SSPC to be the middleware between the object that has a high capacity of calculation and constraint objects.

3 SSPC-IoT

3.1 Main Concept

SSPC-IoT role is to discover services with respecting constraints of objects. It may exist in the form of plugin/software for services management. Each object may integer SSPC inside (virtual object) or outside (exists in another real object). SSPC is composed of two parts: the System Service Provider (SSP) which is responsible of response service for client/customer and the System Service Customer (SSC) which is responsible of service demand. In other way, when Constraint Object sends request service in network, the SSC determine different functionalities dependent to object customer and service or services related the service of Constraint Object, and then SSP role is to discover network and determine these services and constraints that can verify needs of Constraint Object.

The main goal of SSPC is facilitated discover and create services in IoT with considering the IoT challenges (Fig. 1).

SSPC Characteristics:

1. Deployed in constraint networks
2. Multiple SSPCs may exist in network
3. SSPCs are in collaborative between them
4. Can be considered like a root or simple object in network

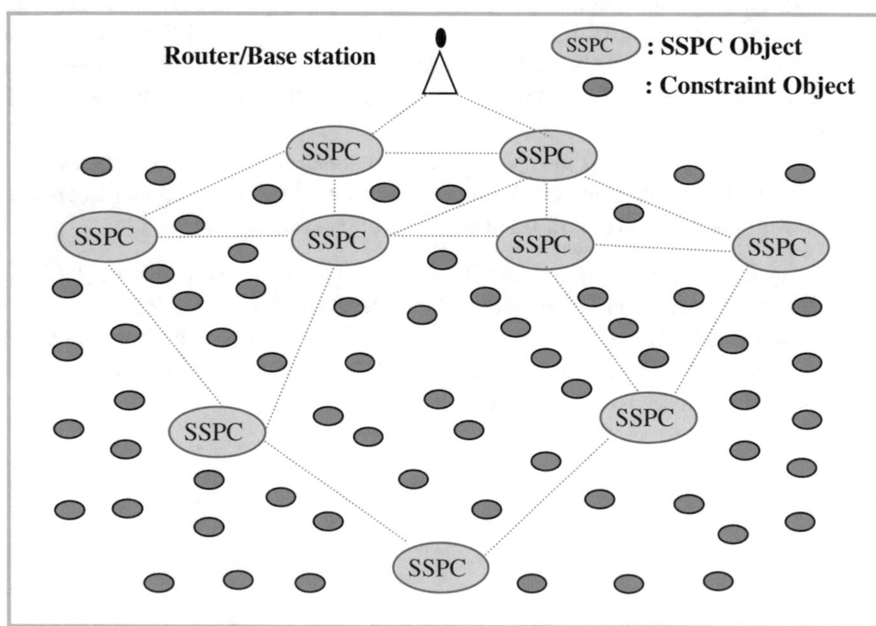

Fig. 1 SSPCs and constraints objects in lossy network

5. Take mission of communication from/to other objects
6. SPPC has high power energy than set of objects.

3.2 SSPC-IoT Workflow

The proposed work is consisting of four phases: (i) SSPC discover (ii) SSPC selection (iii) tree formation, and (iv) data or service transmission.

SSPC discovery: Where a Constraint Object can't get/set service from its network, discover SSPC for helping it to get/set services from other objects network respecting the capability hardware and software of Constraint Object. Each Constraint Object broadcasts a request to discover SSPCs to all neighbors, Algorithm 1. The request discover includes information about device object service, and the constraint of device like cost, reliability, time execution, etc. When SPPC receives the request from Constraint Object, calculates the ratio of correspondence of the Constraint Object. For example, if the time of execution of Constraint Object between 10 and 15 ms, the SPPC verifies if can take mission of data corresponding to this constraint. The SSPC sends the ratio of correspondence to Constraint Object.

SSPC selection: When a Constraint Object receives messages from all SSPCs, the object compares correspondence ratio of each SSPC, the SSPC that has correspondence ratio greater than others, selected like the parent SSPC. Then send a join request message of confirmation to save Constraints Object declared. After that, this object sends data configuration to be saved by SSPC, Algorithm 2.

Algorithm 1: SSPC Discovery

1. *SSPC receives the discovered request*
2. *SSPC calculates the ratio of correspondence*
3. *SSPC sends ratio of correspondence to Constraint Object*
4. *A Constraint Object receives ratio of correspondence of each SSPC*
5. *repeat all steps while there is not response.*

Tree Formation between SPCCs: After SSPC discovery and SSPC selection based on constraints of objects, SSPCs create a logical tree among them by running Algorithm 3. The SSPCs broadcast message contains path from router/base station and number of objects within each SSPC. After receiving messages each SSPC decide the best way to achieve data to the router/base station. Then each SSPC chooses its parent SSPC to achieve to the router.

Algorithm 2: SSPC Selection

1. *A Constraint Object compare all ratios correspondence*
2. *The SPPC has maximum ratio of correspondence added like parent SSPC*
3. *A Constraint Object sends request for selection SSPC*
4. *SSPC sends confirmation of acceptance to Constraint Object.*

Service control and transmission: After period of time the parent SSPC of Constraint Objects, can delete or replace Constraint Object, due to its resources hardware or software or the nature of service presented. Furthermore, each SSPC saves in the memory paths of its objects and of other SSPCs, from the set of paths SSPCs deduce the best length path to get service.

Algorithm 3: Tree Formation Between SSPCs

1. Each SSPC sends to other SSPCs length path from router of constraint objects
2. SSPC save data of other SSPC bridges
3. SSPC bridge chooses the best parent SSPC bridge to be optimization way to sends data to router.

4 Use Case and Functional Comparison

4.1 Use Case

To validate our proposed system components, this section will be described a composite use case. The use case study consists of various applications in different domains like smart home, smart health, smart transportation, and smart civil protection.

Smart home is composed of the following smart objects: fire detector, smoke detector, heat detector, person body detector, alarm device, jet water actuators, and health status of person sensor. All the cited objects are separated physically and connected to SSPCs. As a result, SSPCs manage all services of objects of smart home; control the inside/outside communication and also create and discover services.

In the detect fire scenario, the object sends request service to its SSPC (state: fire detected), when the SPPC received the request data, it checks list of its objects. If it finds services correspondent to object customer, the SSPC can return service desired in case where SSPC has already a cache of service wanted or resends directly the demand to object provider. At the same time, the SSPC running services dependent on fire detectors like smoke detectors, heat detectors, etc.

We point out that if SSPC of detector fire doesn't find all objects related to its object customer, this current SSPC checks its parent SSPC to discover the objects providers to respond to object customers. The same mechanism will be applied by all SSPCs to discover objects providers. Furthermore, discovery of objects often is optimal path length, due to that, when SSPCs deployed objects, they also saved the shorter path that has been discovered.

We assume the fire caused a health disorder for a human, and that person integer an object captured the state of health behavior and can communicate with SSPC. When an object person detects something abnormal with the human body, the process of demand service is beginning. The SSPC take mission, request received by SSPC is processed to determine the domain service wanted. While in this case the service

Table 1 Functional comparison

Characteristics	Our	[15]	[16]	[17]	[18–20]
Mechanism of communication of objects in lossy network	Yes	Non	Non	Yes	Non
Level defined communication services in constraints object	Yes	Partially	Partially	Yes	Partially
Trust object services	Yes	Non	Non	Non	Non
Hierarchy of communication objects	Yes	Non	Non	Yes	Non
Hierarchy of communication service	Yes	Yes	Yes	Yes	Non
Semantic knowledge-based ontologies	Non	Non	Non	Yes	Yes
Service discovery	Yes	Non	Yes	Non	Yes
Self-management of service	Non	Yes	Yes	Non	Non
Security and privacy	Non	Yes	Non	Non	Non

depends on health, the request is directed of objects of smart health. After mining service customers, the SSPC objects of smart health determine each subdomain execute this task with helping of object of smart home. As example the objects of smart health determined the hospitals and doctors can treat the person.

4.2 *Comparison*

To compare our proposed work, we just choose the articles handle mechanism of services management, due to many recent articles take care to developed applications of IoT and platforms, and wireless sensors network (WSN), etc. The comparison of these use case applications proposed solution is not discussed in this table comparison.

Furthermore, we choose some characteristics that we focus in this article: communication in lossy network and constraints devices that are important challengers of IoT, hierarchy of communication objects and services in IoT play an important role for control and manage communication devices and direction of service request/response, semantic knowledge-based ontologies with subdivides each domain service to treat such domain ontology, self-management of service is final goal of all architectures-oriented services, and finally security and privacy are most problems of devices of IoT due low capacity of calculus and low power energy of devices (Table 1).

5 Conclusion and Future Trends

In this paper, a System Service Provider–Customer of IoT for communication services has been proposed. As discussed in previous sections, it is able to facilitate communication between providers and customers in lossy network where devices have constraints. Furthermore, these SSPCs manage services on discovering them deployed objects, also they can get these services in a short time. In our future work, we will focus on the prototype implementation of the proposed system.

References

1. Links, C.: The Internet of things will change our world. ERCIM News **101**(3), 76 (2015)
2. Nikolić, V., Begenešić, N.: The IOT bases: LoRa networks. Zb. Međunarodne Konf. o obnovljivim izvorima električne Energ., **5**(1), 235–238 2017
3. Floarea A.-D., Sgârciu, V.: Smart refrigerator: a next generation refrigerator connected to the IoT. In: 2016 8th International Conference on Electronics, Computers and Artificial Intelligence (ECAI), 2016, pp. 1–6
4. Pacheco, J., Satam, S., Hariri, S., Grijalva, C., Berkenbrock, H.: IoT security development framework for building trustworthy smart car services. In: 2016 IEEE Conference on Intelligence and Security Informatics (ISI), 2016, pp. 237–242
5. Atif, Y., Ding, J., Jeusfeld, M.A.: Internet of things approach to cloud-based smart car parking. Procedia Comput. Sci. **98**, 193–198 (2016)
6. Gubbi, J., Buyya, R., Marusic, S., Palaniswami, M.: Internet of Things (IoT): a vision, architectural elements, and future directions. Futur. Gener. Comput. Syst. **29**(7), 1645–1660 (2013)
7. Ferrández-Pastor, F., García-Chamizo, J., Nieto-Hidalgo, M., Mora-Pascual, J., Mora-Martínez, J.: Developing ubiquitous sensor network platform using internet of things: Application in precision agriculture. Sensors **16**(7), 1141 (2016)
8. Sijun, G.A.O., Yonggang, Z., Xiaolin, Z., Yang, X.U., Bing, F., Lirong, Z.: Design of Internet of things application and service detecting system in agriculture. J. Shanghai Norm. Univ. (Natural Sci.) **44**(1), 51–59 (2015)
9. Castellani, A.P., Bui, N., Casari, P., Rossi, M., Shelby, Z., Zorzi, M.: Architecture and protocols for the internet of things: a case study. In: 2010 8th IEEE International Conference on Pervasive Computing and Communications Workshops (PERCOM Workshops), 2010, pp. 678–683
10. Peng, S., Su, G., Chen, J., Du, P.: Design of an IoT-BIM-GIS based risk management system for hospital basic operation. In: 2017 IEEE Symposium on Service-Oriented System Engineering (SOSE), 2017, pp. 69–74
11. Dong L., Wang, G.: A robust and lightweight name resolution approach for IoT data in ICN. In: 2017 Ninth International Conference on Ubiquitous and Future Networks (ICUFN), 2017, pp. 61–65
12. Truong, T., Dinh, A., Wahid, K.: An IoT environmental data collection system for fungal detection in crop fields. In: 2017 IEEE 30th Canadian Conference on Electrical and Computer Engineering (CCECE), 2017, pp. 1–4
13. Iannacci, J.: Microsystem based energy harvesting (EH-MEMS): powering pervasivity of the Internet of Things (IoT)–a review with focus on mechanical vibrations. J. King Saud Univ. (2017)
14. do Nascimento, N.M., de Lucena, C.J.P.: Fiot: an agent-based framework for self-adaptive and self-organizing applications based on the internet of things. Inf. Sci. (Ny). **378**, 161–176 (2017)
15. Alberti, A.M., Scarpioni, G.D., Magalhaes, V.J., Cerqueira, S.A., Rodrigues, J.J.P.C., da R. Righi, R.: Advancing NovaGenesis architecture towards future Internet of Things. IEEE Internet Things J. (2017)
16. Cirani, S., et al.: A scalable and self-configuring architecture for service discovery in the internet of things. IEEE Internet Things J. **1**(5), 508–521 (2014)
17. Wang, W., Lee, K., Murray, D.: A global generic architecture for the future Internet of Things. Serv. Oriented Comput. Appl. **11**(3), 329–344 (2017)

18. Sarkar, C., Nambi S.N.A.U., Prasad, R.V., Rahim, A., Neisse, R., Baldini, G.: DIAT: a scalable distributed architecture for IoT. IEEE Internet Things J. **2**(3), 230–239 (2015)
19. Nambi, S.N.A.U., Sarkar, C., Prasad, R.V., Rahim, A.: A unified semantic knowledge base for IoT. In: 2014 IEEE World Forum on Internet of Things (WF-IoT), 2014, pp. 575–580
20. Kibria, M.G., Jarwar, M.A., Ali, S., Kumar, S., Chong, I.: Web objects based energy efficiency for smart home IoT service provisioning. In: 2017 Ninth International Conference on Ubiquitous and Future Networks (ICUFN), 2017, pp. 55–60

An Enhanced Moth-Flame Optimizer for Reliability Analysis

Aziz Hraiba, Achraf Touil and Ahmed Mousrij

Abstract In this paper, we devoted the reliability analysis by combining the moth-flame optimizer (MFO) with the first-order reliability method (FORM). To improve the global search ability of MFO, a new position-updated equation is presented according to position update process of accelerated particle swarm (APSO) which can explore the search space quickly and locate the optimal solution efficiently. In the proposed method named as EMFO, FORM is used to evaluate the fitness of each agent. In order to investigate the efficiencies of EMFO in reliability analysis, four classic examples, as well as roof truss model are employed. The results are compared to four well-known heuristic algorithms. The results show that reliability analysis by using EMFO is significantly better than the current heuristic algorithms.

Keywords Reliability analysis · Moth-flame optimizer (MFO) · Accelerated particle swarm optimization (APSO) · First-order reliability method (FORM)

1 Introduction

The structural safety evaluation methods aim to evaluate the likelihood of a violation of the boundary condition by comparing probabilistic models active loads and resistance of a component or structural system. A limit state is a condition beyond which a structure exceeds a specified design requirement expressed in mathematical form by a limit state function $G(X)$. The probability of failure (Pf) is defined as

A. Hraiba (✉) · A. Touil · A. Mousrij
Laboratory of Engineering of Industrial Management and Innovation,
Faculty of Sciences and Technology, Hassan 1st University,
PO Box 577, Settat, Morocco
e-mail: a.hraiba@uhp.ac.ma

A. Touil
e-mail: ac.touil@uhp.ac.ma

A. Mousrij
e-mail: ahmed.mousrij@uhp.ac.ma

V. Bhateja et al. (eds.), *Embedded Systems and Artificial Intelligence*,
Advances in Intelligent Systems and Computing 1076,
https://doi.org/10.1007/978-981-15-0947-6_71

the probability of occurrence of failure ($G(X) \leq 0$), where X is a random variable vector representing the uncertainties of the loads, as well as on the material and geometrical properties of the structure. Although, the uncertainties are quantified in a probabilistic manner and the probability of failure is used as the magnitude used as a basis for the safety measure.

There are several methods in the literature. The most famous is the Monte Carlo simulation (MCS), which represents the reference for all other methods [10]. In a paper by [7] describes that the first-order reliability method (FORM) is more elegant and efficient than simulation methods. However, the previously discussed methods depend on the possibility to calculate the value of $G(X)$ for a vector X. Sometimes these values require the results of other programs (finite element), or the limit state function G is implicit. Or it might be ineffective to link the iteration of the index of reliability with a non-linear dynamic analysis. In addition the computational cost can be very high.

Currently, swarm intelligence algorithms are efficiently used to solve complex optimization problems such as reliability analysis. They work effectively and have many advantages over traditional deterministic methods and algorithms. It has become evident that the researchers concentrated on using single metaheuristics. However, there are some limitations. To overcome this problem, a wide variety of hybrid approaches are proposed in the literature. The main idea of a hybrid with two or more metaheuristics was inspired by the possibility that the new hybridized algorithm combines the strengths of each of these algorithms to provide the following advantages: (i) to produce better solutions, (ii) to provide solutions in less time. In the literature, a wide range of methods has been proposed by combining the generic algorithm and Particle Swarm Optimization for reliability analysis [2]. Recently [15], they proposed a hybrid method based on particle swarm optimization combined with chaotic theory in order to improve the global search of standard PSO. The proposed method was tested on four examples as well as a circular tunnel. The reported results show that the proposed method can identify the design point and compute the corresponding reliability index with high accuracy.

Despite the merits of the above-mentioned works, the problem of local optima entrapment still persist. In addition, there is a theorem in the field of heuristics called No Free Lunch [13] that says there is no optimization algorithm for solving all problems. Since there are differents explicit and implicit state limit functions. Hence, there are possibilities that one algorithm performs well on a state limit set but worse on another. These reasons allow researcher to investigate the efficiencies of new algorithms in enhancing reliability analysis. This is also the contribution of this study, in which the recently proposed moth-flame optimizer (MFO) algorithm [9] and the accelerated particle swarm optimization [6] is proposed to be embedded to reliability analysis. To the best of our knowledge there is no previous work that attempts to use MFO in conjunction with accelerated particle swarm optimization to enhanced exploitation and exploration.

The rest of the paper is organized as follows. Section 2 describes the reliability analysis. Section 3 provides the methodologies utilized in this paper. Section 4 reports the numerical results and discussion. Finally, our conclusions and future work are presented in Sect. 5.

2 Probabilistic Modeling

Probabilistic modeling focuses on the system failure probability, it is not a query to the phenomena that provoke them, but the frequency with which they occur. Therefore, it is not a physical theory, but a theory of probabilities and statistics [3]. Structural reliability is based on the probabilistic model and provides methods to quantify the failure probability. Several important contributions in this area include the work developed by [3, 4, 12].

The reliability is defined as the probability that the performance function $G(X)$ is greater than zero. In other words, the theory of reliability assumes that it is possible to estimate this event using a mathematical model, thus calculating its failure probability [1]. The positive values of $G(X)$ correspond to safety situations and the function negative values give the failure situations. Figure 1, illustrates a general description of reliability analysis.

The reliability experiment is usually expressed in terms of Eq. (1), Where F called failure events. The probability that the event F occurs is given by the fact that the stress exceeds the resistance of the structure.

$$P_f = \Pr\{G(X) < 0\} \tag{1}$$

$G(X)$ represents the performance function, where X random variables noted $X = (X_1, \ldots, X_n)$. These n random variables are called basic variables which represent a physical uncertainty of the model.

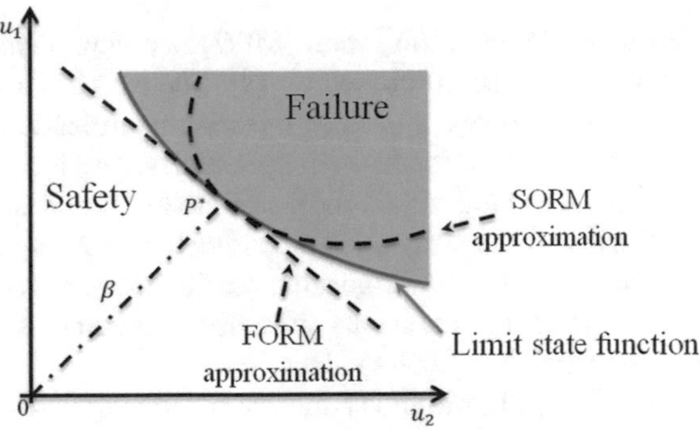

Fig. 1 Probability of failure

Low and Tang [8] proposed a new algorithm for FORM by a new interpret the Hasofer-Lind index, this approach admits the expansion of an ellipsoid in the original space of the basic random variables and minimized the reliability index β as:

$$\beta = \min_{X \in F} \sqrt{\left[\frac{X_i - \mu_i}{\sigma_i}\right]^T [R]^{-1} \left[\frac{X_i - \mu_i}{\sigma_i}\right]} \tag{2}$$

where μ_i and σ_i are respectively the mean and the standard deviation for the random variables X, and R represents the correlation matrix. The probability can be estimated by:

$$P_f = 1 - \phi(\beta) \tag{3}$$

where $\phi(.)$ is the cumulative distribution function of the standard normal variable.

Based on the above assumptions, the following constrained optimization has to be solved:

$$\begin{cases} \text{minimize } \sqrt{[n]^T\, [\,R]^{-1}\, [n]} \\ \text{Subject to:} \\ G(X) = 0 \end{cases} \tag{4}$$

where $G(X)$ is the limited state function.

3 Enhanced Moth-Flame Optimizer

This section describes our proposed moth-flame optimizer and accelerated particle swarm optimization.

3.1 Brief Introduction to Moth-Flame Optimizer (MFO)

The Moth-flame optimization (MFO) is a new population-based stochastic search algorithm recently proposed by [9]. The MFO is a newly developed optimization technique to solve complex engineering optimization problems [5]. It is inspired by the behavior of moths for their special navigation methods in night.

In the MFO algorithm, the set of moths is represented in a matrix M. For all the moths, there is an array OM for storing the corresponding fitness values. The second key components in the algorithm are flames. A matrix F similar to the moth matrix is considered. For the flames, it is also assumed that there is an array OF for storing the corresponding fitness values.

The MFO algorithm is a three-tuple that approximates the global optimal of the optimization problems and defined as follows:

$$\text{MFO} = (I, P, T) \tag{5}$$

I is a function that generates a random population of moths and corresponding fitness values. The methodical model of this function is as follows:

$$I : \varnothing \longrightarrow \{M, OM\} \tag{6}$$

The P function, which is the main function, moves the moths around the search space. This function receives the matrix of M and returns its updated one eventually.

$$P : M \longrightarrow M \tag{7}$$

The T function returns true if the termination criterion is satisfied and false if the termination criterion is not satisfied:

$$T : M \longrightarrow \{\text{True}, \text{False}\} \tag{8}$$

with I, P, and T, the general framework of the MFO algorithm is defined as follows:
$M = I()$;
while $T(M)$ is equal to false
$M = P(M)$;
end

After the initialization, the P function is iteratively run until the T function returns true. The P function is the main function that moves the moths around the search space. As mentioned above the inspiration of this algorithm is the transverse orientation. In order to mathematically model this behavior, we update the position of each moth with respect to a flame using the following equation:

$$M_i = S(M_i, F_j) \tag{9}$$

where M_i indicates the ith moth, F_j indicates the jth flame, and S is the spiral function. A logarithmic spiral is defined for the MFO algorithm as follows:

$$S(M_i, F_j) = D_i \times e^{bt} \times \cos(2\pi t) + F_j \tag{10}$$

where D_i indicates the distance of the ith moth for the jth flame, b is a constant for defining the shape of the logarithmic spiral, and t is a random number in $[-1, 1]$.

D_i is calculated as follows:

$$D_i = |F_j - M_i| \tag{11}$$

where M_i indicate the ith moth, indicates the jth flame, and D_i indicates the distance of the th moth for the jth flame.

Equation (10) describes the spiral flying path of moths. From this equation, the next position of a moth is defined with respect to a flame. The t parameter in the spiral equation defines how much the next position of the moth should be close to the flame ($t = -1$ is the closest position to the flame, while $t = 1$ shows the farthest). A question that may rise here is that the position updating in (10) only requires the moths to move towards a flame, yet it causes the MFO algorithm to be trapped in local optima quickly. In order to prevent this, each moth is obliged to update its position using only one of the flames in (10).

Another concern here is that the position updating of moths with respect to different locations in the search space may degrade the exploitation of the best promising solutions. To resolve this concern, an adaptive mechanism provided the number of flames. The following formula is utilized in this regard:

$$\text{flame–no} = \text{round}\left(N - l * \frac{N-1}{T}\right) \tag{12}$$

where l is the current number of iteration, N is the maximum number of flames, and T indicates the maximum number of iterations.

The gradual decrement in number of flames balances exploration and exploitation of the search space.

3.2 Enhanced Moth Flame Optimizer (EMFO) for Reliability Analysis

In this subsection, we explain the principal phases of the proposed method to obtaining optimal solution as follows.

Enhanced Moth Flame Optimizer (EMFO) It is well-known that the balance between exploration and exploitation are the keys of success of any population-based optimization algorithm, such as GA, PSO, DE and so on. In conventional MFO, it may converge prematurely without enough exploration of search space. In order to increase the diversity of population against premature convergence and accelerate the convergence speed, this paper proposes an enhanced moth-flame optimization (EMFO) algorithm based on the accelerated particle swarm optimization (APSO) developed by [14]. In general, the APSO can explore the search space quickly and locate the optimal solution efficiently. In EMFO, we let each moth take advantage of the information of the another moth to guide the search of candidate moths using Eq. 10 after the position updating, which is formulated as follows:

$$X_i^{t+1} = (1 - \beta)X_i^t + \beta g^* + \alpha r \tag{13}$$

Equation (6) does not contain the velocity, and thus APSO does not need to initialize velocities, and thus simplify the implementation and also avoid any disadvantages associated with velocities in standard PSO. Here, the third term r makes the system more mobile, and do not be trapped in any local solution if a is selected properly, while r can be drawn from a statistical distribution. For the other parameters such as β and α they are chosen according to [6] as follow: $\beta = 0.2 - 0.7$

$$\alpha = \gamma^t \tag{14}$$

where γ is a parameter that can be set 0.1–0.99. Here $t \in [0, t_{\max}]$ and $t_{\max}$ is the maximum of generations.

Evaluation To evaluate the fitness of each agent of EMFO, the constrained problem (4) should be transformed to an unconstrained problem by using the penalty method proposed by [15]. Then the fitness of each agent is given as follows:

$$\text{Fitness} = \sqrt{[n]^T \, [\, R]^{-1} \, [n]} + M \times |G(X)| \tag{15}$$

Framework of EMFO for Reliability Analysis The main steps of the proposed EMFO for reliability analysis are described in Algorithm 1, where N is the population size, $G_{\max}$ is the maximum of generations, and d is the number of decisions' variables.

Algorithm 1 Enhanced Moth Flame Optimizer (EMFO) :

```
Initialize the position of moths and parameters
while (g <= G_max) do
  Update flame no using (12)
  OM= Fitness(M) using (15);
  if (g == 1) then
    F=sort(M);
    OF=sort(OM);
  else
    F=sort(M_t − 1, M_t);
    OF=sort(M_{t−1}, M_t);
  end if
  for i = 1 : N do
    for j = 1 : d do
      Update r and t;
      Calculate D using (11) with respect to the corresponding moth;
      Update M(i, j) using (9) and (10) with respect to the corresponding moth;
    end for
    Update the position of the current search moth M(i, :) as follows:
    M(i, :) = (1 − β) × M(i, :) + β × M(1, :) + α × r
  end for
  Update g = g + 1
end while
Report the best solution
```

4 Results and Discussion

In this section, the proposed algorithm is compared with the chaotic particle swarm based reliability analysis proposed by [15]. The experiments were done using MATLAB R2014a on PC with a 3.30 GHz Intel(R) Core (TM) i5 processor, 4 GB of memory. In the paper by [15] describes a set of state limit functions which are presented in in the following Examples 1–4;

Example 1

$$G(X_1, X_2) = 0.01846154 - 74.76923 \times \frac{X_1}{X_2^3} \tag{16}$$

Example 2

$$G(X_1, X_2) = 2.5 - 0.2357 \times (X_1 - X_2) + 0.00483 \times (X_1 + X_2 - 20)^4 \tag{17}$$

Example 3

$$G(X_1, X_2) = e^{0.4(X_1+2)+6.2} - e^{0.3X_2+5} - 200 \tag{18}$$

Example 4

$$G(X_1, X_2, X_3) = X_1 - \frac{X_2}{X_3} \tag{19}$$

All variables are considered normal distribution with mean, standard deviation and value of penalty coefficient (M) presented in Table 1. For the other parameters are given in Table 2.

For the Example 1, the obtained results are listed in Table 3. It can be seen that the exact failure probability is obtained using MCS with importance sampling is 0.9607×10^2, while as 0.9876×10^2 by using the proposed EMFO which same value as Low and Tang method and better than the obtained with CPSO. In addition, as shown in Fig. 2 the proposed method converges quickly to satisfied reliability index and design points after only a 30 generations which represents a gain in computation time.

For the Example 2, the obtained results are listed in Table 3 and Fig. 3. It can be seen that the proposed EMFO obtained the same value as Low and Tang method and better than the obtained with CPSO.

For the Example 3, according to the obtained results listed in Table 4, the proposed method can compute the reliability index with high efficiency and accuracy. In addition, as shown in Fig. 4 the proposed method converges quickly to satisfied reliability index and design points after only a 40 generations which represents a gain in computation time. Finally, the obtained results for the Example 4 are listed in Table 4 and in Fig. 5. It can be seen that the proposed EMFO are similar to the previous Examples 1–2. This demonstrates that the application of the proposed method to reliability analysis is feasible, efficient, and accurate.

Table 1 Overall data and results

Example	Variables	Distribution	Mean	STD	M
(1)	X_1	Normal	1000	200	1000
	X_2	Normal	250	37.5	
(2)	X_1	Normal	10	3	1000
	X_2	Normal	10	3	
(3)	X_1	Normal	1000	200	0.1
	X_2	Normal	250	37.5	
(4)	X_1	Normal	600	30	1000
	X_2	Normal	1000	33	
	X_3	Normal	2	0.1	

(a) Given data for Examples 1, 2, 3 & 4

Parameter	Value
Population size (NP)	1000
Maximum of generations (Max_Gen)	20
Levy parameter	1.5
Number of rounds	5
APSO parameter (β; γ)	0.5–0.7

(b) Parameters for Examples 1, 2, 3 & 4

Example 1				
Algorithms	Desing Points		RI	$P_f(10^{-2})$
	X_1	X_2		
EMFO	**1118.5655**	**165.4647**	**2.3309**	**0.9879**
CPSO	1125.5766	165.8097	2.3312	0.9871
Low& Tang	1118.5578	165.4640	2.3309	0.9879
MCS	-	-	-	0.9607
Example 2				
Algorithms	Desing Points		RI	P_f (10^{-2})
	X_1	X_2		
EMFO	**15.3034**	**4.6966**	**2.5000**	**0.62**
CPSO	15.3334	4.7267	2.5001	0.62
Low& Tang	15.3034	4.6966	2.5000	0.62
MCS	-	-	-	-

RI: Reliability-Index; P_f: Probability of faillure

(c) Comparison of results for Example 1 & 2

Example 3				
Algorithms	Desing Points		RI	
	X_1	X_2		
EMFO	**−2.5397**	**0.9453**	**2.7099**	
CPSO	−2.5407	0.9427	2.7099	
Low& Tang	−2.5397	0.9454	2.7099	
MCS	-	-	2.685	
Example 4				
Algorithms	Desing Points			RI
	X_1	X_2	X_3	
EMFO	**555.6085**	**1029.0027**	**1.852028**	**2.2697**
CPSO	553.2864	1023.0742	1.84909	2.2784
Low& Tang	555.6091	1029.0028	1.85203	2.2697
MCS	-	-	-	2.2490

RI: Reliability-Index;

(d) Comparison of results for Example 3 & 4

Parameter	Distribution	Mean	CV
$q[N/m]$	Normal	20000	7%
$l[m]$	Normal	12	1%
$A_S[m^2]$	Normal	9, 82Ã 10^{-4}	6%
$A_C[m^2]$	Normal	0,04	12%
$E_S[N/m^2]$	Normal	1×10^{11}	6%
$E_C[N/m^2]$	Normal	2×10^{10}	6%

(e) Variables of Roof truss model structure

Parameters	Algorithms	
	MCS	EMFO
$q[N/m]$	-	20476
$l[m]$	-	12.01
$A_S[m^2]$	-	96.95×10^{-4}
$A_C[m^2]$	-	0.039096
$E_S[N/m^2]$	-	9.4×10^{10}
$E_C[N/m^2]$	-	1.8×10^{10}
$P_f(10^{-3})$	9.38	**9.48**

(f) Comparison of results for Application

Application

Figure 6 shows the roof truss model, this model was considered by [11] in the context of a sensitivity analysis. The limit state function is defined as Eq. (20). The set of parameters are given as follows: q as a distributed load applied to the structure, l is the length between the supports. A_c and A_s are the transverse cross sections, and E_s, E_c are Young's moduli of the steel and concrete beams, respectively. The statistic

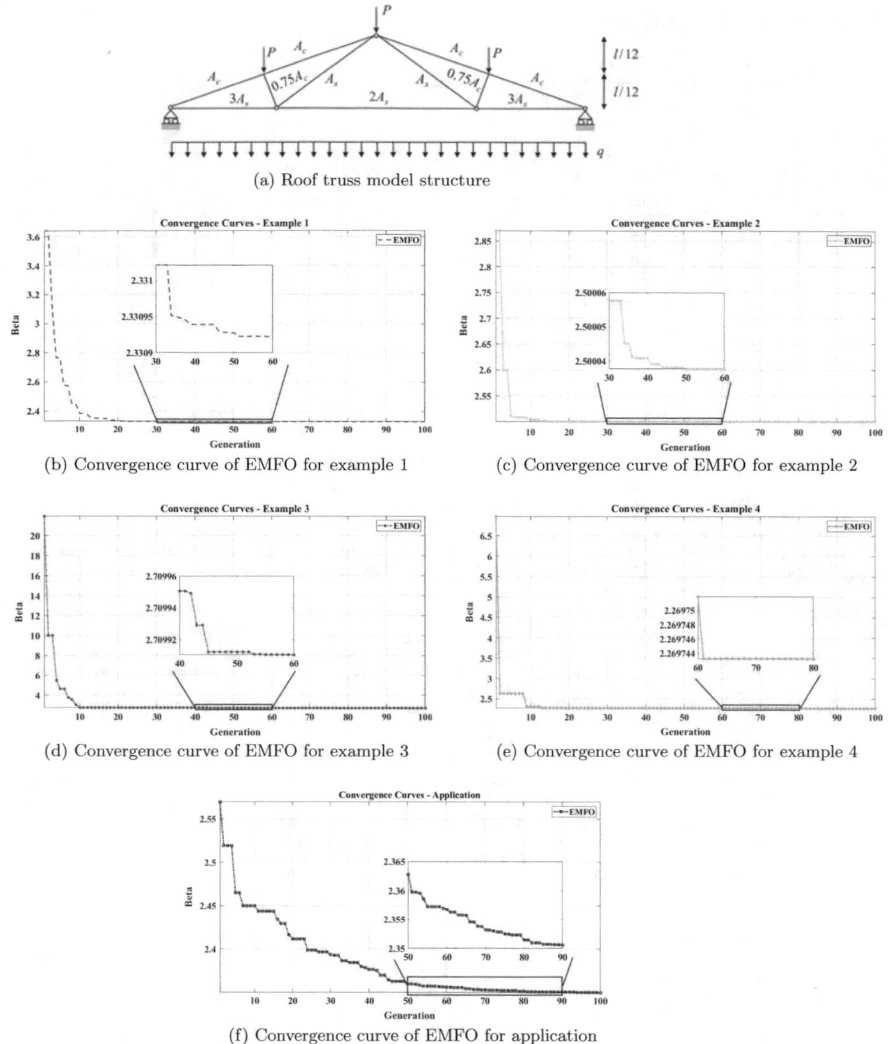

(a) Roof truss model structure

(b) Convergence curve of EMFO for example 1

(c) Convergence curve of EMFO for example 2

(d) Convergence curve of EMFO for example 3

(e) Convergence curve of EMFO for example 4

(f) Convergence curve of EMFO for application

Fig. 2 Overall convergence curves

models are listed in the Table 5. Once more, due to livelihood of the positive parameters, the normal distribution is not an appropriate choice for modeling. However, in order to validate our algorithm, we have proceeded with the identical probabilistic model as in [11].

$$G(q, A_c, A_s, E_c, E_s, l) = 0.03 - \frac{ql^2}{2}\left(\frac{3.81}{A_c E_c} + \frac{1.13}{A_s E_s}\right) \tag{20}$$

For the application, the obtained results are listed in Table 6. It can be seen that the exact failure probability is obtained using MCS with importance sampling is

0.9384×10^2, while as 0.9486×10^2 by using the proposed EMFO. In addition, as shown in Fig. 7 the proposed method converges to satisfied reliability index and design points after 60 generations which represent a gain in computation time.

5 Conclusion

A EMFO-based reliability analysis method was presented. The EMFO algorithm has strong global search capability. The proposed method can identify the design point and compute the corresponding reliability index with high accuracy. It does not require derivative information for the limited state function and is fitted to an implicit limited state function. The method was applied to four classic examples and the reliability of a roof truss mode. The proposed method can be used for reliability analysis in engineering with high efficiency and accuracy.

References

1. Bourinet, J.-M., Deheeger, F., Lemaire, M.: Assessing small failure probabilities by combined subset simulation and support vector machines. Struct. Saf. **33**(6), 343–353 (2011)
2. Cheng, J.: Hybrid genetic algorithms for structural reliability analysis. Comput. Struct. **85**(19–20), 1524–1533 (2007)
3. Deheeger, F.: *Couplage mécano-fiabiliste: 2 SMART-méthodologie d'apprentissage stochastique en fiabilité*. Ph.D. thesis, Université Blaise Pascal-Clermont-Ferrand II (2008)
4. Der Kiureghian, A., Haukaas, T., Fujimura, K.: Structural reliability software at the University of California, Berkeley. Struct. Saf. **28**(1–2), 44–67 (2006)
5. El Aziz, M.A., Ewees, A.A., Hassanien, A.E.: Whale optimization algorithm and moth-flame optimization for multilevel thresholding image segmentation. Expert Syst. Appl. **83**, 242–256 (2017)
6. Gandomi, A.H., Yun, G.J., Yang, X.-S., Talatahari, S.: Chaos-enhanced accelerated particle swarm optimization. Commun. Nonlinear Sci. Numer. Simul. **18**(2), 327–340 (2013)
7. Lemaître, P.: Analyse de sensibilité en fiabilité des structures. PhD thesis, Bordeaux (2014)
8. Low, B.K., Tang, W.H.: New form algorithm with example applications. In: Proceedings of the Fourth Asian-Pacific Symposium on Structural Reliability and its Applications, Hong Kong (2008)
9. Mirjalili, S.: Moth-flame optimization algorithm: A novel nature-inspired heuristic paradigm. Knowl.-Based Syst. **89**, 228–249 (2015)
10. Papadimitriou, D.I., Mourelatos, Z.P., Hu, Z.: Reliability analysis using second-order saddle-point approximation and mixture distributions. J. Mech. Des. **141**(2), 021401 (2019)
11. Papaioannou, I., Breitung, K., Straub, D.: Reliability sensitivity estimation with sequential importance sampling. Struct. Saf. **75**, 24–34 (2018)
12. Sudret, B., Der Kiureghian, A.: Comparison of finite element reliability methods. Prob. Eng. Mech. **17**(4), 337–348 (2002)
13. Wolpert, D.H., Macready, W.G., et al.: No free lunch theorems for optimization. IEEE Trans. Evol. Comput. **1**(1), 67–82 (1997)
14. Yang, X.-S.: Engineering Optimization: An Introduction with Metaheuristic Applications. Wiley, Hoboken (2010)
15. Zhao, H., Zhongliang, R., Chang, X., Li, S.: Reliability analysis using chaotic particle swarm optimization. Qual. Reliab. Eng. Int. **31**(8), 1537–1552 (2015)

Comparative Study on Approaches of Recommendation Systems

Khalid Al Fararni, Badraddine Aghoutane, Jamal Riffi, Abdelouahed Sabri and Ali Yahyaouy

Abstract In the current context, characterized by an information overload, also known as infobesity, it has become essential to design mechanisms that allow users to access what interests them as quickly as possible. Hence, recommender systems have emerged. This article then consists of examining and comparing the different existing recommendation approaches: those content-based filtering, those collaborative filtering, and finally the demographic and social approaches, while indicating, for each approach, the fields of application, some interesting examples, as well as its advantages and limitations. Then, we will indicate the hybridization techniques available to overcome these limitations.

Keywords Recommender systems · Collaborative filtering · Content-Based filtering · Hybrid recommender system

1 Introduction

Frequently, we are faced with making choices. How to dress? Which movie to watch? Which item to buy? Where to travel? The list of possibilities available to us is

K. Al Fararni (✉) · J. Riffi · A. Sabri · A. Yahyaouy
LIIAN Laboratory, Department of Informatics, Faculty of Sciences Dhar El Mahraz, Sidi Mohamed Ben Abdellah University, 1796 Fez-Atlas, Fez, Morocco
e-mail: khalid.alfararni@usmba.ac.ma

J. Riffi
e-mail: jamal.riffi@usmba.ac.ma

A. Sabri
e-mail: abdelouahed.sabri@gmail.com

A. Yahyaouy
e-mail: ali.yahyaouy@usmba.ac.ma

B. Aghoutane
Team of Processing and Transformation of Information, Polydisciplinary Faculty of Errachidia, Moulay Ismaïl University, 11201 Zitoune, Meknes, Morocco
e-mail: b.aghoutane@fpe.umi.ac.ma

V. Bhateja et al. (eds.), *Embedded Systems and Artificial Intelligence*, Advances in Intelligent Systems and Computing 1076,
https://doi.org/10.1007/978-981-15-0947-6_72

generally very large, the evaluation of these possibilities to find what suits us the most is a difficult task and can consume a lot of our time. To address this problem of information overload and choices, recommender systems have emerged in the last decade of the twentieth century.

Recommender systems have been studied in various fields such as the web, e-commerce, and many others.

This paper presents at first a state of the art on the various approaches of recommendation, not only based on preferences of the users, but also on the basis of their behaviors, their demographic profiles, and the judgments of other users (presented on the social networks). A summary table that compares these approaches is provided in Sect. 3. We conclude in Sect. 4, on the perspectives of our work.

2 Recommender Systems

Recommender systems are a specific form of *Filtering Information* to produce to an «active user» (website visitor, a potential buyer of a product, etc.) a personalized list of suggestions (services, products, or links to click, etc.) related to its concerns and expectations. These suggestions are based on what the user has purchased or previously viewed but also on the activity history of other buyers with a similar profile.

Thus, this is how we differentiate *Information Retrieval Systems* for which the user's request to be oriented to the appropriate choices is explicit and recommender systems where the participation of user is non-voluntary, induced in particular by trace engines [1].

Generally, a recommender system makes it possible to compare the profile of a user with certain reference characteristics, and seeks to predict the «opinion» that this user would give to a given item. These characteristics can come from:

- The object itself, we speak about a *Content-based approach.*
- Social environment, we speak about *Collaborative Filtering Approach* or *Social Approach.*
- Demographic data, we speak about a *Demographic Approach.*

2.1 Content-Based Approach

In this approach, also called cognitive filtering, each item is defined by a set of attributes and their value. The choice of the items to recommend is based on a comparison of the topics covered in these items with the topics relevant to the user.

In a content-based recommendation system, the attributes (keywords) that are used to describe the items, and the user profile is designed to indicate the type of item that this user likes. In other words, for the proposal of recommendations to the users, this technique is based on the analysis of the similarities of contents between the items previously consulted by the users (or examine in the present) and those which have not been yet consulted [2].

In particular, various items of candidates are compared with items previously rated by the user and the most relevant items are recommended [3, 4].

This approach has its roots in Information Retrieval (IR) and Artificial Intelligence. In *Information Retrieval*, it is considered that users wanting recommendations are engaged in an information retrieval process. The user expresses a specific need by giving a request (usually a list of keywords). In Information Filtering (IF) Systems, the need is represented by the user's profile.

In *Artificial Intelligence*, the recommendation task can be expressed as a learning problem that exploits past user knowledge. In a simple way, user profiles are in the form of keyword vectors and reflect the long-term interests of the user. Often, it is best for the system to learn the user's profile rather than forcing the user to provide it. This usually involves the application of Machine Learning (ML) techniques. Their purpose is to learn how to categorize new information based on the information previously seen, and which has been implicitly or explicitly labeled as interesting or not by the user. With these labels, Machine Learning's methods can generate a predictive model which, given a new item, will help decide how much the item can interest the user.

There are two types of recommendation based on the content:

- *Keywords-based recommendation*: Consider the characteristics of items that the user has rated in the past to correlate them with their profile. In fact, user and item profiles are represented as weighted keyword vectors. To recommend interesting new items, these SRs attempt to match item attributes with the user's preferences and interests. For a new item, the system performs a cosine similarity calculation between the profile vector and the item vector to predict the user's score on the item.

It should be noted that these keywords are usually either extracted based on automatic indexing or manually assigned.

- *Recommendation based on semantics*: Semantics has been introduced by several methods in the recommendation process. These methods are discussed taking into account several criteria:
  - The type of source of knowledge involved (lexicon, ontology, etc.);
  - The techniques adopted for the annotation or the representation of items;
  - The type of content included in the user profile;
  - The matching strategy between items and profile.

2.2 *Collaborative Filtering*

Draws on the ratings given by a set of users on a set of articles, to generate the predictions. These appreciations, translated into numerical values, can be notes, accounts of purchases made, numbers of visits, etc. The key idea is that a note of the user u for a new item i is likely to be similar to that given by another user v, if u and v have similarly noted other items. Similarly, u is likely to rate two items i and j in the same way, if other users gave similar ratings to these two items. There are two main approaches to collaborative filtering:

- *User-Based Approach* [1]: compare users with each other and find those with common tastes, the notes of a user is then predicted according to his neighborhood.
- *Item-Based Approach* [5]: is to bring together items liked by the same people and to predict user ratings based on articles closest to those they have already noted.

2.3 Demographic Approach

This is a simple recommendation that proposes items relative to the demographic profile of the user [6]. It involves categorizing users into several classes or groups in relation to demographic information such as gender, age, occupation, location, language, country, etc. The principle of this approach is that two users who have evolved in a similar environment share common tastes than two users who have evolved in different environments and therefore do not share the same codes. Many sites use this solution to offer a "personalized" content offer. For example, users are redirected to a particular website based on their language or country. This approach has been very popular in the marketing literature but has received little attention in the field of recommendation algorithms.

2.4 Social Approach

This type of system offers recommendations based on the user's relationships with his friends in social networks, and sometimes this recommendation also depends on the value of user trust in each of his friends. Facebook (2.32 billion users), YouTube (1.9 billion users) are famous examples.

So, the question that arises is how can social models be exploited in information filtering systems?

The basic idea would be to replace the traditional community formation on the basis of votes with that induced by social media (friends and friends of friends). [7] compared the classic collaborative recommendations with those made by friends on three movie recommendation systems (Amazon.com, MovieCritic and Reel.com) and three others for a book recommendation (Amazon.com, RatingZone, and Sleeper). The results showed that users preferred those made by their friends. This can be explained by the fact that the friends are more qualified to advise them since they are supposed to know more about the preferences of the users.

2.5 *Hybrid Approach*

The approaches mentioned previously are not exclusive and different hybrid methods, combining these different types of filtering, have been developed [3]. The use of hybrid approaches improves the relevance of filtering system results by overcoming some of the limitations of the types of filtering presented in Table 1 as Overspecialization problem in content-based filtering; obtaining judgments which is an expensive task for users; etc.

The hybridization takes place in two phases:

1. Independently perform item filtering through collaborative methods or content (or other) to generate candidate recommendations (local scores);
2. Combine these sets of recommendations through hybridization methods such as weighting, mixing, cascading, switching, etc. [2], to produce the final recommendations for users (final scores).

2.5.1 Hybridization Techniques

The difference between hybridization techniques lies in the strategy chosen to combine local scores. Robin Burke summarized hybridization techniques in seven techniques [2], including:

- Weighted: The scores calculated by the different recommendation techniques are combined numerically and weighted to generate a single recommendation.
  The advantage of this method is that the combination is simple to perform and allows to adjust the hybridization depending on performance. The following linear equation can be used:

$$\text{Score}_{\text{finale}} = \alpha \times \text{Score}_1 + (1 - \alpha) \times \text{Score}_2;\ 0 < \alpha < 1 \tag{1}$$

  This strategy is flexible, it is necessary to adjust the parameter α to quantify the contribution of each approach.
- Switching: This technique chooses one of several recommendation approaches, based on several criteria. The system must then define the switching criteria or the cases where the use of another technique is recommended. We can consider that the system tries to predict the note by the first approach and uses the second one in case of failure.

$$\text{Score}_{\text{finale}} = \begin{cases} \text{Score}_1, \text{ si Score}_1 \neq \text{null} \\ \text{Score}_2, \text{ si Score}_1 = \text{null} \end{cases} \tag{2}$$

- Mixed: Recommendations from different techniques are all presented simultaneously to users. The problem that can arise with the use of this technique is the difficulty of calculating the scores to order a recommendation list, when all the techniques recommend the same items but with different notes. To overcome this problem, we can choose for example a list that gathers the five best items according to each technique, aggregate them in a single list, and present the list to the user, or each approach calculates scores of all items; each item has the highest score, the items are then listed according to their scores, and the best ones are recommended.

$$\text{Score}_{\text{finale}} = \max(\text{Score}_1, \text{Score}_2) \tag{3}$$

- Features combination: In a hybrid based on the combination of features, data from collaborative techniques is treated as a feature, and a content-based approach is used on that data.
- Cascade: The cascade involves a step-by-step process. In this case, a recommendation technique is applied first, producing a set of potential candidates. Then, a second technique refines the results obtained in the first step. This method has the advantage that if the first technique generates few recommendations, or if these recommendations are ordered to allow a quick selection, the second technique will no longer be used.
- Features augmentation: is similar to the cascade, but in this case, the results obtained (classification) of the first technique are used by the second as an added feature.
  Although both techniques, in cascade and by augmentation of features, chain two recommenders, with the first one has an influence on the second, they are fundamentally very different. In a hybrid by augmentation, the characteristics used by the second recommender include the output of the first, such as scores for example. In a hybrid by cascade, the second approach does not use the output of the first approach in the production of its ratings, but the results of the two recommenders are combined as a priority.
- Meta-level: In a meta-level hybrid, a first technique is used, but differently than the previous method (feature augmentation), not to produce new features, but to produce a model. And in the second stage, it is the whole model that will serve as input for the second technique [6].

3 Summary Table

In this table, we have reviewed and compared several recommendation approaches, indicating each approach the fields of application and some interesting examples, as well as its advantages and limitations.

Table 1 Summary table on recommendation approaches

	Representation and recommendation techniques		Examples of realization and their fields of application	Advantages	Limitations
Content-based approach	**For user/item profiling**: the vector model of Salton (or the logarithmic model) with a basic weighting such as TF-IDF (term frequency-inverse document frequency), decision trees, ontologies, etc. **For the measure of similarity and the prediction of the votes**: cosine similarity, Sørensen-Dice coefficient, Jaccard index, etc.	Keywords-based filtering	**In the web**: Letizia [8], Syskill and Webert [9], Amalthea [10], Personal WebWatcher [11] **News**: NewT [12], NewsDude [4], YourNews [13] **Books**: Citeseer [14], LIBRA [15] **Music**: Pandora [16] **Movies**17: Movies2GO], INTIMATE [18]	• *User independence*: no need for a large community of users to make recommendations. A list of recommendations can be generated even if there is only one user • *No cold start problem for new items*: valid for the recommendation of new items or unpopular items • *Transparency*: an explanation of how the recommender system works can be provided by explicitly listing the characteristics or descriptions that led to the appearance of an item in a list of recommendations • *The quality increases with time*: the more the user will use the system, the more the relevance of the items that will be proposed to him will be fine	• *Traditional keyword-based profiles are not able to capture the semantics of users' interests*: suffers from polysemy and synonymy problems • *Knowledge of the field is often necessary*: for example, for the recommendation of films, the system needs to know the actors, the directors, etc. • *Overspecialization problem*: if a user is only interested in articles of sports, he will never be offered a political article • *Cold start*[a] *(new user)* • Problem with recommending multimedia documents (images, videos, etc.) in the absence of metadata
		Semantics-based	**Web**: SiteIF [19], SEWeP [20] **News**: Informer [21], News@hand [22] **Academic research papers**: Quickstep and Foxtrot [23] **Other fields**: ITR [24], Informed Recommender [25]		

(continued)

[a]Cold start refers to the lack of information on a new user or a new item that has just been added to the recommendation system.

Table 1 (continued)

	Representation and recommendation techniques		Examples of realization and their fields of application		Advantages	Limitations
Collaborative approach	**Item/user profiling**: rating matrix, models of latent factors, etc. **For the measure of similarity and the prediction of the votes**: Pearson correlation coefficient, cosine similarity, Spearman correlation coefficient, Bayesian clustering, maximum entropy, support-vector machine, singular value decomposition…	Memory-based filtering	User-based	**Movies**: GroupLens Movie (MovieLens) [26] Bellcore Video [27], Firefly [28] **Music**: Ringo [28]	• *Independence from the content*: does not require any knowledge of the content of the item or its semantics • *Works for any type of items*: whose content is either unavailable or difficult to analyze • *No overspecialization problem*: the recommendations are not based on the thematic dimension of the profiles	• *Request enough user to get satisfactory results* • *Sparse matrix* (*sparsity*): difficult to find users who voted the same items • *Cold start* (*new user/item*) • *Scalability*: often, the platforms on which collaborative filters are used have millions of users, products, and content. So it requires a lot of computing power to be able to offer suggestions to users • *Recommends only the most popular items and cannot recommend items not voted*
			Item-based	**TV**: Tivo [29] **E-mails**: Tapestry [30] **E-commerce**: Amazon [31] **Music**: LastFM [32], Mures [6]		
Demogr. approach	**For the measure of similarity and the prediction of the votes**: classification, selection…		**Restaurants**: [32] **E-commerce**: LifeStyle Finder [33], Alambic [34]		• *Simple to implement* • *Requires no history of estimates*	• *Privacy issues* • *The lack of diversity* • *Cold start* (*New item*)

(continued)

Table 1 (continued)

	Representation and recommendation techniques	Examples of realization and their fields of application	Advantages	Limitations
Social approach	**For item/user profiling**: ratings matrix, matrix of trust scores between users, etc. **For the prediction of the votes**: Web of Trust [35], TidalTrust [36]…	**Movies**: FilmTrust [36] **E-commerce**: MoleTrust [35]	• *Adaptability*: as the evaluations database (number of friends) increases, the recommendation becomes more precise	• *Cold start (a new item)* • *The accuracy decreases as the research moves away from the source user*
Hybrid approach	**For the prediction of the votes**: weighting, mixing, switching, features combination, cascade, features increase, and meta-level	**Web**: Fab [37] **News**: P-Tango [38], DailyLearner [39] **Movies**: Netflix, IMDb, CinemaScreen [40], More [41] **Restaurants**: EntreeC [2]	• *Adapts better with some pure filtering problems,* such as cold start, overspecialization, etc.	• *Requires additional settings*: related to the combination of different methods • *Unjustified proposals*: cannot explain why an article was recommended to a user

4 Conclusions

The field of recommendation systems, especially those using a hybrid approach, is still very recent.

In this article, we examined and compared several recommendation approaches, while indicating, for each approach the fields of application and some interesting examples. Then, we recalled that recommendation approaches generally suffer from a set of problems that reduce performance for a subset of users. We rely on hybrid SRs to overcome these problems. We then indicated the available hybridization techniques. However, it should be noted that these techniques do not take into account a lot of useful information to explain why an article was recommended to a user. This requires taking into account different sources of information on the user: his demographic data (Demographic filtering), his social network (Community filtering), etc. and their correlations with the content of the items that are proposed to him.

Thus, current recommender systems still require improvement and thus becoming a rich research area.

References

1. Resnick, P., Varian, H.R.: Recommender systems. Commun. ACM **40**, 56–58 (1997)
2. Burke, R.: Hybrid recommender systems: survey and experiments. User Model. User-Adap. Inter. **12**(4), 331–370 (2002)
3. Pazzani, M.J.: A framework for collaborative, content-based, and demographic filtering. Artif. Intell. Rev. **13**(5–6), 393–408 (1999)
4. Billsus, D., Pazzani, M.: A hybrid user model for news story classification. In: Seventh International Conference on User Modeling, Banff, Canada (1999)
5. Sarwar, B., Karypis, G., Konstan, J., Reidl, J.: Item-based collaborative filtering recommendation algorithms. In Proceedings of the 10th International Conference on World Wide Web (WWW'01), pp. 285–295. New York, NY, USA. ACM (2001)
6. Arnautu, O.R.: Mures: Un système de recommandation de musique. Master's thesis, La Faculté des arts et des sciences Université de Montréal (2012)
7. Sinha, R., Swearingen, K.: Comparing recommendations made by online systems and friends. In: DELOS-NSF Workshop on Personalization and Recommender Systems in Digital Libraries (2001)
8. Liberman, H.: Letizia: an agent that assists web browsing. In: International Joint Conference on Artificial Intelligence (IJCAI-95), pp. 924–929. Morgan Kaufmann publishers Inc., Montreal, Canada (1995)
9. Pazzani, M., Muramatsu, J., Billsus, D.: Syskill and Webert: identifying interesting web Sites. In: Thirteenth National Conference on Artificial Intelligence and the Eighth Innovative Applications of Artificial Intelligence Conference, pp. 54–61. AAAI Press/MIT Press, Menlo Park (1996)
10. Moukas, A.: Amalthaea information discovery and filtering using a multi-agent evolving ecosystem. Appl. Artif. Intell., 437–457 (1997)
11. Mladenic, D.: Machine learning used by personal webwatcher. In: ACAI-99 Workshop on Machine Learning and Intelligent Agents (1999)
12. Sheth, B., Maes, P.: Evolving agents for personalized information filtering. In: 9th Conference on Artificial Intelligence for Applications, pp. 345–352. IEEE Computer Society Press (1993)

13. Ahn, J., Brusilovsky, P., Grady, J., He, D., Syn, S.: Open user profiles for adaptive news systems: help or harm? In: 16th International Conference on World Wide Web, pp. 11–20. ACM (2007)
14. Bollacker, K., Giles, C.: CiteSeer: an autonomous web agent for automatic retrieval and identification of interesting publications
15. Mooney, R.J., Roy, L.: Content-based book recommending using learning for text categorization. In: Proceedings of the fifth ACM conference on digital libraries, pp. 195–204. ACM Press (1999)
16. Bu, J., Tan, S., Chen, C., Wang, C., Wu, H., Zhang, L., He, X.: Music recommendation by unified hypergraph: combining social media information and music content. In: Proceedings of the International Conference on Multimedia, ACM (2010)
17. Mukherjee, R., Jonsdottir, G., Sen S., Sarathi, P.: MOVIES2GO: an online voting based movie recommender system. In: Fifth International Conference on Autonomous Agents, pp. 114–115. ACM Press (n.d.)
18. Mak, H., Koprinska, I., Poon, J.: INTIMATE: a web-based movie recommender using text categorization. In: IEEE/WIC International Conference on Web Intelligence, pp. 602–605. IEEE Computer Society (2003)
19. Magnini, B., Strapparava, C.: Improving user modelling with content-based techniques. In: 8th International Conference of User Modeling, pp. 74–83 (2001)
20. Eirinaki, M., Vazirgiannis, M., Varlamis, I.: SEWeP: using site semantics and a taxonomy to enhance the web personalization process. In: 9th ACM SIGKDD International Conference on Knowledge Discovery and Data Mining, pp. 99–108 (2003)
21. Sorensen, H., O'Riordan, A., O'Riordan, C.: Profiling with the informer text filtering agent. J. Univers. Comput. Sci., 988–1006 (1997)
22. Cantador, I., Bellogin, A., Castells, P.: News@hand: a semantic web approach to recommending news. In: Adaptive Hypermedia and Adaptive Web-Based Systems, pp. 279–283 (2008)
23. Middleton, S., Shadbolt, N., De Roure, D.: Ontological user profiling in recommender systems. In: ACM Trans. Inf. Syst., 54–88 (2004)
24. Degemmis, M., Lops, P., Semeraro, G.: A content-collaborative recommender that exploits WordNet-based user profiles for neighborhood formation. User Model. User-Adap. Inter. J. Personalization Res. (UMUAI), 217–255 (2007)
25. Aciar, S., Zhang, D., Simoff, S., Debenham, J.: Informed recommender: basing recommendations on consumer product reviews. IEEE Intell. Syst. **22**, 39–47 (2007)
26. Konstan, J., Miller, B., Maltz, D., Herlocker, J., Gordon, L., Riedl, J.: GroupLens: applying collaborative filtering to usenet news. Commun. ACM, 77–87 (1997)
27. Hill, W., Stead, L., Rosenstein, M., Furnas, G.: Recommending and evaluating choices in a virtual community of use. In: Human Factors in Computing Systems (1995)
28. Maes, P., Shardanand, U.: Social information filtering: algorithms for automating "Word of Mouth". In: The SIGCHI Conference on Human Factors in Computing Systems. ACM Press, Addison-Wesley Publishing Co., Denver, Colorado, United States (1995)
29. Ali, K., van Stam, W.: TiVo: making show recommendations using a distributed collaborative filtering architecture. In: 10th ACM SIGKDD International Conference on Knowledge Discovery and Data Mining, pp. 394–401 (2004)
30. Goldberg, D., Nichols, D., Oki, B.M., Terry, D.: Using collaborative filtering to weave an information tapestry. Commun. Assoc. Comput. Mach. **35**(12), 61–70 (1992)
31. Linden, G., Smith, B., York, J.: Amazon.com recommendations: item-to-item collaborative filtering. IEEE Internet Comput. **7**, 76–80 (2003)
32. Jaschke, R., Marinho, L., Hotho, A., Schmidt-Thieme, L., Stumme, G.: Tag recommendations in folksonomies. In: Knowledge Discovery in Databases (PKDD 2007), pp. 506–514 (2007)
33. Krulwich, B.: LifeStyle finder: intelligent user profiling using large-scale demographic data. AI Magazine. **18**(2), 37–45 (1997)
34. Aimeur, E., Brassard, G., Fernandez, J.M., Onana, F.S.: Privacy-preserving demographic filtering. In: Proceedings of the ACM Symposium on Applied Computing, pp. 872–878 (2006)
35. Massa, P., Avesani, P.: Trust-aware collaborative filtering for recommender systems. In: On the Move to Meaningful Internet Systems 2004: CoopIS, DOA, and ODBASE. Springer, Berlin, Heidelberg, pp. 3–17 (2004)

36. Golbeck, J., Hendler, J.: FilmTrust: movie recommendations using trust in web-based social networks. Proceedings of the IEEE Consumer Communications and Networking Conference **96**, 2006 (2006)
37. Balabanovic, M., Shoham, Y.: Fab: content-based, collaborative recommendations. Commun. ACM **40**(3), 66–72 (1997)
38. Claypool, M., Gokhale, A., Miranda, T., Murnikov, P., Netes, D., Sartin, M.: Combining content-based and collaborative filters in an online newspaper. In: SIGIR'99 Workshop on Recommender Systems: Algorithms and Evaluation. Berkeley, CA (1999)
39. Billsus, D., Pazzani, M., Chen, J.: A learning agent for wireless news access. In: Proceedings of the 5th International Conference on Intelligent User Interfaces (IUI'00), pp. 33–36. ACM, New York, NY, USA (2000)
40. Salter, J., Antonoupoulos, N.: CinemaScreen recommender agent: combining collaborative and content-based filtering. IEEE Intell. Syst., 35–41 (2006)
41. Lekakos, G., Caravelas, P.: A hybrid approach for movie recommendation. Multimedia Tools Appl. **36**(1–2), 2006 (2006)

A Sentiment-Based Trust and Reputation System in E-Commerce by Extending SentiWordNet

Hasnae Rahimi, Abdellatif Mezrioui and Najima Daoudi

Abstract Trust is a decisive factor in e-services and especially in e-commerce. E-customers usually rely on others' opinions, reviews, recommendations on products, and services to make the right purchase decision. Nevertheless, deceptive reviewers deliberately disseminate fake and dishonest reviews to falsify the products' reputation. Consequently, there is a need for Trust and Reputation Assessment to aggregate these text reviews and compute their related reputation scores. For this purpose, Natural Language Processing cannot be omitted from the process of generating reputation scores. In this paper, we propose a Trust and Reputation System named SentiTrustCom STC which is composed of two subsystems: (1) A Combined Idiomatic Ontology-based Sentiment Orientation System that employs NLP techniques and extends SentiWordNet to analyze Text reviews and compute their related Sentiment orientation scores; (2) Trust and Reputation Engine that proposes algorithms to generate reliable Trust and Reputation scores using the generated Sentiment Polarities as inputs. STC aims to analyze the users' behavioral intention in order to detect any ill-intentioned interventions that could falsify the products' reputation and hence distort the overall trust among reviewers.

Keywords Soft security · Trust and reputation systems · Sentiment mining · Machine learning · Behavioral intention detection · SentiWordNet · Natural language processing · Lexical approach · E-commerce

---

H. Rahimi (✉) · N. Daoudi
Lyrica Research Team, Ecole des Sciences de l'information ESI, Rabat, Morocco
e-mail: Hasnae.rahimi@gmail.com

N. Daoudi
e-mail: daoudinajima@yahoo.fr

H. Rahimi · A. Mezrioui
RAISS Research Team, Institut National des Postes et des Télécommunication INPT, Rabat, Morocco
e-mail: mezrioui@inpt.ac.ma

V. Bhateja et al. (eds.), *Embedded Systems and Artificial Intelligence*, Advances in Intelligent Systems and Computing 1076,
https://doi.org/10.1007/978-981-15-0947-6_73

1 Introduction

The past decades have witnessed tremendous progress in the use of the World Wide Web, the result of innovative advances in Web technologies. Open electronic markets, online collaborative systems, distributed peer-to-peer applications, online social media, and general e-service platforms are the most relevant illustrations of the Web successful accomplishments. In fact, people have increasingly been part of the virtual Web environment, where they have created their content, shared it, interacted with others in various ways, for different purposes, and at a very progressive rate.

Besides, e-commerce applications are frequently required by Web users and e-customers to be well-informed about a target product to purchase. In this constantly evolving environment, customers share an abundant number of opinions which is commonly describing their real-life experiences. They also express their sentiments on specific products or services [1]. As a result, a huge amount of opinions is available in e-commerce applications in the form of overall numeric ratings, text reviews, comments, forum's discussions and posts, recommendations, etc. Nevertheless, text reviews are the most interesting opinions that all customers, sellers, and suppliers rely on to help their reputation as well as to build a reputation image of their target products and services [2].

However, in an e-commerce environment, customers and reviewers can provide fake and dishonest reputation evaluations for different unethical purposes by sharing their reviews and recommendations. This kind of deceitful reviewers who deliberately provide misinformed and fake opinions, threaten and falsify the reliability, the trustworthiness and the reputation of the reviewed products and services [3–5].

Current e-commerce platforms make us poorly prepared for controlling and sanctioning the substantial and increasing number of fraudulent users and service providers holding these unethical and malicious intentions [4, 6]. Therefore, "to Trust or not to Trust", that is the customers' crucial question to be answered.

In Soft Security, Trust and Reputation Systems (TRS) represent the most relevant and intelligent Soft Security mechanisms that aim to assess a mutual trust between service providers and service consumers among e-commerce communities [7, 8]. These engines focus on the improvement of the e-commerce platforms' resistance against fraudulent users and untrustworthy behaviors.

When assuming that these intelligent Trust Management Systems can be equipped with capabilities to reason about trust, risk assessment, and decision making, one can talk about an artificial trust that is enhanced in an e-commerce environment. We believe that robust TRS must accurately formalize Trust as a computational concept to predict future behaviors of reviewers basing on their past actions [6, 8]. As no user in an e-commerce environment is fully trusted, the credibility and trustworthiness of his provided sentiments and opinions must be analyzed and assessed [7, 8].

For this purpose, Sentiment Orientation Systems (SOS) have been widely applied to the customers' text reviews to predict and extract the sentiment orientation that is more likely intentionally expressed, disseminated and communicated by reviewers [1, 2, 8].

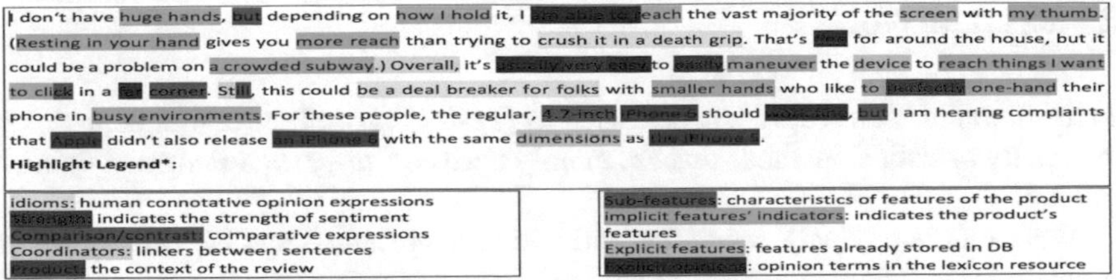

Fig. 1 A highlighted text-based phone review

At this level, a considerable part of the Natural Language Processing (NLP) techniques is enhanced to provide a reliable SOS able to compute a trustful Sentiment Orientation (SO) polarity of the provided text reviews. In fact, Sentiment Mining (SM) approaches are definitely able to mine, analyze and determine the SO of the expressed opinions [2–4]. However, it is important to relate trust and sentiments in a robust system able to exploit sentiment polarities in order to aggregate trust and reputation scores in the context of e-commerce. Through the overall rating, the reviewer only gives a general evaluation of the product. Without providing a specific explicit rating on each feature using preferably text-based reviews, the reader could not have a clear idea on the pros and cons of each feature of the phone as described in Fig. 1.

We highlight in Fig. 1 the main words and phrases to be detected in order to analyze and extract sentiment and opinions. It is necessary to analyze the reviewer's latent rating on each reviewed product's feature. It is also common to use idiomatic expressions to express opinions. How to (1) detect, (2) analyze, and (3) affect SP for idioms?

In this paper, we propose a TRS named SentiTrustCom STC composed of two subsystems:

(1) A Combined Idiomatic Ontology-based Sentiment Orientation System CIOSOS [9] that employs NLP techniques and extends SentiWordNet [10] to analyze Text reviews and compute their related Sentiment orientation scores. CIOSOS has been theoretically presented with no experiment results in previous work [9].
(2) Trust and Reputation Engine that proposes algorithms to generate reliable Trust and Reputation scores using the generated Sentiment Polarities as inputs. STC aims to analyze the users' behavioral intention in order to detect any ill-intentioned reviews that aim to falsify the products' reputation and distort the overall trust among reviewers.

The remainder of this paper is structured as follows. Section 2 presents the related work with a criticizing analysis. Section 3 presents an overall architecture of STC and a brief overview of each of its components. Section 4 exposes the experiment results of STC. Section 5 provides a conclusion and discusses future perspectives of the proposed work.

2 Related Work

In the literature, few research papers such as [11–17] have adopted NLP approaches especially statistical methods that are mainly derived from data mining and opinion mining.

Work on the product's reputation estimation is presented by [11] where authors propose a reputation model that uses opinion mining techniques in order to extract sentiments about the product's features. They provide a method to generate a reputation value for every feature of the product as well as a global product's reputation. They also consider the orientation of the strength of the expressed sentiment. In fact, to calculate the overall product reputation, their proposed system gives more weight to the most important features of the product, having more impact on customers' purchase decisions. The features' impact is calculated by counting the number of occurrences related to every single feature explicitly mentioned in text reviews, assuming that features that are more mentioned by users are more likely important for them.

Authors of [12] analyze an opinionated text data analysis problem called Latent Aspect Rating Analysis (LARA), which aims to analyze opinions expressed about an entity in an online review at the level of topical aspects. The proposed model aims to extract for each review, the latent opinion expressed by the reviewer on each aspect, in addition to the relative emphasis on different aspects when forming the overall judgment of the entity. The authors propose a novel probabilistic rating regression model to solve this text mining problem in a general way. Empirical experiments on a hotel review dataset prove that the proposed model can effectively solve the problem of LARA. Besides, a wide range of application tasks can be supported by the model due to its proposed analysis of opinions at the level of topical aspects.

The work proposed by Lu et al. [13] studies the problem of generating a "rated aspect summary" of short comments. They state that the large scale of the provided information in web services poses the need and challenge of automatic summarization. In e-commerce applications, each of the provided text reviews comes with an overall rating. In fact, this kind of reviews presents a decomposed view of the overall ratings for the major aspects of a target product or entity, so that a user could gain different perspectives towards the product in concern. The proposed methods are quite general and can be used to automatically generate rated aspect summary, given any collection of short comments and their related overall rating. In short comments, opinions on different aspects are usually expressed in concise phrases [13]. Nevertheless, the generation of granular reputation scores for each reviewed feature has not been tackled by the proposed system in [13].

Authors of [14] propose a TRS that generates a user's reputation and exploits this reputation value to calculate weights for the provided ratings. Based on the fact that some users tend to give higher or exaggerative ratings than others, they propose a rating tendency, for each reviewer, which is generated from users' ratings. The proposed system introduces a level of expertise to evaluate the user's reputation based on the user's prediction accuracy-level and his activity frequency. However, the activity frequency of a user does not assess his trustworthiness degree or predict his malicious behavior. The content of his reviews must be analyzed in order to define and predict his behavioral intention.

Besides, authors of [15] introduce the volatility of online ratings in order to reflect the current tendency of users' ratings, relying on the recency of provided ratings. The proposed approach aims to generate a weighted average, where old ratings present less weight than recent ones. The recency factor is defined via a proposed metric representing the Average Rating Volatility, which indicates the level of fluctuation characterizing ratings. Therefore, this level derives a discounting factor involved in the process of weighting old ratings. The recency factor is important but yet insufficient since the relevancy weight of the reviewed features of a product should be combined with the recency factor to attain better performance and effectiveness.

The work in [16] discusses the concept of an interactive reputation system that aims to involve the user in the reputation assessment process to increase the users' ability to detect and analyze malicious behavior. The case study demonstrates the ability of interactive reputation systems to theoretically support the detection of malicious behavior. The authors state that the proposed approach increases the understanding of reputation data and may be applied by the average end-user. So far, this approach has not been experimentally verified. The experiment only focuses on the usage of one specific prototype implementing one interactive visualization. Some specific attacks are involved in the study in order to be detected [16]. Participants who took part in the user study are not a representative sample. In addition, participants who are asked to "solve" cases may have analyzed the seller profiles more critically than in a real-life situation. Adopting machine learning techniques would have been of great interest to (1) train models based on more representative corpora and (2) test the trained model using real-life representative corpora.

Authors of [17] propose a CommTrust system based on the assumption that buyers mostly use free text reviews to express opinions about the target product in e-commerce platforms. CommTrust system is a multidimensional trust model that aims to compute reputation scores from user text reviews. The proposed system employs NLP techniques, opinion mining, and topic modeling to mine text reviews inputted in the weighting process. CommTrust has been used by different websites like Amazon and eBay. However, no details of the opinion mining techniques applied nor the reputation computation approach have been presented or discussed in the paper.

All these research papers mentioned earlier do not tackle the following challenging issues:

1. How to analyze and predict the reviewers' intentions in order to interpret their provided opinions? And What is the appropriate context-dependent approach to detect fraudulent, distorted and colluded reviewers' opinions? How to encounter this fraudulent participation from the beginning?
2. How to tackle the significant lack of human faculties and/or artificial intelligence in TRS to manage idioms in text reviews?

In this paper, we propose a Sentiment-based TRS that handles the abovementioned issues.

3 STC: The Proposed Sentiment-Based Trust and Reputation System

3.1 *Overview of the STC Architecture*

we propose a TRS named SentiTrustCom (STC) which is the combination of the improved version of the Combined Idiomatic Ontology-based Sentiment Orientation System (CIOSOS) [9] and the proposed TrustCom Reputation Engine.

The STC mainly relies on the improved version of the CIOSOS to accurately assess trust and estimate the reputation of the reviewed products in the e-commerce context. The new version of CIOSOS aims to semantically analyze text opinions in order to extract the most likely intended sentiment orientation of different reviewed features and sub-features of the product. Furthermore, the new CIOSOS provides a SO-based classification which is a fine-grained classification of all reviewed entities (reviewers, reviews, products, features, and sub-features). New CIOSOS is a modular system that also adapts and proposes a set of SM and NLP techniques, that are inspired by human cognitive faculties, for the purpose of modeling aspects of the perceived world. Each module of CIOSOS provides an output that is inputted in another module of the STC (cf. Fig. 2).

Besides, TrustCom reputation engine proposes a Trustworthiness Evaluation algorithm that aims to compute the granular reputation score for each reviewed feature, relying on the reputation score of the reviewer and the relevancy weight of the reviewed feature. TrustCom decides on how to involve the numeric rating value in conjunction with the SO polarity of the review, so as to generate the reputation of both the reviewer and the product. TrustCom uses reviews from the knowledge database in its reputation generation process. Reviews of the knowledge base consist of a training set contributing to the STC initialization. The proposed Trustworthiness Evaluation algorithm computes for each review sentence analyzed by the improved CIOSOS a trustworthiness value. This trustworthiness evaluation is based on the distance measurement of the granular SO polarity of the review sentence reviewing a specific feature and the overall SO polarity of the feature. We present in Fig. 2, the STC architecture. In what follows, we call CIOSOS the new version of the Sentiment Orientation System published in [9].

3.2 *The New CIOSOS: The Improved Version of CIOSOS*

As discussed previously, we theoretically presented CIOSOS with no experimental results. Hereafter, we give a detailed overview of the improvement of the modules of CIOSOS. We also shed light on the contribution of CIOSOS in the process of Trust and Reputation assessment.

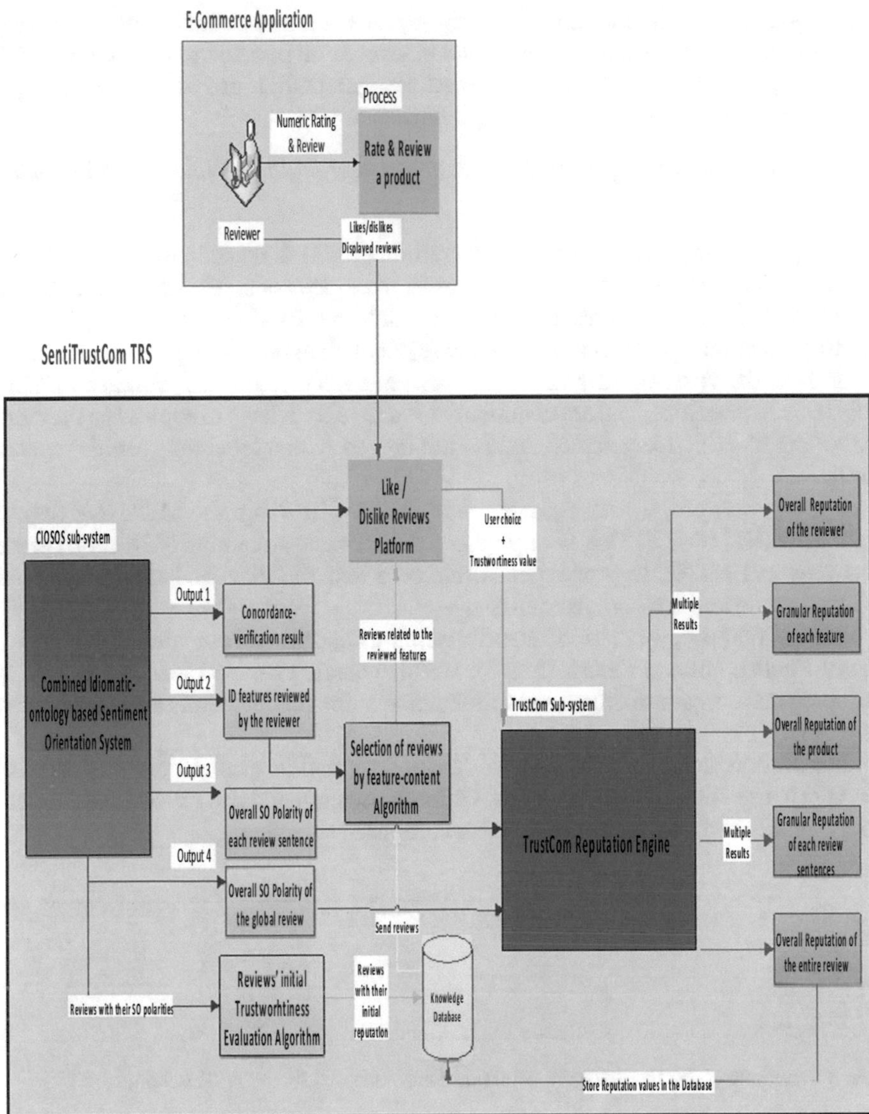

Fig. 2 Architecture of SentiTrustCom: from the review platform to the reputation scores generation

(1) *Preliminary pre-processing: preparing text for sentiment mining*.

We first propose to use a number of pre-processing tasks to prepare the text review for the a posteriori principal sentiment classification modules. These preliminary pre-processing procedures consist of cleansing the text review, which aims to (1) automatically correct the non-word spelling mistakes, (2) automatically correct the real English words spelling errors relying on a context disambiguation process, and

(3) to take into account all consecutively repeated opinion words, opinion phrases or sentences, emoticons, exclamation and interrogation punctuation marks in the SO computation process. After that, we adopt a Segmentation process which attempts to segment paragraph reviews into separated sentences.

(2) *Training and testing of the NER model: detect and tag features and idiomatic expressions.*

We propose to train Named Entity Recognition (NER) to detect and tag:

(a) different forms of features either terms or expressions, (b) sub-features which are characteristics of features themselves, (c) different forms and names of products, (d) English grammatical coordinators and (e) English opinion idiomatic expressions.

In fact, the customized NER model also intends to recognize English opinion idiomatic expressions, that are commonly used in text reviews to express an opinion. All of the identified key named entities are tagged with a label tag specific to each entity.

For this purpose, we have chosen to train the NER from the MIT NLP library called MITIE [18–21]. This library comes with a basic streaming NER tool. So we can only tell MITIE to process each line of a text file independently and output marked up text with a command line.

For this purpose, we created balanced training datasets that contain a considerable set of English opinion idioms (Fig. 3). We also create extended features' ontology database that contains features and sub-features in the form of terms and expressions (Fig. 4).

For the conducted experiments, we selected six mobile phones' reviews datasets in order to train the CIOSOS modules. We have also gathered and created test datasets to granularly evaluate the accuracy of the proposed modules.

```
big cheese
(1,2,idiom)
bird's eye view
(1,4,idiom)
bone of contention
(1,3,idiom)
cock and a bull story
(1,5,idiom)
At the crack of the dawn
```

Fig. 3 A sample of the English opinion idioms dataset labeled for the NER training

```
iPhone 6 S Plus
(1,4,Product)
iPhone 5 S
(1,3,Product)
iPhone 6 s
(1,3,Product)
Samsung Galaxy S 5
(1,4,Product)
Samsung Galaxy S 6
(1,4,Product)
```

Fig. 4 A sample of products and features labeling for the NER training

For the Training process:

1. Collect a set of representative training text reviews containing features, products' names, Brand names, English idioms, exaggerated positive and negative reviews on products.
2. Realize the labeling for each text review.
3. Train a sequence Classifier to predict the labels from the data.

For the Testing process:

1. Create a set of testing text reviews different from the training ones.
2. Run the trained model to recognize, predict and tag each named entity as shown in Fig. 5.

Let's take a text-based review as an input of the customized NER model (Fig. 6).

The output of the customized NER model would be a tagged text-based review (Fig. 7).

For composed named entities, we add underscore for separation since it increases the accuracy level of the dependency parsing in order to determine the right grammatical function of each word of the sentence. Having multiple words in the same sentence falsifies the part of the speech generation process.

(3) *Extending SentiWordNet to SentiWordNet++*

In order to increase the accuracy level of our sentiment mining, we propose to extend the existing lexicon resource SentiWordNet to converge to SentiWordNet++ (swn++) as illustrated in Fig. 15. The extended opinion lexicon resource presents three main novelties:

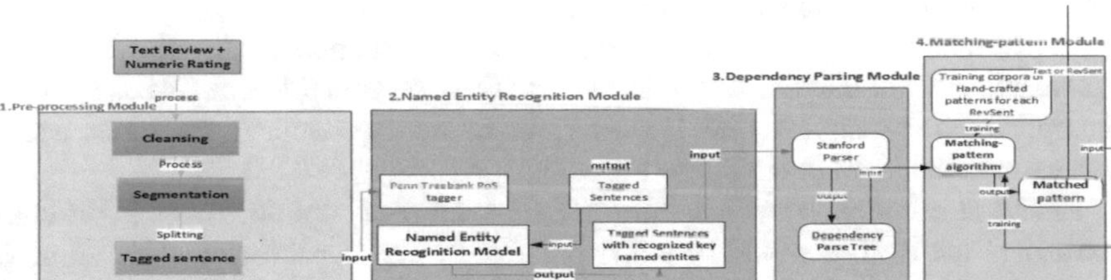

Fig. 5 CIOSOS: module 1–4

"I love the iphone 6S edge camera. The 3d touch screen is just awesome but I hate its edge. I am really caught between a rock and a hard place!"

Fig. 6 A text-based review input of the customized NER model

"I love the iphone_6S_edge/product camera/feature. The 3d_touch_screen/subfeature is just awesome but I hate its edge/feature. I am really caught_between_a rock_and_a_hard_place/idiom !"

Fig. 7 Text-based Output review of the customized NER model

idiom	00981304	0	0.7	**couch_potato#1**	is a lazy person who spends a lot of time sitting down or lying down, watching TV, playing video games, or doing some other activity that doesn't involve physical activity.
idiom	01939984	0.725	0.2	**down_to_earth#1**	A person who is **down_to_earth** is practical, sensible, and realistic. Being down-to-earth is the opposite of being a dreamer, visionary, or "having your head in the clouds."

Fig. 8 Sample of idioms from SentiWordNet++

1. Add connotative opinion expressions in the form of English idioms and phrasal verbs with a new proposed tag "idiom". For each added opinion idiom, we compute positive and negative SO polarities respecting the SentiWordNet threshold which is [0, 1].
2. Update and correct the SO polarities provided by SentiWordNet for some existing idiom phrasal verbs and expressions, in order to increase the effectiveness of the proposed system.
3. Add a specific-label tag which is "idiom" for every idiomatic expression either already stored in SentiWordNet or added by our proposed extension process.

In order to consider idiomatic expressions and phrasal verbs, integral opinion units and proportions of review sentences, we incorporate English opinion idioms with their generated SO polarities in SentiWordNet. First, each idiom would be in a separate line where the Part of speech PoS label would be a specific tag which is the label tag "idiom" as shown in Fig. 5 (Fig. 8).

Concerning the ID of the synset, we choose the most appropriate adjective which is semantically similar and assign its synset ID to the added idiom. We can choose the option to not assign any ID. However, for future improvement and better use of swn++, we should integrate Synstet IDs.

Most English idioms are generally "single-sentiment" opinions even in different contexts. That is, they would be either positive or negative. Nevertheless, some of them are subjected to ironic and sarcastic review form and hence might have both positive and negative SO polarities. As a result, we assign for each idiom a positive score and/or negative score according to SentiWordNet Sentiment orientation policy.

(4) *Dependency parsing and Rules-based Modules.*

After realizing the NER process, we parse the review sentences to generate a typed dependency parse tree for each one of them. In fact, the resulting dependency parse path represents syntactical and grammatical binary relationships between all the words of the sentence in a binary structure. These relationships contain the strength of the expressed sentiment and guide the SM process to the most appropriate SO polarity (Fig. 9).

The new iPhone_6s and 6s_Plus offers a new input feature called 3D_Touch and haptic_feedback.

Fig. 9 A sample treated text-based review

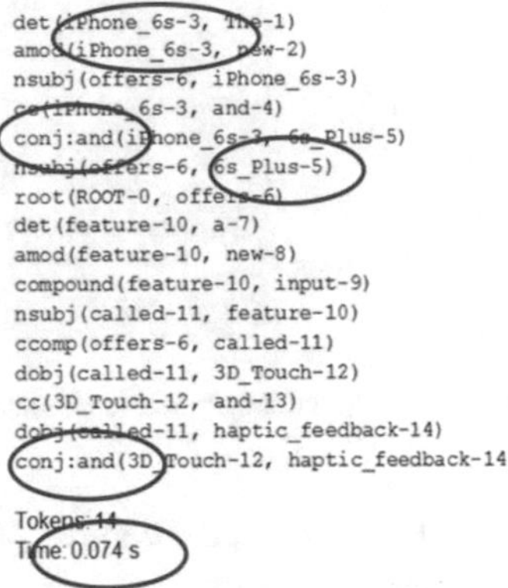

Fig. 10 Output of the dependency parser (with underscore)

We will treat the following text-based review:

Negation forms, semantic connectors, phrasal verbs, and adverb modifiers can also be extracted from obtained relationship dependency trees as shown in Fig. 11.

For this purpose, we propose a pattern-matching algorithm that is trained using a significant number of hand-crafted regular expression patterns, that combine different structures of the syntactical dependency relationships. The trained matching pattern aims to match these patterns with the typed dependency parse path of each review sentence. For each pattern, a number of decision rules are executed in order to associate text opinion to a reviewed feature. Relying on the association-rule, the SO polarity of the text opinion is assigned to the target reviewed feature.

The underscore added to named entities realizes a better time processing and generates more accurate dependencies between words as shown in Fig. 10 compared to the dependency parsing done in Appendix 1.

A rules-based algorithm is then proposed to analyze the matched pattern and its components for the purpose of generating appropriate association-decisions (Fig. 11). In fact, these decisions are about associating the most appropriate opinion text, identified in the matched pattern, to the reviewed feature.

The outputted association intends to assign a SO polarity of the agreed opinion text(s), from the matched pattern to the reviewed feature which is also a part of the pattern in question.

Given a set of review sentences Si, each of them would generate an overall SO polarity SO_Avg for each reviewed feature Fj. For this reason, we are able to generate an overall SO polarity for the target feature from different review sentences. We consider the overall SO polarity of the reviewed feature, the average rating of all of its granular ratings provided in different review sentences.

We propose to constitute a hashmap dictionary for each category of products, which contains lines of Quintuplets values:

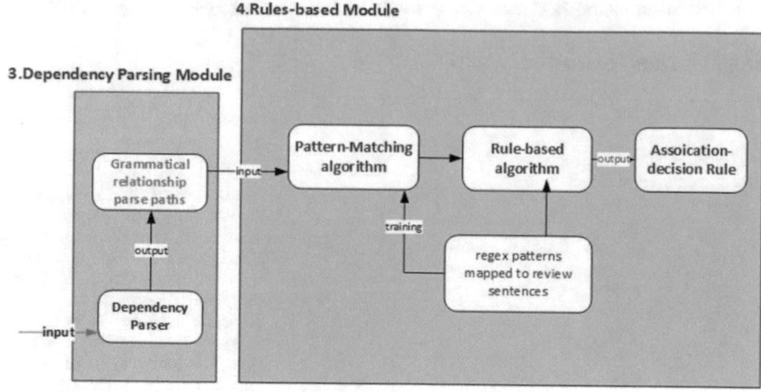

Fig. 11 Dependency parser and rules-based modules

1. ID_Product, 2. Features_Vector, 3. Rev_Sent_Vector, 4. SO_Vector, 5. Overall SO polarity, where:

1. ID_Product: the Id of the product among the set of the product's category.
2. Features_vector: is mapped to the ID_product and presents a vector of IDs of the reviewed features.
3. Rev_Sent_vector: is mapped to ID_feature and presents a vector of review sentences where the ID_feature is reviewed.
4. SO_Vector: is mapped to ID_Rev_Sent and contains SO polarities of the feature ID_feature gathered from the review sentence ID_Rev_Sent.
5. Overall SO polarity vector: is mapped to the ID_feature and each value is then generated from the SO_vector of a specific ID_Rev_Sent but for all review sentences.

Reviews are gathered in text files in the same manner and logic.

We briefly present the algorithm that calculates the overall SO polarity of a specific feature in Fig. 13 (Fig. 12).

Hereafter, we present an example of so computation of features in a text-based review.

This result is obtained from the dependency parse tree presented in Fig. 14.

The generated dependency parse functions as nsubj and conj:nor will match to a regex and then be processed by the So computation algorithm (Fig. 15).

3.3 *TrustCom: The Proposed Solution for the Trust Assessment Major Issues*

The proposed TrustCom reputation engine is initiated when a participant (reviewer) asks to validate his proposed overall numeric rating and text review about a specific product as presented in Fig. 16.

```
Get into the hashmap Dictionary
 For each new review sentence (
   For each ID_product (
      Add a quintuplet line vector to the ID product characterized by the feature ID_feature
      1.ID_Product, 2.Features_Vector, 3.Rev_Sent_Vector, 4.SO_Vector, 5. overall SO polarity
      //map the SO polarities of the ID_feature to the ID of the review sentence
          ID_Product→ ID_feature → ID_RevSent[i] →SO_Vector[j] → SO_polarity_vector[i]
      // Calculate the Average SO polarity of the feature using a loop query
      For each ID_feature (
```

$$Avg_{SO_{Polarity(Fi)}} = AVG\left(\sum_{j=1}^{n} SO_{polarity(fi|Sj)}\right) \quad \text{// where n is the number of sentences}$$

```
                    ))))
```

Fig. 12 Skeleton of gathering granular SO polarities of features of a product

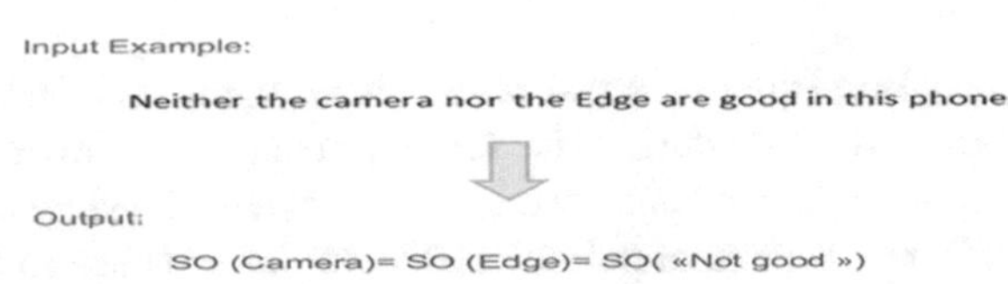

Fig. 13 SO association-based computation of a text-based review

```
cc:preconj(camera-3, Neither-1)
det(camera-3, the-2)
nsubj(good-8, camera-3)
cc(camera-3, nor-4)
det(edge-6, the-5)
conj:nor(camera-3, edge-6)
nsubj(good-8, edge-6)
cop(good-8, are-7)
root(ROOT-0, good-8)
case(phone-11, in-9)
det(phone-11, this-10)
nmod:in(good-8, phone-11)
```

Fig. 14 Dependency parse tree of a text-based review

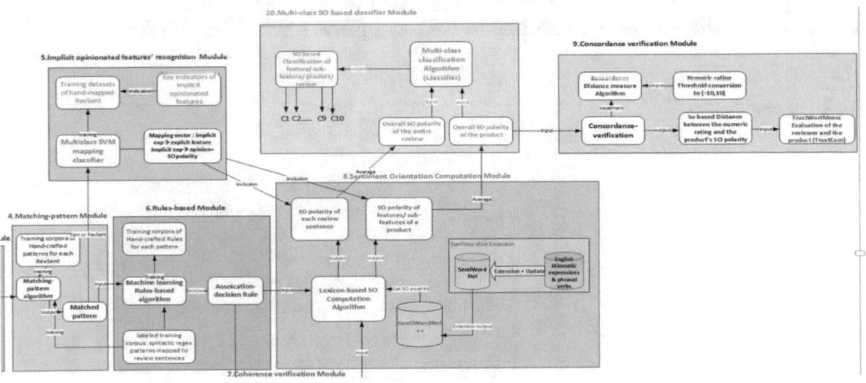

Fig. 15 CIOSOS: extending SWN to SWN++

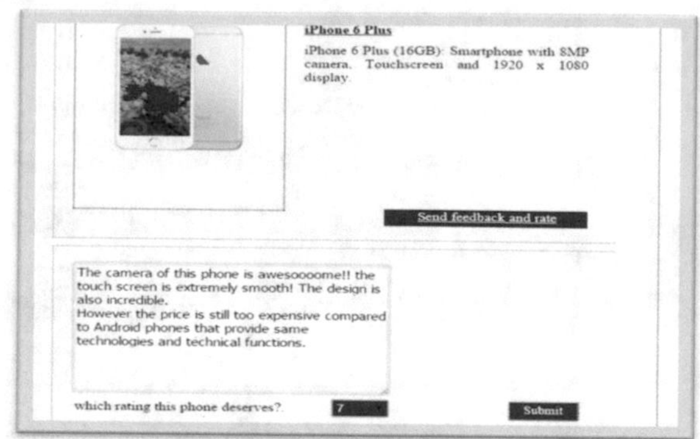

Fig. 16 Reviewing platform

The CIOSOS analyzes the user's text review to compute its overall SO polarity and extract features and sub-features. The TrustCom engine redirects the user to an interface that displays review sentences on the same reviewed features. These review sentences hold different sentiment polarities. The reviewer needs to like or dislike these sentences to validate his opinion (Fig. 17).

Additional pre-fabricated exaggerative negative, positive, and ambiguous short opinions on the reviewed features are displayed to the reviewer. To compute their "initial" reputation scores, a proposed trustworthiness evaluation algorithm is applied to the sentences stored in a knowledge database. The reputation score represents the distance measured between the granular SO polarity of the review sentence on a specific feature and the overall SO polarity of the feature.

(1) *Behavioral intention detection*

The proposed trustworthiness evaluation algorithm takes as input pairs of the user choice (like/dislike) made on the displayed review sentences, in conjunction with the reputation scores of the, respectively, liked or disliked review sentences. The algorithm increments or decrements the reputation score of the reviewer, according to his behavioral attitude towards the displayed set of reviews in the Like/Dislike platform. The displayed pre-fabricated text opinions have different trustworthiness

I want to ONLY like/dislike 4 feedbacks.

"It is a dream coming truuuue!!!!!"

Like Dislike

"So basically i have a complained against apple...and this is the first time i have a complaint for them because they have set the basic starting gb as 16gb and they also have started a 4k video...so if a person buys a 16gb iphone6s ad the user wants to run a 4k video..the video will almost take the whole storage...so after watching the 4k movie they will have to delete it and shizz which will take a long time...so i think apple should have started their base model from 32gb...but still I LOVE U APPLE!!"

Like Dislike

"It's funny because Lenovo K3 Note, beats the crap out of this thing, and it's 10x cheaper."

Like Dislike

"FYI... I have used Android, IOS and Windows... And I love all of them equally for the unique features they offer. Cheers!!!"

Like Dislike

Once you are done with your selection, click HERE to confirm your choices.

Fig. 17 Like/dislike platform

```
//Get the reputation score of the reviewer from the return of the concordance_verification function
// the function returns the updates value of the reputation score of the reviewer
function Trustworthiness_Evaluation_Reviewer (String Array[userchoice][ double feedtrustworth],double Reputation_reviewer) as double
{// according to the applicable case:
Case1: ( (0<feedtrustworth<2) and (userchoice="like")) Or ((-2<feedtrustworth<=0) and (userchoice="dislike"))
Do:Reputation_reviewer+=0.5
Case2: ((2=<feedtrustworth<5) and (userchoice="like")) Or ((-5<feedtrustworth<=-2) and (userchoice="dislike"))
Do:Reputation_reviewer+=1
Case3:((5=<feedtrustworth<7) and (userchoice="like")) Or ((-7<feedtrustworth<=-5) and (userchoice="dislike"))
Do:Reputation_reviewer+=2
Case4:((7=<feedtrustworth<=8) and (userchoice="like"))Or((-8=<feedtrustworth<=-7) and (userchoice="dislike"))
```

Fig. 18 Extract from the trustworthiness evaluation algorithm

degrees ranging from untrustworthy to very trustworthy and are stored in a knowledge base. This knowledge base is initially filled with a data set of gathered as well as prefabricated text reviews with their associated trustworthiness values. While users add reviews in the e-commerce application that solicits STC, the knowledge base grows and would, therefore, contain more text reviews, ready to be displayed depending on reviewers' opinion texts (Fig. 18).

We define the relevancy αi of a feature *fi* as the feature's ranking value, which is computed based on the number of the feature's occurrences among the entire set of review sentences. For instance, among a set of 1000 review sentences, if a feature appears in 20 review sentences, meaning that it is reviewed 20 times. Then, its relevancy weight is computed as the proportion of its occurrences among the number of review sentences. The relevancy weight would then be 20/1000. The most frequently reviewed feature has a relevancy weight equal to "maximum occurrences" out of the "number of review sentences".

$$\alpha = \mathrm{Relevancy}_{\mathrm{weight}}(\mathrm{Idfeature}) = \frac{\sum_{\mathrm{1st\ sentence}}^{\mathrm{last\ sentence}} \mathrm{RevSent}_{\mathrm{Idfeature}}}{\mathrm{sentence\ Total\ number\ of\ review\ sentences}} \tag{1}$$

(2) *Trustworthiness Evaluation algorithm*

A proposed algorithm is applied to generate the average reputation score of all of the reputation scores of the reviewed features *F* including *Fi*. Equation (2) presents the trustworthiness evaluation formula to compute the overall reputation score of the review sentence Id_Rev_Sent.

$$\mathrm{Reputation}_{\mathrm{score}}(\mathrm{Idfeature}) = \frac{\sum_{\mathrm{1st\ rev\ sentence}}^{\mathrm{last\ rev\ sentence}} \alpha i * \mathrm{reputation}_{\mathrm{reviewer}} * \mathrm{SO\ polarity}(\mathrm{Idfeature})}{\sum_{\mathrm{1st\ rev\ sentence}}^{\mathrm{last\ rev\ sentence}} \alpha i * \mathrm{reputation}_{\mathrm{reviewer}}} \tag{2}$$

Finally, the overall reputation score of the product is computed. We propose the formula in equation Eq. (3) that involves the granular reputation score of the reviewed features, the average value of the SO polarity of the review and the provided numeric rating if their semantic concordance is verified.

We present the associated formula in the equation Eq. (3) as follows:

$$\text{Reputation}_{\text{score}}(\text{IdProduct}) = \frac{1}{n} * \sum_{\text{i}=1}^{n} \text{reputation}_{\text{score}}\ (\text{Idfeature } i) + \frac{1}{2} * (\text{SOpolarity}(\text{review}) + \text{rating}) \quad (3)$$

To the best of our knowledge, the SentiTrustCom is a Novel piece of work that analyzes the reviewer's intention for the purpose of computing the most reliable reputation scores. Our word started in 2012 with developing the trustworthiness evaluation algorithms. STC provides a fine-grained SO computation and a granular reputation computation engine. By adopting this approach, STC intends to persuade reviewers to be more honest and truthful while reviewing a product.

4 Related Work's Performance and Experimental Results

For the conducted experiment, we have, respectively, prepared a dataset of opinions for the following phones: the Apple iPhone 6 Plus, iPhone 6S Plus, iPhone 5S, the Samsung S5, the S6, and the S6 edge.

4.1 Experimental Datasets

Datasets of 4800 reviews have been collected from Amazon [22], Reevoo [23] and GSMARENA [24]. Training datasets have also been annotated by trustful moderators who are experts in the studied phones' brand but not working for them. Consequently, they are totally objective and do not tend to support a brand more than another. The training datasets of 3360 (as 70% of total reviews) were employed to train the CIOSOS modules. The trained models have been then applied to the proposed test dataset of 1440 (30% of total reviews) to achieve the SO-based classification.

Training datasets consist of a knowledge base for the TrustCom reputation engine. The SO polarities computed by the CIOSOS were inputted to the TrustCom reputation engine to generate initial reputation scores for all of the review sentences stored in the knowledge base. Relying on the experimental results shown in Table 1, we present the performance graph of the CIOSOS by schematizing the overall precision, recall and f1-measure evaluation metrics of each dataset of the conducted experiments.

4.2 Experimental Results Versus Existing Work Performance

In this experiment, we have assigned to each reviewed feature, the average value of all SO polarities, suggested by the trusted reviewers who assisted in creating the

Table 1 Precision, recall and F1-measure of CIOSOS performance per dataset

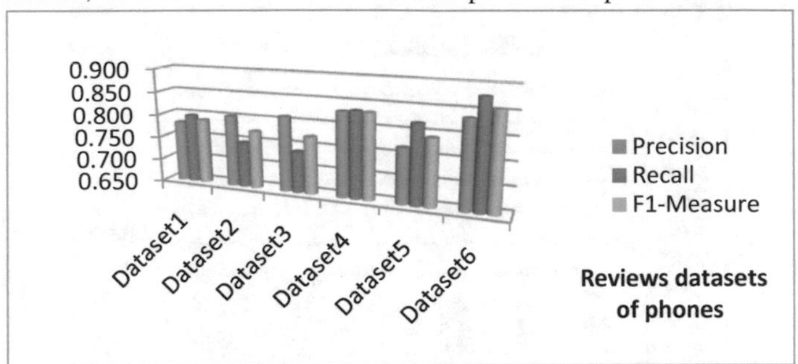

training datasets. The inter-annotator agreement is 87.67%, which is relatively better than the agreement revealed in the reports of existing work in sentiment analysis [25, 26].

Accordingly, we can state that the combination of multiple aspects including the use of an extended opinion lexicon, the features' ontology and the NLP techniques, has provided a better performance compared to SO-based classifiers that only rely on one specific aspect in the analysis such as in [1, 25, 27]. This combination The aspect based reviews' classifier proposed in [25] performed well with respect to the phone domain and showed an accuracy of 67%. The lexicon-based reviews classifier proposed in [1] yielded 69.35% as an accuracy value relying on SentiWordNet as a source of features for a supervised learning algorithm. Besides, the lexicon-based classifier presented in [28] achieved a 70.83% as an overall f-measure. This combination shed light on the necessity of managing both trust and distrust to assess trust and reputation in e-commerce [29].

Besides, recent work in [30] focuses on identifying the features determining a user's trust in online social networks using benchmark datasets. The authors propose a new probabilistic reputation feature model. The enhanced trust prediction framework has been tested and validated on three benchmark datasets namely Wikipedia election dataset, Epinions dataset, and Slashdot dataset. The proposed probabilistic feature enhances the overall accuracy, F1 score, and area under the ROC for the classifier results significantly. We have achieved better overall accuracy. Even for more sophisticated and combined SO-based classifiers such as the one proposed in [26], the experiment results still very comparable to our conducted experiment approach. In fact, the authors of [26] proposed a "boost" framework where the sentiment classification on review text is conducted to boost phrase-level sentiment polarity labeling. The optimal results were generated on only two datasets, where the effectiveness of the framework on the mp3 player dataset was achieved with an F-measure of 82.37%. On the restaurant review data set, it yielded an F-measure of 85.84%, which are still comparable to the results of our experiments conducted on six different phones' datasets.

Concerning the effectiveness evaluation of the TrustCom reputation engine, we measure the similarity between the expected reputation scores and the computed

Table 2 MAE evaluation of TrustCom reputation engine

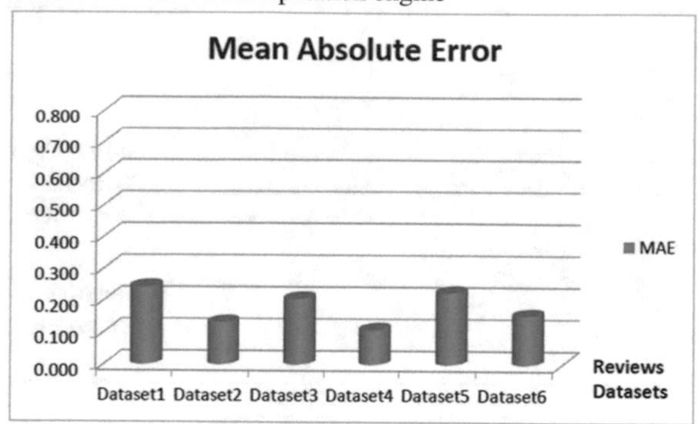

ones. To do this, we calculate the Mean Absolute Errors (MAE) to estimate the Similarity of different ranked lists.

We apply Eq. (4) of the MAE calculation formula to all of the pairs of each dataset(i) which represents the accumulation of the reviews among the training and test datasets (Table 2).

$$\text{MAE} = \frac{\sum_{i=1}^{n} |pi - ri|}{n} \tag{4}$$

We can notice that the MAE evaluation metric value provides a low average value estimated to 0.177 for all of the phones. The lowest MAE value reached 0.108 for the dataset4. The average low MAE estimation indicates a high accuracy level of the TrustCom reputation score prediction.

5 Conclusion and Perspectives

We propose a new sentiment-based TRS named SentiTrustCom, which involves an improved version of the Combined Idiomatic Ontology-based Sentiment Orientation System (CIOSOS) to initiate the TrustCom reputation engine and generates trustful reputation scores for the reviewed products. Experimental Results yielded an overall F-measure of 85%.

We intend to extend the application domain of STC and improve the behavioral intention process by adapting Artificial Intelligence algorithms. We also intend to work on the Contextual disambiguation process in the SO computation Algorithm. In addition, we plan to work on the coherence verification issue where unsuitable (incoherent) opinions are used for features like detection of malicious reviewing. Big data platforms and analytics are to be involved in the analysis process to handle the huge amount of text reviews.

References

1. Taboada, M., Brooke, J., Tofiloski, M., Voll, K., Stede, M.: Lexicon-based methods for sentiment analysis. Proc. Assoc. Comput. Linguist. **37**(2), 267–307 (2011)
2. Sharma, R., Nigam, S., Jain, R.: Mining of product reviews at aspect level. Proc. Int. J. Found. Comput. Sci. Technol. (IJFCST) **4**(3) (2014)
3. Gitari, N.D., Zuping, Z., Damien, H., Long, J.: A lexicon-based approach for hate speech detection. In: Proc. Int. J. Multimedia Ubiquit. Eng. **10**, 215–230. http://dx.doi.org/10.14257/ijmue21. ISSN: 1975-0080
4. Hamouda, A., Rohaim, M.: Reviews classification using SentiWordNet Lexicon. Proc. Online J. Comput. Sci. Inf. Technol. (OJCSIT) **2**(1)
5. Vinodhini, G., Chandrasekaran, R.: Sentiment analysis and opinion mining: a survey. Proc. Int. J. Adv. Res. Comput. Sci. Softw. Eng. Res. Pap. **2**(6) (2012). ISSN: 2277 128X. Available online at: www.ijarcsse.com
6. Kim, H., Song, M.: An ontology-based approach to sentiment classification of mixed opinions in online restaurant reviews. In: The Proceedings of the 5th International Conference, SocInfo 2013, Kyoto, Japan, pp. 95–108, 25–27 Nov 2013
7. Phyu, K., Shein, P.: Ontology based combined approach for sentiment classification. In: The Proceedings of the 3rd International Conference on Communications and information technology, pp. 112–115 (2013)
8. Rahimi, H., El bakkali, H.: State of the art of trust and reputation systems in e-commerce context. Proc. IJCSI Int. J. Comput. Sci. Issues **14**(3) (2017)
9. Rahimi, H., Elbakkali, H.: CIOSOS: combined idiomatic ontology based sentiment orientation system for trust reputation in e-commerce. In: The Proceedings of the International Joint Conference Volume 369 of the series Advances in Intelligent Systems and Computing. International Conference Category B, Springer International Publishing, pp. 189–200, 27 May 2015
10. Esuli, A., Sebastiani, F.: SENTIWORDNET: a publicly available lexical resource for opinion mining. In: Proceedings of the 5th Conference on Language Resources and Evaluation (LREC'06), pp. 417–422
11. Abdel-Hafez, A., Xu, Y., Tjondronegoro, D.: Product reputation model: an opinion mining based approach. In: The Proceedings of the 1st International Workshop on Sentiment Discovery from Affective Data (SDAD 2012), CEUR Workshop Proceedings, Bristol, pp. 16–27
12. Wang, H.., Lu, Y., Zhai, C.: Latent aspect rating analysis on review text data: a rating regression approach. In: The Proceedings of the 16th ACM SIGKDD International Conference on KDD, New York, NY, USA, pp. 783–792 (2010)
13. Lu, Y., Zhai, C., Sundaresan, N.: Rated aspect summarization of short comments. In: The Proceedings of the International World Wide Web Conference Committee (IW3C2). WWW 2009. ACM, Madrid, Spain. 978-1-60558-487-4/09/04. 20–24 Apr 2009
14. Cho, J., Kwon, K., Park, Y.: Q-rater: a collaborative reputation system based on source credibility theory. Proc. Expert Syst. Appl. **36**, 3751–3760 (2009)
15. Leberknight, S., Sen, S., Chiang, M.: On the volatility of online ratings: an empirical study. In: The Proceeding of the E-Life: Web-Enabled Convergence of Commerce, Work, and Social Life, vol. 108, pp. 77–86. Springer Berlin Heidelberg (2012)
16. Saenger, J., Günther, P.: Interactive reputation systems: how to cope with malicious behavior in feedback mechanisms. In: The Proceedings of the Business & Information Systems Engineering. Springer. https://doi.org/10.1007/s12599-017-0493-1 (2017)
17. Firake, V.R., Patil, Y.S.: Survey on CommTrust: multi-dimensional trust using mining e-commerce feedback comments. Proc. Int. J. Innovative Res. Comput. Commun. Eng. IJIRCCE **3**(3). https://doi.org/10.15680/ijircce.2015.03030371640 (2015)
18. https://github.com/mit-nlp/MITIE
19. King, D.E.: Dlib-ml: a machine learning toolkit. J. Mach. Learn. Res. **10**, 1755–1758 (2009)
20. Dhillon, P., Foster, D., Ungar, L.: Eigenwords: spectral word embeddings. J. Mach. Learn. Res. (JMLR) **16** (2015)

21. Joachims, T., Finley, T., Yu, C.N.: Cutting-plane training of structural SVMs. Mach. Learn. **77**(1), 27–59 (2009)
22. http://www.amazon.com/
23. https://www.reevoo.com/products/ratings-and-reviews/
24. http://www.gsmarena.com/
25. Pang, B., Lee, L., Vaithyanathan, S.: Thumbs up? Sentiment classification using machine learning techniques. In: Proceedings of the 2002 Conference on Empirical Methods in Natural Language Processing (EMNLP), pp. 79–86 (2002)
26. Zhang, Y., Zhang, M., Liu, Y., Ma, S.: Do users rate or review? Boost phrase-level sentiment labeling with review-level sentiment classification. In: The Proceedings of the 37th International ACM SIGIR Conference on Research & Development in Information Retrieval, ACM, New York, NY, USA, pp. 1027–1030 (2014)
27. Christiane, F.: WordNet: an electronic lexical database. In: The Proceedings of the MIT Press, Cambridge, MA (1998)
28. Gitari, N.D., Zuping, Z., Damien, H., Long, J.: A lexicon-based approach for hate speech detection. In: Proc. Int. J. Multimedia Ubiquit. Eng. **10**, 215–230. http://dx.doi.org/10.14257/ijmue21. ISSN: 1975-0080 (2015)
29. Lee, S.-J., Ahn, C., Song, K.M., Ahn, H.: Trust and distrust in e-commerce. Sustainability **10**, 1015 (2018)
30. Raj, E.D., Dhinesh Babu, L.D.: An enhanced trust prediction strategy for online social networks using probabilistic reputation features, NeuroComputing Elsevier **219**, 412–421 (2017)

Towards an Ontological Analysis of the Alignment Problems of Fields in the Architecture of an Information System

Lamia Moudoubah, Khalifa Mansouri and Mohammed Qbadou

Abstract The different fields of the company must be described with tools and languages that respond well to its needs and requirements. However, existing architecture techniques are based on limited models, due to a lack of resources to ensure consistency and traceability between the organization's different domains and users. In addition, the governance models of the company's information systems processes are multiple but lack content analysis. To do this, it is necessary to base a good analysis to assess its quality and conformity in order to avoid any form of immensity, imprecision, uncertainty, inconsistency, and fraud. This article attempts to study and compare three approaches used in the architecture of the company's information system to infer the alignment problems encountered in the areas of the latter. These problems mainly concern the integration of languages, the tools for describing the processes gathered in the domains and the automated analysis of its models. This work uses ontologies to align the different fields, through deductive reasoning and inference leading to a particular conclusion in order to successfully analyze existing models.

Keywords Ontology · OWL · Governorship of the information systems · Analyze models · Knowledge engineering · Strategic alignment

1 Introduction

Today, many companies are moving towards using ontology to better manage their processes, and their knowledge to improve the alignment of different processes grouped into different areas, which consists the company's strategy. To stay agile, organizations must be exposed to frequent changes in the evolution of their information systems. However, this evolution is taking place in a dispersed manner, which leads to a rupture in the alignment of these domains and leads to a decrease in

L. Moudoubah (✉) · K. Mansouri · M. Qbadou
SSDIA Laboratory, ENSET of Mohammedia, Hassan II University of Casablanca, Casablanca, Morocco
e-mail: lamiae.modobah@gmail.com

V. Bhateja et al. (eds.), *Embedded Systems and Artificial Intelligence*,
Advances in Intelligent Systems and Computing 1076,
https://doi.org/10.1007/978-981-15-0947-6_74

the company's performance. The complexity of information systems is constantly increasing; the change in the activities that compose business processes, also the change in technologies that require found and developing an evolutive approach and that can be adapted with these changes for sustainable developments. This approach aims to ensure strategic alignment; an approach that consists of redefining the company's overall strategy and technological development strategy in order to be in perfect harmony. The purpose of this work is to redesign the company's architecture by comparing three approaches developed by three different authors, in order to suggest a new approach that will be developed in the future.

2 Previous Works

2.1 Introduction

In this work, ontologies will be used, an ontology is developed at the basis of descriptive approaches, and it aims to define a set of concepts that concern a given domain. However, a descriptive approach must not necessarily respect the constraints of building an ontology. Be careful not to confuse an ontology with a conceptual model. In fact, the latter two agree on the principle of conceptualization and motivation. However, they diverge on the purpose or purpose of modeling; conceptual models are based on approaches that are related to application needs.

Several authors have considered an ontology to be a consensual conceptual model, if you are interested, you can consult the work [1–3]. Ontologies are often used to ensure strategic alignment, which can be summarized as establishing coherence and aligning the company's strategy with its information system. Explicitly, it is a matter of aligning the company's strategy with the business strategy and the business strategy with the IT strategy, this can be presented in the figure below [4] (Fig. 1).

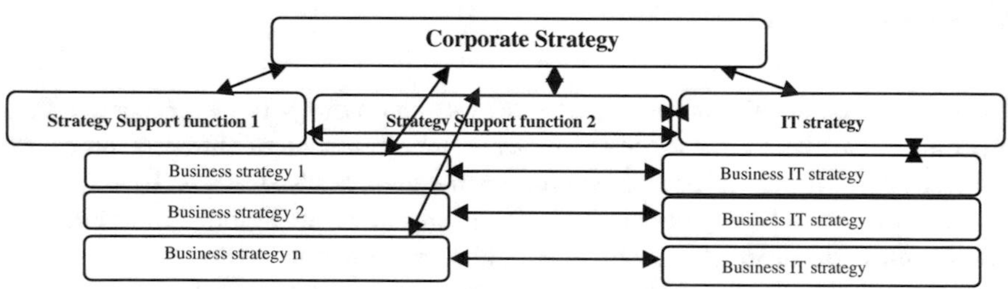

Fig. 1 Alignment diagram of the company's domains [4]. *Source* Cigref

2.2 Approach A: Called the Needs-Based Alignment Documentation Approach

The author of this approach: Bleistein in [5].

The basis of this approach: Bleistein in [5] seeks to align two main entities that are: business strategy and IS, business processes are also mentioned but to a lesser extent. To do this, they propose a requirements engineering approach that defines in the same model (1) the company's strategic objectives and (2) the activities and processes by which these objectives are achieved [6]. Figure 2 illustrates the progression of problem diagrams. The ovals represent the requirements RA, RB, RC, RD, RD, RM which, respectively, refer to the context diagram DA, DB, DC, M where M is the machine (software, hardware, data). RA = corresponds to the company's strategy DA = corresponds to the company level RB = a requirement resulting from an analysis of DA and RA and which refers to DB to satisfy the RA.

Therefore, the entities to be aligned are presented by problem frames, as well as by company-specific models for business processes. The author of this approach to organize the company's strategies [7] also uses the BMM (business motivation model) of the BRG (business rules group).

The positioning of this approach: **Object view** [Number of entities: 3 and Entities: Set {business strategy, business process, system} and Documentation of entities: Set {imposed models (intentional models, problem frames); company-specific models} Link to business components: implicit and Relationship between entities: link typology and Relationship type: mixed]. **Goal View** [Goal: Model, build] **Method View** [Documentation Method: Set {goal model, business model} and Construction Method: top-down and Guidance: undefined and Flexibility: not flexible]. **Tool View**

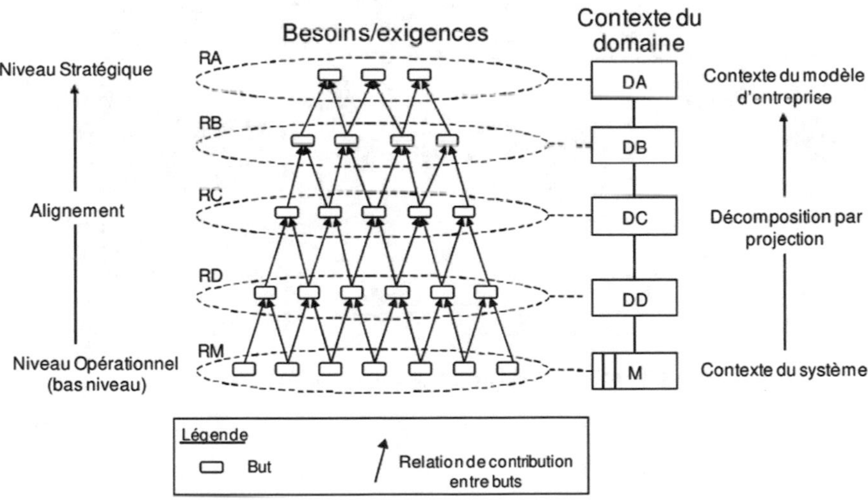

Fig. 2 Integration of a goal model and problem progression [12]

[Documentation tool: Set {goal-oriented language, business model] and Tool to build or evolve derivation rules and Guidance: no tools].

2.3 Approach B: Called the Approach to Evaluate the Degree of Alignment of Business Processes and the IS

The author of this approach: A. Etien in [8].

The basis of this approach: A. Etien in [8], proposes to assess the degree of alignment between two entities: business processes and the system, through the analysis of the degree of alignment between a business process model and a system model. It is, therefore, necessary to define specific metrics for these two models, i.e., between the elements of their metamodels. However, in order to help engineers define the specific metrics associated with the process metamodel used and the system metamodel used, generic metrics are defined between two generic metamodels. Figure 3 shows the approach to defining specific metrics [9].

The positioning of this approach: **Object view** [Number of entities: 2 and Entities: Set {business process, system} and Entity documentation: Set {enterprise-specific templates, imposed models} and Component linkage: implicit Relationship between entities: links and Type of relationship: mixed Purpose view: evaluate]. **Method View** [Evaluation Method: Quantitative and Guidance: set {roadmap, process model} and Flexibility: flexible]. **Tool View** [Evaluation Tools: Metrics and Guidance: Non-Tools] [10].

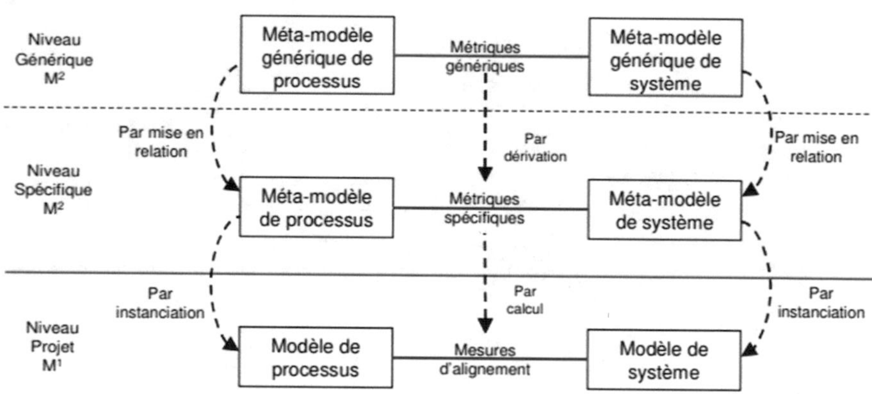

Fig. 3 A system of alignment metrics at three different levels, Etien in [8]

Object view Number of entities: Entire / Entities: Together {business strategy, information technology strategy, values, business model, business process, system, requirement, environment, organization, architecture} / Entity documentation: Set {imposed models; company-specific models; undocumented} / Link to company components: Enum {explicit, implicit, undefined}/ Relationship between entities: Enum {traceability, rules, links, intermediate model, not defined} / Type of relationship: Enum {positive, negative, mixed} / **Goal View** Goal: Together {model, build, evaluate, evolve} / **Method View** Documentation Method: Together {Matrices, Goal Model, Business Model, Value Model} / Construction method: Enum {Top-down, bottom-up, mixed}/ Evaluation method: Enum {quantitative, qualitative}/ Method of evolution: Together {scenario, independence, dependence, double dependence, interdependence, correction Method guidance: Set {good practices, roadmap, process model, undefined} / Flexibility of the method: Enum {flexible, not flexible, not defined} / **Tool View Tool** Documentation or modeling tool: Set {goal-oriented language, dashboards, business model, value-oriented model, undefined} / Evaluation tool: Enum {questionnaire, metrics, framework} / Construction or evolution tool: Enum {dashboards, derivation rules, frame, gaps} / Guidance: Enum {tooled, not tooled} / If equipped with tools: Tool for methodological guidance: Set {roadmaps, fixed processes, patterns}.

Fig. 4 Summary of references terms [12]

2.4 Approach C: Called the Strategic Alignment Assessment and Evolution Approach

The author of this approach: Luftman in [11].

The basis of this approach: Luftman in [11] provides a framework for measuring alignment between two entities: business strategy and IT strategy. This framework is based on the Capability Maturity Model (CMM), created in 1986 by the Software Engineering Institute (SEI) for process improvement. It provides tools to measure progress and maturity in development processes and guides the company in its approach to improving these processes. It has five levels of maturity: initial, managed process, defined process, controlled process and optimization [12]. The proposed approach to achieving and maintaining alignment is based on (1) understanding the maturity of alignment, (2) maximizing success factors and (3) minimizing failure factors. Guidance is provided through a six-step process to be seen in [13].

The positioning of this approach: **Object view** [Number of entities: 2 and Entities: set {business strategy, IT strategy} and Entity documentation: set {business models; undocumented; link to business components: undefined and Relationship between entities: undefined]. **Goal View** [Goal: {evaluate, evolve}]. **Method View** [Evaluation Method: Qualitative and Evolution Method: Correction and Guidance: Process Model and Flexibility: Non-flexible]. **Tool View** [Assessment Tools: Framework and Construction or Evolution Tools: Framework and Guidance: No Tools].

3 The Proposed Approach

There are several interests to model the strategic alignment that is rarely presented or documented, this first allows to check the harmony of the strategy with the information system and the business with the IS as shown in Fig. 1. Many approaches have been designed to establish alignment. It is about establishing a system that is aligned with the company's areas and processes and the organization's requirements. The main objective of these approaches is to build a solid system adapted to the strategic changes often experienced by the organization.

3.1 Reviews: Terms of Reference for the Three Approaches

These approaches are presented according to their positioning in relation to the reference framework, so let's start with an analysis of the reference framework, then move on to compare the three approaches; Laure-Hélène Thevenet proposed in her thesis entitled "Proposition d'une modélisation conceptuelle d'alignement stratégique: La méthode INSTAL" [12]. A framework for studying different aspects of strategic alignment in particular. It consists of four views that correspond to four perspectives from which alignment can be studied (Figs. 4, 5).

In each of these four perspectives are defined attributes that specify the characteristics of the alignment, all attributes contain values defined in a domain. A domain can have a predefined type (integer, Boolean, etc.) that defines the type of the expected value, an enumerated type (noted Enum {value x, value y}) that allows you to choose one and only one value from a predefined list, or a structured type [noted Ensemble (value 1, value 2)] that allows you to choose several values from a predefined list [11].

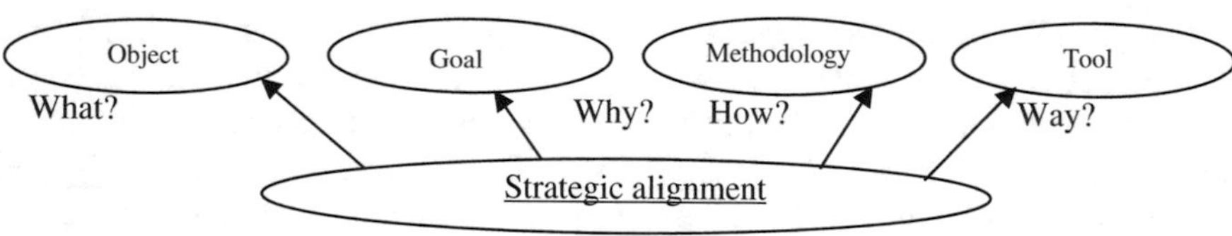

Fig. 5 Key elements of the terms of reference

3.2 Comparison

We compared the three approaches based on their SEO framework (Table 1).

Approach A: focuses on system and business strategy alignment. **Approach B**: Assesses the degree of alignment of the system and business processes. **Approach C**: focuses on the alignment of business and IT strategies.

Table 1 Comparison of the three critical approaches

Les perspectives	Approach A	Approach B	Approach C
Object			
Number of entities	**3**	**2**	**2**
Entities	Business strategy/process/system	Business process/system	Business strategy/IT strategy
Doc. entities	Imposed models (intentional, problem frames), company-specific models	Company-specific models, imposed models	Company-specific models; not documented
Component links	Implicit	Implicit	Not defined
Entity relationship	Type of links	Links	Not defined
Type of relationship	Mixed	Mixed	
Purpose			
Purpose	Model build	Evaluate	Evaluate, evolve
Method			
Of documentation/modeling	Goal model, business model		
Of construction	Top-down		
Of evaluation		Quantitatif	Qualitatif
Of evolution			Correction
Guidance	**Not defined**	Roadmap, process models	Process model
Flexibility	Non-flexible	Flexible	Non-flexible
Tool			
Documentation/modeling	Goal-oriented language, Business model		
Evaluation		Metrics	Framework
Construction and evolution	Derivation rules		Framework
Guidance	No tools	No tools	No tools

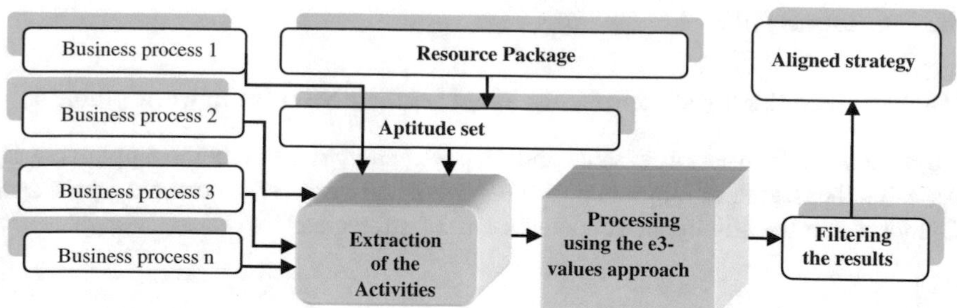

Fig. 6 Model of the proposed approach

Limitations of these approaches: Strategic modeling approaches seek to use conceptual modeling as support for IS engineering by analyzing strategic alignment. However, none explicitly model strategic alignment, with formal ontologies and concepts specific to strategic alignment.

3.3 Presentation of the Proposed Approach

In this section, the proposed approach is explained (Fig. 6).

To achieve alignment of the company's strategy, one must first think about aligning the different areas of the company, these are divided into business processes, and each process represents a set of the company's activities. This during the extraction of its activities requires the use of several resources (human, hardware, software, etc.). After that comes the step of processing these activities, for that, we have opted for the e3-value method, only the e3-value method proposes to model the alignment of activities, to analyze them and to make them evolve. In e3-values, which unlike ACEM and other approaches takes into consideration the strategic level, the evolution process is not very detailed and seems not very flexible.

The positioning of the proposed approach in relation to the reference framework (Fig. 7).

This is why the proposed approach will make it possible, on the one hand, to align the company's strategy with the business lines. On the other hand, to align the business lines with the business IS strategy.

Object view Number of entities: 9 / Entities: business strategy, IT strategy, values, business processes, systems, requirements, environment, organization, architecture / Entity documentation: company-specific templates / Link to company components: explicit, well defined /Relationship between entities: traceability, rules, links, intermediate model / Type of relationship: mixed / **Goal View** Goal: Model, build, evaluate, evolve / **Method View** Documentation Method: Matrices, Goal Model, Business Model, Value Model / Construction method: mixed / Evaluation method: qualitative/ Evolution method: independence / Method guidance: good practices, roadmap, process model / Flexibility of the method: flexible / **Tool View Tool** Documentation tool or modeling: business model, value-oriented model / Evaluation tool: framework / Construction or evolution tool: frame / Guidance: Tooling / If equipped with tools : Tool for methodological guidance: Set {roadmaps, fixed processes, patterns}

Fig. 7 Positioning of the proposed approach

4 Conclusion

The idea behind this paper is based on the process of analyzing and redesigning the strategy for aligning the company's process models. Indeed, the guiding idea of this document is to identify and capture three approaches that model the strategic alignment of an organization, and then to propose a new approach. On the one hand, it will allow the company's strategy to be aligned with the business lines. On the other hand, to align the business lines with the business IS strategy.

Our approach used the techniques of the e3-values approach. The results of our work are not yet well defined, as we still want to detail our approach in order to deduce formal concepts that will be used to build an ontology of the strategic alignment of an organization's process models.

In perspective, we wish to use this research work to diagnose a large structure in order to evaluate the performance of our solution.

References

1. Fankam, C., Bellatreche, L., Dehainsala, H., Ait Ameur, Y., Pierra, G.: SISRO, conception de bases de données à partir d'ontologies de domaine. Technique et Science Informatiques **28**(10), 1233–1261 (2009)
2. Roldan-Garcia, M.M., NavasDelgado, I., Aldana-Montes, J.F.: A design methodology for semantic web database-based systems. In: Third International Conference on Information Technology and Applications (ICITA'05), IEEE, vol. 1, pp. 233–237 (2005)
3. Jean, S.: Langage d'exploitation de bases de données ontologiques. Mémoire pour l'obtention du DEA T3IA, Université de Poiters (2004)
4. Alignement stratégique du système d'information (2002)
5. Bleistein, S., Cox, K., Verner, J., Phalp, K.: B-SCP: a requirements analysis framework for validating strategic alignment. Inf. Softw. Technol. **48**(9) (2006)
6. Casati, F., Ceri, S., Pernici, B., Pozzi, G.: Workflow evolution. In: Proceedings of 15th International Conference on Conceptual Modeling (ER'96). Cottbus, Germany, pp. 438–455 (1996)

7. Yu, E.: Towards modeling and reasoning support for early-phase requirements engineering. In: Proceedings of the 3rd IEEE International Symposium on Requirements Engineering (RE'97), pp. 226 (1997)
8. Etien, A.: L'ingénierie de l'alignement: Concepts, Modèles et Processus. La méthode ACEM pour la correction et l'évolution d'un système d'information aux processus d'entreprise, thèse de doctorat, 13 mars 2006, Université Paris 1 (2006)
9. Wand, Y., Weber, R.: On the ontological expressiveness of information systems analysisand design grammars. J. Inf. Syst. **3**(4), 217–237 (1993)
10. Chakraborty, S.A., Doshi, J.: Reducing query processing time for non-synonymous materialized queries with differed criteria. Int. J. Nat. Comput. Res. (IJNCR) **8**(2), 75–93 (2019)
11. Luftman, J.N.: Assessing business-IT alignment maturity. Commun. Assoc. Inf. Syst. **4**(14), 1–50 (2000)
12. Proposition d'une modélisation conceptuelle d'alignement stratégique: La méthode INSTAL. Laure-Hélène Thevenet
13. Soffer, P., Wand, Y.: Goal-driven analysis of process model validity. In: Proceedings of CAiSE'04, Riga, Latvia (2004)
14. Utilisation conjointe des ontologies et du contexte pour la conception des systèmes de stockage de données. Okba BARKAT

A Multi-Agent Information System Architecture for Multi-Modal Transportation

Jihane Larioui and Abdeltif Elbyed

Abstract Nowadays, we are witnessing significant growth in terms of demand for public transport and the need for travel. This is due to an increase in population density, traffic congestion, lack of information and all the problems that the transport network is facing. Following the reasons cited, the movement of travelers is becoming increasingly difficult. Indeed, as soon as the trip requires the use of one or more services of a transport operator, the planning of the trip requires the traveler to collect the necessary information on the various websites of the transport operators concerned. The objective of this work is to develop an architecture of an information system for multi-modal transport, based on the notions of agents, in order to provide the users with the optimized itinerary to follow and to avoid the consultation of several transportation websites to plan the trip. In addition to this, the system is based on the notion of multi-objective optimization to combine the different modes of transport and satisfy the different criteria such as minimization of the travel time and the number of correspondence between stations.

Keywords Information system · Multi-agent system · Multi-agent architecture · Multi-modal itinerary · Multi-objective optimization · Trip planning

1 Introduction

In multimodal transport, users use at least two different types of transport to reach their destination. Therefore, multimodality refers to the optimal use of different modes of transport. Hence, the need to have a system providing information on departures, itineraries and traffic conditions before and during the journey in real-time [1, 2].

J. Larioui (✉) · A. Elbyed
Laboratory LIMSAD, Hassan II University, Faculty of Science Ain Chock, Casablanca, Morocco
e-mail: jihane.larioui@gmail.com

A. Elbyed
e-mail: abdeltif.elbyed@univh2c.ma

V. Bhateja et al. (eds.), *Embedded Systems and Artificial Intelligence*,
Advances in Intelligent Systems and Computing 1076,
https://doi.org/10.1007/978-981-15-0947-6_75

The traveler wishes to have accurate information according to his personal criteria. He can search for this information himself, but it makes the traveler's orientation more complicated.

Several researchers and industrialists have been involved in the development of multimodal information services and systems not only to improve the comfort of travelers but also to encourage people to use public transport to preserve the environment from pollution and also to address the problems of traffic congestion and urban traffic.

The information system we present is based on a multi-agent architecture to associate user requests with information stored in database of transport operators. It chooses the modes of transport to be combined and provides itineraries responding to the requests while optimizing travel time and number of correspondences. Our multi-agent architecture takes the paradigm of multi-agent systems [2] improving the association of transport operators, route calculation by a multi-objective optimization algorithm and the management of user preferences. The user will avoid consulting several transport Web sites to plan his personal trip [3, 4] since he can express his preferences between different modes of transport and define a decreasing order of priority with several criteria such as time, number of correspondence, cost, and security.

2 Multi-Agent Information System: Adopted Strategy

In literature, agent-based information systems are commonly referred to as Advanced Traveler Information Systems (ATIS) [5, 6]. Indeed, traveler's information is by origin a complicated information. This information usually relates to the itinerary itself and is always relative to the traveler. The latter plans his trip according to his own needs, criteria, and preferences. As a result, the paradigm of multi-agent information systems for personalization, collection, and integration is an effective way to address the problem of multimodal traveler information.

In order to create our multi-modal information system based on a multi-agent architecture, the system to be developed must reach different data sources and integrate the results generated by a multi-objective algorithm for route calculation [7, 8]. We suggest designing an agent-based information system that can find the source of information needed to meet the diverse users' requests. It should also be able to produce a real-time optimized multimodal information and calculate the itinerary.

It would be good to know that the exploitation of the multi-agent approach in information systems favors the collaboration of different entities in order to find the solution which is, in our case, the multimodal information to assist the traveler during his trip. In other words, an optimized multi-modal itinerary.

Our system must access the database of the different transport operators and integrate the results generated by different agents that make up the system to meet the demands of users. In this study, we consider a category of agents that refers to the same transport operator. In this category, each agent represents a transport line of a specific operator and uses his information. This approach will allow us to highlight the effectiveness of the algorithms used during the calculation. As a result, the problem of finding an itinerary between two stations belonging to different operators amounts to identifying the agents concerned and combining several itineraries sought by different operators.

Hence, our system will act on one side as a middleware connecting the client and the different data sources on the other side. It must be able to find the right source of information to query based on the queries of different users and collect information in a meaningful way to provide answers to these queries. It would propose the optimized itinerary to users through different linked agents to different lines of each operator. We are initially forced to find the shortest itinerary on the basis of a single criterion, which is time.

3 Organization of the Multi-Agent Information System

In order to highlight our approach, we propose in this article the first conception of our multi-agent architecture. In this architecture, we have six types of agents. We present in the following the different types of agents that make up this architecture.

3.1 Personal Travel Agent (PTA)

Its main task is to accompany and assist the user, this collaboration and communication between the agent and the user make it possible to better guide the execution of the different delegated tasks. Thus, we call it Agent PTA "Personal Travel Agent". The PTA makes it possible to better formulate the requests and consequently to better orient the field of research from the beginning.

This type of agent is adaptive and has a great learning ability. Based on memorization of the behavior of the user-facing the system, the query history of the latter and the set of preferences required for each connection, this agent manages to integrate into his own knowledge base a model of the user.

3.2 Information Agent (IA)

This agent is able to search, collect, integrate and manipulate information from different data sources. He can adapt his research techniques and criteria according to the needs of the user. Indeed, this agent shows great autonomy. It is able to launch its research or information gathering activities in an autonomous way.

We have used this type of agent for several reasons, including the difficulty of accessing information from different transport operators, so it is better to have access to this information directly from our database.

In this paper, we create a category agent CA^T that groups by operator several information agents (IA) and to associate each information agent (IA) with a carrier line of an operator particular.

For example, as we can see in Fig. 1 we have a category agent for Tramways CA^T that contains two information agents $IA^T_{L_1}$ and $IA^T_{L_2}$. The main role of the agent $IA^T_{L_1}$ is to extract the information relating to line 1 of the tramway, as for the agent $IA^T_{L_2}$ he deals with line 2 of the same operator (Fig. 2).

Indeed, the IA has the protocols and functions that allow it to send requests and receive responses. The agent structure remains the same for all operators. In addition, each agent is responsible for his transportation network, otherwise he can respond to a request for an itinerary if his place of departure and arrival are included in the area administered by the agent. He, therefore, has limited knowledge of the global network. Thus, in response to a request spread over several lines, the information agents responsible for the various operators involved must cooperate to provide the result.

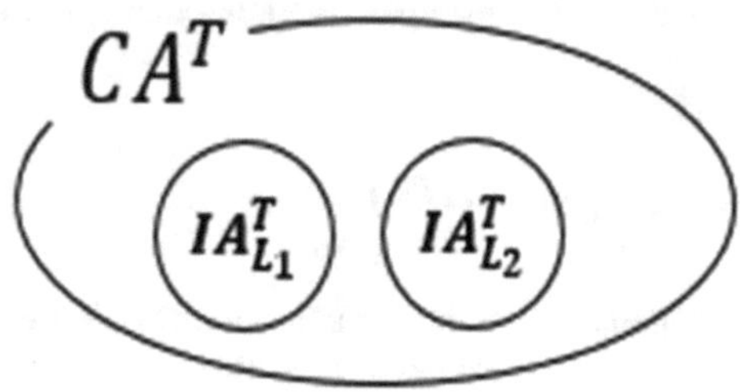

Fig. 1 Category agent for tramways operator

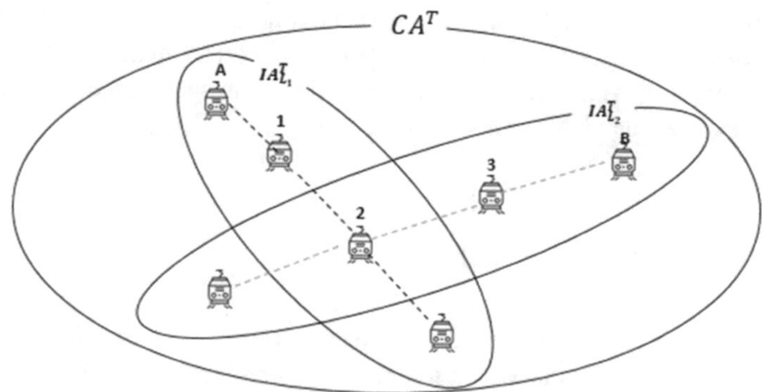

Fig. 2 Associated information agents in a category agent

3.3 Directory Selecting Agent (DSA)

The directory selection agent (DSA) has complete and prior knowledge of all the IAs present in the global multi-agent information system.

Indeed, it has a list of different information agents detailing their crossing stations with other operators. Therefore, this agent has different possibilities of cooperation between the different information agents Fig. 3.

The main role of this agent is to define the search domain by specifying the end agents according to the request of the user (Start/Arrival) and then to suggest a set of groups of agents who will cooperate together to help a selection algorithm based on the shortest path algorithm (Dijkstra). Figure 4 explains the process of the DSA agent in the constitution of cooperative IA agents. Hence, each proposed group is a group of agents representing the path from the departure station to the arrival station.

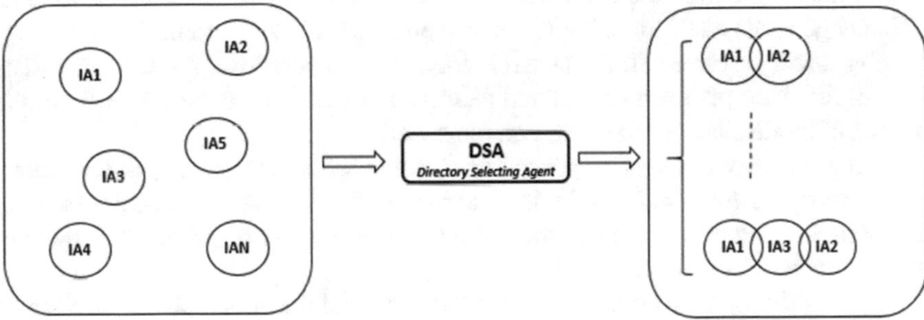

Fig. 3 Set of associated IAs made by the DSA

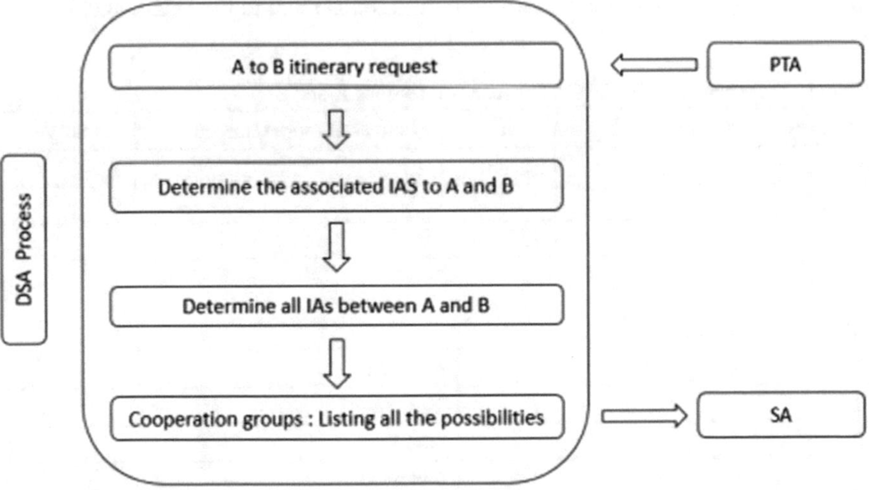

Fig. 4 Directory selecting agent process

3.4 Sorting Agent (SA)

The main role of this sorting agent is to examine the different itineraries proposed by the DSA and to decide how to treat them according to user preferences. For this, the SA delegates the route calculation part to the CA agent and the decision-making part to the DMA agent and then sends the response to the PTA agent.

3.5 Calculating Agent (CA)

The Calculating Agent (CA) is based on the different itineraries that have been proposed by the DSA to calculate the final itinerary based on user preferences.

Indeed, the CA agent calculates for each proposed itinerary the corresponding travel time, the number of mode changes made, the cost to pay and the security rate, to finally generate a table like Table 1 and send it to the DMA agent.

After having received all the possible IA agent cooperations, the CA agent lists them in the form of several cases and examines for each case the travel time, the number of mode changes, the cost, and the security.

Indeed, security is a very important criterion to which the traveler is very interested. We opted for a rating of this criterion under an evaluation scale Table 2. However, in the real world, there are modes of transportation and areas in the city that are much safer than others are.

Thus, according to the modes of transport used and the geographical location of the different stations that compose the itinerary in each case, the CA agent gives a note for the security for each itinerary.

Finally, the CA agent generates a Table 1 that contains the various parameters calculated for each criterion and itinerary and sends them to the DMA agent.

Table 1 Table generated by the CA between two stations A and B

Itinerary	Time (min)	Cost	Number of mode changes	Security
Case 1	30	8	2	2
Case 2	45	5	1	3
Case 3	17	13	3	1

Table 2 Security Scale

Scale	Mentions
1	Very good
2	Good
3	Not good

Table 3 Table generated by DMA Agent

Itinerary	Time (min)	Cost	Number of mode changes	Security	Results (%)
Case 1	30	8	2	2	60
Case 2	45	5	1	3	10
Case 3	17	13	3	1	30

3.6 Decision-Making Agent (DMA)

The decision-making agent is based on the table generated by CA to then apply the TOPSIS "*Technique for Order of Preference by Similarity to Ideal Solution*" method as a MCDM method to facilitate decision-making and ultimately to choose the itinerary that will satisfy the user's preferences.

The results are interpreted in the table above Table 3 as a percentage. Thus, the itinerary with the highest percentage is the most satisfying user's preferences.

4 The Multi-Agent Information System: Suggested Architecture

Figure 5 shows the architecture of the multi-agent system that we propose.

First, the user sends his request, specifies his departure and arrival stations, then, defines his preferences according to a descending order of priority with several criteria such as time, number of correspondence, cost, and security.

The PTA agent receives the request and splits it into two sub-requests: the first sub-query concerns only departure and arrival of the user, this one is sent directly to the DSA agent as for the second concerns his preferences, which is sent to the agent SA.

As soon as the DSA agent receives a request, it queries the different Category Agents to specify the IA agents that will work together to respond to the request of the user. Thus, the Category agents return to the DSA agent the IA agents, respectively, corresponding to departure, arrival and correspondence stations. The DSA agent executes its selection algorithm based on Dijkstra's shortest path algorithm to form several agent groups. Each group is composed of IA agents linking the starting and the ending points. Then, he obtains a set of itineraries and transmits them to the agent SA. The latter bases these itineraries to calculate the final itinerary according to the preferences of the user. Indeed, the CA also calculates for each proposed itinerary the necessary parameters for making decision and sends them to the DMA agent. This one is based on the table generated by the CA agent to then apply the TOPSIS method finally choose the itinerary that satisfies user criteria.

The DMA agent then sends the response (the chosen itinerary) to the SA agent who will send the information to the PTA agent to finally respond to the use.

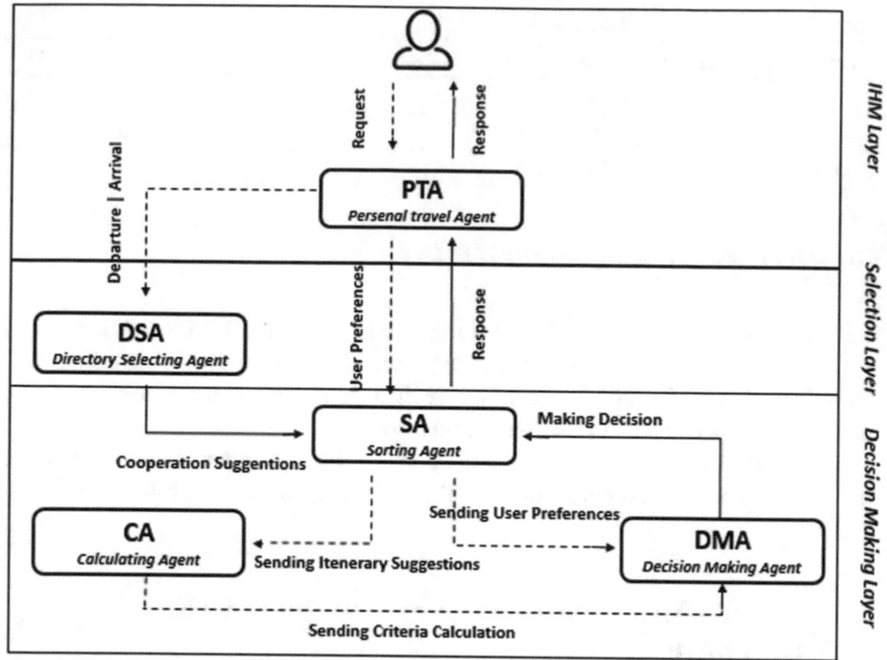

Fig. 5 Multi-agent information system architecture

5 Conclusion

Using a multi-agent technology for the development of an intelligent information system for urban mobility was our approach in this article to address the problems of passenger travel.

The proposed multi-agent architecture is efficient, flexible, designed to easily be adapted and to manage disturbances in the transport network.

In this context, our architecture remains open to include new modules to develop other interesting aspects and can be improved in our future work.

However, this work points the way towards a very promising research perspective, especially, the study of the selection algorithm used by the DSA agent to build the set of information agents that will cooperate, the calculation of the itinerary used by the CA agent as well as the MCDM method used by the DMA agent. These lines of research will be the subject of a rather deep study in our next scientific contributions.

References

1. Ben Khaled, I., Kamoun, M.A, Zidi, K., Hammadi, S.: Vers un Système d'information voyageur multimodal (SIM) à base de système multi-agent (SMA), REE revue de la SEE **1**, 41–47 (2005)

2. Kamoun, M.A., Hammadi, S.: A multi-agent architecture for a transport's multi-modal information system. WSEAS Trans. Syst. **5**(3), 2062 (2004)
3. Kamoun, M.A, Uster, G., Hammadi, S.: Optimisation des systems d'information coopératifs pour l'aide au déplacement multimodal, In Rencontre avec les doctorants des laboratoires ESTAS, LEOST, LIVIC, LTN,Actes INRESTS, INRESTS (2005)
4. Kamoun, M.A., Uster, G., Hammadi, S.: An agent-based cooperative information system for multimodal transport's travelers assistance. In: Proceedings of IMACS 2005 Scientific Computation: Applied Mathematics and Simulation, Paris France (2005)
5. Dotoli, M., Zgaya, H., Russo, C., Hammadi, S.: A multi-agent advanced traveler information system for optimal trip planning in a co-modal framework. IEEE Trans. Intell. Trans. Syst. IEEE, 1524–9050
6. Kamoun, M.A.: Designing an information system for multi-modal travelling assistance: a multi-agent approach for an on-line routes composition. In: PHD Thesis, Ecole Centrale of Lille (2007)
7. Kamoun, M.A., Uster, G., Hammadi, S.: An agent-based cooperative information system for multi-modal travelers route computation. In: Proceedings of IEEE International Conference on Systems, Man and Cybernetics. The Big Island Hawaii USA, 10–12 Oct (2005)
8. Wang, J., Kaempke, K.: Shortest route computation in distributed systems. Comput. Oper. Res. **31**, 1621–1633 (2004)
9. Fayech, B.: Régulation des réseaux de transport multimodal: Systèmes multi-agents et algorithmes évolutionnistes, Thèse de doctorat, Ecolde Centrale de Lille/ Université des Sciences et Technologies de Lille (2003)
10. Danflous, D.: Deployment of national multimodal information systems, Technical report Centre de documentation du CERTU, CE06 5749 (2000)
11. Petit-Roze, C.: Organisation multi-agents au service de la personnalisation de l'information, PHD Thesis, Université de Valenciennes et du Hainaut-Cambré (2003)

Sentiment Analysis Approach Based on Combination of Word Embedding Techniques

Ibrahim Kaibi, El Habib Nfaoui and Hassan Satori

Abstract Sentiment analysis is a field of research that attracts the attention of companies and governments to understand the opinion of the client and citizens, about products, services, policies and more other. With the increased volume of user-generated content on the Web, especially social networks, textual information becomes freely accessible and with a gigantic quantity, which requires powerful automated analysis tools to extract such kind of information (positive or negative sentiment). In this paper, we present sentiment analysis approach depends on pre-trained word embeddings, a frilly high-quality word representation vectors, namely, AraVec and fastText models, and we proposed a combination of the two models, based on vectors concatenation of both models. Sentiment classification was executed employing six different machine learning algorithms, we find that in most of the cases, our proposed method achieves the best results in terms of accuracy, especially with NuSVC classifier which is a type of SVM.

Keywords Sentiment analysis (SA) · Word embedding · Arabic text

1 Introduction

The user's opinion is an important piece of information for decision-making process [1], also called sentiment, this information is extracted using Sentiment Analysis (SA), which refers to the use of many techniques and approaches of several areas, such as machine learning (ML), natural language processing and data mining for the extraction and identification of subjective information from the text [2].

Also known as Opinion Mining, sentiment analysis is a field of research that attracts the attention of businesses and governments to understand how the customer

I. Kaibi (✉) · H. Satori
AICSM Laboratory, Mohammed First University, Oujda, Morocco
e-mail: i.kaibi@ump.ac.ma

E. H. Nfaoui · H. Satori
LIIAN Laboratory, Sidi Mohamed Ben Abdellah University, Fez, Morocco
e-mail: elhabib.nfaoui@usmba.ac.ma

V. Bhateja et al. (eds.), *Embedded Systems and Artificial Intelligence*,
Advances in Intelligent Systems and Computing 1076,
https://doi.org/10.1007/978-981-15-0947-6_76

and citizens think [3], with the increase volume of user-generated content on the Web, especially social networks, textual information becomes freely accessible and with a gigantic quantity. Based on ML, the SA process includes a series of automated tasks, such as data preprocessing, features extraction and sentiment classification to determine whether these emotions are positive or negative.

In the literature, two approaches are used to address the problem of extracting the sentiment, lexicon-based approaches, and machine learning approaches, the ML methods are more accurate than the lexicon-based one [1, 4], this is due to the important number of features used to train ML algorithms. The features are obtained in features extraction phase, in which the well-known method that achieves the best results is word embedding.

Word embeddings, also known as a distributed representation of words, are methods and techniques focused on machine learning for representing each word by a corresponding real number vector, word embedding idea is inspired from distributional hypothesis [5], which is that words used in the same contexts have high proportion of similar meaning. the quality of generated vectors in terms of semantics and syntax, make word embeddings commonly used in sentiment analysis process for extracting the features [6], most of developed models are based on artificial neural networks [7, 8] or count-based models [9, 10] to build word vectors, these distributional representations achieve an important advancement in many NLP tasks, what led to the emergence of a publicly available pre-trained word vectors of many languages [11, 12], because the quality of word vectors directly depends on the amount and quality of data they were trained [11], and the time consuming to train models needs high machine performances.

In this paper, we present the benefit obtained from the use of two different pre-trained word embeddings, AraVec [13] and fastText pre-trained word vectors [11] to represent Arabic tweets for their exploitation in the sentiment classification task. Using six well-known machine learning algorithms, namely, GaussianNB, LinearSVC, NuSVC, LogisticRegression, SGD and RandomForest, the main goal of our work is to improve the accuracy of classifiers comparing to our previews results in which the generated vectors are trained locally, employing word2vec [7] and fastText [8] models on Open Source Arabic Corpora (OSAC) [14]. In this work, we focus on the binary classification of sentiment (positive or negative) in Arabic text.

We recall that AraVec is a pre-trained distributed word representation for Arabic language, the last version provides 16 different word embedding models built on top of two different Arabic content domains; Tweets and Wikipedia Arabic articles, the word vectors are created by the use of word2vec, which is the most popular word embedding model created by Mikolov et al. [7], and is based on two neural network model architectures (Skip-gram and CBOW). Skip-gram takes the word as input and predicts surrounding words as output, conversely, CBOW takes context words and predicts the word. The fastText pre-trained word vectors is a high-quality word representation for 157 languages, two sources of data are used to train fastText pre-trained models: the free online encyclopedia Wikipedia and data from the common crawl project.

The following sections are organized as follows: We discuss some related work in Sect. 2; the methodology is detailed in Sect. 3; Sect. 4 reserved for experiment results and discussion, in Sect. 5 we conclude the paper.

2 Related Works

In the last couple of years, word embedding becomes one of a major technique used to address several NLP tasks, due to the quality of the syntactic and semantic features obtained, especially, word embedding achieve an important improvement in sentiment analysis. In this section, we recall some of the previous works that use pre-trained word embeddings for Sentiment Analysis.

In [4] authors proposed a method based on Word2Vec/GloVe methods, word position algorithm, Part-of-Speech tagging and lexicon-based approaches, named Improved Word Vectors (IWV), they tested the accuracy of their method employing different deep learning models and sentiment datasets, experiment results show that IWV are very effective and increases the accuracy of pre-trained word embeddings in sentiment analysis.

Authors of [15] compared two pre-trained word embeddings, AraVec, and fastText vectors; the models that we use in this work. They work on aspect-based sentiment analysis for Arabic airline tweets, and they formulated the problem as aspect detection step, followed by sentiment polarity classification of the detected aspects, SVM classifier is used in both steps, their experiment results showed that fastText word embeddings archive a little better performance than AraVec-Web, this result is different from that obtained by ourselves, of reason that we use AraVec version trained on tweets data, but they used a version trained on World Wide Web pages Arabic content.

3 Methodology

At the core of our work, we propose to use high quality word representation vectors generated from a very amount dataset, instead of locally training word embeddings models. Tow pre-tainted word embeddings are used (AraVec[1] and fastText[2] vectors) for features extraction phase, after text preprocessing step, we compute the tweet vector representation using simple average word vectors composing this tweet. These vectors carry the important syntactic and semantic information to enhance the training of machine learning algorithms used in sentiment classification step. Figure 1 shows the main architecture of our work.

[1] https://github.com/bakrianoo/aravec.

[2] https://fasttext.cc/docs/en/crawl-vectors.html.

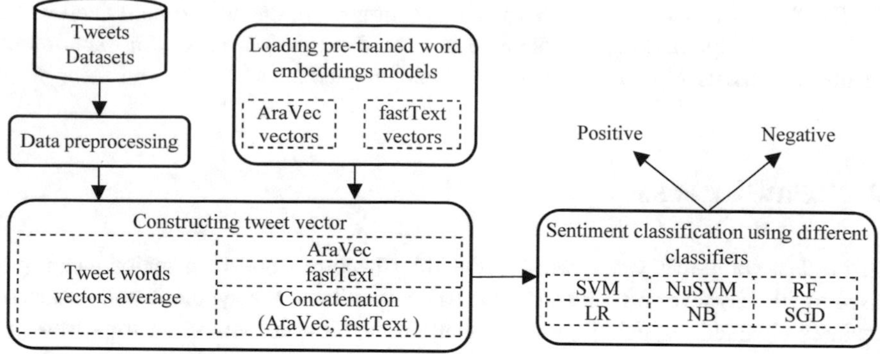

Fig. 1 The main architecture

3.1 Data Collection and Preprocessing

The first step of our work consists of collecting data from several sources. To this end, we have chosen to build our dataset by aggregating four datasets already published for research activities. Each dataset is composed of a well-defined number of tweets belonging to different areas. The collected data is used to train classifiers, for this task, one of the three labels ('1', '0', '−1') is attributed to the tweets, based on the source data, the label 1 for the tweets that contains the positive opinion, 0 for negative tweets and −1 for the rest, more details about the datasets are shown in the following:

- *Arabic Gold Standard Twitter Data for Sentiment Analysis* [16]: Tweets annotated manually using three labels (negative, neutral and positive), the dataset content 6514 Arabic entry;
- *ASTD*: *Arabic Sentiment Tweets Dataset* [17]: A set of 10,000 Arabic tweets labelled as subjective mixed, subjective positive, subjective negative and objective;
- *Twitter Data set for Arabic Sentiment Analysis* [18]: 1000 negative and 1000 positive tweets collected by using a tweet crawler;
- *Twitter corpus of movies in Arabic Language* [19]: 1800 tweets about Arabic movies which are "الحرب العالمية التالتة", "صنع فى مصر", "الفيل الأزرق" and "ميري جوازة".

The total number of tweets is 20,314, and next step consist of cleaning and normalizing these tweets, due to the non-formal nature of tweets text, several text preprocessing operations of Arabic tweets have been applied, mainly, Non-Arabic letters elimination, normalization, and tokenization, for this end, a publicly available python library (pyarabic[3]) is used, from it we apply normalization scripts such as 'tatweel' and 'tashkeel', Finally, each tweet is transformed into a list of words using tokenization, an example of the preprocessing steps is presented in Table 1.

[3]https://pypi.python.org/pypi/pyarabic.

Table 1 Preprocessing steps

Step	Result
Initial tweet	"هـــــــذا خبرررررر ســــــــارّ جدًاااا #nice"
Non-Arabic letters elimination	"هـــــــذا خبرررررر ســــــــارّ جدًاااا"
Normalization	"هذا خبر سار جدا"
Tokenization	"هذا", "خبر", "سار", "جدا"

3.2 *Features Extraction*

One of the primordial phases in the process of sentiment analysis is features extraction, in our work, we focused on pre-trained word embeddings, an important source of syntactic and semantic features [20, 21], we load tow pre-trained models, AraVec, and fastText, with the same configuration parameters except the window size; CBOW as architecture and 300 as vectors dimension, 5 and 3 for window size of fastText and AraVec models, respectively, because AraVec pre-trained vectors are obtained from Twitter dataset in which the maximum length of a tweet is 140 characters.

For loading pre-trained models, we use gensim[4] a free python framework for many text analysis tasks [22]. The vocabulary size of loaded fastText model reach to 2 million words, however, just about 1.3 million words in vocabulary of AraVec model. Each word in the vocabulary is represented by real number vector, which is used to capture tweets features for training classifiers and predict sentiment in the last phase.

3.3 *Tweets Representation*

The tweet is a sequence of words, and for each word, we have a real number vector that represents it, if the word not in vocabulary we ignored it from the tweet, except in case of fastText model, that has ability to generate vectors representation of out of vocabulary words, we well show experimental results of two approaches, ignoring and generating out of vocabulary words.

For representing the tweet, we use the simple average of words vector that composes the tweet. In the Eq. (1), n represents the number of words which form the tweet, and it differs from one tweet to another, V_{tweet} is a generated tweet vector, and $V_{\text{word}(i)}$ is the word vector of the index (i) in the tweet:

$$V_{\text{tweet}} = \frac{\sum_{i=1}^{n} V_{\text{word}_i}}{n} \quad (1)$$

[4]https://radimrehurek.com/gensim/.

```
V_tweet ← 600-dimensional null vector
For each word in tweet
        If word in one of the two vocabulary, then
                If word in AraVec and fastTest vocabulary, then
                        V_AraVec ←Vector of word from AraVec vocabulary
                        V_fastText ←Vector of word from fastText vocabulary
                        V_word ← Concatenate V_AraVec and V_fastText
                If word in one vocabulary and not in the other, then
                        V_exist ←Vector of word from the vocabulary that contains it
                        V_word ← Concatenate V_exist and V_exist
                V_tweet ←V_tweet + V_word
V_tweet ←V_tweet / (number of words in the tweet - number of ignored words)
```

Fig. 2 Pseudo code for pre-trained word embeddings concatenation

We create three different representations of tweets, the first using AraVec model, second using fastText model and third one combining the two models by concatenation of the vectors, in Fig. 2, we describe the combination algorithm, notice that out of vocabulary words are ignored.

3.4 Sentiment Classification

We use tweets vectors obtained from pre-trained word embeddings for training six classifiers, namely, Linear SVC, Random Forest, Gaussian Naive Bayes, NuSVC, Logistic Regression, and Stochastic Gradient Descent Classifier. The tweets are divided into train and test sets, and we compute the accuracy for each classifier. we mention that we use scikit-learn [23] implementation[5] of all classifiers.

4 Experiment Results and Discussion

As we already mentioned, we create three different tweets representation, namely, AraVec vectors, fastText vectors, and the third one is a concatenation of the two models' vectors. For each case we execute binary (positive, negative) sentiment classification, employing six different machine learning algorithms, namely, GaussianNB, LinearSVC, NuSVC, LogisticRegression, SGD, and RandomForest. We point out that of tweets vector dimensionality of AraVec, fastText and the concatenation representation are 300, 300 and 600, respectively. Table 2 shows classifiers accuracy score for each dataset and each type of tweet representation, in the last part of Table 2,

[5]https://scikit-learn.org/stable/supervised_learning.html#supervised-learning.

Table 2 Sentiment classification accuracy results

		Gaussian NB (%)	Linear SVC (%)	NuSVC (%)	Logistic regression (%)	SGD	Random forest (%)
Arabic gold standard twitter data for sentiment analysis	AraVec	76.81	83.33	**86.96**	81.52	78.26	82.97
	fastText	68.84	84.42	85.51	83.70	78.26	81.52
	Concatenation	75.00	78.62	85.51	80.80	82.25	83.70
ASTD: Arabic sentiment tweets dataset	AraVec	68.67	72.69	74.30	71.89	65.86	75.10
	fastText	67.87	78.31	79.12	78.31	79.52	73.09
	Concatenation	73.49	71.08	**80.32**	73.49	73.90	76.31
Twitter data set for Arabic sentiment analysis	AraVec	80.50	80.00	89.00	82.50	77.00	85.00
	fastText	72.50	80.50	88.50	85.50	82.50	81.50
	Concatenation	82.50	82.00	**90.00**	85.00	85.00	85.00
Twitter corpus of movies in Arabic language	AraVec	82.22	88.89	**92.22**	88.89	86.67	90.00
	fastText	81.67	82.78	88.33	83.89	82.78	87.78
	Concatenation	86.11	85.00	**92.22**	87.78	87.22	89.44
All datasets aggregated in one dataset	AraVec	75.25	83.76	87.51	83.98	81.44	85.86
	fastText	70.94	84.20	86.74	84.53	81.88	83.65
	Concatenation	73.92	84.86	**89.28**	85.75	83.31	85.52

we present results of classifiers trained on dataset created bay aggregation of the four datasets.

The best result obtained in terms of accuracy is that of NuSVC classifier trained on tweet vectors generated by a concatenation of AraVec and fastText vectors, except in the case of first dataset, that in which AraVec representation with NuSVC more accurate than concatenation by 1.45%. Even though the vocabulary size of fastText is bigger than AraVec's, we notice that in the average, AraVec outperformed fastText, and that due to the nature of data with which AraVec is trained.

Using another evaluation metrics for checking the classification model's performance; Fig. 3 represents ROC curves of all used classifiers and with different vector representations of tweets, area under ROC curve (auc) turns out that NuSVC trained on concatenated pre-trained vectors, achieve a better capability of distinguishing between positive and negative sentiment.

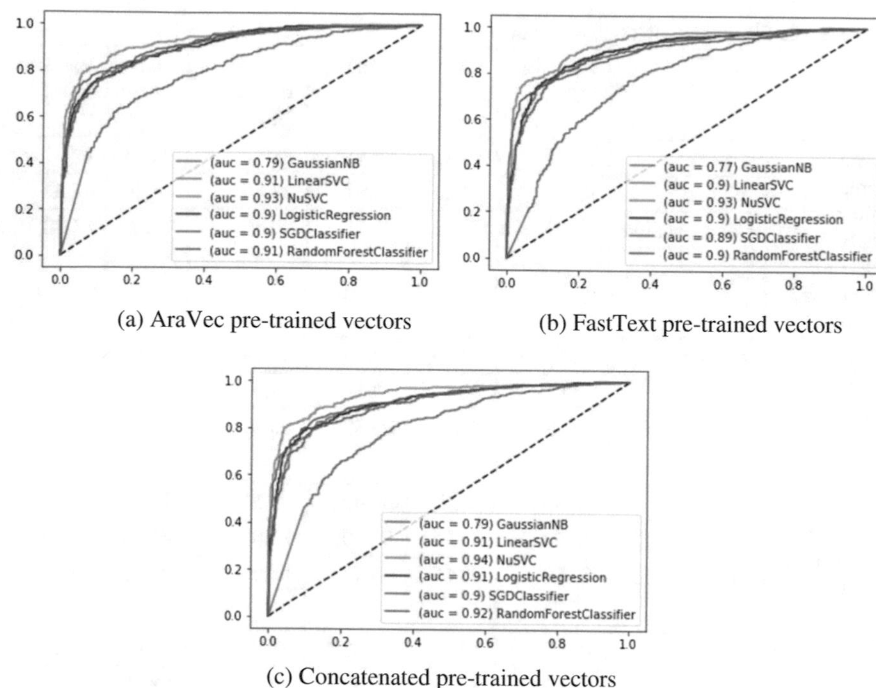

(a) AraVec pre-trained vectors

(b) FastText pre-trained vectors

(c) Concatenated pre-trained vectors

Fig. 3 ROC curves of all used classifiers and with different vector representations of tweets

5 Conclusion

In this paper, we presented an Arabic sentiment analysis based on pre-trained word embeddings, a frilly high-quality word representation vectors, namely, AraVec and fastText models, and we proposed a combination of the two models, based on vectors concatenation of both modes. experimental results show that in most cases, the proposed model achieves the best results in terms of accuracy, especially with NuSVC classifier which is a type of SVM.

References

1. Pang, B., Lee, L.: Opinion mining and sentiment analysis (2008)
2. Boudad, N., Faizi, R., Oulad Haj Thami, R., Chiheb, R.: Sentiment analysis in Arabic: a review of the literature. Ain Shams Eng. J. Elsevier (2017)
3. Tartir, S., Abdul-Nabi, I.: Semantic sentiment analysis in Arabic social media. J. King Saud. Univ. Comput. Inf. Sci. **29**(2), 229–233 (2017)
4. Rezaeinia, S.M., Rahmani, R., Ghodsi, A., Veisi, H.: Sentiment analysis based on improved pre-trained word embeddings. Expert Syst. Appl. **117**, 139–147 (2019)
5. Harris, Z.S.: Distributional Struct. Word **10**(2–3), 146–162 (1954)

6. White, L., Togneri, R., Liu, W., Bennamoun, M.: How well sentence embeddings capture meaning. In: Proceedings of the 20th Australasian Document Computing Symposium on ZZZ—ADCS'15, pp. 1–8 (2015)
7. Mikolov, T., Chen, K., Corrado, G., Dean, J.: Efficient estimation of word representations in vector space. IEEE Trans. Neural Netw. **14**(6), 1–12 (2013)
8. Bojanowski, P., Grave, E., Joulin, A., Mikolov, T.: Enriching word vectors with subword information. Trans. Assoc. Comput. Linguist. **5**(1), 135–146 (2017)
9. Pennington, J., Socher, R., Manning, C.: Glove: global vectors for word representation. In: Proc. Conf. Empirical Methods Nat. Lang. Process. (EMNLP), **28**(1–2), 1532–1543 (2015)
10. Dumais, S.T.: Latent semantic analysis. Annu. Rev. Inf. Sci. Technol. **38**(1), 188–230 (2005)
11. Grave, E., Bojanowski, P., Gupta, P., Joulin, A., Mikolov, T.: Learning word vectors for 157 languages. In: Proceedings of the Eleventh International Conference on Language Resources and Evaluation (LREC-2018), pp. L18–1550 (2018)
12. Fares, M., Kutuzov, A., Oepen, S., Velldal, E.: Word vectors, reuse, and replicability: towards a community repository of large-text resources (2017)
13. Soliman, A.B., Eissa, K., El-Beltagy, S.R.: AraVec: a set of Arabic word embedding models for use in Arabic NLP. Procedia Comput. Sci. **117**, 256–265 (2017)
14. Saad, M., Ashour, W.: OSAC: open source Arabic corpora. In: EEECS'10 6th International Symposium on Electrical Electronics Engineering Computer Science, pp. 118–123 (2010)
15. Ashi, M.M., Siddiqui, M.A., Nadeem, F., Ahmed Siddiqui, M., Nadeem, F.: Pre-trained word embeddings for Arabic aspect-based sentiment analysis of airline. Adv. Intell. Syst. Comput. **845**, 241–251 (2019)
16. Mourad, A., Darwish, K.: Subjectivity and sentiment analysis of modern standard Arabic and Arabic microblogs. In: Proceedings of the 4th Workshop on Computational Approaches to Subjectivity Sentiment and Social Media Analysis, vol. 3, pp. 55–64 (2013)
17. Nabil, M., Aly, M., Atiya, A.: ASTD: Arabic sentiment tweets dataset. In: Proceedings of the 2015 Conference on Empirical Methods in Natural Language Processing, pp. 2515–2519 (2015)
18. Abdulla, N.A., Ahmed, N.A., Shehab, M.A., Al-ayyoub, M.: Arabic sentiment analysis: lexicon-based and corpus-based. Jordan Conf. Appl. Electr. Eng. Comput. Technol. **6**(12), 1–6 (2013)
19. Medhat, W., Hassan, A., Korashy, H.: Corpora preparation and stopword list generation for Arabic data in social network. In: Fourteenth Conference on Language Engineering (ESOLE 2014) (2014)
20. Al-Smadi, M., Qawasmeh, O., Al-Ayyoub, M., Jararweh, Y., Gupta, B.: Deep recurrent neural network versus support vector machine for aspect-based sentiment analysis of Arabic hotels' reviews. J. Comput. Sci. **27**, 386–393 (2018)
21. De Vine, L., Kholghi, M., Zuccon, G., Sitbon, L., Nguyen, A.: Analysis of word embeddings and sequence features for clinical information extraction
22. Rehurek, R., Sojka, P.: Software framework for topic modelling with large corpora. In: LREC Workshop on New Challenges for NLP Frameworks, pp. 45–50 (2010)
23. Pedregosa, F., et al.: Scikit-learn: machine learning in python. J. Mach. Learn. Res. **12**, 2825–2830 (2011)

A Novel Collaborative Filtering Approach Based on the Opposite Neighbors' Preferences

Abdellah El Fazziki, Ouafae El Aissaoui, Yasser El Madani El Alami, Mohammed Benbrahim and Youssouf El Allioui

Abstract Collaborative filtering (CF) has become an effective way to predict useful items. It is the most widespread recommendation technique. It relies on users who share similar tastes and preferences to suggest the items that they might be interested in. Despite its simplicity and justifiability, the collaborative filtering approach experiences many problems, including sparsity, gray sheep and scalability. These problems lead to deteriorating the accuracy of the obtained results. In this work, we present a novel collaborative filtering approach based on the opposite preferences of users. We focus on enhancing the accuracy of predictions and dealing with gray sheep problem by inferring new similar neighbors based on users who have dissimilar tastes and preferences. For instance, if a user **X** is dissimilar to a user **Y** then the user $\neg$**X** is similar to the user **Y**. The Experimental results performed on two datasets including MovieLens and FilmTrust show that our approach outperforms several baseline recommendation techniques.

Keywords Recommender system · Collaborative filtering · Opposite neighbors · Similarity measure

A. El Fazziki (✉) · O. El Aissaoui · M. Benbrahim
University of Sidi Mohammed Ben Abdellah, Fez, Morocco
e-mail: abdellah.elfazziki@usmba.ac.ma

O. El Aissaoui
e-mail: ouafae.elaissaouil@usmba.ac.ma

M. Benbrahim
e-mail: mohammed.benbrahim@usmba.ac.ma

Y. El Madani El Alami
ENSIAS, University of Mohammed V, Rabat, Morocco
e-mail: y.alami@um5s.net.ma

Y. El Allioui
LS3M, FPK, USMS University, B.P.: 145 25000 Khouribga, Morocco
e-mail: yelallioui@gmail.com

V. Bhateja et al. (eds.), *Embedded Systems and Artificial Intelligence*,
Advances in Intelligent Systems and Computing 1076,
https://doi.org/10.1007/978-981-15-0947-6_77

1 Introduction

Recommender systems (RS) are decision support systems used on the web in order to help users to choose useful items [1]. They aim to deal with the information overload problem by predicting useful items based on users' preferences. RS act as a filter that allows passing the relevant item to the user and blocks the irrelevant one [2]. Recommender systems are largely used in various domains such as movies [3], music [4] libraries [5] and e-commerce [6, 7].

In the last decades, various approaches have been proposed for building robust recommender systems. According to [8], recommender systems can be classified into three main categories: collaborative filtering (CF) [9], content-based [10] and hybrid recommender systems [11]. Thanks to its simplicity and justifiability. Collaborative filtering remains the most commonly implemented approach on the web [12]. CF consists in recommending items based on users who share similar tastes and preferences. Despite its strengths, CF encounters many problems which usually lead to deteriorating the accuracy of the recommendations. For instance, Gray sheep is related to users who have unusual tastes and don't share similar preferences with other users [13]. Hence, finding a reliable neighborhood is a hard task. Scalability is another recurrent problem that occurs when computing similarities among all pairs of users. This task is time-consuming, especially in huge datasets. Conjointly, the sparsity of data is caused when users do not provide explicit feedback. Actually, in most cases, users do not rate items in even though they feel an extreme emotion, either satisfaction or discontent [14].

In this paper, we focus to mitigate gray sheep problems and to improve the accuracy of recommendations based on the opposite preferences of users. In other words, we deal with gray sheep problems by generating new users based on dissimilar neighbors. The underlying assumption of our approach is that if a user X has an opposite opinion of a user Y, then, the user $\neg X$ has the same opinion as the user Y. Our approach will increase the number of similar neighbors and then allow building good recommendations.

The rest of this paper is organized as follows:

In Sect. 2 we present an overview of the collaborative filtering baseline approach. Section 3 introduces the proposed approach and the original contribution of this work. In Sect. 4, we investigate the effectiveness of our proposal using an experimental evaluation on several datasets. The conclusions and some perspectives are outlined in Sect. 5.

2 Background

Collaborative filtering techniques are the most used approach in recommender systems thanks to its easiness and efficiency [9]. CF assumes that useful pieces of advice can be predicted from users who share similar tastes and preferences. These preferences can be expressed explicitly by users using ratings regarding their interests in items [15]. They can also be inferred by monitoring users' behavior such as the history of purchases and the time spent on web content called implicit feedback [16]. The set of these triplets forms a matrix called the rating matrix. It is the basic input in collaborative filtering used to build effective prediction models and users' profiles [17].

In CF, model-based and memory-based techniques remain as the main identified categories. The former build or learn models from collected ratings based on machine learning techniques like clustering techniques [18], dimensionality reduction methods [19], support vector machines, neural networks [20]. The latter referred to as neighborhood-based collaborative filtering. Memory-based CF [21] is considered as the earliest CF algorithm. It relies on building recommendations using a similarity-based neighborhood for either users or items. In fact, user-based CF focuses on building a neighborhood for active users in order to make predictions for unseen items. The same reasoning is used for item-based CF. Both use *K*-nearest neighbor classifiers to generate predictions. In what follows, we present the user-based approach.

2.1 *Memory-Based Recommendation Tasks*

As reported by [22], the memory-based approach relies on three steps presented in Fig. 1.

2.1.1 Data Representation

The first task in neighborhood CF consists of building the rating matrix and to fill missing values. In fact, in most cases, the rating matrix is usually sparse since users do not rate items in a regular manner [19]. The most used technique in CF relies on filling missing values with the average user's ratings.

Fig. 1 Memory-based CF process

2.1.2 Neighborhood Formation

In this step, a neighborhood of the most similar users is built based on a similarity metric. The most commonly used formula is the Pearson correlation coefficient. It has values between −1 and +1 where +1 means total positive correlation, −1 a total negative correlation and 0 no association between the two users. It is considered as a standard way of measuring correlation [23]. Thus, the similarity between two users a and b is calculated with the following formula:

$$\mathrm{sim}_{a,b} = \frac{\sum_{j=1}^{n}\left(r_{aj} - \overline{r_a}\right)\left(r_{bj} - \overline{r_b}\right)}{\sqrt{\sum_{j=1}^{n}\left(r_{aj} - \overline{r_a}\right)^2 \sum_{j=1}^{n}\left(r_{bj} - \overline{r_b}\right)^2}} \tag{1}$$

2.1.3 Predictions Generation

The final task in the CF process consists in generating predictions for unseen items. It is computed as an aggregation of similarities between the active user and his neighborhood in addition to their ratings:

$$p_{s,i} = \bar{r}_s + \frac{\sum_{p=1}^{k}\left(r_{p,i} - \overline{r_p}\right) * \mathrm{sim}_{s,p}}{\sum_{p=1}^{k}\left|\mathrm{sim}_{s,p}\right|} \tag{2}$$

K represents the number of closest neighbors. This prediction function uses the KNN technique to estimate the rating of an unseen item i.

Therefore, based on computed predictions, recommender systems can provide top N recommendations as a list of items that the active user has never before shown any interest.

2.2 *Evaluation Metrics*

Many measures have been used in the literature in order to measure the accuracy of a proposed method. MAE (mean absolute error) and RMSE (root mean squared error) remain as the well-known performance metrics which are broadly used in recommender systems. MAE computes the average absolute differences between predicted ratings and real values as presented in the following formula

$$\mathrm{MAE} = \frac{\sum_{(s,i)}\left|p_{s,i} - r_{s,i}\right|}{N} \tag{3}$$

where N is the number of predicted ratings computed during the test phase. $p_{s,i}$ is the predicted rating of user s to item i. $r_{s,i}$ is the real rating value.

RMSE is a quadratic error metric which measures the square root of the average of the squared differences between predicted and actual ratings:

$$\text{RMSE} = \sqrt{\frac{\sum_{(s,i)}(p_{s,i} - r_{s,i})2}{N}} \tag{4}$$

Even though memory-based techniques are easy to implement and provide good recommendations, they encounter many issues such as sparsity, scalability and gray sheep problem which deteriorate the accuracy of predictions. In gray sheep cases, it is hard for a recommender system to find a dense neighborhood with a high number of similar users. In fact, in most cases computed similarities show a low degree of correlation, even negative correlation for some similarity measures like Pearson correlation coefficient.

3 Our Approach

The baseline collaborative filtering approach uses K-nearest neighbors to make new predictions. It relies on selecting useful users who have shown a high positive correlation to the active user. In most cases, computed similarities can be positive or negative that range from -1 to $+1$. Thus, users who have shown a negative correlation are not used in the prediction phase. In addition, in gray sheep cases, the active user seems to be lacking the reliable neighbors since most users have low or negative correlations. Figure 2 below presents an example of a gray sheep situation which occurs in memory-based CF process.

The basic idea behind our approach focuses on dealing with gray sheep problem and then enhancing the accuracy of predictions by increasing the size of the reliable neighborhood.

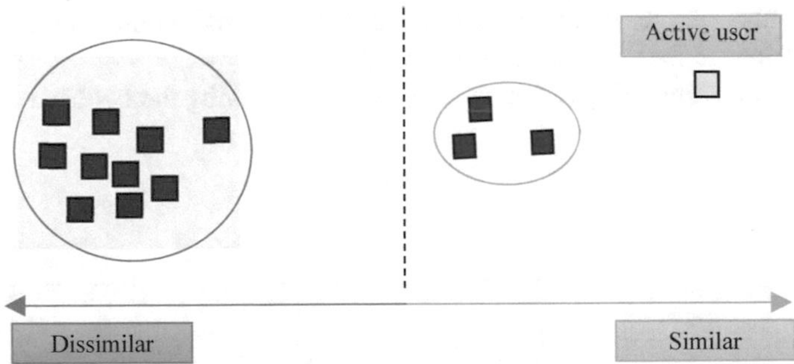

Fig. 2 Example of gray sheep situation in memory-based techniques

Fig. 3 Proposed memory-based CF process

This can be done by exploiting users who have shown a negative correlation or a dissimilarity to the active user in a smart way. To do so, we propose to infer new similar neighbors based on users who have different tastes and preferences for the active user. The underlying assumption of our approach is that if a user X has an opposite interest to a user Y, then, the user $\neg X$ will have the same interest as the user Y. Indeed, new fictive neighbors will be similar to the active user as their similarity values will be close to 1. Therefore, inferred users will enhance the density of the active user neighborhood. Consequently, additional insight will be provided to the recommender engine to make useful recommendations.

The new process includes an additional step (Fig. 3) before forming the active user neighborhood called Rating Matrix augmentation.

Rating matrix augmentation step consists in adding new lines in the rating matrix. Each line will represent an inferred user. He is the opposite user of a real one. This is achieved by deducing the opposite opinion of each rated item using the following formula:

$$\neg r_{\text{aj}} = \text{Max} - r_{\text{aj}} + \text{Min} \tag{5}$$

We denote R the $m \times n$ rating matrix where m is the number of users and n represents the number of items. The entry r_{aj} designates the rating of a user for an item j.

Max and Min represent, respectively, the high and the low value in a given numeric scale.

For instance, in a five-scale rating which ranges from 1 to 5, if a user a provided $r_{\text{aj}} = 5$ as a rating for an item j, then, the inferred rating of user $\neg a$ for the item j will be $\neg r_{\text{ai}} = 1$.

Figure 4 shows an example of opposite ratings on a 5-point scale using the previous formula. As presented, the number 3 has the same value after the opposite transformation. In fact, it represents a neutral opinion.

It lies in inferring ratings of opposite users by providing the opposite opinion on a given user.

Fig. 4 Example of an opposite rating matrix in a 5 point scale

Users \ Items	I_1	I_2	I_3	I_4	I_5	I_6	I_7
a	5		3		2		4
$\neg a$	1		3		4		2

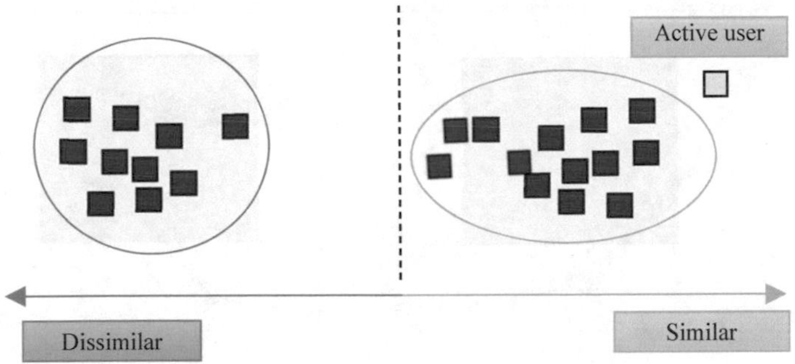

Fig. 5 Example of an active user neighborhood after users inference phase

Figure 5 shows an example of the expected output of the neighborhood formation step. As we can see, blue squares represent the new neighbors based on dissimilar users. The inferred users are likely to be very similar to the active user.

4 Experimentation and Results

We conducted several experiments using MovieLens and FilmTrust datasets to evaluate the effectiveness of our proposed approach. The main objective is to study the performance of the proposed approach over real-world datasets. In this section, we first present a brief description of the used datasets. Second, we present the evaluation procedure and the specification test environment. Then, we summarize our experimental results by comparing the performance of our proposed approach with a well-known baseline CF approach.

4.1 Datasets Collection

The experiments were performed on two commonly used datasets: MovieLens and FilmTrust [31]. Both are academic research projects of web-based movie recommender systems.

MovieLens is a five-point scale rating dataset that ranges from 1 (means bad) to 5 (means excellent). It consists of 1682 movies, 943 users and 100,000 ratings.

FilmTrust dataset consists of 1856 users, 2092 movies and 759,922 ratings. It was collected from a movie recommender systems website based on a social network which includes ratings and reviews. Ratings are numeric values on a 5-point scale between 0.5 and 4 stars.

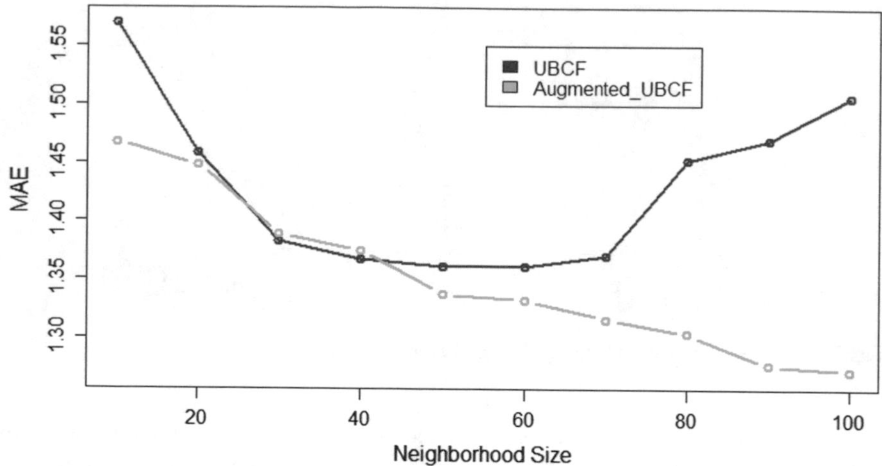

Fig. 6 MAE comparison using FilmTrust dataset

4.2 Experiments

To test our approach, we conducted a set of experiments using MovieLens and FilmTrust datasets. We reported the average results of a 10-fold cross-validation. We ran these experiments on a laptop computer with an Intel i5 at 2.4 GHz and 8 GB RAM.

Figures 6 and 7 show the obtained results of comparing the User-Based Collaborative Filtering approach (UBCF) as a baseline approach, and our proposed approach named Augmented UBCF for each dataset. The figures depict a comparison on MAE where the horizontal axis is the number of users in the neighborhood. It increases from 10 to 100 at an interval of 10. In Fig. 7, we can see the MAE of our approach and the baseline technique, are inversely proportional to the neighborhood size. We can see that our approach has lower MAE than the baseline approach. In Fig. 6 we see that our approach keeps a regular decreasing manner for the MAE while the baseline approach decreases until $N = 30$ then it remains stable until $N = 70$ where MAE starts increasing.

Overall, we can conclude that our approach provides better performance than the baseline approach in both datasets.

5 Conclusion and Perspectives

In this paper, we have proposed a novel collaborative filtering approach based on the opposite preferences of users. We focused on enhancing the accuracy of predictions and dealing with gray sheep problem. Our approach relies on inferring new similar

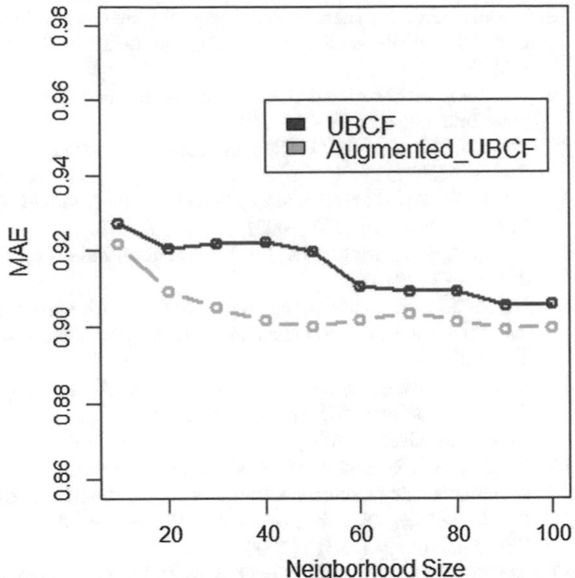

Fig. 7 MAE comparison using MovieLens dataset

neighbors based on users who have shown dissimilar tastes and preferences to the active user. In order to test our algorithm, we compare it with UBCF as a baseline approach. A set of experiments performed on two datasets including FilmTrust and MovieLens datasets show that our proposed approach has achieved good performance while solving gray sheep problem. As future work, we plan to investigate the effectiveness of hybridizing our approach with various machine learning techniques which seem to bring powerful insight to recommender systems.

References

1. Polatidis, N., Georgiadis, C.K.: A dynamic multi-level collaborative filtering method for improved recommendations. Comput. Stand. Interfaces **51**, 14–21 (2017)
2. F. Ortega, B. Zhu, J. Bobadilla, and A. Hernando, "Knowledge-Based Systems CF4J : Collaborative filtering for Java," *Knowledge-Based Syst.*, vol. 0, pp. 1–6, 2018
3. Gomez-uribe, C.A., Hunt, N.: The Netflix recommender system_algorithms. Bus. Value Pdf **6**(4) (2015)
4. Celma, O.: Music recommendation and discovery in the long tail. In: Citeulikeorg, pp. 252 (2008)
5. Callan, J., et al.: Personalisation and recommender systems in digital libraries joint NSF-EU DELOS working group report. Library (Lond) **5**(May), 299–308 (2003)
6. Linden, G., Smith, B., York, J.: Amazon.com recommendations: item-to-item collaborative filtering. IEEE Internet Comput. **7**(1), 76–80 (2003)
7. Linden, G., Smith, B., York, J.: Amazon.com recommendations: Item-to-item collaborative filtering. IEEE Internet Comput. **7**(1), 76–80 (2017)

8. Adomavicius, G., Tuzhilin, A.: Toward the next generation of recommender systems: a survey of the state-of-the-art and possible extensions. IEEE Trans. Knowl. Data Eng. **17**(6), 734–749 (2005)
9. Ekstrand, M.D.: Collaborative filtering recommender systems. Found. Trends® Human–Computer Interact. **4**(2), 81–173 (2011)
10. Pazzani, M.J., Billsus, D.: Content-based recommendation systems. Constr. Build. Mater. **171**, 546–557 (2007)
11. Burke, R.: Hybrid recommender systems: survey and experiments. User Model. User Adap. Interact. **12**(4), 331–370 (2002)
12. Fu, M., Qu, H., Moges, D., Lu, L.: Attention based collaborative filtering. Neurocomputing **311**, 88–98 (2018)
13. Najafabadi, M.K., Mohamed, A., Onn, C.W.: An impact of time and item influencer in collaborative filtering recommendations using graph-based model. Inf. Process. Manag. **56**(3), 526–540 (2019)
14. Vozalis, E., Margaritis, K.G.: Analysis of recommender systems algorithms. In: 6th Hellenic European Conference Computer Mathematics & its Applications (HERCMA), vol. 2003, pp. 1–14. Athens, Greece (2003)
15. Jawaheer, G., Szomszor, M., Kostkova, P.: Comparison of implicit and explicit feedback from an online music recommendation service, pp. 47–51 (2006)
16. Hu, Y., Koren, Y., Volinsky, C.: Collaborative filtering for implicit feedback datasets. Gastroenterology **1**, S415 (2008)
17. Isinkaye, F.O., Folajimi, Y.O., Ojokoh, B.A.: Recommendation systems: principles, methods and evaluation. Egypt. Informatics J. **16**(3), 261–273 (2015)
18. Tsai, C.F., Hung, C.: Cluster ensembles in collaborative filtering recommendation. Appl. Soft Comput. J. **12**(4), 1417–1425 (2012)
19. Paterek, A.: Improving regularized singular value decomposition for collaborative filtering. In: Proceedings of KDD Cup Workshop, pp. 2–5 (2007)
20. Agrawal, S., Agrawal, J.: Survey on anomaly detection using data mining techniques. Procedia Comput. Sci. **60**(1), 708–713 (2015)
21. Breese, J., Heckerman, D., Kadie, C.: Empirical analysis of predictive algorithms for collaborative filtering. In: Proceedings of the 14th Annual Conference on Uncertainty Artificial Intelligence, pp. 43–52 (1998)
22. Bhaidani, S.: Recommender system algorithms. In: Proceedings of International Conference on Weblogs Social Media ICWSM (2007, 2008)
23. Riedl, J., Sarwar, B., Karypis, G., Konstan, J.: Item-based collaborative filtering recommendation algorithms. In: Proceedings of 10th International Conference on WorldWideWeb, pp. 285–295 (200)1

A Review on Data Deduplication Techniques in Cloud

B. Mahesh, K. Pavan Kumar, Somula Ramasubbareddy and E. Swetha

Abstract Day by day the usage of smart devices is increasing and communication between devices which improves data growth in cloud storage. The growth of data affects the performance of storage system. The data deduplication system can optimize performance of data storage by eliminating redundant data, reducing storage cost and improves storage utilization. Deduplication technique classified into few types based on that it produces results. Along with that data deduplication performs on encrypted data will be used to prevent unauthorized users from accessing data. Subsequently, various methods perform deduplication (file or block level) safely over encrypted data in cloud. This paper represents different methods which have used different deduplication technique.

Keywords Cloud computing · Deduplication · Reduction

1 Introduction

There are various advantages in cloud storage including cost savings, availability, and adaptability and so on. Clients around the world tend to move their non-sensitive information to cloud storage. As the data is growing enormously, providing enough cloud storage is a monotonous task for cloud storage providers. Deduplication technique used by cloud storage providers to improve storage efficiency, which says to be saving 90–95% of storage [1, 2]. Initially data Deduplication system developed as a basic storage optimizing system later on it was adapted for large storage systems like cloud storage. Presently, many of the cloud storage providers adapted data deduplication mechanism in order to optimize storage utilization like Dropbox [3], Amazon S3

B. Mahesh · K. Pavan Kumar
CSE, Dr.K.V. Subba Reddy Institute of Technology, Kurnool, Andhra Pradesh, India

S. Ramasubbareddy
IT, VNRVJIET, Hyderabad, Telangana, India

E. Swetha (✉)
CSE, SV College of Engineering, Tirupati, Andhra Pradesh, India
e-mail: swethaevakattu.11@gmail.com

V. Bhateja et al. (eds.), *Embedded Systems and Artificial Intelligence*,
Advances in Intelligent Systems and Computing 1076,
https://doi.org/10.1007/978-981-15-0947-6_78

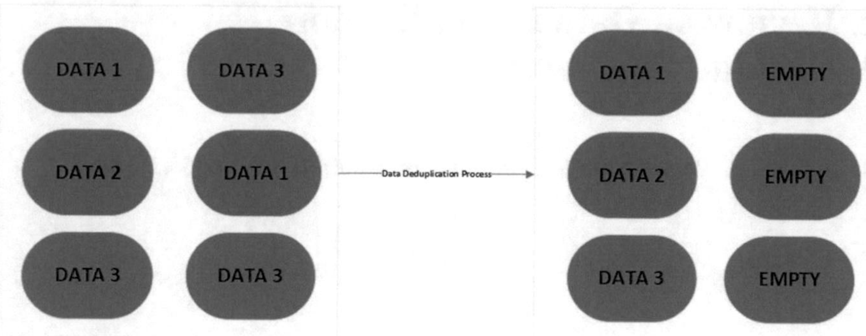

Fig. 1 Data deduplication process

[4], Google Drive [5], etc. data once sent to cloud servers, the data owner does have control on data, most of them tend to store data in encrypted format. Data deduplication is not that efficient to perform deduplication over encrypted data which is done by each individual data owner. i.e., encryption is done by different users with different keys, even similar information, with various keys may produce different ciphertext, making deduplication less possible. Moreover, encryption mechanism is basic technique for providing security for information, on other hand; storage optimization can be achieved by using data deduplication mechanism. Consequently, deduplication and encryption need to work in hand to hand to guarantee secure and optimized storage. Different strategies utilized deduplication techniques over encrypted data are concentrated in this paper. The following Fig. 1 illustrates the process of data deduplication mechanism. In deduplication process, the file is segmented into variable data chunks. Every data chunk is maintained single copy and pointer is used at duplicate data chunks. The deduplication mechanism uses pointer at all duplicate data chunks which free up memory and improves cloud storage utilization.

2 Background

Deduplication

Deduplication is fundamentally a storage optimization strategy for reducing redundant information. Figure 1 illustrates the functioning of data deduplication over data before that can be available into storage. The Deduplication mechanism can be categorized into two types file-level deduplication and block-level deduplication in view of granularity. File-level deduplication considers the whole record, in this manner even little change or alter makes the document unique in comparison to before processed one and so reducing storage optimization. In the case of block-level deduplication, it will divide the entire file into number of chunks and those are considered deduplication. Deduplication can be performed by both client and server-side. Client-side deduplication can save bandwidth by sending hash value if copy is existing [6, 7]. Deduplication is generally is utilized different storage improvement [8].

Convergent Encryption

Convergent encryption mechanism can support deduplication over encrypted data [2]. This converts plaintext to chipper-text by using unique key which can be produced from hash of the data. By applying this technique it generates same ciphertext for identical data, and this is helpful for deduplication to enhance performance.

Proof of Ownership

After deduplication by using hash value provided by user if hash value is matched with existing hash in table the user needs to be authenticated to have access of file for that proof of ownership has been introduced [9]. Without proof of ownership (POW) unauthorized users access the data from storage by stealing hash of the data. Thereby unauthorized user claims that as authorized person for getting access data from storage. To defend such work proof of ownership can be used and various works like [10, 9], etc. Proof of ownership acts as interface between prover and verifier. Prover has to compute M and send it verifier to get access to stored file in storage. [11, 9].

File Level Deduplication

In the case of file-level deduplication [6, 7], the entire file is given as an input to hash algorithm to produce hash value which is used to compare with all existing index in hash table. The file level is more efficient in terms of index computation and takes less time to determine duplicates. It consumes very less computational power because index size is very smaller and reduced number of comparisons.

Block Level Deduplication

In the case of block-level deduplication [6, 7], the entire file is divided into number of blocks or chunks each of these blocks is assigned unique id(hash) by using hash algorithm. Later on each block will be compared with existing index values in hash table. If match found the entire subsequent blocks will not be compared further and pointer to the previously stored data has to be stored. If match not found the entire data has to be stored and new index to be updated in hash table (Fig. 2).

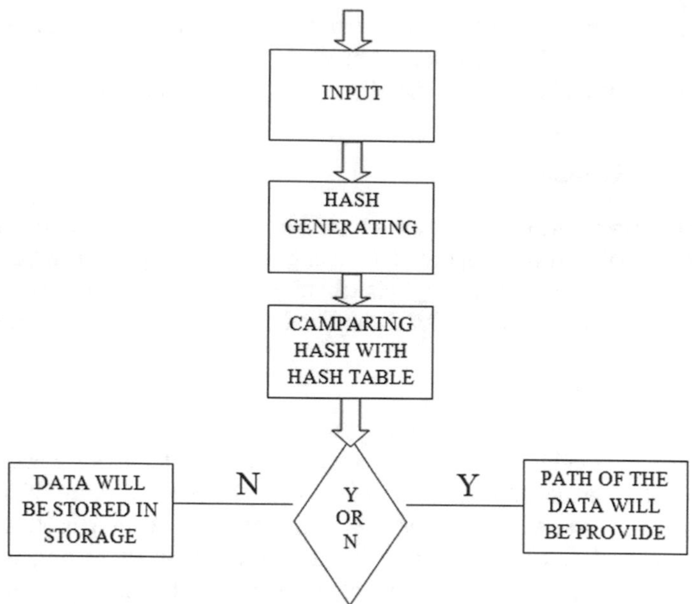

Fig. 2 Flow of data deduplication

3 Related Work

Bellare et al. [12] say the encryption algorithm derives key *K* form message *M* further encrypt the data by using key *k* and then tag T will be generated over ciphertext. Further, this tag will be used for deduplication at server may result in optimizing storage and the key length is comparatively less.

In [13] the key server present at cloud storage provider premises provides key for performing encryption and decryption at client side. In this approach, The key management approach such as Homomorphic encryption is used. encryption key is initially registered by the underlying document uploader and advance circulated subsequent checked uploaded by key server. The Data can be encrypted by using encryption key and later encoded by using hash of record content. Finally, the encrypted data and encrypted keys are sent to corresponding cloud server. HEDup gives guarantee protection on content while deduplication. The disadvantage of this approach is that the server turns into a bottleneck when number of customer's increment if there should be an occurrence of substantial scale sending and a decentralized arrangement of key server can address this issue.

In Bellare et al. [14] assert that Message-locked encryption [15] is liable to Beast compel assault and proposes a new design called DupLess where Beast compel is stood up to. Customer gets message-based keys, for encryption, from key server by means of an oblivious pseudorandom function (OPRF) convention. With OPRF open key for encryption is shared among customers whereas secret key lives with key servers. With this technique assailant's cost of assault expanded and chance is dispensed.

Puzio et al. [16] propose ClouDedup, which provides storage services by using block-level data deduplication approach and also security for stored information using convergent encryption mechanism [7] information secrecy in the meantime utilizing joined key encryption [2] included with square level key administration. This approach can address attacks against convergent encryption by adapting authentication and access control mechanisms. Along these lines, server encryption is connected on top of joined encryption performed by client. For every information portion a mark is connected to it and should be checked for recovering it. The metadata manager (MM) is added as a component to CloudDedup Architecture to manage block-level key management. MM stores various kinds of information such as (meta data about files through file table, storage in signature information through signature table).

Bugiel et al. [17] proposes an approach that for the most part includes two segments—a trusted cloud and commodity cloud. Trusted cloud responsible for all operation on commodity cloud and also responsible for encryption operations over content. Trusted cloud performs security-related operation and commodity cloud will handle outsource data operations. This approach addresses various issues arises against information like loss of information.

In Li et al. [10] proposes a hybrid cloud security mechanism by involving public and private cloud. Private cloud provides tokens in order to access data from cloud storage. For encryption convergent mechanism is adapted for this approach [2] and PoW [11, 9] is ensured to allow only authorized deduplication.

In Storer et al. [18] proposes mechanism to ensure security and space-saving by providing single server and distributed storage systems. The server cannot provide details of which data belongs which user because the encryption information is encrypted with user private key.

The redundancy in information can be reduced (deduplication) securely with two models namely authorized model and anonymous model. Convergent model is very similar to the authenticated model [2]. Anonymous model provides abstraction for readers and writer's details.

In [19], the approach that tends to the issue of uncovering and deduplicating touchy information has been invented. In this approach, several cloud services are provided with information that is split into pieces. A mystery sharing plan that says about the tag consistency and how the information is being classified to share it to the various cloud services opposes the customary deduplication-encryption strategy. Along with this mystery plan, key administration uses another plan that is an Incline mystery sharing plan [20, 21]. A mystery sharing plan in [19] is used in circulating stockpiling framework.

Client Mindful Joined encryption (UACE) and Multilevel key administration (MLK) are used to achieve the goal of [22] that is to propose a savage assault overcoming the issue of substantial keyspace. The reduction of replicated information (deduplication) in two levels cross-client document level and single client square levels can be accomplished through usage of UACE. Server-aided strategy provides document level keys merged encryption keys where a client helped techniques lead to the production of piece keys. When number of sharing clients is improved in count, it decreases the lump keyspace. These level keys are helpful in reducing the keyspace. Besides, to kill the possibility of single purpose of disappointment, this technique utilizes different key servers, outfitted with share-level keys that are created out of document level keys and Shamir's mystery sharing plan [23] is utilized to speak with these circulated servers.

According to the author in [24], the information is classified into pieces in the view of priority where the information with high priority allows less sensitivity in sharing information to various clients and the low priority information allows secured encryption semantically. To address the problem of weak security in [24], a technique has been proposed by author in [19] that is better than cross-client customer side deduplication in working weak leakage-resilient and providing security that does not allow intruders and cloud service providers who are unauthorized.

The concept of DEKEY came into existence overcoming the problem of having huge no. of convergent keys by the author in [25]. The DEKEY concept is applicable to both file-level and block-level. The author in [25], also proposed a concept called master keys to deal with convergent keys in providing security. Security for convergent keys is achieved by DEKEY using secret sharing. Figure 3 describes publication statics on deduplication topic from 1995 to 2019 in various reputed IEEE journals and proceedings. Most of the papers considered for this review from journals (50%), conferences and symposia (50%).

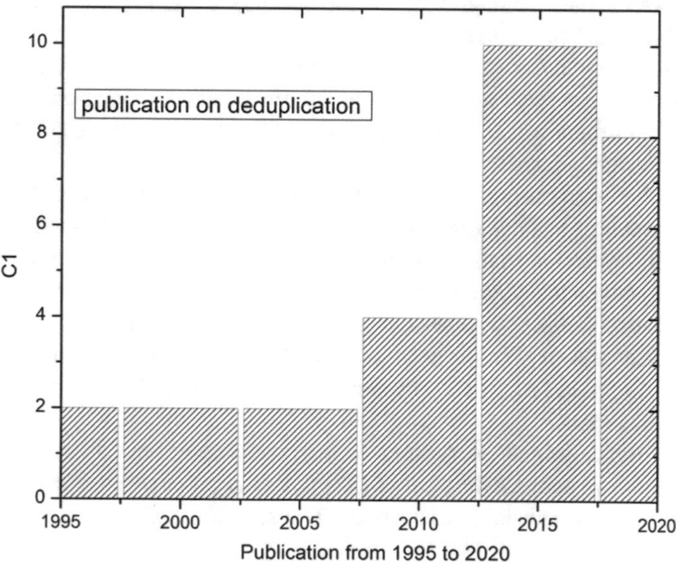

Fig. 3 Publication on deduplication from 1995 to 2019

In Figs. 4 and 5 describe data uploaded by 10 users without and with adaption of duplication mechanism. The storage capacity will be used effectively when more no. of users upload duplicate data (Table 1).

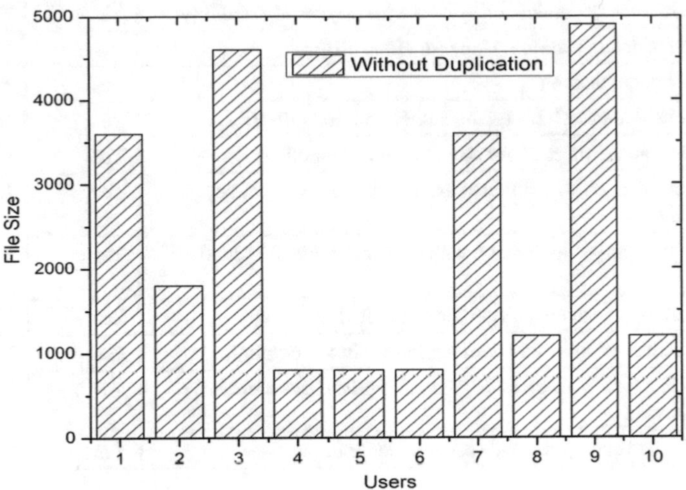

Fig. 4 Without deduplication

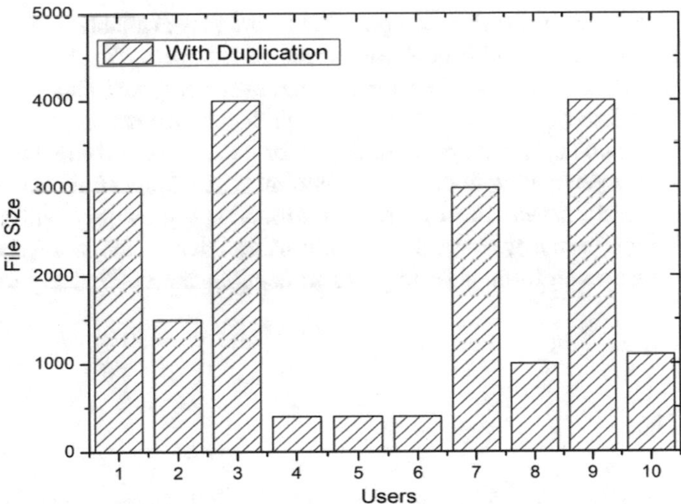

Fig. 5 With deduplication

Table 1 Comparision of various data deduplication approaches

Approach	File-level	Block-level
SecDep: a user aware efficient fine-grained secure deduplication scheme with multilevel key management	Yes	Yes
Secure distributed deduplication system with improved reliability	Yes	No
A secure data deduplication scheme for cloud storage	Yes	No
Secure data deduplication	Yes	No
A hybrid cloud approach for secure authorized deduplication	Yes	No
Twin clouds: an architecture for secure cloud computing	Yes	No
Secure deduplication with efficient and reliable convergent key management	Yes	No
ClouDedup: secure deduplication with encrypted data for cloud storage	Yes	No
DupLess: server-aided encryption for deduplicated storage	Yes	No
HEDup: secure deduplication with homomorphic encryption	Yes	No
BL-MLE: block-level message-locked encryption for secure large file deduplication	Yes	Yes
Message-locked encryption and secure deduplication	Yes	No

4 Conclusion

Deduplication is one of the data compression techniques available in cloud storage for storage optimization and bandwidth reduction. But, deduplication performance is well over normal text and when it performs over encrypted data is difficult to find redundant data because each user store their data by using their own secret key, therefore even same data will appear multiple formats. Most of the methods studied work overconvergent encryption. This method makes deduplication process simple and enhances performance. Day by day the information is growing rapidly; we never compromise both security and storage optimization across the storage area. The techniques need to be formulated to perform deduplication effectively to improve storage.

References

1. OpenDedup: OpenDedup, global inline deduplication for block storage and files. [Online] 2012 Available from: http://opendedup.org/index.php
2. Douceur, J., Adya, A., Bolosky, W., Simon, D., Theimer, M.: Reclaiming space from duplicate files in a server less cloud file system. In: Cloud Computing Systems, 2002. Proceedings of 22nd International Conference on IEEE, pp. 617–624 (2002)
3. Drop Box: http://www.dropbox.com. Tion
4. AmazonS3: http://aws/amazon.com/s3s
5. Google Drive: http://www.drive.google.com

6. SNIA: Advanced deduplication concepts. [Online] 2011. Available from http://www.snia.org/sites/default/education/tutorials/2011/fall/DataProtectionManagement/ThomasRiveria_Advanced_Dedupe_Concepts_FINAL.pdf
7. http://searchdatabackup.techtarget.com/tip/Where-and-how-to-use-data-deduplication-technology-in-disk-based-backup
8. Meyer, D.T., Bolosky, W.J.: A study of practical deduplication. ACM Trans. Storage (TOS) **7**(4), 14 (2012)
9. Yang, C., Ma, J., Ren, J.: Provable ownership of encrypted files in de-duplication cloud storage. Ad Hoc Sens. Wireless Netw. **26**(1-4), 43–72 (2015)
10. Li, J., Li, Y.K., Chen, X., Lee, P.P.C., Lou, W.: A hybrid cloud approach for secure authorized deduplication. Parallel CloudSystems Trans. IEEE **26**(5), 1206–1216 (2015)
11. Halevi, S., Harnik, D., Pinkas, B., Shulman-Peleg, A.: Proofs of ownership in remote storage systems. In: Proceedings of ACM Conference on Computer and Communications Security, pp. 491–500 (2011)
12. Bellare, M., Keelveedhi, S., Ristenpart, T.: Message-locked encryption and secure deduplication. In: Advances in Cryptology–EUROCRYPT 2013, pp. 296–312. Springer, Berlin, Heidelberg (2013)
13. Miguel, R., Aung, K.M.M.: HEDup: secure deduplication with homomorphic encryption. In: International Conference on Networking, Architecture and Storage (NAS), 2015 IEEE, pp. 215–223 (2015)
14. Bellare, M., Keelveedhi, S., Ristenpart, T.: DupLess: server-aided encryption for deduplicated storage. In: Proceedings of the 22nd USENIX Conference on Security. USENIX Association (2013)
15. Chen, R., Mu, Y., Yang, G., Guo, F.: BL-MLE: block-level message-locked encryption for secure large file deduplication. Inf. Forensics Sec. IEEE Trans. **26**(12), 2643–2652 (2015)
16. Puzio, P., Molva, R., Önen, M., Loureiro, S.: ClouDedup: secure deduplication with encrypted data for cloud storage. In: 2013 IEEE 5th International Conference on Cloud Computing Technology and Science (CloudCom), vol. 1, pp. 363–370 (2013)
17. Bugiel, S., Nurnberger, S., Sadeghi, A., Schneider, T.: Twin clouds: an architecture for secure cloud computing. In: Proceedings of Workshop Cryptography Security Clouds, pp. 32–44 (2011)
18. Storer, M.W., Greenan, K., Long, D.D.E., Miller, E.L.: Secure data deduplication. In: Proceedings of 4th ACM International Workshop Storage Security Survivability, pp. 1–12 (2008)
19. Li, J., Chen, X., Huang, X., Tang, S., Xiang, Y., Hassan, M., Alelaiwi, A.H.: Secure cloud deduplication systems with improved reliability. Comput. IEEE Trans. **64**(12), 3569–3579 (2015)
20. Blakley, G.R., Meadows, C.: Security of ramp schemes. Proc. Adv. Cryptol. **196**, 242–268 (1985)
21. Santis, A.D., Masucci, B.: Multiple ramp schemes. IEEE Trans. Inf. Theory **45**(5), 1720–1728 (1999)
22. Zhou, Y., Feng, D., Xia, W., Fu, M., Huang, F., Zhang, Y., Li, C.: SecDep: a user-aware efficient fine- grained secure deduplication scheme with multi-level key management. In: 2015 31st Symposium on Mass Storage Systems and Technologies (MSST), pp. 1–14 (2015)
23. Shamir, A.: How to share a secret. Commun. ACM **22**(11), 612–613 (1979)
24. Stanek, J., Sorniotti, A., Androulaki, E., Kencl, L.: A secure data deduplication scheme for cloud storage. In: Technology Report IBM Research, Zurich, ZUR 1308-022 (2013)
25. Li, J., Chen, X., Li, M., Li, J., Lee, P.P.C., Lou, W.: Secure deduplication with efficient and reliable convergent key management. Parallel CloudSyst. IEEE Trans **25**(6), 1615–1625 (2014)

Analysis of Network Flow for Mitigation of DDoS Attacks in a Cloud Environment

Damai Jessica Prathyusha and K. Govinda

Abstract Cloud computing has increased apparent notoriety and acceptance over the past years. The expediency of this virtual technology accompanies the budgetary issue. One of the essential characteristics of cloud computing, which assistances in decreasing the financial associated issues of the of the cloud providers is the on-demand assessing model. Moreover, malicious users mostly focus on the economic practicality of the cloud customers, as a result, such malicious are proficient to influence the accessibility and availability of cloud facilitated services. These attacks are known as Distributed Denial of service attacks. In this work, we propose a novel methodology dependent on network flow analysis at the targeted side to distinguish and moderate the DDoS attacks against virtual services. Experiments were carried out by using real time datasets to assess the execution of the proposed methodology and the results recommend that our proposed method can identify and alleviate the DDoS attacks with agreeable precision and minimal overhead.

Keywords Cloud environment · DDoS attacks · Attack prevention · Attack detection · Attack mitigation

1 Introduction

Cloud computing is a robust challenger to conventional information technology implementation as it offers minimal effort and "on-demand" based approaches to computing abilities. Many enterprises, moved their entire or the majority their infrastructure into the cloud. Nevertheless, there is a substantial amount of inquiries in customers awareness which are reviewed in the existing work [1]. The majority of the inquiries are explicitly identified with information and their data security [2]. Commonly, there is numerous security associated malicious activities, which are

D. J. Prathyusha · K. Govinda (✉)
School of Computer Science and Engineering, VIT, Vellore, India
e-mail: kgovinda@vit.ac.in

D. J. Prathyusha
e-mail: jessicaprathyusha@gmail.com

V. Bhateja et al. (eds.), *Embedded Systems and Artificial Intelligence*, Advances in Intelligent Systems and Computing 1076,
https://doi.org/10.1007/978-981-15-0947-6_79

very much tended to for the conventional IT frameworks. One of these malicious activities, that has remained a noticeable change is the DDoS attack.

In this work, we offer a network flow examination based novel approach to deal with distinguishing and moderate the event of DDoS attacks. We have performed broad examinations utilizing genuine real time benchmark datasets to test the adequacy of our proposed methodology. The gotten results are empowering and the precision and viability of our methodology.

This work organizes as follows Sect. 2 describes the existing mitigation mechanisms regarding DDoS attacks, the proposed architecture is illustrated in Sect. 3, Sect. 4 defines experimental analysis of this work and finally, we conclude our paper in Sect. 5.

2 Related Work

Authors in [3] used a Software Defined Network (SDN) to refer the DDoS protector module at the SDN manager to recognize the irregular behavior. The projected work gives Deep learning (DL) grounded recognition technique which utilizes the traffic examination tools to channel and advance the packets to the targetted server and can obstruct the contaminated network packets to bring about additional attacks.

A three-way strategy is introduced in [4], in this, each element works out the volumes associating to its preparing abilities. This method entirely uses the prevailing device capacities, for instance, local proxy servers, firewalls, IDS, etc. A masked markov model is applied with funding to disengage DDoS packets from typical entries by [5] their mechanism depended on figuring the probability of the approaching IP address sequence. A neural network, as well as danger hypothesis, could be utilized in SDN to ease the malicious attack [6]. Probably here the related hazard is figured for each host and the equivalent is sent to the very VM which watches out for incoming information stream. On the off casual that the handled hazard factor of the in-stream traffic is more than some default assessment at that point, a few directions and requests are proliferated to the controller of SDN to offer mitigation systems.

3 Proposed Approach

In this section, we present the intended DDoS attack identification and moderation mechanism in a cloud environment. The volume of incoming packets is extreme which permits the malicious users to effectively stow away their quality out of sight. Likewise, malicious users can swindle the location frameworks by sending huge chunks of duplicate incoming traffic. The extreme size of traffic additionally shapes the size of the incoming traffic logs.

It is progressively hard for the defence mechanism to perform a algorithm on the full traffic log. Consequently, so as to lessen the processing weight from the defence

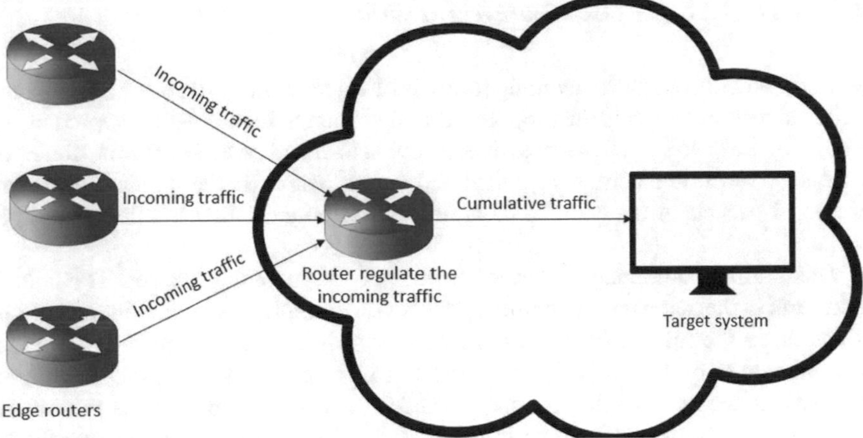

Fig. 1 Collaborative nature of the cloud environment

frameworks inspecting of the incoming traffic is finished. Testing diminishes the measure of information arranges the information that is additionally broke down by the recognition system which empowers the resistance system of self-assertive difficulty to function on traffic along with its size. Nonetheless, sampling offers loss of data which defiantly effects the incoming traffic examination. In our proposed approach, inspecting of traffic is done at different edge routers; hereafter, association among these revelation is vigorous for commanding implementation of our methodology. The routers are measured and managed by the cloud dedicated co-op in the cloud environment. In this manner, our proposed mechanism can be productively executed in the cloud (appeared in Fig. 1).

4 Detection of DDoS Attack Flow in a Cloud Environment

The objective of the DDoS attack is to challenge the budgetary practicality of a cloud-facilitated area. The idea of this attack is essentially not quite the same as the flooding or high rate DDoS attacks. So as to escape from the unfortunate loss's IDS and to have the possibility to proceed the attack for an all-inclusive time period, malicious activity needs to diminish the number of incoming malicious packet in the attack stream.

This makes the recognition of a progressing DDoS attack much progressively confounded, Accordingly, it requires a change in the current DDoS attack position technique to effectively distinguish the low-rate DDoS attack. In this area, we concerning the structure of the proposed DDoS attack identification module.

4.1 DDoS Attack Detection Algorithm

Since the amount of attack incoming traffic is little in low-rate DDoS attacks streams. Below algorithm presents the proposed the novel attack identification approach. To recognize the DDoS attack streams, it is essential to calculate and log the attributes of unaffected traffic and here, we confirm and normally apprise the standard deviation of size of packets of the genuine incoming stream in a catalogue called ostensible shape.

This outline additionally records the rate of the incoming packet. This attack increment in the number of incoming packets is unassuming. A malicious user might attention on the injured individual amid non-top hours. This may keep the most extreme rate of packet amid a given period of time. In this way, to track such kind of activities we separate the timespan into four spaces and observe the rate of the packet for each schedule opening independently. The ostensible profile records the most extreme rate of the packet (P_t) for each timeslot *s*. The present rate of the packet (P_t^1) is observed against the most extreme rate of the packet of that timeslot (P_t) got from the ostensible profile. Henceforth, the malicious activity focusing on the injured individual just amid the non-top periods can be distinguished.

Algorithm: DDoS attack detection algorithm

---

1. Determine the sampling rate (S_a) and threshold (T_h)
2. Isolate incoming packet characteristics [SourceIP, Size of the incoming packet (Packet Size)]
3. Figure the minimal profile 'N_p' throughout the normal period.
4. Observe the packets request rate P_t^1.
5. While $\{(P_t^1 - P_t) > T_h\}$ do
6. Begin sampling the packets at all the routers.
7. Find the collaborative packet flow and calculate the standard deviation.
8. Compute test statistic T_{mon} (monitoring incoming traffic)
9. If $T_{mon} > T_\beta$ (a–1,b–1)

        Consider incoming traffic as attack traffic.

    Else

        Consider incoming traffic as non-attack traffic.

Here (T_{mon}) Traffic monitoring is greater than the significance level or important level (T_β) where T_β $(a-1, b-1)$ is the rate of distribution with $(a-1)$ and $(b-1)$ degree of choice and level of important level. Importance level (β) is a vital parameter that influences the recognition precision. By evolving the estimation of β, we can alteration the positioning between the two streams (attack and non-attack). Now, order speaks to the separation between the T_{mon} and T_β values. The interplanetary between the attack as well as non-attack incoming traffic in the DDoS attack is little. Hence, expanding the interplanetary between both the streams that help in identifying the malicious activity streams whose attributes are in extremely near the non-attack streams.

5 Experiments and Analysis

In this area, we assess our proposed approach by using benchmark world datasets. For the tests, the genuine (attack free) situation is taken from the Intrusion Recognition Assessment dataset (DARPA). The DDoS movement condition is chosen from the NSL KDD real time dataset. These datasets are measured as extremely consistent datasets. We thought about the utilization of either size of the IP packet or source address to figure the of incoming traffic probability amid the given period of time.

To measure the accuracy and viability of this proposed methodology, we assess our methodology on three execution parameters: false positive rate (FP_r), false negative rate (FN_r) and detection rate (D_r). The detection rate is the proportion of a complete number of DDoS attack streams recognized over the absolute number of such streams handled (Fig. 2).

Here FP_r is defined as the proportion of a complete number of an attack free streams being mistakenly treated as attack stream over all out a number of attack free streams handled. The FN_r is the proportion of all out a number of attack streams

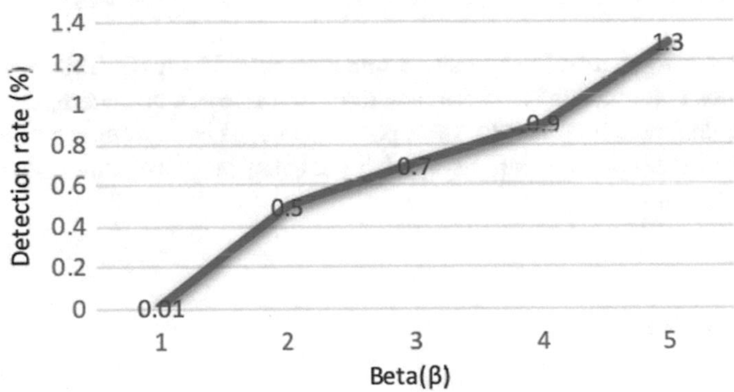

Fig. 2 Detection rate with changing the value of Beta (β)

Table 1 False positive and false negative rate of the DDoS detection mechanism

β	FP_r	FN_r
0.1	0.19	0.10
0.3	0.20	0.09
0.5	0.23	0.08
0.7	0.27	0.07
0.9	0.31	0.07
1.1	0.32	0.06
1.6	0.35	0.04
1.7	0.40	0.04

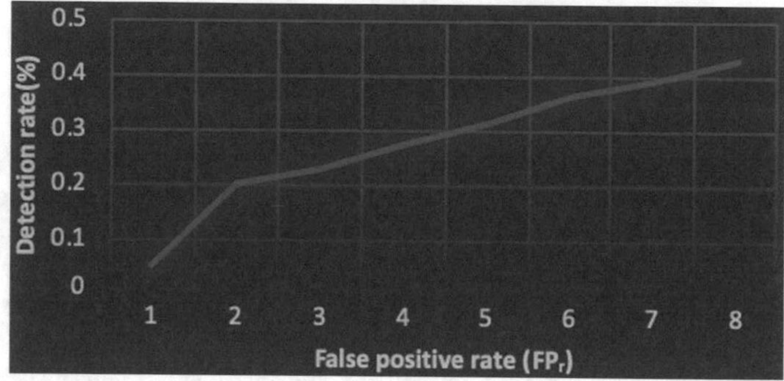

Fig. 3 Detection rate against false positive rate

being inaccurately pronounced as attack free stream over all out a number of attack streams handled. We examined the irregularity recognition rate with change in the value β (Table 1). This connotes that attack recognition rate improves with increment in β.

Detection rate against false positive rate is plotted in Fig. 3. This bend plainly demonstrates that practically 98% attack detection rate can be accomplished, if we are able to deal with around 30% false positive rate. These results recommend that the proposed methodology can distinguish the event of DDoS attack streams with palatable accuracy.

6 Conclusion

This work gives sampling based novel way to deal with distinguish DDoS attack streams in a cloud environment. We propose a new broad sorting of the attack protection system. In this work, an attack stream comprises of both malicious users and legitimate users. In this way, to isolate attack customers from the non-attack clients we have given a system stream examination technique. The obtained results show that these methodologies work adequately and steadily with slight computational overhead. Taking everything into account, our proposed methodologies can viably and effectively recognize the DDoS attack streams and can able to categorize the attack customers also.

References

1. Kandukuri, B.R., Rakshit, A: Cloud security issues. In: 2009 IEEE International Conference on Services Computing, pp. 517–520. IEEE (2009, September)
2. Kaufman, L.M.: Data security in the world of cloud computing. IEEE Secur. Priv. **7**(4), 61–64 (2009)
3. Priyadarshini, R., Barik, R.K.: A deep learning based intelligent framework to mitigate DDoS attack in fog environment. J. King Saud University-Comput. Inf. Sci (2019)
4. Zhou, L., Guo, H., Deng, G.: A fog computing based approach to DDoS mitigation in IIoT systems. Comput. Secur (2019)
5. Xu, X., Sun, Y., Huang, Z.: Defending DDoS attacks using hidden Markov models and cooperative reinforcement learning. In: Pacific-Asia Workshop on Intelligence and Security Informatics, pp. 196–207. Springer, Berlin, Heidelberg (2007, April)
6. Mihai-Gabriel, I., Victor-Valeriu, P.: Achieving DDoS resiliency in a software defined network by intelligent risk assessment based on neural networks and danger theory. In: 2014 IEEE 15th International Symposium on Computational Intelligence and Informatics (CINTI), pp. 319–324. IEEE (2014, November)

Critical Review on Course Recommendation System with Various Similarities

A. Revathi, D. Kalyani, Somula Ramasubbareddy and K. Govinda

Abstract A user-based Course Recommendation System is developed in this paper. The recommender system is constructed as an online website/application which is capable of producing a personalized list of courses which a user can take. Modern versions of traditional recommender system, such as collaborative filtering are considered to be efficient in this domain of user-based recommendation. On the basis of collaborative filtering principle, the recommendation process is divided into three steps, representation of student data, generation of neighbor user (student) and the generation of course recommendations. In order to compute the likenesses amid each user, the Cosine method is adopted during the process of the generation of neighbors. Then the course recommendations are generated according to the ratings of the other students and his personal ratings.

Keywords Recommendation systems · GQL · Cosine similarity · DBMS

1 Introduction

This undertaking will give a presentation of extemporized proposals framework in the area of instruction and online training sites which at first pass by prevalence pooling and the main personalization they give is as for the classification choices. The enhancements my methodology will offer needs to do with exploiting online life relations into record, the suggestions obviously will be founded on what your companions have done or are doing also. This is the place the database part hits in, the established DBMS show we manage relations in a fairly unintuitive manner,

---

A. Revathi · D. Kalyani · S. Ramasubbareddy (✉)
IT, VNRVJIET, Hyderabad, Telangana, India
e-mail: svramasubbareddy1219@gmail.com

K. Govinda
SCOPE, VIT UNIVERSITY, Vellore, Tamilnadu, India

V. Bhateja et al. (eds.), *Embedded Systems and Artificial Intelligence*,
Advances in Intelligent Systems and Computing 1076,
https://doi.org/10.1007/978-981-15-0947-6_80

putting them away in lines and segments. Another methodology is utilizing Graph Databases like Neo4j, which give GQL and demonstrate these relations in a fairly more instinctive way. Anyway, I will not utilize any of these in my venture however will develop it from ground level. Site will have every one of the highlights of a customary removed training site in addition to this course proposal framework with joining of Collaborative Filtering model which will result to community oriented and enhanced learning background consequently.

2 Literature Survey

2.1 Method Comparison

The development of the Internet usually made it very hard to viably separate useful Data from all the accessible information in online [1]. The amazing amount of data requires tools for effective data isolating. Communitarian classification is one of the frameworks used for dealing with this issue [2].

The inspiration for community separating originates from the possibility that individuals regularly get the best proposals from somebody with tastes like themselves. Shared separating incorporates strategies for coordinating individuals with comparative interests and making recommendations on this premise [8, 9].

Cooperative separating calculations regularly require (1) clients' dynamic support, (2) a simple method to speak to clients' interests, and (3) calculations that can coordinate individuals with comparative interests.

Normally, the work process of a communitarian categorizing framework is:

1. Clients can communicate their inclinations by ranking things (for example books, films or CDs) of the framework. These rankings can be seen as an estimated portrayal of the client's interest for the specific area [3].
2. The framework coordinates this current client's ranking against other clients' and identifies the individuals with maximum "comparative" tastes [4].
3. With the same sort of clients, the framework prescribes things that the comparative clients have ranked exceedingly however not ranked by this particular client (probably the absence of ranking is regularly considered as the newness of a thing).

A fundamental issue of collective classification is the means by which to consolidate as well as weight the inclinations of client neighbors. Once in a while, clients can quickly rank the suggested things. Thus, the framework picks up an undeniably accurate portrayal of client inclinations [5, 6].

2.2 Methodology

Collaborative categorizing in Recommender frameworks.

Collaborative categorizing systems have numerous methods, however regular frameworks can be compact to two steps:

1. Scrutinize the clients who will communicate similar ranking patterns with another active client (the client whom the expectation is for).
2. Utilize the rankings from the clients who has similar kind of taste noticed in first step to figure a prediction for the other client.

This can come under the classification of client based collaborative categorizing. This is one of the applications of client based Nearest Neighbor mechanism.

On the other hand, thing-based community classification (clients who purchased x and also purchased y), continues in a thing driven way [10]

1. Build a thing framework that governs connections between sets of things.
2. Infer the preferences of the present client by analyzing the matrix and coordinating that user's information.

For instance, the Slope Individual thing-based community categorizing group.

The same sort of communitarian filtering can be established on comprehended impression of a regular customer lead (rather than the false conduct constrained by a ranking errand). These structures perceive what a customer has performed collaboratively through what other customers have performed (what tune they are checked out, what items they obtained) and use that statistics to predict the customer's lead in formative work, or to foresee how a customer may grow a great of the opportunity to continue given the shot. These gauges should be isolated through commercial logics to choose in what way they could impact the exercises of a business structure. For example, it isn't useful to proposal to present individual a particular accumulation of song in the occurrence till this point have demonstrated that they guarantee that music. Depending on a scoring or rating framework which is found the middle value of overall clients disregards particular requests of a client, and is especially poor in assignments where there is huge variety in enthusiasm (as in the proposal of music). Not with standing, there are different techniques to battle data blast, such as web search and data bunching. Contingent upon a ranking system which is noticed the center value of over all customers dismisses specific solicitations of a customer, and is particularly insignificant in assignments where there is enormous assortment in energy (as in the proposition of music). Nevertheless, there are distinctive strategies to fight information impact, for example, information batching and web searching.

3 Client-Client Collaborative Categorizing

It is an analytical system for collaborative categorizing by Konstan, Herlocker, Riedl and Borch-ers (Proc. SIGIR 1999). The compute grade for a client and thing is specified as:

$$s(u,i) = \left[\sum \text{Rating}(vi)\right] \Big/ |C|$$

where C is referred as number of clients, and by this ranking can be calculated by that client. We are as yet intrigued by score. Along these, we will complete one major advance to attempt this move towards client collaborative categorizing. In this way, despite everything we're keen on the score that we anticipate for client and a thing regardless we're keen on the possibility that we should need to calculate other clients in our framework.

However, we're currently moving to take the ranking and increase it counts a type of a load. And now, presently the condition we have will be

$$s(u,i) = \left[\sum \text{Rating}(v,i) \cdot \text{Weight}(u,v)\right] \Big/ \sum \text{Weight}(u,v)$$

Okay, that's our first step, but there's a problem with this first step. And that problem starts with the fact that people rate on a very different scale.

In what way we represent that? Normally, when we were in this non-customized positioning, we could improve this calculation more advanced! When we standardized it to the client's evaluating measure then, we will figure S (u, i) which get as the average ranking that client provides for all the items in addition to some deviation. Additionally, that deviation will be how much this thing is preferred or more deteriorate then normal. Also, the ranking that client gives the thing subtracted by that client's normal ranking.

The normal rating that client provides for each thing in addition to some deviation. Deviation will be at what amount this thing is preferable or more regrettable over normal thing. Moreover, average of the clients, the rating that client offers the thing minus that client's average ranking.

We will entirety up how distant from every client versus own normal ranking they graded this thing that by the quantity of clients we're calculating together and after that, count that to the normal score for this particular client. Most frameworks choose they will cluster the evaluations that one will not function below the base score or over the best score in one's expectation. No individual could organize that at the calculation.

Taking a step forward we now have

$$s(u, i) = \text{Avg}(\text{Rating}(u)) + \left(\left[\sum \text{Rating}(v, i) - \text{Avg}(\text{Rating}(v)).\text{Weight}(u, v)\right] \Big/ \sum \text{Weight}(u, v)\right)$$

Now we are much closer to achieve the kind of collaborative filtering we wanted, we have removed the problem of variable scaling with this algorithm. But next thing is how we get these weights, well for that we can use Pearson Co-relation, but that is the most basic one, other more impressive one is to compute what we call Cosine Similarities.

So, formalizing everything, we reach at our final algorithm:

//Given a set of things, and a set of clients C and sparse matrix //of rankings R, we compute the calculation S (u, i) as follows:
for us in Clients, where v! = u:
Calculate Weight (u, v) //Comparation metric (Cosine similarity)
Select neighborhood of V in User with highest Weight(u,v)
ComputePerdiction() //With the equation above

3.1 Bottlenecks

Given number of users is *M* and number of items is *N*, then computation can be a very heavy bottleneck.

1. Corrclation between two users is linear, O(N)
2. All correlation for a user will be like *O*(*M***N*)
3. All pairwise Correlations will be *O*(*M*^2 * *N*)
4. Best case complexity of the recommendations will be *O*(*M***N*)

To improve this lets first understand why it even works:

Assumptions: our past agreement predicts our future agreement.

1. Our tastes are either independently steady or move in sync with one another.
2. Our framework is perused inside a field of contract.

4 Thing-Thing Collaborative Categorizing (Improvement)

We begin with the impediments of client shared collaborative classification that persuaded the expansion of this thing method. Client community-oriented categorizing was extraordinary . It gave great outcomes. Individuals were content with the expectations and proposals that it created, however it had two or three issues. The quantity of items that any given client would have ranked was excessively little. Furthermore, that sparsity prompted the issue were frequently focuses where no proposal might be formulated. Sometimes, a client just does not have common in anything with any other person. In some cases, a thing that was no real way to prescribe to an expansive group of clients on particular site [7].

That insufficiency could be tended to in a variety of courses. There was effort done on classification that would put in many false methodical evaluations. Another major problem was computational performance.

To tackle all this, we came up with a new model, known as item-item collaborative filtering. The item-item model is stable. This is dependent on having more users than items. Average item has many more ratings than a user. Intuitively, items don't generally change rapidly—at least not in ratings space. Thing resemblance is a way to computing a forecast of clients thing preferences.

ITEM-ITEM collaborative filtering look for items that are like the articles that user has already rated and recommend most similar articles. But what does that mean when we say item-item similarity? In this case we don't mean whether two items are the same by attribute like Fountain pen and pilot pen are similar because both are pen. Instead, what similarity means is how people treat two items the same in terms of like and dislike.

Thing-thing collaborative classification search for things that resemble the articles which client has ranked and prescribe most comparative articles. However, concerning thing-thing similarity, one does not mean whether two things are the equivalent by considering attribute like Fountain pen as well as pilot pen are comparative in light of the fact that two are pens. Rather, what similitude implies how individuals treat two things the equivalent as far as like and dislike things.

This strategy is very steady when contrasted with Client based collective categorizing due to the average or normal thing has significantly a larger number of evaluations than the normal client. In this way, an individual ranking cannot be affected to such an extent.

To compute comparability between two things, we examine the arrangement of things the objective client has ranked such that the objective thing I and after that chooses k most comparative things. Similitude among two things is determined by taking the evaluations of the clients who have ranked two things and from that point utilizing the cosine comparison stated below.

When we have the similitude between the things, the forecast is then registered by taking average of client's evaluations on these comparative things. The equation to compute ranking is fundamentally the same as the client based cooperative categorizing with the exception of among things rather than weights among clients. Furthermore, we utilize the present clients ranking for particular thing and for different things, in-stead of different clients ranking for the present things.

$$w_{ij} = \text{sim}(i, j) = \frac{\sum_{u \in U_i \cap U_j} \hat{r}_{ui} \hat{r}_{uj}}{\sqrt{\sum \hat{r}_{uj}} \sqrt{\sum \hat{r}_{uj}}}$$

Using rating algorithm in this way makes it computationally way better

$$s(i, u) = \mu_i + \frac{\sum_{j \in I_u} \left(r_{uj} - \mu_j\right) w_{ij}}{\sum_{j \in I_u} \left|w_{ij}\right|}$$

5 Experimental Results

The Fig. 1 gives the mean absolute error of recommendation based on the user and the items, whereas the product/tem based recommendation gives minimum mean absolute error compared to user based recommendation.

The Tables 1 and 2 shows the different similarities for user based recommendation for six users and the corresponding bar charts are shown in Figs. 2 and 3. It is comparatively much simpler as it is basically just the time it took two different algorithms to make predictions, lower is better.

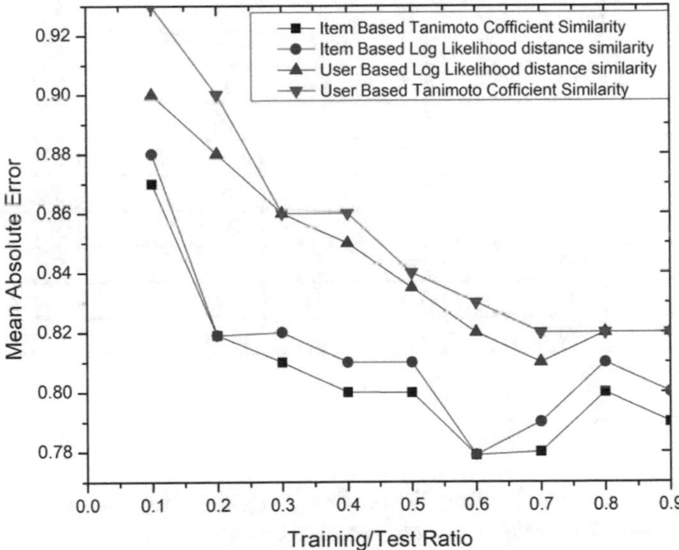

Fig. 1 Impact of training/test ratio on item-based and user based algorithms by using mean absolute error

Table 1 Average recommendation times for six user based similarities in milliseconds

User based similarities				
Euclidean distance	Log likelihood	Pearson correlation	Unentered cosine	Spearman correlation
547	579	515	469	532
547	610	547	531	547
547	640	515	500	562
507	625	548	547	531
563	640	547	516	595
516	626	515	516	609
516	656	531	516	578
516	656	500	500	563
547	625	532	532	578
516	672	532	500	594
532.2	**632.9**	**528.2**	**512.7**	**568.9**

Table 2 Average recommendation times for five item based similarities in milliseconds

Item based similarities				
Euclidean distance	Log likelihood	Pearson correlation	Tanimoto coefficient	Uncentered cosine
500	531	501	500	500
500	563	578	485	531
536	578	531	537	536
516	562	516	500	548
532	563	547	516	539
562	547	516	524	531
563	547	531	516	532
534	563	547	524	581
547	562	547	516	575
547	531	531	522	532
533.7	**554.7**	**534.5**	**511.6**	**540.5**

6 Conclusion

Recommender frameworks offer clients a few things that they may want to purchase from a business. This framework utilizes client databases for taking enhancement esteem for business. Recommender frameworks help client things that they might want to purchase from business. In like manner, these frameworks assist the commercial by happening more deals. Recommender frameworks are turning into a basic device in online business on the Web. New advances are required that can build

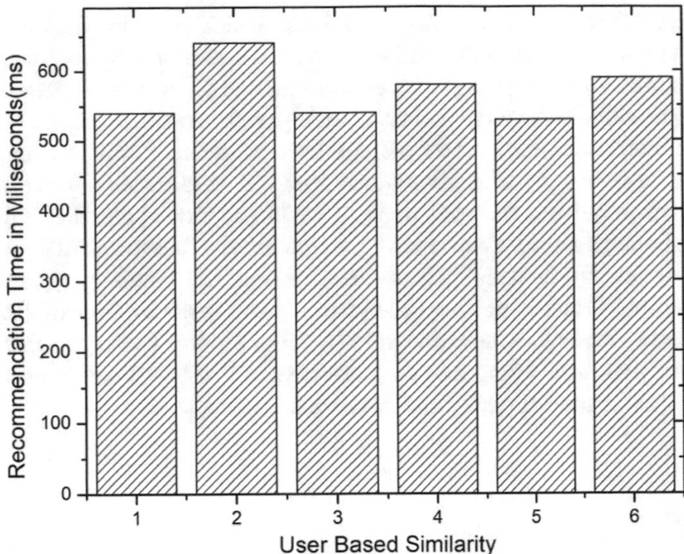

Fig. 2 Impact of item based algorithms based on recommendations times

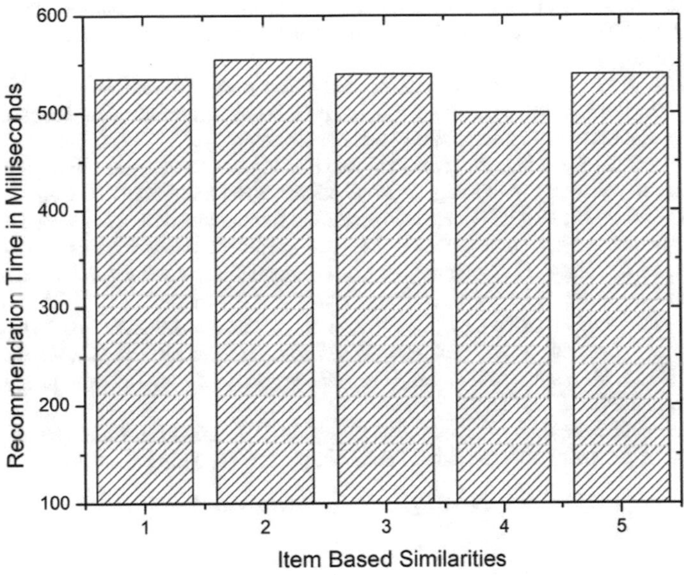

Fig. 3 Impact of item based algorithms based on recommendations times

up the adaptability of recommender frameworks which are being underlined by the immense volume of client information in current databases. Synergistic sifting is another approach to separating information that can choose from database. Community separating frameworks gather client's past information around a thing, for example, films, book, music, thoughts, feeling, and items. For prescribing the best thing, there are numerous calculations that depend on various methodologies. As indicated by Collaborative Filtering Systems, there is a common point that is foundation of similitude among clients and things. Communitarian based calculation stretches out to enormous informational indexes likewise bolster astounding suggestions. We look at User-based and Item-based calculations with various comparability file. By utilizing these calculations, and discovered that our proposed change, which is Item-Item based synergistic sifting through performed client based community oriented separating both computationally and as far as consistency.

References

1. Adomavicius, G., Tuzhilin, A.: Toward the next generation of recommender systems: a survey of the state-of-the-art and possible extensions. IEEE Trans. Knowl. Data Eng
2. Basu, C., Hirsh, H., Cohen, W.: Recommendation as classification: using social and content-based information in recommendation. In: Proceedings of the National Conference on Artificial Intelligence
3. Bell, R.M., Koren, Y., Volinsky, C.: All together now: a perspective on the Netflix prize. Chance
4. Billsus, D., Pazzani, M.J.: Learning collaborative information filters. In: Proceedings of the Fifteenth International Conference on Machine Learning
5. Aggarwal, C.C., Wolf, J.L., Wu, K.L., Yu, P.S: Horting hatches an egg: a new graph-theoretic. Approach to collaborative filtering. In: Proceedings of the ACM KDD'99 Conference (1999)
6. Basu, C., Hirsh, H., Cohen, W.: Recommendation as classification: using social and content-based. Information in recommendation. In: Recommender System Workshop '98 (1998)
7. Pazzani, M.J., Billsus, D.: Content-based recommendation systems. In: The Adaptive Web, pp. 325–341. Springer, Berlin, Heidelberg (2007)
8. Chang, P.C., Lin, C.H., Chen, M.H.: A hybrid course recommendation system by integrating collaborative filtering and artificial immune systems. Algorithms **9**(3), 47 (2016)
9. Lü, L., Medo, M., Yeung, C.H., Zhang, Y.C., Zhang, Z.K., Zhou, T.: Recommender systems. Phys. Rep. **519**(1), 1–49 (2012)
10. Amatriain, X., Jaimes, A., Oliver, N., Pujol, J.M.: Data mining methods for recommender systems. In: Recommender Systems Handbook, pp. 39–71. Springer, Boston, MA (2011)

Web Scraping for Unstructured Data Over Web

G. Naga Chandrika, Somula Ramasubbareddy, K. Govinda and E. Swetha

Abstract The need and significance for extracting information from the web is rising up with an increasing trend. Almost every day, we end up in a circumstance, where we need to extract information from the web. This is not always about finding new courses, but we also have to prone for reviews and data for providing a brief about them. Mostly, the issue is how we scrap the data and store it, but not on the strategy we use to perform it. The project done here is a proof of concept for how we could integrate it with php to make the updating by running python on the server. Here, we are going to use the python script to extract the info from a website and parse it to get the required information regarding our needs. Then, the data is send to the webserver hosted on the internet, the php running on the server will get the data from the python script. The data is been store on the mysql server using the php script and every time, we run the python, the data in the website changes.

Keywords Bellcore attack · CRT · Fault attacks · RSA · Embedded platforms

1 Introduction

Web scraping is a concept, which can be used to extract data from a website, which is been hosted online. Web scraping tools access the internet through the protocols (i.e., http/https) or through the web browsers which we commonly use. Most the data extraction is done by the software or the script, and it is an automated process. Hence, it can also be referred as web crawling. The actions done by the web scrapper can be considered as an act of copying the data, but the end product is in the required form, which we need.

G. Naga Chandrika · S. Ramasubbareddy
IT, VNRVJIET, Hyderabad, Telangana, India

K. Govinda
SCOPE, VIT University, Vellore, Tamilnadu, India

E. Swetha (✉)
SV College of Engineering, CSE, Tirupati, Andhra Pradesh, India
e-mail: swethaevakattu.11@gmail.com

V. Bhateja et al. (eds.), *Embedded Systems and Artificial Intelligence*,
Advances in Intelligent Systems and Computing 1076,
https://doi.org/10.1007/978-981-15-0947-6_81

Most of the web scraper fetches the html page of the website or download the html document. Once the download is completed, it will start the parsing and processing of the data. Therefore, webscraper is a vital component for scrapping as it helps in fetching, processing and storing. The data from the scrapper can be stored in the desired format (example .xml .pdf .txt .csv). Normally web scraper will extract a data to make some other form of data. An example would be scaping the data of name of the movie and its year of release into an excelsheet. It can also be used in data mining, web indexing, price comparison and weather monitoring. Most of these applications are data oriented and hence, it also requires processing of the data before making the required format (Fig. 1).

Website pages are build from the languages called mark-up languages (HTML and XHTML) and are having adequate data for the users in the text form. Most of the automated tools cannot use these information as they are intended for the humans. This is the reason behind the origin of web scraper. Most of the websites does not have a proper API for pulling the data out. Hence, we deploy web scraper. Only a few sites like Amazon, Twitter and Google provide API to the users with free of cost. Some web scrapers listen to the website for a particular change (example, stock market and weather monitoring).

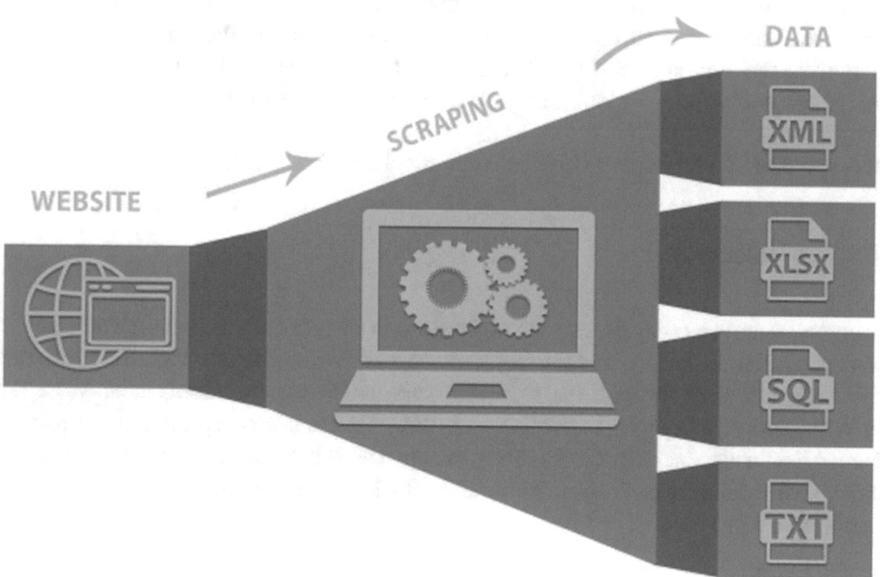

Fig. 1 Scrapping process

2 Literature Survey

In this paper, we find a tool that consequently gets, downloads and solidifies atmosphere information from an Internet database where the information are contained on numerous Site pages. The instrument is known as the Canadian Atmosphere Information Scratching Apparatus (CCDST) and was created to improve get to and streamline examination of atmosphere information from Canada's National Atmosphere Information and Data Chronicle (NCDIA). The CCDST deconstructs a URL for a specific atmosphere station in the NCDIA and after that iteratively changes the date parameters to download expansive volumes of information, expel singular record headers, and union information documents into one useful document. This computerized arrangement improves access to atmosphere information by considerably minimising the time expected to physically download information from numerous Site pages. Without the CCDST, the time engaged with physically downloading atmosphere information limits the research people from working on the climate trends. The device is coded as a Microsoft Excel macro and is accessible to specialists and understudies with free of cost. The primary idea and structure of the apparatus can be altered for other Web databases facilitating geophysical information [1].

Recommendation Frameworks have been around for over 10 years now. Picking what book to study next has dependably been an tedious for some people. Most students feel uncomfortable to always search for the books, instead of spending time in studying and getting a good book is always an question mark. In this paper, a model for an online customized book recommender framework which used various metrics for getting the preferred book and helps in filtering the unwanted based on the content. Recommendations can be done to different age, gender, sex and county people using the statistical parameters. Scratching data from the web and utilizing the data acquired from this procedure can be similarly helpful in making proposals [2].

There is a lot of meteorological and air quality information accessible on the web. Regularly, extraordinary sources give straying and useless repudiating information for the same land region and time. This makes the clients to go through it and find the reliable and the accurate data among them. It displays a novel information combination strategy that unions the information from various hotspots for a given zone and time, guaranteeing the best information quality. The strategy is an interesting blend of land-use regression systems, statistical air quality demonstrating and an outstanding information combination calculation. It demonstrates tests where a mixed temperature estimate predicts singular temperature figures from a few suppliers. Additionally, it exhibits that the nearby hourly NO_2 fixation can be assessed precisely with our combination technique while a more traditional extrapolation strategy may miss the mark. The strategy frames some portion of the online services PESCaDO, intended to provide customized natural data to clients [3].

RCrawler is a contributed R bundle for space based web scraping and content extraction. As the principal usage of a parallel web crawler in the R domain, RCrawler can extract, parse, store pages, remove substance, and create information that can be straightforwardly utilized for web content mining applications. It is additionally

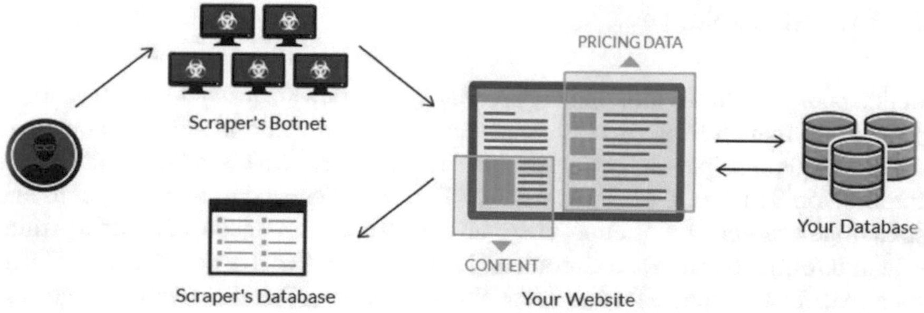

Fig. 2 Architecture

adaptable, and could be adjusted to different applications. The fundamental highlights of RCrawler are multi-threaded functionality, content extraction, and copy content recognition. The crawler has a highly improved framework, and can download a substantial number of pages with in a every second, while being robust against specific accidents and bug traps. In this paper, It portrays the outline and usefulness of RCrawler, and provide details regarding our experience of using it in a R domain, including optimizations that handle the constraints of R [4] (Fig. 2).

3 Proposed Method

We all know that python comes under open source license. Numerous libraries are available to perform the single action. Hence, we must choose the leading among them. I prefer using Beautiful Soup (python library) as it is easy to use and fun to work on. The other important modules are.

Urllib2: Data can be fetched by providing the URL as a parameter. It can also perform basic actions like authentication, redirections, cookies. kindly, refer the documentations for more information.

BeautifulSoup: It is an efficient tool for extracting or parsing the useful information for the website fosted online. We use this to fetch the html tags and to extract the data from it. In this project, we are using Beautiful Soup-4. kindly, refer the documentations for more information.

Beautiful Soup cannot fetch the website through http or https. Therefore, we use urllib2 with BS4 to do the work for us.

The other alternatives for BS4 are,

- mechanize
- scrapy
- scrapemark.

We are going to use the python script to extract the info from a website and parse it to get the required information regarding our needs. Then, the data is send to the

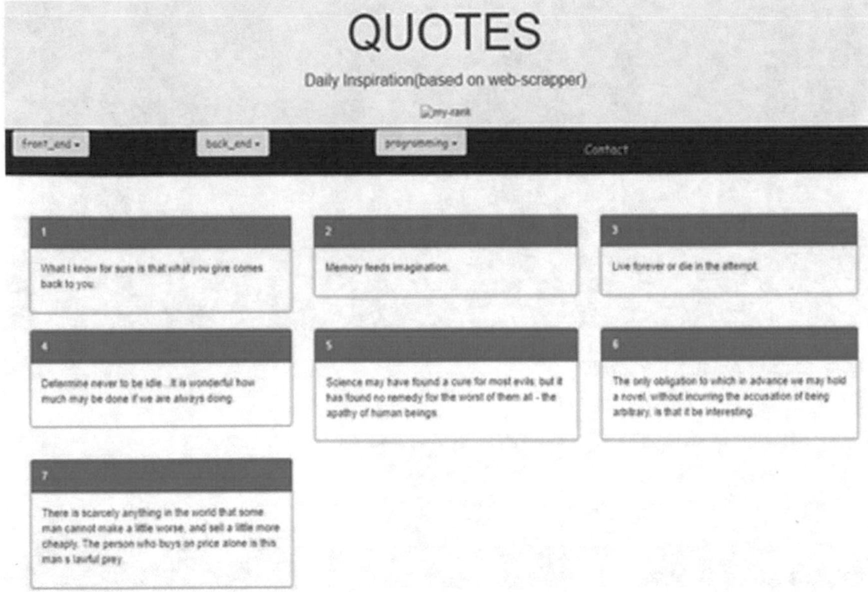

Fig. 3 Before running the script

webserver hosted on the internet by post request, the php running on the server will get the data from the python script. The data is been store on the mysql server using the php script and everytime, we run the python, the data in the website changes. The stored data will be displayed in the website created.

4 Experimental Results

The picture down shows that the picture of the website before and after the script runs. The updation happens whenever the script runs, the running process can be automated by hosting it on the cloud. The data can also be extracted from the website by setting the *limits to the number of content* (Figs. 3, 4 and 5).

5 Conclusion

We took a look at web scratching strategies utilizing "BeautifulSoup" and "urllib2" in Python and furthermore, how it is been applied to the database in the server utilizing php. Additionally, we looked at the basics of web scratching and how it tackles the issues. The vast majority of the utilization cases lie in the Web based business, Statistical surveying and weather monitoring. Since we're in the enormous

```
he quote is : Perhaps too much of everything is as bad as too little.
he status code is : 200
tatus Text : OK
ction :  Records inserted successfully....


he quote is : The most merciful thing in the world, I think, is the inability of the human mind to correlate all its contents.
he status code is : 200
tatus Text : OK
ction :  Records inserted successfully....


he quote is : I have noticed that the people who are late are often so much jollier than the people who have to wait for them.
he status code is : 200
tatus Text : OK
ction :  Records inserted successfully....


he quote is : I guess sometimes you have to lie to find the truth.
he status code is : 200
tatus Text : OK
ction :  Records inserted successfully....
```

Fig. 4 During running the script

Fig. 5 After running the script

information business where the vast majority of our answers rotate around web scraping and extraction. To say a couple of clear and important ones, there are some huge statistical surveying houses that straightforwardly manage the producer marks and might want to nearly monitor client assessments for particular high-esteem items. For this, they might want to extract item audits from different sources and have the capacity to break down it for processing. At that point there are other value examination engines which might want to total the item costs from around the web and run some investigation over it. There are people who want to build an engine for a particular use and need as much more publically existing data. So basically, there's no restriction to how you can utilize the available data in the online to build our system efficiently because we cannot do much with the limited data sets, we have. A few organizations utilize it to enhance their client encounters while some utilization it for look into and different investigations that prompts more current business advancements.

References

1. https://www.sciencedirect.com/science/article/pii/S009830041400243X-CCDST:%20A%20free%20Canadian%20climate%20data%20scraping%20tool
2. https://ieeexplore.ieee.org/document/7012952?arnumber=7012952-Web-basedpersonalized%20hybrid%20book%20recommendation%20system
3. https://www.sciencedirect.com/science/article/pii/S1364815214003478-Fusion%20of%20meteorological%20and%20air%20quality%20data%20extracted%20from%20the%20web%20for%20personalized%20environmental%20information%20services
4. https://www.sciencedirect.com/science/article/pii/S2352711017300110?via%3Dihub-RCrawler:An%20R%20package%20for%20parallel%20web%20crawling%20and%20scraping

'Customer Satisfaction', Loyalty and 'Adoption' of E-Banking Technology in Mauritius

Vani Ramesh, Vishal Chandr Jaunky, Randhir Roopchund and Heemlesh Sigh Oodit

Abstract Small Island, Mauritius is adopting drastic changes with reference to technology. The prime focus of extant study is exploratory in nature and trying to examine the relationship between the factors that are inspiring Mauritian consumers and customers to go with technology and technological advances that are user friendly. E-banking services are one among them. Also, try to test which of the dimensions has the strongest potential in influencing 'e-customer'. This study focuses mainly on perception of the customer, e-banking adoption motif, 'e-banking satisfaction' and 'quality' of 'e-banking services' in Mauritius. Primary data is collected with the help of drop-off survey and were statistically analysed with the help of Structural Equation Modelling (SEM) and ordered probit and logit regressions. The empirical evidences of this study reveals, reliability, responsiveness, secured transaction, comfort, efficiency, dependence and ease-of-use has a striking impact on the 'customer satisfaction' and e-loyalty. With these findings, for Mauritian content, 'electronic banking' would be indispensable 'banking service' that can be, provided, well executed to improve 'e-customer satisfaction' and 'e-customer loyalty' to sustain goodwill of the e-banking customers. Relative importance and service quality of e-banking technology at Maurtius would support the banking industry to motivate the customers.

Keywords E-customer · Goodwill · Ease of use · Significance · Efficiency

V. Ramesh (✉)
REVA University, Bengaluru, India
e-mail: drvanisarada@gmail.com

V. C. Jaunky
LTU, Southfield, Sweden
e-mail: vishal.jaunky@ltu.sc

R. Roopchund
UDM, Roches-Brunes, Mauritius
e-mail: rroopchund@udm.ac.mu

H. S. Oodit
UOM, Moka, Mauritius
e-mail: hoodit@gmail.com

V. Bhateja et al. (eds.), *Embedded Systems and Artificial Intelligence*,
Advances in Intelligent Systems and Computing 1076,
https://doi.org/10.1007/978-981-15-0947-6_82

1 Introduction

Information Technology has been very successful in construction of innumerable characteristics of human life easier for recent societies [1]. Customer Satisfaction is one of the main responsibilities of any business or service. Providing quality service, creating a trust among the customers and captivating their satisfaction and loyalty is a fundamental duty of any business or service industry. Same as applicable for 'Electronic-Banking Services' (**e-services**) [2].

'e-banking' services are basically count on the information exchange among the bank employees and bank customers, through technology adoption rather traditional modes of 'head-on' interaction [3]. Mauritian banking sector has recognized 'e-banking' technology and tangled enthusiastically in improving its services and business operations constantly. Since then, quality of service is available 24/7, with secured transaction, customer's retention and customer satisfaction is very much practical. Also, the cost effective of their transactions with reference to time, energy and value addition added to strengthen their e-loyalty.

There are ample number of studies that have investigated on e-banking technology adoption and service quality. Considering few of them, those are directly related to the present study, which are measured for 'internet-banking service quality' and 'internet-banking customer satisfaction', the study being first and prime most in Mauritius tries to identify the gap in literature. Similarly, also the aims at filling the gap in existing literature on the area of research, by addressing the 'e-banking adoption', testing empirically the considering all relevant determinants for a small and developing Island, like Mauritius. To our knowledge, so far there is no evidence of such study that has been undertaken earlier. Perhaps, the core findings of the study highlights, importantance of 'Internet-Banking Service Quality' on 'Internet-Banking Customer Satisfaction', and reliability, quality of service and perception of 'e-banking customers'.

The structure of the paper is organized as, introduction followed by literature review, methodology, analysis and findings, interpretation and discussion, finally conclusion which recaps the study and presents relevant limitation and commendations for imminent research.

2 Literture Review

2.1 e-banking

The first ATM (Automated Teller Machine) was invented by Shephered-Barron, a Scottish, on June 27th, 1967 at Barclays bank in Enfiled, North London. The first ATM was installed on July 200th, 2017 at Mauritius Commercial Bank Ltd (MCB) by BIRGER. Though it is recent phenomenon in the country, recently it has been taken momentum in diversification skills and experiencing a rapid economic growth.

Mauritian banking sector plays a key role in contributing for GDP, providing more facilities for the transactions through internet and encouraging the flow of foreign currency.

As on now, the largest networking banks such as, Mauritius Commercial Bank (MCB), Hong Kong Shanghai Bank (HSBC) and the State Bank of Mauritus Ltd are offering the internet banking services. The main services include, inter account transfer, fund transfer to credit card account, transfers to other accounts, SWIFT payments, foreign transfers… etc. These e-banking services are spread among the customers so rapidly due to technological advances [4]. According to Lustsik [5], the banking services with the help of technology assistance has numerous option for providing good quality services. Due to drastic changes in lifestyle and vast exposure to global networks, the consumers and customers likes to operate their banking transactions at their convenience to save time and energy in cost effective way. The following Table 1 summarizes the most important definitions of 'Internet-banking' (e-banking) services.

Shaikh and Karjaluot [6] has thrown some insights on e-banking with reference to mobile applications. Followed by Tieto [7] and Arcand et al. [8], defining e-banking as a technology which is user friendly and unique in offering customers a quality service.

Table 1 Definitions of "e-banking services"

Researcher	Definition
Jane et al. (2004)	Provide banking services to customers via the Internet starting from the ATM service and the end of the transfer of funds electronically
Riedl et al. (2009)	Provide banking services to customers through the Internet and self-service centers
Ahmad and Al Zubi (2011)	The bank gives the opportunity to its customers to gain access to their accounts and the implementation of their transactions and purchase products directly via the Internet
Ala' Eddin et al. (2011)	A portal through which customers can use all kinds of banking services, from paying bills and end of the financial investments
Asli (2011)	System that enables customers to access their accounts and obtain general information about the services offered by the bank via the Internet
Jazani et al. (2012)	Banking services to provide a high degree of convenience that enables customers to access their accounts at time and place that suits them
Aliyu (2012)	Using technology by the ban k by to connect with customers, and presentation of financial services from its path
Salhieh et al. (2013)	Provide banking services to customers and allow them to carry out financial operations of (withdrawals and deposits, transfer funds, pay bills, request loans, make sure the account balances…. and others) 24 h through the Internet

Source Authors Computation

Maurtius is still developing on updating the technology, yet inadequate to accessibility of internet and mobile technology comparatively, may be due to deliberate progress of Information Technology and infrastructural facilities that are needed for the safe and smooth functioning of e-commerce and e-services. Having considering these hurdle, this study outlines the concept as the e-banking adoption, satisfaction of customer and e-loyalty of e-customer.

2.2 *The Concept-'Customer Satisfaction'*

The concept, 'Customer satisfaction', in broader terms, the process of concluding the purchase with post purchase phenomena, right from the attitude, purchase intention, repeat purchase and loyalty to the brand [9]. As rightly pointed out by Churchill and Surprenanat [10], the post purchase satisfaction that the customer derives out of the product or service shows a vital part in creating brand doppelgänger and brand loyalty. Generally, sense of gratification ascends once customers starts comparing their insight about the product or service, that they have actually perceived and developed the taste and preference towards it, with same attitude and expectation fro them [11]. Many definitions define 'customer satisfaction' in different ways, but the generally accepted and understood by a lay man, in simple terms is given by Tse and Wilton [12]. According to their definition, "The term, customer satisfaction is a understood as response to the assessment of the apparent variation among expectations and concluding consequence of subsequent consumption". Kotler and Kelle [13], describes the satisfaction as "A response on the post purchase valuation and expectations from prior-purchasing stage".

There are some diverse opinion from some studies. Few researchers pragmatic about the effect on the experience in purchasing and consuming phase. This can also have a significant consequence on the customers decisions towards satisfaction [14]. So, the term 'customer satisfaction' can be defined as a is a sensation of liking or disliking that the consumer/customer after utilising the product or service of their anticipation [15].

Concluding, for the purpose of study, it remains tacit, concept 'customer satisfaction' is a perception and insolence of 'e-customer'/'e-consumer' on 'e-banking technology' and quality of 'e-banking service' at Mauritius.

2.3 *Affiliation of Internet-Banking (e-banking) Quality Service with e-customer Satisfaction*

Basic intention of this study is to understand relationship among 'Quality' of e-banking technology, 'services' and 'satisfaction' of customer. There are number of variables that are tested by many studies to examine, which are important motifs behind approval, acceptance of e-banking technology. Table 2 summarises some of components that are tested by different studies.

Referring to Parasuraman et al. [16], the "service quality of e-banking" and "satisfaction of e-customer" seems very robustic. During 2018 Asiyanbi and Ishola [9], in their empirical evidence, points out, "The unit of satisfaction of customers in the banking sector surges while consuming services of e-banking". According to our view, and the relevant literature review reveals that, the important dimensions to test the e-customer satisfaction and loyalty can be well situated to cluster further down as, an 'efficiency', 'reliability', 'privacy', 'security', 'ease of use', 'responsiveness' and 'good communication' [17].

3 Methodology

Extensive literature review recommends that, secured transaction, reliability, privacy, comfort, 'responsiveness' and 'communication' networks remain very significant dimensions need to be tested for the country. The concept, 'customer satisfaction', 'e-loyalty' with 'e-banking service quality' and 'satisfaction' is tested with the help of empirical evidence in Mauritian context. A survey was conducted at the major commercial points of the country to collect the feed back with the help of well structured questionnaire, consisting '5-point' likert scale [18]. Convenience sampling techniques is used to data collection. The reliant variable is 'customer satisfaction', loyalty, adoption of e-banking in Mauritius. The data is analyzed with the help of AMOS (21) for structural equation modelling and soft ware STATA for ordered logit and probit regressions.

The general hypothesis of the study projected a constructive and substantial rapport among service quality of e-banking, e-customer satisfaction and loyalty, as explained below:

- H (1): The 'Efficiency' of 'Internet-banking services' at Maurititus has constructive outcome on 'e-customer satisfaction'.
- H (2): The 'Reliability' of 'Internet-banking services' at Mauritius has constructive outcome on 'customer satisfaction'.
- H (3): The 'secured transaction' of 'Internet-banking services' at Mauritius has positive effect on satisfaction of customer.
- H (4): The 'responsiveness' and 'communication' of 'e-banking' services at Mauritius has constructive outcome on 'customer satisfaction'.

4 Analysis and Findings

4.1 Descriptive Statistics

The sample is normally distributed among male respondents and female respondents across different locations of Mauritius. Male respondents constitute 62% and female respondents constitute 38%. Most of the respondents are below middle age, between 25 and 40 years of age group, indicating these groups are more technology usable and having affirmative attitude towards 'e-banking services'. Majority of them are working professionals, which shows the technology is very convenient ant comfortable for them to have safe transactions 24/7. Also, majority of them are earning relatively good pay. About 46% of the respondents are earning more than 30,000 MRU. The Tables 2 and 3 explains the logit and probit regressions along with e-satisfaction and e-loyalty.

Linear Modelling Technique ('Ordinary Least Squares', OLS) is computed for testing the variables of e-banking technology. The findings of this analysis demonstrates that, there is an encouraging influence of e-satisfaction of consumer which is a dependent variable, by the independent variables, such as quality of e-banking services, reliability, secured transaction, responsiveness and efficiency. The results shows that, most of the variables are significant at 1, 5 and 10% which is impacting the e-satisfaction and e-loyalty. Also, it is observed that pseudo R^2 is 0–1, which is quite acceptable for the study.

4.2 Structural Equation Modelling (SEM)-Measurement Model

SEM, a measurement model is used to evaluate the relationship among the latent construct and dependent variables. Factor Analysis (FA) and Confirmatory Factor Analysis (CFA) is computed, finally path diagram is drawn. This has been demonstrated in a path diagram, Fig. 1.

The measurement model shows standard model fit indices and found adequate. These are explained in a Table 4.

There are exogenous variables (quality of e-banking services, reliability, secured transaction, efficiency) and one endogenous Moreover, t-test for specified value of all the independent and dependent variables are at significate level which is displayed in Table 6. Therefore, and e-satisfaction of the customer is acceptable.

Table 2 Descriptive analysis of variables

Variables	Mean	Standard Deviation	Minimum	Maximum
Dependent Variable:	332.24865	0.8423639	1	5
Consumer Satisfaction with Internet banking in Mauritius.				
Age	0.951351	9.192554	18	63
Gender:				
Male				
Female				
Residential Area:				
Urban	0.389189	0.4888895	0	1
Rural				
Educational Achievement:				
Primary				
Secondary	0.6108108	0.4888895	0	1
Diploma				
Undergraduate	0.189189	0.392721	0	1
Others	0.459459	0.499706	0	1
Income:				
Level 1: Below Rs 10000				
Level 2: Rs 10000-Rs 20000	0.248648	0.499706	0	1
Level 3: Rs 20000- Rs 30000	0.102702	0.304394	0	1
Level 4: Above Rs 30000	0.183783	0.388358	0	1
Family Size	0.194594	0.396962	0	1
Marital Status:	3.627027	1.357877	0	8
Single				
Married				
Divorced	0.421621	0.495158	0	1
Others	0.016216	0.126648	0	1
e-trust	0.005405	0.073521	0	1
Reliability				
Efficiency	3.88468	0.672016	1	5
Comfort	3.81982	0.723128	1	5
Confidence	3.68108	0.654060	1	5
Secure transaction	3.76576	0.760056	1	5
Dependence	3.11351	0.905497	1	8.75
	3.33693	0.795510	1	5

Source Authors own computation

Table 3 Ordered logit regression: e-trust

Variables	Ordered logit	Outcome 1	Outcome 2	Outcome 3	Outcome 4	Outcome 5
Age	−0.0241412 (0.0267292)	0.000219 (0.0002831)	0.0011146 (0.0012498)	0.0017443 (0.0019476)	−0.0000622 (0.0004046)	−0.0030157 (0.0033113)
Gender: (baseline female) Male Educational achievement	−0.3451426 (0.3703387) −1.089145 (0.6390763)*	0.0031304 (0.0037981) 0.0098783 (0.0081134)	0.0159359 (0.0174679) 0.0502881 (0.032009)	0.0249374 (0.0269976) 0.0786934 (0.047082)*	−0.0008887 (0.0057248) −0.0028043 (0.017925)	−0.043115 (0.0461423) −0.1360555 (0.0788037)*
(Baseline secondary)	0.874405 (−0.6095082)	0.0079307 (0.0072847)	0.0403731 (0.0297689)	0.0631779 (0.0447184)	−0.0022514 (0.0144572)	−0.1092303 (0.0751946)
Diploma						
Undergraduate others	−0.1626591 (0.6347552)	−0.0026655 (0.0035072)	0.0075103 (0.0293919)	0.0117525 (0.0459039)	−0.0004188 (0.0031178)	−0.0203193 (0.0793048)
Residential area						
(Baseline rural) Urban	0.2938893 (0.3367441)	0.0050501 (0.0068247)*	−0.0135695 (0.0158642)	−0.0212342 (0.0243755)	0.0007567 (0.0048892)	0.0367125 (0.04193)
Monthly income (Baseline above Rs. 30,000)	−0.5568011 (0.6743996)	0.010865 (0.0081415)	0.0257087 (0.031715)	0.0402302 (0.0490637	−0.0014336 (0.0093253)	−0.0695553 (0.0839034)
Below Rs. 10000	−1.197936 (0.56877)*	0.0085974 (0.0065394)	0.0553112 (0.0296822)**	0.0865538 (0.0425089)*	−0.0030844 (0.0196998)	−0.1496456 (0.069832)
Rs. 10,000						
Rs. 20,000	−0.9479106	0.0008642	0.043767	0.0684888	−0.0024407	−0.1184125
	(0.48105)*	(0.0011761)	(0.0251075)	(0.0353908)*	(0.0155643)	(0.0594076)*
Rs. 20,000						
Rs. 30,000	−0.0952824	0.0026659	0.0043994	0.0068844	−0.0002453	−0.0119026
	(0.1218743)	(0.003989)	(0.0057626)	(0.0088914)	(0.0015997)	(0.015191)

(continued)

Table 3 (continued)

Variables	Ordered logit	Outcome 1	Outcome 2	Outcome 3	Outcome 4	Outcome 5
Family size	+					
	−0.2939363	−0.1459896	0.0135716	0.0212376	−0.0007568	−0.0367184
Civil status	(0.4235663)	(8.116549)	(0.0198478)	(0.0308718)	(0.0048959)	(0.0530136)
(Baseline single)						
Married	16.09623	−0.002118	−0.7431962	−1.162992	0.0414442	2.010733
Divorce	(894.8483)	(0.0190263)	(41.31742)	(64.65523)	(2.319066)	(111.784)
Others						
	0.2335176	−0.0000519	−0.010782	−0.0168722	0.0006013	0.0291709
	(2.088246)	(0.0029729)	(0.0954102)	(0.1508215)	(0.006504)	(0.2608949)
e-trust:	0.0057199	−0.0052278	−0.0002641	−0.0004133	0.0000147	0.0007145
Reliability	(0.3278752)	(0.0042055)	(0.0151402)	(0.0236886)	(0.000847)	(0.0409595)
Efficiency	0.5763975	−0.0106083	−0.0266135	−0.0416461	0.0014841	0.0720033
Comfort	(0.3316944)	(0.0070772)	(0.0163688)*	(0.0248006)*	(0.0094878)	(0.0408177)*
Confidence	1.169632	−0.0025635	−0.0540043	−0.0845088	0.0030115	0.1461099
Secure	(0.4259147)	(0.003289)	(0.0232697) *	(0.0325553)*	(0.0190801)	(0.0529318)*
Transaction	0.2826459	0.0016727	−0.0130503	−0.0204219	0.0007278	0.035308
	(0.3221561) +	(0.0021545)	(0.0150883)	(0.023455)	(0.0046709)	(0.0402463)
Dependence	0.1844275	−0.0045389	0.0085154	0.0133253	−0.0004749	−0.0230386
	(0.2127307)	(0.0036447)	(0.01012)	(0.0154903)	(0.003109)	(0.0263928)
	0.5004434	0.0221065	−0.0231065	−0.0361582	0.0012885	0.0625151
/cut1	(0.2778108)*	(0.0141747)	(0.0141747	(0.0207767)	(0.0082942)	(0.0339362)*

(continued)

Table 3 (continued)

Variables	Ordered logit	Outcome 1	Outcome 2	Outcome 3	Outcome 4	Outcome 5
	0.5712443					
/cut2|	(1.806576)					
	3.13675					
/cut3|	(1.74242)					
	4.602404					
/cut4|	(1.745298)					
	8.03364					
	(1.805758)					

Source Authors own computation
Note: *, + and # denote 1%, 5% and 10% significance level respectively

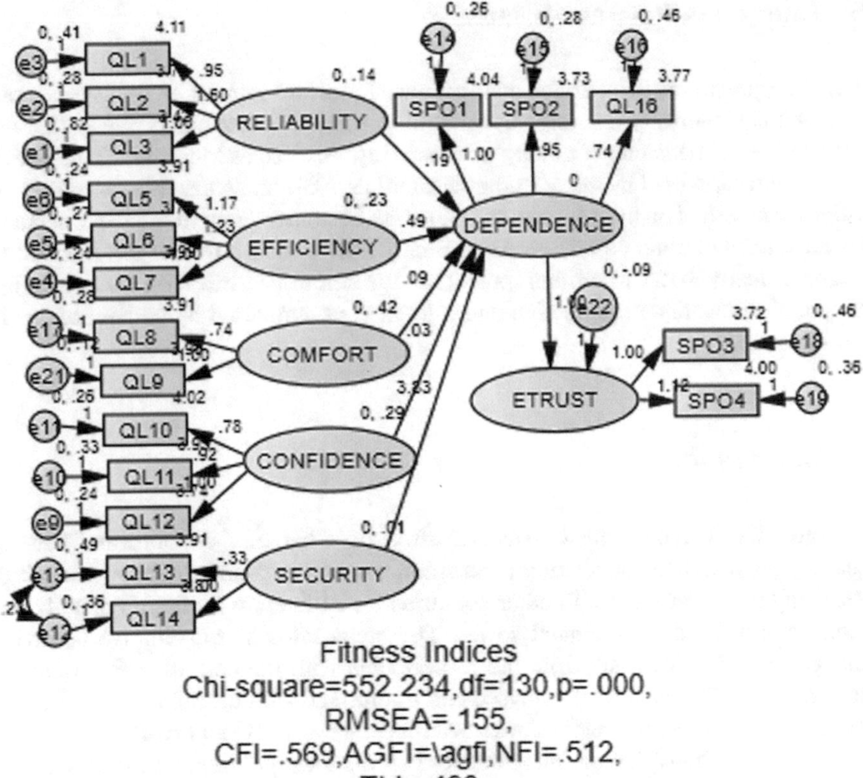

Fig. 1 Path Diagram-E-satisfaction and e-loyalty. *Source* Author's computation; *Note* ** denotes significant at 1% level

Table 4 Modelfit indices (*t*-test and *p* value)

Statement on Environmental Perceptions	Mean	SD	*t* value	*P* value
Dependency	3.90	1.07	16.776	< 0.001**
Security	4.10	1.03	21.291	< 0.001**
Confidence	3.99	1.14	17.472	< 0.001**
Comfort	3.90	1.21	14.878	< 0.001**
Reliability	3.80	1–20	13.878	< 0.001**

Source Author's computation
** denotes significant at 1% level

5 Interpretation and Discussion

The study identifies the 'service quality' has significant rapport on e-customer satisfaction and e-loyalty. Among all, reliability, ease of use, secured transaction are the sturdiest dimensions of quality of e-banking services at Mauritius. This is evidenced and supported by many studies earlier [19]. The study also identifies there is a significant effect on the e-loyalty of Mauritian customers with reference to secured transactions and good communication. Finally, the dimensions that are selected to test empirically for the Mauritian context has the noteworthy effect on the 'e-banking adoption', e-customer satisfaction and e-loyalty, as supported by earlier studies in the literature.

6 Conclusion

The present research intendeds to test the effect of 'e-banking' adoption and 'service quality' on customer satisfaction (e-satisfaction) and customer loyalty (e-loyalty). Referring to earlier studies in other countries (ref. literature section), no study has been done at Mauritian context so far. The study adopted quantitative approach and gathered the feedback from the respondents with the help of well structured questionnaire. The data were analysed and interpreted with the help of both STATA (ordered logit and probit regressions of e-trust and e-loyalty) and AMOS (SEM and path diagram). Study findings recommends that the set hypotheses for this research is satisfactorily reinforced for empirical evidence (primary data). The key impact of this study is, all the selected variables for testing, such reliability, secured transaction, ease of use, service quality, good communication are the main interpreter of e-satisfaction and e-loyalty.

Though the e-banking services are taking a shape and trying to stabilize, comparatively, it is still developing and need to upgrade the technology and infrastructure in Mauritius. The country need to upgrade the infrastructural facilities for the banks to perform better and provide quality service. One among them may be a good governing with reference to strict law and security for the customer to have more trust on e-technology.

The study recommends for further study, based on the limitations encountered would be, most importantly, since the perception and understanding differs from person to person, deep investigation need to be done in testing socio-cultural, economic, and environmental contexts of different parts of the country carefully.

Acknowledgements Sincere thanks gratitude goes to Dr. M. Dhanamjaya, Registrar, REVA University, Dr. Ramesh, Director Planning, REVA University and Dr. S. Ramesh, Dean and Director, MCC.

References

1. Rust, R.T., Oliver, R.W.: The death of advertising. J. Advertising **23**(4), 71–77 (1994)
2. Stone, G., Joseph, M.: An empirical evaluation of US bank customer perceptions of the impact of technology on service delivery in the banking sector. Int. J. Retail Distrib. Manage. **31**, 190–202 (2003)
3. Lakhtaria, K.I. et al.: The impact of the new Web 2.0 technologies in communication, development, and revolutions of societies. J. Adv. Inf. Technol. **2**, 204–216 (2011)
4. Khadem, P., Mousavi, S.: Effects of self-service technology on customer value and customer readiness: the case of banking industry. Manag. Sci. Lett. **3**, 2107–2112 (2013)
5. Lustsik, O.: Can e-banking services be profit-able? (University of Tartu Economics and Business Adminis-tration Working Paper No.30-2004) (2004, November). https://doi.org/10.2139/ssrn.612762
6. Shaikh, A.A., Karjaluoto H.: Mobile banking adoption. A literature review. Telematics Inform. **32**(1) 129–42 (2015)
7. Tieto.: History of Banking. http://www.tieto.com/sites/default/files/atoms/files/26nov_history_of_banking-reddad_v3.pdf (2016). Accessed 28 Feb 2017
8. Arcand, M., Prom Tep, S., Brun, I., Rajaobelina, L.: Mobile banking service quality and customer relationships. Int. J. Bank Mark. **35**(7), 1068–1089 (2017)
9. Asiyanbi, H., Ishola, A.: E-Banking services impact and customer satisfaction in selected bank branches in Ibadan metropolis, Oyo state, Nigeria. Accounting **4**(4), 153–160 (2018)
10. Churchill, G.A., Surprenant, C.: An investigation into the determinants of customer satisfaction. J. Mark. Res. **19**, 491–504 (1982)
11. Oliver, R.L.: Measurment and evaluation of satisfaction processes in retail settings. J. Retail. **57**(3), 25–48 (1981)
12. Tse, D.K., Wilton, P.C.: Models of consumer satisfaction formation: an extension. J. Mark. Res. **25**, 204–212 (1988)
13. Kotler, P., Keller, K.: Marketing Management, 14th edn. Pearson, New York, NY (2011)
14. Koschate, N., et al.: The role of cognition and affect in the formation of customer satisfaction: a dynamic perspective. J. Mark. **70**(3), 21–31 (2006)
15. Keller, K.L., Lehmann, D.R.: Brands and branding: research findings and future priorities. Mark. Sci. **25**, 740–759 (2006)
16. Parasuraman, et al.: SERVQUAL: a multiple-item scale for measuring consumer perceptions of service quality. Retail. Crit. Concepts **64**(1), 140–161 (2002)
17. Tan, M., Teo, T.S.H.: Factors influencing the adoption of internet banking. J. AIS **1**(1) (2017)
18. Hunain, M., et al.: The impact of e-banking on customer satisfaction: evidence from banking sector of Pakistan. J. Bus. Administration Res. **5**(2), 27–40 (2016)
19. Vanijikovan.: Customer acceptance of online banking: an extension of the technology acceptance model. Internet Res. **14**(3), 224–235 (2017)

Factor's Persuading 'Online Shopping' Behaviour in Mauritius: Towards Structural Equation Modelling

Vani Ramesh, Vishal C. Jaunky and Christopher Lafleur

Abstract Recent years, the idea of shopping online has become a necessity than a luxury. The singularities of online shopping are budding considerably. The players of this inclination show that there is still a huge possible demand for these shopping trends. There are many contributing factors that are influencing these behaviours. This study intends to scrutinize the contributing factors that are persuading the customers and consumers to opt for online shopping in Mauritius. The sample size is from the respondents having exposure to online shopping or e-stores. The primary data collected through survey with the help of well-designed questionnaire (5-point Likert scale), with SERVQUAL, CDMM and OSAM models. The data is analysed to test the hypothesis and model fit using AMOS. From the findings and conclusion of this study, it is understood that there is a significant influence of personality traits (SERVQUAL), Internet knowledge, purchase intention and experience, followed by post-purchase satisfaction (OSAM, CDMM). The contribution of this study for corporate academics and existing literature is very stimulating for further research, being first and pioneer in Mauritius.

Keywords Trend · Information technology · Online shopping · Loyalty · Personality

---

V. Ramesh (✉)
REVA University, Bengaluru, India
e-mail: sarada889@yahoo.in

V. C. Jaunky
LTU, Luleå, Sweden
e-mail: vishal.jaunky@ltu.se

C. Lafleur
University of Mauritius, Moka, Mauritius
e-mail: chrislafleur1234@yahoo.com

V. Bhateja et al. (eds.), *Embedded Systems and Artificial Intelligence*,
Advances in Intelligent Systems and Computing 1076,
https://doi.org/10.1007/978-981-15-0947-6_83

1 Introduction

The term 'Online Shopping' (e-shopping) in 'Digital Age' is a 'i-Generation' or 'i-Gen', 'Post-Millennials', where *technology has permitted* electronic commerce (e-commerce) to have business dealings and providing information via 'computer networks' [1]. More than 50% of the business communications are now tangled in or supporter of three elementary 'e-commerce applications, i.e. e-commerce business-to-consumer, business-to-business and consumer-to-consumer' [2].

As rightly said by Lee et al. [3], online shopping behaviour, online purchase intention, perceived usefulness and subjective norm have an important direct affiliation on re-purchase. The advances in technology have stimulated; e-shopping convention is no longer restricted to schmoozing for advertising, also, for sales and marketing led to the growth of online shopping with comfort. Further, 50% of Mauritians are habituated to Internet and social networking sites [4]. Bearing in mind these changes, the present study antedates to examine the model that will govern the factors persuading consumer online shopping choices. The outcome of this study is anticipated to be beneficial in serving e-commerce's sites owners and business entities associated with e-commerce and e-trade in Mauritius.

1.1 Theories Adopted

1.1.1 The Consumer Decision-Making Model (CDMM)

"Nicosia Model" (CDMM), projected by Nicosia [5], exhibits how consumer purchase is very complex process. Right from need recognition to post-purchase behaviour, the model demonstrates the online shopping behaviour [6]. Considering positive point, the model aims at inspiring the users and guides the marketers in designing marketing strategies at each stage of online shopping decision-making process. Also, the model deliberates the element of customers' devotion in searching the information, assessing different alternatives that are available until they choose the end product/service to mollify their expectations.

1.1.2 Online Shopping Acceptance Model (OSAM)

According to OSAM, though positive relationship exists among usage of Internet and e-shopping intention [7], in some of the recent studies, the relationship was not found to be significant [8]. This study adopts and examines this model with the help of empirical evidence, whether it has been accepted and what is the acceptance level in Mauritian context.

1.1.3 The Technology Acceptance Model (TAM)

TAM is adopted to test the reasoned action [9] and planned behaviour of the online customers [10]. Since technology acceptance model is used to understand the user behaviour, the same has been considered to test the online shopping behaviour of Mauritians [11].

1.1.4 Freud's Theory of Motivation

People behaviour is very unpredictable, and this theory is tested to understand the online consumer behaviour of Mauritians [12].

2 Based on the Review of Previous Studies, the Following Hypothesis Was Set for the Present Study

Since the determination is to discover and to test relevant factors influencing 'Online Shopping Technology' adoption and 'Online Shopping Satisfaction' of Mauritians. The study also extends in evaluating the factors influencing to motivating in accepting online shopping which leads to 'e-loyalty' and 'e-trust'.

Hypothesis I Personality traits significantly influence online shopping intentions in Mauritius.

Hypothesis II Online shopping orientation has significant positive impact on online shopping intension in Mauritius.

Hypothesis III Information available has significant positive impact on online shopping motivation.

Hypothesis IV Online shopping satisfaction has significant influence on e-loyalty of Mauritian consumers.

2.1 Methodology

Demographic, psychological, normative beliefs, purchase decision, satisfaction, e-trust and e-loyalty are tested. The questionnaire was constructed codified into 6 sections are distributed to online shoppers across the island, administered as a drop-off survey. The response rate was 87.3% (443 out of 500). The data is annualized, 'Path analysis', 'multiple regressions' are computed. The results show that the Mauritian e-consumers/e-customers are really fascinated to accept online shopping technology.

3 Literature Review

Online buying intentions are directly influenced by trust and consumer attitudes [13]. With the practical direction, expediency, worth, wide product choice, and earnings also have robust [14]. As rightly whispered by Shergill and Chen [15] in his investigative factor analysis, it is initiated that the 'site design', [16] 'Reliability of online sites', 'Service rendered by the sites' and 'safety of online sites' are leading in influencing online consumer perceptions. Similarly, Heijden et al. [17] identified online purchase intentions with the help of 'Technology-oriented Perspective' and 'Trust-oriented Perspective'.

3.1 Evolution of Online Shopping

During late 1981, Thomson Holidays invented online shopping to release a first product, the world's first direct business-to-business (B2B), where the buyers is connected interactively with the seller's electronic device directly without third party's intervention. There is a theoretical evidence; in the 1980s, the first B2C concept was introduced by the pilots to pay for the social security funding and used to investigate the potential for IT to reduce social disadvantage by offering in the home services to vulnerable and physically challenged citizens. Amazon started eBay bidding in the year 1995; 2015 is given increasingly boost for online shopping with mobile shopping 'staggering $1.55 trillion" in America and later expanded across the globe. China that avowals the major economy, worth $562.66 billion.

3.1.1 Evolution of Online Shopping—Mauritius

Mauritius has witnessed remarkable growth in information technology (IT)-based business activities. So far, there is no evidence in the literature that any study has focussed on examining the factors that are influencing the adoption of online shopping by the customers and consumers. Moreover, from the point of manufacturers and sellers, this innovation has hiked the turnovers with huge margins, since there is no much overhead costs of showroom rents, labour wages, and others. The present study looks into possible motivating factors for Mauritian context and swotted that there are over 280,000 online customers in Mauritius buying products and services [18]. 'We are in a new era where every visitor can enjoy a product or service, 'lepoint.mu' or 'tantebazar.com'. Making a 'bazaar' in two plays, Secured payments? Are some of the active online shopping portals that are playing very vital role in Mauritius. E-secure with SBML and MCB offers secured online transactions. Winners launched online purchase during 2012 and also one of the online shopping innovations in Mauritius.

Table 1 KMO and Bartlett's test

Kaiser–Meyer–Olkin measure of sampling adequacy		0.868
Bartlett's test of sphericity	Approx. Chi-Square	132.621
	df	84
	Sig.	0.000

Source Authors Computation

4 Analysis and Discussions

To test the acceptability and reliability of the questionnaire, preliminary test was done with the sample size of 50 respondents from different parts of the country and has goodness of fit at 0.80 [19]. Accordingly, the Cronbach's alpha is computed, where the results are within the acceptable range (0.824 and 0.912) [20] to proceed with the study. The total sample collected for the study is 432 (300 female and 132 male). Goodness-of-fit indices were generated and modified the model with repeated computation to improve the results as per the acceptance range for the study [21]. The female respondents are more in number; it is presumed that the females in Mauritius are more dependent and comfortable to do online shopping.

The respondents are from diverse family structure, different economic backgrounds, educational qualification, regions, type of products purchased, and online shopping portals used.

4.1 Exploratory Factor Analysis (EFA)

With the sample size of 432, the factor analysis is extracted [22] (Tables 1, 2, and 3; Fig. 1).

Regression weights are positive and moderate values. Squared multiple correlation R^2 value test reliability is also positive and low. As the final conclusion, model considered is moderately acceptable. Analysis of variance indicated that overall satisfaction is dependent on the set of independent variables such as connectivity, accessibility, and details (Table 4; Figs. 2, and 3).

The factors are divided into personality traits, purchase decision, satisfaction and e-loyalty. The hypothesis is sustained by *p*-value (***).

5 Discussion and Conclusion

The findings showed the hypothesis was supported to the maximum extent of online buyers' behaviours towards online shopping in assessment for satisfaction, purchase decision, e-trust and e-loyalty [6]. Consequence of purchase intention towards online

Table 2 Model summary of multiple regression

Model	*R*	*R* Square	Adjusted *R* Square	Std. error of the estimate	Change statistics				
					R Square change	*F* change	df1	df2	Sig. *F* change
1	0.484[a]	0.234	0.217	0.56427	0.234	13.576	3	133	0.000

[a]Predictors: (constant), personality traits, shopping decision, satisfaction and e-loyalty. *Source* Authors Computation

Table 3 'Measurement model fit indices' for first output-modified CFA

Absolute fit indicators						Incremental fit indicators					
Model fit index	Chi- Square ($\chi 2$)	df	*p*-value of $\chi 2$	CMIN/df ($\chi 2$/df)	RMSEA	CFI	GFI	AGFI	NFI	TLI	$\Delta\chi 2$
Acceptable value	Small		<0.05	<is 5	<0.05 is good, <0.08 is acceptable	>is 0.95 great, >is 0.7 tolerable	Same	Same	>is 0.90 great, >is 0.7 tolerable	>is 0.95 great, >is 0.7 tolerable	
First output	132.62	84	0	2.47	0.65	0.934	0.81	0.77	0.843	0.918	
Modified	82.95	57	0.004	2.17	0.058	0.960	0.832	0.795	0.887	0.946	117.9

Source Primary data collected from questionnaire
Source Authors Computation

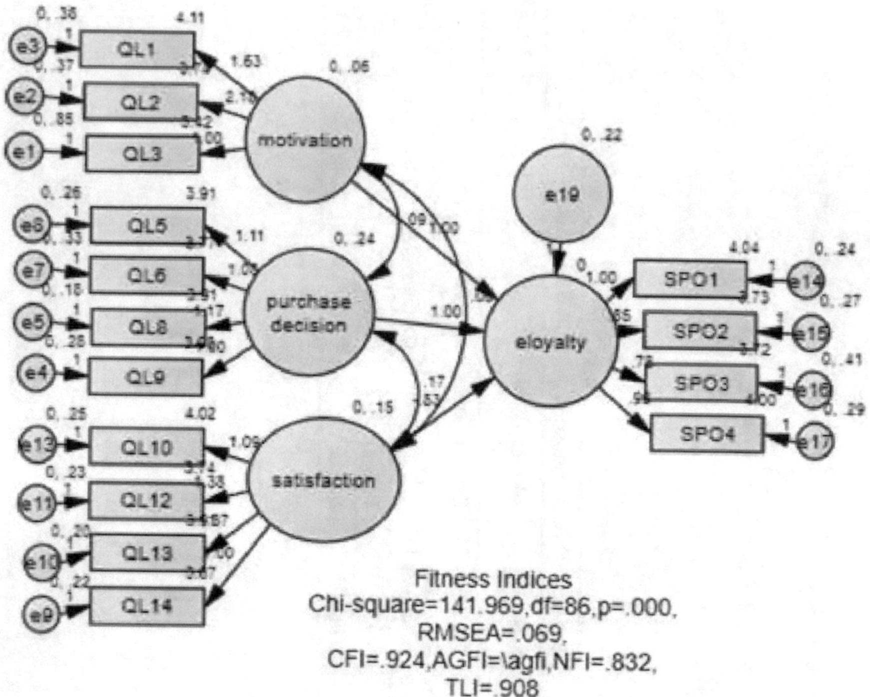

Fig. 1 Structural equation modelling with modified CFA. *Source* Authors Computation

Table 4 Coefficients[a]

Model		Unstandardized coefficients		Standardized coefficients	*T*	Sig.
		B	Std. error	Beta		
1	(Constant)	1.640	0.389		4.214	0.000
	Personality traits	0.097	0.092	0.092	1.059	0.292
	Purchase decision	0.419	0.107	0.407	3.918	0.000
	e-loyalty	0.046	0.118	0.040	0.391	0.696

[a]Dependent Variable: Overall_Satisfaction. *Source* Authors Computation

shopping by the Mauritian consumers and customers is reliable as per previous studies [23]. The findings evidenced that Mauritians purchase intention is influenced by personality traits, socio-economic, cultural and law. Though there is a scarcity of resources and infrastructure, still Mauritians are trying to cope up with the technological advances across the globe, by updating themselves. The present study has revealed that there is an amplified illuminating power of Mauritian online shopping intention and also provides guidelines for further research to go in depth to test more dimensions. Thus, it is suggested for the further study.

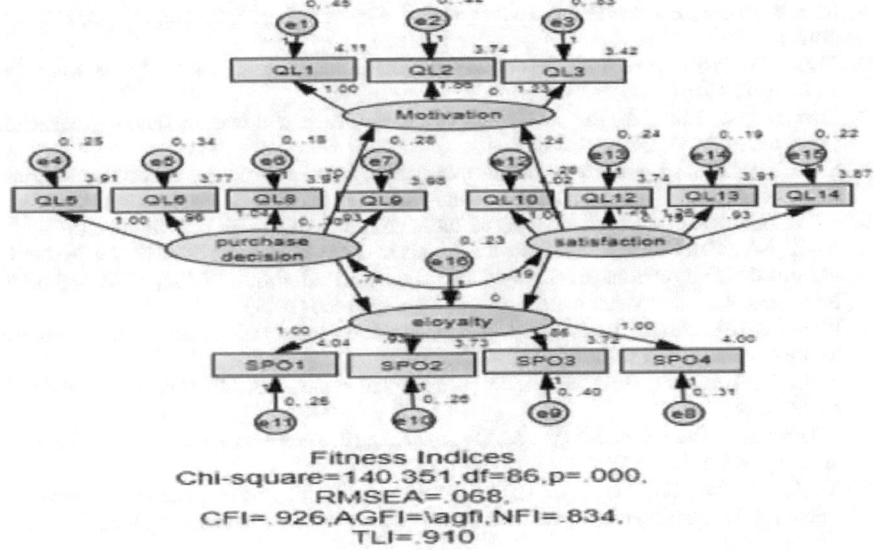

Fig. 2 SEM model modification. *Source* Authors Computation

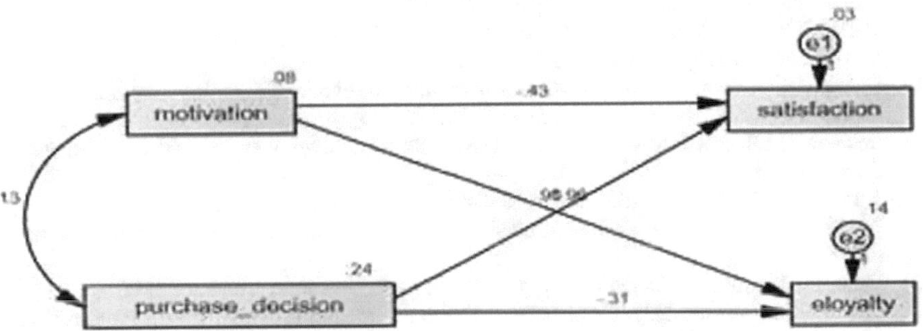

Fig. 3 Multiple regression modified SEM. *Source* Authors Computation

References

1. Satapathy, S.C., et al.: Smart computing and informatics. In: Proceedings of the First International Conference on SCI, vol. 1 (2016); O'Brien, J.A., Marakas, G.M.: Management Information Systems, 7th edn. McGraw Hill, New York. Occupation and the organization. J. Organ. Behav. **15**(6), 535–547 (2006). O'Brien, J.A.: Introduction to Information Systems, 12th edn. McGraw Hill, New York
2. Law, J.: 11 Street Officially launches in Malaysia. Retrieved June 2018, from hardwarezone. http://www.hardwarezone.com.my/tech-news (2018)
3. Lee, et al.: Analyzing key determinants of online repurchase intentions. Asia Pac. J. Market. logistics **23**(2), 200–221 (2011)
4. Clemes, M.D., Gan, C., Zhang, J.: An empirical analysis of online shopping adoption in Beijing, China. J. Retail. Consum. Serv. **21**(3), 364–375. Prentice-Hall, Cliffs, NJ (2014)

5. Nicosia, F.: Consumer behavior toward sociology of consumption. J. Consum. Res. **5**, 121–133 (1976)
6. Bhasin, H.: Factors Affecting Consumer Buying Behavior. www.marketing91.com. Accessed 26 Jan 2017 (2010)
7. Bhatnagar, A., Misra, S., Rao, H.R.: On risk, convenience, and internet shopping behavior. Commun. ACM **43**(11), 98–105 (2000)
8. Cho, J.: Likelihood to abort an online transaction: influences from cognitive evaluations, attitudes, and behavioral variables. Inf. Manag. **41**, 827–838 (2004)
9. Ajzen, I., Fishbein, M.: Understanding attitudes and predicting social behavior (1980)
10. Jamil, N.A., Mat, N.K.: To investigate the drivers of online purchasing behavioral in Malaysia based on the theory of planned behavior (TPB): a structural equation modeling (SEM) approach. In: International Conference on Management, pp. 453–460 (2017)
11. Pavlou, P.A.: Understanding and predicting electronic commerce adoption: an extension of the theory of planned behavior. **30**(1), 115–143 (2016)
12. Kotler, P., Armstrong, G.: Atmospherics as a marketing tool. J. Retail. Consum. Serv. **49**(4), 48–64 (2005)
13. Delafrooz, N., Paim, L.H., Khatibi, A.: Understanding consumer's internet purchase intention in Malaysia (2011)
14. Kwek, C.L., Tan, H.P., Lau, T.C.: The effects of shopping orientations, online trust and prior online purchase experience toward customers' online purchase intention. Int. Bus. Res. **3**(3), 63–76 (2010)
15. Shergill, G.S., Chen, Z.: Web-based shopping: consumers' attitudes towards online shopping in New Zealand. J. Electron. Commer. Res. **6**(2), 79–94 (2005)
16. Satapathy, S.C., et al. (eds.): Computer communication, networking and internet security. In: Proceedings of IC3T 2016, vol. 5. Springer (2017)
17. van Heijden, H., Verhagen, T., Creemers, M.: Understanding Online Purchase Intentions: Contribution from Technology and Trust (2003)
18. Clemes, M.D., Gan, C., Zhang, J.: An empirical analysis of online shopping adoption in Beijing, China. J. Retail. Consum. Serv. **21**(3), 364–375 (2014)
19. Sekaran, U.: Research Methods for Business: A Skill Building Approach, 4th edn. Wiley, New Jersey (2003)
20. Nunnally, J.C.: Psychometric Theory. McGraw-Hill, Michigan (1978)
21. Agresti, A., Finlay, B.: Statistical Methods for the Social Sciences. Prentice Hall, New Jersey (1997)
22. Field, A.: Discovering Statistics Using SPSS. Sage Publications, London (2005)
23. He, et al.: Empirical Study of Consumers' Purchase Intentions in C2C Electronic Commerce. **13**(3), 287–292 (2009)

Ensemble Attribute Collection and Conformable Leader Election for Data Clustering

Anuradha Yarlagadda

Abstract Clustering is one among the outstanding information mining approach extensively accustomed to categorize the information supported by their similarity. High-dimensional information analysis always presents a new era in the territory of data mining. Many variable selection techniques like density-based, model-based, and grid-based exist in the literature. However, high-dimensionality cluster analysis is lagging behind in accuracy and requires high processing time. To handle the dimensionality curse, efficient dimensionality reduction technique with appropriate feature selection is to be identified yielding clustering quality with less processing time. The proposed approach effectively handles the dimensionality by identifying most representative and sufficient subset of the attributes using ensemble approach along with conformable leader algorithm to cluster the data objects. Experimental evaluation proves that the quality of the clusters is improved with conformable leader approach and the execution time is reduced with dimensionality reduction.

Keywords Feature selection · Data clustering · Ensemble feature selection · Dimensionality reduction

1 Introduction

Features selected for analysis have impressive significance in pattern recognition, data analysis, information retrieval, machine learning, and data mining application [1]. Feature determination is vital for a few reasons; importantly, the noisy attributes can reduce the efficiency of the information mining algorithms. In supervised learning, it is realized that choice of the attribute has great influence in improving the efficiency of classifiers [2], prompting progressively affordable (both in storage space and calculation) classifiers, and, as a rule, it might prompt interpretable models. Feature determination has been broadly contemplated with regard to supervised learning [3], where a definitive objective is to choose attributes that can accomplish the most

A. Yarlagadda (✉)
Department of Information Technology, GVPCE, Visakhapatnam, India
e-mail: anuradhayarlagaddag@gmail.com

V. Bhateja et al. (eds.), *Embedded Systems and Artificial Intelligence*,
Advances in Intelligent Systems and Computing 1076,
https://doi.org/10.1007/978-981-15-0947-6_84

astounding accuracy on concealed data. Advancement in equipment and programming innovation made high-dimensional information available. Cluster analysis of high-dimensional information is bringing about less accuracy with high execution time. To address these issues, dimensionality reduction techniques are presented. In such scenario, the questions that arise are firstly the choice of features to be considered for analysis, secondly the choice of appropriate measures to be presented and finally to assess the cluster quality dependent on these selected features and measures.

For some real-world issues, which perhaps include much feature interaction, we need a dependable and sensibly competent technique to take out the less influencing features. Most of learning strategies do not carry on well in these conditions as, with numerous insignificant, yet noisy, attributes give very little information. A feature subset selection concentrates on identifying a smaller group of features that are essential and adequate for analyzing the data. To settle on a choice on which features to hold and which to dispose of, we need a trustworthy and proficient strategy for evaluating their significance to the target objective. Considering above concerns, the authors propose a new feature subset selection algorithm for data with high dimensions based on clustering. The algorithm includes

(i) Ignoring unessential attributes,
(ii) Building a minimum spanning tree (MST), and
(iii) Choosing most relevant attributes by partitioning MST.

Employing any single attribute selection method to find the feature subset for analysis may confine on generating local optima. Ensembles of strategies for attribute selection endeavor to join various attribute selection techniques as opposed to utilizing a single one. This paper concentrates on presenting an algorithm for structuring ensemble of feature selection approach. This ensemble approach finds a versatile and reliable solution for finding attributes appropriate for different information mining assignments, such as classification, clustering, and so forth.

The organization of the remaining paper is as follows. The related work is depicted in Segment 2. The proposed methodology is described in Sect. 3. Exploratory outcomes and comparisons of the outcomes with different techniques are talked about in Sect. 4, and conclusions are given in Sect. 5.

2 Related Work

Kira et al. in 1992 [4] proposed initially the algorithm Relief. It is an algorithm propelled by occurrence-based algorithms depending upon weights. It appraises the attributes quality based on how their qualities well recognize the equivalent and distinctive classes that are close to one another [5]. Some of the algorithms for attribute subset selection used in preprocessing are Relief, extension of Relief is ReliefF, and its extension is RReliefF [6]. The Relief arbitrarily samples an occasion and finds its closest neighbor from the equivalent and inverse class. The weights of the attributes are updated based on their closeness to a class. This procedure is carried for required number of occasions. The reason is that intra-similarity should be high than inter-similarity [7]. Relief has two limitations: One is it could not handle efficicntly redundant attributes, and secondly, it suits for only binary classes. ReliefF, an extension of Relief, handles the above problem by avoiding closest hits and misses as the predicted value is continuous and is used for regression problems. Additional knowledge is needed to check whether two instances are in the same class or not, and therefore, the concept of probability is considered to check the status of the predicted values of two instances [8]. This probability is estimated by considering the relative dissimilarity between instance-associated class values. Identifying more influential representatives of the entire data repository for further analysis using clustering or classification is the fundamental idea of (kNNModel) [9] technique.

3 Proposed Approach

A novel approach is proposed which is an ensemble technique for selecting most relevant attributes. In this model, attribute subsets are obtained by utilization of multiple feature selection strategies. After choice of feature subsets from various techniques either greater part convergence or election of all component subsets to get increasingly appropriate features will be chosen for next learning strategies.

Advantages:

1. Ensemble techniques result in getting best relevant feature subsets for analysis.
2. Ensures in obtaining the best choice of feature subsets that are highly correlated with class and uncorrelated with each other.
3. It productively and viably manages both not relevant and repeated features and gets apt subset of attributes (Fig. 1).

Feature selection (FS) algorithms result is combined to get feature merge function (FMF) which in turn results in the final feature subset (FSS). Feature selection algorithms called as ReliefF + kNN and ReliefF + Parzen are used to build the proposed ensemble approach. RELIEF was initially anticipated as an online attribute selection algorithm based on some heuristic intuitions [10]. Later, it was extend as Relief-F

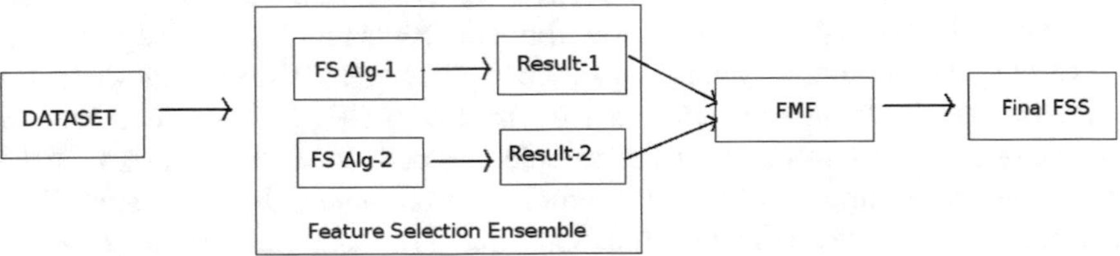

Fig. 1 Proposed architecture

by Kononenko, 1994. Relief deals with nominal and numerical attributes. However, it cannot handle unprocessed data and problems with multiple classes. ReliefF is the extension addressing the above-said problems.

Good results for feature subset selection are obtained from the composition of Relief-F + kNN. This model arbitrarily selects instances from given training dataset and later discovers k nearest neighborhood set called Ns and k closest Miss set called Qs. Weights of the each feature are updated considering theses sets basing on the average of the difference between nearest neighborhood set (Ns) and Miss set (Qs) with the existing instance and current feature using Euclidian distance as similarity measure. The algorithm is described as follows:

Algorithm 1—ReliefF + kNN

Feature values vector and the training instance associated class is the input

Vector Wt of the behavior of features is the output

1. Initialize weights Wt[O]: = 0;
2. For j = 1 to n begin
   a. arbitrarily identify an instance $\mathrm{I_j}$
   b. get k nearest neighbours $\mathrm{Ns_j}$;
   c. for each class $\mathrm{C} \neq \mathrm{class(I_j)}$
      get k misses Qs(C) from class C;
   d. for O = 1 to all the features

$$\mathrm{Wt[O]} : = \mathrm{Wt[O]} - \sum_{\mathrm{j=1}}^{\mathrm{k}} \mathrm{diff}\big(O, I_{i,} Ns_J\big)/(q, k)$$
$$+ \sum_{\mathrm{c \neq class(Ii)}} [p(C)/(1 - p(\mathrm{class}(I_i)))$$
$$+ \sum_{\mathrm{j=1}}^{\mathrm{k}} \mathrm{diff}\,(O, I_i, Qs_J(C))]/(q, k)$$

end;

Second algorithm "Relief-F + Parzen" estimator [11] is used as Gaussian RBF kernel. This model is slightly modified version of previous Relief-F + kNN by replacing kNN with Parzen density estimation. Emanuel Parzen invented this approach in the early 1960s, providing a rigorous mathematical analysis. Parzen windowing is an approach to estimate the *probability density function* (PDF), $P(X)$ for some random sample or instance x from which that sample was derived. Its major task includes monitoring each sample by placing a kernel function to read the observation. This observation is used to predict the probability that how many observations of x_i can contribute in the current window. Thus, the PDF $P(x)$ is the total of contribution from the overall observations to the given window. It is defined as follows:

$$P(x) = \frac{1}{n}\sum_{i=1}^{n}\frac{1}{h_n^d}K\left(\frac{x - x_1}{h_n}\right) \tag{1}$$

where kernel function is termed as $K(x)$ and $h_n > 0$ is the window width and is generally estimated depending on number of observations. Gaussian PDF is used most of the time for Parzen window density estimation. In that case above Eq. 1 can be updated as

$$P(x) = \frac{1}{n}\sum_{i=1}^{n}\frac{1}{(h\sqrt{2\pi})_n^d}\exp\left(-\frac{1}{2}\left(\frac{x - x_i}{h_n}\right)^2\right) \tag{2}$$

where h_n is the standard deviation for each dimension. In the proposal, this kernel can be estimated for both Hit and Miss sets to evaluate the difference between each kernel where the average results for both Hit and Miss sets are, respectively, $K(H)$ and $K(M)$. The weight of each attribute is greedily calculated to get the optimal quality of the feature subset. It is represented as

$$\begin{aligned} &\text{Max} \sum_{n=1}^{N} w^t m_n^p \\ &\text{s.t}\|w\|_2^2 = 1, w >= 0 \end{aligned} \tag{3}$$

where m_n^p is the margin and is defined as difference between $K(H)$ and $K(M)$. The above Eq. 3 is called as "Parzen + Relief". For simplicity, it can be referred as PRF(A) for attribute A. Now, the algorithm 1 can be modified as follows:

Algorithm 2—Parzen + ReliefF
Input: A feature value vector and training instance associated Class

Output: the vector W of weights of the attribute Qualities

3. Assign weights WT[A]: = 0;
4. For j: = 1 to n begin
    a. R_i, a random occurrence is selected
    b. HS, the hit set is found
    c. MS, the Miss set is found
    d. for A: = 1 to a do

$$\text{WT}[A] := \text{WT}[A] - \text{PRF}(A,R,\text{HS}) + \text{PRF}(A,R,\text{MS})$$

5. end;

The proposed ensemble attribute collection method is explained below:

Algorithm 3—Ensemble Attribute Collection (EFS)
Data repository *D* for evaluation is considered as input,

Attribute subset ASS is the resultant output

1. AS1 = kNNReliefF(D, p)
2. AS2 = PRF(D)
3. Compute ASS = {AS1}∩{AS2}
4. Return D

Initially, some subset of attributes (AS1) is chosen from first algorithm and other subset of attributes.

(AS2) is chosen from other algorithms. A final attribute set is obtained by our proposed approach which considers the common attributes obtained from both the algorithms. This set obtained is utilized as the attribute set for information mining tasks.

In the present paper, clustering technique is employed on the dataset which consists of features obtained from the proposed ensemble feature subset selection. According to leader algorithm [12], a leader is created arbitrarily and each element of the dataset is assigned to a particular cluster whose dissimilarity between the leader and the data object is within the minimum threshold. Otherwise, a new cluster is created with the data point as a new leader. The point that is noteworthy in this methodology is each data occurrence is appointed to closest cluster if the dissimilarity measure between data object and leader is within the threshold. Already allocated data objects have no knowledge about new leaders that are formed later, and there is no check to see if those data points are much closer to new leader. The proposed approach takes care of this point and finds the distance between already assigned objects to the new leader, and if it is less, then re-allocation of that data point to the new cluster is made by removing it from the older one. The extension of the leader algorithm considering the above fact is given below.

Algorithm: Conformable Leader Algorithm

1. Input: Data repository and a threshold value
2. Output: Clusters C
3. **FS = EFS(Dataset)**
4. Arbitrarily consider an element as a leader from dataset and attach to leaders' list
5. For all data points except the one considered as leader
6. {
    a. compute the dissimilarity between the data object and all leaders
    b. identify the closest cluster center
    c. if (dissimilarity measure to closest cluster center < threshold)
    d. {
        i. allocate the instance to the closest leader
        ii. cluster number is to be noted

iii. append it to element list of this cluster
iv. cluster elements count is to be incremented
e. }
f. else
g. {
i. Add instance to leader list
ii. leader count is to be incremented
h. }

7. }
8. Evaluate again distance between all the information points with the cluster centers ignoring leader instances and cluster centers obtained before existing instance to search out for the new nearest leader.
9. If the distance between the data point to new leader is less then that data point is assigned to that leader removing it from existing cluster.
10. Quality of the Cluster is evaluated.

If all already assigned data points are to be compared with all the new leaders, then computational complexity of the algorithm increases. Hence, in the proposed algorithm, every data instance is matched with the leaders created in a window of data points. The proposed strategy can be well suited for dynamic information, and it can deal with concept drift in data streams where leaders are created dynamically. Subsequently, the proposed approach is appropriate for both online and offline analysis. The proposed algorithm employs preprocessing at step 3 by capturing the relevant subset suitable for evaluation. Because of dimensionality reduction, the time complexity of the proposed algorithm is reduced and works efficiently for all high-dimensional datasets. The performance of the proposed approach is evaluated by finding the cluster quality. The quality of the clusters is evaluated using Silhouettes [13] and Davies and Bouldin [14] measures.

4 Experimental Evaluations

Experiments are carried to evaluate the viability of the newly presented algorithm on few datasets from UCI Machine Learning Repository. The dataset for analysis is considered in such a way that few instances with less dimensions, huge data size, and some with high dimensions, few number of data points and finally large data size with more number of dimensions. Experiments are carried on test datasets like KDDCup99, 9Tumeors, and Breast Cancer.

The cluster quality is evaluated for conformable leader algorithm, and time complexity is reduced remarkably after application of ensemble attribute collection algorithm. Figure 2 shows the cluster quality of leader follower algorithm. Figure 3 shows the cluster quality of the conformable leader algorithm. Experimental evaluation reveals that best-quality clusters are visualized at a threshold of 0.695–0.748.

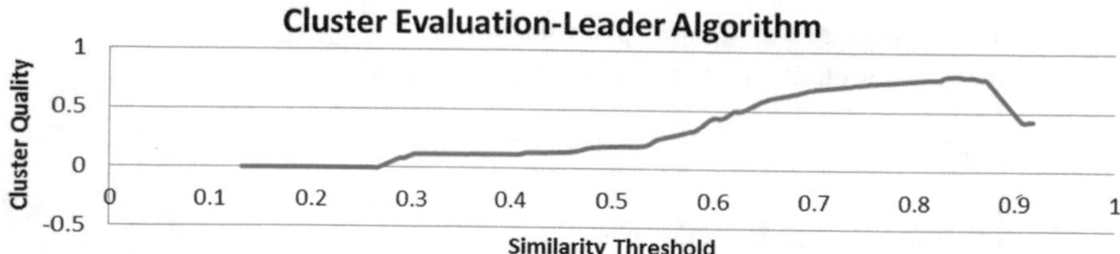

Fig. 2 No. of dimensions = 50, no. of thresholds = 500, no. of clusters = 32, quality = 0.7, and max clusters = 60

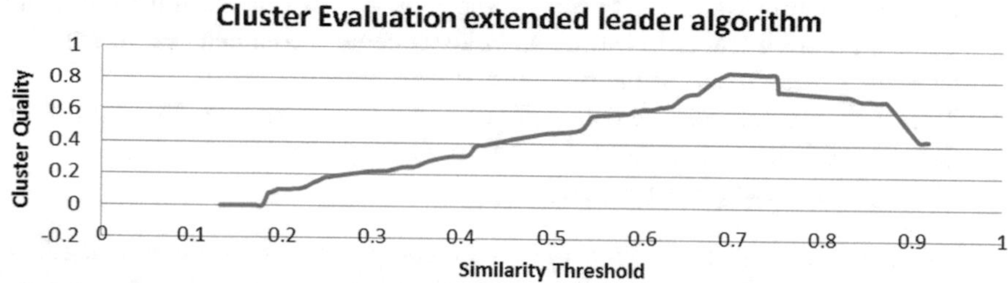

Fig. 3 No. of dimensions = 50, no. of thresholds = 500, no. of clusters = 15, quality = 0.89, and max clusters = 60

Figure 4 shows the comparison of cluster qualities with and without implementing ensemble attribute collection algorithm for different threshold values. Figures 5 and 6 give the computational complexity of the proposed ensemble attribute collection algorithm.

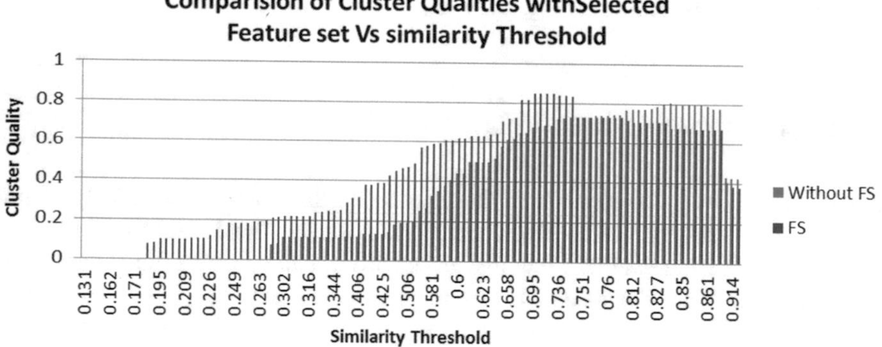

Fig. 4 Comparing cluster quality with different similarity thresholds

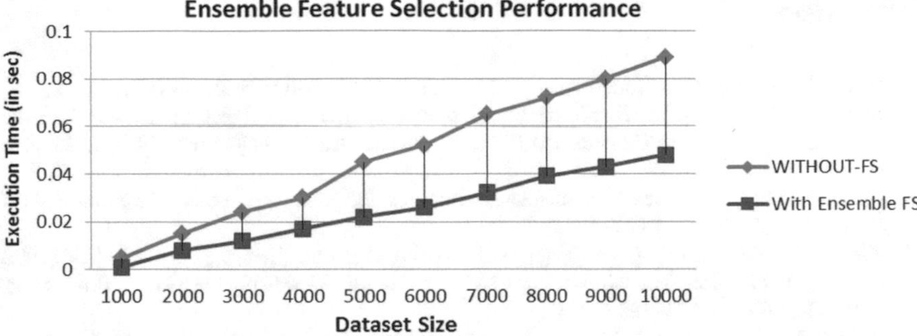

Fig. 5 Performance evaluation graph

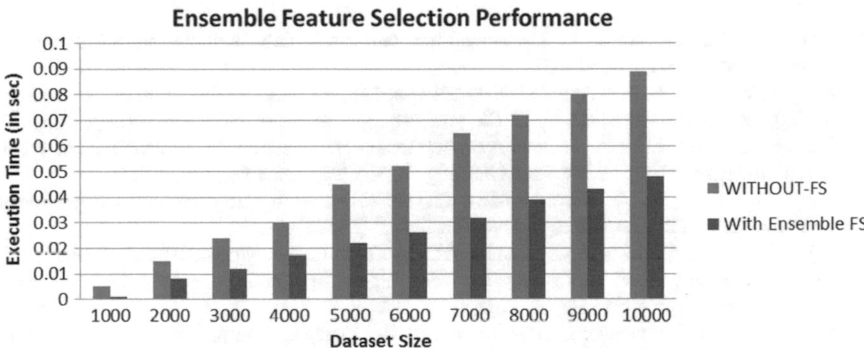

Fig. 6 Performance of feature ensemble FS

5 Conclusion and Future Work

In this work, limitations of the well-known clustering technique leader follower algorithm for large datasets and the limitations of feature selection algorithms are highlighted and details of the proposed clustering method conformable leader algorithm with ensemble attribute collection method are presented. Our investigational outcomes on numerical datasets show that the proposed algorithms perform well than the leader follower algorithm. Thus, the proposed ensemble approach effectively handles high-dimensional online and offline data. In order to solve the problem that results in feature selection fluctuated with trained instances, a novel ensemble attribute collection algorithm is proposed. Finally, the performance in target recognition is significantly increased by the feature subsets generated by proposed algorithm. Experimental results revel that computational complexity is reduced with good cluster quality. In future, our plan is to process these algorithms concurrently on multicore systems.

References

1. Hsu, C.W., Chang, C.C., Lin, C.J.: A practical guide to support vector classification
2. Raudys, S.J., Jain, A.K.: Small sample size effects in statistical pattern recognition: recommendations for practitioners. IEEE Trans. Pattern Anal. Mach. Intell. **13**(3), 252–264 (1991)
3. Blum, A., Langley, P.: Selection of relevant features and examples in machine earning. Artif. Intell. **97**(1–2), 245–271 (1997)
4. Kira, K., Rendell, L.A.: A practical approach to feature selection. Mach. Learn. 249–256 (1992)
5. Liu, H., Yu, L., Dash, M., Motoda, H.: Active feature selection using classes. In: Proceedings of PAKDD'03, pp. 474–485 (2003)
6. Robnik, M., Kononenko, I.: Machine Learning. Kluwer Academic Publishers, vol. 53, pp. 23–69 (2003)
7. Huang, Y., McCullagh, P.J., Black, N.D.: Feature selection via supervised model construction. In: Proceedings of the Fourth IEEE International Conference on Data Mining, pp. 411–414 (2004)
8. Sikonja, M.R., Kononenko, I.: Theoretical and empirical analysis of ReliefF and RReliefF. Mach. Learn. J. **53**, 23–69 (2003)
9. Guo, G., Wang, H., Bell, D., et al.: kNN model-based approach in classification. In: Proceedings of CoopIS/DOA/ODBASE 2003, LNCS, vol. 2888, pp. 986–996. Springer-Verlag (2003)
10. Kira, K., Rendell, L.A.: A practical approach to feature selection. In: Proceedings of Ninth International Workshop on Machine Learning (ICML '92), pp. 249–256 (1992)
11. Yang, S.-H., Hu, B.-G.: Discriminative feature selection by nonparametric Bayes error minimization. IEEE Trans. Knowl. Data Eng. **24**(8) (2012)
12. Vijaya, P.A., Narasimha Murty, M., Subramanian, D.K.: Leaders-subleaders: an efficient hierarchical clustering algorithm for large data sets. Pattern Recogn. Lett. archive **25**(4), 505–513 (2004). Elsevier Science Inc., New York, NY, USA
13. Rousseeuw, P.: Silhouettes: a graphical aid to the interpretation and validation of cluster analysis. J. Comput. Appl. Math. **20**(1), 53–65 (1987)
14. Davies, D., Bouldin, D.: A cluster separation measure. IEEE PAMI **1**(2), 224–227 (1979)

A Comparative Study on Effective Approaches for Unsupervised Statistical Machine Translation

B. Tarakeswara Rao, R. S. M. Lakshmi Patibandla and M. Ramakrishna Murty

Abstract Although Machine Translation has historically trusted on huge amounts of parallel corpora, the latest analysis has accomplished to prepare each Neural and Statistical Machine Translation system using monolingual corpora only. In spite of the prospective of this methodology for low-resource settings, obtainable structures square measure way outstanding their supervised counterparts, restraining their concrete interest. In this paper, Sect. 1 contains numerous deficiencies of existing unsupervised SMT approaches by exploiting subword information. Section 2 consists of another methodology established on phrase-based statistical machine translation that significantly cessations the gap with supervised structures. Principled Unsupervised Statistical Machine Translation in Sect. 3. Results and discussions in Sect. 4 and conclusion in Sect. 5.

Keywords Subword · Machine · Translation · Unsupervised

1 Introduction

Neural Machine Translation (NMT) has as of late turned into the predominant worldview in machine interpretation [1]. As opposed to progressively unbending Statistical Machine Translation (SMT) architectures (Koehn et al. 2003), NMT models are prepared to start to finish, abuse persistent portrayals that alleviate the sparsity issue, and beat the territory issue by utilizing unconstrained settings. On account of this extra flexibility, NMT can all the more successfully abuse huge parallel corpora, despite the fact that SMT is as yet unrivaled when the preparation corpus isn't huge

B. Tarakeswara Rao (✉)
Department of CSE, KHIT, Chowdavaram, Andhra Pradesh, India
e-mail: tarak7199@gmail.com

R. S. M. L. Patibandla
Department of IT, VFSTR, Vadlamudi, Andhra Pradesh, India
e-mail: patibandla.lakshmi@gmail.com

M. R. Murty
Department of CSE, ANITS, Visakhapatnam, India

V. Bhateja et al. (eds.), *Embedded Systems and Artificial Intelligence*,
Advances in Intelligent Systems and Computing 1076,
https://doi.org/10.1007/978-981-15-0947-6_85

enough (Koehn and Knowles 2017). Fairly incomprehensible, while most machine interpretation looks into has concentrated on asset-rich settings where NMT has for sure supplanted SMT, [2] are penny profession has figured out how to prepare an NMT framework with no supervision, depending on monolingual corpora alone [2, 3]. Given the shortage of parallel corpora for most language sets, including less-resourced dialects yet, in addition, numerous mixes of real dialects, this examination line opens energizing chances to bring successful machine interpretation to a lot more situations. All things considered, existing arrangements are still a long way behind their directed partners, extraordinarily constraining their pragmatic convenience. For example, existing unsupervised NMT frameworks acquire between 15 and 16 BLEU focuses on WMT 2014 English–French interpretation, though a best in class NMT framework gets around 41 [2, 3] (Yang et al. 2018). In this paper, we investigate whether the inflexible and secluded nature of SMT is increasingly appropriate for these unsupervised settings and a novel unsupervised SMT framework that can be prepared on monolingual corpora alone [4, 5]. The staying of this paper is composed as pursues. Section 2 at that point depicts our Phrase-based Statistical Machine Translation, while Sect. 3 talks about Principled Unsupervised Statistical Machine Translation. We then present the correlations of analyses that were done and the outcomes acquired in Sect. 4, and Sect. 5 is a Conclusion of this study.

2 Phrase-Based Statistical Machine Translation

While initially propelled as an uproarious channel model (Brown et al. 1990), express based SMT is currently defined as a log-direct blend of a few factual models that score interpretation competitors (Koehn et al. 2003). The parameters of these scoring capacities are evaluated autonomously dependent on recurrence checks, and their loads are then tuned in a different approval set [6]. At surmising time, a decoder attempts to find the interpretation hopeful with the most elevated score as per the subsequent consolidated model. The specific scoring models found in a standard SMT framework are as per the following:

- **Phrase-table**: The expression table is an accumulation of source language n-grams and a rundown of their potential interpretations in the objective language alongside various scores for each of them. In order to interpret longer groupings, the decoder consolidates these fractional n-gram interpretations and positions the subsequent competitors as indicated by their relating scores and the remainder of scoring capacities [1, 3]. So as to manufacture the expression table, SMT processes word arrangements in the two bearings from a parallel corpus, symmetrizes these arrangements utilizing various heuristics [7], removes the arrangement of steady expression matches, and scores them dependent on recurrence tallies. For that reason, standard SMT utilizes 4 scores for each expression table passage: the immediate and backward lexical weightings, which are gotten from word-level arrangements, and the immediate and opposite expression interpretation probabilities, which are registered at the expression level.

- **Language model**: The language model doles out a likelihood of a word grouping in the objective language [7, 8]. Customary SMT utilizes n-gram language models for that, which utilizes straightforward recurrence checks over a vast monolingual corpus with back-off and smoothing.
- **Reordering model**: The reordering model records for various words arranges crosswise over dialects, scoring interpretation hopefuls as indicated by the situation of each deciphered expression in the objective language. Standard SMT consolidates two such models: a separation based bending model that punishes deviation from a monotonic order, and a lexical reordering model that fuses expression introduction frequencies from a parallel corpus [9].
- **Word and expression punishments**: The word and expression punishments appoint a fixed score to each produced word and state and are helpful to control the length of the yield content and the inclination for shorter or longer expressions [10, 11].

Having prepared all these various models, a tuning procedure is connected to advance their loads in the subsequent log-straight model, which normally augments some assessment metric in a different approval parallel corpus [12]. A typical decision is to improve the BLEU score through Minimum Error Rate Training (MERT) [7].

3 Principled Unsupervised Statistical Machine Translation

Expression-based SMT is defined as a log-straight blend of a few factual models: an interpretation model, a language model, a reordering model, and a word/expression punishment. In that capacity, fabricating an unsupervised SMT framework requires taking in these various segments from monolingual corpora [13]. Things being what they are, this is direct for a large portion of them: the language model is found out from monolingual corpora by definition; the word and expression punishments are parameterless; and one can drop the standard lexical reordering model at a little expense and do with the bending model alone, which is likewise parameterless. Thus, the principle challenge left is learning the interpretation model that is, building the expression table.

3.1 Preliminary Expression Table

So as to manufacture our underlying expression table, we pursue [5] and learn n-gram embeddings for every language freely, map them to a mutual space through self-learning, and utilize the subsequent cross-lingual embeddings to concentrate and score state sets. All the more solidly, we train our n-gram embeddings utilizing phrase2vec1, a basic expansion of skip-gram that applies the standard negative inspecting loss of [14] to bigram context and trigram-setting sets notwithstanding the typical word-setting pairs. Having done that, we map the embeddings to a cross-lingual space utilizing VecMap3 with indistinguishable statement [4], which fabricates an underlying arrangement by adjusting indistinguishable words and iteratively improves it through self-learning [15]. At long last, we remove interpretation applicants by taking the 100 closest neighbors of each source expression, and score them by applying for the delicate max work over their cosine similitudes:

$$\varphi(\bar{f}|\bar{e}) = \frac{\exp(\cos(\bar{e}, \bar{f})/\tau)}{\sum_{\bar{f}'} \exp(\cos(\bar{e}, \overline{f'})/\tau)}$$

φ where the temperature τ is estimated using maximum likelihood estimation over a dictionary induced in the reverse direction. In addition to the phrase translation probabilities in both directions, the forward and reverse lexical weightings are also estimated by aligning each word in the target phrase with the one in the source phrase most https://github.com/artetxem/phrase2vec. So as to keep the model size within a reasonable limit, we restrict the vocabulary to the most frequent 200,000 unigrams, 400,000 bigrams, and 400,000 trigrams. Likely generating it, and taking the product of their respective translation probabilities. The reader is referred to [5] for more details.

3.2 *Adding Sub Word Information*

A natural constraint of existing unsupervised SMT frameworks is that words are taken as nuclear units, making it difficult to misuse character-level data. This is reflected in the known difficulty of these models to decipher named substances [16], as it is exceptionally testing to segregate among related formal people, places or things dependent on distributional data alone, respecting interpretation blunders like "Sunday Telegraph" $\rightarrow$ "The Times of London" [17]. In order to defeat this issue, we suggest fusing subword data once the underlying arrangement is done at the word/expression level [17, 18]. For that reason, we add two extra loads to the underlying expression table that are practically equivalent to the lexical weightings, however, utilize a character-level likeness work rather than word interpretation probabilities:

$$\text{score}(\bar{f}|\bar{e}) = \prod_{i} \max(\varepsilon, \max j \ \text{sim}(\overline{f_i}, \overline{e_j}))$$

where $\varepsilon = 0.3$ ensures a base comparability score, as we need to support interpretation applicants that are comparable at the character level without exorbitantly punishing those that are most certainly not. For our situation, we utilize a basic comparability work that standardizes the Levenshtein separate lev(·) [18] by the length of the words len(·):

$$\mathrm{sim}(f, e) = 1 - \frac{\mathrm{lev}(f, e)}{\max(\mathrm{len}(f), \mathrm{len}(e))}$$

3.3 Unsupervised Tuning

Having prepared the hidden measurable models freely, SMT tuning expects to change loads of their subsequent log-straight mix to streamline some assessment metrics like BLEU in a parallel approval corpus, which is commonly done through Minimum Error Rate Training or MERT [7]. Obviously, this is impossible in carefully unsupervised settings, however, we contend that it would, in any case, be attractive to streamline some unsupervised paradigm that is required to connect well with test execution. Shockingly, neither of the current unsupervised SMT frameworks do as such: Artetxe et al. [5] utilize a heuristic that manufactures two introductory models in inverse ways, utilizes one of them to produces an engineered parallel corpus through back-interpretation (Sennrich et al. 2016), and applies MERT to tune the model in the switch bearing, emphasizing until combination, though [17] do not play out any tuning whatsoever. In what pursues, we propose an increasingly principled way to deal with tuning that defines an unsupervised foundation and an advanced method that is ensured to combine to a neighborhood ideal of it. Motivated by the past work on CycleGANs (Zhu et al. 2017) and double learning (He et al. 2016), our technique takes two introductory models in inverse ways, and defines an unsupervised advancement target that joins a cyclic consistency misfortune and a language model misfortune over the two monolingual corpora E and F:

$$L = L_{\mathrm{cycle}}(E) + L_{\mathrm{cycle}}(F) + L_{\mathrm{lm}}(E) + L_{\mathrm{lm}}(F)$$

The cyclic consistency misfortune catches the instinct that the interpretation of an interpretation ought to be near the first content [14, 19]. In order to measure this, we take a monolingual corpus in the source language, make an interpretation of it to the objective language and back to the source language, and process its BLEU score accepting the first content as reference:

$L_{\mathrm{cycle}}(E) = 1 \quad \mathrm{BLEU}(T_{F\to E}(T_{E\to F}(E)), E)$. In the meantime, the language model misfortune catches the instinct that machine interpretation should deliver the fluent message in the objective language. For that reason, we gauge the per-word entropy in the objective language corpus utilizing an n-gram language model, and punish higher per-word entropies in machine interpreted content as follows[4]: $L_{\mathrm{lm}}(E) = LP \cdot \max(0, H(F) - H(T_{E\to F}(E)))^2$ where the length punishment $LP = LP(E) \cdot LP(F)$

$$LP(E) = \max\left(1, \frac{\mathrm{len}(T_{F\to E}(T_{E\to F}(E)))}{\mathrm{len}(E)}\right)$$

punishes unreasonably long translations[5]: limit the consolidated adversity work, we adjust MERT to mutually enhance the parameters of the two models. In its fundamental structure, MERT approximates the scan space for each source sentence through an n-best rundown and plays out a type of organize plunge by registering the ideal incentive for every parameter through an efficient line look technique and voraciously making the stride that prompts the biggest addition. The procedure is rehashed iteratively until combination, increasing an n-best rundown with the refreshed parameters at every emphasis in order to get a superior guess of the full inquiry space. Given that our improvement target consolidates two interpretation frameworks $T_{F \to E}(T_{E \to F}(E))$, this would require creating an n-best rundown for $T_{E \to F}(E)$ first and, for every passage on it, producing another n-best rundown with $T_{F \to E}$, yielding a joined n-best rundown with N2 sections.

4 Results and Discussions

In order to make comparisons to previous work, we use the French-English and German-English datasets from the WMT 2014 shared task. More concretely, training data consists of the concatenation of all News Crawl monolingual corpora from 2007 to 2013, which make a total of 749 million tokens in French, 1,606 million in German, and 2,109 million in English, from which we take a random subset of 2,000 sentences for tuning Preprocessing is done using standard Moses tools, and involves punctuation normalization, tokenization with aggressive hyphen splitting, and true casing. SMT implementation is based on Moses8, and we use the KenLM [20] tool included in it to estimate 5-gram language model with modified Kneser-Ney smoothing. Unsupervised tuning implementation is based on Z-MERT (Zaidan 2009), and we use FastAlign [12] for word alignment within the joint refinement procedure. Table 1 reports the consequences of the hybridized framework in contrast with past work. As it very well may be seen, our full framework acquires the best-distributed outcomes in all cases, beating the past cutting edge by 5–7 BLEU focuses on all datasets and interpretation bearings. A generous piece of this improvement originates from our increasingly principled unsupervised SMT approach, which beats all past SMT based frameworks by around 2 BLEU focuses. In any case, it is the NMT hybridization that brings the biggest additions, improving the consequences of this underlying SMT framework by 5–9 BLEU focuses.

As appeared in Table 2, our supreme additions are impressively bigger than those of past hybridization techniques, regardless of whether underlying SMT framework is generously better and in this way more difficult to enhance.

Table 3 reports the consequences of various managed frameworks in the same WMT 2014 test set. All the more solidly, we incorporate the best outcomes from the common undertaking itself, which reflect the best in class in machine translation back in 2014; those of Vaswani et al. (2017), who presented the now overwhelming transformer engineering; and those of [11], who apply back-interpretation at a vast scale and hold the present best outcomes in the test set.

Table 1 Results comparison

		NMT				SMT					NMT + SMT				
		Artetxe et al. [2]	Lample et al. [16]	Yang et al. (2018)	Lample et al. [17]	Artetxe et al. [5]	Lample et al. [17]	Marie and Fujita [19]	detok. Sacre BLEU*	HUSMT	Lample et al. [17]	Marie and Fujita [19]*	Ren et al. (2019)	Detok. Sacre BLEU*	HUSMT
WMT-14	fr-en	15.6	14.3	15.6	24.2	25.9	27.2	–	27.9	28.4	27.7	–	28.9	33.2	33.5
	en-fr	15.1	15.1	17.0	25.1	26.2	28.1	–	27.8	30.1	27.6	–	29.5	33.6	36.2
	de-en	10.2	–	–	–	17.4	–	–	19.7	20.1	–	–	20.4	26.4	27.0
	en-de	6.6	–	–	–	14.1	–	–	14.7	15.8	–	–	17.0	21.2	22.5
WMT-16	de-en	–	13.3	14.6	21.0	23.1	22.9	20.2	24.8	25.4	25.2	26.7	26.3	33.8	34.4
	en-de	–	9.6	10.9	17.2	18.2	17.9	15.5	19.4	19.7	20.2	20.0	21.7	26.4	26.9

Table 2 Results of different unsupervised MT for neural MT hybridization

		Initial SMT + NMT hybrid					
		Lample et al. [17]		Marie and Fujita [19]*		HUSMT	
WMT-14	fr-en	27.2	27.7(+0.5)	–	–	28.4	33.5(+5.1)
	en-fr	28.1	27.6(−0.5)	–	–	30.1	36.2(+6.1)
WMT-16	de-en	22.9	25.2(+2.3)	20.2	26.7(+6.5)	25.4	34.4(+9.0)
	en-de	17.9	20.2(+2.3)	15.5	20.0(+4.5)	19.7	26.9(+7.2)

Table 3 Results of USMT hybridization to different supervised systems

		WMT-14			
		fr-en	en-fr	de-en	en-de
Unsupervised	HUSMT	33.5	36.2	27.0	22.5
	detok. SacreBLEU*	33.2	33.6	26.4	21.2
Supervised	WMT best	35.0	35.8	29.0	20.6
	Vaswani et al. [1]	–	41.0	–	28.4
	Edunov et al. [11]	–	45.6	–	35.0

Subjective outcomes Table 4 shows some translation examples from hybridized framework in contrast with those revealed by Artetxe et al. [5]. We pick precisely the same sentences detailed by Artetxe et al. [5], which were arbitrarily taken from the news test 2014, so they should be illustrative of the general conduct of the two frameworks. While not flawless, the hybridized framework delivers for the most part fluent interpretations that precisely catch the importance of the first content. Just in accordance with quantitative outcomes, this recommends unsupervised machine interpretation can be a usable option in pragmatic settings. Contrasted with Artetxe et al. [5], interpretations are commonly more fluent, which isn't amazing given that they are delivered by an NMT framework as opposed to an SMT framework.

5 Conclusions

In this paper, we identify several deficiencies in previous unsupervised SMT systems and recommend a more principled approach that addresses them by incorporating subword information, using a theoretically well-founded unsupervised tuning method. In addition to that, we use improved SMT approach to initialize a dual NMT model that is further improvement through on-the-fly back translation. The comparison shows improvement of previous state-of-the-art in unsupervised machine translation by 5–7 BLEU points in French–English and German–English WMT 2014 and 2016.

Table 4 Comparison of randomly selected examples from French → English news test 2014 reported by Artetxe et al. [5]

Source	Reference	Artetxe et al. [5]	HUSMT
D'autres révélations ont fait état de documents divulgués par Snowden selon lesquels la NSA avait intercepté des données et des communications émanant du téléphone portable delachancelièreallemande Angela Merkel et de ceuxde34autreschefsd'État	Other revelations cited documents leaked by Snowden that the NSA monitored German Chancellor Angela Merkel's cellphone and those of up to 34 other world leaders	Other disclosures have reported documents disclosed by Snowden suggested the NSA had intercepted communications and data from the mobile phone of German Chancellor Angela Merkel and those of 32 other heads of state	Other revelations have pointed to documents disclosed by Snowden that the NSA had intercepted data and communications emanating from German Chancellor Angela Merkel's mobile phone and those of 34 other heads of state
La NHTSA n'a pas pu examiner la lettre d'information aux propriétaires en raison de l'arrêtde16joursdesactivités gouvernementales, ce qui a ralentilacroissancedesventes de véhicules en octobre	NHTSA could not review the owner notification letter due to the 16-day government shutdown, which tempered auto sales growth in October	The NHTSA could not consider the letter of information to owners because of halting 16-day government activities, which slowed the growth in vehicle sales in October	NHTSA said it could not examine the letter of information to owners because of the 16-day halt in government operations, which slowed vehicle sales growth in October
Le M23 est né d'une mutinerie, en avril 2012, d'anciens rebelles, essentiellement tutsi, intégrés dans l'armée en 2009 après un accord de paix	The M23 was born of an April 2012 mutiny by former rebels, principally Tutsis who were integrated into the army in 2009 following a peace agreement	M23 began as a mutiny in April 2012, former rebels, mainly Tutsi integrated into the national army in 2009 after a peace deal	M23 was born into a mutiny in April 2012, of former rebels, mostly Tutsi, embedded in the army in 2009 after a peace deal
Tunks a déclaré au Sunday Telegraph de Sydney que toutela familleétait "extrêmement préoccupée" du bienêtre de sa fille et voulait qu'elle rentre en Australie	Tunks told Sydney's Sunday Telegraph the whole family was "extremely concerned" about his daughter's welfare and wanted her back in Australia	Tunks told The Times of London from Sydney that the whole family was "extremely concerned" of the welfare of her daughter and wanted it to go in Australia	Tunks told the Sunday Telegraph in Sydney that the whole family was "extremely concerned" about her daughter's well-being and wanted her to go into Australia

References

1. Vaswani, A., Knight, K., Dyer, C.: Unifying bayesian inference and vector space models for improved decipherment. In: Proceedings of the 53rd Annual Meeting of the Association for Computational Linguistics and the 7th International Joint Conference on Natural Language Processing, vol. 1, Long Papers, pp. 836–845. Association for Computational Linguistics, Beijing, China (2015)

2. Artetxe, M., Labaka, G., Agirre, E., Cho, K.: Unsupervised neural machine translation. In: Proceedings of the 6th International Conference on Learning Representations (ICLR 2018) (2018c)
3. Conneau, A., Lample, G., Ranzato, M.A., Denoyer, L., Jégou, H.: Word translation without parallel data. In: Proceedings of the 6th International Conference on Learning Representations (ICLR 2018) (2018); Dou, Q., Knight, K.: Large scale decipherment for out-of-domain machine translation. In: Proceedings of the 2012 Joint Conference on Empirical Methods in Natural Language Processing and Computational Natural Language Learning, pp. 266–275 (2012)
4. Artetxe, M., Labaka, G., Agirre, E.: A robust self-learning method for fully unsupervised cross-lingual mappings of word embeddings. In: Proceedings of the 56th Annual Meeting of the Association for Computational Linguistics, vol. 1, Long Papers, pp. 789–798. Association for Computational Linguistics (2018a)
5. Artetxe, M., Labaka, G., Agirre, E.: Unsupervised statistical machine translation. In: Proceedings of the 2018 Conference on Empirical Methods in Natural Language Processing, pp. 3632–3642. Association for Computational Linguistics, Brussels, Belgium (2018b)
6. Artetxe, M., Labaka, G., Agirre, E.: Learning bilingual word embeddings with (almost) no bilingual data. In: Proceedings of the 55th Annual Meeting of the Association for Computational Linguistics, vol. 1, Long Papers, pp. 451–462. Association for Computational Linguistics, Vancouver, Canada (2017)
7. Och, F.J.: Minimum error rate training in statistical machine translation. In: Proceedings of the 41st Annual Meeting of the Association for Computational Linguistics, pp. 160–167. Association for Computational Linguistics, Sapporo, Japan (2003)
8. Ott, M., Edunov, S., Grangier, D., Auli, M.: Scaling neural machine translation. In: Proceedings of the Third Conference on Machine Translation: Research Papers, pp. 1–9. Association for Computational Linguistics, Belgium, Brussels (2018)
9. McCallum, A., Bellare, K., Pereira, F.: A conditional random field for discriminatively-trained finite-state string edit distance. In: Proceedings of the Twenty-First Conference on Uncertainty in Artificial Intelligence, pp. 388–395 (2005)
10. Dou, Q., Knight, K.: Dependency-based decipherment for resource-limited machine translation. In: Proceedings of the 2013 Conference on Empirical Methods in Natural Language Processing, pp. 1668–1676. Association for Computational Linguistics, Jeju Island, Korea, Seattle, Washington, USA (2013)
11. Edunov, S., Ott, M., Auli, M., Grangier, D.: Understanding back-translation at scale. In: Proceedings of the 2018 Conference on Empirical Methods in Natural Language Processing, pp. 489–500. Association for Computational Linguistics, Brussels, Belgium (2018)
12. Dyer, C., Chahuneau, V., Smith, N.A.: A simple, fast, and effective reparameterization of IBM model 2. In: Proceedings of the 2013 Conference of the North American Chapter of the Association for Computational Linguistics: Human Language Technologies, pp. 644–648. Association for Computational Linguistics, Atlanta, Georgia (2013)
13. Hassan, H., Aue, A., Chen, C., Chowdhary, V., Clark, J., Federmann, C., Huang, X., Junczys-Dowmunt, M., Lewis, W., Li, M., et al.: Achieving human parity on automatic Chinese to English news translation (2018)
14. Mikolov, T., Sutskever, I., Chen, K., Corrado, G.S., Dean, J.: Distributed representations of words and phrases and their compositionality. In: Advances in Neural Information Processing Systems, vol. 26, pp. 3111–3119 (2013)
15. He, D., Xia, Y., Qin, T., Wang, L., Yu, N., Liu, T.-Y., Ma, W.-Y.: Dual learning for machine translation. In: Advances in Neural Information Processing Systems, vol. 29, pp. 820–828 (2016). arXiv:1803.05567
16. Lample, G., Conneau, A., Denoyer, L., Ranzato, M.A.: Unsupervised machine translation using monolingual corpora only. In: Proceedings of the 6th International Conference on Learning Representations (ICLR 2018) (2018a)
17. Lample, G., Ott, M., Conneau, A., Denoyer, L., Ranzato, M.A.: Phrase-based & neural unsupervised machine translation. In: Proceedings of the 2018 Conference on Empirical Methods in Natural Language Processing, pp. 5039–5049 (2018b)

18. Levenshtein, V.I.: Binary codes capable of correcting deletions, insertions, and reversals. In: Soviet physics doklady, vol. 10, pp. 707–710. Association for Computational Linguistics, Brussels, Belgium (1966)
19. Marie, B., Fujita, A.: Unsupervised neural machine translation initialized by unsupervised statistical machine translation (2018). arXiv:1810.12703
20. Heafield, K., Pouzyrevsky, I., Clark, J.H., Koehn, P.: Scalable modified Kneser-ney language model estimation. In: Proceedings of the 51st Annual Meeting of the Association for Computational Linguistics, vol. 2, Short Papers, pp. 690–696. Association for Computational Linguistics, Sofia, Bulgaria (2013)
21. Post, M.: A call for clarity in reporting bleu scores. In: Proceedings of the Third Conference on Machine Translation: Research Papers, pp. 186–191. Association for Computational Linguistics, Belgium, Brussels (2018)

Correction to: Automatic Synthesis Approach for Unconstrained Face Images Based on Generic 3D Shape Model

Hamid Ouanan, Mohammed Ouanan and Brahim Aksasse

Correction to:
Chapter "Automatic Synthesis Approach for Unconstrained Face Images Based on Generic 3D Shape Model" in: V. Bhateja et al. (eds.), *Embedded Systems and Artificial Intelligence*, Advances in Intelligent Systems and Computing 1076, https://doi.org/10.1007/978-981-15-0947-6_43

The original version of the book was inadvertently published with an incorrect affiliation of the co-authors of the chapter "Automatic Synthesis Approach for Unconstrained Face Images Based on Generic 3D Shape Model", which has now been corrected as the following:

M. Ouanan and B. Aksasse

Department of Computer Science, M2I Laboratory, Faculty of Science and Techniques, Moulay Ismail University, BP 509 Boutalamine 52000 Errachidia, Morocco

The chapter and book have been updated with the changes.

The updated version of this chapter can be found at https://doi.org/10.1007/978-981-15-0947-6_43

V. Bhateja et al. (eds.), *Embedded Systems and Artificial Intelligence*, Advances in Intelligent Systems and Computing 1076,
https://doi.org/10.1007/978-981-15-0947-6_86

Author Index

V. Bhateja et al. (eds.), *Embedded Systems and Artificial Intelligence*, Advances in Intelligent Systems and Computing 1076,
https://doi.org/10.1007/978-981-15-0947-6

Y

Z

Printed in the United States
By Bookmasters